AF538479

SWITCHED CAPACITOR CIRCUITS

SWITCHED CAPACITOR CIRCUITS

Phillip E. Allen

Associate Professor
Department of Electrical Engineering
Texas A&M University
College Station, TX

Edgar Sánchez-Sinencio

Professor and Head
Department of Electronics
National Institute of Astrophysics, Optics and Electronics
Puebla, Pue.
Mexico

Van Nostrand Reinhold
Electrical/Computer Science and Engineering Series

VNR VAN NOSTRAND REINHOLD COMPANY
NEW YORK CINCINNATI TORONTO LONDON MELBOURNE

Library of Congress Catalog Card Number: 83–10507
ISBN: 0–442–20873–1

Manufactured in the United States of America

Published by Van Nostrand Reinhold Company Inc.
135 West 50th Street,
New York, N.Y. 10020

Van Nostrand Reinhold Company Limited
Molly Millars Lane
Wokingham, Berkshire RG11 2PY, England

Van Nostrand Reinhold
480 Latrobe Street
Melbourne, Victoria 3000, Australia

Macmillan of Canada
Division of Gage Publishing Limited
164 Commander Boulevard
Agincourt, Ontario M1S 3C7, Canada

15 14 13 12 11 10 9 8 7 6 5 4 3 2 1

Library of Congress Cataloging in Publication Data

Allen, P. E. (Phillip E.)
Switched capacitor circuits.
(Van Nostrand Reinhold electrical/computer science and engineering series)
Includes index.
1. Switched capacitor circuits. I. Sanchez-Sinencio, Edgar. II. Title. III. Series.
TK7868.S88A38 1983 621.3815'3 83–10507
ISBN 0–442–20873–1

To Our Wives
Margaret and Jamy
and Our Children
Kurt and Liza
Cheryl Zenia
Paul

Van Nostrand Reinhold
Electrical/Computer Science and Engineering Series
Sanjit Mitra, Series Editor

HANDBOOK OF ELECTRONIC DESIGN AND ANALYSIS PROCEDURES USING PROGRAMMABLE CALCULATORS, by Bruce K. Murdock

COMPILER DESIGN AND CONSTRUCTION, by Arthur B. Pyster

SINUSOIDAL ANALYSIS AND MODELING OF WEAKLY NONLINEAR CIRCUITS, by Donald D. Weiner and John F. Spina

APPLIED MULTIDIMENSIONAL SYSTEMS THEORY, by N. K. Bose

MICROWAVE SEMICONDUCTOR ENGINEERING, by Joseph F. White

INTRODUCTION TO QUARTZ CRYSTAL UNIT DESIGN, by Virgil E. Bottom

DIGITAL IMAGE PROCESSING, by William B. Green

SOFTWARE TESTING TECHNIQUES, by Boris Beizer

LIGHT TRANSMISSION OPTICS, Second edition, by Dietrich Marcuse

REAL TIME COMPUTING, edited by Duncan Mellichamp

HARDWARE AND SOFTWARE CONCEPTS IN VLSI, edited by Guy Rabbat

MODELING AND IDENTIFICATION OF DYNAMIC SYSTEMS, by N. K. Sinha and B. Kuszta

COMPUTER METHODS FOR CIRCUIT ANALYSIS AND DESIGN, by Jiri Vlach and Kishore Singhal

HANDBOOK OF SOFTWARE ENGINEERING, edited by C. R. Vick and C. V. Ramamoorthy

SWITCHED CAPACITOR CIRCUITS, by Phillip E. Allen and Edgar Sánchez-Sinencio

Preface

The objective of the book is to provide sufficient background and understanding to enable its readers to design and apply switched capacitor circuits whether these are to be implemented as discrete circuits or by MOS technology. Since this is the first book devoted entirely to the subject of switched capacitor circuits, it has no pattern to follow. Fortunately, it was developed in an environment where many of the circuits and concepts it discusses could actually be integrated as MOS integrated circuits. It is hoped that this environment has created a selection process that has enhanced the contents.

Switched capacitor circuits provide an example of the influence that technology can have on the field of electrical engineering. Only seven years ago, the problem of building analog circuits and systems using standard MOS technology was still unsolved. Although analog circuits and systems were implemented by means of integrated circuit technology, they were neither economical nor competitive. The act of combining analog sampled data techniques with MOS technology has solved this difficulty. As a result, the field of switched capacitor circuits has developed into maturity in a relatively short period of time.

This book makes no apology for not including todays idea's that will become tomorrow's practice. The field is simply growing too fast to make such a claim. Instead, the principles of the subject are given, along with examples of these principles, so that anyone not familiar with switched capacitor circuits can quickly come to a point of general understanding of them and not only apply this new circuit technique but even use integrated circuit technology to build them if desired.

One of the problems with any book of a pioneering kind is what to include and what to leave out. Because of the wide applicability of switched capacitor techniques, the authors decided to include both filter and nonfilter circuits.

This book was developed from notes for a graduate course on Analog Sampled Data Circuits taught by the authors at Texas A&M University. The material in Chapter 8 was used to supplement Gray and Meyer's *Analysis and Design of Analog Integrated Circuits,* the text assigned for an A&M graduate course on Analog Integrated Circuit Design. A preliminary version of the present text was widely circulated in industry, both within and outside the United States, and it has also been used as the text in five four-day short courses taught primarily to engineers from industry. Exposure of this kind helped bring this book to its present state.

Switched Capacitor Circuits should be useful both as a reference book in industry and as a textbook in senior or graduate courses on the subject. The engineer in industry will find the book useful in bringing him up to date on a subject that may not have been available during his formal training. The student will find the book useful as a text that places emphasis on developing design capability in the area of analog circuits. Students may well find that they do not need the material of Chapter 1, whereas the practicing engineer may need to review it carefully. This situation may be reversed in Chapter 8.

The first chapter addresses the concepts and background needed to study this subject. It presents pertinent definitions concerning analog sampled data signals. It also includes a thorough presentation of the interrelationships between continuous time and discrete time frequency domains. These relationships will be used at every point along the way in this study.

In Chapter 2, the reader is introduced to circuits that implement basic signal processing operations by using switched capacitor circuits. Most of the circuits presented here and in the remainder of the book can be built only by using capacitors and MOS transistors. Of course, since the MOS transistor can be used as a switch or combined with other transistors to form an op amp, the basic components of switched capacitor circuits are capacitors, switches and op amps. Most of the circuits presented in this chapter will be used in later chapters. The methods of analyzing switched capacitor circuits presented in this chapter form the basis for the computer-assisted methods discussed in the following chapter.

Chapter 3 is the first of three chapters devoted to the use of switched capacitor circuits for implementing filters. It provides a background of continuous time active filter design since this is the primary starting point for switched capacitor filter designs. Fig. 3.1–16 identifies and categorizes the many filter synthesis techniques that follow. Whenever possible, the synthesis technique under study is referred back to this figure. Next, the chapter addresses the methods used to characterize discrete time filters. This discussion is followed by techniques that use switched capacitor simulations of resistors. These filters are easily understood from their continuous time counterparts. The last section of the chapter presents several methods for analyzing switched capacitor circuits by using a computer. An available means of analyzing complex switched capacitor circuits in the frequency domain is a very important capability in the design of switched capacitor circuits.

Chapter 4 presents a simple method of realizing higher order filters, starting with passive RLC prototype filters. This method is very general and permits low-pass, bandpass, high-pass, band-elimination, and general RLC filter realizations. A unique feature of this presentation is its avoidance of signal flow graphs. Although signal flow graphs can be very useful in achieving realizations, many manipulations are required, and these manipulations can become very complex, particularly for high-pass filters. In the present approach, all manipulations are done at the equation level using the concept of state variables. The result is hopefully a simpler and more general method. For most of the filters presented in the first part of Chapter 4, it is assumed that the clock frequency is much higher than the filter passband frequencies. The last section of this chapter addresses this problem and shows how to design filters when the clock frequency is not much greater than the filter passband frequencies.

The last chapter on switched capacitor filters is Chapter 5. This chapter presents filters that are designed in the discrete time domain. The method adopted transforms the requirements of continuous time domain filters into the discrete time domain (z-domain). Realizations of the filter are developed in the z-domain and designed to meet the transformed requirements. The result is a filter that is not influenced by the clock frequency. This method yields filters that have a higher frequency capability but are typically more complex to design. Several schemes to permit the realization of higher order filters are presented. Chapter 5 also examines the nonideal performance of general switched capacitor filters caused by switches, parasitics, and the dynamics of the operational amplifier. In sum, Chapters 3, 4, and 5 represent a general introduction to the design of switched capacitor filters.

Nonfilter applications of switched capacitor circuits are the concern of Chapter 6. It is important to realize that switched capacitor techniques are useful for the design of general analog circuits and systems. This chapter addresses the subjects of comparators, rectification, waveshaping, modulation, multiplication and division, oscillators, phase shifting and equalization networks, phase locked loops, and programmable switched capacitor circuits. It has been written with the intent of showing that analog circuits in general are compatible with MOS technology.

Digital-to-analog and analog-to-digital converters are the subject of Chapter 7. Since switched capacitor circuits have established analog signal processing circuits as a viable means of signal processing that uses MOS technology implementation, the system designer can select either a digital or analog approach to system design depending upon his particular requirements. As a consequence, the need to convert between analog and digital signals has increased. The prime objective of this chapter is to present the techniques of conversion between analog and digital signals that are compatible with MOS technology.

The last chapter presents the details of analog MOS circuit design. This information is very important if the switched capacitor circuit is to be implemented with MOS technology. Even if the designer of a system is not involved in the design of its various components, it is necessary that he understand the influence of the components on switched capacitor circuits. Chapter 8 first presents a brief review of MOS technology and MOS modelling. Next, it considers the MOS switch. Finally, it presents design techniques for analog MOS circuits that lead to the design of MOS operational amplifiers.

Over the past two years, the material in this book has been classroom tested not only at Texas A&M, but at the National Institute of Astrophysics, Optics and Electronics in Puebla, Mexico, at California State University at San Luis Obsipo, and at Georgia Institute of Technology. Since there is more material than can be taught in a one-semester senior or graduate level course on the subject, it can be handled in one semester only by skipping some of the topics or otherwise reducing the coverage. A one-semester course emphasizing filters, for example, would include Chapters 1 through 5 and perhaps pertinent portions of Chapter 8. A one-semester course emphasizing switched capacitor circuits would include all of Chapters 1 and 2, selected portions of Chapters 3 through 5, all of chapter 6, and pertinent sections of Chapters 7 and 8.

If appropriate fabrication facilities are available, the material could be taught as a two-semester course, with design, simulation, fabrication, and testing made a part of the sequence. In this case, the instructor should quickly move into the more advanced topics so that the design of the project to be fabricated can begin during the first semester; he can then come back and fill in the details during the second semester while the project is being fabricated. A reasonable approach for the first semester would be a brief presentation of Chapters 1 and 2, the first section of Chapter 3, and all of Chapters 4 and 8. The second semester would then incorporate a review of Chapter 2, the remainder of Chapter 3, and all of Chapters 5 through 7.

A reasonably large number of problems and examples have been included to help illuminate important concepts. Their presence should make the text attractive for self-study as well as for a classroom environment. The prerequisite for the text is a basic course on electronics and circuit theory. A background in sampled data systems and MOS technology will be helpful, but it is not necessary.

The reader should be encouraged to supplement his reading with the construction of various circuits using op amps, precision capacitors, and analog switches. Many lessons that may not be altogether clear in the text will become immediately obvious on the bench. This book contains sufficient information to allow one to design and fabricate switched capacitor circuits wherever such fabrication facilities are available. Although a discrete switched capacitor circuit represents an integrated switched capacitor circuit reasonably well, there are some significant differences (mostly because of smaller capacitances), and these cannot be appreciated until the circuits have been tried.

The authors gratefully acknowledge the generosity and support of Texas Instruments, Inc., who made available at no cost a consistent and quick turn-around fabrication capability that has inspired the completion of this text. The dedication and patience of Mr. Herman van Beek of Texas Instruments, Inc., in coordinating the fabrication efforts were invaluable. The support and encouragement of Dr. W. B. Jones, Jr., Head of the Department of Electrical Engineering at Texas A&M University, and of colleagues in the Department of Electronics at the Institute of Astrophysics, Optics and Electronics in Puebla, is gratefully acknowledged. One of the authors (ESS) would like to thank the Organization of American States for partial financial support of research in this field.

The authors would also like to express their appreciation and gratitude to all the other persons who have contributed to the development of this book. These include many graduate students who suffered through early versions of the text at Texas A&M University, the Institute of Astrophysics, Optics, and Electronics, and Georgia Institute of Technology. Particular thanks must go to Prof. J. A. Connelly, Prof. Gustav Wassel, and Jose Silva-Martinez. The authors would also like to acknowledge the typing assistance of Keri Ridge, Debbie Hemphill, Margarita Gomez-Sarabia, and Patricia Sánchez-Garcia, as well as the excellent artwork by Selma Campos-Garcia. Finally, our special thanks go to our wives and families for their patience and encouragement during the preparation of the manuscript.

Phillip E. Allen
Edgar Sánchez-Sinencio

Contents

SWITCHED CAPACITOR CIRCUITS

1
Analog Sampled Data Concepts

1.1 INTRODUCTION

This book is about the analysis and design of analog circuits using periodic sampling techniques. This class of circuits is called *analog sampled data circuits*. The periodic sampling of an analog signal results in a sequence of pulses. The amplitudes of these pulses correspond to the amplitude of a continuous signal at the time it was sampled. Processing of analog signals by analog sampled data circuits amounts to the processing of each pulse in the sequence of the sampled analog signal.

The periodic sampling of analog signals has been used for many years. Such functions as transversal filters, shift registers, analog-to-digital converters, and control systems are among the many examples of analog sampled data circuit applications. The first known record of sampling analog signals is found on pages 420–425 of *Treatise on Electricity and Magnetism* by James Clerk Maxwell (see Bibliography) where he developed Eq. (20) of Sec. 2.2 in his discussion on the equivalent resistance of a periodically switched capacitor. The theory of sampling analog signals was well developed by the late 1950s. In the 1960s several schemes were proposed that used switches and capacitors to simulate filters.[1] It was shown that the filter transmittances had the important property of depending only on capacitor ratios. However, the importance and significance of many of the ideas proposed for analog sampled data signal processing had to wait until technology provided the means of turning ideas into practical reality. In 1972, it was suggested that MOS technology would be applicable for the construction of analog sampled data filters.[2] This suggestion was then followed by a rapid development in the implementation of analog signal processing circuits using analog sampled data techniques and MOS technology.[3] As a result, almost every analog signal

---

[1] A. Fettweis, "Realization of General Network Functions using the Resonant-Transfer Principle," *Proc. Fourth Asilomar Conf. on Circuits and Systems,* Pacific Grove, CA, Nov. 1970, pp. 663–666.

[2] D. L. Fried, "Analog Sample-Data Filters," *IEEE J. of Solid-State Circuits,* Vol. SC-7, No. 4, August 1972, pp. 302–304.

[3] D. A. Hodges, P. R. Gray, and R. W. Brodersen, "Potential of MOS Technologies for Analog Integrated Circuits," *IEEE J. of Solid-State Circuits,*" Vol. SC-13, No. 3, June 1978, pp. 285–294.

processing circuit built by continuous methods can also be built using analog sampled data techniques combined with MOS technology. The result is an economical means of mass producing analog signal processing functions, such as amplification, addition, subtraction, multiplication, division, nonlinear waveshaping, oscillation, filtering, and many other functions.

This chapter is an introduction to the basic concepts and background for studying the analysis and design of analog sampled data networks. Pertinent definitions concerning analog sampled data signals are presented. Emphasis will be placed on the interrelationships between the continuous time frequency domain and the discrete time frequency domain. These relationships will be used throughout our study of analog sampled data circuits.

1.2 SAMPLED SIGNALS

A signal is a detectable value of voltage, current, or charge and is a key quantity in our study. In this section we wish to define several types of signals that are pertinent to analog sampled data circuits. A *continuous time* signal is a function $y(t)$ that is defined over a continuous range of time, although the amplitude might contain discontinuities. An *analog signal* is both continuous in time and amplitude. Under this definition, an analog signal could be considered as a particular case of a continuous time signal. Commonly, the terms *continuous time* and *analog* are frequently interchanged in usage.[4] Both terms are often used to mean the same thing.

A *discrete time signal* is a function that is defined only over a particular set of values of time. There are two important subcases of a discrete time signal. The first case occurs when the amplitude of a discrete time signal is permitted to assume a continuous range of values. This function is called a *sampled data signal.* The second case occurs when the amplitude is defined only at a particular set of values. This function is said to be a *digital signal.* A digital signal is represented by a sequence of numbers in which each number has a finite number of digits, which corresponds to a finite number of amplitude states. For example, an analog signal may be sampled every T seconds and converted to a digital signal with a given number of bits, say N. Consequently there are only 2^N possible binary numbers that can represent the analog signal amplitude at any point in time. By sampling, we mean that the value of $y(t)$ for values of $t = nT$ is measured or obtained in some manner.

Figure 1.2–1 illustrates the concepts presented above. Figure 1.2–1(a) shows a continuous time signal, and Fig. 1.2–1(b) is a periodic signal, $p(t)$, with a sampling frequency of $f_s = 1/T$. A sampled data signal can be developed

[4] L. R. Rabiner et al., "Terminology in Digital Signal Processing," *IEEE Trans. Audio and Electroacoustics,* Vol. AU-20, December 1972, pp. 323–337.

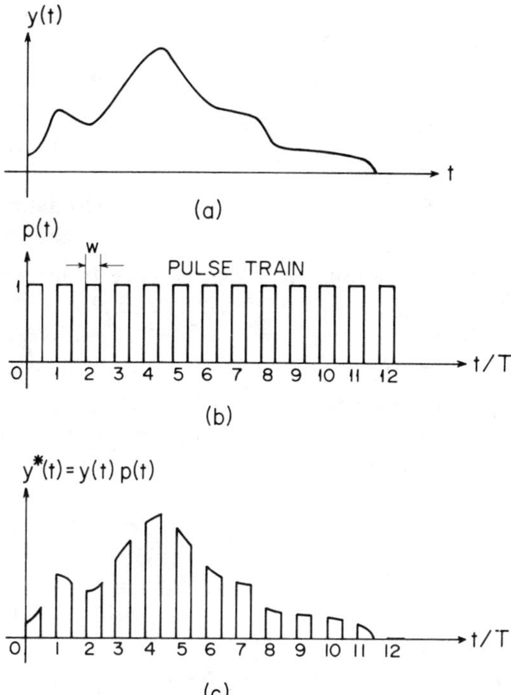

Fig. 1.2–1. (a) Continuous time signal $y(t)$, (b) periodic signal, (c) sampled-data signal $y^*(t)$.

by sampling a continuous time signal at periodic intervals of time T. This process is illustrated in Fig. 1.2–1(c): $y^*(t)$ represents a sampled data signal and is the product of $y(t)$ and a pulse train $p(t)$. The multiplication of two signals in the time domain is shown by the block diagram in Fig. 1.2–2. In a communication system, this particular form of a sampled data signal is called a *pulse amplitude modulated (PAM) signal.* It is assumed that each sample has a width W, so that the resulting signal $y^*(t)$ consists of a series of relatively narrow pulses whose amplitudes are modulated by the original analog signal.

A similar development for a digital signal using a zero width pulse sampling can be derived. This case is illustrated in Fig. 1.2–3, where $N = 3$ bits have been used, which correspond to 2^3 or 8 possible states. The dotted curve is included for the purpose of comparison. Note that amplitudes of the impulses do not necessarily fall on the original curve.

In many cases the sampled signal in an analog sampled data circuit has a piecewise constant amplitude, rather than the form of Fig. 1.2–1. The physical reason for this is that a signal stored at one location (such as charge on a capacitor) is often completely transferred to another location (another

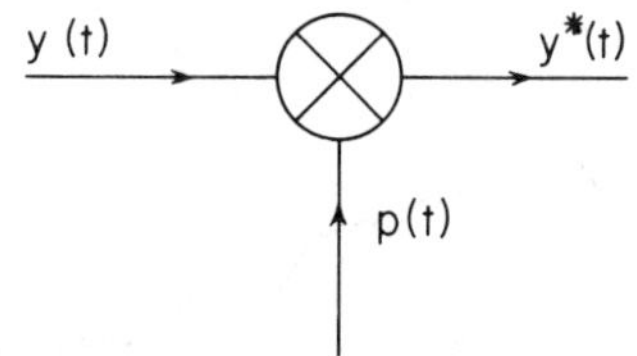

Fig. 1.2–2. Symbol for the multiplication of two signals, $y^*(t) = y(t) \cdot p(t)$.

capacitor) each sampling period. In most cases this transfer occurs in much less time than W of Fig. 1.2–1. The resulting sampled data signal $y_H(t)$, is shown in Fig. 1.2–4(a). In some cases the waveform is sampled, and that sample value is held until the next sample. Such a waveform is called a *zero-order sample-and-hold* signal and is illustrated in Fig. 1.2–4(b). In our study of analog sampled data circuits, most analog sampled data signals will either have the form of Fig. 1.2–1(c) or Fig. 1.2–4(a) with $W \leq T/2$.

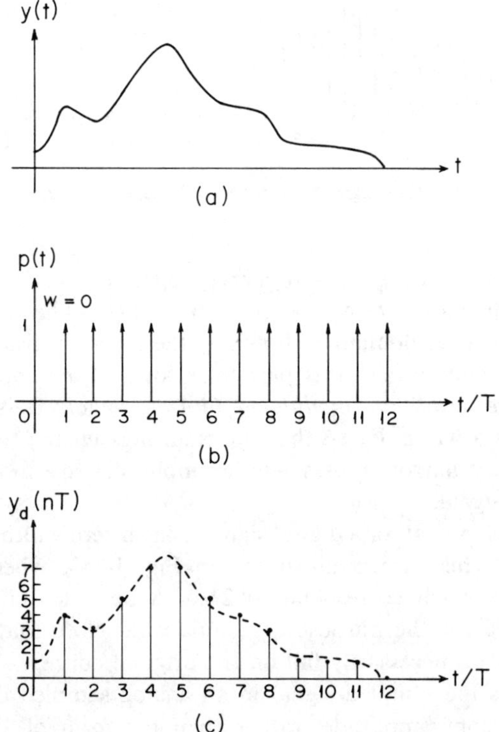

Fig. 1.2–3. (a) Analog signal $y(t)$, (b) periodic signal $p(t)$, (c) digital signal obtained as the product $y(t) \cdot p(t)$.

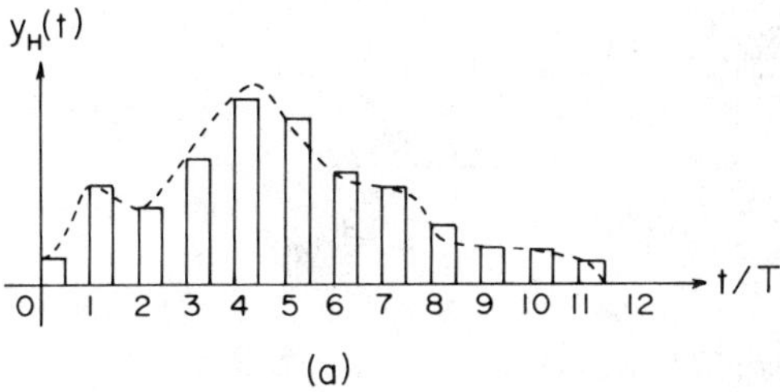

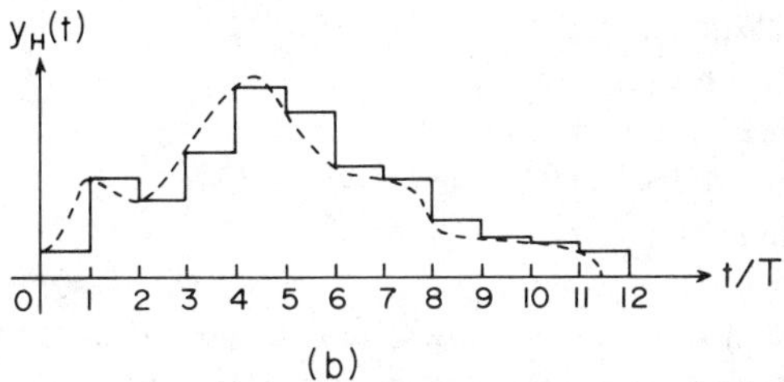

Fig. 1.2–4. Examples of sampled data signal. (a) Constant amplitude and (b) zero-order sample-and-hold.

When analog sampled data signals have the form of Fig. 1.2–4(b), the time the sample is transferred to the next storage location becomes important. Figure 1.2–5 shows a sinusoid that has been sampled and is a zero-order sample-and-hold signal. Because the amplitude of the sample is available during the entire clock period, the transfer to the next storage location could occur at any time during this period. However, a constant delay in time can occur, and it might produce undesirable results if the sampling period is not much smaller than the period of the sinusoid (or any periodic signal). For example, if the sample is transferred at the end of the period, as shown by the right-hand sinusoid, it has a significant phase shift with respect to

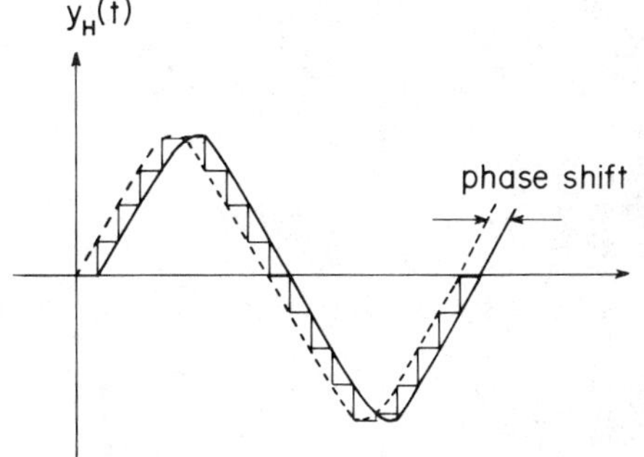

Fig. 1.2–5. An illustration of phase shift in a sample-and-hold signal.

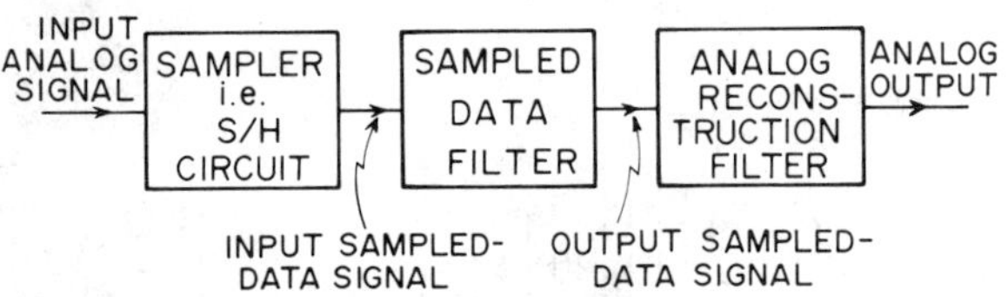

Fig. 1.2–6. Processing of analog signal by a sampled data filter.

the leftmost sinusoid. The amount of phase shift depends upon the frequency. This effect can be ignored if the sampling period is much smaller than the period of the sampled signal.

To process an analog signal by sampled data techniques, the signal generally must first be band-limited through the use of a low-pass filter, then sampled. Next, the sampled data signal can be processed by a sampled data system. After the desired processing has been accomplished, the signal may be converted back into analog form. A simplified block diagram illustrating the sampled data processing of analog signals is shown in Fig. 1.2–6.

Because sampled data signals are often derived from analog signals $y(t)$ by periodic sampling, it is important to understand how the sample data signals $y^*(t)$ are related to the original analog signal $y(t)$. Let us consider Fig. 1.2–7, where $y(t)$ is the analog signal with a sampling frequency f_s, and the relation between f_s and f_0 is $f_s \cong f_0/4$. It can be observed from Fig. 1.2–7 that the corresponding $y^*(t)$ has taken the identity of a lower frequency than that of the original $y(t)$. If the $y^*(t)$ is processed as shown in the block diagram of Fig. 1.2–6, it is impossible to recover the correct output signal having the original frequency f_0. Therefore we can observe that, in general, a signal cannot be uniquely recovered from its sampled values. The process shown in Fig. 1.2–7 is sometimes called *aliasing*. In general aliasing will not occur if the following conditions are met:

1. The analog signal is frequency-limited to a finite range.
2. The sampling rate is much higher than the highest frequency of the analog signal.

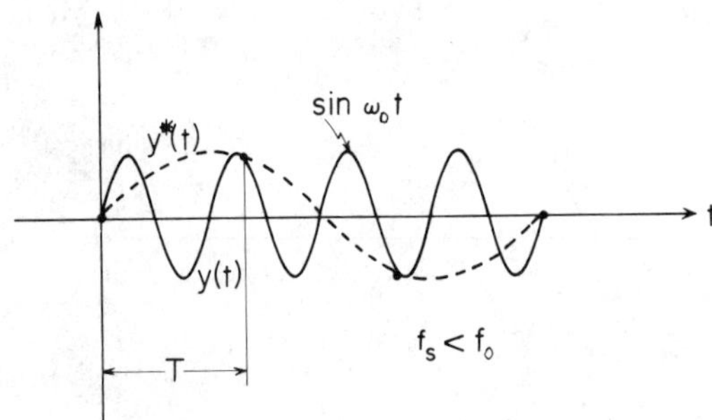

Fig. 1.2–7. Input analog signal and its corresponding sampled data signal.

In practice, condition 1 is difficult to satisfy exactly. However, the spectra of most practical signals can be assumed to be band-limited. As mentioned before, in many sampled data systems it is common practice to filter the analog signal before sampling. This guarantees that the frequency-limited condition is satisfied for all practical purposes.

To avoid aliasing, it is required that $f_s - f_h \geq f_h$, where f_s and f_h are the sampling frequency and the highest frequency contained in the analog signal, respectively. This results in the following condition

$$f_s \geq 2 f_h \tag{1}$$

Equation (1) is known as the *sampling theorem,*[5,6] which requires that the analog signal must be sampled at a rate at least twice the highest frequency in the spectrum.

It is of interest to compare the frequency spectrum of an analog signal and its sampled data representation. Let us recall that the sampled-data signal $y^*(t)$ can be expressed as

$$y^*(t) = y(t) \cdot p(t) \tag{2}$$

The Fourier series expression for $p(t)$ is given as

$$p(t) = \sum_{m=-\infty}^{\infty} C_m e^{j\omega_s mt} \tag{3}$$

where m is an integer defining the order of the harmonic. Therefore Eq. (2) becomes

$$y^*(t) = \sum_{m=-\infty}^{\infty} C_m y(t) e^{j\omega_s mt} \tag{4}$$

The spectrum of $y^*(t)$ can be found by taking the Fourier transform of $y^*(t)$ to get

$$Y^*(\omega) = \sum_{m=-\infty}^{\infty} C_m Y(\omega - m\omega_s) \tag{5}$$

---

[5] W. D. Stanley, *Digital Signal Processing,* Reston Publishing Co., Inc., 1975.

[6] L. R. Rabiner and B. Gold, *Theory and Application of Digital Signal Processing,* Prentice-Hall, Inc., Englewood Cliffs, NJ, 1975.

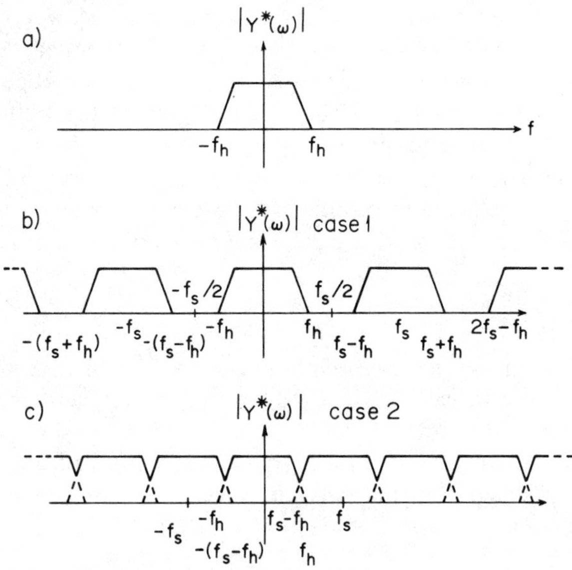

Fig. 1.2–8. Spectra of an analog signal and the sampled data signal obtained from it. (a) Analog signal, (b) sampled data signal satisfying the Nyquist frequency, (c) sampled data signal not satisfying the Nyquist frequency condition $f_s < 2\,f_h$.

The process discussed above is illustrated in Fig. 1.2–8 for the case of an impulse train. It should be noticed that $|Y^*(\omega)|$, which is the spectrum of a sampled data signal, is periodic. It consists of the original spectrum plus an infinite number of translated versions of the original spectrum. For the case of an impulse train, the magnitudes are multiplied by the C_m coefficients, which follow a $2\sin(m\omega_s t)/m\omega_s t$ frequency weighting; $2\,f_h$ is often called the *Nyquist frequency.*

This section has presented some of the basic concepts of analog sampled data circuits to be used in our study of this subject. Examples of the various signals and how they are generated will be found as the reader progresses. The next section will focus on concepts dealing with further relationships and properties of analog and sampled data signals and systems.

1.3 LAPLACE TRANSFORM TECHNIQUES

The linear analysis of an analog signal can be easily accomplished through the use of the Laplace transform. In this section we shall briefly review the Laplace transform and its application. The concepts presented here will be useful in developing analytical methods for analog sampled data circuits, which also use the transform method.

Conventionally, the Laplace transform is considered as an operational method for solving a linear differential equation. After the Laplace transformation, the differential equation becomes an algebraic equation that can be solved by common algebraic operations. The roots of the algebraic equation give information about the stability of the system. The definition of the one-sided Laplace transform is given by

$$Y(s) \triangleq \mathscr{L}[y(t)] = \int_0^\infty y(t)\, e^{-st}\, dt \tag{1}$$

The inverse Laplace transform is defined as

$$y(t) \triangleq \mathscr{L}^{-1}[Y(s)] = \frac{1}{2\pi j} \int_C Y(s)\, e^{st}\, ds \tag{2}$$

where C is a contour in the s-plane, chosen to include all singularities of $Y(s)$. The following property

$$\begin{aligned} \mathscr{L}[y^{(r)}(t)] &= \int_0^\infty \left(\frac{d^r y(t)}{dt^r} \right) e^{-st}\, dt \\ &= s^r Y(s) - \sum_{n=0}^{r-1} s^{r-1-n} \frac{d^n y(t)}{dt^n} \bigg|_{t \to 0} \end{aligned} \tag{3}$$

makes the Laplace transform very useful for solving linear differential equations with constant coefficients. Some common Laplace transforms are given in Table 1.3–1.

Some properties involving analog systems can be developed by the use of the Laplace transforms. Consider the block diagram of Fig. 1.3–1. An excitation, $x_{in}(t)$, is applied to a linear analog system resulting in a response, $y_o(t)$. The transfer function $H(s)$, characterizing the system is given by

$$H(s) = \frac{Y_o(s)}{X_{in}(s)} = \frac{b_m s^m + b_{m-1} s^{m-1} + \cdots + b_0}{a_n s^n + a_{n-1} s^{n-1} + \cdots + a_0}, \qquad m \le n \tag{4}$$

or

$$H(s) = K \frac{\prod_{i=1}^{m} (s - s_{oi})}{\prod_{i=1}^{n} (s - s_{pi})} \tag{5}$$

Table 1.3–1 Some common Laplace transforms

	$Y(s)$	$y(t)$
1	1	$\delta(t)$
2	k/s	k
3	k/s^n	$kt^{n-1}/(n-1)!\ (n=1, 2, \ldots)$
4	$1/(s-a)^n$	$t^{n-1}e^{at}/(n-1)!\ (n=1, 2, \ldots)$
5	$\dfrac{a}{s^2+a^2}$	$\sin at$
6	$\dfrac{s}{s^2+a^2}$	$\cos at$
7	$\dfrac{1}{(s-a)^2+b^2}$	$\dfrac{1}{b} e^{at} \sin bt$
8	$\dfrac{s-a}{(s-a)^2+b^2}$	$e^{at} \cos bt$
9	$\dfrac{1}{s(s^2+a^2)}$	$\dfrac{1}{a^2}(1-\cos at)$
10	$\dfrac{1}{s^2(s^2+a^2)}$	$\dfrac{1}{a^3}(at-\sin at)$

where K is a real gain factor. s_{0i} and s_{pi} are called the *zeros* and the *poles* of the system, respectively. Observe that any s_{pi} of the transfer function $H(s)$ is a natural frequency of the corresponding $y_0(t)$, and the corresponding impulse is a linear combination of the exponentials $e^{s_{pi}t}$. In general, s_{pi} are complex and given as

$$s_{pi} = \sigma_i \pm j\,\omega_i \tag{6}$$

A given σ_i represents the damping factor for an exponential (when it is negative), and ω_i represents the radian frequency of a sinusoidal oscillation. The natural response of the system involves only the poles of $H(s)$. The forced response is characterized by the poles of the excitation, $X_{in}(s)$. Furthermore, the locations of the poles s_{pi} determine the stability of the system. Thus a system is stable if the poles of the transfer function $H(s)$ are in the

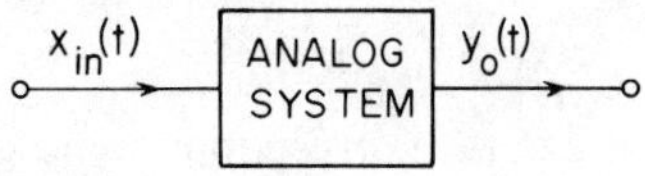

Fig. 1.3–1. Block diagram of an analog signal processing system.

left-hand complex frequency plane, i.e., $\sigma_i < 0$, and the terms $e^{-s_{pi}t}$ eventually vanish as t approaches infinity. Zeros of $H(s)$ are permitted anywhere in the complex frequency plane for most transfer functions. Some examples illustrating these properties follow.

Example 1.3–1. *Stability of a transfer function.* Consider the transfer function of an analog system given as

$$H(s) = \frac{10\, s^2 + 40\, s + 80}{s^2 + 2\, s + 17} \tag{7}$$

The zeros and poles of this transfer function are found to be

$$\begin{aligned} s_{01}, s_{02} &= -2 \pm j2 \\ s_{p1}, s_{p2} &= -1 \pm j4 \end{aligned} \tag{8}$$

Since the poles, s_{p1} and s_{p2} are in the left half of the complex frequency plane, the system is stable. The natural response will correspond to the type having an exponential decay.

Example 1.3–2. *An RLC circuit.* The voltage transfer function of the RLC passive circuit shown in Fig. 1.3–2 can be obtained by considering the circuit to be a voltage divider consisting of two impedances given as

$$Z_1(s) = 1 \qquad \text{and} \qquad Z_2(s) = \frac{4s}{s^2 + 4}$$

Therefore,

$$H(s) = \frac{V_0(s)}{V_{\text{in}}(s)} = \frac{Z_1}{Z_1 + Z_2} = \frac{s^2 + 4}{s^2 + 4\, s + 4} \tag{9}$$

The corresponding poles and zeros are

$$\begin{aligned} s_{01} &= 2_j \\ s_{02} &= -2j \\ s_{p1} = s_{p2} &= -2 \end{aligned}$$

The frequency response of Eq. (9) corresponds to a notch (band-elimination) filter, where $\omega = 2$ is the notch frequency at which there is no signal transmis-

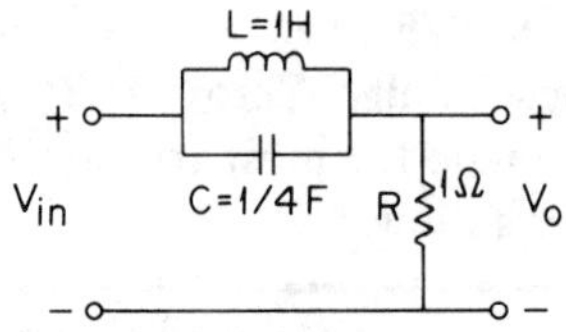

Fig. 1.3–2. RLC passive circuit.

sion through the filter. Again in this case, because the poles are located in the left half of the s-plane, the system is stable. This can be confirmed by obtaining the impulse response, by using entry 4 of Table 1.3–1 to get

$$h(t) = \delta(t) + 4\, e^{-2t}\,(2t - 1) \tag{10}$$

which goes to zero as t approaches infinity.

1.4 Z-TRANSFORM TECHNIQUES

In the previous section, we presented an operational method of solving a continuous time system. In this section we shall develop a similar method of solving a continuous system that is sampled. The resulting operational method will be known as the *z-transform.* As in the previous presentation, our discussion is limited to the one-sided z-transform. The definition of the z-transform is given by

$$Y(z) = \sum_{n=0}^{\infty} y(nT)z^{-n} \tag{1}$$

for all z for which the series $Y(z)$ converges. Basically all the properties of the Laplace transform apply, with some minor modifications, to the z-transform. This is illustrated in Table 1.4–1. Note that when a complex variable, z, is expressed in polar form as $z = re^{j\omega T}$, where ω is given in radians per second and T in seconds, then Eq. (1) can be expressed as

$$Y(re^{j\omega T}) = \sum_{n=0}^{\infty} y(nT)(re^{j\omega T})^{-n} \tag{2}$$

For the particular case when $r = 1$, Eq. (2) yields

$$Y(e^{j\omega T}) = \sum_{n=0}^{\infty} y(nT)\, e^{-j\omega nT} \tag{3}$$

Table 1.4–1 Summary of Some z-transform Properties

	z-TRANSFORM	SEQUENCE
1	$a\,X(z) + b\,V(z)$	$ax(n) + bv(n)$
2	$z^{-n_1}\,Y(z)$	$y(n - n_1)$
3	$Y(z/b)$	$b^n\,y(n)$
4	$-z\dfrac{dY(z)}{dz}$	$n\,y(n)$
5	$Y(z^{-1})$	$y(-n)$
6	$X(z)\,V(z)$	$x(n) * v(n)$†
7	$\dfrac{1}{2\pi j}\displaystyle\int_c X(u)V\left(\frac{z}{u}\right)u^{-1}\,du$	$x(n)\,v(n)$

† The symbol * denotes convolution.

which can be seen as the *Fourier transform* of the sequence $y(nT)$. Furthermore from Eq. (1) we can observe that if the sequence $y(nT)$[7] is the coefficient of z^{-n}, then the z-transform can be directly obtained.

Example 1.4–1. *Obtaining a sequence from a z-transform.* Consider the z-transform

$$Y(z) = \frac{1}{(1 - bz^{-1})^2}, \quad |z| > |b|$$

Carrying out the division, we obtain

$$\begin{aligned} Y(z) &= 1 + 2bz^{-1} + 3b^2z^{-2} + 4b^3z^{-3} + \cdots \\ &= \sum_{n=0}^{\infty} (n+1)\,b^n z^{-n} \end{aligned}$$

Therefore

$$y(n) = (n+1)\,b^n\,u(n)$$

Several common z-transforms are summarized in Table 1.4–2.

---

[7] For simplicity we will use $y(n) = y(nT)$.

Table 1.4–2 Simple z-transforms

	$Y(z)$	$y(n)$
1	1	$\delta(n)$
2	$\dfrac{kz}{z-a}$	$k\,u(n)\,a^n$
3	$\dfrac{Tz}{(z-1)^2}$	nT
4	$\dfrac{z}{z-e^{-bt}}$	e^{-nbt}
5	$\dfrac{z \sin bT}{z^2 - 2z \cos bT + 1}$	$\sin nbT$
6	$\dfrac{z^2 - z \cos bT}{z^2 - 2z \cos bT + 1}$	$\cos nbT$

Next, it is shown that the operator z^{-1} corresponds to a unit delay in the sequence of discrete samples in the time domain. Multiplying Eq. (1) by z^{-1} results in

$$z^{-1}Y(z) = z^{-1} \sum_{n=0}^{\infty} y(n)z^{-n} \tag{4}$$

which can be modified to

$$z^{-1}Y(z) = \sum_{n=0}^{\infty} y(n)z^{-(n+1)} \tag{5}$$

Let us start the summation with $n = 1$ and change accordingly $y(n)$ and $z^{-(n+1)}$ so that

$$z^{-1}Y(z) = \sum_{n=1}^{\infty} y(n-1)z^{-n} \tag{6}$$

or

$$z^{-1}Y(z) = \sum_{n=0}^{\infty} y(n-1)z^{-n} - y(-1) = Y[y(n-1)] - y(-1) \tag{7}$$

where $Y[y(n)]$ means the z-transform of the sequence $y(n)$. Furthermore, if $y(n) = 0$ for $t < 0$, then $y(-1) = 0$, and we obtain

$$z^{-1}\,Y(z) = z^{-1}Y[y(n)] = Y[y(n-1)] \tag{8}$$

Therefore, each sample in the sequence, $Y(z)$, has been delayed by T seconds. In a similar manner we can derive the general relation for delay as

$$z^{-m}Y(z) = Y[y(n-m)] \tag{9}$$

Let us now consider the *stability* of a second-order homogeneous difference equation given as

$$y(n) + a_1 y(n-1) + a_2 y(n-2) = 0 \tag{10}$$

The characteristic equation is given by

$$(1 + a_1 z^{-1} + a_2 z^{-2})\,Y(z) = 0 \tag{11}$$

or

$$z^2 + a_1 z + a_2 = (z - z_{p1})(z - z_{p2}) = 0 \tag{12}$$

where

$$z_{p1},\ z_{p2} = -\frac{a_1}{2} \pm \sqrt{(a_1/2)^2 - a_2} \tag{13}$$

If the roots are complex, i.e., $(a_1/2)^2 < a_2$, then the poles can be conveniently expressed in polar form as

$$z_{p1},\ z_{p2} = x \pm jy = re^{\pm j\theta} \tag{14}$$

where

$$r = (x^2 + y^2)^{1/2} = \sqrt{a_2} \tag{15}$$

and

$$\cos\theta = \frac{x}{r} = \frac{-a_1}{2\sqrt{a_2}} \tag{16}$$

The characteristic equation can be rewritten as

$$z^2 - 2r\cos\theta\, z + r^2 = 0 \tag{17}$$

The natural response for a system with complex roots has the following form

$$y(n) = k_1 r^n \cos(n\theta) + k_2 r^n \sin(n\theta) \tag{18}$$

or

$$y(n) = Kr^n \cos(n\theta + \phi) \tag{19}$$

where r and θ are determined by the parameters a_1 and a_2. K and ϕ are determined by the initial conditions.

From Eqs. (18) and (19), the system will be stable if and only if $r < 1$. The region of stability in the plane (a_1, a_2) is shown in Fig. 1.4–1. In general, a discrete system is stable if all its poles lie inside the unit circle.

A sampled data system, such as that shown in Fig. 1.4–2, can be characterized by its transfer function, $H(z)$. The general transfer function of a sampled data system can be described by

$$H(z) = \frac{Y(z)}{X(z)} = \frac{b_m z^m + b_{m-1} z^{m-1} + \cdots + b_0}{a_n z^n + a_{n-1} z^{n-1} + \cdots + a_0} \tag{20}$$

or

$$H(z) = K \frac{\prod_{i=1}^{m} (z - z_{0_i})}{\prod_{i=1}^{n} (z - z_{pi})}, \qquad m \leq n \tag{21}$$

where K is the gain factor. z_{0i} and z_{pi} are the zeros and poles of the system, respectively. Stability can be determined by the poles of the transfer function.

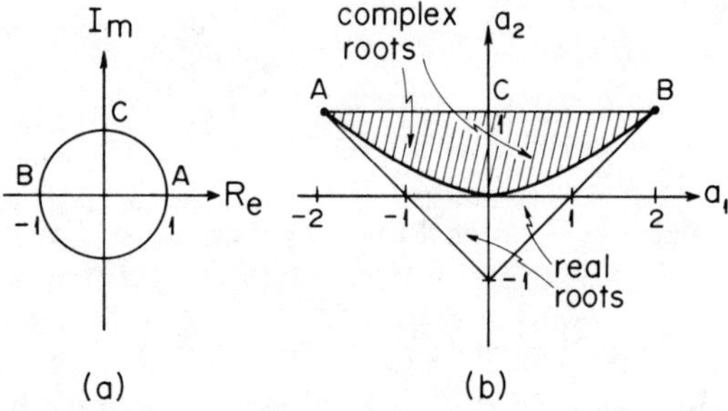

Fig. 1.4–1. Relationship between the poles in (a) the z-plane and (b) coefficients a_1 and a_2.

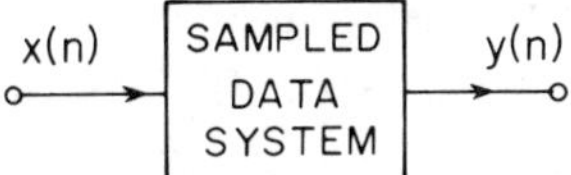

Fig. 1.4–2. Block diagram of a sampled data system.

In general the impulse response of the system involves a linear combination of terms of the form z_{pi}^n. Therefore, the only condition for the impulse response to eventually vanish (when $t \to \infty$) and make the system stable is that $|z_{pi}| < 1$. Thus a discrete system is stable if all its poles lie inside the unit circle. The location of the zeros, in general, is not critical regarding stability and can lie anywhere in the z-plane.

The frequency spectrum $H_d(e^{j\omega T})$ of a discrete impulse response $h(nT)$ can be found through the use of the Fourier transform and can be written as

$$H_d(e^{j\omega T}) = \frac{1}{T}\sum_{l=0}^{\infty} H(j\omega + jl\omega_c) \tag{22}$$

where the subscript d denotes discrete and $H(j\omega)$ is the frequency response[8] of the continuous signal $h(t)$. Furthermore, $H_d(e^{j\omega T})$ can be expressed through the Fourier sum as

$$H_d(e^{j\omega T}) = \sum_{n=0}^{\infty} h(nT)e^{-jn\omega T} \tag{23}$$

It can be observed that $H_d(e^{j\omega T})$ is a periodic function of frequency ω_c. This property follows from the fact that $e^{j(\omega+\omega_c)nT} = e^{j\omega nT}$. In the z-plane, a transfer function is evaluated for physical frequencies by substituting $z = e^{j\omega T}$. Thus, the function is evaluated over the unit circle. Note that in practice $H(z)$ can be evaluated from $z = 1$ to $z = -1$, which is equivalent for frequencies from 0 to $\omega_c/2$. After $\omega > \omega_c/2$ the frequency spectrum is mirrored about $\omega_c/2$. Note that both ω_c and ω_s will be used throughout the text to mean the clock or sampling frequency.

The following examples illustrate the concepts presented in this section.

Example 1.4–2. *Stability of a sampled data system.* Consider a system described by the difference equation

$$y(n) - 1.2728\,y(n-1) + 0.81\,y(n-2) = x(n) + x(n-1) \tag{24}$$

[8] A. S. Sedra and P. O. Brackett, *Filter Theory and Design: Active and Passive,* Matrix Publishers, Portland, OR, 1978.

The transfer function $H(z)$ is determined by taking the z-transform of Eq. (24) and forming the ratio of $Y(z)/X(z)$ to get

$$H(z)=\frac{1+z^{-1}}{1-0.9\sqrt{2}\,z^{-1}+0.81\,z^{-2}} \tag{25}$$

or

$$H(z)=\frac{z(z+1)}{(z-0.9\,e^{-j45^\circ})(z-0.9\,e^{j45^\circ})} \tag{26}$$

The poles are located at $0.9\,\underline{/\pm 45^\circ}$, which are inside the unit circle. Therefore the system is stable.

Example 1.4–3. *Impulse response of a sampled data system.* Consider the transfer function

$$H(z)=\frac{Az(1-e^{-\alpha_1})}{z^2+[(A-1)-(A+1)e^{-\alpha_1}]z+e^{-\alpha_1}} \tag{27}$$

and find the impulse response, $h(n)$, of this system. The transfer function may be expressed as

$$H(z)=\frac{Az(1-e^{-\alpha_1})}{(z+\alpha+j\beta)(z+\alpha-j\beta)} \tag{28}$$

Expanding $H(z)$ in partial fractions results in

$$H(z)=\frac{A(1-e^{-\alpha_1})}{2j\beta}\left[\frac{-z}{(z+\alpha+j\beta)}+\frac{z}{z+\alpha-j\beta}\right] \tag{29}$$

Using Table 1.4–2, we obtain

$$h(n)=\frac{A(1-e^{-\alpha_1})}{2j\beta}[(-(\alpha-j\beta))^n-(-(\alpha+j\beta))^n]$$

If $\alpha \pm j\beta = -re^{\pm j\omega}$, then the impulse response of $H(z)$ is

$$\begin{aligned} h(n)&=\frac{A(1-e^{-\alpha_1})}{\beta}(r)^n\,\frac{e^{-j\omega n}-e^{j\omega n}}{2j} \\ &=\frac{A(1-e^{-\alpha_1})}{\beta}(r^n)\sin n\omega \end{aligned} \tag{30}$$

Assume that $A = 10$ and $\alpha_1 = 1/10$. The impulse response becomes

$$h(n) = -1.156(-0.95123)^n \sin 5.23n$$

Example 1.4–4. *Obtaining the z-domain transfer function.* Consider the sequence defined by

$$x(n) = \begin{cases} a^n \sin(nb) & n \geq 0 \\ 0 & n < 0 \end{cases} \tag{31}$$

Applying Eq. (1) to $x(n)$ gives

$$\begin{aligned} X(z) &= \sum_{n=0}^{\infty} a^n \sin(nb)\, z^{-n} \\ &= \sum_{n=0}^{\infty} \frac{a^n}{2j} (e^{jnb} - e^{-jnb})\, z^{-n} \\ &= \left\{ \sum_{n=0}^{\infty} (ae^{jb} z^{-1})^n - \sum_{n=0}^{\infty} (ae^{-jb} z^{-1})^n \right\} \frac{1}{2j} \end{aligned} \tag{32}$$

Both series converge if $|ae^{jb} z^{-1}| = |ae^{-jb} z^{-1}| < 1$ or $|z| > |a|$. Therefore

$$X(z) = \frac{1}{2j} \left[\frac{1}{1 - ae^{jb} z^{-1}} - \frac{1}{1 - ae^{-jb} z^{-1}} \right] \tag{33}$$

Thus we obtain

$$X(z) = \frac{a \sin(b)\, z^{-1}}{(1 - ae^{jb} z^{-1})(1 - ae^{-jb} z^{-1})}, \quad |z| > |a| \tag{34}$$

Example 1.4–5. *Obtaining the z-domain transfer function of a sequence.* Assume a sequence is given as

$$y(n) = \begin{cases} n\, k^n & n \geq 0 \\ 0 & n < 0 \end{cases} \tag{35}$$

and obtain $Y(z)$. Applying Eq. (1) to $y(n)$ gives

$$Y(z) = \sum_{n=0}^{\infty} n\, k^n z^{-n} = -z \frac{d}{dz} \left(\sum_{n=0}^{\infty} k^n z^{-n} \right) \tag{36}$$

Since

$$\sum_{n=0}^{\infty} k^n z^{-n} = \sum_{n=0}^{\infty} (k z^{-1})^n = \frac{1}{1 - kz^{-1}} \tag{37}$$

we have

$$\frac{d}{dz}\sum_{n=0}^{\infty} k^n z^{-n} = \frac{d}{dz}\left(\frac{1}{1-kz^{-1}}\right) = \frac{-kz^{-2}}{(1-kz^{-1})^2} \tag{38}$$

Therefore substituting this in Eq. (36) results in

$$Y(z) = \frac{kz^{-1}}{(1-kz^{-1})^2} \tag{39}$$

In this section we have introduced the z-transform and illustrated its application. The z-transform provides a convenient operational method of analyzing sampled data circuits. Some of the properties of the z-domain pertinent to our study were presented. These properties include stability, roots, the relationship between the sequence and its z-domain, and the evaluation of the z-transfer function. In the next section we examine the relationship between the continuous domain and the discrete domain.

1.5 RELATIONSHIPS BETWEEN THE CONTINUOUS AND DISCRETE DOMAINS

In the previous two sections, concepts of the continuous time and discrete time domains have been presented. In this section we wish to develop the relationships between these domains. The motivation for this study is found in fact that analog signal processing circuits are frequently specified in the continuous time domain, but realized using discrete time domain techniques. For example, many analog signal processing functions can be described very succinctly in the continuous frequency domain. It therefore becomes important to develop relationships between the continuous frequency domain and the sampled data frequency domain. To avoid confusion, we shall use the term *continuous* or *analog frequency* when the corresponding time domain is continuous and the term *discrete* or *sampled data frequency* when the corresponding time domain is discrete.

One technique of converting a continuous system to a discrete system is to replace the differential equations by difference equations that approximate the differential equations. Consider a set of differential equations given as

$$\sum_{i=0}^{n} \frac{d^i y(t)}{dt^i} b_i = \sum_{i=0}^{m} \frac{d^i x(t)}{dt^i} a_i, \qquad n \geq m \tag{1}$$

Equation (1) can be written in discrete form as

$$\sum_{i=0}^{n} \Delta^i[y(n)]\, b_i = \sum_{i=0}^{m} \Delta^i\, [x(n)]\, a_i \tag{2}$$

where

$$\Delta^0\, [y(n)] = y(n) \tag{3}$$

$$\Delta^1\, [y(n)] = \frac{1}{T}\, [y(n) - y(n-1)] \qquad \text{(backward difference)} \tag{4}$$

or

$$\Delta^1\, [y(n)] = \frac{1}{T}\, [y(n+1) - y(n)] \qquad \text{(forward difference)} \tag{5}$$

and

$$\Delta^i\, [y(n)] = \Delta^1[\Delta^{i-1}[y(n)]] \tag{6}$$

Observe that $\Delta^1\,[y(n)]$ can be obtained by either the *backward difference* or the *forward difference.* Taking the z-transform of each side of Eq. (2), we obtain

$$H(z) = \frac{Y(z)}{X(z)} = \frac{\sum_{i=0}^{m} a_i\, W^i(z)}{\sum_{i=0}^{n} b_i\, W^i(z)} \tag{7}$$

where

$$W(z) = \frac{1}{T}\,(1 - z^{-1}) \qquad \text{(backward difference)} \tag{8}$$

or

$$W(z) = \frac{1}{T}\left(\frac{1 - z^{-1}}{z^{-1}}\right) \qquad \text{(forward difference)} \tag{9}$$

In a similar manner, taking the Laplace transform of Eq. (1), we have

$$H(s) = \frac{\sum_{i=0}^{m} a_i s^i}{\sum_{i=0}^{n} b_i s^i} \tag{10}$$

Comparing Eqs. (7) and (10), results in the following relations between the continuous complex frequency variable, s, and the discrete complex variable z. For the backward difference, we get

$$s = \frac{1 - z^{-1}}{T} \tag{11}$$

or

$$z = \frac{1}{1 - Ts} \tag{12}$$

For the forward difference, we have

$$s = \frac{1}{Tz^{-1}}(1 - z^{-1}) \tag{13}$$

or

$$z = 1 + Ts \tag{14}$$

Pictorically we can represent these mappings as shown in Fig. 1.5–1. We note that the forward mapping is capable of transforming stable poles in the analog domain into unstable poles in the discrete domain.

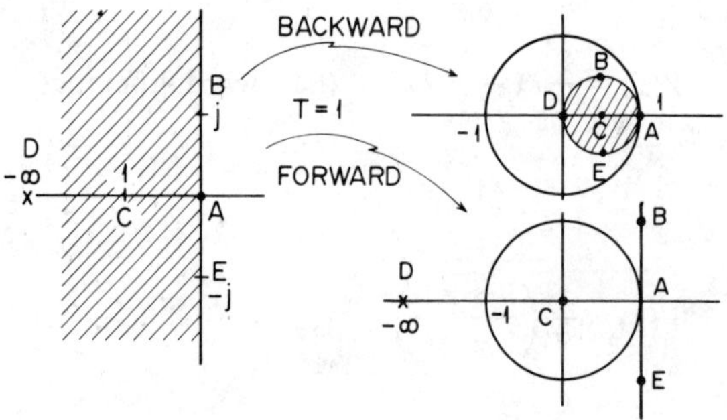

Fig. 1.5–1. Illustration of the backward and forward mappings.

A third transformation, which uses a combination of the forward and backward transformations, is the *bilinear transformation.* This transformation is based on integrating the differential equation and then approximating the integrals numerically by the area of the trapezoidal whose bases are $y(n-1)T$ and $y(n)T$ and whose height is T. The resulting bilinear transformation is

$$s = \frac{2}{T}\frac{1-z^{-1}}{1+z^{-1}} \tag{15}$$

or

$$z = \frac{1+(T/2)s}{1-(T/2)s} \tag{16}$$

This bilinear mapping is illustrated in Fig. 1.5–2.

Comparing the bilinear transformation with the backward or forward transformation shows that only the bilinear transformation maps the $j\omega$ axis of the continuous frequency domain onto the unit circle of the discrete frequency domain. We note that $H(s)\big|_{s=j\omega}$ for $-\infty \leq \omega \leq \infty$ is compressed into the interval $-\pi/T \leq \omega \leq \pi/T$ by the bilinear transformation. This mapping may be developed by replacing z in Eq. (15) by $e^{j\omega T}$ to obtain

$$s = \frac{2}{T}\left[\frac{1-e^{-j\omega T}}{1+e^{-j\omega T}}\right] = \frac{2}{T} j \tan\left(\frac{\omega T}{2}\right) = \sigma + j\Omega_a \tag{17}$$

Since $\sigma = 0$, then

$$\omega = \omega_d = \frac{2}{T}\tan^{-1}\left(\frac{\Omega_a T}{2}\right) \tag{18}$$

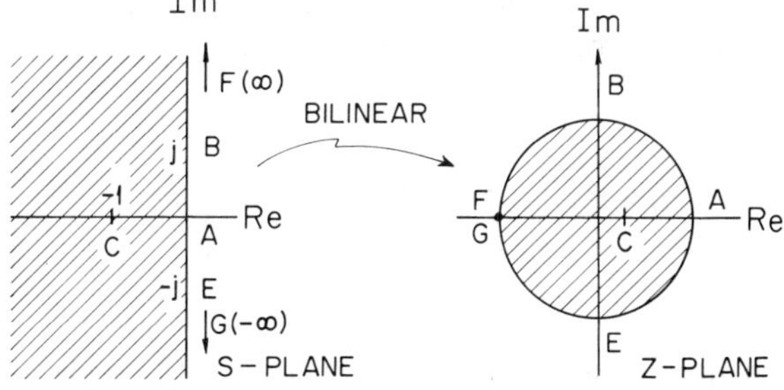

Fig. 1.5–2. Illustration of the bilinear mapping.

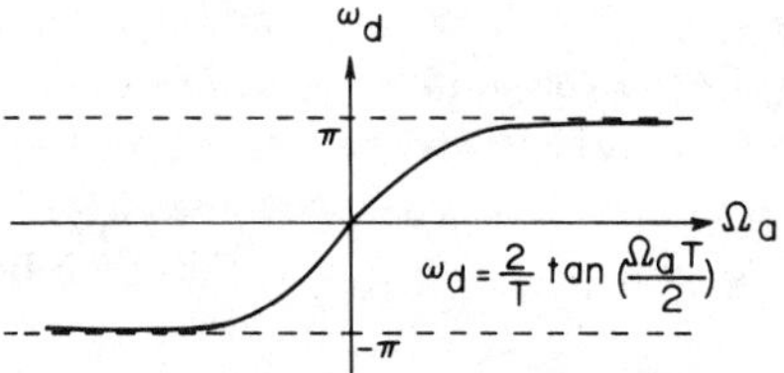

Fig. 1.5–3. Mapping of the analog frequency axis Ω_a onto the unit circle using the bilinear transformation.

This relationship is illustrated in Fig. 1.5–3, where ω_d and Ω_a are the sampled data and analog frequencies, respectively. It is seen that the bilinear transformation introduces a distortion between the analog and discrete frequencies. Equation (18) will be useful as we begin to consider the problem of realizing in one domain specifications that are given in the other domain.

A fourth transformation is called the *impulse invariant* or *impulse invariance transformation.* This transformation is based on the fact that

$$h_d = h(n) = h_a(nT) \tag{19}$$

It can be shown that

$$z = e^{sT} \tag{20}$$

or

$$s = \frac{1}{T} \ln z \tag{21}$$

which is the impulse invariant transformation. The validity of Eq. (20) or Eq. (21) may be demonstrated by making a partial fraction expansion of the analog transfer function, $H_a(s)$, to obtain the corresponding $h_a(nT)$. Using Eq. (19) results in the unit-sample response, $h(n)$. The corresponding $H(z)$ may be obtained, resulting in Eq. (20) or Eq. (21). We can express $H(z)$ as

$$H(z) = \frac{1}{T} \sum_{k=-\infty}^{\infty} H\left(s + jk\frac{2\pi}{T}\right)\Bigg|_{s=\frac{1}{T}\ln z} \tag{22}$$

Therefore, each horizontal strip of $2\pi/T$ width in the s-plane is overlayed onto the z-plane to form the discrete function from the analog function.

Unfortunately, this transformation is not a simple one-to-one algebraic mapping from the s-plane to the z-plane.

The concepts of the previous four transformations are illustrated in the following example. Additional relationships between $H(s)$ and $H(z)$ via the different transformations are left as a problem (see Problem 1.12).

Example 1.5–1. *Illustration of the four types of transformations.* Given the poles $p_{1,2}$ in the s-plane, obtain the corresponding poles in $z_{1,2}$ in the z-domain, using the four transformations discussed in this section. First consider the RHP poles

$$p_1 = 1 - j$$

$$p_2 = 1 + j$$

If we assume $T = 1$, then the poles in the z-domain are

1. backward transformation

$$z_{1,2} = \frac{1}{1 - p_{1,2}} = \frac{1}{1 - 1 \pm j} = \pm j \tag{23}$$

2. forward transformation

$$z_{1,2} = 1 + p_{1,2} = 1 + 1 \pm j = 2 \pm j \tag{24}$$

3. bilinear transformation

$$z_{1,2} = \frac{1 + (½)p_{1,2}}{1 - (½)p_{1,2}} = \frac{3 \pm j}{1 \pm j} = 1 \pm 2j \tag{25}$$

4. impulse invariant transformation

$$z_{1,2} = e^{1 \pm j} \tag{26}$$

These results are illustrated in Fig. 1.5–4.

Next consider the LHP poles

$$p_1 = -1 - j$$

$$p_2 = -1 + j$$

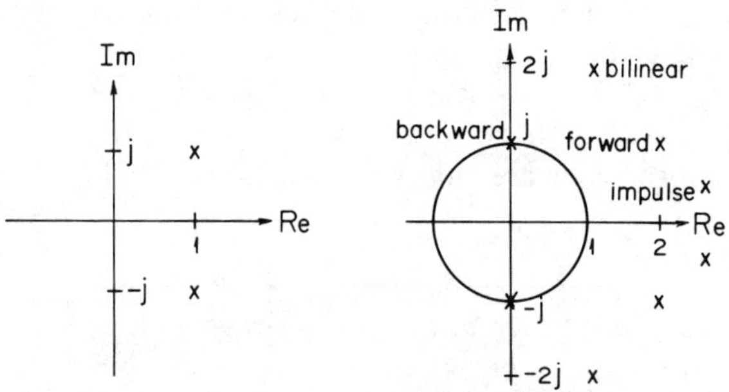

Fig. 1.5–4. Example of mapping right-hand poles from the s- to the z-plane.

Again consider $T = 1$; then the poles in the z-domain are:

1. backward transformation

$$z_{1,2} = \frac{1}{1 + 1 \pm j} = \frac{1}{5}(2 \pm j) \tag{27}$$

2. forward transformation

$$z_{1,2} = 1 - 1 \pm j = \pm j \tag{28}$$

3. bilinear transformation

$$z_{1,2} = \frac{1 + \frac{1}{2}(-1 \pm j)}{1 - \frac{1}{2}(-1 \pm j)} = \frac{1}{5}(1 \pm 2j) \tag{29}$$

4. impulse invariant transformation

$$z_{1,2} = e^{-1 \pm j} \tag{30}$$

Fig. 1.5–5 illustrates the results of the LHP poles.

Example 1.5–2. *Computation of* H(s) *and* H(z). Given the following differential equation

$$a_1 \frac{dx(t)}{dt} + a_0 x(t) = b_0 y(t)$$

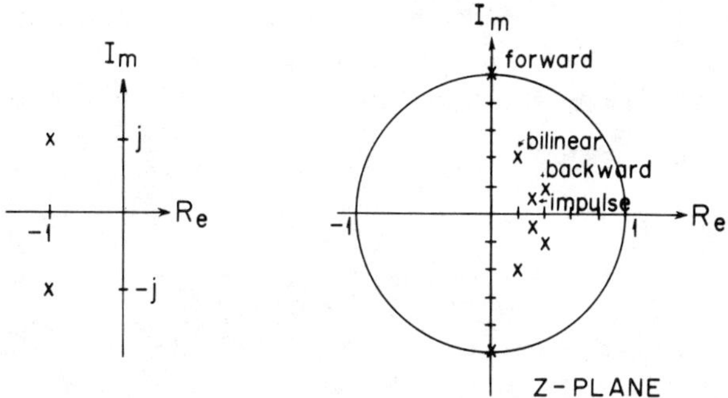

Fig. 1.5–5. Example of mapping left-hand poles from the s-plane to the z-plane.

1. Find the transfer functions $H(s) = X(s)/Y(s)$, and compute its magnitude and phase.
2. Using the bilinear transformation, compute the magnitude and phase of $H(z) = X(z)/Y(z)$.

The solution to this example is as follows.

1. From the differential equation and using the Laplace transform, we obtain

$$H(s) = \frac{b_0}{a_1 s + a_0}$$

and its magnitude and phase are given by

$$|H(s)| = \frac{b_0}{\sqrt{a_0^2 + (\omega a_1)^2}}$$

$$\underline{/H(s)} = -\tan^{-1}\frac{\omega a_1}{a_0}$$

2. Transforming $H(s)$ to $H(z)$ through the bilinear mapping results in

$$H(z) = \frac{b_0 T(1 + z^{-1})}{2a_1 + a_0 T + z^{-1}(a_0 T - 2a_1)}$$

or

$$H(z) = \frac{b_0 T(z + 1)}{(a_0 T - 2a_1) + (2a_1 + a_0 T)z}$$

In the z-plane, a transfer function is evaluated for physical frequencies by substituting $z = e^{j\omega T}$. Observe that the transfer function is periodic and evaluated over the unit circle, that is, $0 \leq \omega T \leq \pi$ and that the point $z = -1$ in the z-plane corresponds to $\omega = \dfrac{\pi}{T} = \dfrac{\omega_c}{2}$.

$$|H(e^{j\omega})| = \frac{\sqrt{\sin^2\omega T + (1+\cos\omega T)^2}}{\sqrt{[a_0T - 2a_1 + (2a_1 + a_0T)\cos\omega T]^2 + [(2a_1 + a_0T)\sin\omega T]^2}}\, b_0T$$

$$\underline{/H(e^{j\omega})} = \theta_z - \theta_p$$

$$\theta_z = \text{arc tan}\left(\frac{\sin\omega T}{1+\cos\omega T}\right)$$

$$\theta_p = \text{arc tan}\left(\frac{(2a_1 + a_0T)\sin\omega T}{a_0T - 2a_1 + (2a_1 + a_0T)\cos\omega T}\right)$$

1.6 RELATIONSHIPS BETWEEN THE S- AND Z-DOMAINS FOR HIGH SAMPLING RATES

The problem of obtaining a function $H(s)$ given $H(z)$ is not straightforward if the original transformation from the s- to the z-plane is not known. However, when the sampling rate is much greater than the frequencies of interest, various assumptions can be made to obtain approximate relationships between the two planes. In this section we shall develop these approximate relationships and illustrate their application.

Consider the s-plane pole, s_p. The corresponding pole in the z-plane, z_p, can be obtained through any of the four conventional transformations. The z-plane pole is expressed in terms of the s-plane pole using the four transformations as

$$z_1 = \frac{1}{1 - s_pT} \text{ (backward)} \tag{1}$$

$$z_2 = 1 + s_pT \text{ (forward)} \tag{2}$$

$$z_3 = \frac{1 + s_pT/2}{1 - s_pT/2} \text{ (bilinear)} \tag{3}$$

$$z_4 = e^{s_pT} \text{ (impulse)} \tag{4}$$

Now for small $|s_p|T$, i.e., $|s_p|T < 1$, the previous expressions can be written as

$$z_1 = 1 + s_pT + (s_pT)^2 + (s_pT)^3 + \cdots \tag{5}$$

$$z_2 = 1 + s_pT \tag{6}$$

$$z_3 = \left(1 + \frac{s_pT}{2}\right)(1 + s_pT/2 + (s_pT/2)^2 + (s_pT/2)^3 + \cdots$$

or

$$z_3 = 1 + s_pT + \frac{(s_pT)^2}{2} + \frac{(s_pT)^3}{4} + \cdots \tag{7}$$

and

$$z_4 = 1 + s_pT + \frac{(s_pT)^2}{2} + \frac{(s_pT)^3}{6} + \cdots \tag{8}$$

For high sampling rates, the second-order effects can be ignored, so all the above four transformations can be roughly approximated as

$$z_p \cong z_1 \cong z_4 \cong z_3 \cong z_2 = 1 + s_pT \tag{9}$$

It can be noticed that as the sampling period T is decreased, z_p approaches the point $z = 1$. Figure 1.6–1 illustrates the relationships between s_p and z_p in both planes for the case of a pair of complex poles.

Next the relationships between both planes are considered. The poles in the s-domain are characterized by

$$s^2 + \frac{\omega_n}{Q}s + \omega_n^2 = (s + a)^2 + b^2 \tag{10}$$

where a and b are shown in Fig. 1.6–1 and expressed as

$$a = \frac{\omega_n}{2Q} \tag{11}$$

and

$$b = \frac{\omega_n}{2Q}\sqrt{4Q^2 - 1} \tag{12}$$

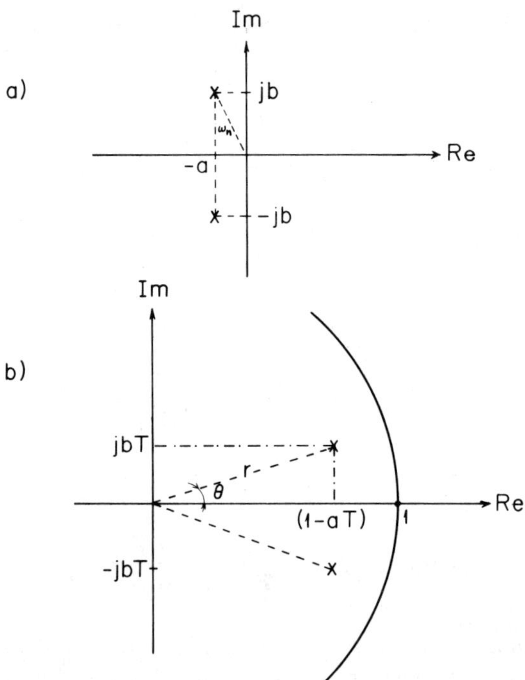

Fig. 1.6–1. Relations between the s- and z-plane for small sT; (a) s-plane, (b) z-plane.

Thus, the locations of the poles are

$$s_{p_1}, s_{p_2} = -a \pm jb = -\frac{\omega_n}{2Q} \pm j\frac{\omega_n}{2Q}\sqrt{4Q^2 - 1} \tag{13}$$

The poles in the z-domain are characterized by the equation

$$z^2 + b_1 z + b_2 = 0 \tag{14}$$

or

$$z^2 - 2r\cos\theta\, z + r^2 = 0 \tag{15}$$

Using the approximation of Eqs. (9) and (13) and for $Q \geq 1$, one gets

$$z_{p_i} = 1 + s_{p_i}T \tag{16}$$

or

$$z_{p_1}, z_{p_2} = (1 - aT) \pm jbT \tag{17}$$

The poles in the z-domain can also be expressed in polar coordinates as

$$z_{p_1}, z_{p_2} = r\, e^{\pm j\theta} \tag{18}$$

where

$$r = \sqrt{(1 - aT)^2 + (bT)^2} \tag{19}$$

or

$$r = \sqrt{1 - \frac{\omega_n T}{Q} + (\omega_n T)^2} \tag{20}$$

From Eq. (20), one can obtain the expression for Q, which is

$$Q = \frac{\omega_n T}{1 - r^2 + (\omega_n T)^2} = \frac{\omega_n T}{(1 - r)(1 + r) + (\omega_n T)^2} \tag{21}$$

θ can be given in terms of ω_n and Q by Eqs. (13), (18), and (19) and is

$$\theta = \tan^{-1} \frac{bT}{1 - aT} = \tan^{-1} \frac{\dfrac{\omega_n T}{2Q}\sqrt{4Q^2 - 1}}{1 - \dfrac{\omega_n T}{2Q}} \tag{22}$$

$$= \tan^{-1} \frac{\omega_n T \sqrt{4Q^2 - 1}}{2Q - \omega_n T}$$

Note that Eqs. (20), (21), and (22) are approximations for the case of high sampling rate of the backward, bilinear, and impulse invariant transformations; however, no approximation is made regarding the forward transformation. For the forward mapping, the transformation of Eq. (9) is not an approximation, and expressions (21) and (22) are exact. Clearly if we do not use the forward transformation, one can get different expressions for r and θ.

A summary of the above results is given in Table 1.6–1. θ was obtained for the backward and forward[9] difference, assuming that $\tan\theta \approx \theta$, i.e.,

---

[9] The expressions of f_n and Q for the forward difference were obtained assuming $r^2 \cong 1 - \dfrac{\omega_n T}{Q}$ and $\theta \cong \dfrac{\omega_n T\sqrt{4Q^2 - 1}}{2Q}$. Similarly for the backward difference $r^2 \cong 1/(1 + \omega_n T/Q)$ and $\theta \cong \dfrac{\omega_n T\sqrt{4Q^2 - 1}}{2Q}$.

Table 1.6–1 High Sampling Rate Comparison

PARAMETER	TRANSFORMATIONS FORWARD	BACKWARD	IMPULSE INVARIANCE
r	$[1+(\omega_n T)^2-\omega_n T/Q]^{1/2}$	$[1+(\omega_n T)^2+\omega_n T/Q]^{-1/2}$	$\exp(-\omega_n T/2Q)$
θ	$\dfrac{\omega_n T[4Q^2-1]^{1/2}}{2Q-\omega_n T}$	$\dfrac{\omega_n T[4Q^2-1]^{1/2}}{2Q+\omega_n T}$	$\dfrac{\omega_n T[4Q^2-1]^{1/2}}{2Q}$
f_n	$\dfrac{f_s}{2\pi}\left[\theta^2+\dfrac{(1-r^2)^2}{4}\right]^{1/2}$	$\dfrac{f_s}{2\pi}\left[\theta^2+\left(\dfrac{1}{r^2}-1\right)^2/4\right]^{1/2}$	$\dfrac{f_s}{2\pi}[\theta^2+\ln^2 r]^{1/2}$
Q	$\dfrac{\omega_n T}{1-r^2+(\omega_n T)^2}$	$\dfrac{\omega_n T}{\dfrac{1}{r^2}-1-(\omega_n T)^2}$	$-\dfrac{\omega_n T}{2\ln r}$

see Eq. (22). In practice, all the expressions in Table 1.6–1 are good approximations for small θ and for r approaching unity. The expressions for the bilinear transformation are left as an exercise (see Problem 1.13).

Another useful approximation, not just for high sampling rate where $\omega_n T << 1$, but also assuming Q greater than unity (i.e., r approaching unity), is given as follows. It can be observed from Table 1.6–1 that for $Q >> 1$, all the cases reduce to

$$\theta \cong \omega_n T \tag{23}$$

If r approaches unity, Eq. (21) can be written as

$$Q \cong \frac{\omega_n T}{2(1-r)+(\omega_n T)^2} \tag{24}$$

Now if the term $(\omega_n T)^2$ in Eq. (24) can be neglected, i.e., $2(1-r) >> (\omega_n T)^2$, Eq. (24) yields

$$Q \cong \frac{\omega_n T}{2(1-r)} = \frac{\theta}{2(1-r)} \tag{25}$$

Observe that Eq. (25) is an approximation of Eq. (21). Equation (25) also could be derived from a different expression for Q obtained from the impulse invariance case where the approximation $\ln r \cong -(1-r)$ can be used to obtain Eq. (25).

As we will see in the following chapters, relations between b_1 and b_2 of Eq. (14) with θ and Q of Eqs. (23) and (25) will be useful in the design of sampled data circuits. Thus from Eq. (25) we can write

$$r \cong 1 - \frac{\theta}{2Q} \tag{26}$$

and cos θ can be approximated as

$$\cos \theta \cong 1 - \frac{\theta^2}{2} \tag{27}$$

Thus making use of Eqs. (26), (27), (14), and (15), we can write the following expressions

$$b_1 = -\left(2 - \frac{\theta}{Q} - \theta^2\right) \tag{28}$$

$$b_2 = \left(1 - \frac{\theta}{2Q}\right)^2 \tag{29}$$

These expressions are valid at high sampling rates and for large values of Q. If b_2 is approximated by

$$b_2 \cong 1 - \frac{\theta}{Q} \tag{30}$$

then we can write the following expressions

$$\theta^2 = 1 + b_2 + b_1 \tag{31}$$

and

$$Q = \frac{\theta}{1 - b_2} \tag{32}$$

The relationships developed in this section will be useful in the later chapters of the book where, in many practical problems, the sampling rate is high and the roots are close to $r = 1$. These relationships will provide considerable simplification.

1.7 SUMMARY

This chapter has presented the basic concepts of analog sampled data circuits and has developed the necessary tools to be used in the analysis and design of analog sampled data circuits. We have focused on the relationships between the continuous and discrete domains, because they are very pertinent to our study. Other concepts necessary for the following material will be developed as required. Much more detailed results are available in many existing books that deal with the subject of sampled data control systems or digital signal processing. A list of these texts is found in the bibliography.

PROBLEMS

1.1 (Sec. 1.2). Determine the transfer function $H(j\omega)$ for the zero-order sample-and-hold circuit. Assume that the impulse response is given by

$$y(t) = \frac{1}{T}[u(t) - u(t - T)]$$

Plot magnitude and phase of $H(j\omega)$ vs. ω.

1.2 (Sec. 1.3). Find the mathematical expression for the quantized, continuous time signal shown in Fig. P1.2 and its corresponding Laplace transforms.

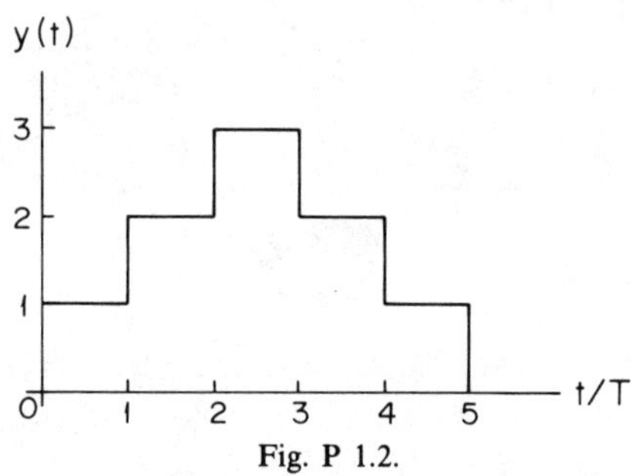

Fig. P 1.2.

1.3 (Sec. 1.3). Plot the magnitude of $H(s)$ as a function of ω ($s = j\omega$) for Eq. (7) of Sec. 1.3. Find the impulse and step response of this transfer function.

1.4 (Sec. 1.3). Obtain a differential equation relating i_1 and i_2 in the circuit of Fig. P1.4. Find the transfer function $I_2(s)/I_1(s)$.

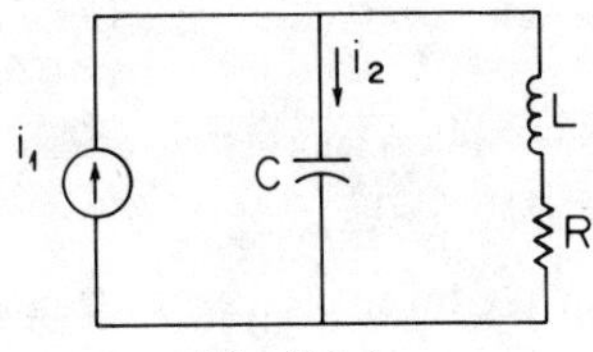

Fig. P 1.4.

1.5 (Sec. 1.3). The initial and final value theorems for the s-domain are

$$\lim_{t \to 0+} h(t) = \lim_{s \to \infty} sH(s)$$

and

$$\lim_{t \to \infty} h(t) = \lim_{s \to 0} sH(s)$$

respectively. Evaluate the initial and final values of

$$H(s) = \frac{Ks}{s^2 + \dfrac{\omega_0}{Q}s + \omega_0^2}$$

1.6 (Sec. 1.4). Find the inverse z-transform of

a. $H(z) = \left(\dfrac{T}{2}\right)\dfrac{1 + z^{-1}}{1 - z^{-1}}$

b. $H(z) = \dfrac{z(z+1)}{z^2 - 1.2z + 0.72}$

c. $H(z) = \dfrac{0.5}{(1 - z^{-1})(1 - 0.5z^{-1})}$

d. $H(z) = \dfrac{z^2}{z^2 - 2r\cos\theta\, z + r^2}$

1.7 (Sec. 1.4). The initial and final value theorems for the z-domain are

$$\lim_{t \to 0} y^*(t) = \lim_{z \to \infty} Y(z)$$

and

$$\lim_{t \to \infty} y^*(t) = \lim_{z \to 1} (1 - z^{-1})Y(z)$$

respectively. Evaluate the initial and final time values of

$$H(z) = \frac{k(z+1)(z-1)}{z^2 - (2r\cos\theta)z + r^2}$$

1.8 (Sec. 1.4). Prove the initial and final value theorems given in Prob. 1.7.

1.9 (Sec. 1.4). Apply the final value theorem given in Prob. 1.7 to $H(z)$ of Example 1.4–3 to find $h(\infty)$. Plot $h(n)$ vs. n to verify your results.

1.10 (Sec. 1.4). Prove that the system given by Eq. (8) of Sec. 1.4 is stable by obtaining the impulse response and using the final value theorem.

1.11 (Sec. 1.4). Obtain the z-transforms $Y(z)$ of the following discrete time signals:

1. $y(n)=\begin{cases} k & \text{for} \quad 0 \le n \le m \\ 0 & \text{otherwise} \end{cases}$

2. $y(n)=-y(n+1)+T\,y(n-1)$

3. $$y(n)=\begin{cases} 0 & \text{for} \quad n<0 \\ 1 & 0<n \le T/2 \\ 2 & T/2<n \le T \\ 4 & n>T \end{cases}$$

4. $y(n)=u(n)-u(n-1)$

1.12 (Sec. 1.5). Given the transfer function $H(s)$

$$H(s)=\frac{k}{s^2+\dfrac{\omega_0}{Q}s+\omega_0^2}$$

a. Obtain the corresponding $H(z)$ through the use of the forward, backward, bilinear, and impulse invariant transformations

$$H(z)=\frac{\text{numerator}}{1+b_1z^{-1}+b_2z^{-2}}$$

b. Prove that the relationships between ω_0, Q, and b_1 and b_2 for the different transformations are given by Table P 1.12.

1.13 (Sec. 1.5). The transfer function $H(s)$ is given as

$$H(s)=\frac{\dfrac{\omega_0}{Q}s}{s^2+\dfrac{\omega_0}{Q}s+\omega_0^2}$$

where $\omega_0=2\,\pi$ and $Q=10$.

a. Obtain the corresponding $H(z)$ using the forward, backward, bilinear, and impulse invariant transformations if $T=0.5$ seconds. Plot the magnitude of the sampled data frequency response for each case.
b. Repeat part (1) if $T=0.1$ seconds.
c. Compare the results of parts (1) and (2).

1.14 (Sec. 1.6). Develop the expressions for the bilinear transformation similar to those in Table 1.6–1.

Table P 1.12

	ω_0	Q	b_1	b_2
Forward	$\frac{\sqrt{(1+b_1+b_2)}}{T}$	$\frac{\sqrt{(1+b_1+b_2)}}{b_1+2}$	$-2+\frac{\omega_0 T}{Q}$	$1-\frac{\omega_0 T}{Q}+(\omega_0 T)^2$
Backward	$\frac{\sqrt{\frac{1+b_1+b_2}{b_2}}}{T}$	$-\frac{\sqrt{b_2(1+b_1+b_2)}}{b_1+2b_2}$	$-\frac{2+\frac{\omega_0 T}{Q}}{1+\frac{\omega_0 T}{Q}+(\omega_0 T)^2}$	$\frac{1}{1+\frac{\omega_0 T}{Q}+(\omega_0 T)^2}$
Bilinear $a=2/T$	$a\sqrt{\frac{1+b_1+b_2}{1-b_1+b_2}}$	$\frac{\sqrt{(1+b_1+b_2)(1-b_1+b_2)}}{2(1-b_2)}$	$\frac{2(\omega_0^2-a^2)}{a^2+\frac{\omega_0}{Q}a+\omega_0^2}$	$\frac{a^2-\frac{\omega_0}{Q}a+\omega_0^2}{a^2+\frac{\omega_0}{Q}a+\omega_0^2}$
Impulse invariant	$\frac{\sqrt{\left(\cos^{-1}\frac{-b_1}{2\sqrt{b_2}}\right)^2+\frac{1}{4}(\ln b_2)^2}}{T}$	$\frac{\sqrt{\left(\cos^{-1}\frac{-b_1}{2\sqrt{b_2}}\right)^2+\frac{1}{4}(\ln b_2)^2}}{-\ln b_2}$	$-2e^{-\frac{\omega_0 T}{2Q}}\cos\left(\frac{\omega_0 T}{2Q}\sqrt{4Q^2-1}\right)$	$e^{-\frac{\omega_0 T}{Q}}$

1.15 (Sec. 1.6). Given a pair of complex poles in the s-domain located at $p = -1 \pm j1$, find the corresponding complex poles in the z-domain using the backward and bilinear mapping for

a. $T = ½$ seconds, and
b. $T = 0.01$ seconds.

Obtain the approximate values of Q and f_n for the different transformations and different sample periods T.

2
Implementation of Basic Signal-Processing Operations

In this chapter we introduce some of the analog sampled data realizations for basic analog signal-processing functions. These functions include amplifiers, summers, delays, sample-and-hold, integration, and differentiation. Examples of some simple first-order circuits are also presented. Also we shall introduce methods of analyzing analog sampled data circuits and show how the transformations of the previous chapter are related to the realizations of the basic signal-processing operations.

In this chapter we shall develop switch capacitor networks that simulate continuous analog circuits. The operating concept of these circuits and their analyses will be presented. The op amp is combined with switches and capacitors to complete the repertoire of components necessary for analog sampled data circuits. The circuits developed point out the need for general methods of analysis.

One of the primary advantages of analog sampled data circuits is that they provide a means of economically and accurately implementing analog circuit functions with existing integrated circuit technology. Continuous analog circuits contain resistors, capacitors, and active devices. Unfortunately, the performance of these circuits depends upon the accuracy of the resistors and capacitors. In filters this becomes a serious problem, because the RC product must be accurately defined for satisfactory performance. Present integrated circuit technology cannot provide sufficient absolute value accuracy of the resistors and capacitors without trimming. An additional problem concerns the fact that integrated resistors have poor linearity and temperature characteristics. Besides these undesirable properties, large values of time constants require large values of resistance, which in turn require large areas of the integrated circuit. RC active filters have been built using integrated circuit technology, but they require careful trimming techniques and are generally implemented as hybrid circuits. Such techniques can be too costly and have discouraged widespread development of integrated analog signal-processing functions.

Analog sampled data techniques provide a unique solution to the above problems. It will be shown below that the resistor can be replaced by switches and capacitors. This results in the important fact that the circuit performance is determined by *capacitor ratios.* Ratios of elements are always easier to control. Besides this, the capacitor is more suitable to integrated circuit technology than the resistor. The result is a technique of realizing analog signal-processing circuits that will be in use for a long time.

2.1 NOTATION

Before developing an equivalent resistor from switches and capacitors, it is necessary to establish some notation. Figure 2.1–1 shows a network with all switches external to the network. The switch symbol will always be drawn open. It will be assumed that the switches are controlled by the application of various clocks. The number of clocks applied to a system will be designated by n. Typically we say that a system has an n-*phase clock,* i.e., a two-phase clock. Let us take a period of time T and divide it into n equal segments as illustrated in Fig. 2.1–2(a). The period T is called the *clock period.* The segments will be called *phase periods.* Each switch in the system will be designated by the symbol

$$\phi_{k,l,m,} \ldots \tag{1}$$

where $k, l, m, \ldots$ are numbers designating which phase period that particular switch is closed. Typically a switch is closed only once during the clock period, so that ϕ would have only a single numerical subscript. It will also be assumed that the phase periods are separated by a finite period of time in which all the switches are open. This leads to a situation called *nonoverlapping clocks,* which is a very important property for switched capacitor networks. Typically the actual width of the clock waveform that controls the switch is slightly less than the phase period. This ensures that the nonoverlap-

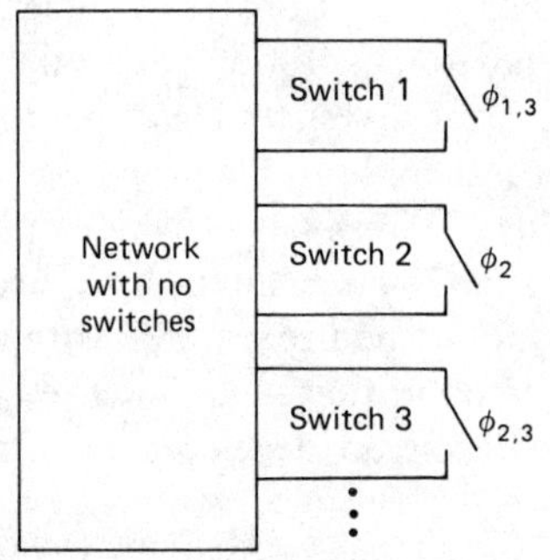

Fig. 2.1–1. Schematic representation of switches.

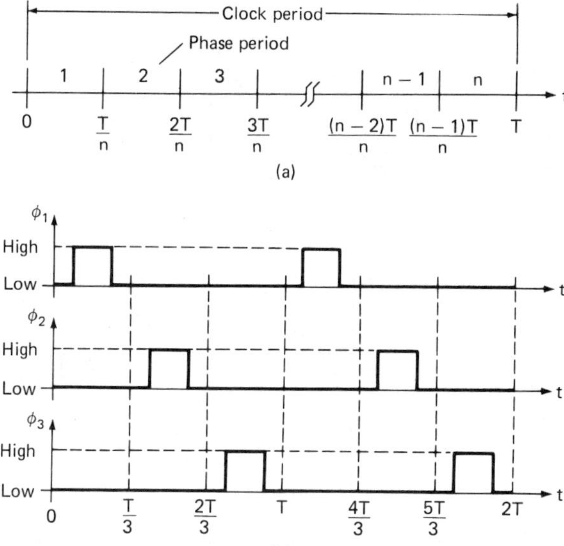

Fig. 2.1–2. (a) Illustration of phase periods of a clock sequence. (b) Waveforms for a typical three-phase clock.

ping property is satisfied. In the following material it will be assumed that when a clock waveform is drawn, such as in Fig. 2.1–2(b), that the switch is closed when the waveform is high and open when the waveform is low. This convention is not always followed by manufacturers of discrete MOS switches.

In general, the networks considered in this textbook contain switches, capacitors, and operational amplifiers. As the switches open and close, a typical biphase network[1] changes alternately between two topologies. We can consider that the time-varying[2] analog sampled data realizations correspond to one topology during the ϕ_2(or ϕ_1) clock phase and to a second topology during the ϕ_1(or ϕ_2) clock. A useful complementary notation for the clock phases is denoted by *even* and *odd,* which can be associated to ϕ_2 and ϕ_1 by definition. Thus, the sampled data waveforms can be expressed as the sum of their even and odd components. To illustrate graphically, consider the zero-order, sample-and-hold waveform in Fig. 1.2–4(b); this is reproduced again with

[1] All the networks considered in this textbook will use biphase switches unless otherwise specified.

[2] C. F. Kurth and G. S. Moschytz, "Two-Port Analysis of Switched-Capacitor Networks using Four-Port Equivalent Circuits in the z-Domain," *IEEE Trans. on Circuits and Systems,* Vol. CAS-26, No. 3, March 1979, pp. 166–179.

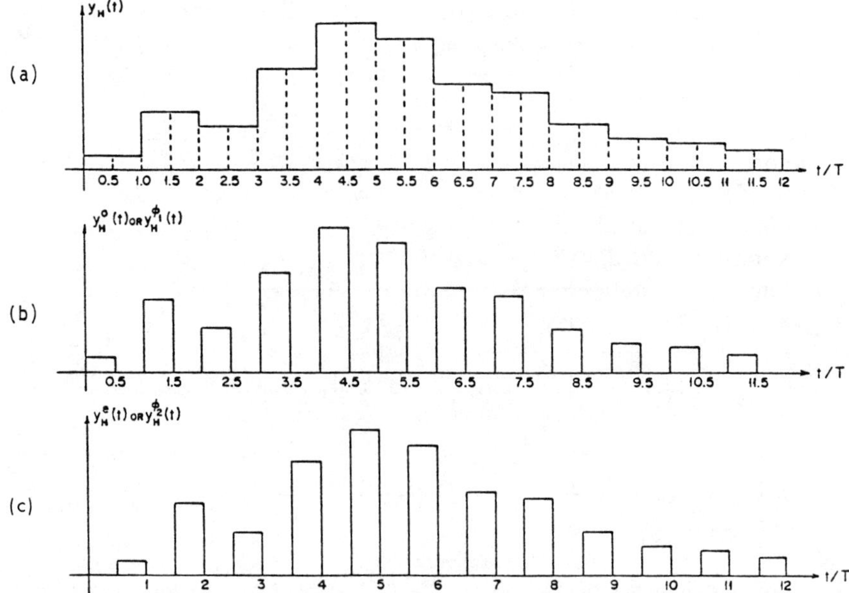

Fig. 2.1–3. (a) Sample-and-hold waveform and its respective (b) odd and (c) even components.

its even and odd components in Fig. 2.1–3. Mathematically it can be expressed as

$$y_H(t) = y_H^o(t) + y_H^e(t) \tag{2}$$

or in the z-domain

$$Y_H(z) = Y_H^o(z) + Y_H^e(z) \tag{3}$$

where

$$Y_H^o(z) = z^{-1/2}\, Y_H^e(z) \tag{4}$$

In fact, the input and output voltages of a time-varying sampled data network can be expressed as

$$V_{in}(z) = V_{in}^e(z) + V_{in}^o(z) \tag{5a}$$

$$V_O(z) = V_O^e(z) + V_O^o(z) \tag{6a}$$

This can be made equivalent to

$$V_{in}(z) = V_{in}(z)\Big|_{\phi_2} + V_{in}(z)\Big|_{\phi_1} \tag{5b}$$

$$V_O(z) = V_O(z)\Big|_{\phi_2} + V_O(z)\Big|_{\phi_1} \tag{6b}$$

Therefore, at least four possible transfer functions are possible. In practice, usually $V_{in}^{o}(z) = 0$ or $V_{in}^{e}(z) = 0$ and similarly for $V_O(z)$. Although, if $V_O(z)$ is sampled at all times, then the effects during the ϕ_1 and ϕ_2 clock phases must be added.

A convenient notation for the transfer functions is

$$H^{ij}(z) = \frac{V_O^j(z)}{V_{in}^i(z)} \tag{7}$$

where i and j can be either e or o. For example, $H^{oe}(z)$ represents $V_O^e(z)/V_{in}^O(z)$. Also, a transfer function $V_O(z)/V_{in}(z)$ can be defined as $(V_O^e(z) + V_O^o(z))/(V_{in}^e(z) + V_{in}^o(z))$.

2.2 SWITCHED CAPACITOR EQUIVALENT RESISTORS

In this section we shall develop realizations of continuous resistors using switched capacitor networks.[3,4] Our objective is to be able to replace the resistors of an RC active network resulting in a sampled data equivalent network. However, we shall find that while the switched capacitor resistors in this section are exactly equivalent to resistors *by themselves,* such an equivalence may not hold true when the realizations of this section are used to replace resistors of an RC active network.[5] We shall learn in the next few sections of this chapter how to determine the performance of a network containing switched equivalent resistors and other components.

Let us begin by considering the network of Fig. 2.2–1(a). Networks such as this, which contain switches, capacitors, and dependent and independent sources, are called *switched capacitor networks.* It is our objective to show

---

[3] J. T. Caves, M. A. Copeland, C. F. Rahim, and S. D. Rosenbaum, "Sampled Analog Filtering Using Switched Capacitors as Resistor Equivalents," *IEEE J. of Solid-State Circuits,* Vol. SC-12, December 1977, pp. 592–599.

[4] B. J. Hosticka, R. W. Brodersen, and P. R. Gray, "MOS Sampled-Data Recursive Filters Using Switched Capacitor Integrators," *IEEE J. of Solid-State Circuits,* Vol. SC-12, December 1977, pp. 600–608.

[5] Y. P. Tsividis, "Analytical and Experimental Evaluation of a Switched Capacitor Filter and Remarks on the Resistor/Switched-Capacitor" (Correspondence), *IEEE Trans. on Circuits and Systems,* Vol. CAS-26, No. 2, February 1979, pp. 141–144.

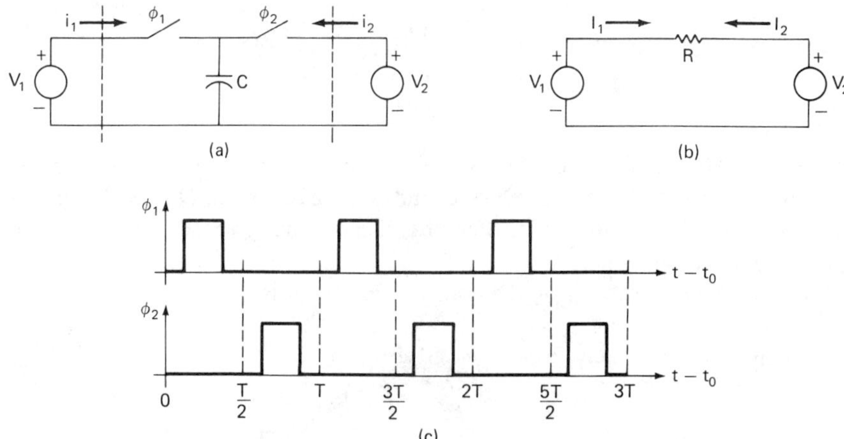

Fig. 2.2–1. (a) Parallel switched capacitor realization of a continuous resistor. (b) Continuous resistor. (c) Clock waveforms for the switched capacitor realization.

that Fig. 2.2–1(a) is equivalent to the circuit of Fig. 2.2–1(b). We begin by assuming that the entire system is inactive. That is, both switches are open and the capacitor C is completely discharged. For convenience sake, we assume that V_1 and V_2 are two independent DC voltage sources. This assumption represents no loss of generality, because V_1 and V_2 will typically be sampled signals similar to those in Fig. 1.2–4, which are constant during most or all of the phase period. The clock waveforms are shown in Fig. 2.2–1(c).

At time $t_0 = nT$, we apply the clocks to Fig. 2.2–1(a). The first clock pulse, ϕ_1, which will occur during the first phase period, will close switch 1. At this time, C will be charged to V_1. In practice, a finite resistance, R, is associated with the switch. This resistance is connected in series with the capacitor and prevents instantaneous charging of C. Obviously, the RC time constant must be much less than the width of ϕ_1 for the charge to be transferred. When C has charged to V_1, we may say that the charge which flowed into C, across the left dotted line in the direction designated by i_1, is equal to

$$Q(t_0 + T/2) = C\,V_1 \tag{1}$$

The time $t_0 + T/2$ is used because we shall assume that, from t_0 to $t_0 + T/2$, switch 1 was closed long enough to charge C to V_1 and is now open.

Next, we progress to the second phase period. When ϕ_2 becomes high, switch 2 will close, and C will become charged to V_2, assuming that the

RC time constant is very small. However, this time the only charge flowing across the right dotted line in the direction designated by i_2 is equal to

$$Q(t_0 + T) = C\,V_2 - C\,V_1 = C(V_2 - V_1) \tag{2}$$

which is the difference between the charge that is placed on C by V_2 and the previous charge that was placed on C by V_1 during the first phase period. We note that final charge on the capacitor during a phase period is not necessarily equal to the charge flowing past the dotted lines during that phase period.

Finally, let us consider the next phase period, which is a repeat of the first case, except this time C is charged to $C\,V_2$ during the previous phase period. When ϕ_1 causes switch 1 to close for the second time, the charge flow across the left dotted line in the direction designated by i_1 can be written as

$$Q\left(t_0 + \frac{3T}{2}\right) = C\,V_1 - C\,V_2 = C(V_1 - V_2) \tag{3}$$

This sequence of events will continue indefinitely, and we have now reached a steady-state condition. Equations (2) and (3) describe the charge flow during the appropriate phase period. Let us define the resistance of Fig. 2.2–1(b) as

$$R = \frac{V_1 - V_2}{I_1} = \frac{V_2 - V_1}{I_2} \tag{4}$$

We are assuming that V_1 and V_2 in Figs. 2.2–1(a) and (b) are constant in this analysis. The current flow past a point in a circuit can be expressed as

$$i = \frac{dq}{dt} \tag{5}$$

Let us consider the flow of charge across the left dotted line of Fig. 2.2–1(a) under steady-state conditions. The total charge flow past this point Q_1, can be expressed from Eq. (5) as

$$Q_1 = \int_{t_0+T}^{t_0+3T/2} i_1\,dt \tag{6}$$

However, the limits of Eq. (6) can be extended to that of Eq. (7) because $i_1 = 0$ during the previous phase period.

$$Q_1 = \int_{t_0+T/2}^{t_0+3T/2} i_1\,dt \tag{7}$$

We see that the charge given by Eq. (7) must equal the charge that flows across the left dotted line given by Eq. (3). Equating (3) to (7) and dividing through by T results in

$$\frac{Q\left(t_0 + \dfrac{3T}{2}\right)}{T} = \frac{1}{T}\int_{t_0+T/2}^{t_0+3T/2} i_1\,dt = I_1\,(\text{aver.}) \tag{8}$$

Replacing $Q(t_0 + 3T/2)$ by Eq. (3) and rearranging terms gives

$$\frac{V_1 - V_2}{I_1(\text{aver.})} = \frac{T}{C} \tag{9}$$

Comparing Eq. (4) with Eq. (9) yields the desired relationship that

$$R = \frac{T}{C} \tag{10}$$

where we have assumed that $I_1 = I_1(\text{aver.})$. This assumption is exact if both V_1 and V_2 are constant during the clock period, T. If we designate the clock period as T_c, rather than T, then we may rewrite Eq. (10) as

$$R = \frac{T_c}{C} = \frac{1}{f_c C} \tag{11}$$

where f_c is the clock frequency in Hertz. The switched capacitor resistor of Fig. 2.2–1(a) will be called the *parallel switched capacitor resistor realization.* This name will help distinguish this realization from several others that follow.

A second switched capacitor realization of the continuous resistor is given in Fig. 2.2–2. This configuration is called the *series switched capacitor resistor realization* of the continuous resistor. Its operation can be developed in a manner similar to the parallel resistor realization. Using the same assumptions as for the parallel resistor realization, it may be shown (see Problem 2.1a) that the equivalent resistance of Fig. 2.2–2 is given by Eqs. (10) and (11). There is a distinction between the series and parallel realizations that will

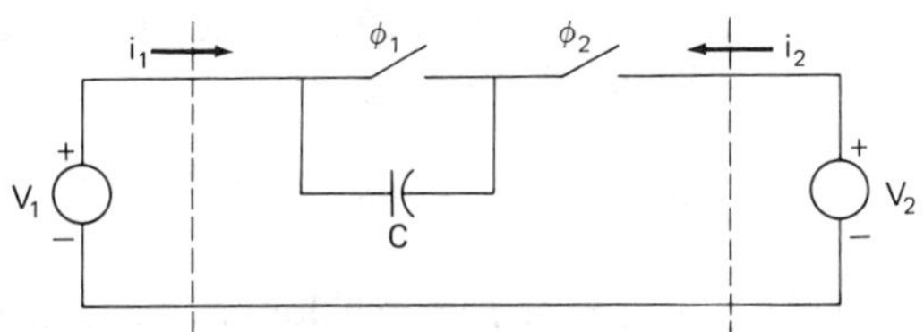

Fig. 2.2–2. Series switched capacitor realization of a continuous resistance.

become apparent in later work. This distinction has to do with the fact that in the parallel case, V_1 and V_2 are never simultaneously coupled, though in the series case V_1 is connected to V_2 through C for a portion of the clock period. Also i_1 is always equal to $-i_2$. The series SC[6] resistor realization of Fig. 2.2–2 is valid only at ϕ_2.

A third realization is a combination of the parallel and series configuration and is shown in Fig. 2.2–3.[7] This configuration will be called the *series-parallel realization* of a resistor. This circuit may be analyzed using similar techniques as for the previous two realizations. Assume that both capacitors

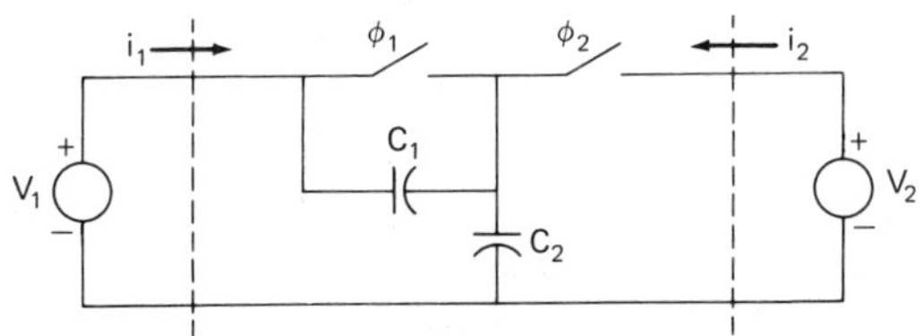

Fig. 2.2–3. Series-parallel switched capacitor realization of a continuous resistance.

are discharged and the two-phase clock of Fig. 2.2–1(c) is applied at $t = t_0 = nT$. At ϕ_1 or $t = t_0 + T/2$, the charge which flowed across the left dotted line in the direction of i_1 is

$$Q_1\left(t_0+\frac{T}{2}\right)=C_2\,V_1 \tag{12}$$

At (ϕ_2) $t = t_0 + T$, the charge which flowed across the right dotted line in the direction of i_2 is

$$Q_2(t_0+T)=C_1(V_2-V_1)+C_2(V_2-V_1) \tag{13}$$

---

[6] We shall abbreviate "switched capacitor" by SC for simplicity.

[7] C. F. Rahim, et al., "A Functional MOS Circuit for Achieving the Bilinear Transformation in Switched-Capacitor Filters," *IEEE J. of Solid State Circuits,* Vol. SC-13, No. 6, December 1978, pp. 906–909. This realization is only valid at ϕ_1.

Also at this time (ϕ_2), the charge which flowed across the left dotted line in the direction of i_1 has occurred and is given by

$$Q_1(t_0 + T) = C_1(V_1 - V_2) \tag{14}$$

Actually the portion of $Q_2(t_0 + T)$ which flowed through C_1 is equal to $-Q_1(t_0 + T)$. Again at ϕ_1, that is at $t = t_0 + 3T/2$, the charge which has just flowed across the left dotted line in the direction of i_1 is

$$Q_1\left(t_0 + \frac{3T}{2}\right) = C_2(V_1 - V_2) \tag{15}$$

At this point we have reached steady state and can solve for the net charge flowing into either side of Fig. 2.2–3 during a clock period. Integrating the expression

$$i_1 = \frac{dq}{dt} \tag{16}$$

over one full clock period, i.e., $t_0 + T/2$ to $t_0 + 3T/2$, results in

$$Q_1 = \int_{t_0+T/2}^{t_0+3T/2} i_1\,dt = \int_{t_0+T/2}^{t_0+T} i_1\,dt + \int_{t_0+T}^{t_0+3T/2} i_1\,dt$$
$$= Q_1(t_0 + T) + Q_1\left(t_0 + \frac{3T}{2}\right) \tag{17}$$

The average current can be found by dividing Eq. (17) by T to get

$$\frac{(V_1 - V_2)(C_1 + C_2)}{T} = \frac{1}{T}\int_{t_0+T/2}^{t_0+3T/2} i_1\,dt = I_1(\text{aver.}) \tag{18}$$

Comparing Eq. (18) with Eq. (4) results in

$$R = \frac{T}{C_1 + C_2} = \frac{T}{2C} \tag{19}$$

if $C_1 = C_2 = C$. The series-parallel switch is seen to have half the resistance of the previous realizations. The reason is because twice the charge is being transferred across the dotted boundaries of Fig. 2.2–3. We shall see later that this switch has some attractive properties not found in the prior two

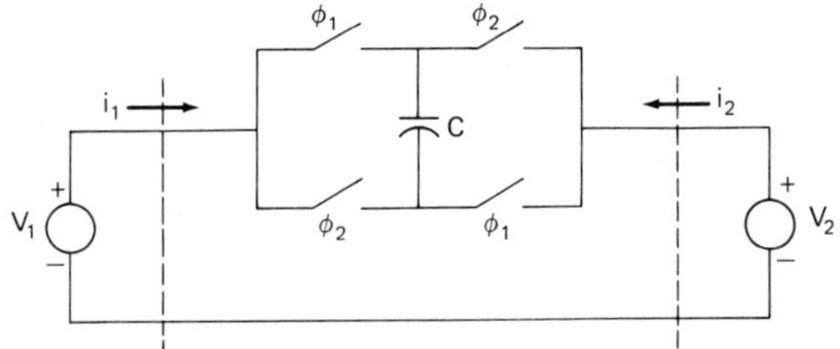

Fig. 2.2–4. Bilinear switched capacitor realization of a continuous resistance.

realizations that make the additional capacitor worthwhile. Note that the resistance values of Eqs. (10) and (19) are identical if $C_1 = C_2 = C/2$.

A fourth switched capacitor resistor realization is shown in Fig. 2.2–4.[8,9] This configuration is called the *bilinear switched capacitance realization* of a resistor. It can be shown that this realization results in a resistor given below (see Problem 2.1b).

$$R = \frac{T_c}{4C} \tag{20}$$

Though the bilinear realization has basically the same performance as the series-parallel realization, there are some practical differences that are important. When each of the above realizations is embedded in a switched capacitor network each will be found to possess different properties. This will be shown in the following sections.

The advantage of the switched capacitor realization of resistors can be appreciated by comparing the RC product of a resistor designated as R_1 and a capacitor designated as C_2. Let us assume that the product of R_1 and C_2 forms the time constant τ, given as

$$\tau = R_1 C_2 \tag{21}$$

Let us further assume that it is important to control τ as accurately as possible. The dependence of the accuracy of τ upon R_1 and C_2 can be found by the following expression

$$\frac{d\tau}{\tau} = \frac{dR_1}{R_1} + \frac{dC_2}{C_2} \tag{22}$$

---

[8] G. C. Temes and I. A. Young, "An Improved Switched-Capacitor Integrator," *Electronics Letters,* Vol. 14, No. 9, April 27, 1978, pp. 287–288.

[9] C. K. Sutton, W. K. Jenkins, and T. N. Trick, "New Structures for Switched Capacitor Sampled Data Filters," *Proc. 1978 Midwest Symp. Circuits and Systems,* August 1978, pp. 169–173.

where dx/x is interpreted as the accuracy of x. The worst-case accuracy of τ will be the sum of the absolute accuracies of R_1 and C_2, which is very poor if R_1 and C_2 are implemented on an integrated circuit.

Now let us replace R_1 with a switched capacitor equivalent. Assuming that R_1 is replaced by a series or parallel switch capacitor equivalent, Eq. (21) becomes

$$\tau = \frac{1}{f_c} \frac{C_2}{C_1} = T_c \frac{C_2}{C_1} \tag{23}$$

The accuracy of τ can be expressed as

$$\frac{d\tau}{\tau} = \frac{dT_c}{T_c} + \frac{dC_2}{C_2} - \frac{dC_1}{C_1} \tag{24}$$

Assuming that T_c is perfectly accurate gives

$$\frac{d\tau}{\tau} = \frac{dC_2}{C_2} - \frac{dC_1}{C_1} \tag{25}$$

Because the two capacitors C_1 and C_2 are built close together, using the same technology, the accuracy of Eq. (25) is much improved over that given in Eq. (22). Typical values of $d\tau/\tau$ compatible with standard MOS technologies are in the neighborhood of 0.1%. This is indeed a very satisfying result when τ must be carefully controlled. Furthermore, because the capacitors are similar in many respects, such properties as linearity and temperature coefficients are well behaved. More details on the technological aspects will be considered in Chapter 8. It may be helpful to refer to the technological information presented in Table 8.0–1 as the reader progresses through this text.

2.3 SWITCHED CAPACITOR EQUIVALENT RC NETWORKS

The next step in our study of switched capacitor networks is to develop realizations for continuous RC passive networks. The realizations of such networks amount to the straightforward replacement of the continuous realizations developed in the last section. However, the resulting circuits furnish excellent examples for developing the methods of analyzing switched capacitor networks. In a sense, the objective of this section is to develop methods of analysis for switched capacitor networks.

Let us consider several realizations of the continuous RC circuit shown

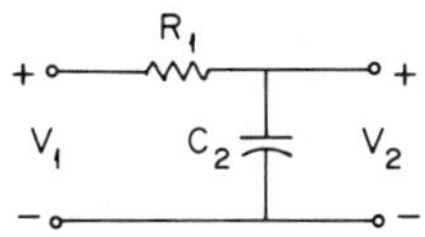

Fig. 2.3–1. Continuous RC circuit.

in Fig. 2.3–1. The continuous frequency domain voltage transfer function of this circuit is given as

$$H(s)=\frac{\dfrac{1}{R_1C_2}}{s+\dfrac{1}{R_1C_2}}=\frac{1}{s\,\tau_1+1}=\frac{1}{s/\omega_1+1} \tag{1}$$

where $\tau_1 = R_1C_2$ and $\omega_1 = 1/\tau_1$. The frequency response can be found by replacing s by $j\omega$. The magnitude is

$$|H(j\omega)|=[1+(\omega\,R_1C_2)^2]^{-1/2} \tag{2}$$

and the argument or phase shift is

$$\text{Arg } H(j\omega)=-\tan^{-1}(\omega\,R_1C_2) \tag{3}$$

The frequency response of Fig. 2.3–1 is given in Fig. 2.3–2(a) and (b). We see that this circuit has a low-pass frequency response. The roots of Fig. 2.3–1 are shown in Fig. 2.3–2(c).

An analog sampled data realization of Fig. 2.3–1 can be obtained by replacing the resistor with one of the switched capacitor equivalent resistances developed in the last section. Figure 2.3–3(a) shows a switched capacitor RC realization using the parallel SC resistor equivalent of Fig. 2.2–1(a). To analyze this circuit, we need to specify the clock sequence. Figure 2.3–3(b) is a shorthand method of illustrating the clock sequence. ϕ_1 and ϕ_2 specify the phase periods during which switches designated as ϕ_1 and ϕ_2 close and will be denoted as the odd and even phase clocks.

It is necessary in the following analysis to have a more precise definition of the phase period. Although each phase period is $T/2$ in length, we must specify whether or not the end points are included. In Fig. 2.3–3(b), we arbitrarily define the odd phase period, ϕ_1, of a two-phase clock to be the period of time where

$$(n+i)\le\frac{t}{T}<\left(n+i+\frac{1}{2}\right)\qquad\text{odd }(\phi_1)\text{ phase period} \tag{4}$$

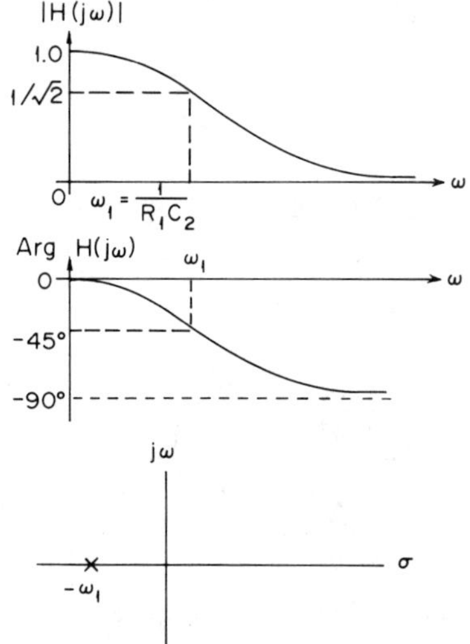

Fig. 2.3–2. Characteristics of Fig. 2.3–1. (a) Magnitude of the frequency response. (b) Phase response. (c) Root locations in the complex frequency plane.

and the even phase period, ϕ_2, of a two-phase clock to be the period of time where

$$\left(n+i+\frac{1}{2}\right) \leq \frac{t}{T} < (n+i+1) \qquad \text{even } (\phi_2) \text{ phase period} \tag{5}$$

where $i = \{\ldots, -2, -1, 0, 1, \ldots\}$ corresponds to the ith clock period. Observe that we have defined the phase period to include the left end point only. This convention assures that the nonoverlapping property is preserved.

In the analysis of Fig. 2.3–3(a), we shall assume that $v_1(t)$ is constant during phase periods. Thus $v_1(t)$ as a sampled data signal would have a waveform similar to the zero-order sample-and-hold of Fig. 2.1–3. Let us consider first the odd phase period where $(n-1) \leq t/T < (n-½)$, when the ϕ_1 switch is closed. In our analysis we are assuming that the

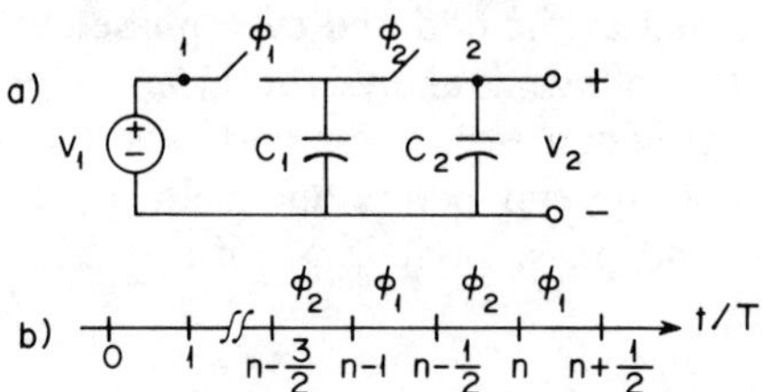

Fig. 2.3–3. (a) Switched capacitor realization of Fig. 2.3–1 using the parallel configuration. (b) Clock phasing.

switch ϕ_1 closes immediately after $t = (n - 1)T$ and that C_1 is instantaneously charged to $v_1^o[(n - 1)T]$. In practice, the time required for v_1 to charge C_1 to $v_1^o[(n - 1)T]$ should be small compared with $T/2$. Also note that, once C_1 has been charged to $v_1^o[(n - 1)T]$, the ϕ_1 switch can reopen, ideally without causing any change in the circuit. Our comments given here for the ϕ_1 switch also apply to the ϕ_2 switch during its phase period. It is fortunate that the only concern is that the switches must be closed long enough to transfer the charge. Otherwise the clock circuits would face very severe timing requirements.

During the odd phase period ϕ_1 at $(n - 1) \leq t/T < (n - ½)$, we may redraw Fig. 2.3–3(a) as shown in Fig. 2.3–4(a). From this figure we see that

$$v_{C_1}(t) = v_1^o[(n - 1)T] = v_1^o(n - 1) \tag{6}$$

and

$$v_{C_2}(t) = v_2^o[(n - 1)T] = v_2^o(n - 1) \tag{7}$$

We have dropped the clock period T in Eqs. (6) and (7) because it adds no useful information and simplifies the notation. We will continue to follow this convention where no misinterpretation is likely.

In the next even phase period $(n - ½) \leq t/T < n$, the ϕ_1 switch is open, and the ϕ_2 switch closes. Figure 2.3–4(b) represents Fig. 2.3–3(a) during this phase period. During this phase period C_1 and C_2 are paralleled, resulting in a new value of v_2. We may convert the output portion of the circuits in

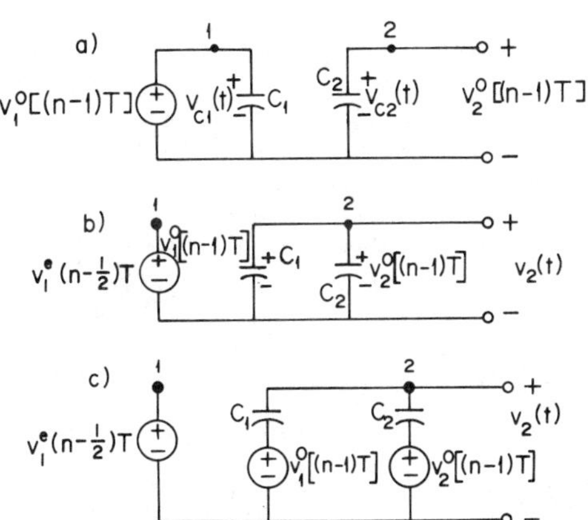

Fig. 2.3–4. (a) Equivalent odd circuit of Fig. 2.3–3(a) when ϕ_1 switch is closed. (b) Equivalent even circuit of Fig. 2.3–3(a) when ϕ_2 switch is closed. (c) Alternative form of (b).

Fig. 2.3–4(b) to that shown in Fig. 2.3–4(c) with uncharged capacitors. The voltage sources representing the initial voltages on the capacitors are assumed to be multiplied by a unit step function that starts at $t = (n - 1)T$. Using superposition techniques, we can solve for the voltage v_2 to get

$$v_2(t) = \frac{C_1}{C_1 + C_2} v_1^o(n-1) + \frac{C_2}{C_1 + C_2} v_2^o(n-1) \tag{8}$$

If we evaluate $v_2(t)$ at $t = (n - ½)$, we get

$$v_2^e\left(n - \frac{1}{2}\right) = \frac{C_1}{C_1 + C_2} v_1^o(n-1) + \frac{C_2}{C_1 + C_2} v_2^o(n-1) \tag{9}$$

Looking forward at the next phase period, we observe that

$$v_2^o(n) = v_2^e\left(n - \frac{1}{2}\right) \tag{10}$$

because the voltage on C_2 has not changed since it was charged to $v_2^e(n - ½)$ at $t = (n - ½)$. Using Eq. (10) and the relationships of Table 1.4–1, we can write Eq. (9) as

$$V_2^o(z) = \frac{C_1 z^{-1}}{C_1 + C_2} V_1^o(z) + \frac{C_2 z^{-1}}{C_1 + C_2} V_2^o(z) \tag{11}$$

Solving for $V_2^o(z)/V_1^o(z)$ results in

$$H^{oo}(z) = \frac{V_2^o(z)}{V_1^o(z)} = \frac{1}{1+\alpha} \frac{z^{-1}}{1 - \dfrac{\alpha}{1+\alpha} z^{-1}} \tag{12}$$

where $\alpha = C_2/C_1$. Observe from Eq. (10) that $V_2^o(z) = z^{-1/2}\, V_2^e(z)$; then we can write

$$H^{oe}(z) = \frac{V_2^e(z)}{V_1^o(z)} = \frac{1}{1+\alpha} \frac{z^{-1/2}}{1 - \dfrac{\alpha}{1+\alpha} z^{-1}} \tag{13}$$

thus

$$H^{oo}(z) = z^{-1/2}\, H^{oe}(z) \tag{14}$$

The previous analysis of Fig. 2.3–3(a) used conventional network analysis methods. Let us consider another, called *charge conservation.*[10,11] This approach is basically an application of Kirchoff's current law, where charge is used instead of current. At one particular clock phase, charges are instantaneously redistributed, maintaining charge conservation at every node in the circuit. Thus for the biphase circuits, two distinct, but coupled, nodal charge equations can be obtained, which characterize the charge conservation condition at a particular node for all time instants of one period T. We can write the two nodal charge equations for one node, for the even clock phase as

$$q_L^e(t') = q_m^o(t) + q_c^{o,e}(t),\ t' > t \tag{15}$$

and for the odd clock phase

$$q_L^o(t') = q_m^e(t) + q_c^{o,e}(t), \qquad t' > t \tag{16}$$

where t' is a time reference, $q_L(t')$ is the charge *left* at one particular node at equilibrium, $q_m(t)$ is the charge at that particular node from the previous phase period, referred to as the *memory* charge, and $q_c(t)$ is the charge injected at that particular node, designated as the *contribution* charge. Note[11] that even though the reference time t' is not explicitly indicated in the symbol $q_c(t)$, it will always be clear what is the reference time. The superscripts of $q_c^{o,e}(t)$ imply that the contribution charge can be from the even or odd clock phases or both. The relationship between the even and odd topologies can be interpreted as topologically decoupled, with the states of one determining the initial conditions for the other.[12] Thus, a biphase time-varying SC network can be considered as two interrelated time-invariant networks. Let us consider again the analysis of the circuit of Fig. 2.3–3(a); from Fig. 2.3–4(b), we identify the components of Eqs. (15) as

$$q_L^e\left(n - \frac{1}{2}\right) = (C_2 + C_1)\, v_2^e\left(n - \frac{1}{2}\right) \tag{17a}$$

$$q_m^o(n-1) = C_2\, v_2^o\,(n-1) \tag{17b}$$

$$q_c^o(n-1) = C_1\, v_1^o\,(n-1) \tag{17c}$$

[10] C. F. Kurth and G. S. Moschytz, "Nodal Analysis of Switched-Capacitor Networks," *IEEE Trans. on Circuits and Systems,* Vol. CAS-26, No. 2, February 1979, pp. 93–105.

[11] Y. P. Tsividis, "Analysis of Switched Capacitive Networks," *Proc. of the 1979 Int. Symp. on Circuits and Systems ISCAS,* July 1979, pp. 752–755.

[12] K. R. Laker, "Equivalent Circuits for the Analysis and Synthesis of Switched Capacitor Networks," *Bell Syst. Tech. J.,* March 1979, pp. 729–769.

and looking at the next odd phase period, we have

$$q_L^o(n) = C_2 v_2^o(n) \tag{18a}$$

$$q_m^e\left(n-\frac{1}{2}\right) = C_2 v_2^e\left(n-\frac{1}{2}\right) \tag{18b}$$

$$q_c^{o,e}(t) = 0 \tag{18c}$$

Using the above equations and taking the z-transform, we can obtain Eqs. (12) and (13). The charge equation in Eqs. (15) and (16) can be generalized to include any number of nodes. Further discussion of this method is included in Section 3.6.

The switched capacitor RC realization of Fig. 2.3–3(a) has demonstrated a general property of sampled data SC networks. This property is that the z-domain transfer function depends upon which phase period the output is sampled. We have developed two transfer functions for Fig. 2.3–3(a), given in Eqs. (12) and (13). This situation is illustrated on a general basis in Fig. 2.3–5 for a sampled data system having a two-phase clock.

Let us confirm whether or not Fig. 2.3–3(a) is an equivalent realization of Fig. 2.3–1(a). In discrete time systems, the steady-state frequency response can be found by replacing z by $e^{j\omega T}$, where ω is the frequency of the applied sinusoid, and T is the clock period. Therefore, Eq. (12) can be written as

$$H^{oo}(e^{j\omega T}) = \frac{V_2^o(e^{j\omega T})}{V_1^o(e^{j\omega T})} = \frac{1}{(1+\alpha)\cos\omega T - \alpha + j(1+\alpha)\sin\omega T} \tag{19}$$

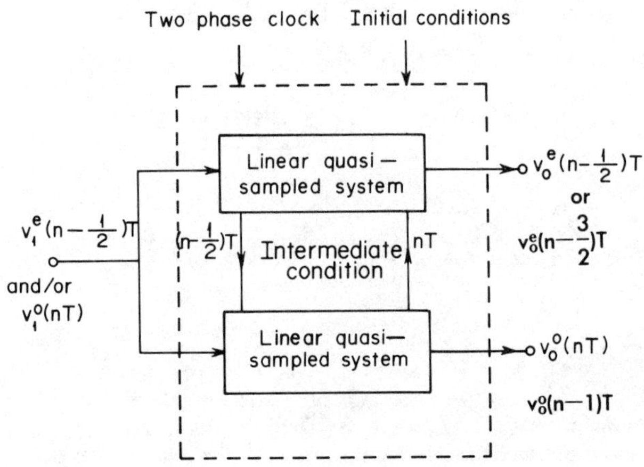

Fig. 2.3–5. Time domain model of switched capacitor networks.

where Euler's formula has been used to simplify $e^{j\omega T}$. The magnitude of Eq. (19) is

$$|H^{oo}(e^{j\omega T})| = \frac{1}{[1 + 2\alpha(1+\alpha)(1 - \cos \omega T)]^{1/2}} \tag{20}$$

and the phase shift of Eq. (19) is

$$\operatorname{Arg}[H^{oo}(e^{j\omega T})] = -\tan^{-1}\left[\frac{\sin \omega T}{\cos \omega T - \dfrac{\alpha}{1+\alpha}}\right] \tag{21}$$

The question now arises of how to select α to meet a prescribed frequency response. We shall assume that ω_1 of Eq. (1) defines the desired frequency response of this first-order system. One answer to this question would be to apply the forward transformation of Section 1.5 to Eq. (1) to get[13]

$$H(s) \xrightarrow[s=\frac{1}{T}(z-1)]{} H(z) = \frac{\omega_1 T z^{-1}}{1 - (1 - \omega_1 T)z^{-1}} \tag{22}$$

Comparing Eqs. (12) and (22) shows that

$$\alpha = \frac{1}{\omega_1 T} - 1 = \frac{1}{2\pi}\frac{\omega_c}{\omega_1} - 1 \tag{23}$$

This expression would allow α to be designed when ω_c/ω_1 was specified.

Another method of answering this question would be to start with Eq. (12) and assume that $\omega_1 << \omega_c$, so that z can be approximated by $1 + sT$. This result gives

$$H(z) = \xrightarrow[z=1+sT]{} H(s) = \frac{1}{sT(1+\alpha) + 1} \tag{24}$$

Comparing Eq. (24) with Eq. (1) gives

$$\frac{1}{\omega_1} = T(1+\alpha) \tag{25}$$

[13] We have dropped the superscripts of the $H(z)$ and will continue to follow this convention where no misinterpretation is likely.

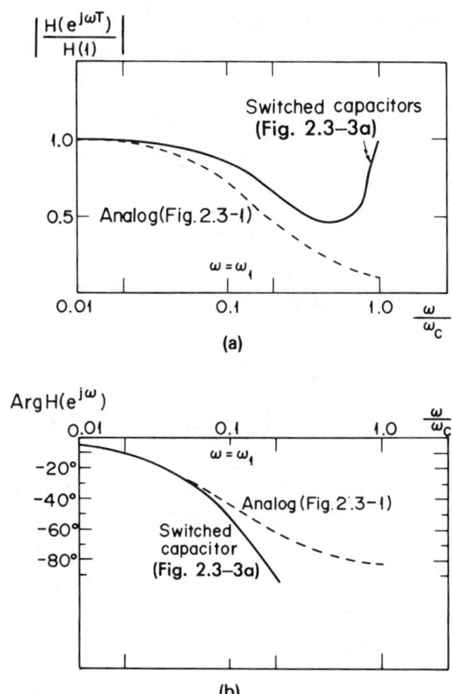

Fig. 2.3–6. (a) Magnitude and (b) phase response of Fig. 2.3–3(a).

which results in Eq. (23). The frequency response of Eq. (19) is plotted in Fig. 2.3–6 for $\omega_c/\omega_1 = 10$, which corresponds to $\alpha = 0.5915$. For frequencies of ω less 0.02 ω_c, the switched capacitor circuit of Fig. 2.3–3(a) is a good approximation of the analog circuit of Fig. 2.3–1. We note that the switched capacitor is a very poor approximation for $\omega > 0.1\ \omega_c$. At $\omega/\omega_c = 0.5$, the lowest attenuation occurs, and at $\omega = \omega_c$ the magnitude is at the starting value, and the phase shift will be $-360°$. The roots of Eq. (12) are shown in Fig. 2.3–7 and consist of a pole located at $\alpha/(1 + \alpha)$ on the positive real axis.

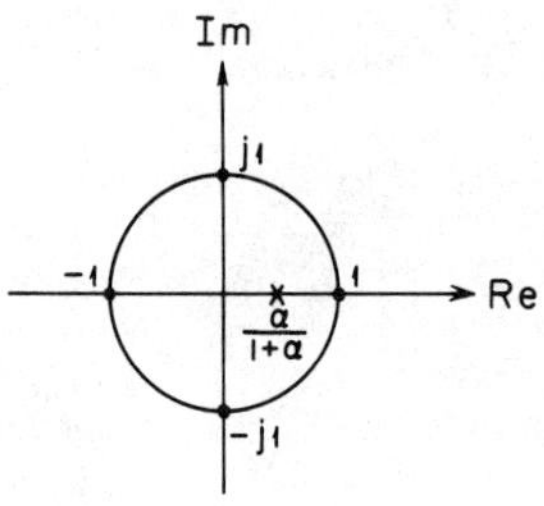

Fig. 2.3–7. z-plane roots of Fig. 2.3–3(a).

To improve the switched capacitor realization of Fig. 2.3–1, it is necessary to increase ω_c or alternatively reduce ω_1. This is considered in the following example.

Example 2.3–1. *Frequency response of Fig. 2.3–3(a) with $\omega_1 = 0.02\omega_c$.* Let us find the frequency response of Fig. 2.3–3(a) if $\omega_1 = 0.02\omega_c$. Equation (23) gives

$$\omega_1 = \frac{1}{\tau} = \frac{1}{(1+\alpha)T} = .02\omega_c \tag{26}$$

Because $T = 1/f_c$, we find

$$\alpha = \frac{25}{\pi} - 1 = 6.9577 \tag{27}$$

Using this value, we replot the magnitude and phase of Fig. 2.3–1 and Fig. 2.3–3(a) in Fig. 2.3–8. We observe much better correlation in the frequency range around ω_1.

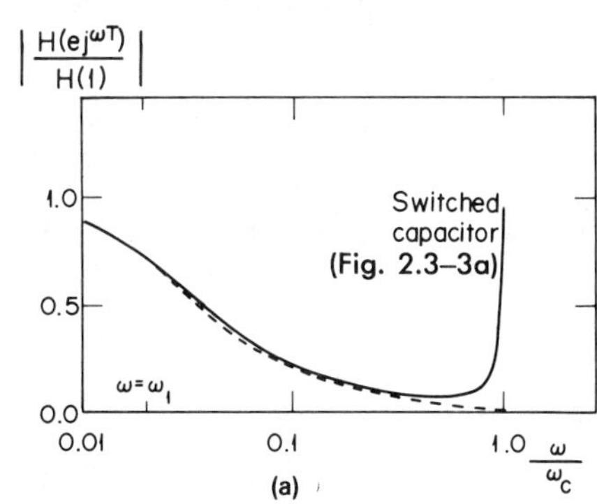

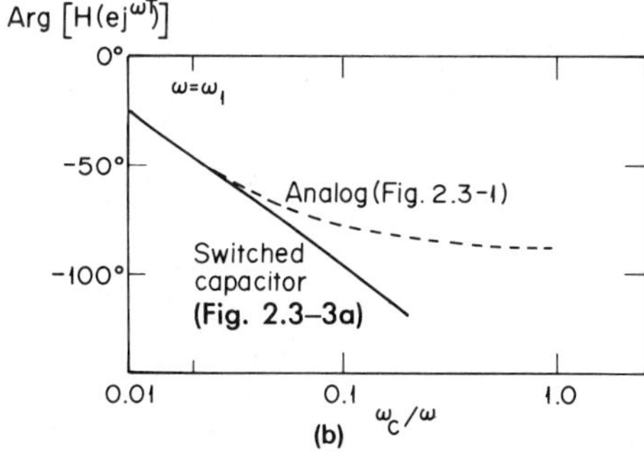

Fig. 2.3–8. (a) Magnitude and (b) phase response of Example 2.3–1. $\omega_c/\omega_1 = 50$.

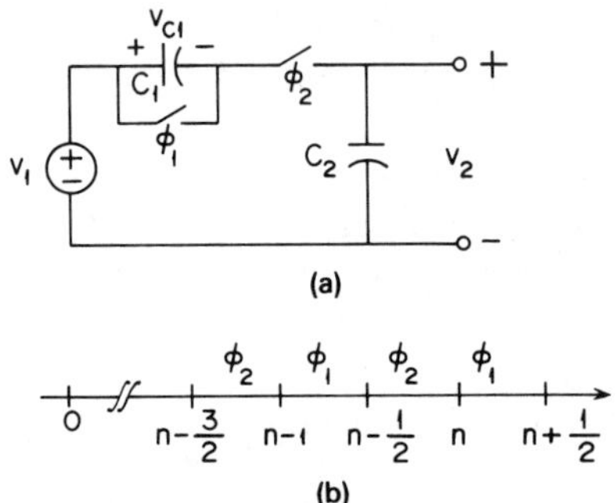

Fig. 2.3–9. (a) RC equivalent network using the series switched capacitor resistor realization. (b) Clock phasing.

Next, we consider another RC equivalent. A switched capacitor network with a series SC resistor equivalent is shown in Fig. 2.3–9. To illustrate the two approaches for analysis, we will discuss them both again for this example. First using conventional analysis, for $(n - ½) \leq t/T < n$, i.e., for the even phase period where ϕ_1 is off and ϕ_2 is on, we have the situation illustrated in Fig. 2.3–10. By superposition we obtain

$$v_2^e\left(n - \frac{1}{2}\right) = \frac{C_2}{C_1 + C_2}\, v_2^o(n-1) + \frac{C_1}{C_1 + C_2}\, v_1^e\left(n - \frac{1}{2}\right) \tag{28}$$

When ϕ_1 closes, we have for the next odd phase period

$$v_2^e\left(n - \frac{1}{2}\right) = v_2^o(n) \tag{29}$$

Combining Eqs. (28) and (29) and taking the z-transform, we get the corresponding transfer function

$$H^{eo}(z) = \frac{V_2^o(z)}{V_1^e(z)} = \frac{z^{-1/2}}{1 - \dfrac{\alpha}{1+\alpha} z^{-1}} \frac{1}{1+\alpha} \tag{30}$$

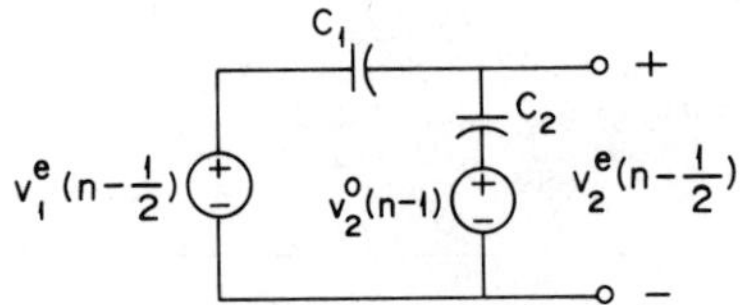

Fig. 2.3–10. Equivalent circuit of Fig. 2.3–9(a) for even ϕ_2 period.

From Eqs. (29) and (30), we can also write

$$H^{ee}(z) = \frac{V_2^e(z)}{V_1^e(z)} = \frac{1}{1 - \dfrac{\alpha}{1+\alpha} z^{-1}} \frac{1}{1+\alpha} \tag{31}$$

Now using the charge conservation approach, illustrated by Eq. (15), we can write for ϕ_2, the even phase period

$$C_2 \, v_2^e\left(n - \frac{1}{2}\right) = C_2 \, v_2^o(n-1) + C_1\left[v_1^e\left(n - \frac{1}{2}\right) - v_2^e\left(n - \frac{1}{2}\right)\right] \tag{32a}$$

and for the odd phase period

$$C_2 \, v_2^o(n) = C_2 \, v_2^e\left(n - \frac{1}{2}\right) \tag{32b}$$

By using Eqs. (32a) and (32b) we can again obtain Eqs. (30) and (31). The roots of Fig. 2.3–9 are obtained from Eq. (30) and are illustrated in Fig. 2.3–11. In comparing Fig. 2.3–2(c), 2.3–7, and 2.3–11, we note that the series switch possesses a zero at the origin. Although this zero influences the phase response, it has no effect on the magnitude of the frequency response.

The frequency response of Eq. (31) can be found by using $z = e^{j\omega T}$ to get

$$H^{ee}(e^{j\omega T}) = \frac{1}{1 + \alpha - \alpha \cos \omega T + j\alpha \sin \omega T} \tag{33}$$

The magnitude and phase response of Eq. (33) is

$$|H^{ee}(e^{j\omega T})| = \frac{1}{[1 + 2\alpha(1+\alpha)(1 - \cos \omega T)]^{1/2}} \tag{34}$$

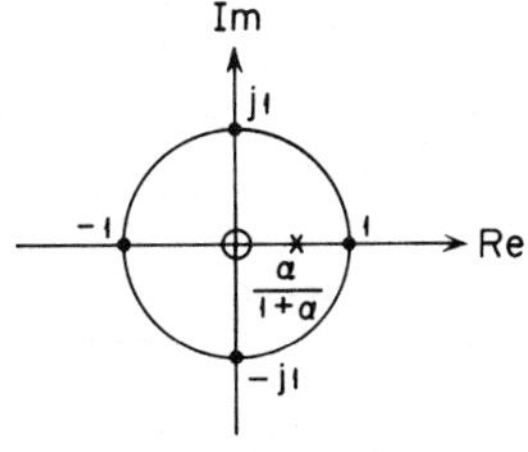

Fig. 2.3–11. *z*-plane roots of Fig. 2.3–9(a).

and

$$\text{Arg } H^{ee}(e^{j\omega T}) = -\tan^{-1}\left[\frac{\sin \omega T}{\dfrac{1+\alpha}{\alpha} - \cos \omega T}\right] \tag{35}$$

Though the magnitude is identical to the parallel resistor RC realization, the phase shift is different due to the presence of the zero.

To design the frequency response of Fig. 2.3–9(a), it is necessary to relate α to ω_1 of Fig. 2.3–1 and Eq. (1). Applying the backward transformation of Section 1.5 to Eq. (1) results in

$$H(s) \xrightarrow[s = \frac{1+z^{-1}}{T}]{} H(z) = \frac{\dfrac{\omega_1 T}{1+\omega_1 T}}{1 - \dfrac{1}{1+\omega_1 T} z^{-1}} \tag{36}$$

Comparing Eqs. (31) and (36) gives

$$\alpha = \frac{1}{\omega_1 T} = \frac{1}{2\pi}\frac{\omega_c}{\omega_1} \tag{37}$$

We may also assume that $\omega << \omega_c$, so that z in Eq. (31) may be replaced by $1 + sT$ and compared to Eq. (1). This approach results in

$$H(z) \xrightarrow[z \approx 1+sT]{} H(s) = \frac{1+sT}{sT(1+\alpha)+1} \approx \frac{1}{sT(1+\alpha)+1} \tag{38}$$

Equating Eq. (38) to Eq. (1) gives

$$\alpha = \frac{1}{\omega_1 T} - 1 = \frac{1}{2\pi}\frac{\omega_c}{\omega_1} - 1 \tag{39}$$

which is different from Eq. (37). The difference is due to the presence of the zero and assuming ωT is much less than 1 but not $\omega T(1 + \alpha)$. We shall choose Eq. (37) where α is given by the backward transformation rather than the approximation of Eq. (39) in the following work.

Example 2.3–2. *Frequency response of Fig. 2.3–9.* Let us evaluate and plot the phase response of Eq. (33) for the case of ($\omega_1 = 0.1\omega_c$) and ($\omega_1 =$

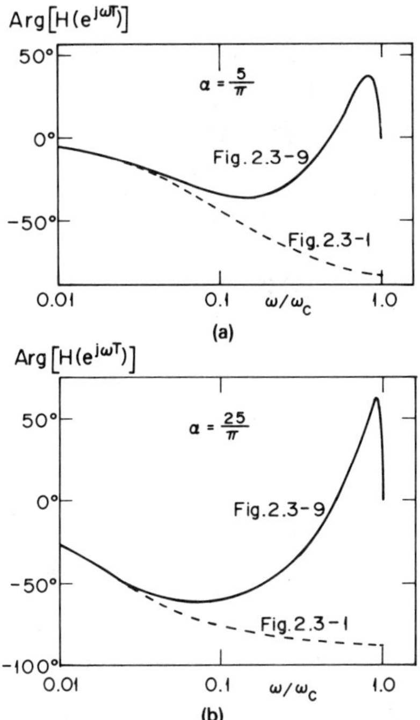

Fig. 2.3–12. Phase response of Fig. 2.3–9 considered in Example 2.3–2. (a) $\omega_1 = 0.1\omega_c$, $\alpha = 5/\pi$. (b) $\omega_1 = 0.02\omega_c$, $\alpha = 25/\pi$.

$.02\omega_c$). The results of the two cases are plotted in Fig. 2.3–12. Also included is the response of Fig. 2.3–1. We see that again the sampling rate has a strong influence on the results. The phase shift does not have the 2π lag at ω_c, due to the zero. In fact, the phase shift becomes positive above the Nyquist frequency.

The transfer function of sampled data filters can be represented as signal flow diagrams in the same form as digital filters.[14] For instance, consider the transfer function of Eq. (12) for the parallel switched capacitor RC realization, which can be rewritten as

$$V_2^o(z) = \frac{\alpha}{1+\alpha} V_2^o(z)\, z^{-1} + V_1^o(z)\, z^{-1}/(1+\alpha) \tag{40}$$

[14] A. V. Oppenheim and R. W. Schafer, *Digital Signal Processing*, Prentice-Hall, Inc., Englewood Cliffs, NJ, 1975.

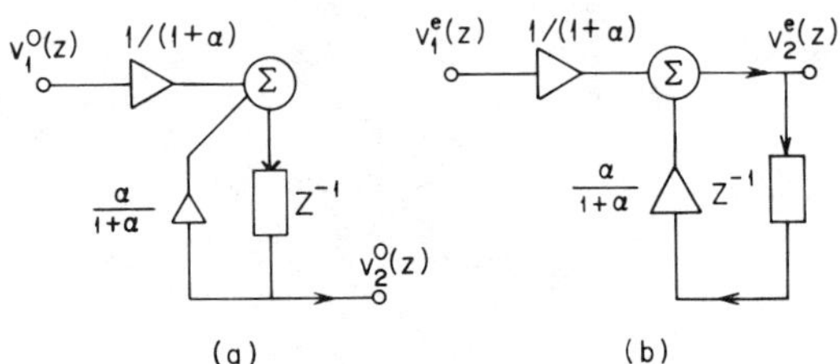

Fig. 2.3–13. Flow diagrams of (a) Fig. 2.3–3(a) and (b) Fig. 2.3–9(a).

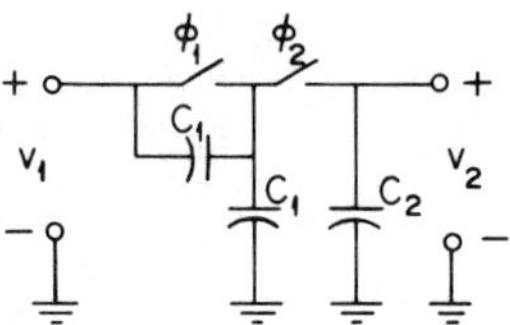

Fig. 2.3–14. Realization of Fig. 2.3–1 using series-parallel equivalent resistor.

The corresponding equation for the series-switched capacitor RC realization can be written as

$$V_2^e(z) = \frac{\alpha z^{-1}}{1+\alpha} V_2^e(z) + \frac{1}{1+\alpha} V_1^e(z) \tag{41}$$

These flow diagrams are also called z-domain block diagrams. The flow diagrams for Eqs. (40) and (41) are given in Fig. 2.3–13.

Next let us consider the bilinear SC resistor equivalent used to replace the resistor of Fig. 2.3–1(a).[15] One possible implementation of Fig. 2.3–1 is shown in Fig. 2.3–14. This implementation uses the series-parallel SC resistor of Fig. 2.2–3. The equivalent representations of Fig. 2.3–14 for the odd and even phases clocks are shown in Fig. 2.3–15. The charge conservation equation for ϕ_2 is

$$\begin{aligned}(C_1 + C_2)\, v_2^e\left(n - \frac{1}{2}\right) &= C_2\, v_2^o(n-1) + C_1\, v_1^o(n-1) \\ &\quad + C_1\left[v_1^e\left(n - \frac{1}{2}\right) - v_2^e\left(n - \frac{1}{2}\right)\right]\end{aligned} \tag{42a}$$

[15] C. F. Rahim et al., "A Functional MOS Circuit for Achieving the Bilinear Transformation in Switched-Capacitor Filters," *IEEE J. of Solid-State Circuits,* Vol. SC-13, No. 6, December 1978, pp. 906–909.

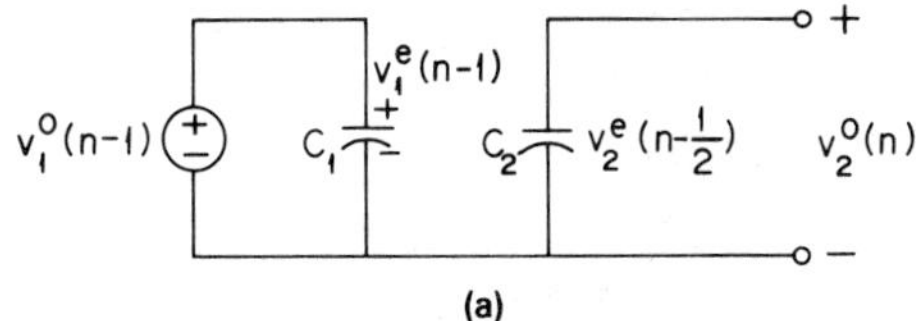

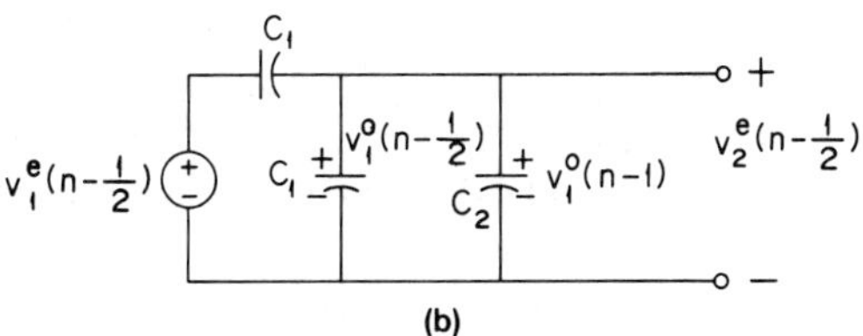

Fig. 2.3–15. Equivalent representations of Fig. 2.3–14(a) when ϕ_1 is closed and (b) when ϕ_2 is closed.

Looking at the next odd phase clock, we observe that

$$v_2^e\left(n-\frac{1}{2}\right)=v_2^o(n) \tag{42b}$$

and assuming a sampled and held signal

$$v_1^e\left(n-\frac{1}{2}\right)=v_1^o(n) \tag{42c}$$

Using the relationship of Table 1.4–1 and $C_1 = C/2$, $\alpha_1 = C_2/C$ results in

$$H^{oo}(z)=\frac{V_2^o(z)}{V_1^o(z)}=\frac{1}{2(1+\alpha_1)}\,\frac{1+z^{-1}}{1-\dfrac{\alpha_1}{1+\alpha_1}z^{-1}} \tag{43}$$

The z-domain block diagram representation of Fig. 2.3–14 is shown in Fig. 2.3–16. We note that the series-parallel configuration is indeed a combination of the series and parallel z-domain block diagrams.

The roots of Eq. (43) are shown in Fig. 2.3–17. Besides the anticipated pole at $\alpha_1/(1+\alpha_1)$, we see that there is a zero at $z = -1$. This zero will result in a notch or zero magnitude at half the sampling frequency.

To plot the frequency response and compare it with the analog response

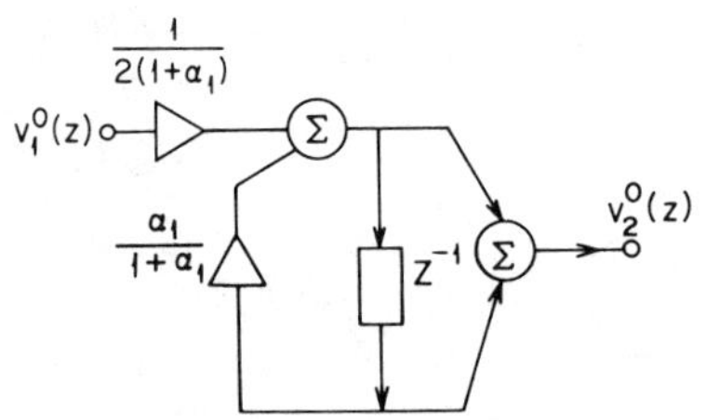

Fig. 2.3–16. Flow diagram for Fig. 2.3–14(a).

of Fig. 2.3–2, it is necessary to develop a relationship between α_1 of Eq. (43) and ω_1 of Eq. (1). Let us examine two possible approaches to this objective. The first approach is to apply the bilinear transformation given by Eq. (15) in Section 1.5 to $H(s)$ given by Eq. (1) of this section. The result will be equated to Eq. (43). Following this approach gives

$$H(s) \xrightarrow[s=\frac{2}{T}\left(\frac{1-z^{-1}}{1+z^{-1}}\right)]{} H(z) = \frac{1+z^{-1}}{1-z^{-1}\dfrac{1-\dfrac{T\omega_1}{2}}{1+\dfrac{T\omega_1}{2}}} \; \frac{T\omega_1/2}{1+\dfrac{T\omega_1}{2}} \tag{44}$$

In this passive RC case, the series-parallel SC resistor equivalent corresponds exactly to the bilinear mapping if

$$2\alpha_1 = \frac{2}{\omega_1 T} - 1 \tag{45}$$

Example 2.3–3. *Frequency response of Fig. 2.3–14.* Let us evaluate and plot the frequency response of Eq. (43) for the cases where ω_c/ω_1 is equal to 10 and to 50. The results of these two cases are plotted in Fig. 2.3–18. Also included is the frequency response of Fig. 2.3–1. We note that Fig.

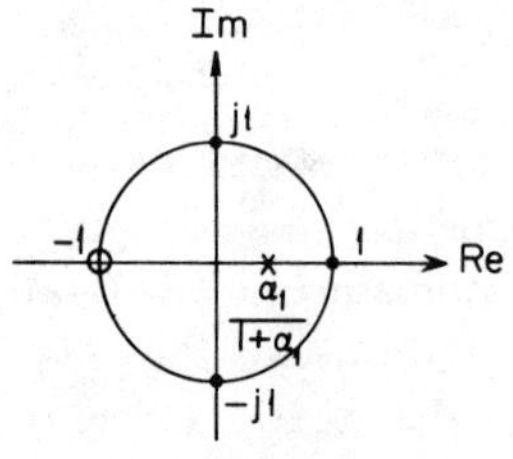

Fig. 2.3–17. Roots of Eq. (43).

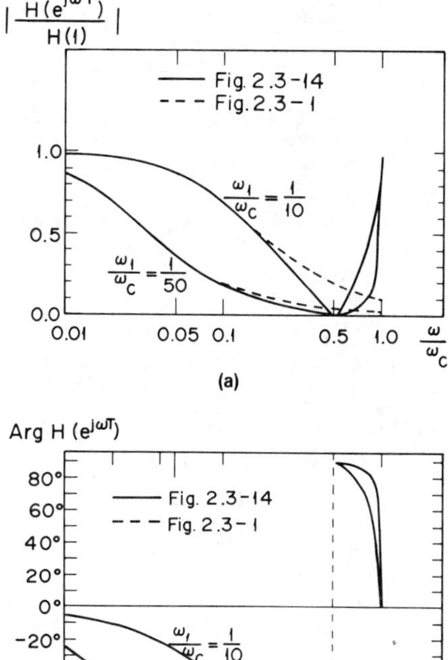

Fig. 2.3–18. Frequency response of Example 2.3–3 for two different values of ω_1. (a) Magnitude and (b) phase response.

2.3–14 is a good approximation of Fig. 2.3–1 for frequencies less than $0.1\omega_c$ for the values of ω_c/ω_1 considered in this example.

The last SC realization of Fig. 2.3–1(a) to be considered is shown in Fig. 2.3–19 and uses the bilinear SC resistor simulation of Fig. 2.2–4.[16] If the even phase period ϕ_2 occurs between $(n - 1/2)T$ and nT, then Fig. 2.3–20(a) shows an equivalent circuit for Fig. 2.3–19. We note that C_1 is charged to the voltage

$$v_{C1}^e\left(n - \frac{1}{2}\right) = v_1^e\left(n - \frac{1}{2}\right) - v_2^e\left(n - \frac{1}{2}\right) \tag{46}$$

[16] J. A. McKinney and C. A. Holijak, "The Periodically Reversed Switched Capacitor," *IEEE Trans. on Circuit Theory,* Vol. CT-15, 1968, pp. 288–290.

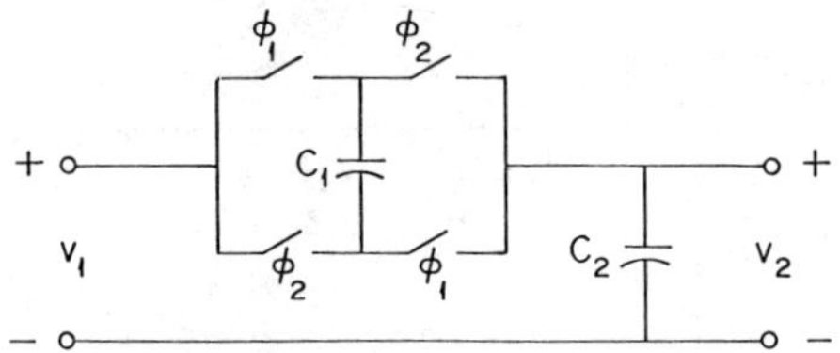

Fig. 2.3–19. Bilinear SC resistor stimulation of Fig. 2.3–1.

and that C_2 is charged to

$$v_{c2}^{e}\left(n-\frac{1}{2}\right)=v_{2}^{e}\left(n-\frac{1}{2}\right) \tag{47}$$

Fig. 2.3–20(b) shows the equivalent circuit for Fig. 2.3–19 at the beginning of the next odd phase ϕ_1, $n \le t/T < (n + ½)$. Using the charge conservation approach, we identify

$$q_L^o(n)=C_2\, v_2^o(n) \tag{48a}$$

$$q_m^e\left(n-\frac{1}{2}\right)=C_2\, v_2^e\left(n-\frac{1}{2}\right) \tag{48b}$$

$$q_c^{o,e}=C_1\, v_{c1}^e\left(n-\frac{1}{2}\right)+C_1\, v_{c_1}^o(n) \tag{48c}$$

where

$$v_{c_1}^o(n)=v_1^o(n)-v_2^o(n) \tag{49}$$

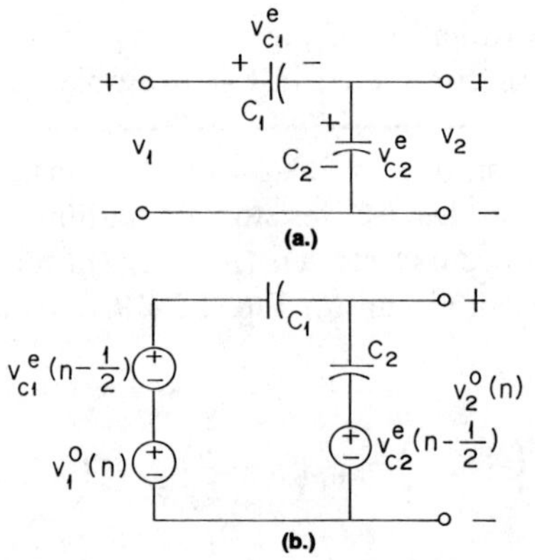

Fig. 2.3–20. (a) An equivalent circuit for Fig. 2.3–19 during the ϕ_2 phase period. (b) Equivalent circuit of Fig. 2.3–19 at the beginning of the ϕ_1 period.

Combining Eqs. (48) and (49) and using the z-transform, we can write Eq. (16) as

$$(C_2 + C_1)V_2^o(z) = z^{-1/2}(C_2 - C_1)V_2^e(z) + C_1(V_1^e(z)z^{-1/2} + V_1^o(z)) \quad (50)$$

In a similar way we can obtain the following equation

$$(C_2 + C_1)V_2^e(z) = z^{-1/2}(C_2 - C_1)V_2^o(z) + C_1[V_1^o(z)z^{-1/2} + V_1^e(z)] \quad (51)$$

during the even phase period. By summing Eqs. (50) and (51) we can obtain the transfer function

$$H(z) = \frac{V_2(z)}{V_1(z)} = \frac{V_2^e(z) + V_2^o(z)}{V_1^e(z) + V_1^o(z)} = \frac{1 + z^{-1/2}}{(1 + \alpha) + (1 - \alpha)z^{-1/2}} \quad (52)$$

where $\alpha = C_2/C_1$.

In the bilinear SC resistor simulation, a complete clock period is really $T_c/2$, rather than T, because all switches change position twice every T_c seconds. If a new period is defined as

$$T' = \frac{T_c}{2} \quad (53)$$

then Eq. (52) can be written

$$H(z') = \frac{z' + 1}{(1 - \alpha) + (1 + \alpha)z'} = \frac{1 + z'^{-1}}{1 - \dfrac{\alpha - 1}{\alpha + 1} z'^{-1}} \frac{1}{1 + \alpha} \quad (54)$$

Equation (54) can be exactly obtained by applying the bilinear transformation to Eq. (1) where

$$\alpha = \frac{4}{\omega_1 T'} = \frac{C_2}{C_1} \quad (55)$$

The frequency response of Eq. (54) can be written as

$$|H(e^{j\omega T'})| = \left[\frac{1 + \cos\omega T'}{(1 + \cos\omega T') + \alpha^2(1 - \cos\omega T')}\right]^{1/2} = \left[\frac{1}{1 + \alpha^2 \tan^2\left(\dfrac{\omega T'}{2}\right)}\right]^{1/2} \quad (56)$$

and

$$\text{Arg } H(e^{j\omega T'})$$
$$= \tan^{-1}\left(\frac{\sin \omega T'}{1+\cos \omega T'}\right) - \tan^{-1}\left(\frac{(1+\alpha)\sin \omega T'}{(1-\alpha)+(1+\alpha)\cos \omega T'}\right) \quad (57)$$
$$= -\tan^{-1}\left[\frac{\alpha \sin \omega T'}{1+\cos \omega T'}\right] = -\tan^{-1}\left[\alpha \tan \frac{\omega T'}{2}\right]$$

The frequency response of Fig. 2.3–19 can be plotted using the relationship of eq. (55), as illustrated in the following example.

Example 2.3–4. *Frequency response of Fig. 2.3–19.* Plot the phase and magnitude of Fig. 2.3–19 for ω_c/ω_1 equal to 10 and to 50. Compare this result with the actual frequency response of Fig. 2.3–1. Figure 2.3–21 shows the frequency response of Fig. 2.3–19 plotted as a function of ω/ω_c (rather than ω/ω_c'). The actual clock frequency ω_c' is twice the clock frequency (ω_c) for all the previous realizations for Fig. 2.3–1.

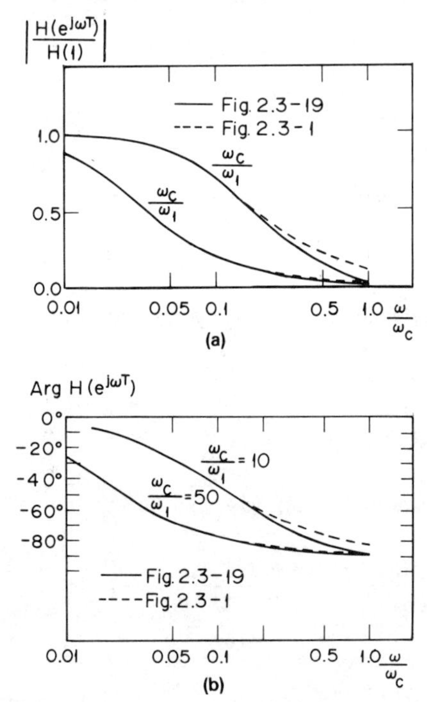

Fig. 2.3–21. (a) Magnitude and (b) phase response for Fig. 2.3–19 for $\omega_c/\omega_1 = 10$ and 50.

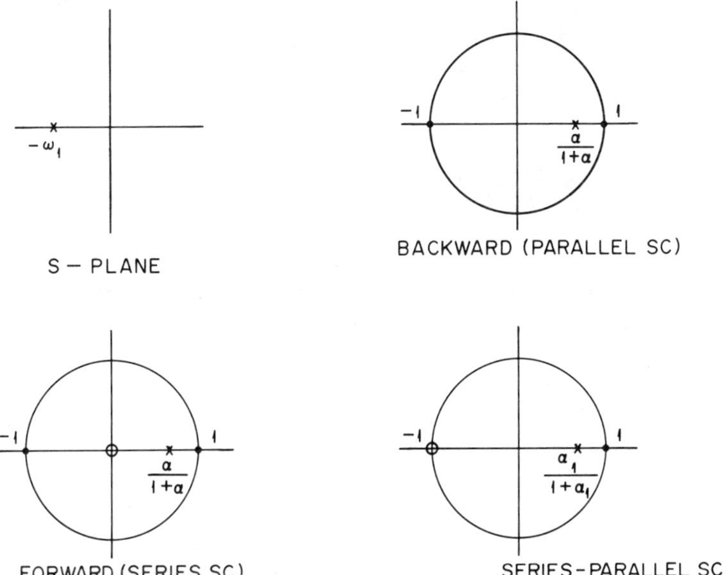

Fig. 2.3–22. Zero-pole diagram of the RC circuit of Fig. 2.3–1 and its SC equivalents.

In this section we have examined the use of the SC resistor simulations of Section 2.2 in realizing the passive low-pass, first-order network of Fig. 2.3–1. The zero-pole diagrams for the RC prototype and the first three RC-SC realizations are shown in Fig. 2.3–22.

In this particular case, of a passive RC low-pass, it was shown that the series, parallel, and bilinear SC resistor equivalents of Fig. 2.3–1 and shown in Figs. 2.3–3(a), 2.3–9(a), and 2.3–19 imply the backward, forward and bilinear mappings. The series-parallel (SP) combination of Fig. 2.3–14 also implies the bilinear mapping.

Care should be exercised when the SC resistor equivalents are used. For instance: (1) the simulation of a conventional grounded resistor using a parallel SC will not involve a backward mapping, but rather a forward mapping; (2) The series-parallel SC resistor equivalent in a passive RC circuit under certain circumstances does not involve a bilinear mapping, although, as will be shown in Section 2.5, an inverting active-RC integrator, the SP-SC resistor equivalent, involves exactly bilinear mapping. This section illustrated the importance of having the sampling frequency higher than the signal frequency. Also in this section, we have shown how the frequency response of an analog sampled data system may be calculated and plotted. The techniques used in this section will be applied throughout this text to other types of switched capacitor networks.

So far, our considerations have included only passive networks. In the next section we introduce the use of the op amp in switched capacitor networks. In many ways the op amp will simplify the methods of analysis and provide performance not possible in passive networks.

2.4 USE OF THE OP AMP IN SWITCHED CAPACITOR NETWORKS

Up to this point we have considered analog sampled data networks containing only switches and capacitors. These circuits have several disadvantages that limit their usefulness. One of these, for example, is the inability to completely transfer the charge on one capacitor to another. In this section we introduce the op amp into our repertoire of components. The op amp provides the needed flexibility to accomplish the design of practically all analog signal processing circuits using analog sampled data techniques.

The op amp is an extensively used component in analog circuit design. We shall consider briefly some of the ideal characteristics of op amps and then turn our attention to SC circuits containing op amps. In Chapter 8, the practical characteristics of the op amp will be presented in more detail. In Section 5.5, the influence of the practical characteristics of the op amp on the performance of analog sampled data circuits will be considered. Further information concerning op amps can be found in numerous texts, some of which are included in the bibliography.

Let us consider the voltage-controlled voltage-source (VCVS) in Fig. 2.4–1(a). This VCVS has a differential input. The difference in the voltage V_1 and V_2 controls the voltage source having a voltage gain of A_v. The output voltage, V_o, is referenced to AC ground and is equal to $A_v(V_1 - V_2)$ if the load resistance is much larger than R_o. R_i represents the differential input resistance seen between terminals 1 and 2. R_o is the output resistance in series with the controlled voltage source.

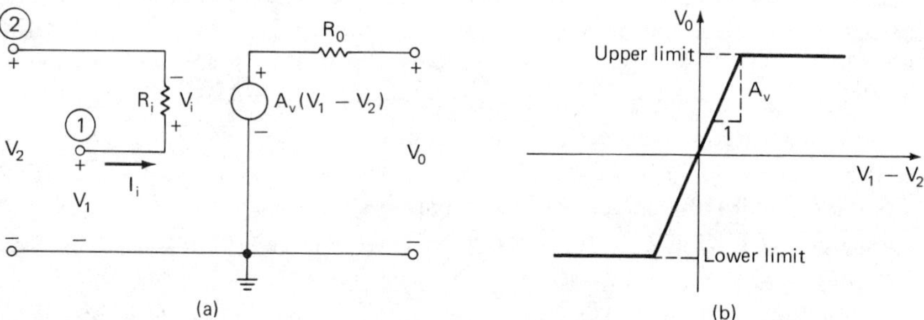

Fig. 2.4–1. (a) Model for a differential-input, voltage-controlled, voltage-source amplifier. (b) Output-input characteristics for (a).

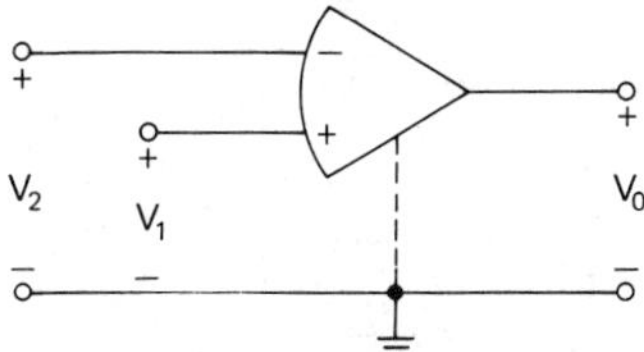

Fig. 2.4–2. Symbol for the op amp.

Figure 2.4–1(b) shows the transfer characteristics of Fig. 2.4–1(a) when $R_o = 0$ (or the load resistor is infinity). We see that for $V_1 - V_2$ close to zero that the output voltage is related to $V_1 - V_2$ by the voltage gain A_v. However, as $V_1 - V_2$ continues increasing or decreasing, the output voltage is limited. This limitation occurs because of finite power supplies.

In terms of Fig. 2.4–1, the ideal op amp is defined as

$$A_v = \infty \tag{1}$$

$$R_i = \infty \tag{2}$$

and

$$R_o = 0 \tag{3}$$

The symbol for the op amp is shown in Fig. 2.4–2. The ground connection shown by the dotted line is usually omitted. In reality, the gain of an ideal op amp is not infinity, though it is so large that one cannot tell whether it is infinity or not. Although op amps can be made from any controlled source whose forward gain becomes infinite, we shall assume in this work that the op amp is the VCVS type.

We next wish to show that the input terminals of an op amp can be modeled as a null port. A *null port* can be defined from Fig. 2.4–3 as a pair of terminals where both

$$V = 0 \quad \text{and} \quad I = 0 \tag{4}$$

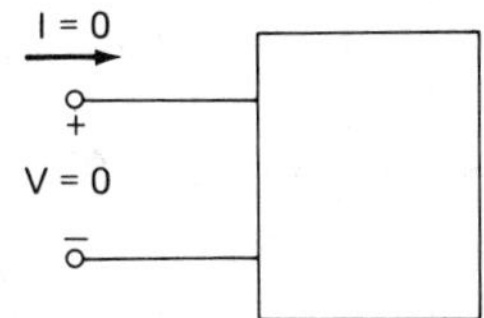

Fig. 2.4–3. Illustration of a null port.

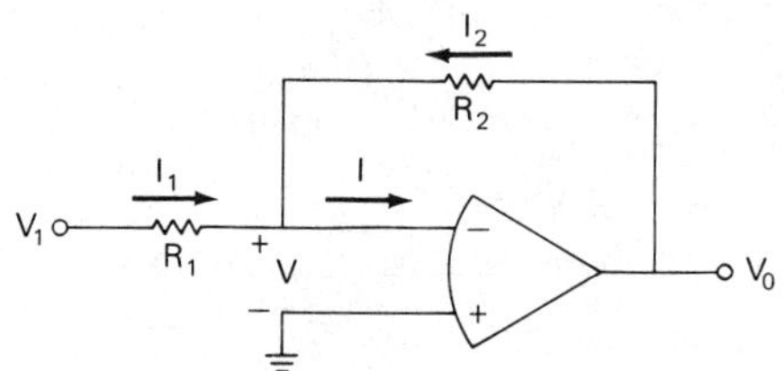

Fig. 2.4–4. Use of the null port to solve for the gain of an inverting amplifier.

The null port is often called a *virtual ground* when one of the terminals is at AC ground. It is easy to show that, if the gain of the VCVS realizing the op amp goes to infinity, the input terminals form a null port if some form of feedback exists. If the output voltage and current are finite, the transmission parameters of Fig. 2.4–1(a) are found to be

$$\begin{aligned} V_i = V_1 - V_2 &= (1/A_v)V_o - (R_o/A_v)I_o \\ I_i &= (1/A_vR_i)V_o - (R_o/A_vR_i)I_o \end{aligned} \tag{5}$$

If V_o and I_o are finite, then V_i and I_i approach zero as A_v approaches infinity.

The importance of the null port concept in analyzing op amp circuits can be illustrated in Fig. 2.4–4. This configuration is called the *inverting configuration.* Note that since the voltage and current of a null port are both zero, their polarity is not important.[17] The currents flowing in R_1 and R_2 are given as

$$I_1 = \frac{V_1 - V}{R_1} = \frac{V_1}{R_1} \tag{6}$$

and

$$I_2 = \frac{V_0 - V}{R_2} = \frac{V_0}{R_2} \tag{7}$$

since $V = 0$. At the junction of R_1, R_2, and the negative or inverting terminal of the op amp, the following nodal equation can be written

$$I_1 + I_2 = I = 0 \tag{8}$$

[17] One must keep track of the polarity of the op amp to obtain the desired feedback and to avoid instability.

since $I = 0$. Therefore $I_1 = -I_2$, and Eqs. (6) and (7) can be combined to give

$$\frac{V_0}{V_1} = -\frac{R_2}{R_1} \tag{9}$$

Thus, Fig. 2.4–4(a) is an inverting amplifier with a gain of R_2/R_1. Under ideal conditions the input resistance of Fig. 2.4–4(a) is

$$R_{\text{in}} = \frac{V_1}{I_1} = R_1 \tag{10}$$

and the output resistance is zero. In practice the input resistance is still approximately R_1, and the output resistance is given by

$$R_{\text{out}} = \frac{R_0}{1 + A_v \dfrac{R_1}{R_1 + R_2}} \tag{11}$$

where R_o and A_v are given in Fig. 2.4–1(a).

Example 2.4–1. *An inverting integrator.* Let us design a realization for an inverting integrator whose s-domain transfer function is given as

$$\frac{V_0(s)}{V_1(s)} = -\frac{\omega_0}{s} \tag{12}$$

Replacing R_1 and R_2 in Eq. (9) by $Z_1(s)$ and $Z_2(s)$, we get

$$\frac{V_0(s)}{V_1(s)} = -\frac{Z_2(s)}{Z_1(s)} \tag{13}$$

If we are restricted to resistors and capacitors, then an obvious choice for $Z_2(s)$ is a capacitor C_2 and for $Z_1(s)$ is a resistor R_1. Thus Eq. (13) becomes

$$\frac{V_0(s)}{V_1(s)} = -\frac{1}{sR_1C_2} \tag{14}$$

where $\omega_0 = 1/R_1C_2$. Figure 2.4–5 shows the resulting inverting integrator.

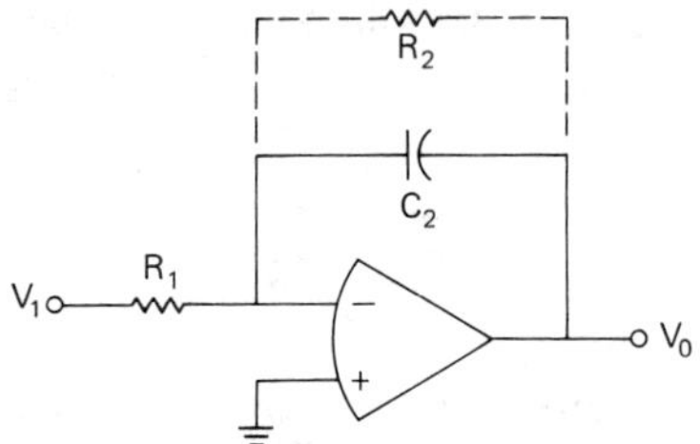

Fig. 2.4–5. Design of an inverting integrator from Example 2.4–1.

Example 2.4–1 brings out an important problem that can occur in the use of the op amp. This problem is that at $\omega = 0$, there is no feedback around the op amp. Consequently the gain $V_0(0)/V_1(0)$ becomes equal to A_v, which is approaching infinity. If $V_1(0)$ is finite, then $V_0(0)$ will be extremely large and in practice will be at either the upper or lower limits. Consequently, we always want to have some form of DC feedback around the op amp to prevent this circumstance. In Fig. 2.4–5, the addition of R_2 where $R_2 >> R_1$ will solve this problem.

Figure 2.4–6 illustrates the noninverting configuration of an op amp. We may write

$$V_1 - V = I_1 R_1 \tag{15}$$

$$I_1 + I = I_2 \tag{16}$$

and

$$V_0 = I_1 R_1 + I_2 R_2 \tag{17}$$

But using the null port concept, Eqs. (15) and (16) simplify to

$$V_1 = I_1 R_1 \tag{18}$$

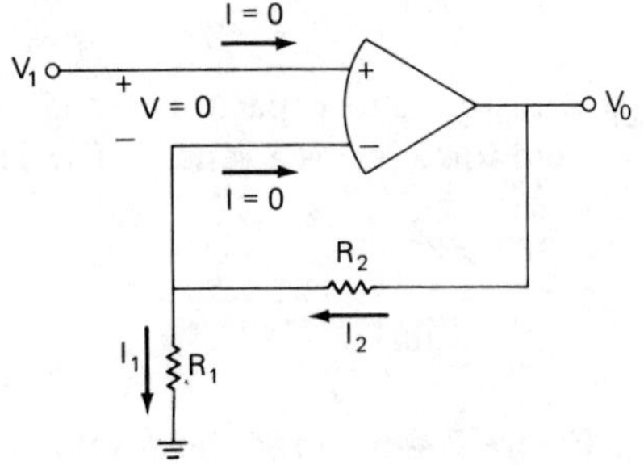

Fig. 2.4–6. Use of the null port to solve for the gain of a noninverting amplifier.

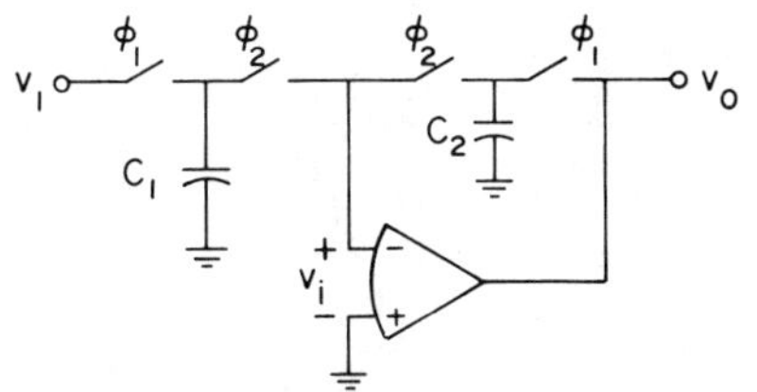

Fig. 2.4–7. An impractical realization of Fig. 2.4–4.

and

$$I_1 = I_2 \tag{19}$$

Substituting Eqs. (18) and (19) into Eq. (17) gives

$$\frac{V_0}{V_1} = \frac{R_1 + R_2}{R_1} = 1 + \frac{R_2}{R_1} \tag{20}$$

We observe that the gain of Fig. 2.4–6 can never be less than unity. Under ideal conditions, the input resistance of Fig. 2.4–6 is infinity, and the output resistance is zero. Under practical conditions, the input resistance is

$$R_{\text{in}} = R_{\text{i}}\left(1 + \frac{A_v\, R_1}{R_1 + R_2}\right) \tag{21}$$

where R_i and A_v are shown in Fig. 2.4–1(a). The output resistance for the noninverting configuration is the same as the inverting configuration and is given by Eq. (11).

In this text we are constraining ourselves to circuits that use only capacitors, switches, and op amps. Consequently let us consider the replacement of the resistors in Fig. 2.4–4 and 2.4–6 by SC equivalents. Fig. 2.4–7 shows an inverting amplifier that does not use resistors. The resistors of Fig. 2.4–4 have been replaced by the parallel SC resistor realization of Fig. 2.2–1(a). Let us first see if the circuit will work. Unfortunately, we see that for both the ϕ_1 and ϕ_2 phase periods the op amp has no feedback path. Under this condition no null port can exist. Therefore, v_i is likely to have a nonzero value of voltage. Because this voltage is amplified by A_v, the output will be at one of its limits. Although this circuit can be easily fixed by several modifications, let us find a different realization that avoids the above problem.

In Fig. 2.4–8(a), we replace R_1 of Fig. 2.4–4 with the parallel SC resistor realization and R_2 with a modified series SC resistor realization. The series switch of this series SC resistor realization has been removed because it is

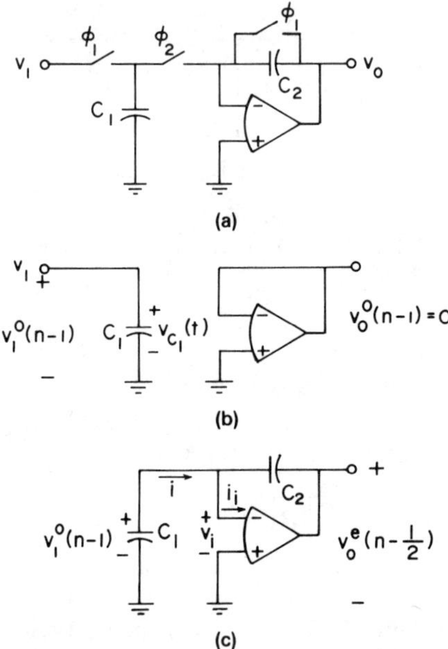

Fig. 2.4–8. (a) A practical realization of Fig. 2.4–4. (b) Equivalent circuit of (a) when ϕ_1 is closed. (c) Equivalent circuit of (a) when ϕ_2 is closed.

not necessary and would cause the op amp to have no feedback during one of the phase periods. The operation of Fig. 2.4–8(a) is developed as follows. Let us start with the odd phase period $(n-1) \le t/T < (n-\frac{1}{2})$, which will correspond to ϕ_1. Figure 2.4–8(b) shows the equivalent circuit of Fig. 2.4–8(a) during this phase period. We see that C_1 is charged to

$$v_{c_1}^o(n-1) = v_1^o(n-1) \tag{22}$$

and that C_2 is discharged, resulting in

$$v_0^o(n-1) = 0 \tag{23}$$

During the next even phase period, $(n-\frac{1}{2}) \le t/T < n$, the ϕ_2 switch closes. The conditions when the ϕ_2 switch closes at $t=(n-\frac{1}{2})$ are illustrated in Fig. 2.4–8(c). First, we will derive the transfer function $H^{oe}(z)$ using conventional circuit analysis. Because C_1 is across the null port of the op amp ($v_i = 0$), it discharges instantaneously, assuming no series resistance. This discharge creates a current i that cannot flow into the op amp because of the null port ($i_i = 0$). Therefore i flows through C_2 and places the charge previously on C_1 all on C_2. A useful relationship for this analysis is

$$i = C\frac{dv}{dt} \tag{24}$$

where i flows through the capacitor C from the + terminal to the − terminal. Due to the discharge of C_1, i of Fig. 2.4–8(c) is given as

$$i = -C_1\left[\frac{v_1^e\left(n-\frac{1}{2}\right) - v_1^o(n-1)}{T/2}\right] = \frac{C_1\, v_1^o(n-1)}{T/2} \tag{25}$$

Applying Eq. (24) to C_2 gives

$$i = -C_2\left[\frac{v_0^e\left(n-\frac{1}{2}\right) - v_0^o(n-1)}{T/2}\right] = -\frac{C_2 v_0^e\left(n-\frac{1}{2}\right)}{T/2} \tag{26}$$

Equating Eqs. (25) and (26) gives

$$v_0^e\left(n-\frac{1}{2}\right) = -\frac{C_1}{C_2}\, v_1^o(n-1) \tag{27}$$

Applying Table 1.4–1 to Eq. (27) results in

$$H^{oe}(z) = \frac{V_0^e(z)}{V_1^o(z)} = -\frac{C_1}{C_2}\, z^{-1/2} \tag{28}$$

We now see that Fig. 2.4–8 acts like an inverting amplifier with a gain of C_1/C_2 and a delay of $T/2$ seconds. Note that voltage gain can be achieved if C_1 is larger than C_2. Also we note that at no time is the op amp without some form of feedback. In the even phase period, C_2 is charged to whatever voltage is given by Eq. (27) and will hold that voltage indefinitely under ideal conditions. Therefore, as long as $v_1^o(n-1)$ does not cause $v_o^e(n-½)$ to limit, the circuit performs as expected.

Let us also use the charge conservation approach to analyze Fig. 2.4–8(b). Equation (16) of Sec. 2.3 applied to C_2 results in identifying $q_m^o(n-1)$ as zero and $q_c^o(n-1)$ as $-C_1\, v_1^o(n-1)$. Therefore

$$q_c^o(n-1) = -C_1\, v_1^o(n-1) \tag{29}$$

but since

$$q_L^e\left(n-\frac{1}{2}\right)=C_2\,v_0^e\left(n-\frac{1}{2}\right) \tag{30}$$

we arrive at the same result as Eq. (27) by equating Eqs. (29) and (30).

The block diagram of this circuit in the z-domain is given by Fig. 2.4–9. We see that the signal $V_1(z)$ is multiplied (or amplified) by $-C_1/C_2$ and delayed by $T/2$. In most applications where the sampling rate is high, the realization approaches ideally Fig. 2.4–4. This circuit will be very useful in our later studies.

An alternative form of Fig. 2.4–8 may be developed by using the series SC resistor equivalent of Fig. 2.2–3 to replace R_1 of Fig. 2.4–4. The result is shown in Fig. 2.4–10(a). In the ϕ_2 phase period, both capacitors are discharged, so that in the ϕ_1 phase period the equivalent circuit is given in Fig. 2.4–10(b). This circuit can be analyzed by continuous methods to get

$$H^{oo}(z)=\frac{V_0^o}{V_1^o}=-\frac{C_1}{C_2} \tag{31}$$

which holds only during the odd phase period. We note that Fig. 2.4–10(a) does not have the delay of Fig. 2.4–8(a).

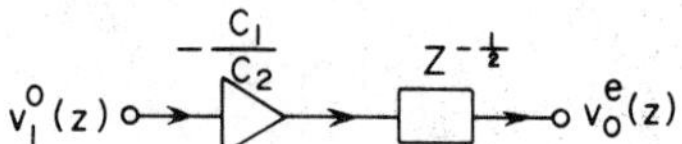

Fig. 2.4–9. z-domain block diagram of Fig. 2.4–8(a).

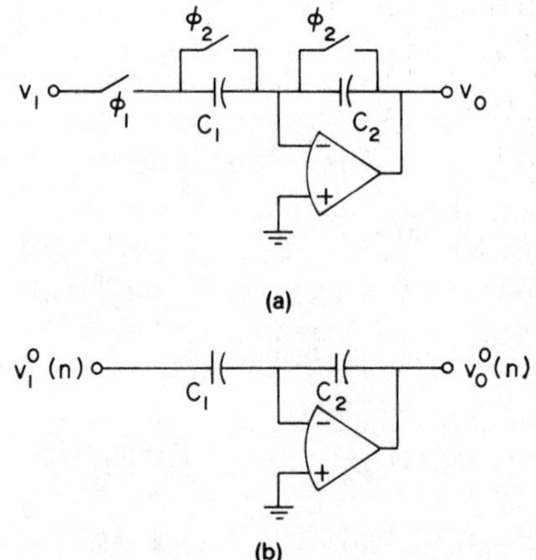

Fig. 2.4–10. (a) An SC realization of Fig. 2.4–4 using series configuration. (b) Equivalent circuit of (a) during the odd phase period.

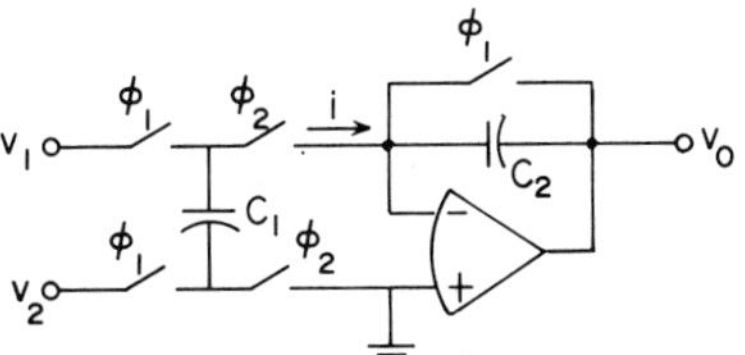

Fig. 2.4–11. SC realization of a differential amplifier.

Figure 2.4–8(a) is probably the most useful form of the inverting amplifier. Because of the flexibility of switches, it turns out that Fig. 2.4–8(a) can be easily converted to a noninverting amplifier. Let us consider the circuit of Fig. 2.4–11. Here, instead of grounding C_1 during ϕ_1 phase period, we connect it between two voltages, $v_1(t)$ and $v_2(t)$. Therefore Eq. (29) becomes

$$q_c^o(n-1) = -C_1[v_1^o(n-1) - v_2^o(n-1)] \tag{32}$$

As in Eq. (30) we have

$$q_L^e\left(n-\frac{1}{2}\right) = C_2\, v_o^e\left(n-\frac{1}{2}\right) \tag{33}$$

Using the charge conservation equation, we get

$$\begin{aligned} v_o^e\left(n-\frac{1}{2}\right) &= -\frac{C_1}{C_2}\, v_1^o(n-1) + \frac{C_1}{C_2}\, v_2^o(n-1) \\ &= \frac{C_1}{C_2}\,[v_2^o(n-1) - v_1^o(n-1)] \end{aligned} \tag{34}$$

where we have now developed a differential amplifier. If a noninverting amplifier only is desired, then we simply ground v_1 to get

$$v_o^e\left(n-\frac{1}{2}\right) = \frac{C_1}{C_2}\, v_2^o(n-1) \tag{35a}$$

Thus

$$H^{oe}(z) = \frac{C_1}{C_2}\, z^{-1/2} \tag{35b}$$

Because it is more desirable to use the op amp with the noninverting terminal grounded, we shall use Fig. 2.4–11 to obtain noninverting amplifiers rather than develop an SC realization for Fig. 2.4–6.

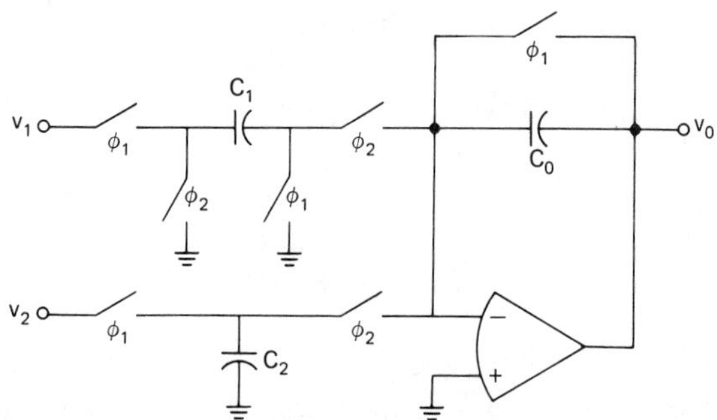

Fig. 2.4–12. Realization for Example 2.4–2.

Example 2.4–2. *Design of an SC amplifier.* Find an SC realization of the continuous function $v_0(t) = 4v_1(t') - 2v_2(t')$, $t > t'$. Figure 2.4–12 is seen to have the response

$$v_o^e\left(n-\frac{1}{2}\right) = \frac{C_1}{C_o}\, v_1^o\,(n-1) - \frac{C_2}{C_o}\, v_2^o\,(n-1) \tag{36}$$

Selecting $C_1/C_o = 4$ and $C_2/C_o = 2$ gives the desired realization.

The SC equivalent resistors of Figs. 2.2–3 and 2.2–4 can also be used to realize Fig. 2.4–4. The series-parallel SC realization has the same problem as Fig. 2.4–7 and offers no advantage over Figs. 2.4–8(a) or 2.4–10(a). The bilinear SC resistor equivalent does not have the problem of causing the op amp output to limit and can essentially be placed anywhere in an RC op amp circuit without problems. An inverting amplifier realization using the bilinear SC resistor is shown in Fig. 2.4–13. It can be shown (see Problem 2.6c) that the z-domain transfer function of Fig. 2.4–13 is

$$H(z) = -\frac{C_1}{C_2} \tag{37}$$

It should be noted that the clock phases are not important in this configuration as long as each capacitance is reversed once during each clock period.

Figure 2.4–14(a) shows a resistive summer that is very useful in continuous analog circuits. This circuit can be implemented by any of the previous techniques to provide an SC realization of Fig. 2.4–14(a). Figure 2.4–14(b) gives an example of an SC summer extending the concepts of Fig. 2.4–8(a). Note that, unlike the continuous resistive summer of Fig. 2.4–14(a), the analog

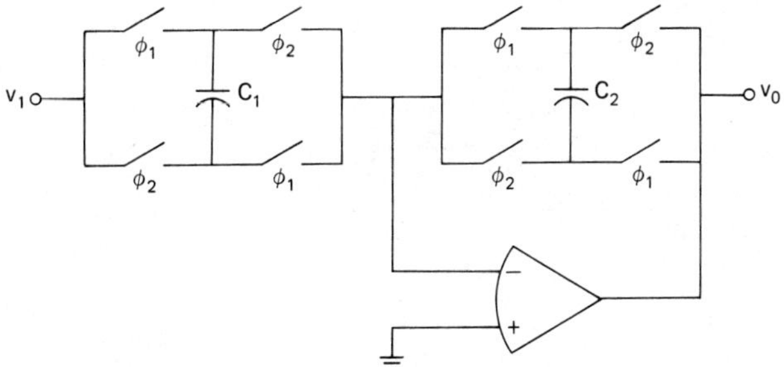

Fig. 2.4–13. SC realization of Fig. 2.4–4 using bilinear configuration.

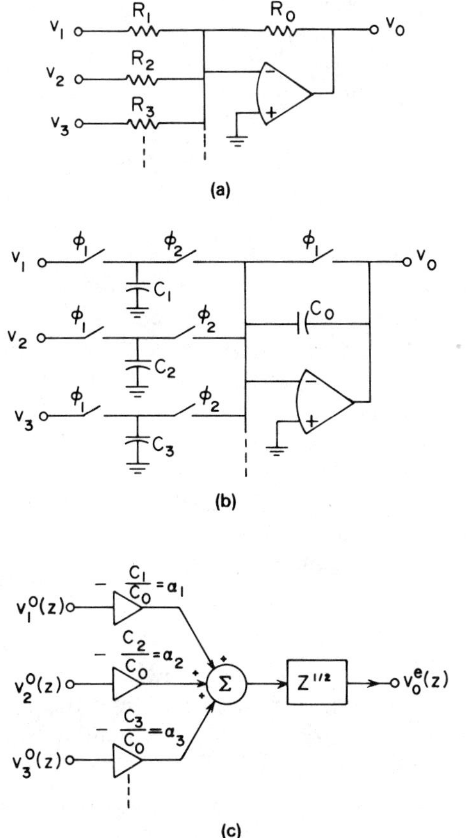

Fig. 2.4–14. (a) Continuous inverting summer. (b) SC realization of inverting summer. (c) z-domain block diagram of (b).

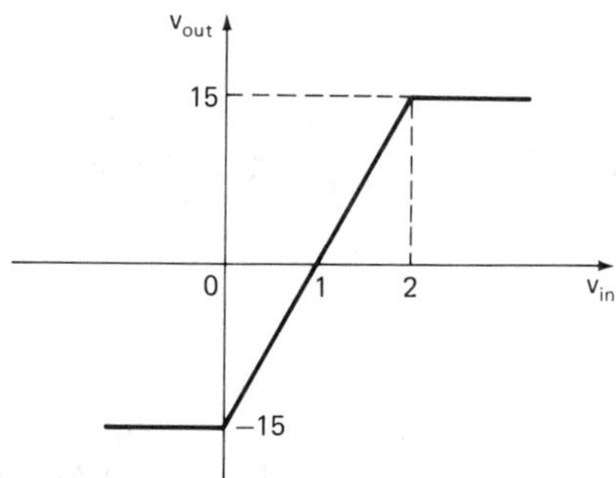

Fig. 2.4–15. Desired transfer characteristic for Example 2.4–3.

sampled data version is capable of summing signals of either polarity using only one op amp, with the input configuration of Fig. 2.4–11 with $v_1 = 0$. The z-domain block diagram of Fig. 2.4–14(b) is given in Fig. 2.4–14(c). This circuit will become a useful building block in Chapter 5.

Example 2.4–3. *Design of an analog sampled data amplifier with DC offset.* An amplifier is to be built using the techniques of this section and having the transfer characteristics shown in Fig. 2.4–15. Assume that the op amp is powered from $\pm$ 15 volts and that the output is also limited at $\pm$ 15 volts. We see from Fig. 2.4–15 that the slope of the linear portion of the transfer characteristics is 15. Therefore the signal gain must be 15. However, at $v_{in} = 0$, v_{out} must be -15 volts. Consequently a DC potential must be summed with the signal input. In Example 2.4–2, a circuit was given in Fig. 2.4–12 that is capable of performing the desired function if $v_1 = v_{in}$ and $v_2 = 15$ volts. Therefore

$$v_o^e\left(n-\frac{1}{2}\right)=\frac{C_1}{C_o}\,v_{in}^o\,(n-1)-\frac{C_2}{C_o}\,15 \tag{38}$$

If we let $C_1/C_o = 15$ and $C_2/C_o = 1$, then the characteristics of Fig. 2.4–15 will be realized.

Some circuits built from capacitors, switches, and op amps were in use long before such components were used to design the circuits shown in this text. One such case is the sample-and-hold circuit shown in Fig. 2.4–16(a). In this circuit the switch is closed connecting the capacitor and unity gain buffer[18] to the input signal v_1. When the switch opens, say at $t = t_o$, then

[18] *Buffer* is a term for a circuit with unity voltage gain, very high input impedance, and very low output impedance.

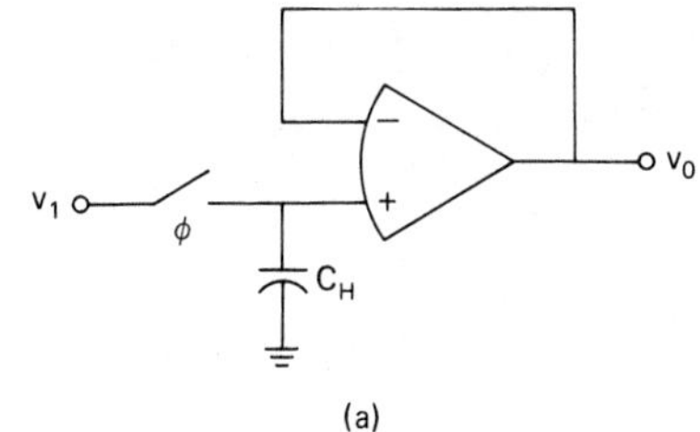

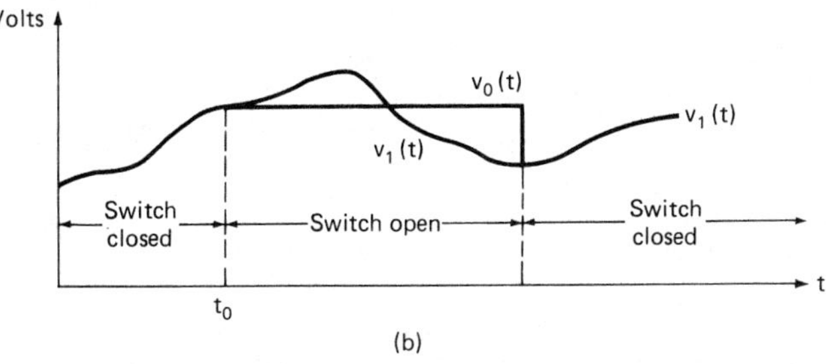

Fig. 2.4–16. (a) Simple sample-and-hold circuit. (b) Waveforms illustrating the operation of (a).

the value of $v_1(t_o)$ is stored on C and is available at v_o. The waveforms describing this circuit are shown in Fig. 2.4–16(b). One useful application of Fig. 2.4–16(a) is found in most of the previous SC amplifier realizations. It was observed that the z-domain transfer function was multiplied by a half-delay, $z^{-1/2}$. This $z^{-1/2}$ can be changed to a full delay, z^{-1}, by the use of a sample-and-hold circuit. Figures 2.4–8(a) and 2.4–16(a) can be combined to result in Fig. 2.4–17. Here we see that during the ϕ_2 phase $v_o(n)$ is stored in the holding capacitor C_H. During the next phase period, when ϕ_1 closes, $v_o(n)$ is available at v_{OH}. Therefore

$$v_{OH}^{o}\left(n+\frac{1}{2}\right)=v_o^e(n) \tag{39}$$

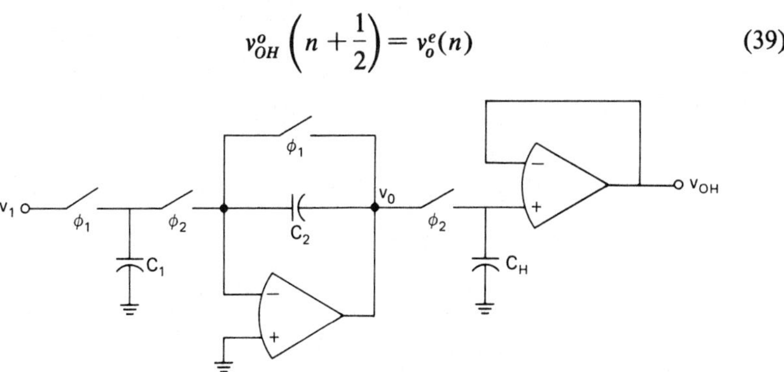

Fig. 2.4–17. Use of the sample-and-hold to obtain an inverting amplifier with a full delay.

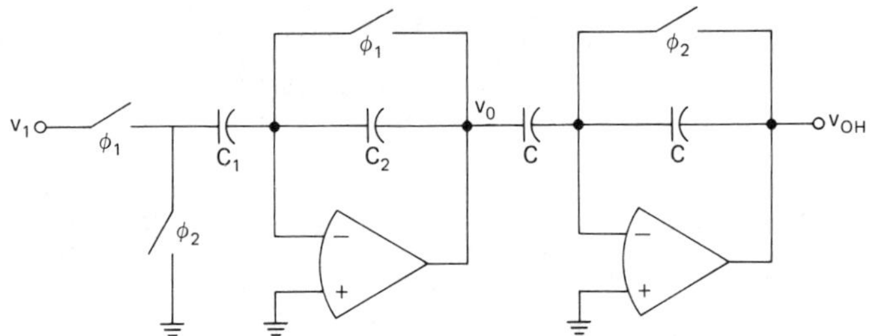

Fig. 2.4–18. Implementation of a sample-and-hold arrangement to achieve an amplifier with a full delay.

or

$$V_{OH}^o(z) = V_o^e(z)\, z^{-1/2} = -\frac{C_1}{C_2}\, z^{-1}\, V_1^o(z) \tag{40}$$

where $V_o^e(z)$ has been replaced by Eq. (28). Illustrations of Eqs. (28) and (39) are found in Figs. 1.2–4(a) and (b), respectively. Figure 2.4–18 is also an equivalent realization of Fig. 2.4–17. Other forms of sample-and-hold circuits can be used in place of Fig. 2.4–16(a), such as the circuit shown in Fig. 2.4–19. The operating principles are the same as in Fig. 2.4–14(a), and the objective of the additional complexity is to increase the performance.

In many situations it is not necessary to use switches and to still implement continuous functions without using resistors. Such an example is the charge amplifier of Fig. 2.4–20(a). The gain is seen to be given as

$$\frac{v_O}{v_1} = -\frac{C_1}{C_2} \tag{41}$$

Unfortunately, a resistor such as R_2 is required for the reasons given following Example 2.4–1. Consequently, while Fig. 2.4–20(a) or the summer of Fig. 2.4–20(b) may be a convenient realization, if a DC feedback path does not

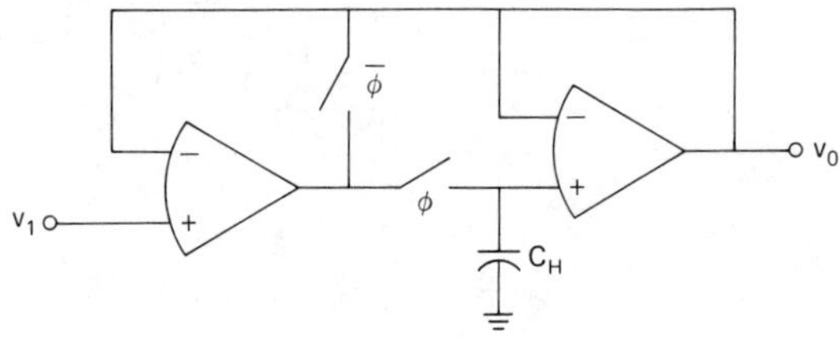

Fig. 2.4–19. A two-op amp sample-and-hold circuit.

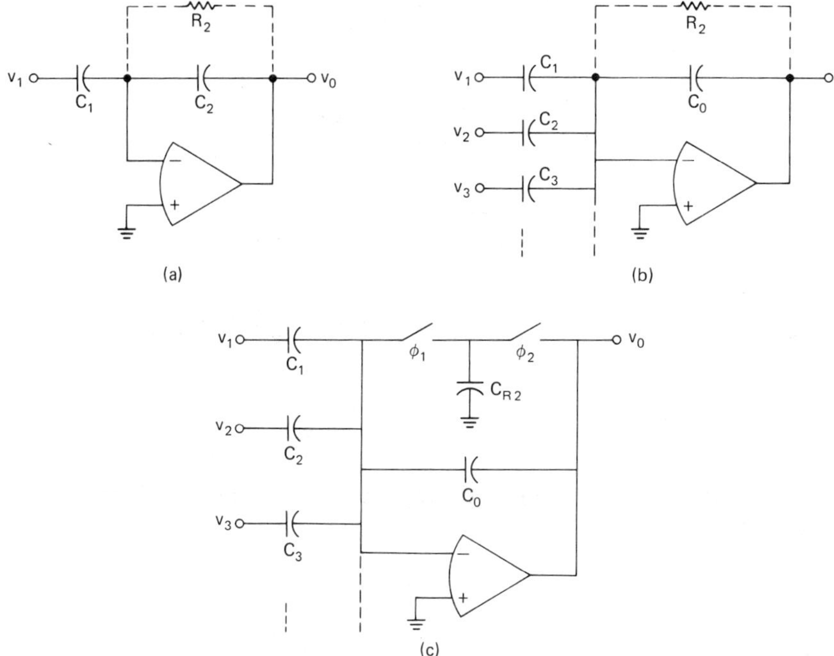

Fig. 2.4–20. (a) Inverting amplifier using capacitors and op amps. (b) Summer using only capacitors and op amps. (c) Practical realization of (b) using SC techniques.

exist, then an SC resistor must be included, as shown in Fig. 2.4–20(c). However, in many cases there will be external DC feedback paths around the amplifier, so that R_2 is not necessary.

2.5 INTEGRATORS

One of the more important circuits in analog signal processing systems is the integrator. Its primary use is in the realization of active filters, although it is used in many other applications. In this section we examine methods of simulating this very useful building block using analog sampled data circuits. The approach presented will replace the resistances of the integrator by the SC resistor equivalent circuits presented in Section 2.2. Each of the resulting SC integrators will be characterized with respect to its z-domain transfer function, frequency response, and z-domain block diagram.

An inverting analog integrator is shown in Fig. 2.5–1. We shall ignore the fact that in practice the op amp will require some form of DC feedback. The transfer function can be easily found using the null port concept intro-

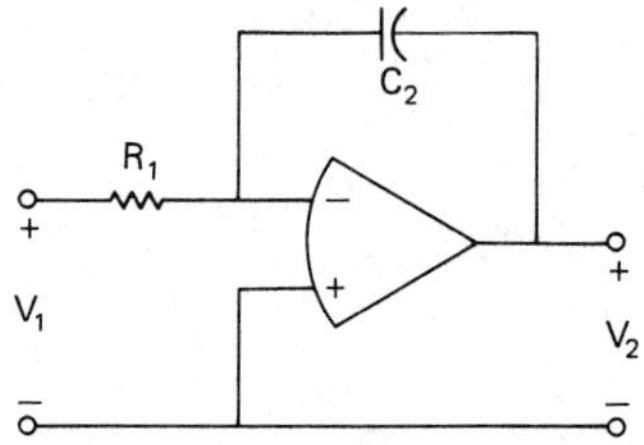

Fig. 2.5–1. Inverting analog integrator.

duced in the previous section. From Eq. (13) of Section 2.4, we find that

$$H(s)=\frac{V_2(s)}{V_1(s)}=-\frac{1}{sR_1C_2}=-\frac{1}{s\tau}=-\frac{\omega_o}{s} \tag{1}$$

where τ is the time constant of the integrator. The magnitude of the inverting integrator is

$$|H(j\omega)|=\frac{\omega_o}{\omega} \tag{2}$$

and the phase shift is

$$\text{Arg H}(j\omega)=\frac{\pi}{2} \tag{3}$$

The frequency response of the ideal inverting integrator of Fig. 2.5–1 is shown in Fig. 2.5–2.

A noninverting analog integrator is more complex than Fig. 2.5–1. Although an inverter can be cascaded with Fig. 2.5–1, we shall consider the

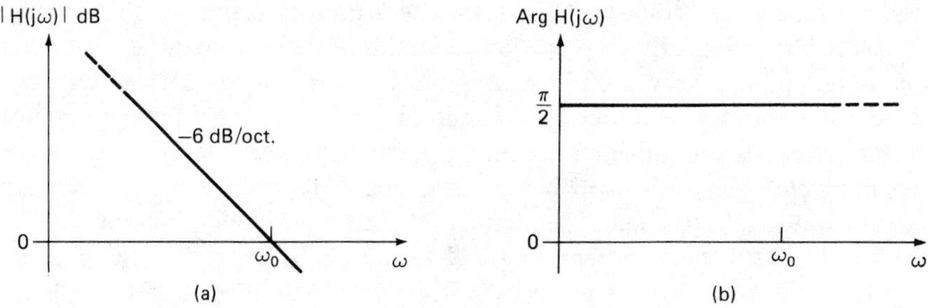

Fig. 2.5–2. (a) Magnitude and (b) phase response of Fig. 2.5–1.

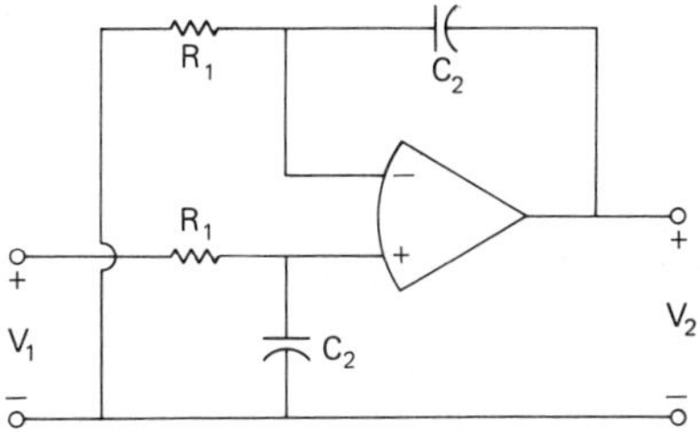

Fig. 2.5–3. Noninverting analog integrator.

circuit of Fig. 2.5–3. It can be shown using the null port concept that

$$H(s) = \frac{V_2(s)}{V_1(s)} = \frac{1}{sR_1C_2} = \frac{1}{s\tau} = \frac{\omega_o}{s} \tag{4}$$

The magnitude of this noninverting integrator is given by Eq. (2). The phase shift is given as

$$\text{Arg } H(j\omega) = -\frac{\pi}{2} \tag{5}$$

Except for the phase shift, Figs. 2.5–1 and 2.5–3 are alike.

The performance described above for the inverting and noninverting analog integrators will be simulated by analog sampled data circuits. Each analog sampled data realization will be compared with the performance of these two circuits. In Chapter 5 we shall consider nonideal integrators and shall examine some of the influences of the op amp finite DC gain and finite gain-bandwidth product.

The first SC realization of the analog integrator will replace the resistor R_1 by a parallel SC resistor equivalent. This realization will be called the *parallel SC integrator.* R_1 of Fig. 2.5–1 is replaced by the parallel SC resistor equivalent of Fig. 2.2–1(a). The resulting SC integrator is shown in Fig. 2.5–4(a). Fig. 2.5–4(b) shows the clock sequence for the SC integrator. During the odd phase periods we note that the charge across C_2 is constant. This charge cannot change because to do so would require current flow into the null port of the op amp. An equivalent circuit of Fig. 2.5–4(a) is shown in Fig. 2.5–4(c) for the *even phase period.* Using the charge conservation at node A (Fig. 2.5–4(c)) and identifying

$$q_L^e\left(n - \frac{1}{2}\right) = C_2\, v_2^e\left(n - \frac{1}{2}\right) \tag{6}$$

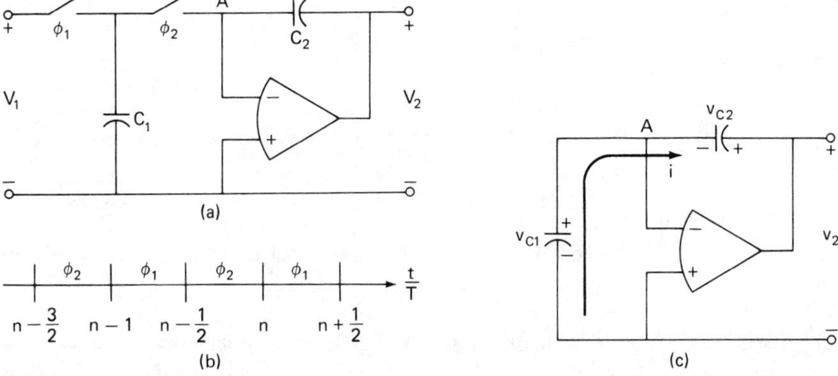

Fig. 2.5–4. (a) Parallel SC inverting integrator. (b) Clock sequence. (c) Equivalent circuit during the ϕ_2 phase period.

$$q_m^o(n-1) = C_2\, v_2^o(n-1) \tag{7}$$

and

$$q_c^o(n-1) = -\, C_1\, v_1^o(n-1) \tag{8}$$

Thus we can write

$$C_2\, v_2^e\left(n-\frac{1}{2}\right) = C_2\, v_2^o(n-1) - C_1\, v_1^o(n-1) \tag{9}$$

After taking the z-transform and multiplying by $z^{1/2}$, we can write Eq. (9) as

$$C_2\, V_2^e(z) = C_2\, V_2^o(z)z^{-1/2} - C_1\, V_1^o(z)z^{-1/2} \tag{10}$$

During the *odd phase,* the charge on C_2 does not change; then we can write

$$v_2^o(n-1) = v_2^e\left(n-\frac{3}{2}\right) \tag{11a}$$

or

$$V_2^o(z) = z^{-1/2}\, V_2^e(z) \tag{11b}$$

Combining Eqs. (11b) and (10), we obtain

$$H^{oe}(z) = \frac{V_2^e(z)}{V_1^o(z)} = -\frac{C_1}{C_2}\frac{z^{-1/2}}{1-z^{-1}} \tag{12}$$

Using (11b), Eq. (12) can be also expressed as

$$H^{oo}(z) = \frac{V_2^o(z)}{V_1^o(z)} = -\frac{C_1}{C_2}\frac{z^{-1}}{1-z^{-1}} \tag{13}$$

The integrator defined by $H^{oe}(z)$ has a delay of $T/2$ in the forward path. An integrator with this half-delay property is known as a *Type I Lossless Discrete Integrator* (LDI)[19,20] Note that the integrator defined by $H^{oo}(z)$ has a full delay in the forward path of the integrator. This integrator is known as a *Type I Direct-Transform Discrete Integrator* (DDI).

The reason this integrator is called the direct-transform discrete integrator can be seen as follows. In Eq. (1) we can replace R_1 by Eq. (10) of Section 2.2 to get

$$H(s) = -\frac{C_1}{s\,TC_2} \tag{14}$$

Equating Eqs. (13) and (14) gives

$$s = \frac{z-1}{T} = \frac{1-z^{-1}}{Tz^{-1}} \tag{15}$$

which is the *forward transformation* described in Chapter 1. Consequently, we can say that using the Type I DDI to replace Fig. 2.5–1 is equivalent to applying the forward transformation. This is very useful information, for now we know the proper mapping between the s- and z-domains for SC circuits where all analog integrators have been replaced by Type I DDI equivalents and the proper sampling of the input and output signals.

Let us examine the frequency response of the parallel SC integrator of Fig. 2.5-4(a). Replacing z by $e^{j\omega T}$ in Eq. (12) results in

---

[19] L. T. Bruton, "Low Sensitivity Digital Ladder Filters," *IEEE Trans. on Circuits and Systems,* Vol. CAS-22, No. 3, March 1975, pp. 168–176.

[20] R. W. Brodersen, P. R. Gray, and D. A. Hodges, "MOS Switched-Capacitors Filters," *Proc. of IEEE,* Vol. 67, No. 1, January 1979, pp. 61–75.

$$H^{oe}(e^{j\omega T}) = -(C_1/C_2)\left(\frac{e^{-j\omega T/2}}{1 - e^{-j\omega T}}\right) \tag{16}$$

Multiplying the numerator and denominator by $e^{j\omega T/2}$ gives

$$H^{oe}(e^{j\omega T}) = -\frac{C_1}{C_2}\frac{1}{e^{-j\omega T/2} - e^{-j\omega T/2}} \tag{17}$$

Application of Euler's formula to (17) gives the desired result expressed as

$$H^{oe}(e^{j\omega T}) = -\frac{C_1}{j\omega TC_2}\left(\frac{\dfrac{\omega T}{2}}{\sin\dfrac{\omega T}{2}}\right) = -\frac{\omega_o}{j\omega}\left(\frac{\dfrac{\omega T}{2}}{\sin\dfrac{\omega T}{2}}\right) \tag{18}$$

where $\omega_o = C_1/TC_2$, which is consistent with the replacement of R_1 in Eq. (1) with T/C_1. We see that the magnitude of (18) is

$$|H^{oe}(e^{j\omega T})| = \frac{\omega_o}{\omega}\left(\frac{\omega T/2}{\sin\dfrac{\omega T}{2}}\right) \tag{19}$$

and the phase shift is

$$\text{Arg } H^{oe}(e^{j\omega T}) = \frac{\pi}{2} \tag{20}$$

It is seen that the magnitude is equal to the analog magnitude given by Eq. (2) except for the term in parentheses, and the phase shift is exactly that of Eq. (3). The term in parentheses is called a *gain error.* The error it causes in the magnitude as a function of the ratio of the applied frequency to the clock frequency is given in Table 2.5–1. We see that when $f < .05\ f_c$ this magnitude error is negligible.

The frequency response of Eq. (13) can be found in a similar manner and is given as

$$H^{oo}(e^{j\omega T}) = -\frac{\omega_o}{j\omega}\left(\frac{\dfrac{\omega T}{2}}{\sin\dfrac{\omega T}{2}}\right)\exp\left(-j\frac{\omega T}{2}\right) \tag{21}$$

Table 2.5–1 Magnitude and phase (delay) errors in switched capacitor integrators versus normalized frequency (Type I DDI)

NORMALIZED FREQUENCY $\frac{f}{f_c}$	ERROR IN GAIN CONSTANT MAGNITUDE	ERROR IN PHASE FROM IDEAL 90°
0.00	0.00%	0°
0.05	0.41%	9°
0.10	1.66%	18°
0.15	3.80%	27°
0.20	6.90%	36°
0.25	11.07%	45°
0.30	16.50%	54°
0.35	23.41%	63°
0.40	32.13%	72°
0.45	43.13%	81°
0.50	57.08%	90°

Therefore the magnitude of Eq. (21) is

$$|H^{oo}(e^{j\omega T})| = \frac{\omega_o}{\omega}\left(\frac{\frac{\omega T}{2}}{\sin\frac{\omega T}{2}}\right) \tag{22}$$

and the phase shift of Eq. (21) is

$$\text{Arg } H^{oo}(e^{j\omega T}) = \frac{\pi}{2} - \frac{\omega T}{2} \tag{23}$$

The primary difference between the Type I LDI and Type I DDI integrators is the additional phase lag term shown in Eq. (23). The effects of this additional delay are also shown in Table 2.5–1.

The noninverting analog integrator, Fig. 2.5–3, can be realized by simply providing a means of reversing C_1 of Fig. 2.5–4. Using the technique illustrated in Fig. 2.4–11 (with $v_1 = 0$) results in Fig. 2.5–5. The transfer function of this circuit is exactly equivalent to that of Fig. 2.5–4(a) except that the integrator is noninverting. Therefore the z-domain transfer function of Fig. 2.5–5 used as a Type I LDI is

$$H^{oe}(z) = \frac{C_1}{C_2}\frac{z^{-1/2}}{1 - z^{-1}} \tag{24}$$

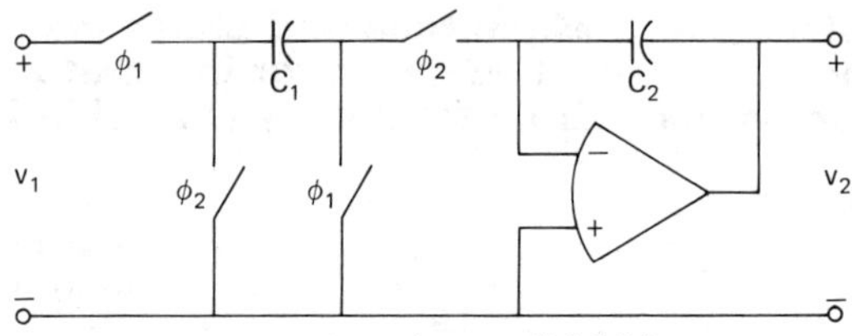

Fig. 2.5–5. A noninverting parallel SC integrator.

and as a Type I DDI is

$$H^{oo}(z) = \frac{C_1}{C_2} \frac{z^{-1}}{1 - z^{-1}} \tag{25}$$

The magnitude of the frequency response is given by Eqs. (19) and (22), the phase shift for the Type I LDI is given by Eq. (5), and the phase shift for the Type I DDI is given by

$$\text{Arg } H^{oo}(e^{j\omega T}) = -\frac{\pi}{2} - \frac{\omega T}{2} \tag{26}$$

The results of this parallel SC integrator are summarized in Table 2.5–2, where the magnitude and the phase shift of the frequency response are given.

Example 2.5–1. *Error in frequency response of a Type I DDI.* Find the actual magnitude and phase of an inverting Type I DDI if the frequency applied to the integrator is 10 kHz, the clock frequency is 100 kHz, and ω_0 is 20π krps. The ideal integrator will have a magnitude of unity at 10 kHz and a phase shift of $\pi/2$. Because $f/f_c = 0.1$, Table 2.5–1 shows that the gain will be 1.0166 rather than unity, and the phase shift will be 72° rather than 90°.

The analog and SC integrator frequency responses for the circuits previously considered are shown in Fig. 2.5–6. The z-domain block diagram of the parallel SC integrators can be developed as follows. Consider Eq. (13) rewritten as

$$V_2^o(z) = z^{-1}\, V_2^o(z) - \frac{C_1}{C_2}\, z^{-1}\, V_1^o(z) \tag{27}$$

The block diagram can be developed using the methods illustrated in Section 2.3. The result is shown in Fig. 2.5–7(a). Figure 2.5–7(b) shows the block

Table 2.5–2 Comparison of integrators

TYPE OF INTEGRATOR	MAGNITUDE, $\lvert H(e^{j\omega T})\rvert$	PHASE, $\text{Arg}\, H(e^{j\omega T})$
Analog RC, inverting (noninverting)	$\frac{\omega_o}{\omega}$	$\frac{\pi}{2}\left(-\frac{\pi}{2}\right)$
Inverting, parallel SC		
a. Type I LDI	$\frac{\omega_o}{\omega}\frac{\omega T/2}{\sin(\omega T/2)}$	$\frac{\pi}{2}$
b. Type I DDI	$\frac{\omega_o}{\omega}\frac{\omega T/2}{\sin(\omega T/2)}$	$\frac{\pi}{2}-\frac{\omega T}{2}$
Noninverting parallel SC		
a. Type I LDI	$\frac{\omega_o}{\omega}\frac{\omega T/2}{\sin(\omega T/2)}$	$-\frac{\pi}{2}$
b. Type I DDI	$\frac{\omega_o}{\omega}\frac{\omega T/2}{\sin(\omega T/2)}$	$-\frac{\pi}{2}-\frac{\omega T}{2}$
Inverting series SC		
a. Type II LDI	$\frac{\omega_o}{\omega}\frac{\omega T/2}{\sin(\omega T/2)}$	$\frac{\pi}{2}$
b. Type II DDI	$\frac{\omega_o}{\omega}\frac{\omega T/2}{\sin(\omega T/2)}$	$\frac{\pi}{2}+\frac{\omega T}{2}$
Bilinear SC and series-parallel SC	$\frac{2\omega_o}{\omega}\frac{\omega T/2}{\tan(\omega T/2)}$	$\frac{\pi}{2}$

diagram for the Type I LDI. If the multiplier constants of $-C_1/C_2$ are changed to C_1/C_2, then the block diagrams of Fig. 2.5–7 correspond to the noninverting SC integrators.

The second realization of the analog integrator uses the series SC resistor equivalent. This realization will be called the *series SC integrator.* R_1 of Fig. 2.5–1 is replaced by the series SC resistor equivalent of Fig. 2.2–2 to obtain the resulting SC integrator of Fig. 2.5–8(a). Another version of the series SC integrator is shown in Fig. 2.5–8(b). Although both realizations are equivalent on an ideal basis, it will be shown in Chapter 8 that Fig. 2.5–8(b) is less sensitive to parasitics. The clock sequence is shown in Fig. 2.5–8(c). Let us consider the odd phase (Fig. 2.5–8c) at which

$$q_L^o\left(n-\frac{1}{2}\right)=C_2v_{c_2}^o\left(n-\frac{1}{2}\right)=C_2\,v_2^o\left(n-\frac{1}{2}\right) \tag{28}$$

$$q_m^e(n-1)=C_2v_{c_2}^e(n-1)=C_2\,v_2^e\,(n-1) \tag{29}$$

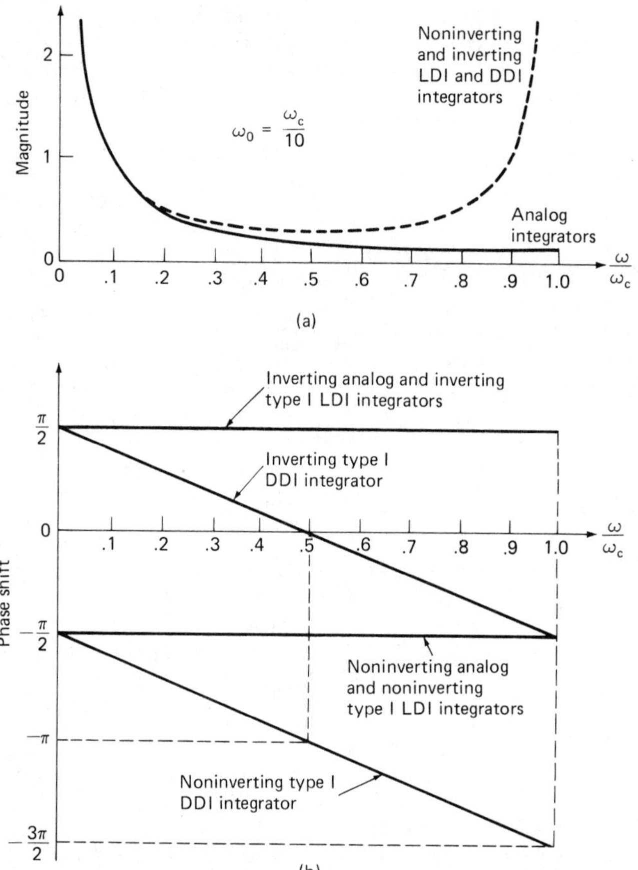

Fig. 2.5–6. Frequency response of various integrators. (a) Magnitude response for $\omega_0 = 0.1$ ω_c and (b) phase response.

$$q_c^o\left(n-\frac{1}{2}\right) = -C_1 v_{c_1}^o\left(n-\frac{1}{2}\right) = -C_1\left[v_1^o\left(n-\frac{1}{2}\right)\right] \tag{30}$$

Using charge conservation analysis, rearranging terms and taking the z-transform, we get

$$C_2\, V_2^o(z) - C_2\, V_2^e(z)z^{-1/2} = -C_1\, V_1^o(z) \tag{31}$$

Looking at the next even phase period, we observe that

$$V_2^e(z) = z^{-1/2}\, V_2^o(z) \tag{32}$$

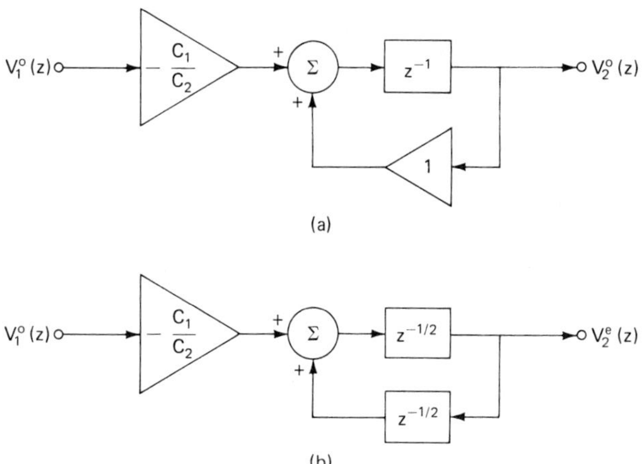

Fig. 2.5–7. A z-domain block diagram of (a) the Type I DDI inverting integrator and (b) the Type I LDI inverting integrator.

Equation (32) is a result that the charge on C_2 does not change from the odd to the even phase period. Thus combining Eqs. (31) and (32), we obtain

$$H^{oo}(z) = \frac{V_2^o(z)}{V_1^o(z)} = -\frac{C_1}{C_2}\frac{1}{1-z^{-1}} \tag{33}$$

We see from Eq. (33) that there is no delay in the forward path between the input and the output signals. This type of SC integrator is called the *Type II Direct-Transform Discrete Integrator* (DDI) and will be very useful in later work. The Type II DDI has only an inverting realization, because the flexibility of reversing capacitors is lost in the series SC resistor equivalent. Comparing Eqs. (14) and (33) gives

$$s = \frac{1-z^{-1}}{T} \tag{34}$$

This is recognized as the *backward transformation* described in Chapter 1. Therefore, the Type II DDI implements the backward mapping from the s-domain to the z-domain. We can also obtain $H^{oe}(z)$ from Eqs. (31) and (32), that is

$$H^{oe}(z) = \frac{V_2^e(z)}{V_1^o(z)} = \frac{-\dfrac{C_1}{C_2}z^{-1/2}}{1-z^{-1}} \tag{35}$$

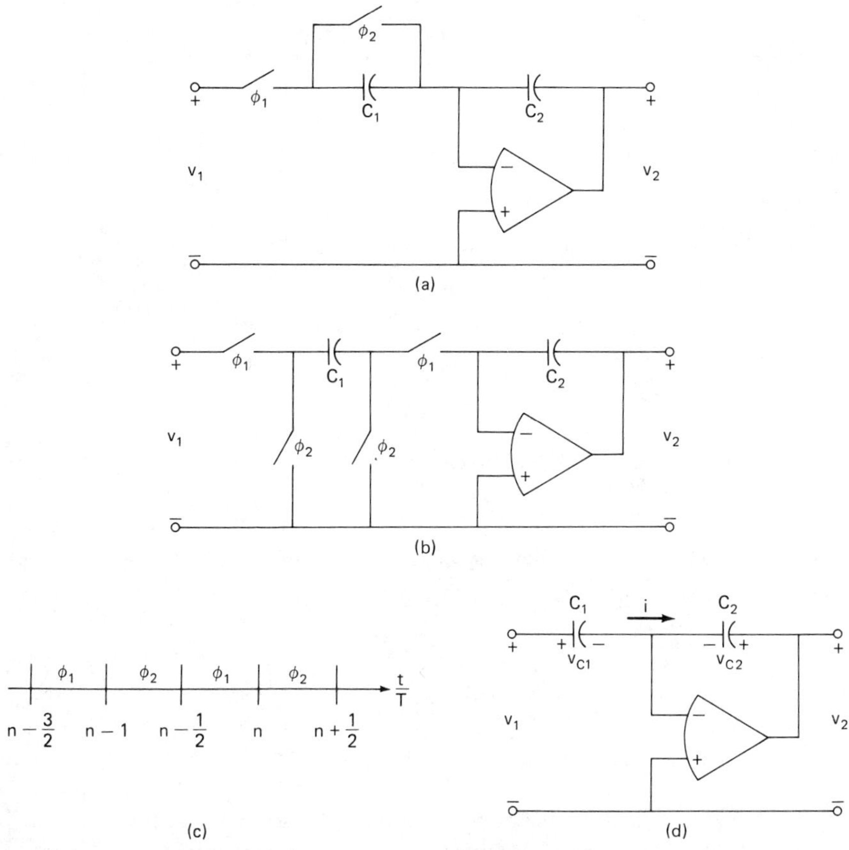

Fig. 2.5–8. (a) Series SC inverting integrator. (b) Parasitic insensitive version of (a). (c) Clock sequence. (The phase periods are arbitrarily chosen.) (d) Equivalent circuit during ϕ_1 period.

This SC integrator is seen to be identical to the inverting Type I LDI. Thus we define the SC integrator described by Eq. (35) as the *Type II Lossless Discrete Integrator* (LDI).

The z-domain block diagram of the Type II DDI realization is shown in Fig. 2.5–9. We observe that the Type II DDI has no delay in the forward path. The z-domain block diagram of the Type II LDI is identical to that of the Type I LDI shown in Fig. 2.5–7(b).

The frequency response of the Type II LDI is identical to that of the inverting Type I LDI given in Eq. (16) and following. However, the Type II DDI has a slightly different frequency response from the Type I DDI and is given as

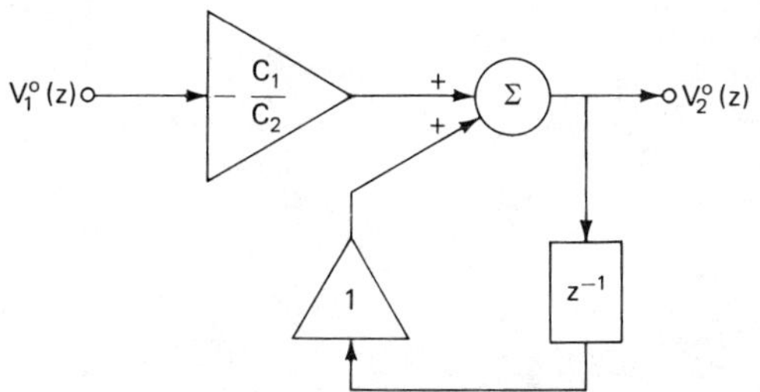

Fig. 2.5–9. z-domain block diagram of an inverting Type II DDI.

$$H^{oo}(e^{j\omega T}) = \frac{-\dfrac{C_1}{C_2}}{1 - e^{-j\omega T}} = -\frac{C_1}{C_2}\,\frac{e^{j\omega T/2}}{e^{j\omega T/2} - e^{-j\omega T/2}}$$

$$= -\frac{\omega_o}{j\omega}\left(\frac{\dfrac{\omega T}{2}}{\sin\dfrac{\omega T}{2}}\right)\exp\left(\frac{j\omega T}{2}\right) \tag{36}$$

where $\omega_o = C_1/(TC_2)$. The magnitude of Eq. (36) is identical to Eq. (22), but the phase shift is given as

$$\text{Arg } H^{oo}(e^{j\omega T}) = \frac{\pi}{2} + \frac{\omega T}{2} \tag{37}$$

which is a leading phase shift. An interesting feature of this result is that the cascade of an inverting Type I DDI with an inverting Type II DDI should cancel the phase shift due to the sampling rate.

Table 2.5–2 summarizes the magnitude and phase expressions for the series SC integrators. The frequency response of the Type II LDI is the same as the Type I LDI shown in Fig. 2.5–6. The magnitude of the Type II DDI is also the same as the magnitude of the Type I DDI of Fig. 2.5–6(a). The phase response of the Type II DDI is shown in Fig. 2.5–10.

The series-parallel SC resistor equivalent of Fig. 2.2–3 may be used to replace R_1 of Fig. 2.5–1, resulting in the series-parallel SC integrator shown in Fig. 2.5–11(a). Considering the even phase and Fig. 2.5–11(c), we have

$$q_L^e(n) = C_2\, v_2^e(n) \tag{38a}$$

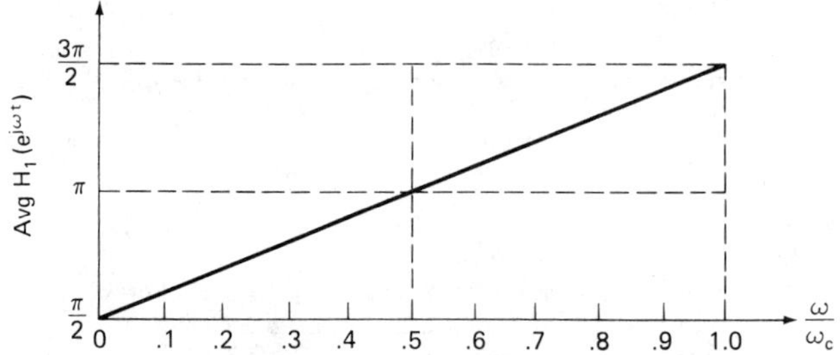

Fig. 2.5–10. Phase response of the inverting Type II DDI.

$$q_m^o\left(n-\frac{1}{2}\right)=C_2\,v_2^o\left(n-\frac{1}{2}\right) \tag{38b}$$

$$q_c^{e,o}=-\,C_1\,v_1^e(n)-C_1\,v_1^o\left(n-\frac{1}{2}\right) \tag{38c}$$

Combining Eqs. (38a) through (38c) and taking the z-transform, we can write

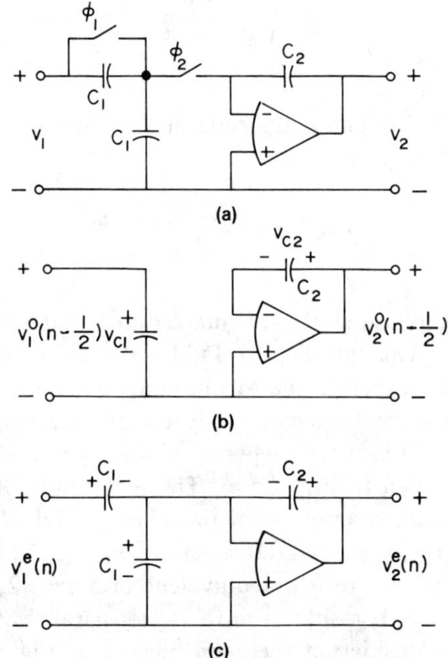

Fig. 2.5–11. (a) Series-parallel SC inverting integrator. (b) Equivalent circuit of (a) during the ϕ_1 phase period. (c) Equivalent circuit of (a) during the ϕ_2 phase period. The phase periods are arbitrarily chosen.

$$C_2 V_2^e(z) = C_2 V_2^o(z) z^{-1/2} - C_1 V_1^e(z) - C_1 z^{-1/2} V_1^o(z) \tag{39}$$

Looking at the next odd phase, we notice that

$$V_2^o(z) = z^{-1/2} V_2^e(z) \tag{40}$$

and

$$V_1^o(z) = z^{-1/2} V_1^e(z) \tag{41}$$

Thus, by combining (39) through (41), we can write

$$H^{ee}(z) = \frac{V_2^e(z)}{V_1^e(z)} = -\frac{C_1}{C_2} \frac{1 + z^{-1}}{1 - z^{-1}} \tag{42}$$

and

$$H^{eo}(z) = -\frac{C_1}{C_2} \frac{(1 + z^{-1}) z^{-1/2}}{1 - z^{-1}} \tag{43}$$

The frequency response of $H^{ee}(z)$ is given as

$$H^{ee}(e^{j\omega T}) = -\frac{C_1}{C_2} \frac{e^{j\omega T/2} + e^{-j\omega T/2}}{e^{j\omega T/2} - e^{-j\omega T/2}} = -\frac{2\omega_o}{j\omega} \frac{\omega T/2}{\tan \dfrac{\omega T}{2}} \tag{44}$$

where $\omega_o = C_1/(TC_2)$. The magnitude of $H^{ee}(e^{j\omega T})$ is

$$|H^{ee}(e^{j\omega T})| = \frac{\omega_o T}{\tan \dfrac{\omega T}{2}} \tag{45}$$

and the phase shift is

$$\text{Arg } H^{ee}(e^{j\omega T}) = \frac{\pi}{2} \tag{46}$$

We see that, although the phase shift is ideal, the magnitude has an error of $(\omega T/2)/(\tan(\omega T/2))$. Figure 2.5–12 shows the magnitude response of Fig. 2.5–11(a). We note that the magnitude has small error at low frequencies.

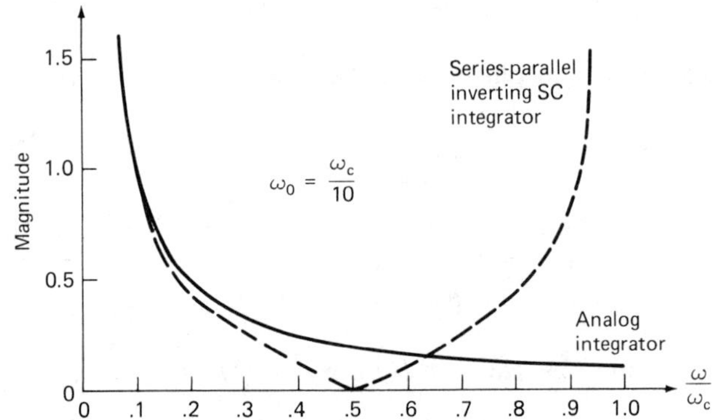

Fig. 2.5–12. Magnitude response of the inverting series-parallel integrator.

The frequency response of the series-parallel SC integrator is summarized in Table 2.5–2. The z-domain block diagram corresponding to Eq. (42) is shown in Fig. 2.5–13. The summation of the signal paths at the output leads to the numerator expression of Eq. (42).

The last SC integrator to be presented here uses the bilinear SC resistor

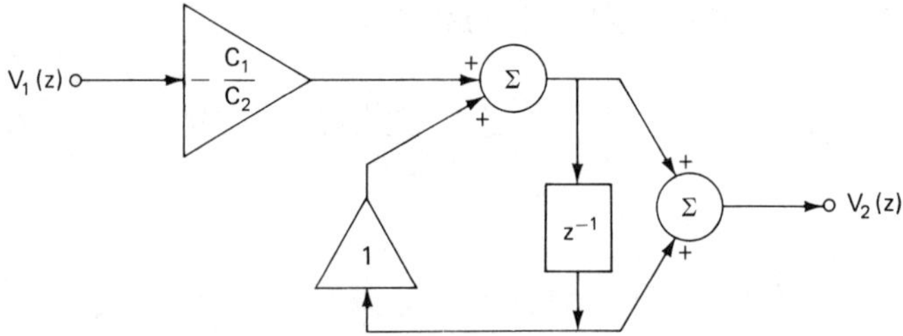

Fig. 2.5–13. z-domain block diagram of the inverting series-parallel integrator.

equivalent of Fig. 2.2–4 in place of R_1 in Fig. 2.5–1. The resulting bilinear SC integrator is shown in Fig. 2.5–14(a). Fig. 2.5–14(b) shows an equivalent circuit of Fig. 2.5–14(a) during an odd phase period that occurs in the interval $(n - 1/2) \leq t/T < n$. We can identify that

$$q_L^o\left(n - \frac{1}{2}\right) = C_2\, v_2^o\left(n - \frac{1}{2}\right) \tag{47a}$$

$$q_m^e(n-1) = C_2\, v_2^e(n-1) \tag{47b}$$

$$q_c^{o,e} = -C_1\, v_{c_1} = -C_1\, v_1^e(n-1) - C_1\, v_1^o\left(n - \frac{1}{2}\right) \tag{47c}$$

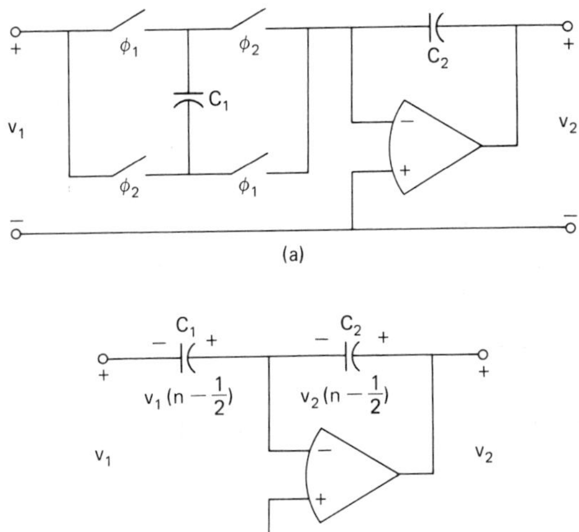

Fig. 2.5–14. (a) Bilinear SC integrator realization. (b) An equivalent circuit during the phase period of $(n - 1/2) \leq t/T < n$.

Combining Eqs. (47a) through (47c) and taking the z-transform, we can obtain

$$C_2\, V_2^o(z) = C_2\, V_2^e(z) z^{-1/2} - C_1\, V_1^e(z) z^{-1/2} - C_1\, V_1^o(z) \tag{48}$$

During the even phase, in a similar way, we can obtain

$$C_2\, V_2^e(z) = C_2\, V_2^o(z) z^{-1/2} - C_1\, V_1^o(z) z^{-1/2} - C_1\, V_1^e(z) \tag{49}$$

If the output V_2 is sampled at both phases, the corresponding transfer function can be obtained by adding Eqs. (48) and (49); that is

$$H(z) = \frac{V_2(z)}{V_1(z)} = \frac{V_2^o + V_2^e}{V_1^o + V_1^e} = -\frac{C_1}{C_2}\frac{1 + z^{-1/2}}{1 - z^{-1/2}} \tag{50}$$

It should be noticed from Fig. 2.5–14 that all switches change position twice every T seconds, thereby doubling the effective period. This can be also observed by comparing Eqs. (42) and (50). Thus, we can rewrite Eq. (50) as

$$H(z) = -\frac{C_1}{C_2}\frac{1 + z'^{-1}}{1 - z'^{-1}} \tag{51}$$

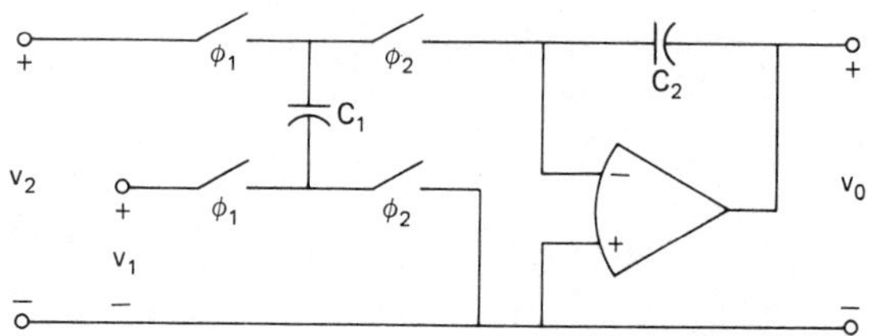

Fig. 2.5–15. A differential SC integrator using the parallel switch configuration.

if a new period is defined $T_c = 2T'$ (see Eq. (53) of Section 2.3).

We may show that the series-parallel or bilinear SC integrator implements the bilinear transform. If R_1 in Eq. (1) is replaced by Eq. (19) of Section 2.2, we get

$$H(s) = -\frac{2C_1}{sTC_2} \tag{52}$$

Now if Eq. (52) is equated to Eq. (42), we find that

$$s = \frac{2}{T}\left(\frac{1-z^{-1}}{1+z^{-1}}\right) \tag{53}$$

which is the bilinear transform of Chapter 1. Thus the series-parallel SC or bilinear SC integrator realizations implement the bilinear mapping from the s-domain to the z-domain.

Finally, in this section we consider *SC differential integrator circuits.* One possible implementation is shown in Fig. 2.5–15. This configuration corresponds to the forward transformation. The corresponding output is

$$V_0^o(z) = \frac{C_1}{C_2}\frac{z^{-1}}{1-z^{-1}}[V_1^o(z) - V_2^o(z)] \tag{54}$$

This integrator will be a useful building block in SC filters.

Table 2.5–2 summarizes the results of this section. The variations for the magnitude of the SC integrators with respect to the conventional analog RC integrator are relatively small and are equivalent to a small error in the capacitor ratios. The magnitude error decreases as the sampling frequency is increased. Therefore, this error can be reduced at the expense of increased capacitor ratios. The problem regarding phase is more serious. The LDI and bilinear realizations have ideal phase characteristics, and the DDI realizations contribute a phase shift error that increases if the sampling rate is decreased. The effects of these errors in the frequency response of SC integrators will be examined in the various applications of these circuits.

2.6 FIRST-ORDER CIRCUITS

In this section we shall apply analog sampled data methods to simple first-order circuits. The typical circuit examined may have one pole, one zero, or both. We shall also use this section to introduce some of the methods by which one may design analog sampled data circuits to realize a given specification.

A very general first-order circuit is shown in Fig. 2.6–1(a). The s-domain transfer function of this circuit can be found to be given by

$$H(s) = \frac{V_2(s)}{V_1(s)} = -\frac{C_3}{C_4}\left(\frac{s + \dfrac{1}{R_1C_3}}{s + \dfrac{1}{R_2C_4}}\right) = -\frac{C_3}{C_4}\left(\frac{s - s_{0i}}{s - s_{pi}}\right) \tag{1}$$

where s_{0i} and s_{pi} are the zero and pole locations, respectively, of Fig. 2.6–1(a). The gain of this circuit at DC is given by $-(R_2/R_1)$, and the gain as

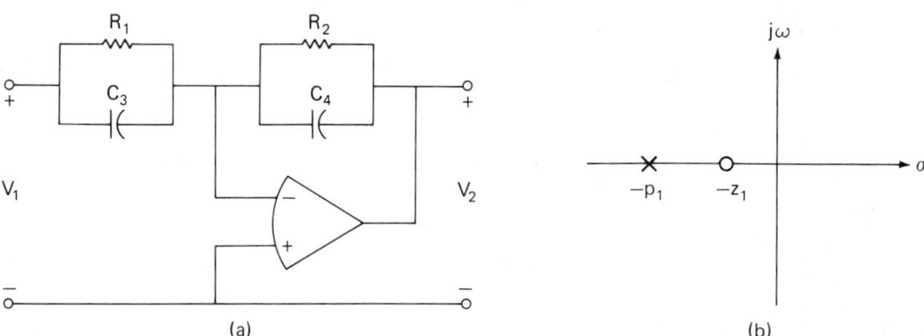

Fig. 2.6–1. (a) Continuous first-order circuit. (b) Possible root locations.

ω approaches infinity is $-(C_3/C_4)$. By adjusting these components, the circuit can easily be made to emphasize high frequencies or low frequencies. Figure 2.6–1(b) shows the roots of Eq. (1) in the s-domain.

An SC simulation[22] of Fig. 2.6–1(a) may be obtained by replacing the

[22] We shall use the term *simulation* because, in most cases, the SC circuit only approximates the desired continuous circuit due to the fact that the SC resistors of Section 2.2 may no longer be equivalent to an analog resistor. The equivalence depends upon many other factors.

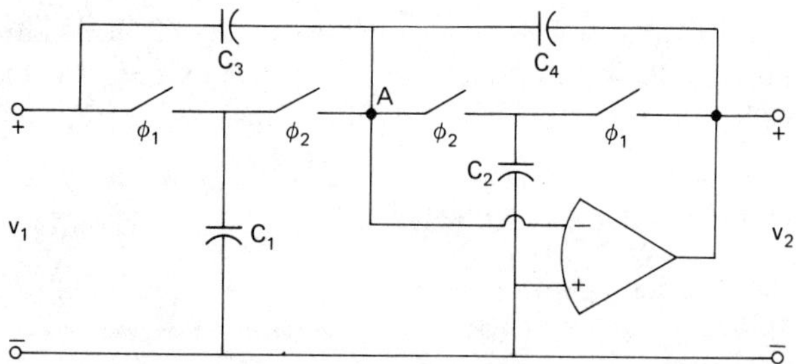

Fig. 2.6–2. SC realization of Fig. 2.6–1(a) using parallel SC equivalent resistors.

resistors by SC equivalent resistors. Assuming C_3 and C_4 are present, then we can select any of the SC resistor equivalents without too much concern for the op amp. For example, if we select the parallel SC resistor equivalent, then Fig. 2.6–2 results. During the ϕ_2 phase period the nodal charge equilibrium conditions at A become

$$C_4\, v_{C4} + C_1\, v_{C1} + C_2\, v_{C2} + C_3\, v_{C3} = 0 \tag{2}$$

where

$$v_{C4} = v_2^e(n) - v_2^o\left(n - \frac{1}{2}\right) \tag{3a}$$

$$v_{C3} = v_1^e(n) - v_1^o\left(n - \frac{1}{2}\right) \tag{3b}$$

$$v_{C2} = v_2^o\left(n - \frac{1}{2}\right) \tag{3c}$$

$$v_{C1} = v_1^o\left(n - \frac{1}{2}\right) \tag{3d}$$

Combining Eqs. (2) and (3), and taking the z-transform results in

$$C_4 V_2^e(z) - C_4 V_2^o(z) z^{-1/2} + C_2 V_2^o(z) z^{-1/2} = -C_1 V_1^o(z) z^{-1/2} - C_3 V_1^e(z) + C_3 V_1^o(z) z^{-1/2} \tag{4}$$

We can observe from the odd phase that

$$V_1^o(z) = V_1^e(z) z^{-1/2} \tag{5}$$

$$V_2^o(z) = V_2^e(z)z^{-1/2} \tag{6}$$

where we have assumed that the input is sampled and held. The transfer function for $H^{ee}(z)$ can be written from Eq. (4) as

$$H^{ee}(z) = \frac{V_2^e(z)}{V_1^e(z)} = -\frac{C_3}{C_4}\frac{z - z_{o_1}}{z - z_{p_1}} = -\frac{C_3}{C_4}\frac{1 - z_{o_1}z^{-1}}{1 - z_{p_1}z^{-1}} \tag{7}$$

where the z-domain zero and pole are

$$z_{o_1} = 1 - \frac{C_1}{C_3} \tag{8}$$

$$z_{p_1} = 1 - \frac{C_2}{C_4} \tag{9}$$

Also, the transfer function $H^{eo}(z)$ can be expressed as

$$H^{eo}(z) = \frac{V_2^o(z)}{V_1^e(z)} = -\frac{C_3}{C_4}\frac{1 - z_{o_1}z^{-1}}{1 - z_{p_1}z^{-1}}z^{-1/2} \tag{10}$$

We note that the pole or zero can be anywhere on the real axis from $-\infty$ to $+1$. For a stable realization the pole must be within the range of -1 to $+1$.

It would be very useful to be able to visualize the influence of the roots in the z-domain upon the frequency response of the system. This can be done by starting with the general z-domain transfer function given in Eq. (20) of Section 1.4 and repeated here as

$$H(z) = \frac{b_m z^m + b_{m-1}z^{m-1} + \cdots + b_0}{a_n z^n + a_{n-1}z^{n-1} + \cdots + a_0}, \qquad m \le n \tag{11}$$

An equivalent form of Eq. (11) was given in Eq. (21) of Section 1.4 and is

$$H(z) = K\frac{\prod_{i=1}^{m}(z - z_{oi})}{\prod_{i=1}^{n}(z - z_{pi})}, \qquad m \le n \tag{12}$$

Replacing z by $e^{j\omega T}$ to evaluate the frequency response gives

$$H(e^{j\omega T}) = K \frac{\prod_{i=1}^{m} (e^{j\omega T} - z_{0i})}{\prod_{i=1}^{n} (e^{j\omega T} - z_{pi})} \tag{13}$$

If we examine any of the products of the numerator, $(e^{j\omega T} - z_{0i})$, or any of the products of the denominator, $(e^{j\omega T} - z_{pi})$, we see that these products are each complex numbers in the z-domain. We will designate the products in the numerator of Eq. (13) as $\mathbf{W}_i(\omega)$ and those in the denominator as $\mathbf{U}_j(\omega)$. Therefore, Eq. (13) could be written as

$$H(e^{j\omega T}) = K \frac{\mathbf{W}_1(\omega)\,\mathbf{W}_2(\omega)\,\cdots\,\mathbf{W}_{m-1}(\omega)\,\mathbf{W}_m(\omega)}{\mathbf{U}_1(\omega)\,\mathbf{U}_2(\omega)\,\cdots\,\mathbf{U}_{n-1}(\omega)\,\mathbf{U}_n(\omega)} \tag{14}$$

The magnitude of $H(e^{j\omega T})$ becomes

$$|H(e^{j\omega T})| = K \frac{W_1(\omega)\,W_2(\omega)\,\cdots\,W_{m-1}(\omega)\,W_m(\omega)}{U_1(\omega)\,U_2(\omega)\,\cdots\,U_{n-1}(\omega)\,U_n(\omega)} \tag{15}$$

where $W_i(\omega)$ is the magnitude of $\mathbf{W}_i(\omega)$, etc. The phase shift of $H(e^{j\omega T})$ is

$$\begin{aligned} \operatorname{Arg} H(e^{j\omega T}) = {} & \operatorname{Arg} \mathbf{W}_1(\omega) + \operatorname{Arg} \mathbf{W}_2(\omega) + \cdots + \operatorname{Arg} \mathbf{W}_m(\omega) \\ & - \operatorname{Arg} \mathbf{U}_1(\omega) - \operatorname{Arg} \mathbf{U}_2(\omega) - \cdots - \operatorname{Arg} \mathbf{U}_n(\omega) \end{aligned} \tag{16}$$

As ω is varied from 0 to ω_c, the magnitude and phase response can be found by evaluating Eqs. (15) and (16).

Example 2.6–1. *Evaluation of frequency response from* $H(z)$. Assume in Fig. 2.6–2 that $C_2 = C_4 = C_3$, $C_1 = 2C_3$. Find the frequency response of $H^{ee}(z)$. Equation (7) becomes

$$H(z) = -\left(\frac{z+1}{z}\right)$$

These roots are illustrated in Fig. 2.6–3(a). We start with $\omega_0 = 0$ and pick values up to $\omega_0 = \omega_c$ and use Eqs. (15) and (16) to evaluate the frequency

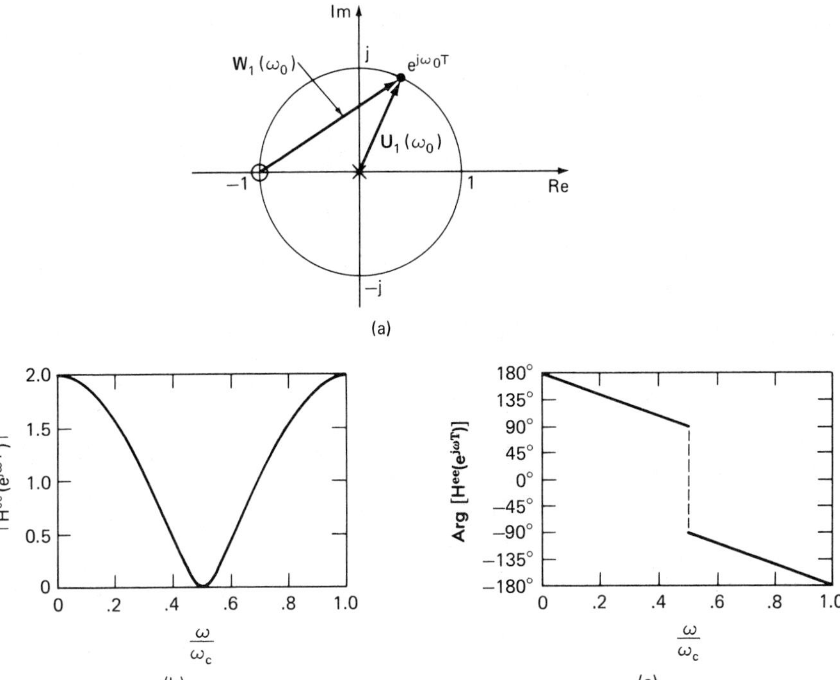

Fig. 2.6–3. (a) Roots of Example 2.6–1. (b) Magnitude and (c) phase response of Example 2.6–1.

response. Table 2.6–1 shows the details of the frequency response calculations. Figure 2.6–3(b) and Fig. 2.6–3(c) give the magnitude and phase of the frequency response. We see that under the conditions of the example, Fig. 2.6–2 acts as a low-pass circuit.

The additional flexibility of SC circuits over continuous circuits can be illustrated by replacing the parallel SC resistor equivalents C_1 of Fig. 2.6–2 by the inverting parallel SC resistor equivalent used in Fig. 2.5–5. If the parallel SC resistor equivalent C_1 is replaced by the inverting type, then Fig. 2.6–2 is converted to Fig. 2.6–4. Calculations of $H(z)$, when replacing one form of parallel SC resistor equivalent with another, can be done very conveniently by noting in Fig. 2.6–5 that such a replacement is equivalent to replacing a positive value capacitor by a negative capacitor having the same value. For example, in Fig. 2.6–2, the parallel SC resistor equivalent using C_1 has been replaced by the inverting parallel SC resistor equivalents

Table 2.6–1

ω/ω_c	$W_1(\omega)$	$U_1(\omega)$	$\lvert H^{ee}(e^{j\omega T})\rvert$	arg $[W_1(\omega)]$	arg $[U_1(\omega)]$	arg $[H(e^{j\omega T})]$
0.0	2	1	2.000	0°	0°	180°
0.1	1.902	1	1.902	18°	36°	162°
0.2	1.618	1	1.618	36°	72°	144°
0.3	1.176	1	1.176	54°	108°	126°
0.4	0.618	1	0.618	72°	144°	108°
0.5	0	1	0.000	±90°	180°	±90°
0.6	0.618	1	0.618	−72°	216°	−108°
0.7	1.176	1	1.176	−54°	252°	−126°
0.8	1.618	1	1.618	−36°	288°	−144°
0.9	1.902	1	1.902	−18°	324°	−162°
1.0	2	1	2.000	0°	360°	−180°

to get Fig. 2.6–6a. Thus, the transfer function of Fig. 2.6–6a can be found from Eq. (7) by replacing C_1 by $-C_1$ to get

$$H^{ee}(z) = \frac{V_2^e(z)}{V_1^e(z)} = -\frac{C_3}{C_4}\left(\frac{z - z_{01}}{z - z_{p1}}\right) \tag{17}$$

where

$$z_{01} = 1 + \frac{C_1}{C_3} \tag{18}$$

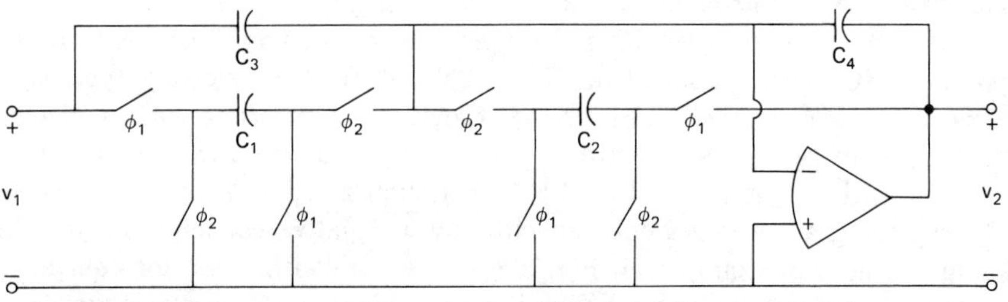

Fig. 2.6–4. Replacement of the noninverting parallel SC equivalent resistors in Fig. 2.6–2 with the inverting parallel SC equivalent resistor.

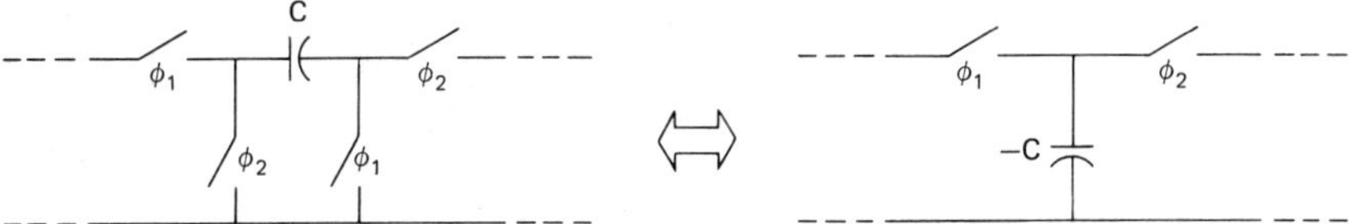

Fig. 2.6–5. Equivalent circuits between the noninverting and inverting parallel SC resistor equivalent circuits.

and

$$z_{p_1} = \left(1 - \frac{C_2}{C_4}\right) \tag{19}$$

We note in this realization that the zero is outside the unit circle.

Example 2.6–2. *First-order high-pass realization.* Let us find a realization for the roots, z_{01} and z_{p1} shown in Fig. 2.6–6(a). Using Fig. 2.6–6(b) with $C_1 = 0$ and $C_2 = C_4 = C_3$ results in the desired roots. Prob. 2.16 examines the details of the frequency response calculations. The frequency response of this realization is shown in Fig. 2.6–6(c) and 2.6–6(d), which are equivalent to those of Example 2.6–1, and demonstrates high-pass behavior.

The z-domain block diagram of the parallel SC resistor realizations of Fig. 2.6–1(a) can be developed from Eq. (7). The results are shown in Fig. 2.6–7. z_{01} and z_{p_1} depend upon the actual configuration selected for the SC resistor equivalents. The series, series-parallel, and bilinear SC resistor equivalents can also be used to achieve different realizations of Fig. 2.6–1(a).

Another useful first-order circuit is a particular case of Fig. 2.6–2 where C_3 is removed. In this case we have what is called a *damped integrator,* and the transfer function is found from Eq. (1) by letting $C_3 = 0$. The result is

$$H(s) = \frac{V_2(s)}{V_1(s)} = \frac{-\dfrac{1}{R_1C_4}}{s + \dfrac{1}{R_2C_4}} = -\frac{R_2}{R_1}\frac{\dfrac{1}{R_2C_4}}{s + \dfrac{1}{R_2C_4}} = -\frac{R_2}{R_1}\left(\frac{p_1}{s + p_1}\right) \tag{20}$$

We see that a damped integrator is equivalent to a low-pass, first-order system having a gain of R_2/R_1 at low frequencies. SC realizations of Eq. (20) can be developed by removing C_3 from Fig. 2.6–2 or Fig. 2.6–4. The noninverting

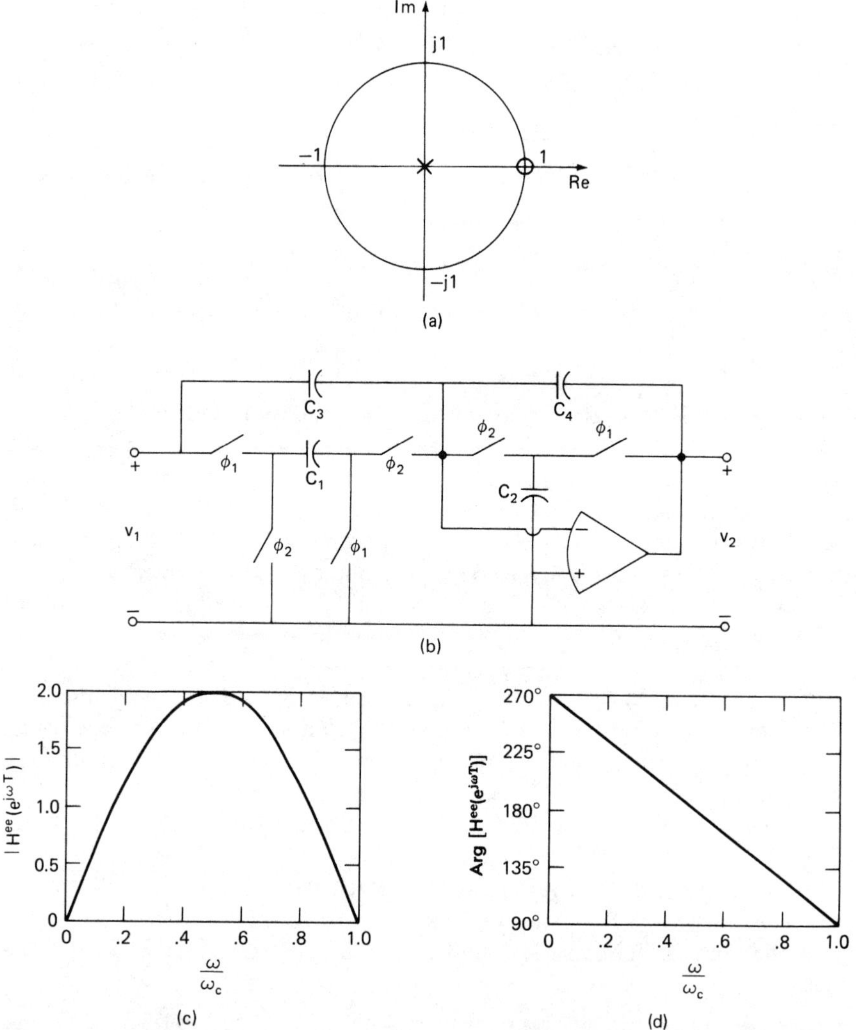

Fig. 2.6–6. (a) Roots of Example 2.6–2. (b) Realization for Example 2.6–2. (c) Magnitude response of Fig. 2.6–6(a). (d) Phase response of Fig. 2.6–6(a).

realization is developed from Fig. 2.6–4 and is shown in Fig. 2.6–8. The transfer function can be found from Eq. (7) as

$$H^{ee}(z) = \frac{V_2(z)}{V_1(z)} = \frac{\dfrac{C_1}{C_4} z^{-1}}{1 - z^{-1} + \dfrac{C_2}{C_4} z^{-1}} = \frac{\dfrac{C_1}{C_4}}{z - \left(1 - \dfrac{C_2}{C_4}\right)} \tag{21}$$

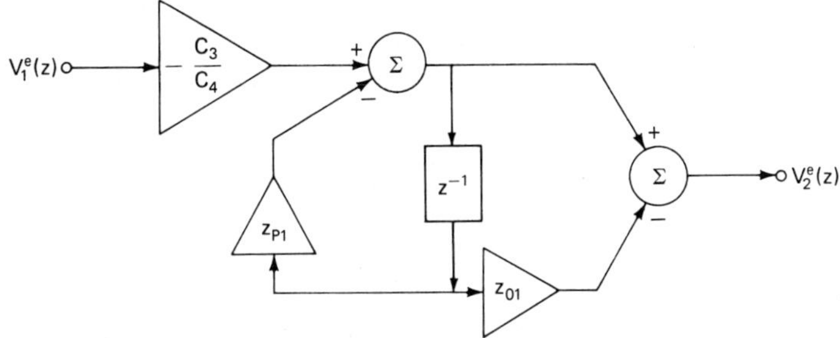

Fig. 2.6–7. General z-domain block diagram for first-order realization.

which has a pole at

$$z_1 = \left(1 - \frac{C_2}{C_4}\right) \tag{22}$$

For a stable system, the pole can be located anywhere on the real axis between −1 and +1. The polarity of Eq. (21) can be changed from plus to minus by replacing the noninverting parallel SC resistor equivalent by the inverting

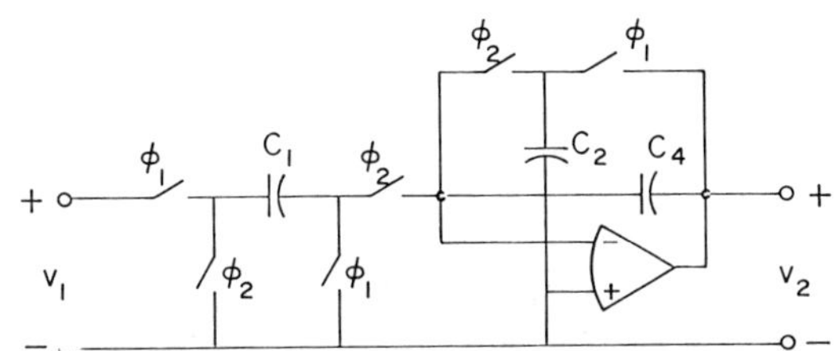

Fig. 2.6–8. Noninverting, low-pass, first-order SC realization.

parallel SC resistor equivalent illustrated in Fig. 2.6–5. Furthermore, we can obtain a low-pass, amplifier, or a high-pass realization by simply changing the capacitor ratio C_2/C_4, that is

$$0 \le \frac{C_2}{C_4} < 1, \qquad \text{LP: } 0 < z_{p1} \le 1 \tag{23}$$

$$\frac{C_2}{C_4} = 1, \qquad \text{amplifier and delay: } z_{p1} = 0 \tag{24}$$

$$1 < \frac{C_2}{C_4} \le 2, \qquad \text{HP: } -1 < z_{p1} < 0 \tag{25}$$

If clock phases for the parallel SC in the feedback are alternated, that is, $\phi_1(\phi_2)$ becomes $\phi_2(\phi_1)$, then the transfer function $H^{oo}(z)$ becomes[22]

$$H^{oo}(z) = \frac{C_1}{C_4} \frac{\left(1 - \frac{C_2}{C_4}\right)}{z - \left(1 - \frac{C_2}{C_4}\right)} \tag{26}$$

It is seen that the DC gain for the odd component of the output has been modified by the factor $1 - (C_2/C_4)$. For the conditions in Eq. (24), $H^{oo}(z) = 0$. The transfer function $H^{oe}(z)$ is given by

$$H^{oe}(z) = \frac{\frac{C_2}{C_4} z^{-1/2}}{1 - \left[1 - \frac{C_2}{C_4}\right] z^{-1}} \tag{27}$$

Example 2.6–3. *Application of the first-order system to frequency domain design.* It is desired to use Fig. 2.6–8 to realize the frequency response shown in Fig. 2.6–9. The frequency ω_0 is to be equal to $\omega_c/20$. A general form of Eq. (26) is given as

$$H^{oo}(z) = \frac{a_1}{z - b_1} \tag{28}$$

Thus we have only two variables, a_1 and b_1, that can be used to satisfy the specification. We shall therefore choose a_1 and b_1 in order to satisfy the magnitude at $\omega = 0$ and ω_0. The frequency response of Eq. (28) is

$$H(e^{j\omega T}) = \frac{a_1}{e^{j\omega T} - b_1} \tag{29}$$

At $\omega = 0$ we have

$$10 = \frac{a_1}{1 - b_1} \tag{30}$$

---

[22] K. R. Laker, "Equivalent Circuits for the Analysis and Synthesis of Switched Capacitor Networks," *Bell Syst. Tech. J.*, Vol. 58, No. 10, October 1979, pp. 2235–2269.

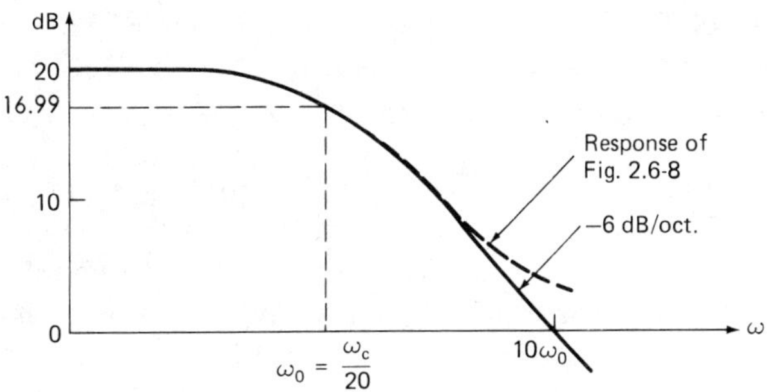

Fig. 2.6–9. Frequency response specification for Example 2.6–3.

or

$$a_1 = 10(1 - b_1) \tag{31}$$

The square of the magnitude of the frequency response of Eq. (28) is

$$|H(e^{j\omega T})|^2 = \frac{a_1^2}{1 - 2b_1 \cos \omega T + b_1^2} = \frac{100(1 - b_1)^2}{1 - 2b_1 \cos \omega T + b_1^2} \tag{32}$$

At $\omega = \omega_0$, Eq. (32) is equal to 50, which gives the following equation

$$b_1^2 - 2(2 - \cos \omega_0 T)\, b_1 + 1 = 0$$

The only stable root corresponds to the value

$$b_1 = 0.7323$$

From Eq. (31) we have

$$a_1 = 2.6773$$

Equating Eq. (26) to Eq. (28) results in the desired capacitor ratios of

$$\frac{C_2}{C_4} = 0.2677$$

and

$$\frac{C_1}{C_4} = 3.6560$$

The response of Fig. 2.6–8 with these values is plotted on Fig. 2.6–9 for comparison purposes. The realization is seen to be very good at frequencies below ω_0, but to become poor at frequencies above ω_0.

Fig. 2.6–1 is a very general first-order realization. Its generality is illustrated in Table 2.6–2, where various transfer functions in the complex frequency domain have been given. Many SC realizations can be developed by replacing the resistors of these configurations by various types of SC resistor equivalents. Polarity reversal may be accomplished if Fig. 2.6–5 is used for one of the resistors R_1 or R_2 of each realization. The pole or zero can be switched from the positive real axis to the negative real axis or vice versa by using the appropriate parallel SC equivalent resistor in place of the resistor forming the RC time constant associated with the pole or zero.

Let us consider another first-order low-pass filter by substituting R_2 with

Table 2.6–2 First-order analog circuits and their transfer functions

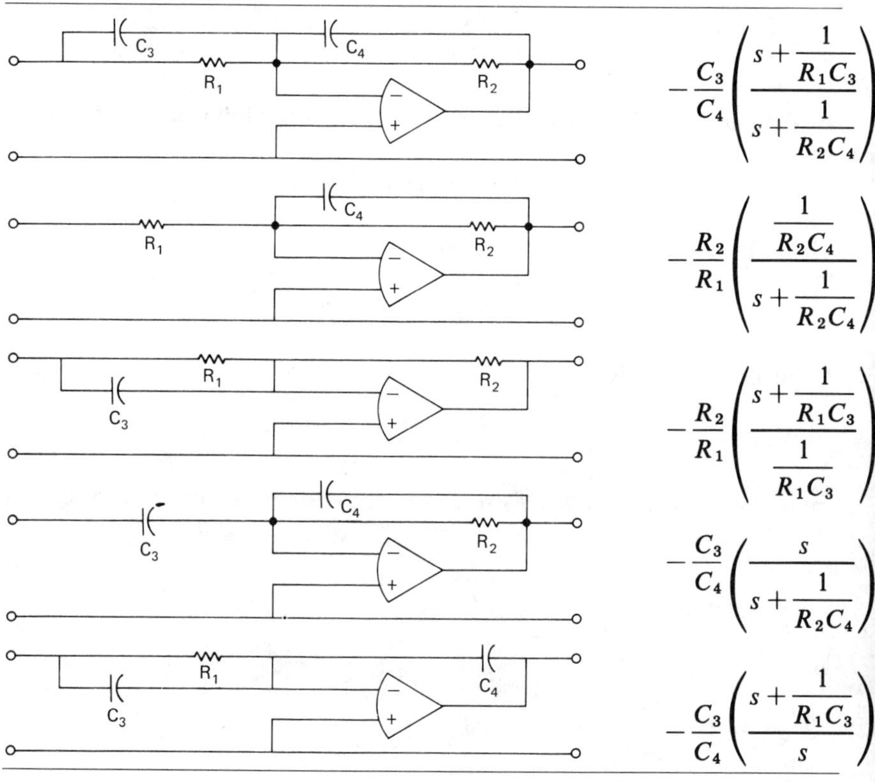

$$-\frac{C_3}{C_4}\left(\frac{s+\frac{1}{R_1C_3}}{s+\frac{1}{R_2C_4}}\right)$$

$$-\frac{R_2}{R_1}\left(\frac{\frac{1}{R_2C_4}}{s+\frac{1}{R_2C_4}}\right)$$

$$-\frac{R_2}{R_1}\left(\frac{s+\frac{1}{R_1C_3}}{\frac{1}{R_1C_3}}\right)$$

$$-\frac{C_3}{C_4}\left(\frac{s}{s+\frac{1}{R_2C_4}}\right)$$

$$-\frac{C_3}{C_4}\left(\frac{s+\frac{1}{R_1C_3}}{s}\right)$$

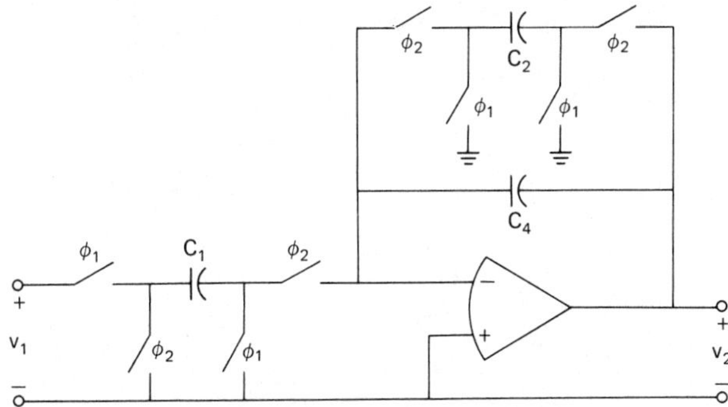

Fig. 2.6–10. A stray insensitive, first-order, low-pass SC filter.

a series SC resistor equivalent and R_1 with an inverting parallel SC resistor. This is shown in Fig. 2.6–10. The transfer functions can be obtained as

$$H^{oe}(z) = \frac{\dfrac{C_1}{C_4} z^{-1/2}}{1 + \dfrac{C_2}{C_4} - z^{-1}} \tag{33}$$

and

$$H^{oo}(z) = \frac{\dfrac{C_1}{C_4} z^{-1}}{1 + \dfrac{C_2}{C_4} - z^{-1}} = \frac{\dfrac{C_1}{C_4}}{z\left(1 + \dfrac{C_2}{C_4}\right) - 1} \tag{34}$$

The pole of these transfer functions is located at $1/(1 + C_2/C_4)$. Even though this structure is not as versatile as the one shown in Fig. 2.6–4, it will be shown (in Chapter 8) to be insensitive to stray capacitances.

Next we shall consider some simple first-order circuits that illustrate further the use of analog sampled data methods for the realization of analog circuits. An RC differentiator is shown in Fig. 2.6–11. The transfer function in the s-domain is found as

$$H(s) = \frac{V_2(s)}{V_1(s)} = -sC_1R_2 \tag{35}$$

An SC realization of this circuit requires replacement of R_2. Although we should be careful not to cause the op amp to operate in open loop, let us

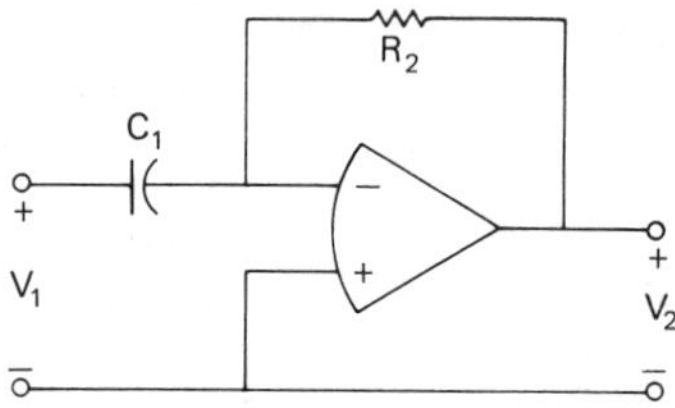

Fig. 2.6–11. Analog differentiator.

ignore this aspect at the present and consider the SC realization of Fig. 2.6–11 shown in Fig. 2.6–12(a). During the ϕ_1 phase period, the circuit performs as a charge amplifier. We shall assume that the odd phase period occurs during the interval of $(n - 1) \le t/T < (n - 1/2)$. During the even phase period, v_{C_2} is set to zero, and v_{C_1} remains constant because it cannot charge or discharge due to the lack of a current path. At the next odd (ϕ_1) phase period, the charge relationship associated with C_2 is

$$q_L^o\left(n - \frac{1}{2}\right) = C_2\, v_2^o\left(n - \frac{1}{2}\right) \tag{36a}$$

$$q_m^e = 0 \tag{36b}$$

and

$$q_c^{o,e} = -\,C_1\, v_{c_1} = -C_1\, v_1^o\left(n - \frac{1}{2}\right) + C_1\, v_1^e(n - 1) \tag{36c}$$

Using the charge conservation method with Eqs. (36a) through (36c), we find that the z-domain transfer function is

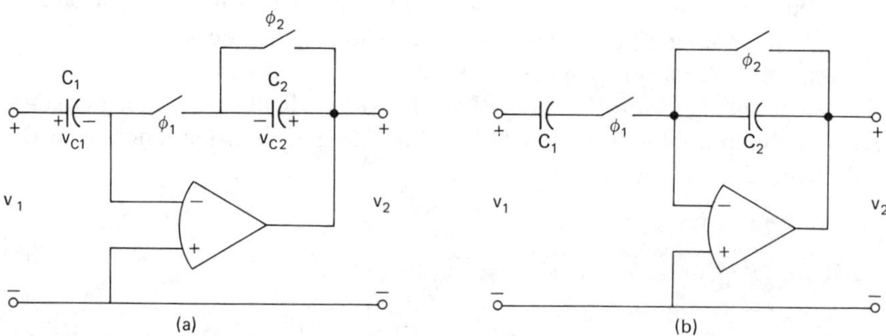

Fig. 2.6–12. (a) An SC realization of a differentiator. (b) A practical realization of Fig. 2.6–11.

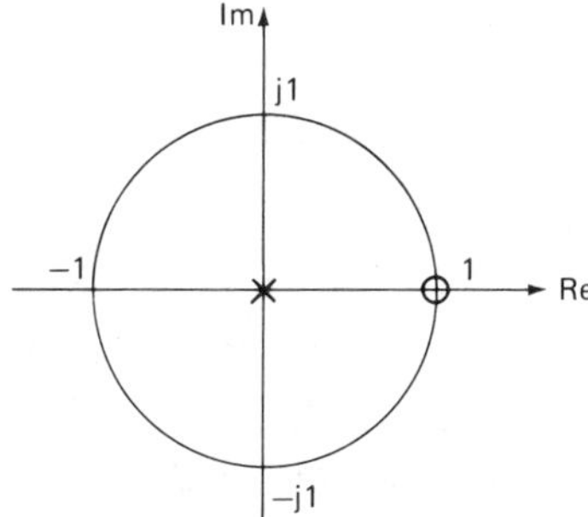

Fig. 2.6–13. Roots of Fig. 2.6–12.

$$H^{oo}(z) = \frac{V_2^o(z)}{V_1^o(z)} = -\frac{C_1}{C_2}(1 - z^{-1}) = -\frac{C_1}{C_2}\frac{z-1}{z} \tag{37}$$

assuming that $V_1^o(z) = V_1^e(z)z^{-1/2}$. We may make the circuit practical by moving the ϕ_1 switch in series with C_1, as shown in Fig. 2.6–12(b). With this modification we do not change the transfer function, but we do avoid causing the op amp to be open loop during a portion of the clock cycle.

The roots of Eq. (37) are shown in Fig. 2.6–13. The SC differentiator using the series SC resistor equivalent has a pole at the origin and a zero at 1. To see if this circuit simulates Eq. (35), we substitute Eq. (9) of Section 1.6 into Eq. (37) to get

$$H(s) = \frac{V_2(s)}{V_1(s)} \approx -s\frac{TC_1}{C_2} \tag{38}$$

Comparing Eq. (38) with Eq. (35) confirms that the series SC differentiator of Fig. 2.6–12 will simulate a continuous differentiator when $\omega << \omega_c$.

The actual frequency response of Eq. (37) can be expressed as

$$H^{oo}(e^{j\omega T}) = -\frac{C_1}{C_2}(1 - \cos\omega T - j\sin\omega T) \tag{39}$$

We note that, unlike the analog differentiator, the SC differentiator does not approach a magnitude larger than $2C_1/C_2$ as frequency increases and therefore avoids the unstable tendencies of an increasing closed-loop gain as frequency increases. Figure 2.6–14 shows the z-domain block diagram of the series SC differentiator. Other types of SC equivalent resistors can be used to replace R_2 of Fig. 2.6–11. These SC differentiators will exhibit slightly different properties (see Problem 2.13).

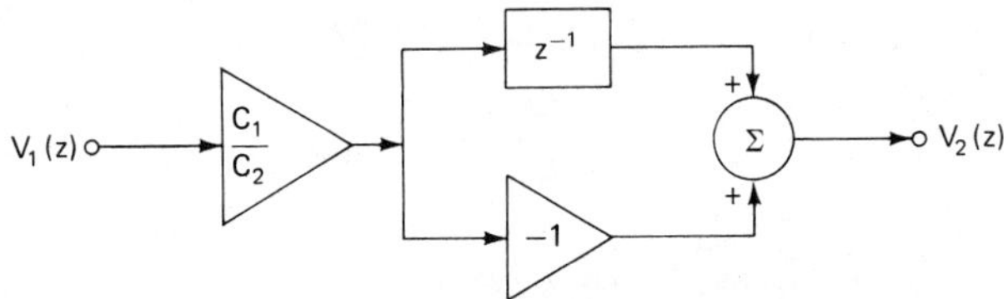

Fig. 2.6–14. z-domain block diagram of the series SC differentiator.

Example 2.6–4. *Series SC differentiator design.* Design an SC differentiator using the circuit of Fig. 2.6–12(b). Let the time constant of the differentiator be $20/\omega_c$. The time constant of the differentiator is given from Eq. (38) as

$$\tau = T\frac{C_1}{C_2} \tag{40}$$

Therefore C_1/C_2 becomes equal to $10/\pi$. The frequency response of the series SC differentiator is shown in Fig. 2.6–15 and is compared to an analog differentiator for reference.

Let us consider how SC techniques might be used to realize an inductor. The defining equation of an inductance L can be expressed as

$$v = L\frac{di}{dt} \tag{41}$$

In ideal SC circuits the currents that charge capacitors do not flow continuously but occur in impulses. Therefore, Eq. (41) may be replaced by an analogous charge equation

$$v(t) = \frac{L}{T/2}[i(t) - i(t - T/2)] \tag{42}$$

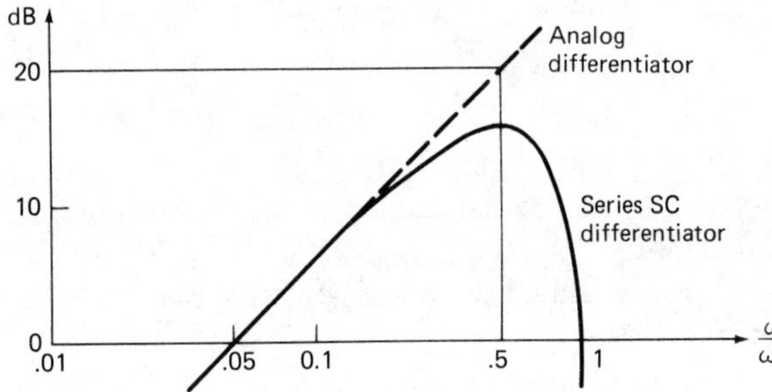

Fig. 2.6–15. Frequency response of the series SC differentiator of Example 2.6–4.

Equation (42) can be simulated by the circuit in Fig. 2.6–16(a). The voltage across the capacitor may be expressed as

$$v(t) = \frac{1}{C}\int_0^{T/2} i\,dt = \frac{1}{C}\left[i\left(\frac{T}{2}\right) - i\,(0)\right] \tag{43}$$

The quantity in the brackets is the area under the $i(t)$ curve. However, because $i(t)$ consists of two impulses, one at $t = (n - 1)T$ and the other at $t = nT$, it is necessary to define $i(T/2) = [i(t)]\ T/2$ and $i(0) = [i(t - T/2)]\ T/2$ to maintain the proper dimensions. These definitions result in

$$v(t) = \frac{T}{2C}\,[i(t) - i(t - T/2)] \tag{44}$$

Comparing Eqs. (42) and (44) gives the value of SC simulated inductance as

$$L = \frac{T^2}{C} \tag{45}$$

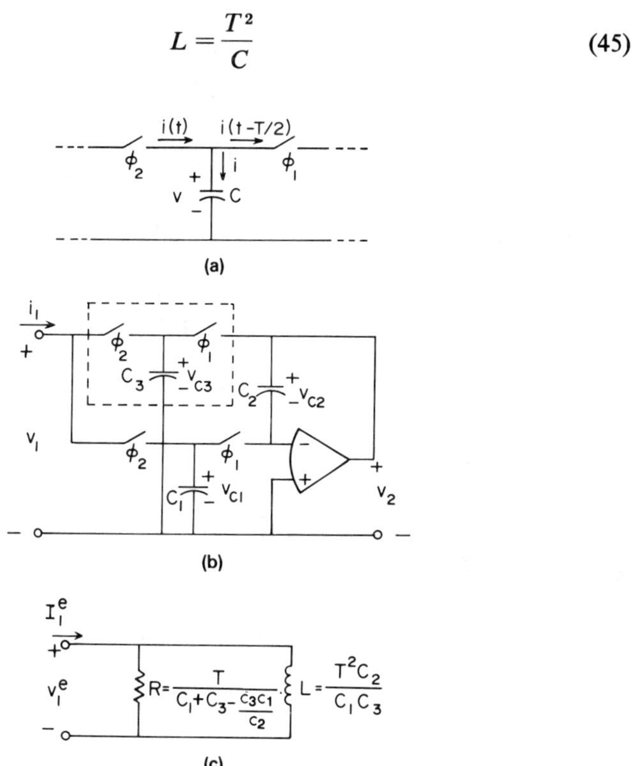

Fig. 2.6–16. (a) Simulation of inductance using SC methods. (b) Practical realization of (a). (c) Equivalent circuit for (b).

A realization of Fig. 2.6–16(a) is given in Fig. 2.6–16(b), where the equivalent portion of Fig. 2.6–16(a) is shown within the dotted box. We now analyze Fig. 2.6–16(b) to find an expression for the z-domain input admittance $Y_1(z)$. Let us consider the analysis for the even phase ϕ_2, i.e., $(n - 1/2) \leq t/T < (n)$. By Kirchoff's current law we can write

$$i_1 = i_{C_3} + i_{C_1} \tag{46}$$

and

$$i_{C_3} = \frac{C_3 v_{C_3}}{T} \tag{47}$$

$$i_{C_1} = \frac{C_1 v_{C_1}}{T} \tag{48}$$

where

$$v_{C_3} = v_1^e\left(n - \frac{1}{2}\right) - v_2^o(n-1) \tag{49}$$

$$v_{C_1} = v_1^e\left(n - \frac{1}{2}\right) \tag{50}$$

Thus Eq. (46) becomes

$$Ti_1^e\left(n - \frac{1}{2}\right) = (C_3 + C_1)\, v_1^e\left(n - \frac{1}{2}\right) - C_3 v_2^o(n-1) \tag{51}$$

Also note that

$$v_2^e\left(n - \frac{1}{2}\right) = v_2^o(n-1) \tag{52}$$

During the ϕ_1 (odd phase) we see that

$$C_1 v_{C_1} + C_2 v_{C_2} = 0 \tag{53}$$

that is

$$-C_1 v_1^e\left(n - \frac{1}{2}\right) = C_2\left[v_2^o(n) - v_2^e\left(n - \frac{1}{2}\right)\right] \tag{54}$$

Substituting Eq. (52) into Eq. (54) results in

$$v_2^o(n-1) = \frac{C_1}{C_2} v_1^e\left(n - \frac{1}{2}\right) + v_2^o(n) \tag{55}$$

Combining Eqs. (55) and (51), assuming a sampled and held input, i.e., $v_1^e(n - 1/2) = v_1^o(n)$, and taking the z-transform, we obtain

$$Y_1^e(z) = Y_{en}^e = \frac{I_1^e}{V_1^e} = \left[C_3 + C_1 - \frac{C_3C_1}{C_2} + \frac{C_3C_1}{C_2}\frac{1}{z-1}\right]/T \tag{56}$$

Because the Type I DDI was used in this realization, we should be able to use Eq. (15) of Section 2.5 to transform from the z-plane back to the s-plane replacing $z - 1$ by sT. The result is

$$Y_1(s) = \left(\frac{C_1}{T} + \frac{C_3}{T}\right) - \frac{C_3C_1}{C_2} + \frac{C_1C_4}{sT^2C_2} = G + \frac{1}{sL} \tag{57}$$

where

$$G = \frac{1}{R} = \frac{C_1}{T} + \frac{C_3}{T} - \frac{C_3C_1}{C_2T} \tag{58}$$

and

$$L = \frac{T^2C_2}{C_1C_3} \tag{59}$$

Therefore, Fig. 2.6–16(b) is equivalent to a resistor R in parallel with an inductor L, as illustrated in Fig. 2.6–16(c).

Let us illustrate the use of the circuit of Fig. 2.6–16(b) to realize the simple RL circuit of Fig. 2.6–17(a). The transfer function of this circuit is a high-pass type and is given as

$$\frac{V_{\text{out}}(s)}{V_{\text{in}}(s)} = \left(\frac{R}{R + R_0}\right)\left(\frac{s}{s + \dfrac{R_0R}{L(R_0 + R)}}\right) \tag{60}$$

A realization of this circuit using Fig. 2.6–16(b), together with a parallel SC resistor equivalent, is shown in Fig. 2.6–17. The transfer function is ob-

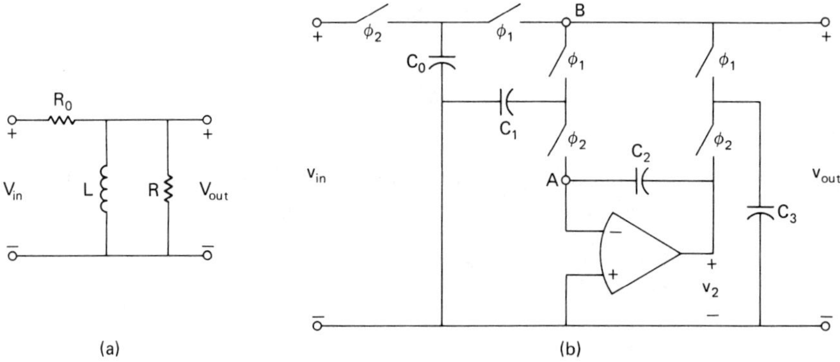

Fig. 2.6–17. (a) Simple *RL* circuit. (b) SC realization of (a).

obtained using the charge conservation technique. We can consider two nodal charge conservation equations at A and B of Fig. 2.6–17(b), one set for the even period ϕ_2 and one for the odd period ϕ_1. Thus, for the even period at node A, we see that

$$C_1 v_{c_1} + C_2 v_{c_2} = 0 \tag{61}$$

At node B no charge exists. For the odd period, we obtain

$$\text{Node } A\text{:}\quad C_2 v_{c_2} = 0 \tag{62}$$

$$\text{Node } B\text{:}\quad C_0 v_{c_0} + C_1 v_{c_1} + C_3\, v_{c_3} = 0 \tag{63}$$

Equations (61) through (63) are expressed, respectively, as

$$C_1 v_{\text{out}}^o(n-1) + C_2 v_2^e\left(n - \frac{1}{2}\right) - C_2 v_2^o(n-1) = 0 \tag{64}$$

$$C_2\left[v_{\text{out}}^o(n) - v_2^e\left(n - \frac{1}{2}\right)\right] = 0 \tag{65}$$

$$C_0\left[v_{\text{out}}^o(n) - v_{in}^e\left(n - \frac{1}{2}\right)\right] + C_1 v_{\text{out}}^o(n) + C_3\left[v_{\text{out}}^o(n) - v_2^e\left(n - \frac{1}{2}\right)\right] = 0 \tag{66}$$

Taking the z-transform of Eqs. (64) through (66), we can obtain

$$H^{eo}(z)=\frac{C_0 z^{-1/2}(1-z^{-1})}{(C_0+C_1+C_3)\left[1-z^{-1}\left(1-\frac{C_3C_1}{C_2(C_0+C_1+C_3)}\right)\right]} \tag{67}$$

It is evident that Fig. 2.6–17(b) could have been developed from the analog circuit of Fig. 2.6–18. The dotted portion of this figure simulates a parallel *RL* circuit.

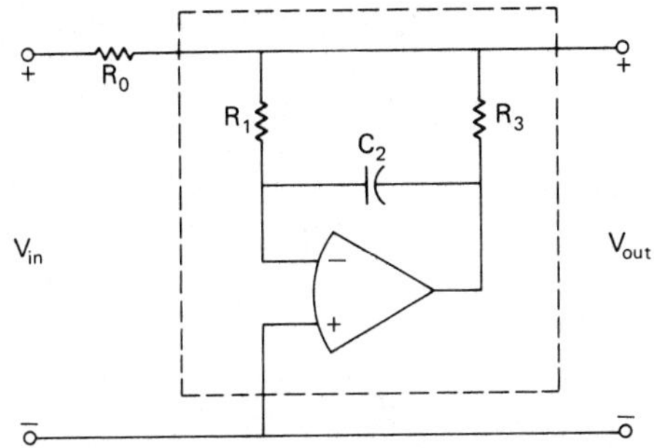

Fig. 2.6–18. RC active realization of Fig. 2.6–17(a).

The z-domain block diagram of Fig. 2.6–17(b) is shown in Fig. 2.6–19. The multiplier constants of k_1 and k_2 are given as

$$k_1=\frac{C_0}{C_0+C_1+C_3} \tag{68}$$

and

$$k_2=1-\frac{C_3C_1}{C_2(C_0+C_1+C_3)} \tag{69}$$

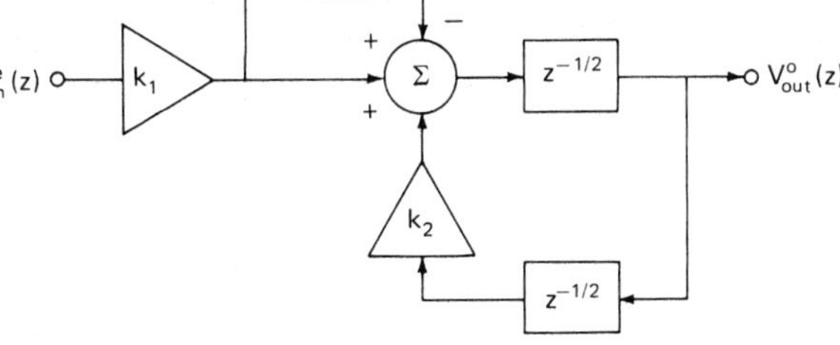

Fig. 2.6–19. z-domain block diagram of Fig. 2.6–17(b).

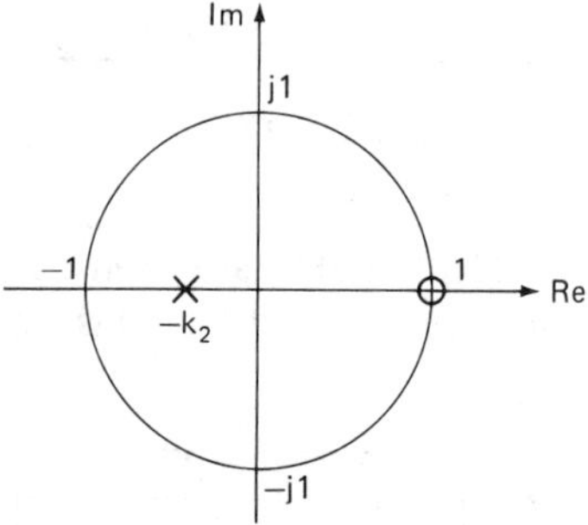

Fig. 2.6–20. Roots of Fig. 2.6–17(b).

Figure 2.6–20 shows the roots of Fig. 2.6–17(b). The high-pass behavior is clearly evident, for

$$1 < \frac{C_3 C_1}{C_2(C_0 + C_1 + C_3)} < 2$$

Though there are many more first-order circuits that could be developed here, we have constrained ourselves to those that are typical and demonstrate the principles of such networks. We continue to see that SC circuits offer much more flexibility than their analog counterparts. More examples will be considered in Chapter 6, which presents applications of analog sampled data networks.

2.7 SUMMARY

This chapter has introduced the basic methods of realizing analog continuous circuits with analog sampled data circuits. We have examined SC equivalent circuits for resistors and RC circuits. The introduction of the op amp led to a means of completely removing the charge from one capacitor and placing it all on another. This result led to the integrator and other first-order networks. It was observed that the SC realization of a continuous analog circuit often had much more flexibility than its continuous counterpart. Many of the ideas and circuits presented here will be used in following chapters.

PROBLEMS

2.1 (Sec. 2.2). Show that:

a. The series-switched capacitor realization of Fig. 2.2–2 has a resistance value of $R = 1/Cf_c$, defined at the even phase ϕ_2.

b. The bilinear switched capacitor realization of Fig. 2.2–4 of the conventional resistor has a resistance value of

$$R = \frac{1}{4Cf_c}$$

Under what conditions are the above equivalent resistors valid?

2.2 (Sec. 2.3). Obtain the exact frequency response at ϕ_1 of the switched capacitor circuit shown in Fig. P 2.2. Define $\alpha = C_1/C_2$, and solve for the -3-dB frequency in Hertz when $\alpha = 3$ and the clock frequency is (1) 64 kHz and (2) 128 kHz. The clock phases are nonoverlapping, and v_1 is a sample-and-hold signal.

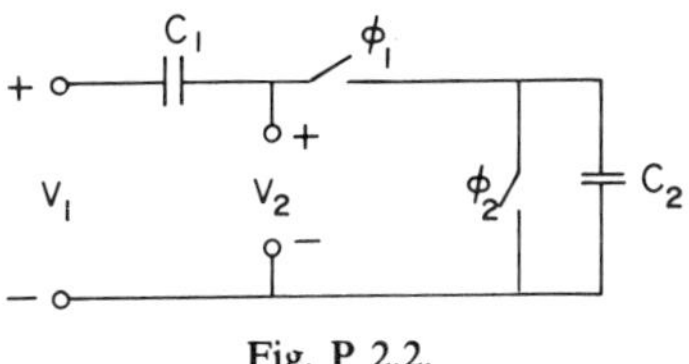

Fig. P 2.2.

2.3 (Sec. 2.3). Find the transfer function $H(s)$ of the circuit shown in Fig. P 2.3. Obtain the exact transfer function $H^{oo}(z)$ when R_1 and R_2 are (parallel) SC resistor equivalents.

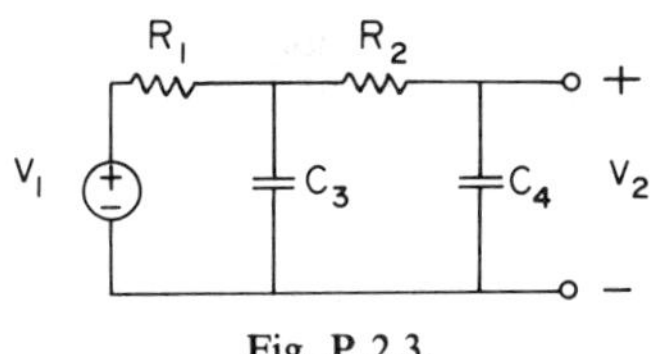

Fig. P 2.3.

2.4 (Sec. 2.3). Plot magnitude and phase of the transfer functions $H(z)$ and $H(s)$ obtained in Prob. 2.3 for

a. $R_1 = R_2 = 100\ \mathrm{k}\Omega$, $C_3 = C_4 = 200$ pF, and $f_s = 128$ kHz.
b. $R_1 = 100\ \mathrm{k}\Omega$, $R_2 = 270\ \mathrm{k}\Omega$, $C_3 = 200$ pF, $C_4 = 150$ pF, and $f_s = 64$ kHz.

2.5 (Sec. 2.3). Plot magnitude and phase of the transfer functions of Problem 2.2.

a. Obtain the equivalent R_2C_1 of Fig. P 2.2.
b. Determine the corresponding values of R_2C_1 for the data of Problem 2.2b.

2.6 (Sec. 2.4). Given the prototype of Fig. 2.4–4, obtain the exact transfer function $H(z)$ when

a. R_1 becomes a (series) SC resistor equivalent and R_2 becomes a (bilinear) SC resistor equivalent.
b. R_1 becomes an inverting (parallel) SC resistor equivalent and R_2 becomes a (series) SC resistor equivalent.
c. R_1 and R_2 become bilinear SC resistors.

2.7 (Sec. 2.4). Plot the results of Problem 2.6 when $R_1 = 6.8$ kΩ and $R_2 = 68$ kΩ, when $f_s = 64$ kHz.

2.8 (Sec. 2.4). Obtain the transfer functions $H^{eo}(z)$ and $H^{ee}(z)$ of the circuit shown in Fig. P 2.8, and plot the results when

$$\frac{C_2}{C_3} = 1.99, \frac{C_1}{C_3} = 10, \text{ and } T = 1/(128 \times 10^3)$$

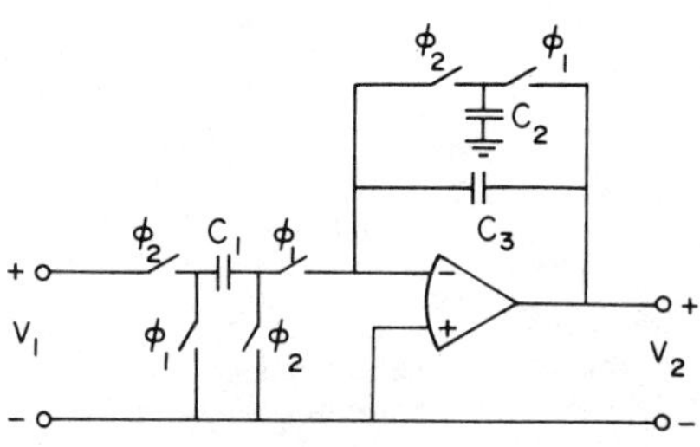

Fig. P 2.8.

2.9 (Sec. 2.5). Given the noninverting analog integrator of Fig. 2.5–3, obtain $H^{oo}(z)$ when the resistors are substituted by series SC resistor equivalents. Generate a table similar to Table 2.5–1.

2.10 (Sec. 2.5). Show that the circuit of Fig. P 2.10 has the transfer functions

$$H^{oe}(z) = -\frac{\dfrac{\dfrac{C_2}{C} z^{-1/2}}{1 + \dfrac{C_2}{C}}}{1 - \dfrac{1}{1 + \dfrac{C_2}{C}} z^{-1}}$$

$$H^{oo}(z) = -z^{-1/2} H^{oe}(z)$$

2.11 (Sec. 2.6). Given the RC prototype shown in Fig. 2.4–5, determine $H(s)$ and

a. Determine $H(z)$ when R_1 and R_2 are replaced by parallel SC resistor equivalents. Identify the type of filters when (1) $1 < C_2/C_3 < 2$ and (2) $C_2 = C_3 = C$.

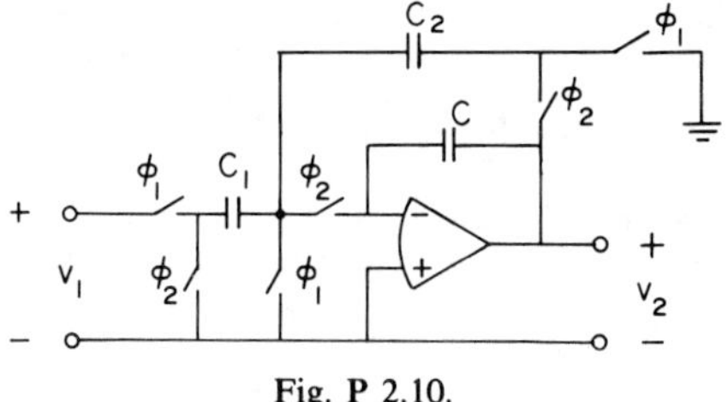

Fig. P 2.10.

b. Plot $H(s)$ and $H_{00}(z)$ when $R_1 = 1$ kΩ, and $C_3 = 10^{-6}/2\pi$ for (1) $T = 10^{-4}$ and (2) $T = 1/(128 \times 10^3)$.

2.12 (Sec. 2.6). Determine the exact frequency response $H^{oo}(z)$, $H^{oe}(z)$ of the lossy integrator SC circuit shown in Fig. P 2.12. Solve for the −3-dB frequency in Hertz if

a. $\alpha_1 = 2 \qquad \alpha_2 = 1$
b. $\alpha_1 = 2 \qquad \alpha_2 = \pi$

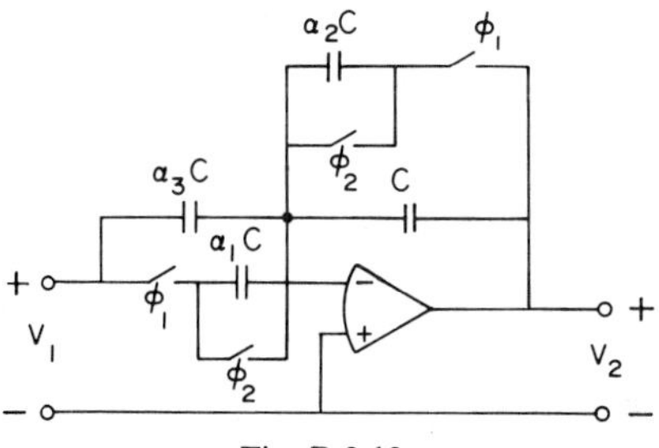

Fig. P 2.12.

2.13 (Sec. 2.6). Given the prototype of the RC analog differentiator, replace R_2 by a bilinear SC equivalent, and plot both $H(j\omega)$ and $H(e^{j\omega t})$ when $R_2 = 120$ kΩ, $C_1 = 100/2\pi$ pF, and $f_c = 64$ kHz.

2.14 (Sec. 2.6). For the circuit of Fig. P 2.14, obtain $H(s)$ and $H(z)$ for the series SC resistor equivalent. Plot the results for $R_1 = 11.255$ kΩ, $R_2 = 22.505$ kΩ, $C_3 = C_4 = 0.1$ μF for (1) $f_c = 1$ kHz and (2) $f_c = 7.072$ kHz.

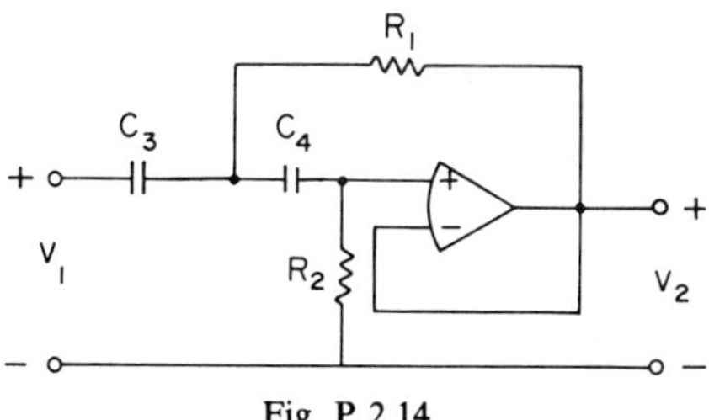

Fig. P 2.14.

2.15 (Sec. 2.6). Obtain the transfer function $H(s)$ of Fig. 2.6–18 and compare with $H(z)$ when the resistors are replaced by series SC resistor equivalents. Design the circuit for a cutoff frequency of 3.4 kHz for $f_c = 128$ kHz.

2.16 (Sec. 2.6). Develop a table similar to Table 2.6–1 for Example 2.6–2. From this table, verify that Fig. 2.6–6(c) and Fig. 2.6–6(d) are correct.

3
Introduction to Switched Capacitor Filters

Our purpose in this chapter and the next is to present the basic design techniques of filters that are implemented using analog sampled data methods. The largest application of analog sampled data techniques to date has been in the area of filtering. The reasons for this situation have been discussed in earlier chapters. With the material of this and the next chapters, coupled with the background of Chapter 8, the reader will be in a position to design integrated circuit filters using analog sampled data techniques. Many aspects of IC technology will influence the material presented on filters.

In this chapter we shall briefly review the important concepts of continuous filter design. Also the performance characterization of SC filters are presented. Next we shall try to generalize the approach to designing an SC filter and how this approach is related to the specifications and design of a continuous filter. The first SC filter technique presented will be that of second-order blocks where the continuous integrators are replaced by SC integrators. In this area we ignore the problem of s- to z-domain mappings by assuming that the sampling rate is high. The next method of designing SC filters uses a direct bilinear replacement of the resistor. In this section the influence of the bilinear mapping is discussed. Scaling and frequency prewarping techniques necessary for analog sampled data filters will be presented at the point in this chapter where they are needed. A similar technique that uses the backward transformation to simulate passive components is given. Some comments regarding advantages and disadvantages for each SC resistor simulation are given. Finally, the analysis and simulation techniques of SC networks are presented.

3.1 CONTINUOUS FILTER DESIGN

This section presents the background necessary for design of filters. This background includes types of filters, transformations and normalizations useful in design, determination of complexity, and use of tabulated data for filters. Continuous and analog sampled data filter design approaches will be compared.

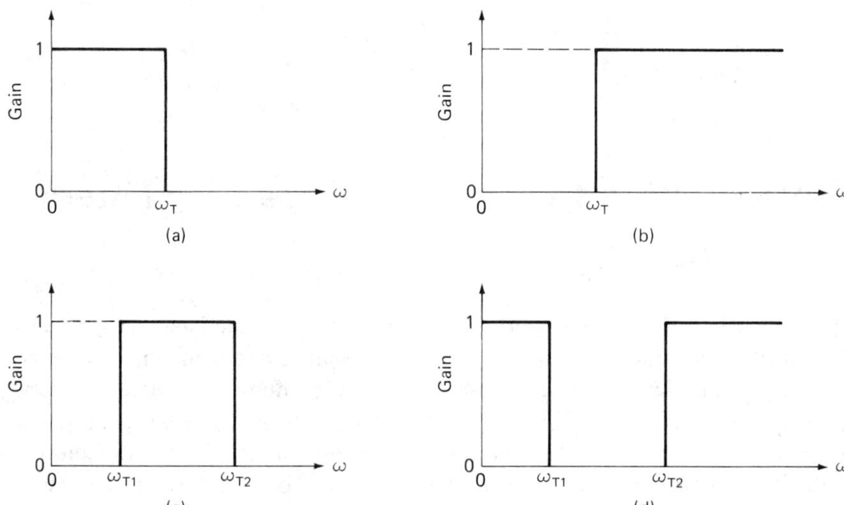

Fig. 3.1–1. Types of filters. (a.) Low pass. (b.) High pass. (c.) Bandpass. (d.) Band elimination.

A brief review of continuous filter design follows. More detail and information can be found in the references listed in the bibliography. Filters can be classified into four different groups, depending on what ranges of frequencies are transmitted and what ranges of frequencies are rejected. Ideally we can assume that the gain of the filter is unity in the *passband* where frequencies are transmitted and zero in the *stopband* where frequencies are rejected. Furthermore, we shall assume that under ideal conditions the passband and stopband are adjacent to each other and that ω_T is the transition frequency from one band to the other.

Using these definitions, we show four types of filters in Fig. 3.1–1. Figure 3.1–1(a) shows a *low-pass filter,* which has the ideal frequency response

$$H_{LP}(j\omega) = \begin{cases} 1, & 0 \le \omega \le \omega_T \\ 0, & \omega_T < \omega \le \infty \end{cases} \tag{1}$$

Fig. 3.1–1(b) illustrates a *high-pass filter,* described as

$$H_{HP}(j\omega) = \begin{cases} 0, & 0 \le \omega < \omega_T \\ 1, & \omega_T \le \omega \le \infty \end{cases} \tag{2}$$

A *bandpass filter* is shown in Fig. 3.1–1(c). The ideal frequency response of this filter is given as

$$H_{BP}(j\omega) = \begin{cases} 0, & 0 \le \omega < \omega_{T1} \\ 1, & \omega_{T1} \le \omega \le \omega_{T2} \\ 0, & \omega_{T2} < \omega \le \infty \end{cases} \tag{3}$$

Finally, a *band-elimination filter* is shown in Fig. 3.1–1(d). The ideal frequency response of this filter is

$$H_{BE}(j\omega) = \begin{cases} 1, & 0 \le \omega \le \omega_{T1} \\ 0, & \omega_{T1} < \omega < \omega_{T2} \\ 1, & \omega_{T_2} \le \omega \le \infty \end{cases} \tag{4}$$

Equations (1) through (4) represent four types of ideal filter responses based on the transmission of a signal. The phase shift of each of the filters of Fig. 3.1–1 should be linear and can be defined as

$$\text{Arg } H(j\omega) = -T_d\,\omega, \qquad 0 \le \omega \le \infty \tag{5}$$

Because the time delay, T_d, of the filter is equal to the negative of the derivative of the phase shift, then Eq. (5) defines a system that has a constant time delay of T_d seconds for each frequency transmitted by the filter.

In practice the ideal filter characteristics of Fig. 3.1–1 cannot be realized. It is not possible to have abrupt changes in transmission as a function of frequency or to have zero transmission over a region or band of frequencies. Consequently we can only approximate the ideal filter characteristics of Fig. 3.1–1. Let us consider how we can modify the ideal filter specifications of Fig. 3.1–1 so that they can be realized in practice. We shall focus our efforts on the low-pass filter. The considerations that result are easily extended to the other three types of filters. Let us consider the low-pass filter specifications of Fig. 3.1–2. The frequency range has been divided into three parts. From zero to the passband frequency, ω_{PB}, the gain must be within G_{PB} and unity. From the *stopband frequency,* ω_{SB}, to infinity, the gain must be within G_{SB} and zero. The frequency range between ω_{PB} and ω_{SB} is called the *transition region.* The gain of the filter is undefined in this region, although a good approximation would be expected to follow closely a straight line drawn from point A to point B. Another way of interpreting Fig. 3.1–2 is to say that the filter response must fall in the shaded regions. The reasons for the transition region and a nonzero value of G_{SB} are clear from the ideal consideration. However, it is not clear why G_{PB} is not unity. The answer is that there is a tradeoff between how close G_{PB} is to unity and the width of the transition region. We shall see that permitting G_{PB} to be less than unity allows smaller values of ω_{SB} and thus a smaller transition region.

There are several types of filter approximations that satisfy Fig. 3.1–2. These filter types are based on the rational polynomial transfer function given as

$$H(s) = \frac{b_0 + b_1 s + b_2 s^2 + \cdots + b_{n-1} s^{n-1} + b_n s^n}{a_0 + a_1 s + a_2 s^2 + \cdots + a_{n-1} s^{n-1} + a_n s^n} \tag{6}$$

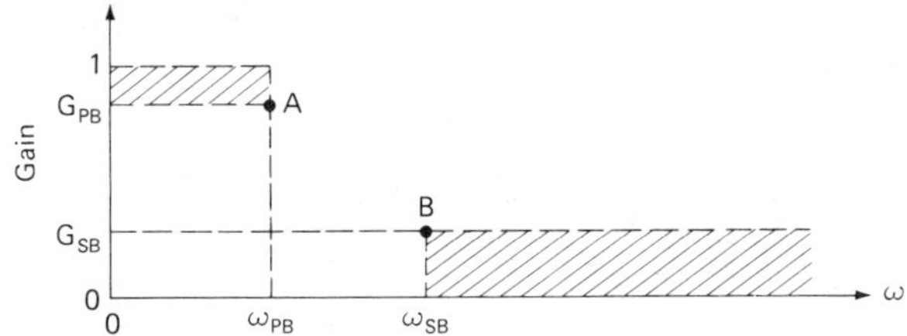

Fig. 3.1–2. Practical low pass filter specifications.

The order of this function is given as *n.* The types of filter approximations are generally classified by some aspect of their frequency response. For example, the *Butterworth filter approximation* has a magnitude response that is maximally flat in the passband and is monotonically decreasing in the transition and passband regions. For the Butterworth filter approximation, all $b_i = 0$ except for $i = 0$ in Eq. (6). An example of several Butterworth filter approximations for various values of n are given in Fig. 3.1–3. It is customary to normalize the passband frequency ω_{PB} to 1 rps. G_{PB} is normally $1/\sqrt{2}$ in the Butterworth filter approximation. The phase shift of the Butterworth filter approximation may be found in the references given in the bibliography (see Zverev).

Several other filter approximations are often used in filter design. One of these is called the *Chebyshev filter approximation* and is illustrated in Fig. 3.1–4 for various values of *n.* We see that the magnitude of the Chebyshev filter approximation ripples in the passband and is monotonically decreasing in the transition region and the stopband. The advantage of the Chebyshev filter approximation is that for a given *n,* the slope of the magnitude response in the vicinity of ω_{PB} is steeper than that of a Butterworth filter approximation, resulting in a smaller transition region if G_{PB} and G_{SB} are identical.

A third filter approximation is called the *elliptic* filter approximation and is shown in Fig. 3.1–5. It is seen that the elliptic filter approximation ripples

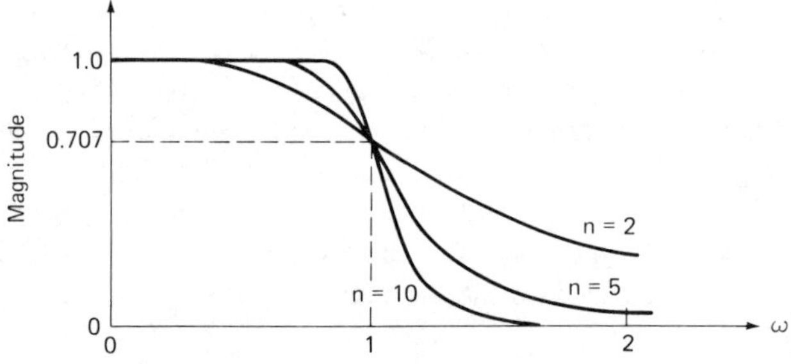

Fig. 3.1–3. Magnitude characteristics of a Butterworth approximation for orders of n = 2, 5, and 10.

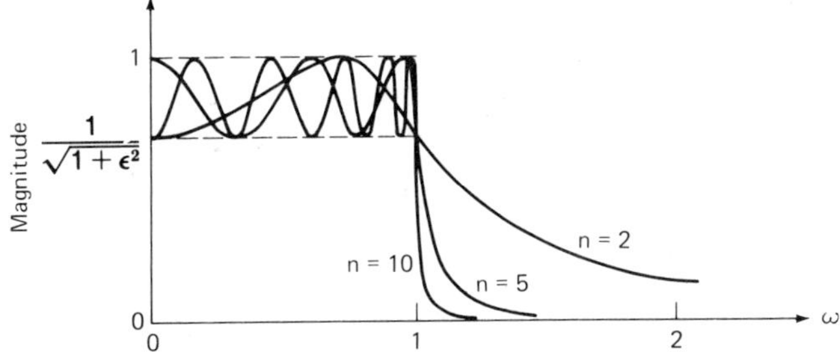

Fig. 3.1–4. Magnitude characteristics of a Chebyshev approximation for an arbitrary ϵ and for orders of n = 2, 5, and 10.

both in the stopband and in the passband, but is monotonically decreasing in the transition region. The elliptic filter approximation has the narrowest transition region of any type of filter characteristic given the same value of n, G_{PB}, and G_{SB}. Other filter approximations have been tabulated and feature other frequency characteristics, such as linear phase shift.

Fortunately, there is considerable tabulated information available on standard filter approximations. The objective of the filter designer in using this tabulated information is to obtain a polynomial, such as Eq. (6) and its roots, or a passive RLC ladder network. Both cases are the starting point in active filter design. Given the polynomial and its roots, the designer can use one of many methods to arrive at an active filter realization of the filter approximation. Another method involves the simulation of the passive RLC ladder network by RC active networks.

To use the tabulated information, it is necessary to be able to determine the order n from the filter specification. Though the filter specification can be given as shown in Fig. 3.1–2, it is more customary to convert the vertical axis into decibels. Figure 3.1–6(a) shows the gain specification of a low-pass filter with the vertical axis in decibels and the frequency axis normalized

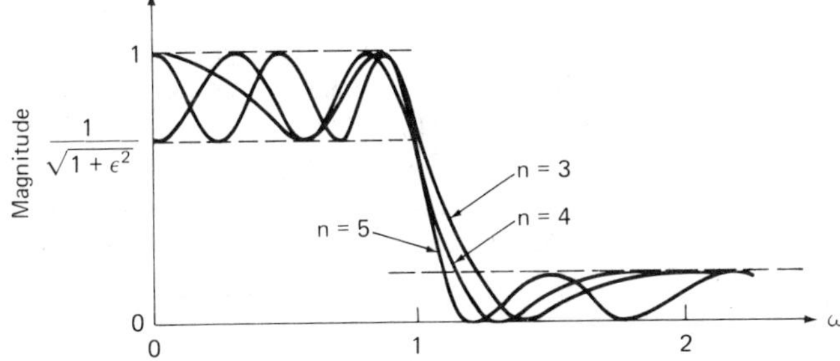

Fig. 3.1–5. Magnitude characteristics of an Elliptic approximation with arbitrary ϵ and for orders of n = 3, 4, and 5.

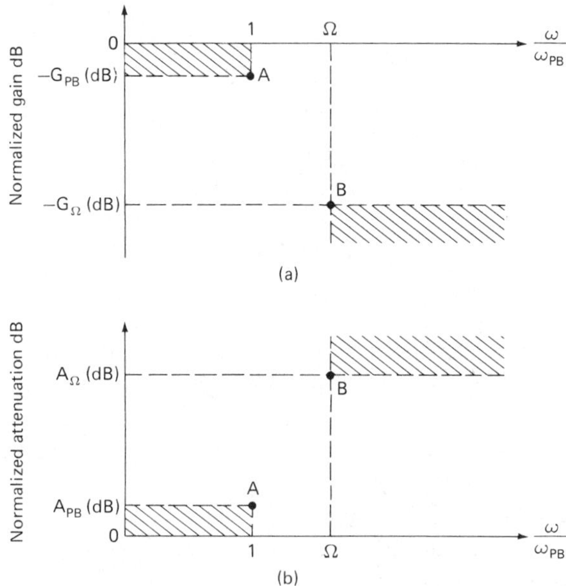

Fig. 3.1–6. (a.) Specifications for a low pass filter in terms of gain. (b.) Specifications for a low pass filter in terms of attenuation.

to ω_{PB}. We have defined $\Omega = \omega_{SB}/\omega_{PB}$ and $G_\Omega = G_{SB}$. As the gains of most filters are less than unity, attenuation rather than gain is used. Figure 3.1–6(b) shows the equivalent specification in terms of attenuation, A, plotted in decibels.

From the parameters A_Ω, A_{PB}, and Ω, we can determine the order of the filter using a set of nomographs.[1] Figure 3.1–7 is a nomograph for determining the order of the Butterworth filter approximation. The use of these nomographs is illustrated in Fig. 3.1–8. A straight line is drawn through the specified values of A_{PB} and A_Ω, shown as points 1 and 2 in Fig. 3.1–8. The intersection of this line with the ordinate of the graph determines point 3. This intersection is then extended horizontally until it meets a line drawn vertically from point 4, giving the normalized stopband frequency Ω. The resulting intersection at point 5 establishes the required order of the filter. If point 5 is between two of the "order loci," the higher one must be used.

Example 3.1–1. *Determination of the order of a low-pass Butterworth filter approximation.* Find the order of a Butterworth filter approximation to Fig.

[1] M. Kawakami, "Nomographs for Butterworth and Chebyshev Filters," *IEEE Trans. Circuit Theory,* Vol. CT-10, June 1963, pp. 288–298.

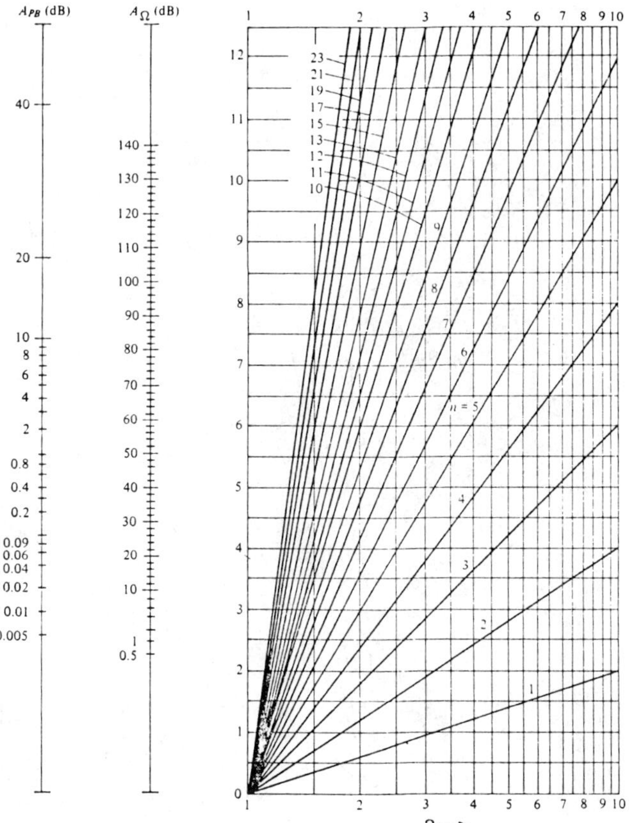

Fig. 3.1–7. A nomograph for determining the order of a maximally flat magnitude function (© 1963, IEEE).

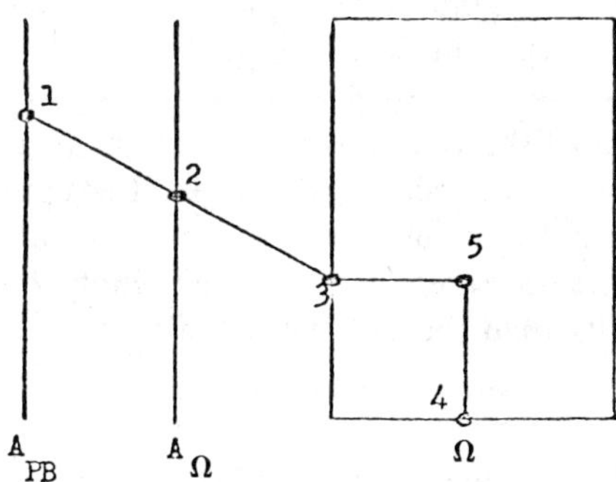

Fig. 3.1–8. The method for using the nomograph of Fig. 3.1–7.

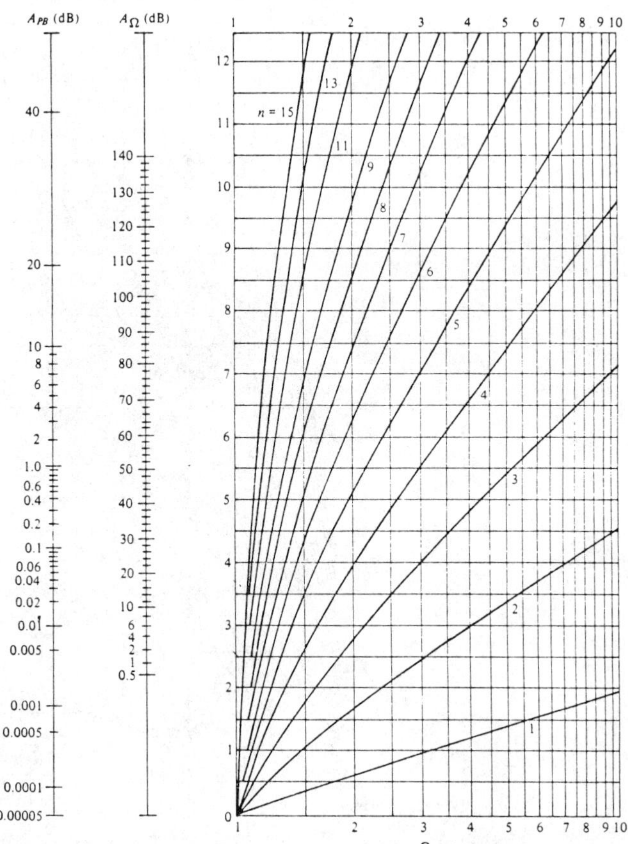

Fig. 3.1–9. A nomograph for determining the order of an equal-ripple magnitude function (© 1963, IEEE).

3.1–6(b) where $A_\Omega = 40$ dB, $A_{PB} = 3$ dB, and $\Omega = 2$. Using the nomograph of Fig. 3.1–7, we find that point 5 lies between $n = 6$ and $n = 7$. Therefore the order of the filter approximation will be 7.

Figure 3.1–9 and 3.1–10 give the corresponding nomographs for determining the complexity of a Chebyshev and elliptic filter approximation, respectively. These nomographs are used in the same manner as the one in Fig. 3.1–7. Let us illustrate how the Chebyshev and elliptic filter approximations require less complexity than the Butterworth filter approximation in the following example.

Example 3.1–2. *Determination of the order of low-pass Chebyshev and elliptic filter approximations.* Repeat the previous example to find the order of the Chebyshev and elliptic filter approximations that will meet the specifications.

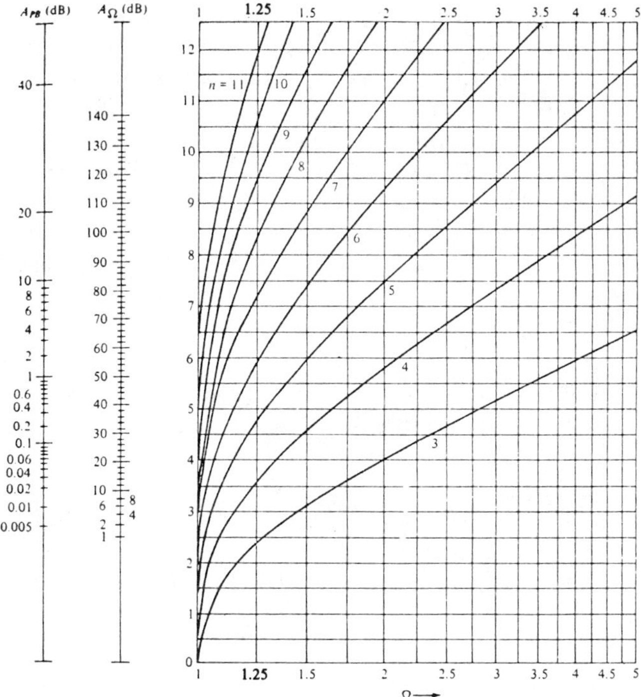

Fig. 3.1–10. A nomograph for determining the order of an elliptic magnitude function (© 1963, IEEE).

From Fig. 3.1–9, we see that $n = 4$ for the Chebyshev filter approximation to satisfy the specifications in Example 3.1–1. From Fig. 3.1–10, we see that $n = 3$ is sufficient for the elliptic filter approximation to satisfy the specifications.

Once the order of the filter approximation is known, then the designer must decide how the filter is to be realized. We shall consider two approaches. The first approach starts with Eq. (6) in the form of a second-order products (one first-order product will be necessary if n is odd). This information can be found in the many sources of tabulated filter characteristics. For example, Table 3.1–1 shows the denominator coefficients of Eq. (6) with $b_0 = 1$ and gives the pole locations and quadratic functions of the denominator of Eq. (6) for the Butterworth filter approximation. The general form of the ith second order (quadratic) function will be

$$H(s) = \frac{\pm H_{0i}\ \omega_{pi}^2}{s^2 + \dfrac{\omega_{pi}}{Q_i}\ s + \omega_{pi}^2} \tag{7}$$

where H_{0i} is the gain at $\omega = 0$, ω_{pi} is the undamped natural frequency, and Q_i is the quality factor of the pole p_i.

Typically the information tabulated on filter design is normalized so that ω_{PB} is unity. However, the filter specifications are usually in the neighborhood of thousands of cycles per second. To convert the *normalized frequency* values of the design tables to the actual filter frequency will require a *frequency denormalization.* This denormalization will involve a change of the complex-frequency variable. If we consider p as the normalized complex-frequency variable and s as the denormalized (actual) one, then the frequency denormalization is defined as

$$s = \Omega_n p \tag{8}$$

where Ω_n is a unitless frequency denormalization constant. Table 3.1–2 shows how this denormalization will act on the normalized values of R, L, and C.

Table 3.1–1(a) Denominator Coefficients of Maximally Flat Magnitude (Butterworth) Functions of the Form: $s^n + a_{n-1}s^{n-1} + a_{n-2}s^{n-2} + \cdots + a_2s^2 + a_1s + 1$ with Passband 0 – 1 rad/s

n	a_1	a_2	a_3	a_4	a_5
2	1.414214				
3	2.000000				
4	2.613126	3.414214			
5	3.236068	5.236068			
6	3.863703	7.464102	9.141620		
7	4.493959	10.097835	14.591794		
8	5.125831	13.137071	21.846151	25.688356	
9	5.758770	16.581719	31.163437	41.986386	
10	6.392453	20.431729	42.802061	64.882396	74.233429

(By permission from L. P. Huelsman and P. E. Allen, *Introduction to the Theory and Design of Active Filters,* McGraw-Hill Book Co., New York, N.Y., 1980.)

Table 3.1–1(b) Pole Locations and Quadratic Factors ($s^2 + a_1 s + 1$) of Maximally Flat Magnitude (Butterworth) Functions with Passband 0 – 1 rad/sec.; Note: All Odd-order Functions Also have a Pole at $s = 1$

n	POLES	a_1
2	$-0.70711 \pm j0.70711$	1.41421
3	$-0.50000 \pm j0.86603$	1.00000
4	$-0.38268 \pm j0.92388$	0.76536
	$-0.92388 \pm j0.38268$	1.84776
5	$-0.30902 \pm j0.95106$	0.61804
	$-0.80902 \pm j0.58779$	1.61804
6	$-0.25882 \pm j0.96593$	0.51764
	$-0.70711 \pm j0.70711$	1.41421
	$-0.96593 \pm j0.25882$	1.93186
7	$-0.22252 \pm j0.97493$	0.44504
	$-0.62349 \pm j0.78183$	1.24698
	$-0.90097 \pm j0.43388$	1.80194
8	$-0.19509 \pm j0.98079$	0.39018
	$-0.55557 \pm j0.83147$	1.11114
	$-0.83147 \pm j0.55557$	1.66294
	$-0.98079 \pm j0.19509$	1.96158
9	$-0.17365 \pm j0.98481$	0.34730
	$-0.50000 \pm j0.86603$	1.00000
	$-0.76604 \pm j0.64279$	1.53208
	$-0.93969 \pm j0.34202$	1.87938
10	$-0.15643 \pm j0.98769$	0.31286
	$-0.45399 \pm j0.89101$	0.90798
	$-0.70711 \pm j0.70711$	1.41421
	$-0.89101 \pm j0.45399$	1.78202
	$-0.98769 \pm j0.15643$	1.97538

(By permission from L. P. Huelsman and P. E. Allen, *Introduction to the Theory and Design of Active Filters,* McGraw-Hill Book Co., New York, N.Y., 1980.)

Table 3.1–2 Effect of Frequency and Impedance Denormalization on Network Elements

DENORMALIZED	R	C	L
$s = \Omega_n p$	R	$\frac{C}{\Omega_n}$	$\frac{L}{\Omega_n}$
$Z = z_n Z_n$	$z_n R$	$\frac{C}{z_n}$	$z_n L$
$Z(s) = z_n Z_n(p)$	$z_n R$	$\frac{C}{\Omega_n z_n}$	$\frac{z_n}{\Omega_n} L$

$$Z(s) = z_n \, Z_n(s) \tag{9}$$

A second type of denormalization that is often used is called *impedance normalization.* It permits an arbitrary scaling to be applied on the passive elements of a filter in order to get more desirable values. A normalized impedance $z_n(s)$ can be denormalized to the impedance $z(s)$ by the relation where z_n is a unitless impedance denormalization constant. Table 3.1–2 shows how this denormalization affects the passive elements. Also the combined effects of the denormalizations of Eqs. (8) and (9) are illustrated.

Figure 3.1–11 shows three possible realizations of Eq. (7). Figure 3.1–11(a) uses a finite-gain amplifier and an RC feedback network. This circuit was originally proposed by Sallen and Key in 1955[2] and was one of the first active RC filter structures. This circuit does not have the flexibility of realizing Eq. (7) if H_{0i} is specified for H_{0i} will be equal to $3 - 1/Q_i$. In Fig. 3.1–11(a), one may choose a suitable value for R and then ω_{pi} and Q_i define the rest of the elements. Alternatively one may choose a suitable value for the capacitors and then the resistors are defined as $(\omega_{pi}C)^{-1}$. Figure 3.1–11(a) realizes Eq. (7) with a plus sign.

Figure 3.1–11(b) is called an *infinite-gain realization.* It has the capability of realizing simultaneously H_{0i}, ω_{pi}, and Q_i. Figure 3.1–11(b) realizes Eq. (7) with a minus sign. Figure 3.1–11(c) is called the Tow-Thomas[3,4] circuit and consists of the cascade of a damped inverting integrator, an inverting integrator, and an inverter. This circuit has a great deal of flexibility and is very easy to tune. If the output is taken at V_2, then Fig. 3.1–11(c) realizes

[2] R. P. Sallen and E. L. Key, "A Practical Method of Designing RC Active Filters," *IRE Trans. Circuit Theory,* Vol. CT-2, March 1955, pp. 74–85.

[3] J. Tow, "Design Formulas for Active RC Filters Using Operational Amplifier Biquad," *Electronics Letters,* July 24, 1969, pp. 339–341.

[4] Lee C. Thomas, "The Biquad: Part I—Some Practical Design Considerations," *IEEE Trans. Circuit Theory,* Vol. CT-18, May 1971, pp. 350–357.

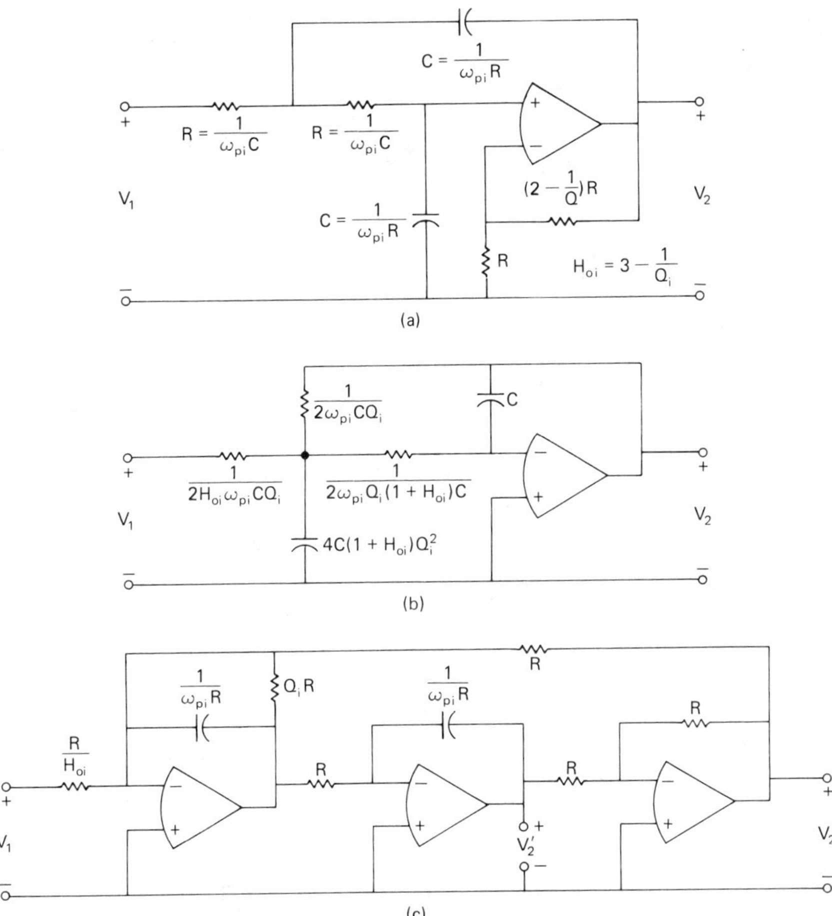

Fig. 3.1–11. Three possible realizations of eq. (7). a.) Sallen and Key (finite gain) structure. b.) Infinite gain structure. c.) Thomas-Tow (resonator) structure.

Eq. (7) with a minus sign. If the output is taken at V_2', then Fig. 3.1–11(c) realizes Eq. (7) with a plus sign. Many other realizations of Eq. (7) exist; however, the circuits of Fig. 3.1–11 are representative.

Example 3.1–3. *Design of a low-pass Butterworth filter approximation.* A low-pass Butterworth filter approximation is to be designed for the specifications of $A_\Omega = 30$ dB, and $\omega_{PB} = 2000\ \pi$ and $\omega_{SB} = 4000\ \pi$. Thus if we normalize to ω_{PB}, we get $\Omega = 2$. From Fig. 3.1–7 we see that $n = 5$ will satisfy the filter specification. From Table 3.1–1 we have two second-order stages cascaded with one first-order stage. Figure 3.1–12(a) shows a block diagram of how the stages may be cascaded. The order is arbitrary, although

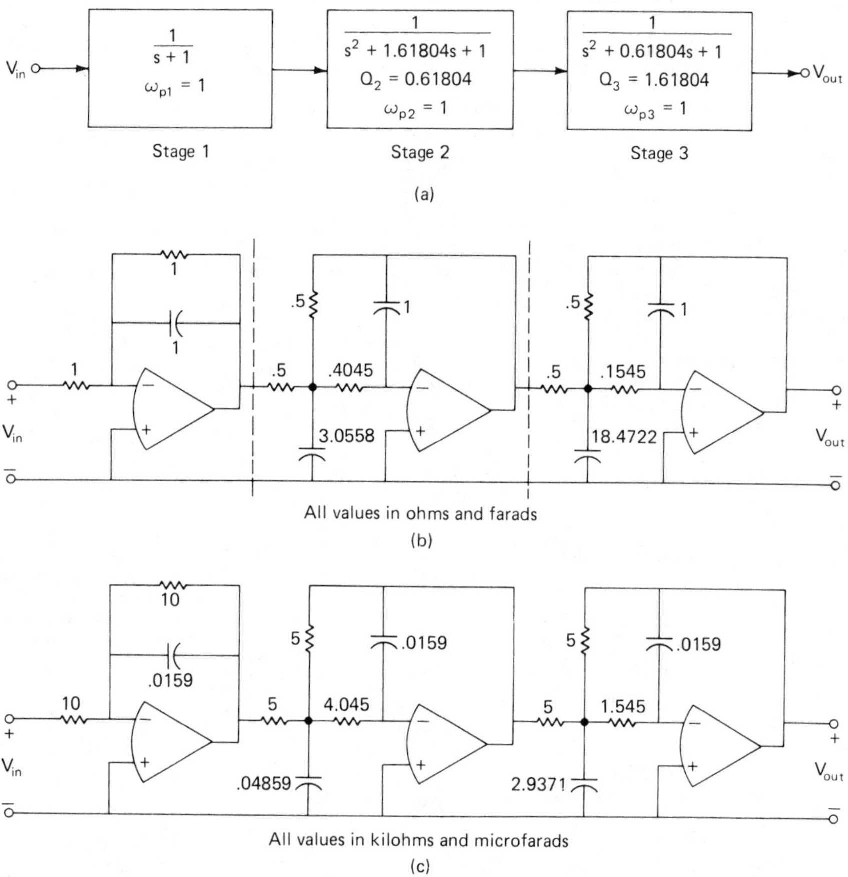

Fig. 3.1–12. Realization for Ex. 3.1–3. a.) Stage ordering. b.) Normalized realization. c.) Unnormalized realization.

one typically chooses the high Q stages as the last stages. The normalized transfer function for each of the stages is shown, as well as Q_i and ω_{pi}. Note that stage 1 is a simple first-order circuit so that Q is meaningless. Stage 1 may be realized simply with a damped integrator, whereas one of the circuits in Fig. 3.1–11 can be used for stages 2 and 3. Selecting Fig. 3.1–11(b) results in Fig. 3.1–12(b), where we have used the formulas of Fig. 3.1–11(b) with $\omega_{pi} = 1$ $(i = 1, 2, 3)$, $Q_2 = 0.61804$, and $Q_3 = 1.61804$. The last step is to frequency-denormalize using Eq. (8). We want to denormalize from a frequency of 1 radian/second to 2000 π radians/second. Therefore Ω_n is 2000 π. We shall also impedance-denormalize to avoid 1-ohm resistors. Selecting z_n as 10^4 results in Fig. 3.1–12(c). If a Butterworth filter approximation had been used for any value of A_{max} other than -3 dB, then normalized passband would not be unity. This must be taken into account when finding the proper Ω_n.

Filters other than low-pass can be designed using the first approach illustrated above. If tabular information giving the quadratic factors is not available, then one must use transformations on the complex frequency variable. These transformations allow one to take the tabulated low-pass information and to transform it to high-pass, bandpass, or band-elimination. The transformed filter information can then be realized in a cascaded manner using the appropriate RC active stages.

A low-pass to high-pass transformation can be defined as

$$s = \frac{\omega_0^2}{p} \tag{10}$$

where s is the low-pass complex frequency variable, p is the high-pass complex frequency variable, and ω_0 is a constant (normally unity). Table 3.1–3 illustrates how Eq. (10) transforms an ideal low-pass filter to an ideal high-pass filter. The poles of a high-pass filter can be found by substituting the low-pass poles for s in Eq. (10). One must also remember that a high-pass realization of nth-order has n zeros at the origin. Alternatively, one can use Eq. (10) with $\omega_{pi} = 1$ to replace s in expressions such as Eq. (7) to get

$$H_i(p) = \frac{\pm H_{0i}\, p^2}{p^2 + \dfrac{\omega_0}{Q_i}\, p + (\omega_0)^2} \tag{11}$$

which is a high-pass quadratic form of Eq. (7). Second-order RC active realizations of Eq. (11) similar to those shown in Fig. 3.1–11 can be found in the literature on RC active filters. Figure 3.1–13 gives three realizations using the same structures shown in Fig. 3.1–11.

A low-pass to bandpass transformation can be defined as

$$s = p + \frac{\omega_r^2}{p} \tag{12}$$

where s is the low-frequency complex variable, p is the bandpass complex frequency variable, and ω_r is defined in Table 3.1–3 as the geometric center frequency, given as

$$\omega_r = \sqrt{\omega_{T1}\, \omega_{T2}} \tag{13}$$

Table 3.1–3 Changes of Network Elements Under Frequency Transformations

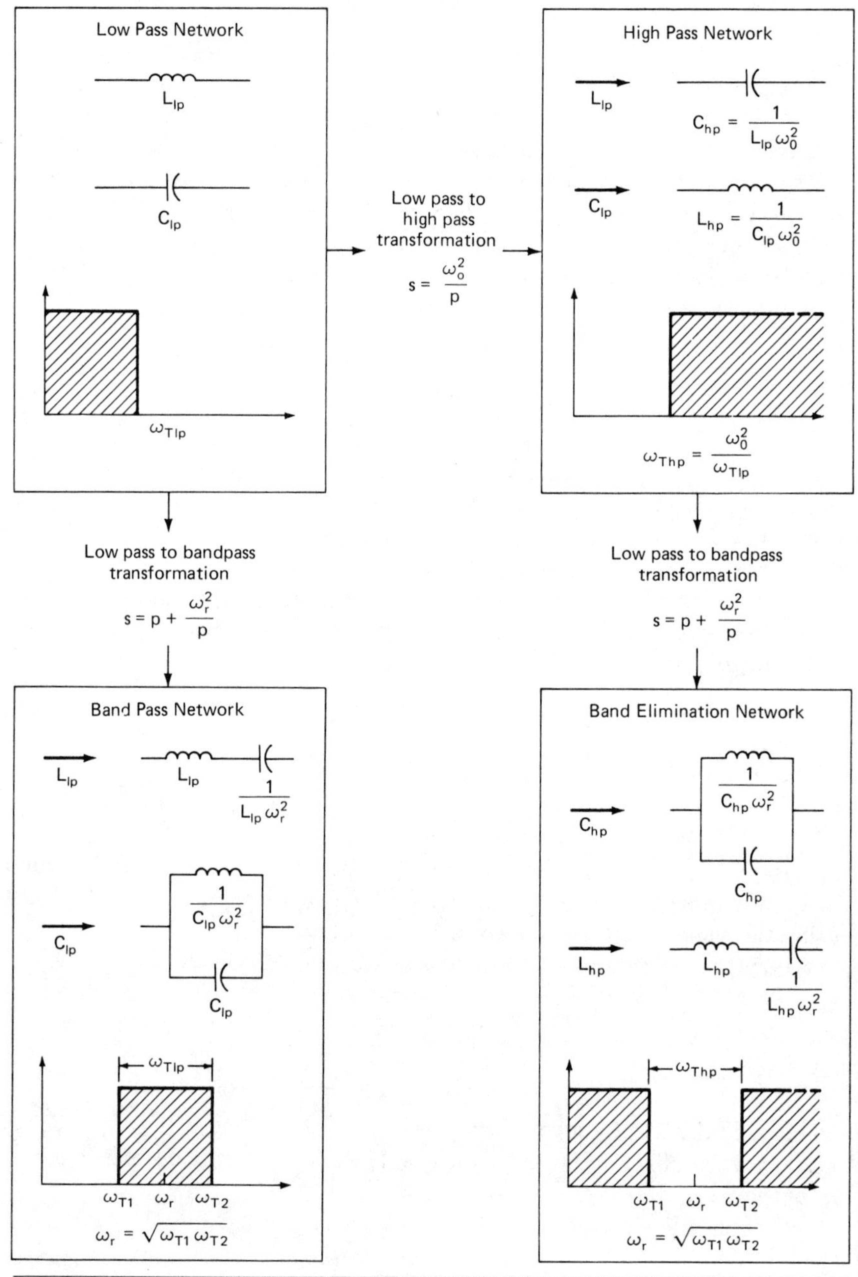

(By permission of L. P. Huelsman and P. E. Allen, *Introduction to the Theory and Design of Active Filters*, McGraw-Hill Book Co., New York, N.Y., 1980.)

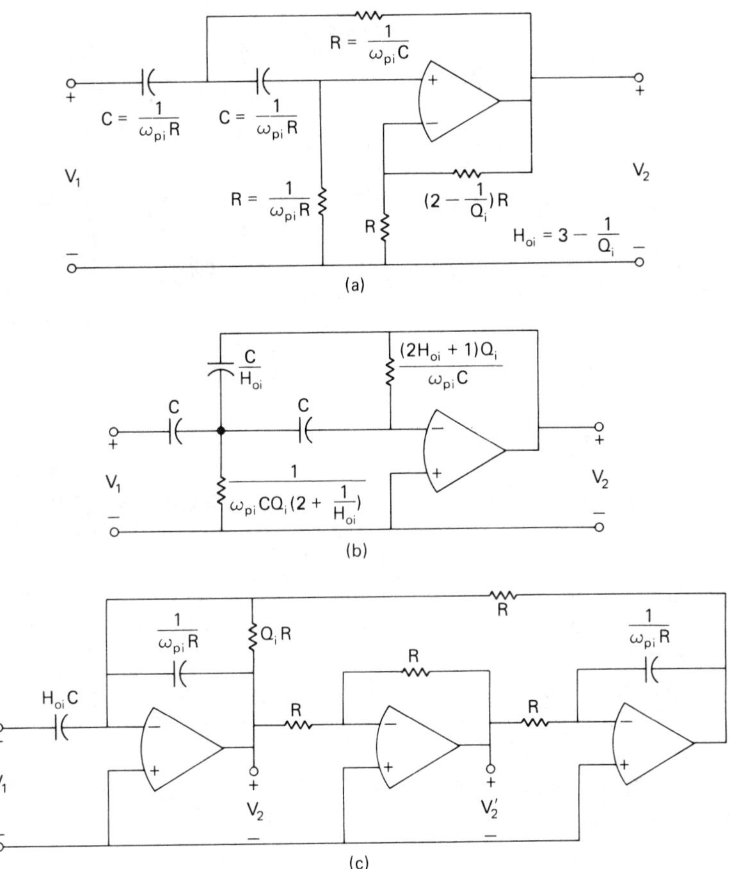

Fig. 3.1–13. Three realizations of eq. (11). a.) Finite gain or Sallen and Key structure. b.) Infinite gain structure. c.) Thomas-Tow or resonator structure.

We note that the bandwidth is defined as $\omega_{T2} - \omega_{T1}$ and is equal to the original passband of the low-pass filter, ω_{T1p} (or ω_{PB}). The transformation of Eq. (12) is used in a similar manner as the transformation of Eq. (10) to get either the new bandpass roots or the bandpass transfer function. The transformation of Eq. (12) doubles the order of the low-pass filter. For example, if we had a first-order low-pass filter given as

$$H_{LP}(s) = \frac{H_0\, \omega_1/Q}{s + \omega_1/Q} \tag{14}$$

then application of the low-pass to bandpass transformation would result in

$$H_{BP}(p) = \frac{\pm H_0 \dfrac{\omega_1}{Q} p}{p^2 + \dfrac{\omega_1}{Q} p + \omega_r^2} \tag{15}$$

We see that the second-order bandpass structure has two complex poles and a zero at the origin and at infinity in the complex frequency plane. A more general form of the second-order bandpass transfer function is

$$H_{BP}(s) = \frac{\pm H_{0i} \dfrac{\omega_{pi}}{Q_i} s}{s^2 + \dfrac{\omega_{pi}}{Q_i} s + \omega_{pi}^2} \tag{16}$$

where we have reverted back to the notation of s for the complex frequency variable. Figure 3.1–14 shows three realizations of Eq. (16) using the same structures used in Figs. 3.1–11 and 3.1–13.

The band-elimination transformation can be achieved by first using Eq. (10) followed by Eq. (12). The second-order transfer functions resulting from this transformation will contain zeros on the $j\omega$ axis. The realizations of these functions are more complex than those previously considered and will not be included here. Further information on the band-elimination filter will be presented in Section 4.5.

The second approach to active filter design, once the order of the filter is known, is to go directly to a passive realization. Passive RLC realizations for normalized low-pass filter approximations have also been tabulated. Such a tabulation can be found in Appendix A. To illustrate the use of this tabulation and to demonstrate the second approach, we repeat Example 3.1–3.

Example 3.1–4. *Realization of a passive RLC filter.* Design a passive RLC filter that will satisfy the specification of Example 3.1–3. This filter is to be driven from a voltage source having zero resistance and will be terminated in a 1000-ohm load. From Example 3.1–3, we know that $n = 5$. From Appendix A, we obtain the realization shown in Fig. 3.1–15. This realization is normalized so that $\omega_{PB} = 1$ rps and $R_L = 1$ ohm. It is now necessary to denormalize using both Eqs. (8) and (9). Because ω_{PB} of Fig. 3.1–15(a) is 1, and we desire an $\omega_{PB} = 2000\ \pi$, then $\Omega_n = 2000\ \pi$. Because R_L of Fig. 3.1–15(a) is 1 Ω, and we want $R_L = 1000\ \Omega$, then $z_n = 1000$. Applying Table 3.1–2 to Fig. 3.1–15(a) results in the final realization shown in Fig. 3.1–15(b). Figure 3.1–15(b) should be equivalent to Fig. 3.1–12.

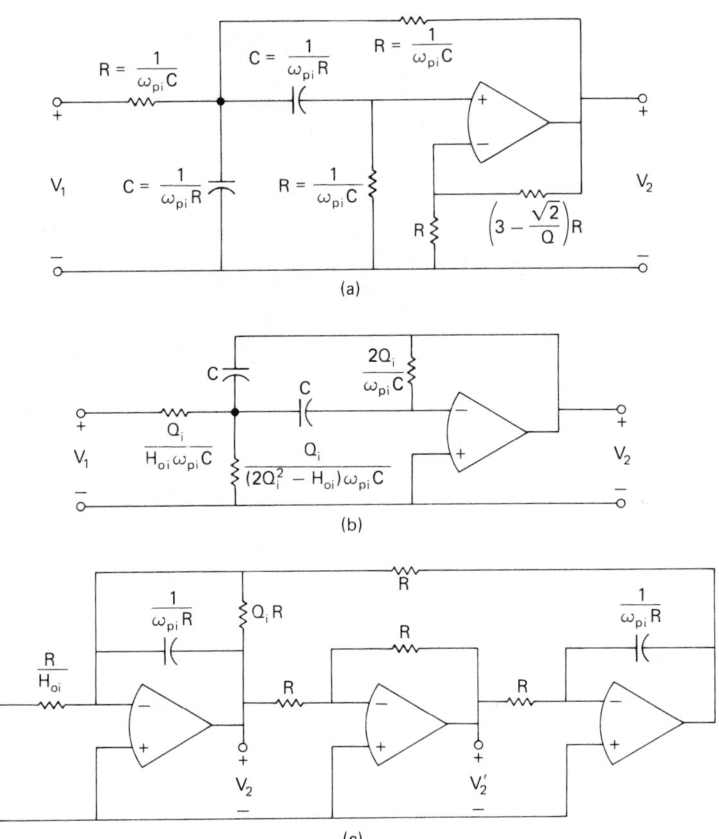

Fig. 3.1–14. Three realizations of eq. (15). a.) Finite gain or Sallen and Key. b.) Infinite gain. c.) Thomas-Tow or resonator.

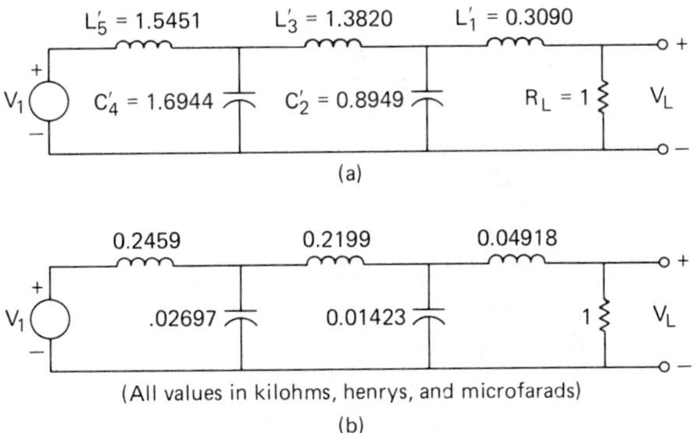

Fig. 3.1–15. a.) Passive RLC prototype for Ex. 3.1–4. b.) Unnormalized realization of a.).

The passive RLC low-pass realizations may be easily converted to high-pass, band-pass, or band-elimination using Table 3.1–3. The passive RLC realizations are also used as the starting point in certain types of active filter design. We shall present a technique based upon this approach in the next chapter.

In this section we have presented a brief overview of active filter design techniques. Most of the SC filter techniques developed in this text will start from the active filter design. Figure 3.1–16 shows the various forms of SC filter techniques that will be covered. Typically the filter design starts with specifications in the frequency domain. Figure 3.1–6 is an example. Next we have two possible paths to choose from. One path is to develop a continuous filter realization using the principles outlined in this section. Then one can either simulate the resistances of the active filter realization or the reactances of the passive filter realization. There are three basic techniques in each case, depending on which transformation, forward, backward, or bilinear is used. If the sampling rate is high, then the SC realization will closely approxi-

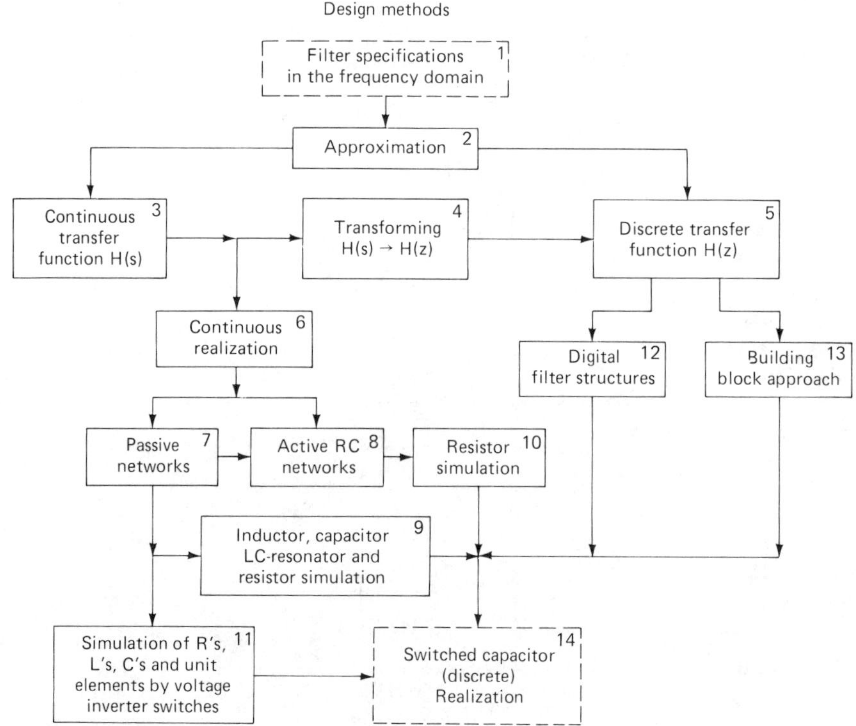

Fig. 3.1–16. Illustration of switched capacitor design methods.

mate the frequency response of the continuous realization. Under certain circumstances, it may be necessary to prewarp the specifications to account for the effects of the SC simulation of the continuous filter components.

The second path does not develop a continuous realization. Instead the continuous domain specifications are translated to the z-domain, resulting in an $H(z)$. This translation can be accomplished by various transformations between the s- and z-domains or may be developed by techniques that synthesize an $H(z)$ to meet a given frequency specification. Once $H(z)$ has been determined, then it may be realized by one of two methods. The first method uses building blocks to implement a z-domain block diagram of $H(z)$. The second method uses a direct implementation and takes advantage of the large amount of information and structures used in digital filters. The latter path will be developed in Chapter 5.

3.2 PERFORMANCE CHARACTERIZATION OF SC FILTERS

In this section, several measures are developed to aid the designer in comparing various SC filter realizations that satisfy a set of design specifications. It is common to find filter realizations that meet the same frequency response specification. However, these filter realizations may differ in parameter sensitivities, dynamic range, total chip area required, etc. A group of measures are needed that are easy to calculate and meaningful with respect to measurements in the frequency domain. Also in this section, as in the previous one, the z-domain transfer functions are described, for the popular generic forms designed as: low-pass (LP), high-pass (HP), bandpass (BP) and band-elimination (BE), or notch.

Sensitivity is a good measure of circuit performance. *Sensitivity* is a measure of the deviation in some performance characteristic of the circuit due to some change in the nominal value of one or more of the elements of the network. Low-sensitivity circuits are naturally preferred over high-sensitivity circuits. The *relative sensitivity* is defined as

$$S_p^f \triangleq \frac{\partial f}{\partial p}\frac{p}{f} = \frac{\partial \ln (f)}{\partial \ln (p)} \tag{1}$$

where f is the performance characteristic function, and p is the parameter or network element that is causing the deviation. f can be any appropriate performance characteristic, such as the pole (or zero) locations, i.e., $f = r$, $f = \theta$, or the quality factor Q of the network.

Let us first develop S_p^Q and S_p^θ for a second-order system for a high sampling rate, i.e., $\omega T << 1$. A complex pair of poles at $re^{\pm j\theta}$ can be characterized as

$$(z - re^{-j\theta})(z - re^{j\theta}) = z^2 - 2r\cos\theta z + r^2 \tag{2}$$

The denominator $D(z)$ of the transfer function for this pair of poles can be written as

$$D(z) = 1 - 2r\cos\theta\, z^{-1} + r^2 z^{-2} \tag{3}$$

where z^2 has been factored out. In general

$$D(z) = 1 + b_1 z^{-1} + b_2 z^{-2} \tag{4}$$

From Section 1.6, Eqs. (31) and (32), the pole frequency ω_0 and the quality factor Q of the continuous time domain are related to the pole locations in the discrete z-domain under high-sampling conditions as follows

$$\theta^2 \cong (\omega_0 T)^2 = 1 + b_2 + b_1 \tag{5}$$

$$Q = \frac{\theta}{1 - b_2} \tag{6}$$

or

$$Q = \frac{[1 + b_2 + b_1]^{1/2}}{1 - b_2} \tag{7}$$

and

$$b_2 = r^2 = \left(1 - \frac{\theta}{2Q}\right)^2 \cong 1 - \frac{\theta}{Q} \tag{8}$$

In practice b_1 and b_2 are functions of more than one capacitor ratio and not in all cases will the sensitivities be small. In fact, the sensitivities with respect to the different capacitor ratios depend on each SC structure. Thus, the sensitivities of the parameters given in Eqs. (5), (7), and (8), with respect to a capacitor ratio α_i, are

$$S^{\theta}_{\alpha_i} = \frac{\alpha_i}{2\theta^2}\left[\frac{db_2}{d\alpha_i} + \frac{db_1}{d\alpha_i}\right] \tag{9}$$

$$S^{Q}_{\alpha_i} = S^{\theta}_{\alpha_i} + \frac{\alpha_i Q}{\theta}\frac{db_2}{d\alpha_i} \tag{10}$$

$$S^{r}_{\alpha_i} = \frac{\alpha_i}{2\left(1 - \dfrac{\theta}{Q}\right)}\frac{db_2}{d\alpha_i} \tag{11}$$

Therefore, the S_p^f evaluation is a very adequate measurement to compare several possible realizations that satisfy the same frequency response.

Sometimes the measurement of high Q's and their sensitivities is difficult. Another alternative is the sensitivity of the maximum value, H_{max}, of a transfer function, where

$$H_{max} = H_m = \max |H(e^{j\omega T})| \tag{12}$$

Again consider a second-order transfer function of an all pole filter of the form

$$H(z) = \frac{kz^{-2}}{1 - 2r\cos\theta\, z^{-1} + r^2 z^{-2}} \tag{13}$$

The angle ϕ where H_{max} occurs, assuming $H_{max} > H(1)$, is given by

$$\phi = \cos^{-1}\left(\frac{1 + r^2}{2r}\cos\theta\right) \tag{14}$$

and the corresponding value of H_{max} is

$$H_{max} = \frac{k}{(1 - r^2)\sin\theta} \tag{15}$$

The sensitivity of H_{max} with respect to a capacitor ratio α_i, can be expressed as a function of $S^r_{\alpha_i}$ and $S^\theta_{\alpha_i}$. That is

$$S^{H_m}_{\alpha_i} = \frac{2r^2}{1 - r^2} S^r_{\alpha_i} - \frac{\theta}{\tan\theta} S^\theta_{\alpha_i} \tag{16}$$

Notice that if Q, as in Table 1.6, is defined as

$$Q = \frac{\theta}{-2\ln r} \tag{17}$$

then the sensitivity of Q can be expressed as

$$S^Q_{\alpha_i} = -\frac{1}{\ln r} S^r_{\alpha_i} + S^\theta_{\alpha_i} \tag{18}$$

Comparing expressions (16) and (18), we notice that for $S^\theta_{\alpha_i} << S^r_{\alpha_i}$ they are similar, that is, large (small) sensitivities of H_m correspond to large (small) sensitivities of Q. Several other alternatives of Eq. (18) exist. Let us illustrate the application of these sensitivities with an example.

Example 3.2–1. *Second-order filter.* Consider a sampled data filter whose transfer function[5] is given by

$$H(z) = \frac{\alpha_1 \alpha_2 z^{-2}}{1 - (2 - \alpha_2) z^{-1} + (1 + \alpha_1 \alpha_2 - \alpha_2) z^{-2}} \tag{19}$$

where α_1 and α_2 are capacitor ratios. Determine the sensitivities for two cases: (a) high sampling rate $\omega T << 1$, $r = 0.99$, and $\theta = 8°$, (b) moderate sampling rate, $r = 0.99$, and $\theta = 22.5°$. A simple set of design equations for case (a) can be obtained. Using Eq. (8) and the approximations of $\cos\theta \cong 1 - \theta^2/2$, the following expressions result

$$2 - \alpha_2 = 2 - \theta^2 - \frac{\theta}{Q} \tag{20}$$

$$1 + \alpha_1 \alpha_2 - \alpha_2 \cong 1 - \frac{\theta}{Q} \tag{21}$$

or

$$(\omega_0 T)^2 = \theta^2 = \alpha_1 \alpha_2 \tag{22a}$$

$$\alpha_2 = \theta^2 - 2 \ln r \tag{22b}$$

$$Q = \left(\frac{\alpha_1}{\alpha_2}\right)^{1/2} \frac{1}{1 - \alpha_1} \tag{23}$$

The following sensitivities of θ and Q with respect to α_1 and α_2 can be derived using Eqs. (9) through (11)

$$S^{\theta}_{\alpha_1} = S^{\theta}_{\alpha_2} = \frac{1}{2} \tag{24}$$

$$S^{Q}_{\alpha_1} = \frac{1}{2} + \theta Q = \frac{1}{2} + \frac{\theta^2}{1 - r^2} \tag{25}$$

$$S^{Q}_{\alpha_2} = -\frac{1}{2} \tag{26}$$

---

[5] B. J. Hosticka et al., "MOS Sampled Data Recursive Filters Using Switched Capacitor Integrators," *IEEE J. of Solid-State Circuits,* Vol. SC-12, No. 6, December 1977, pp. 600–608.

$$S^{r}_{\alpha_1} = \frac{\theta^2}{2r^2} \tag{27}$$

$$S^{r}_{\alpha_2} = S^{r}_{\alpha_1} - \frac{\theta^2 - 2 \ln r}{r^2} \tag{28}$$

For case (b), we want to evaluate Eqs. (16) and (17). Observing that

$$r^2 = 1 - \alpha_2 (1 - \alpha_1) \tag{29}$$

and

$$\cos \theta = \frac{2 - \alpha_2}{2\sqrt{1 - \alpha_2 (1 - \alpha_1)}} \tag{30}$$

then

$$S^{r}_{\alpha_1} = \frac{1 + r^2 - 2r \cos \theta}{2r^2} \tag{31}$$

$$S^{r}_{\alpha_2} = -\frac{1 - r^2}{2r^2} \tag{32}$$

$$S^{\theta}_{\alpha_1} = \frac{1 + r - 2r \cos^2\theta}{2r^2\, \theta \tan \theta} \tag{33}$$

$$S^{\theta}_{\alpha_2} = -\frac{1 - r^2}{2r^2\, \theta \tan \theta} + \frac{1 - r \cos \theta}{r\theta \sin \theta} \tag{34}$$

Thus from Eqs. (16) and (17)

$$S^{H_m}_{\alpha_1} = \frac{1 + r^2 - 2r \cos \theta}{1 - r^2} - \frac{1 + r^2 - 2r \cos \theta}{2r^2 \tan^2 \theta} \tag{35}$$

$$S^{H_m}_{\alpha_2} = -1 - \frac{r^2(1 - 2 \cos \theta) + 2r - 1}{2r^2 \tan^2 \theta} \tag{36}$$

$$S^{Q}_{\alpha_1} = -\frac{1 + r^2 - 2r \cos \theta}{2r^2 \ln r} + \frac{1 + r^2 - 2r \cos \theta}{2r^2\theta \tan \theta} \tag{37}$$

$$S^{Q}_{\alpha_2} = \frac{1 - r^2}{2r^2 \ln r} + \frac{r^2(1 - 2 \cos \theta) + 2r - 1}{2r^2\, \theta \tan \theta} \tag{38}$$

The numerical results for case (a), using Eqs. (24) through (28), are

$$S^{\theta}_{\alpha_1} = S^{\theta}_{\alpha_2} = -S^{Q}_{\alpha_2} = 0.5$$

and

$$S^{Q}_{\alpha_1} = 1.474, \ S^{r}_{\alpha_1} = 0.0099, \ S^{r}_{\alpha_2} = -0.071$$

Equations (24) through (28) are simple, but they are valid only for a high sampling rate. Equations (31) to (38), although more complicated, are more general. They can be also used for case (a); then

$$S^{r}_{\alpha_1} = 0.0098$$
$$S^{r}_{\alpha_2} = -0.1015$$
$$S^{\theta}_{\alpha_1} = 0.5035$$
$$S^{\theta}_{\alpha_2} = 0.4796$$
$$S^{Hm}_{\alpha_1} = 0.4730$$
$$S^{Hm}_{\alpha_2} = -1.47$$
$$S^{Q}_{\alpha_1} = 1.48$$
$$S^{Q}_{\alpha_2} = -0.50$$

We observe that the simplified expressions for a high sampling rate are very accurate, and that they differ very little in comparison with the precise expressions.

For case (b) only the more exact expressions can be used to get

$$S^{r}_{\alpha_1} = 0.077$$
$$S^{r}_{\alpha_2} = -0.010$$
$$S^{\theta}_{\alpha_1} = 0.473$$
$$S^{\theta}_{\alpha_2} = 0.402$$
$$S^{Hm}_{\alpha_1} = 7.13$$
$$S^{Hm}_{\alpha_2} = -0.381$$
$$S^{Q}_{\alpha_1} = 8.13$$
$$S^{Q}_{\alpha_2} = -0.593$$

We can consider particular expressions of Q and ω_0 based on different s- to z-transformations given in Table 1.6–1. Consider for instance the impulse invariance transformation without any approximation. The expressions are

$$f_0 = \frac{\sqrt{\theta^2 + 2 \ln r}}{2\pi T} \tag{39}$$

$$Q = -\frac{\omega_0 T}{2 \ln r} \tag{40}$$

It can be proved (see Problem 3.5) that[6]

$$S_{\alpha_i}^{f_0} = \frac{1}{(\omega_0 T)^2} \left[\frac{\ln r + \theta \cot \theta}{2} S_{\alpha_i}^{b_2} - \theta \cot \theta S_{\alpha_i}^{b_1} \right] \tag{41}$$

$$S_{\alpha_i}^{Q} = \frac{1}{(\omega_0 T)^2} \left[\frac{-\theta^2 + \theta \ln r \cdot \cot \theta}{2 \ln r} S_{\alpha_i}^{b_2} - \theta \cot \theta \, S_{\alpha_i}^{b_1} \right] \tag{42}$$

It should be noticed from Eqs. (9) through (11) and (41) through (42) that the computation of sensitivities requires essentially the determination of $S_{\alpha_i}^{b_1, b_2}$ for each particular SC filter topology where b_2 and b_1 are algebraic expressions of the capacitor ratios.

To compare different SC structures, it is appropriate to adjust all the output voltage levels. If one or more than one output voltage level is too high, overloads may occur, or if it is very low, additional noise might be involved. This voltage-adjusting process is sometimes called *scaling*. During a given clock phase, an active SC structure has some switches open and others closed and becomes simply a network with op amps and capacitors. Let us consider, without loss of generality, the structure of Fig. 3.2–1(a). The gain voltage of this structure is

$$\frac{V_3}{V_{\text{in}}} = -D_1(-F_1)(-G_1) = -D_1 F_1 G_1 \tag{43}$$

Suppose D_1 and therefore the output voltage V_1 is very small and the voltage level V_1 is to be adjusted without affecting V_3, i.e., keeping the total voltage gain of Eq. (43) constant. To accomplish this, an adjustment of V_1 by an amount x is required indicated as

$$V_1 \rightarrow x V_1 \tag{44}$$

Next, it is necessary to scale $F_1 C_2$ and C_1 as

$$(C_1, F_1 C_2) \rightarrow \left(\frac{1}{x} C_1, \frac{1}{x} F_1 C_2 \right) \tag{45}$$

---

[6] D. E. Hocevar and W. K. Jenkins, "Sensitivity Analysis of Switched Capacitor Circuits," *23rd Midwest Symposium on Circuits and Systems,* Toledo, 1980, pp. 439–443.

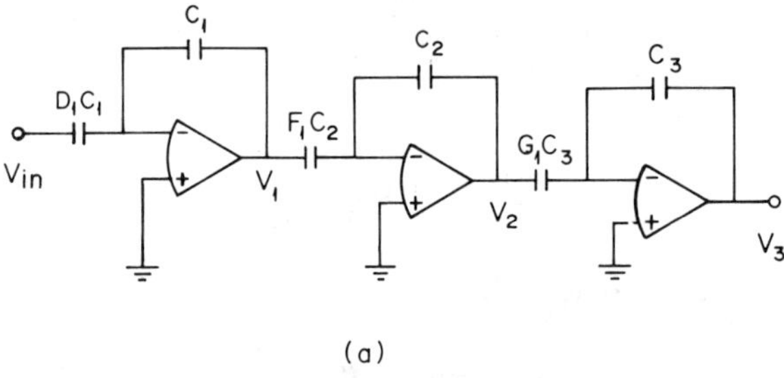

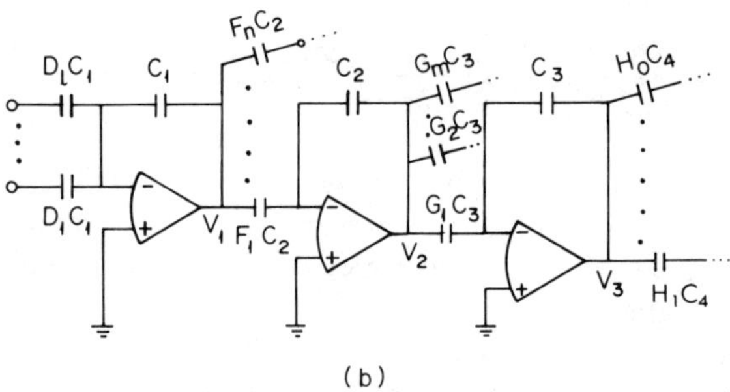

Fig. 3.2–1. Networks illustrating scaling. (a) Capacitive network. (b) A more general capacitive network.

The total voltage gain associated with Eq. (43) remains constant, that is

$$\frac{V_3}{V_{in}} = -D_1 x \left(-\frac{F_1}{x}\right)(-G_1) = -D_1 F_1 G_1 \tag{46}$$

Now assume that the output V_2 also needs to be adjusted by an amount y. Then in a similar form, scale C_2 and G_1C_3 as

$$(C_2,\ G_1C_3) \rightarrow \left(\frac{1}{y}\ C_2, \frac{1}{y}\ G_1C_3\right) \tag{47}$$

Again expression (43) is unaffected, i.e.

$$\frac{V_3}{V_{in}} = -D_1 x \left(\frac{-F_1 y}{x}\right)\left(\frac{-G_1}{y}\right) = -D_1 F_1 G_1 \tag{48}$$

This procedure can be applied for any output voltage. Consider the circuit of Fig. 3.2–1(b). Suppose that the output voltage V_1 requires adjustment by a quantity x; then the following capacitors need to be scaled

$$(C_1,\ F_1C_2, \ldots F_nC_2) \rightarrow \left(\frac{C_1}{x}, \frac{F_1C_2}{x}, \ldots \frac{F_nC_2}{x}\right) \tag{49}$$

Notice that $F_2C_2, \ldots, F_nC_2$ are capacitors not necessarily connected to the second op amp. In exactly the same way any output voltage can be adjusted, i.e.

$$V_3 \rightarrow rV_3 \tag{50}$$

Thus, it is necessary to scale only the following capacitors

$$(C_3,\ H_1C_4,\ \ldots,\ H_0C_4) \rightarrow \left(\frac{1}{r}C_3, \frac{1}{r}H_1C_4, \ldots, \frac{1}{r}H_0C_4\right) \tag{51}$$

As the reader has observed, the adjustment of an output voltage level amounts to scaling all the capacitors connected to the output voltage node involved, independently of where the other terminal of the capacitors are connected.

Typically, this capacitor value scaling procedure is going to modify the *total capacitance area* needed to realize the IC version. Let us consider the problem of calculating the total capacitance required to implement an SC network. Two ways to compute the total capacitance in an SC network are presented. The first approach simply requires one to scale so that the minimum capacitance value in the network becomes unity, C_u.[7] Thus, all the capacitor values are a function of C_u, and the total capacitance is just the sum of all the capacitors. The second approach can be readily applied to any SC network where capacitors can be combined in different groups. The capacitors in each group are identified by observing that all of them are connected to the same input node of one of the op amps. This can be illustrated by considering Fig. 3.2–1(a), where the three groups can be easily identified

$$\{D_1C_1,\ C_1\},\ \{F_1C_2,\ C_2\},\ \{G_1C_3,\ C_3\} \tag{52}$$

Now for each group we can define a minimum unit capacitance.

Example 3.2–2. *Total capacitance.* Given the circuit of Fig. 3.2–1(a), where

$$D_1C_1 = 0.2\ pF, \qquad C_1 = 1\ pF$$
$$F_1C_2 = 1.2\ pF, \qquad C_2 = 0.6\ pF$$
$$G_1C_3 = 0.25\ pF, \qquad C_3 = 0.8\ pF$$

[7] The unit capacitance C_u in practice is typically $0.5\ \text{pF} \le C_u \le 2.0\ \text{pF}$.

calculate the total capacitance. The first approach of calculating the total capacitance ignores the concept of a grouping; therefore the capacitance values yield

$$D_1C_1 = 1\ C_u, \qquad C_1 = 5\ C_u$$

$$F_1C_2 = 6\ C_u, \qquad C_2 = 3\ C_u$$

$$G_1C_3 = 1.25\ C_u, \qquad C_3 = 4\ C_u$$

and the total capacitance becomes $C_T = 20.25\ C_u$. Now, the second approach gives us the following set of values

$$D_1C_1 = 1\ C_u, \qquad C_1 = 5\ C_u$$

$$F_1C_2 = 2\ C_u, \qquad C_2 = 1\ C_u$$

$$G_1C_3 = 1\ C_u, \qquad C_3 = 3.2\ C_u$$

and the C_T is reduced to $13.2\ C_u$.

In general, the second approach of calculation of the C_T is optimal whenever it is possible to identify the different groups of capacitors. In most practical SC circuits it is possible to identify the different groups of capacitors.

As discussed in Section 3.1, filters can be classified into four basic different groups, illustrated in Fig. 3.1–1. For simplicity, without loss of generality, consider the case of second-order systems. Figure 3.1–16 illustrates several SC design approaches. One of them consists of, given the filter specifications in the s-domain, obtaining the corresponding z-domain specifications and then obtaining a realization by means of building blocks or direct realization. Typically to transform a continuous transfer function $H(s)$ to a sampled data transfer function $H(z)$ requires an s- to z-mapping. These mappings were discussed in Section 1.5. In what follows we discuss how, given an $H(s)$, a corresponding $H(z)$ is obtained under different mappings (see also Table P. 1.12). The problem of prewarping involved in the mappings is not considered here. The following sections will include a discussion about prewarping. Initially let us concentrate on the case of a low-pass filter. The transfer function in the s-domain can be characterized by

$$H(s) = \frac{\pm H_0\, \omega_p^2}{s^2 + \dfrac{\omega_p}{Q} s + \omega_p^2} \tag{53}$$

Assume a backward substitution, i.e., $s \rightarrow (1 - z^{-1})/T$. Then Eq. (53) becomes

$$H(z) = \frac{\pm K_0}{1 - 2r\cos\theta\, z^{-1} + r^2 z^{-2}} \tag{54}$$

where

$$2r\cos\theta = \frac{2 + \omega_p T/Q}{1 + \omega_p T/Q + (\omega_p T)^2} \tag{55}$$

$$r^2 = 1/[1 + (\omega_p T)/Q + (\omega_p T)^2] \tag{56}$$

$$K_0 = \pm\, H_0(\omega_p T)^2\, r^2 \tag{57}$$

Now using a forward substitution, i.e., $s \rightarrow (1 - z^{-1})/Tz^{-1}$, we obtain

$$H(z) = \frac{K_0 z^{-2}}{1 - 2r\cos\theta\, z^{-1} + r^2 z^{-2}} \tag{58}$$

where

$$2r\cos\theta = 2 - \frac{\omega_p T}{Q} \tag{59}$$

$$r^2 = 1 - \frac{\omega_p T}{Q} + (\omega_p T)^2 \tag{60}$$

$$K_0 = \pm\, H_0(\omega_p T)^2 \tag{61}$$

Finally, consider a bilinear transformation, i.e., $s \rightarrow \frac{2}{T}\,\frac{1 - z^{-1}}{1 + z^{-1}}$ Equation (53) yields

$$H(z) = \frac{K_0(1 + z^{-1})^2}{1 - 2r\cos\theta\, z^{-1} + r^2 z^{-2}} \tag{62}$$

where

$$2r\cos\theta = \frac{4 - \dfrac{\omega_p T}{Q} + (\omega_p T)^2}{4 + \dfrac{\omega_p T}{Q} + (\omega_p T)^2} \tag{63}$$

$$r^2 = \frac{4 - \dfrac{\omega_p T}{Q} + (\omega_p T)^2}{4 + \dfrac{\omega_p T}{Q} + (\omega_p T)^2} \tag{64}$$

and

$$K_0 = \frac{\pm H_0(\omega_p T)^2}{4 + \dfrac{\omega_p T}{Q} + (\omega_p T)^2} \tag{65}$$

A summary of these results is shown pictorially in Fig. 3.2–2. The transfer functions of the other type of filters can be obtained in a similar form. Table 3.2–1 describes the different types of conventional filters. The denominator,

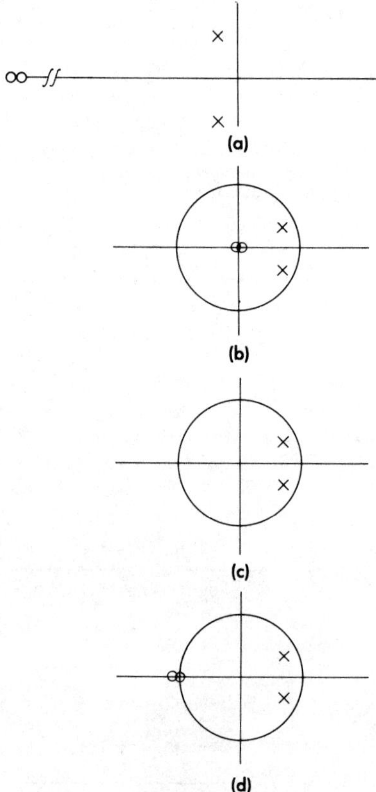

Fig. 3.2–2. Low-pass characteristics. (a) S-plane, (b) Backward mapping, (c) Forward mapping, (d) Bilinear mapping.

Table 3.2–1–$H(z)$ second-order transfer functions in the z-plane

TYPE OF FILTER	NUMERATOR
LP 20 (bilinear)	$K_0(1 + z^{-1})^2$
LP 02 (forward)	$K_0 z^{-2}$
LP 00 (backward)	K_o
LP 11	$K_0 z^{-1}(1 + z^{-1})$
LP 10	$K_0(1 + z^{-1})$
LP 01	$K_0 z^{-1}$
BP 20 (bilinear)	$K_0(1 - z^{-1})(1 + z^{-1})$
BP 01 (forward)	$K_0 z^{-1}(1 - z^{-1})$
BP 00 (backward)	$K_0(1 - z^{-1})$
HP	$K_0(1 - z^{-1})^2$
Notch (symmetric)	$K_0[1 + 2(1 - \cos\theta_0)z^{-1} + 2(1 - \cos\theta_0)z^{-2}]$ where $\cos\theta_0 = \dfrac{2r\cos\theta}{1 + r^2}$; θ_0 and θ are the angles of the zero and pole locations, respectively
Low-pass notch	Same as above with $\cos\theta_0 > \dfrac{2r\cos\theta}{1 + r^2}$
High-pass notch	Same as above with $\cos\theta_0 < \dfrac{2r\cos\theta}{1 + r^2}$
All pass	$K_0(r^2 - 2r\cos\theta\ z^{-1} + z^{-2})$

which is not shown, is of the type given in Eqs. (54), (58), or (62). The basic differences among them are involved in the numerator of the transfer functions. It can be seen in Table 3.2–1 that the LP and BP functions contain a family of different transfer functions with different numbers of zeros. Some of the transfer functions do not correspond to a particular mapping. These forms are referred to in Table 3.2–1[8] as LP_{ml}, where m denotes the number of $(1 + z^{-1})$ factors, and l denotes the number of z^{-1} terms.

The criterion to choose a particular type of transfer among the different forms for the LP or BP depends on different factors. It can be noticed that the zeros at $z = -1$ are involved only when the bilinear transform is used.

[8] P. E. Fleischer and K. R. Laker, "A Family of Active Switched Capacitor Biquad Blocks," *Bell System J.* Vol. 58, No. 10, December 1979, pp. 2235–2269.

These zeros in the z-plane are related to zeros at infinity in the s-plane, and their effects are negligible as the sampling frequency increases, i.e., small θ's. Sometimes ignoring those zeros will result in more economical realizations. The final decision about the form of the transform function will depend on the design requirements, cost, and complexity of the chosen realization.

3.3 RESISTOR SIMULATION USING THE PARALLEL SC EQUIVALENT RESISTOR

In this section SC filter structures are obtained based on conventional active RC structures by substituting the resistors by parallel SC equivalent resistors. See boxes 1, 2, 3, 6, 8, 10, and 14 of Fig. 3.1–16.

Let us first consider a single amplifier topology. A low-pass, infinite-gain, RC multiple-feedback filter is shown in Fig. 3.1–11(b) and repeated here in Fig. 3.3–1(a). The resistors are simulated by parallel SC equivalent resistors as shown in Fig. 3.3–1(b). Notice that R_1 is simulated by a reverse (inverting)

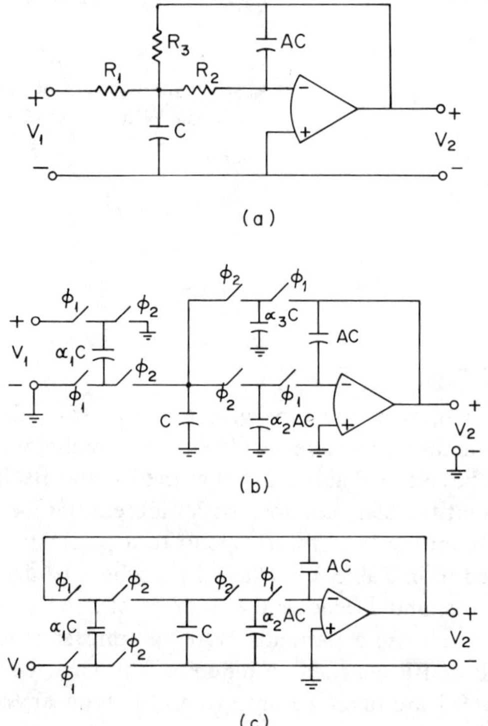

Fig. 3.3–1. (a) Low pass infinite gain SC multiple-feedback filter. (b) Noninverting parallel SC equivalent. (c) Simplified SC version.

parallel SC of the form discussed in Chapter 2. The transfer function is given by

$$H^{ee}(z) = \frac{Kz^{-1}}{1 - 2r\cos\theta\, z^{-1} + r^2 z^{-2}} \tag{1}$$

where

$$r^2 = \frac{1}{1 + \alpha_3 + \alpha_1 + \alpha_2 A} \tag{2}$$

$$2r\cos\theta = r^2(2 + \alpha_1 + \alpha_3 + \alpha_2 A - \alpha_2\alpha_3) \tag{3}$$

$$K = \alpha_1\alpha_2\, r^2 \tag{4}$$

Now, if we set $\alpha_1 = \alpha_3$, the resulting circuit is shown in Fig. 3.3–1(c). Furthermore, if we make $\alpha_2 A = \alpha_1$, then the design equations become

$$\alpha_1 = \frac{1 - r^2}{3r^2} \tag{5a}$$

and

$$A = \frac{-\alpha_1{}^2}{2r\cos\theta(1 + 3\alpha_1) - 2 - 3\alpha_1} \tag{5b}$$

The largest to smallest capacitor ratio (LSC) is approximately $9Q^2$ for Q larger than 1 and small f_0/f_s, where f_0 and f_s are the cutoff and sampling (clock) frequencies, respectively. The sensitivities of Q and θ are smaller than 1 (see Problem 3.7), and for the range of Q from 0.5 to near 3 the LSC is reasonable.

It should be noticed that given the $H(s)$ of Fig. 3.3–1(a), the resulting $H(z)$ obtained by substituting $s = (1 - z^{-1})/Tz^{-1}$ and $R_i = T/C_i$ is different from Eq. (1). Thus, the analysis of the circuit of Fig. 3.3–1(b) cannot be obtained by the simple substitutions mentioned above. It requires a charge conservation analysis. Furthermore, if the clock phases associated with the capacitor $\alpha_2 AC$ are interchanged the transfer function of Eq. (1) becomes[9]

[9] B. J. Hosticka, R. W. Brodersen, and P. R. Gray, "MOS Sampled Data Recursive Filters Using Switched Capacitor Integrators," *IEEE J. Solid-State Circuits,* Vol. SC-12, No. 6, December 1977, pp. 600–608.

$$K = k\alpha_1\alpha_2 \tag{6}$$

$$k = \frac{1}{(1+\alpha_1)(1+A\alpha_2)} \tag{7}$$

$$2r\cos\theta = k + 1 \tag{8}$$

$$r^2 = k(1+\alpha_1\alpha_2) \tag{9}$$

Because of the charge sharing between all the capacitors in this circuit the design of this filter is somewhat more complicated. Let us consider the design of this filter for the practical case of a high sampling rate (i.e., $\omega T << 1$). Thus, we can use the approximations developed for $\omega T << 1$ in Section 1.6 (see Eqs. (26) and (27) of Section 1.6). For this case we find that

$$k + 1 \cong 2 - \frac{\theta}{Q} - \theta^2 \tag{10}$$

and for high Q's

$$k(1+\alpha_1\alpha_2) \cong \left(1 - \frac{\theta}{2Q}\right)^2 \cong 1 - \frac{\theta}{Q} \tag{11}$$

The design equations are developed from Eqs. (10) and (11) and can be expressed as

$$\theta^2 \cong k\alpha_1\alpha_2 \tag{12}$$

$$Q \cong \frac{\sqrt{k\alpha_1\alpha_2}}{1 - k(1+\alpha_1\alpha_2)} \tag{13}$$

$$k \cong 1 - \frac{\theta}{Q} - \theta^2 \tag{14}$$

A design procedure can be developed as follows. Given θ and Q, determine k. Next, choose a value of α_2, and then determine α_1. Finally, from Eq. (7) determine A. The computation of the filter sensitivities for high sampling rates is left as an exercise (see Problem 3.8).

Example 3.3–1. *Low-pass infinite gain.* Design an SC version Fig. 3.3–1(c) of a multiple-feedback filter characterized by Eqs. (6) through (9) with $Q = 73$ and $\theta \cong 13.18°$ (i.e., $f_0 = 3.75$ kHz, and $f_c = 102.4$ kHz). Using Eq. (14)

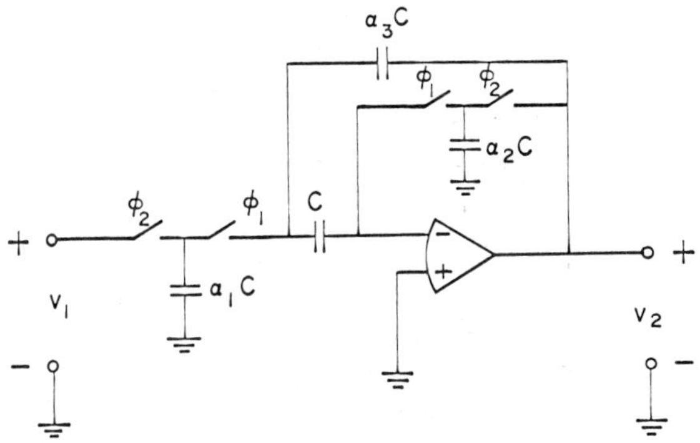

Fig. 3.3–2. Friend bandpass SC circuit.

$$k = 0.944$$

Thus from Eq. (12), and choosing $\alpha_2 = 1.99$, α_1 is obtained

$$\alpha_1 = 0.028$$

Finally, from Eq. (7) A can be calculated as

$$A = 0.015$$

Letting AC be equal to the unit capacitance, C_u, gives

$$\{\alpha_1 C, C, \alpha_2 AC, AC\} \rightarrow \{1.87, 66.67, 1.99, 1\} C_u$$

The total capacitance is approximately 72 C_u. Note that the filter characterized by Eqs. (1) through (5) for the same design specifications would require more than 47,705 C_u (see Problem 3.9a).

A single amplifier biquad circuit that realizes a bandpass response is shown in Fig. 3.3–2. This SC circuit is based on the RC prototype of Fig. 3.1–14(b) with the grounded resistor having a value of infinity. The transfer function is given by

$$H^{ee}(z) = \frac{-\alpha_1(1 - z^{-1})z^{-1}/\alpha_3}{1 - \dfrac{2\alpha_3 - \alpha_2 - \alpha_1\alpha_2 - \alpha_2\alpha_3}{\alpha_3} z^{-1} + \dfrac{\alpha_3 - \alpha_2(1 + \alpha_3)}{\alpha_3} z^{-2}} \tag{15}$$

If $\alpha_3 = 1$, the design equations become

$$\alpha_2 = \frac{1 - r^2}{2} \tag{16}$$

$$\alpha_1 = \frac{2}{\alpha_2}(1 - r\cos\theta - \alpha_2) \tag{17}$$

Although the sensitivities are very low, the major disadvantage of this circuit is that the LSC is very large, approximately $4Q^2$, for small ratios f_0/f_s.[10]

In what follows, several possible SC structures using state-variable design techniques that can be used to realize a pair of complex conjugate poles are presented. The name *state-variable* is derived from the fact that state-variable methods of solving differential equations are used in the development of the realization. Consider a second-order high-pass transfer function characterized by

$$\frac{d^2v_2(t)}{dt^2} + \frac{\omega_0}{Q}\frac{dv_2(t)}{dt} + \omega_0^2\, v_2(t) = K\frac{d^2v_1(t)}{dt^2} \tag{18}$$

where $v_2(t)$ and $v_1(t)$ are the output and input of the system, respectively. We can rewrite Eq. (18) as

$$v_H(t) = \frac{d^2v_2(t)}{dt^2} = K\frac{d^2v_1(t)}{dt^2} - \frac{\omega_0}{Q}\frac{dv_2(t)}{dt} - \omega_0^2 v_2(t) \tag{19}$$

An analog computer block for solving Eq. (19) is shown in Fig. 3.3–3. Observe that there are three possible outputs. If negative and positive integrators are available, there are four possible combinations of signs for the coefficients and the integrators designated as I_1 and I_2, that is

I_1	I_2	ω_0/Q	ω_0^2	
+	+	−	−	(20a)
−	−	+	−	(20b)
+	−	−	+	(20c)
−	+	+	+	(20d)

and each loop involves negative feedback. $v_{BP}(t)$ and $v_{LP}(t)$ are the outputs of the first and second integrator and correspond to the bandpass and low-pass functions. The three functions are simultaneously available with the same poles.

---

[10] D. E. Hocevar and W. K. Jenkins, op. cit.

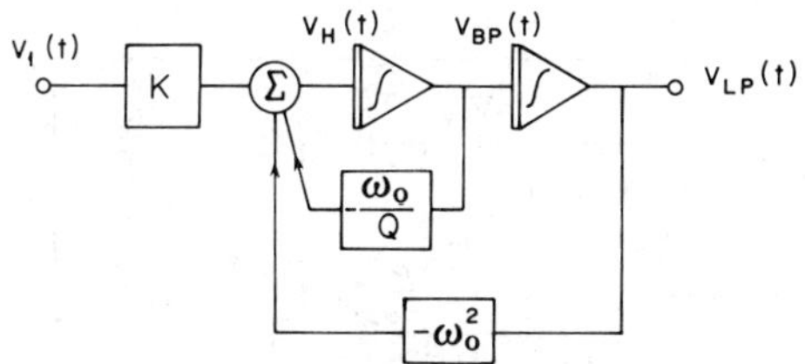

Fig. 3.3–3. Computer block diagram of Eqn (19).

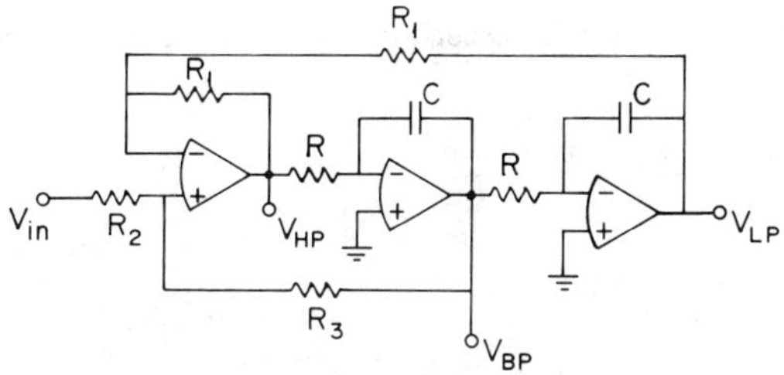

Fig. 3.3–4. KHN State variable filter.

In active RC filters, the most common integrator is the inverting Miller integrator shown in Fig. 2.5–1. The first basic implementation of Fig. 3.3–3 using both inverting integrators (Eq. (20b)) is the KHN state-variable structure.[11] This circuit is shown in Fig. 3.3–4. Also note that the time constant RC associated with the integrators is defined as

$$RC = 1/\omega_0 \tag{21}$$

The output can be expressed as

$$V_{\mathrm{HP}} = \frac{2R_3}{R_2 + R_3} V_{in} + \frac{2R_2}{R_2 + R_3} V_{\mathrm{BP}} - V_{\mathrm{LP}} \tag{22}$$

According to the signs given by Eq. (20b), the corresponding equation for Eq. (22) is

$$V_{\mathrm{HP}} = KV_{in} + \frac{1}{Q} V_{\mathrm{BP}} - V_{\mathrm{LP}} \tag{23}$$

The resistor ratios become

[11] W. J. Kerwin et al., "State Variable Synthesis for Insensitive I.C. Transfer Functions," *IEEE J. Solid-State Circuits,* Vol. SC-2, September 1967, pp. 87–92.

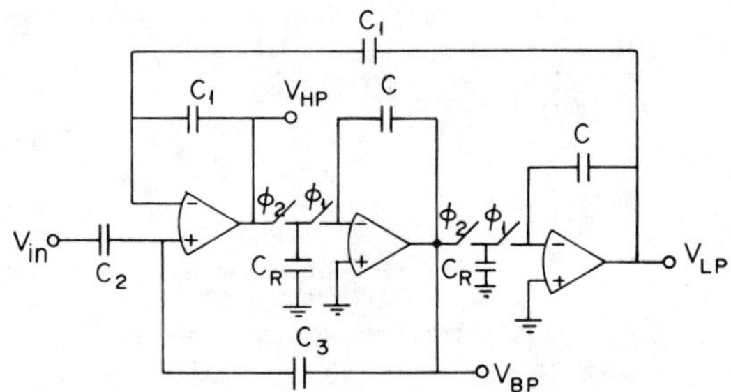

Fig. 3.3–5. A KHN state variable SC version using parallel SC resistors and inverting integrators.

$$\frac{R_3}{R_2} = 2Q - 1 \tag{24}$$

and

$$K = 2(1 - 1/2Q) \tag{25}$$

An SC version of the KHN filter using the parallel-SC resistor equivalent and a capacitive summer[12] is shown in Fig. 3.3–5. The corresponding equations for this structure are

$$V_{\mathrm{HP}} = \frac{2C_2}{C_3 + C_2} V_{\mathrm{in}} + \frac{2C_3}{C_3 + C_2} V_{\mathrm{BP}} - V_{\mathrm{LP}} \tag{26}$$

Because we are using Type I DDI integrators,[13] we can write

$$V_{\mathrm{BP}} = \frac{-\dfrac{C_R}{C} z^{-1}}{1 - z^{-1}} V_{\mathrm{HP}} \tag{27}$$

and

$$V_{\mathrm{LP}} = \frac{-\dfrac{C_R}{C} z^{-1}}{1 - z^{-1}} V_{\mathrm{BP}} \tag{28}$$

---

[12] An SC resistor across the feedback capacitor C_1 and the C's can be connected to provide a DC feedback. Each SC resistor requires two switches and a very small capacitor, such as the parasitic capacitance involved with the switches.

[13] For the state-variable SC filters, the superscripts e and o are omitted for simplicity. The inputs and outputs of the integrators are sampled at the same clock phase.

and therefore

$$V_{LP} = \frac{\left(\frac{C_R}{C}\right)^2 z^{-2}}{(1 - z^{-1})^2} V_{HP} \tag{29}$$

Solving the above equations, we can write the high-pass transfer function as

$$H_{HP}(z) = \frac{V_{HP}}{V_{in}} = \frac{-\frac{2C_2}{C_2 + C_3}(1 - z^{-1})^2}{1 - z^{-1}\left[2 - \frac{2C_3\alpha}{C_2 + C_3}\right] + z^{-2}\left[1 - \alpha\left(\frac{2C_3}{C_3 + C_2} - \alpha\right)\right]} \tag{30}$$

where $\alpha = C_R/C$. The bandpass transfer function is

$$H_{BP}(z) = \frac{V_{BP}}{V_{in}} = \frac{\frac{2C_2\alpha}{C_2 + C_3} z^{-1}(1 - z^{-1})}{1 - z^{-1}\left[2 - \frac{2\alpha C_3}{C_2 + C_3}\right] + z^{-2}\left[1 - \alpha\left(\frac{2C_3}{C_3 + C_2} - \alpha\right)\right]} \tag{31}$$

and the low-pass transfer function is

$$H_{LP}(z) = \frac{-\frac{2C_2\alpha^2}{C_2 + C_3} z^{-2}}{1 - z^{-1}\left[1 - \frac{2\alpha C_3}{C_2 + C_3}\right] + z^{-2}\left[1 - \alpha\left(\frac{2C_3}{C_3 + C_2} - \alpha\right)\right]} \tag{32}$$

The pole locations, r and θ, are given by

$$r = \left[1 - \alpha\left(\frac{2C_3}{C_3 + C_2} - \alpha\right)\right]^{1/2} \tag{33}$$

and

$$\cos\theta = \frac{1 - \frac{\alpha C_3}{C_2 + C_3}}{\left[1 - \alpha\left(\frac{2C_3}{C_3 + C_2} - \alpha\right)\right]^{1/2}} \tag{34}$$

or

$$\alpha^2 = 1 + r^2 - 2r\cos\theta \tag{35}$$

Next we consider another of the several possible implementations of the state-variable sampled data filter. In this case we use both noninverting integrators; this corresponds to option (20a) of the analog computer block diagram of Fig. 3.3–3. Besides, we assume that both integrators may have different time constants α_1 and α_2. This is illustrated in Fig. 3.3–6. Thus, the equations characterizing this filter are given by

$$V_{HP} = \alpha_5 V_{in} - \alpha_4 V_{BP} - \alpha_3 V_{LP} \tag{36}$$

$$V_{BP} = \frac{\alpha_1 z^{-1}}{1 - z^{-1}} V_{HP} \tag{37}$$

$$V_{LP} = \frac{\alpha_2 z^{-1}}{1 - z^{-1}} V_{BP} \tag{38}$$

or

$$V_{LP} = \frac{\alpha_1 \alpha_2 z^{-2}}{(1 - z^{-1})^2} V_{HP} \tag{39}$$

Solving these equations results in

$$\frac{V_{HP}}{V_{in}} = \frac{\alpha_5(1 - z^{-1})^2}{1 - z^{-1}(2 - \alpha_1\alpha_4) + (1 - \alpha_1\alpha_4 + \alpha_1\alpha_2\alpha_3)z^{-2}} \tag{40}$$

$$\frac{V_{BP}}{V_{in}} = \frac{\alpha_1\alpha_5(1 - z^{-1})z^{-1}}{1 - z^{-1}(2 - \alpha_1\alpha_4) + (1 - \alpha_1\alpha_4 + \alpha_1\alpha_2\alpha_3)z^{-2}} \tag{41}$$

$$\frac{V_{LP}}{V_{in}} = \frac{\alpha_1\alpha_2\alpha_5\, z^{-2}}{1 - z^{-1}(2 - \alpha_1\alpha_4) + (1 - \alpha_1\alpha_4 + \alpha_1\alpha_2\alpha_3)z^{-2}} \tag{42}$$

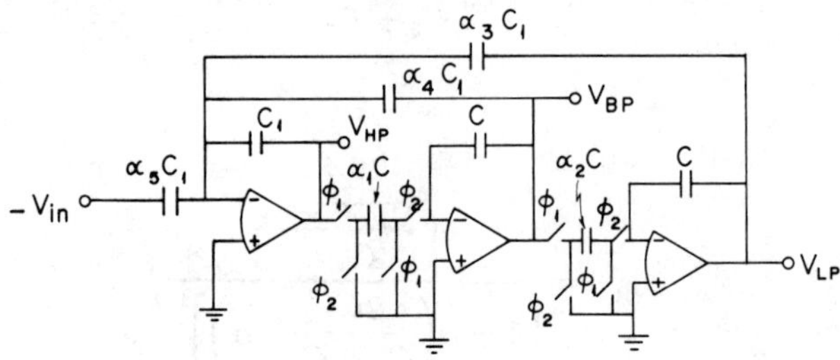

Fig. 3.3–6. A state variable SC filter using noninverting integrators.

The design equations as functions of r and θ are

$$\alpha_1\alpha_4 = 2 - 2r\cos\theta \tag{43}$$

$$\alpha_1\alpha_2\alpha_3 = 1 + r^2 - 2r\cos\theta \tag{44}$$

Also we observe that

$$\left.\frac{V_{\mathrm{LP}}}{V_{\mathrm{in}}}\right|_{\omega=0} = \frac{\alpha_5}{\alpha_3} \tag{45}$$

The center frequency f_0 and the selectivity Q can be expressed as functions of θ and r. This topic was discussed in Section 1.6. Furthermore, for a high sampling rate and small values of $(1 - r)$, practically any of the approximations discussed in Section 1.6 can be used, i.e., $Q \cong \theta/2(1 - r)$ and $\theta = \omega_0 T$. We should emphasize that there exist several other possible SC implementations of Fig. 3.3–3 besides those indicated by Eqs. (20a) to (20d). In addition, the resistor of each integrator can be replaced with either a forward, backward, or bilinear type, providing more alternatives. Structures involving stray-insensitive integrators will be discussed in Chapter 5.

Next, we consider SC implementations using only two op amps.[14] There are several advantages of SC filters in comparison with conventional RC active filters. One of them is the ability to invert the signal without requiring additional inverting amplifiers. Thus, for example consider the Tow-Thomas biquad[15,16] shown in Fig. 3.3–7(a). This biquad has the advantages that each op amp has a grounded input and that Q is directly determined by the resistor QR. Unfortunately, it does not provide a high-pass output, which is lost in the desire to have all negative feedback loops. A Tow-Thomas SC version is shown in Fig. 3.3–7(b). The inverting amplifier has been eliminated, and the summation is carried out at the negative terminal (virtual ground) of the first op amp of Fig. 3.3–7(b). The equations characterizing this structure are given by

$$V_{\mathrm{BP}} = \frac{-\alpha z^{-1}}{1 - z^{-1}} V_{\mathrm{in}} - \frac{\alpha_3 z^{-1}}{1 - z^{-1}} V_{\mathrm{BP}} + \frac{\alpha_1 z^{-1}}{1 - z^{-1}} V_{\mathrm{LP}} \tag{46}$$

$$V_{\mathrm{LP}} = \frac{-\alpha_2 z^{-1}}{1 - z^{-1}} V_{\mathrm{BP}} \tag{47}$$

[14] B. J. Hosticka et al, "MOS Sampled Data Recursive Filters Using Switched Capacitor Integrators," *IEEE J. Solid-State Circuits,* Vol. SC-12, December 1977, 600–608.

[15] L. C. Thomas, "The Biquad: Part I—Some Practical Design Considerations," *IEEE Trans. Circuit Theory,* Vol. CT-18, May 1971, pp. 350–357.

[16] J. Tow, "Active RC Filters—A State-Space Realization," *Proc. IEEE,* Vol. 56, 1968, pp. 1137–1139.

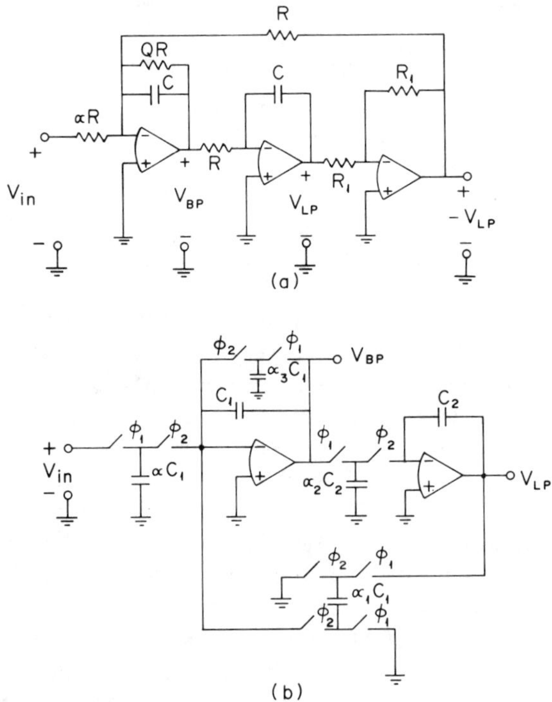

Fig. 3.3–7. a) Tow-Thomas RC-Biquad. $\omega_0 = 1/RC$ b) Tow-Thomas state variable SC version.

Therefore we can write the following transfer equations

$$\frac{V_{BP}}{V_{in}} = \frac{-\alpha z^{-1}(1 - z^{-1})}{1 - (2 - \alpha_3)z^{-1} + (1 + \alpha_1\alpha_2 - \alpha_3)z^{-2}} \tag{48}$$

$$\frac{V_{LP}}{V_{in}} = \frac{-\alpha\alpha_2 z^{-2}}{1 - (2 - \alpha_3)z^{-1} + (1 + \alpha_1\alpha_2 - \alpha_3)z^{-2}} \tag{49}$$

If $\alpha = \alpha_1$, then two switches and one capacitor can be saved. The modified structure is shown in Fig. 3.3–8. It should be noticed that Eq. (48) and Eq. (49) can also be directly obtained by using the charge conservation analysis for each feedback capacitor and each phase ϕ_1 and ϕ_2 (see Problem 3.14).

Let us consider another possible structure, the corresponding case of Fig. 3.3–6 with two noninverting integrators. A possible realization with only two op amps is shown in Fig. 3.3–9. To obtain the transfer function the charge conservation equations can be used. For both feedback capacitors C and during the ϕ_1 and ϕ_2 phase periods, we can obtain the charge conservation equations for Fig. 3.3–9. After taking the z-transform, we can write

$$\frac{V_{\text{BP}}}{V_{\text{in}}} = \frac{\alpha z^{-1}(1 - z^{-1})}{1 - (2 - \alpha_3)z^{-1} + (1 + \alpha_1\alpha_2 - \alpha_3)z^{-2}} \tag{50}$$

$$\frac{V_{\text{LP}}}{V_{\text{in}}} = \frac{\alpha\alpha_2 z^{-2}}{1 - (2 - \alpha_3)z^{-1} + (1 + \alpha_1\alpha_2 - \alpha_3)z^{-2}} \tag{51}$$

A simple set of design equations for a high sampling rate and high Q of a bandpass filter is, from Section 1.6, the following

$$2 - \alpha_3 \cong 2 - \theta^2 - \frac{\theta}{Q} \tag{52}$$

$$1 + \alpha_1\alpha_2 - \alpha_3 \cong 1 - \frac{\theta}{Q} \tag{53}$$

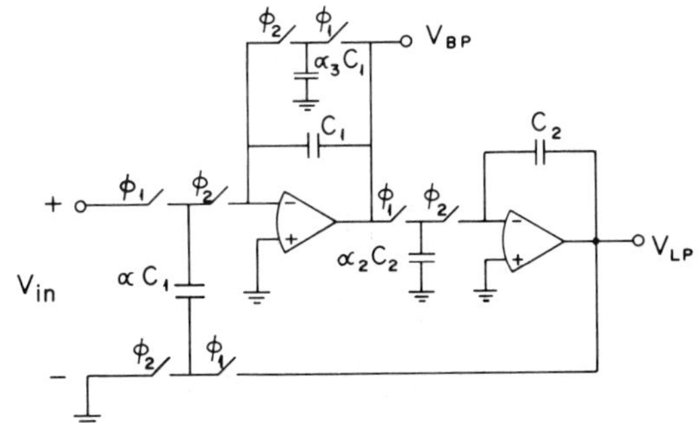

Fig. 3.3–8. Simplified state-variable SC filter using inverting integrators.

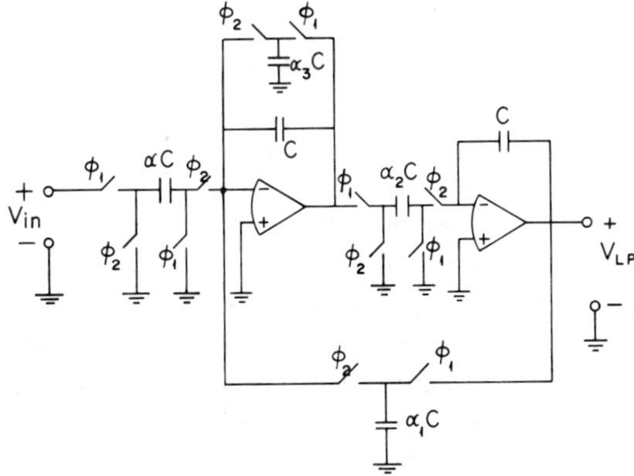

Fig. 3.3–9. A state variable SC filter using two noninverting integrators.

or

$$\alpha_1 \alpha_2 = \theta^2 \tag{54}$$

$$\alpha_3 = \theta^2 + \frac{\theta}{Q} \tag{55}$$

$$\alpha = K = \alpha_1 H_{DC} = \alpha_1 \; H(1) \tag{56}$$

The sensitivity expressions for Q and ω_0 can be obtained, for

$$\theta^2 \cong \alpha_1 \alpha_2 \tag{57}$$

$$Q \cong \frac{(\alpha_1 \alpha_2)^{1/2}}{\alpha_3 - \alpha_1 \alpha_2} \tag{58}$$

Thus

$$S_{\alpha_1}^{\theta} = S_{\alpha_1}^{\omega_0} = \frac{1}{2} \tag{59}$$

$$S_{\alpha_2}^{\omega_0} = \frac{1}{2} \tag{60}$$

$$S_{\alpha_3}^{\omega_0} = 0 \tag{61}$$

$$S_{\alpha_1}^{Q} = \frac{1}{2} + Q\theta \tag{62}$$

$$S_{\alpha_2}^{Q} = \frac{1}{2} + Q\theta \tag{63}$$

$$S_{\alpha_3}^{Q} = -1 - \theta Q \tag{64}$$

Observe that for this structure with high Q, that a very high f_c is required in order to obtain low Q sensitivity. Other structures with better sensitivity properties are discussed in other sections.

For a better understanding in the z-domain of structures of Figs. 3.3–9 and 3.3–7, Fig. 3.3–10 shows their flow diagrams using delays and multipliers. Notice that they are topologically the same and the common basic requirement is for all the loops to be negative. Furthermore, we have two loops around the delays, that is $(1 - \alpha_3)$ and 1, and if we interchange them we will obtain exactly the same low-pass output but the bandpass output will be modified. The original pole-zero pattern of the V_{BP} is shown in Fig. 3.3–10(c), and the zeros are located at

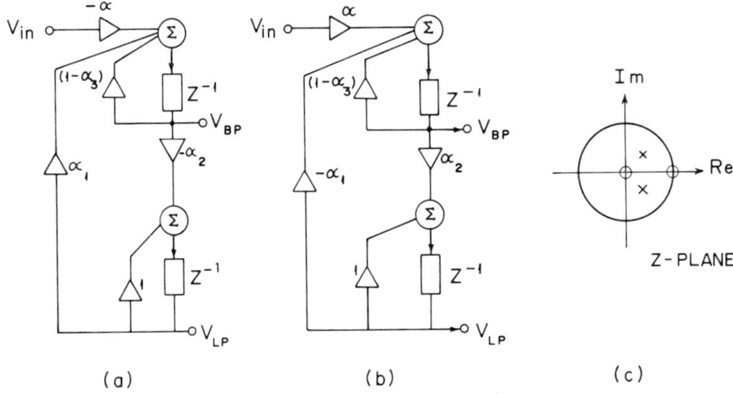

Fig. 3.3–10. Flow diagram of the Tow-Thomas state variable SC filter. (a) Inverting integrators corresponding to Fig. 3.3–7, and (b) Noninverting type corresponding to Fig. 3.3–9. (c) Pole-zero pattern for V_{BP}.

$$z_1 = 0 \tag{65}$$

$$z_2 = 1 \tag{66}$$

When the loops are interchanged, the zeros are relocated at

$$z_1 = 0 \tag{67}$$

$$z_2 = 1 - \alpha_3 \tag{68}$$

The physical implementation to obtain Eq. (68) involves moving the capacitor $\alpha_3 C$ (and switches) from the first op amp to the second op amp.

A few comments regarding practical implementations are due. The two-op amp structures use the inverting, forward integrator which has a top-plate parasitic of C_p.[17] C_p is a nonlinear voltage dependent capacitance. However, for small constant signals, C_p is directly added to the parallel SC; this is illustrated in Fig. 3.3–11.

Ideally we would like to have inverting and noninverting forward integrators that are insensitive to stray capacitance. Unfortunately we have available only stray-insensitive, noninverting forward and inverting backward integrators. We can combine these two types of stray-insensitive integrators to develop other structures that are not strictly state-variable, but have good properties (see Chapter 5).[18,19] Other practical considerations of SC filters are also presented in more detail in Chapter 5.

[17] D. J. Allstot, "MOS Switched Capacitor Ladder Filters," Ph.D. thesis, University of California, Berkeley, 1979.

[18] K. Martin and A. S. Sedra, "Strays-Insensitive Switched Capacitor Filters Based on Bilinear Z-Transform," *Electronics Letters,* Vol. 15, June 1979, pp. 365–366.

[19] P. E. Fleisher, A. Ganesan, and K. R. Laker, "Parasitic Compensated Switched Capacitor Circuits," *Electronics Letters,* Vol. 17, November 1981, pp. 929–931.

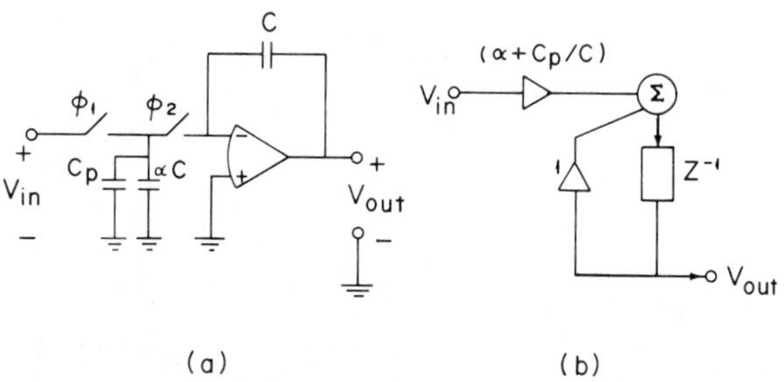

Fig. 3.3–11. a) Inverting integrator with top-plate parasitic capacitance. b) Corresponding flow-diagram

---

Example 3.3–2. *Tow-Thomas SC filter realization.* Design a bandpass filter with a center frequency $f_0 = 2041$ Hz, a quality factor $Q = 16$, and $H_0 = 1$ for a sampling frequency of 100 kHz. Because $\theta = \omega_0 T = 7.3476°$ is small and $Q = 16$, we can use the approximate design equations (54) through (56). Then the following relations are obtained

$$\alpha_1\alpha_2 = 0.016445,\ \alpha_1 = \alpha_2 = 0.128240$$

$$\alpha_3 = 0.02446$$

$$\alpha = 1$$

There are two groups of capacitors, in Fig. 3.3–7(b), associated with each op amp that can be calculated independently.

$$\text{Group 1:}\quad \{\alpha C_1,\ \alpha_1 C_1,\ \alpha_3 C_1,\ C_1\}$$

$$\text{Group 2:}\quad \{\alpha_2 C_2,\ C_2\}$$

The capacitor values are obtained so that the minimum capacitance value becomes the unit capacitance C_u. Thus

$$\text{Group 1:}\quad \{5.25,\ 5.25,\ 1,\ 40.88\}\ C_u$$

$$\text{Group 2:}\quad \{1.0,\ 7.8\}\ C_u$$

and the total capacitance is 62 C_u. Aside from a dynamic range point of view, the voltage V_{BP} may need to be scaled. Scaling, as discussed in the previous section, can be obtained by simply modifying $\alpha_3 C_1$, C_1 and $\alpha_2 C_2$. This scaling does not change the overall transfer function for V_{LP}. If V_{BP} is to be modified to render xV_{BP}, then it is necessary to scale only

$$(C_1, \alpha_3 C_1, \alpha_2 C_2) \text{ to } \left(\frac{C_1}{x}, \frac{\alpha_3 C_1}{x}, \frac{\alpha_2 C_2}{x}\right)$$

In a similar manner if the gain associated with V_{LP} is to be scaled by yV_{LP}, the following capacitors are modified

$$(C_2, \alpha_1 C_1) \text{ to } \left(\frac{C_2}{y}, \frac{\alpha_1 C_1}{y}\right)$$

It should be noticed that the scaling alters the total capacitance required for the implementation. Thus, in general, a compromise between total capacitance, sensitivity, and dynamic range should be achieved. The frequency response of the bandpass output is shown in Fig. 3.3–12. The total capacitance also depends on the design specifications. For instance for $Q = 0.7071$, $H_0 = 1$, $\theta \cong 7.2°$, the capacitor ratios are $\{\alpha C_1, \alpha_1 C_1, \alpha_3 C_1, C_1\} = \{1,1, 1.54, 7.96\}$ and $\{\alpha_2 C_2, C_2\} = \{1, 7.96\}$, and the total capacitance is only $20C_u$.

The low-pass filters considered in this section showed that in general for high Q the one-op amp structure of Fig. 3.3–1(b) and (c) based on the multiple-

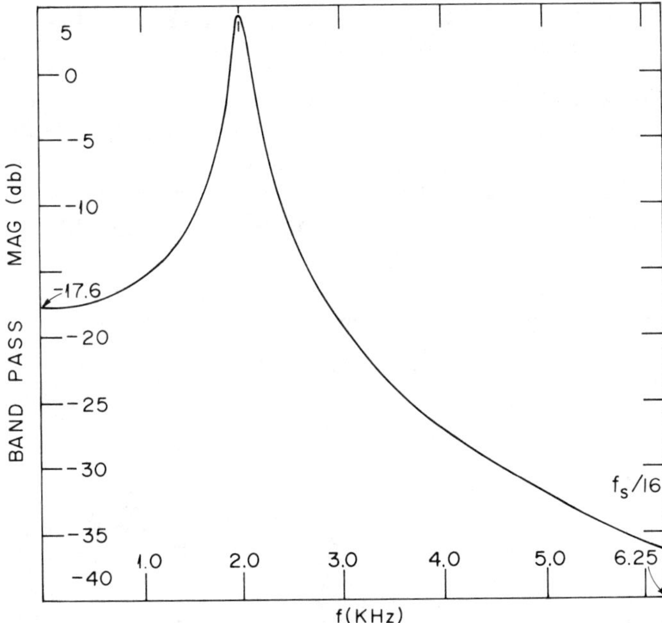

Fig. 3.3–12. Frequency response of the bandpass filter of Fig. 5.3–7(b) with a clock frequency of 100 KHz.

feedback RC prototype has low Q sensitivity magnitude; however, the total capacitance is proportional to Q^2. A variation of Fig. 3.3–1(c) when the clock switches associated with $\alpha_2 AC$ are interchanged gives a structure with higher sensitivities but smaller total capacitance, and a certain tradeoff is possible for the two-op amp circuits; the Q sensitivity is proportional to Q and has low LSC values. A different circuit shown in Fig. P 3.16 has low Q sensitivity magnitude, although the LSC is proportional to Q. Similar comparisons can be made regarding other types of filters. Clearly, for a specific application and design choice, a more specific comparison can be made.

The SC design approach of this section based on active RC-prototypes is convenient and practical when the sampling rate is high, that is for $\theta Q < 1$. Furthermore, the inverting and noninverting configurations can be easily implemented. However, because a forward mapping is involved, problems of stability may arise. The reader should remember from Chapter 1 that the forward transformation is not always a stable mapping. Thus for large θ, it is preferable to use other structures obtained from unconditionally stable mapping or to design directly in the z-plane for particular zero and pole locations (r, r_0, θ and θ_0), rather than to make approximations of equivalent resistors.

3.4 SC FILTERS USING THE BILINEAR Z-TRANSFORM FOR COMPONENT SIMULATION

This section deals with an SC filter design approach based on an element-by-element replacement of all resistors in analog RC filters.[20] The approaches taken use boxes 1, 2, 3, 6, 8, 10, and 14 or 1, 2, 3, 6, 7, 9, and 14 of Fig. 3.1–16. The difference between the first approach of this section and Section 3.3 is found in box 10. The bilinear z-transform rather than the forward z-transform is used to develop a resistor simulation.

Consider the linear active RC filter shown in Fig. 3.4–1(a). For a resistor R_m in the time domain, we can write the following relation

$$q_m(t) = \int_{-\infty}^{t} i_m(\tau)\, d\tau = G_m \int_{-\infty}^{t} v_m(\tau)\, d\tau \tag{1}$$

and hence in the s-domain

$$Q_m(s) = \frac{G_m}{s} V_m(s) \tag{2}$$

[20] G. C. Temes, "The Derivation of Switched-Capacitor Filters from Active RC Prototypes," *Electronics Letters*, Vol. 14, June 1978, pp. 361–362.

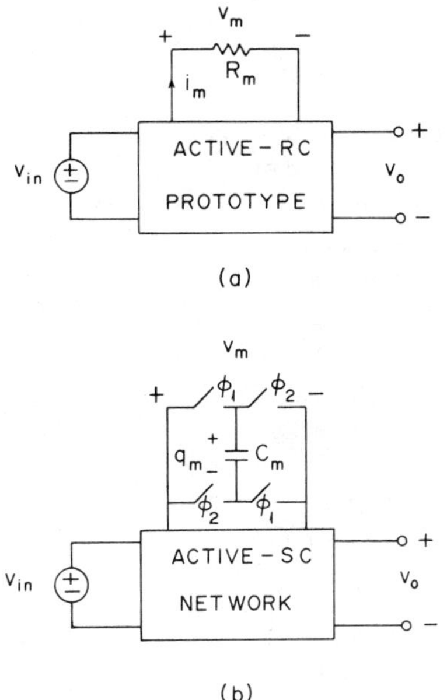

Fig. 3.4–1. (a) A linear active-RC filter. (b) Corresponding SC filter.

Now consider the corresponding SC filter of Fig. 3.4–1(b). This circuit was derived by replacing each branch resistor R_m by a two-terminal circuit containing a single capacitor C_m and four switches (two toggle switches). Assuming that all switches in the circuit of Fig. 3.4–1(b) operate once every T seconds, and if the same switching arrangement is used in every branch, we can write the charge conservation equation as

$$q_m(n) - q_m(n-1) = C_m[v_m(n-1) + v_m(n)] \tag{3}$$

Assuming that the arms of the switches make only momentary contact at sample times, Eq. (3) can be written in the z-domain as

$$Q_m(z) = C_m \frac{1+z^{-1}}{1-z^{-1}} V_m(z) \tag{4}$$

A comparison of Eqs. (2) and (4) shows that if $C_m = G_m T/2$, then the replacement of the resistor R_m by the SC C_m is equivalent to the bilinear mapping

$$s \leftrightarrow \frac{2}{T}\frac{1 - z^{-1}}{1 + z^{-1}} \tag{5}$$

where $T = T_c/2$, because all switches change position twice every T_c seconds. The properties of this bilinear mapping have been discussed in Chapter 1. Due to this bilinear transformation, a stable RC active filter is transformed into a stable SC active network. Thus, except for some frequency warping and the expected differences between the analog and sampled data circuits, the two circuits ideally behave identically.

Equation (3) was derived, as mentioned before, assuming that the switches make only momentary contact at sample times. If the switches make contact at all times, except (momentarily) at $t = nT$, $n = 0, \pm 1, \pm 2, \ldots$, we obtain a different expression for Eq. (3). The branch voltage and charge waveforms of the equivalent resistor in Fig. 3.4–1(b) are shown in Fig. 3.4–2. Thus from Fig. 3.4–2, at times $t_n = nT - \tau$ where $\tau << T$, omitting T in the equation, we can write

$$\begin{aligned} q_m(t_n) - q_m(t_{n-1}) = C_m[v_m(n-1) - v_m(t_{n-1}) + 2v_m(n-1) + v_m(t_n) \\ - v_m(n-1)] \cong C_m[v_m(t_n) + v_m(t_{n-1})] \end{aligned} \tag{6}$$

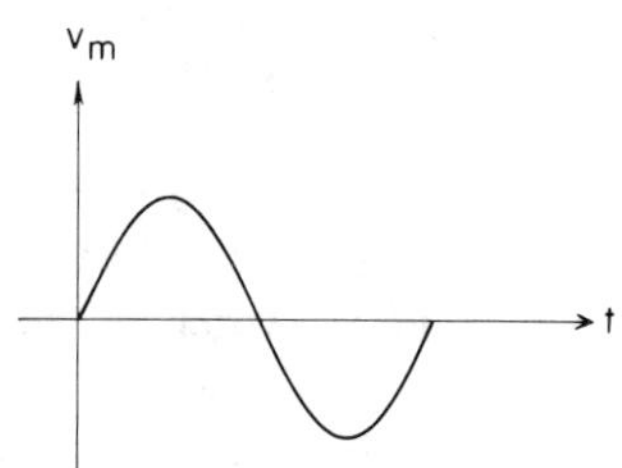

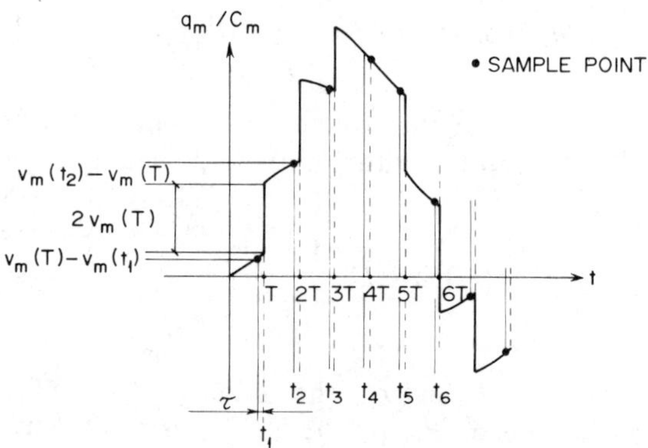

Fig. 3.4–2. Bilinear branch voltage and charge waveforms

Equation (6) agrees well with Eq. (3). The relation between the continuous-time waveforms is, from Fig. 3.4–2

$$q_m(t) = C_m v_m(t) + 2C_m \sum_{n=-\infty}^{t} v_m(n)u(t-n) \tag{7}$$

In Eq. (7), the first term is a charge passing through the branch due to the presence of C_m between the nodes at all times, not only at the sampling instances t_n. This undesirable leakage can be fortunately eliminated by cascading an S/H circuit with the SC filter. The resulting filter response, $H_{\mathrm{RE}}(\omega)$, combines the bilinear transform of the RC filter characteristic with the spectrum of a zero-order hold circuit, that is

$$H_{\mathrm{RE}}(\omega) = H\left(\frac{2}{T}\tan\frac{\omega T}{2}\right)\frac{\sin \omega T/2}{\omega T/2} \tag{8}$$

If $\omega T << 1$, the S/H does not affect the output. Otherwise the output will be weighted by the zero-order hold transfer function.

The design of the active SC filter based on an element-by-element resistor replacement in analog RC filters via the bilinear transform can be described as follows.[21]

1. The desired passband and stopband limit frequencies, ω_d's, of the SC filter are prewarped to obtain the corresponding ω_a's of the active RC prototype filter, using the relation

$$\omega_a = \frac{2}{T}\tan\left(\frac{\omega_d T}{2}\right) \tag{9}$$

2. The active RC prototype is designed from the transform frequencies ω_a's.
3. Each resistor R_m in the active RC circuit is replaced by an SC simulation of value

$$C_m = \frac{G_m T}{2} = \frac{T}{2R_m} \tag{10}$$

Because all switches change position twice every T_c seconds, $T = T_c/2$, Eq. (10) yields

[21] G. C. Temes, H. J. Orchard, and M. Jahanbegloo, "Switched Capacitor Filter Design Using the Bilinear Z-Transform," *IEEE Trans. Circuits and Systems*, Vol. CAS-25, December 1978, pp. 1039–1044.

$$C_m = \frac{T_c}{4R_m} \tag{11}$$

4. An S/H circuit, operating at twice the sampling rate, is added at the input. At the output of the SC filter, as discussed in Section 1.2, usually a simple smoothing low-pass filter is connected.

Next an example illustrating this approach is considered.

Example 3.4–1. *Single-amplifier, infinite-gain filter.* Given the low-pass prototype shown in Fig. 3.4–3, design the equivalent bilinear SC version. The desired center frequency, ω_0, and quality factor, Q, of the SC circuit are 2π krad/s and 0.7071, respectively. Consider two cases of sampling frequency

a. $T_c = 1/50$ ms

b. $T_c = 1/8$ ms

The transfer function of the prototype is given by

$$H(s) = \frac{-\dfrac{1}{R_1R_3C_4C_5}}{s^2 + \left(\dfrac{1}{R_1} + \dfrac{1}{R_2} + \dfrac{1}{R_3}\right)\dfrac{s}{C_4} + \dfrac{1}{R_2R_3C_4C_5}} \tag{12}$$

Thus

$$\omega_0^2 = \frac{1}{R_2R_3C_4C_5} \tag{13}$$

$$\frac{\omega_0}{Q} = \frac{1}{C_4}\left(\frac{1}{R_1} + \frac{1}{R_2} + \frac{1}{R_3}\right) \tag{14}$$

$$H(0) = -R_2/R_1 \tag{15}$$

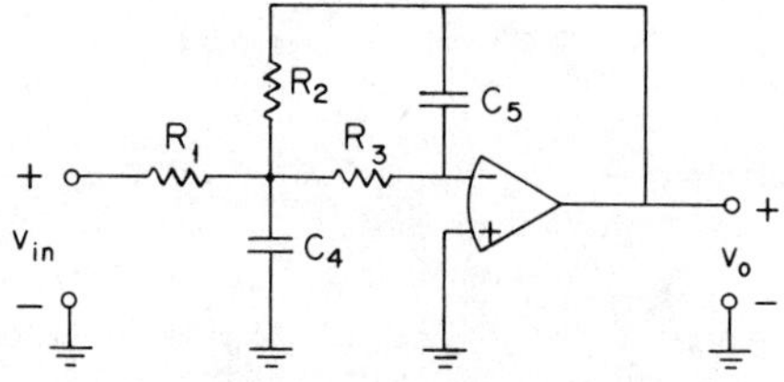

Fig. 3.4–3. An active-RC low-pass filter.

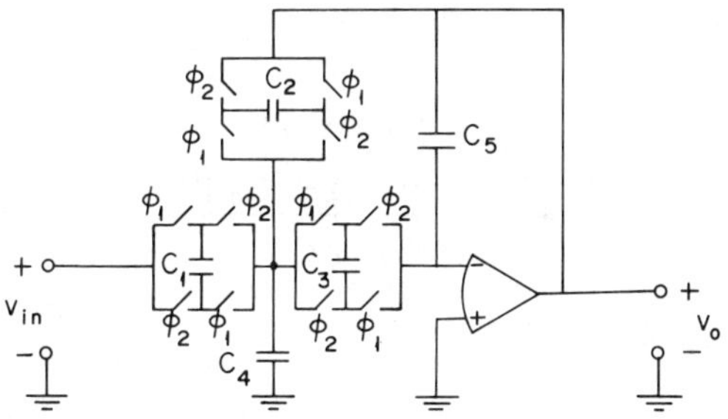

Fig. 3.4–4. An active-SC low-pass filter.

The equivalent SC filter is shown in Fig. 3.4–4.

Now, following the four-step design procedure described before, we have

$$\omega_{0_a} = \frac{2}{T} \tan\left(\frac{\omega_{0_d} T}{2}\right) \tag{16}$$

Thus ω_0 in Eqs. (13) and (14) is replaced by Eq. (16), and the R's and C's are calculated accordingly. Next, the resistors are replaced by Eq. (11). The corresponding $H(z)$ is given by

$$H(z) = \frac{-\dfrac{4C_1C_3(1+z^{-1})^2}{C_4C_5\left[1+\dfrac{2}{C_4}(C_1+C_2+C_3)+\dfrac{4C_2C_3}{C_4C_5}\right]}}{1-\dfrac{2\left(1-\dfrac{4C_2C_3}{C_4C_5}\right)z^{-1}+\left[1+\dfrac{4C_2C_3}{C_4C_5}-\dfrac{2}{C_4}(C_1+C_2+C_3)\right]z^{-2}}{1+\dfrac{4C_2C_3}{C_4C_5}+\dfrac{2}{C_4}(C_1+C_2+C_3)}} \tag{17}$$

For (a) we can obtain

$$\omega_{0_a} \cong 2\pi \times 1.0003291 \times 10^3$$

First the design is carried out in the s-domain,[22] for $H_0 = 1$. Choose $C_4 =$ 240 pF. Calculate

$$C_5 = mC_4, \text{ where } \qquad m \leqq \frac{1}{4Q^2(1+H_0)} \cong 0.25 \tag{18}$$

---

[22] L. P. Huelsman and P. E. Allen, *Introduction to the Theory and Design of Active Filters*, Chap. 5, McGraw-Hill Book Co., New York, 1980.

Next, set arbitrarily $m = 0.2$. Thus

$$C_5 = 48 \text{ pF}$$

$$R_2 = \frac{1 \pm [1 - 4mQ^2(1 + H_0)]^{1/2}}{2\omega_{0_a} C_5 Q} = 1.3 \text{ M}\Omega \tag{19}$$

Using Eq. (11) we obtain $C_2 = 3.85$ pF

$$R_1 = \frac{R_2}{H_0} = R_2 \tag{20}$$

Therefore $C_1 = 3.85$ pF

$$R_3 = \frac{1}{\omega_{0_a}^2 R_2 m C_4^2} \tag{21}$$

$$R_3 = 1.69 \text{ M}\Omega$$

Again, after using Eq. (11), C_3 becomes

$$C_3 = 2.96 \text{ pF}$$

The capacitor values can be summarized as

$$C_1 = 3.85 \text{ pF} \rightarrow 1.3\ C_u$$

$$C_2 = 3.85 \text{ pF} \rightarrow 1.3\ C_u$$

$$C_3 = 2.96 \text{ pF} \rightarrow C_u$$

$$C_4 = 240 \text{ pF} \rightarrow 81.08\ C_u$$

$$C_5 = 48 \text{ pF} \rightarrow 16.216\ C_u$$

and the total capacitance, TC, is about $101 C_u$, and if $C_u = 2.96$ pF, then TC is 299 pF. The frequency response is shown in Fig. 3.4–5. The corresponding pole angle θ is about $(2\pi f_0/f_c) = 7.2°$.

Now for (b) where $f_c = 8$kHz, we have

$$\omega_{0_a} = 2\pi \times 1.0548 \times 10^3$$

Therefore, $R_2 = 1.23 \times 10^6\ \Omega$, and $C_2 = C_1 = 25.4$ pF. $R_3 = 1.60 \times 10^6\ \Omega$, and the corresponding $C_3 = 19.5$ pF. C_4 and C_5 are as before. Thus

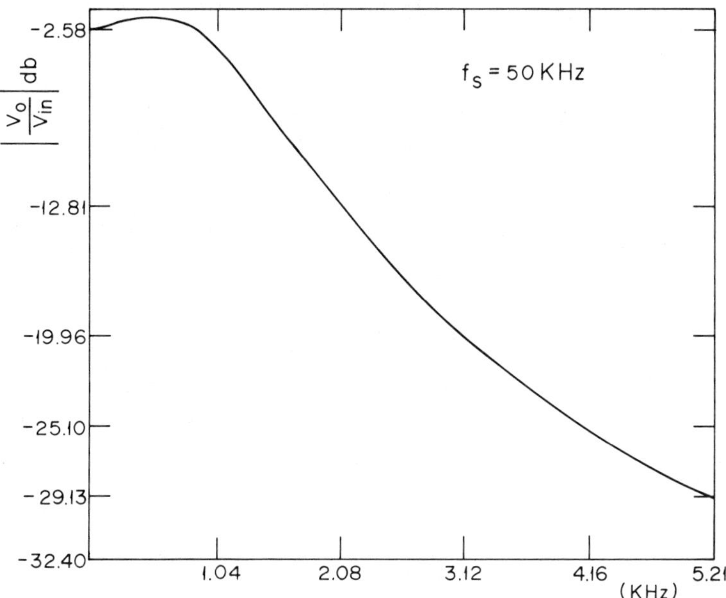

Fig. 3.4–5. Frequency response of low-pass SC active filter of Fig. 3.4–4, case a)

$[C_1, C_2, C_3, C_4, C_5] = [1.3, 1.3, 1, 12.31, 2.47]\ C_u$. The total capacitance is about 19 C_u. The frequency response for case (b) is shown in Fig. 3.4–6.

Comparing both cases, we see that the largest capacitor ratios are nearly 82 and 13 for cases (a) and (b), respectively. Case (b) with θ approximately 45° needs less area (about ⅕) to be implemented, although the antialising filter associated with it involves more stringent requirements than the one associated with case (a).

The following expressions are based on Fig. 3.4–4 and Eq. (17)

$$2r\cos\theta = 2\frac{1-\dfrac{4C_2C_3}{C_4C_5}}{1+\dfrac{4C_2C_3}{C_4C_5}+\dfrac{2}{C_4}(C_1+C_2+C_3)} \tag{22}$$

$$r^2 = \frac{1+\dfrac{4C_2C_3}{C_4C_5}-\dfrac{2(C_1+C_2+C_3)}{C_4}}{1+\dfrac{4C_2C_3}{C_4C_5}+\dfrac{2(C_1+C_2+C_3)}{C_4}} \tag{23}$$

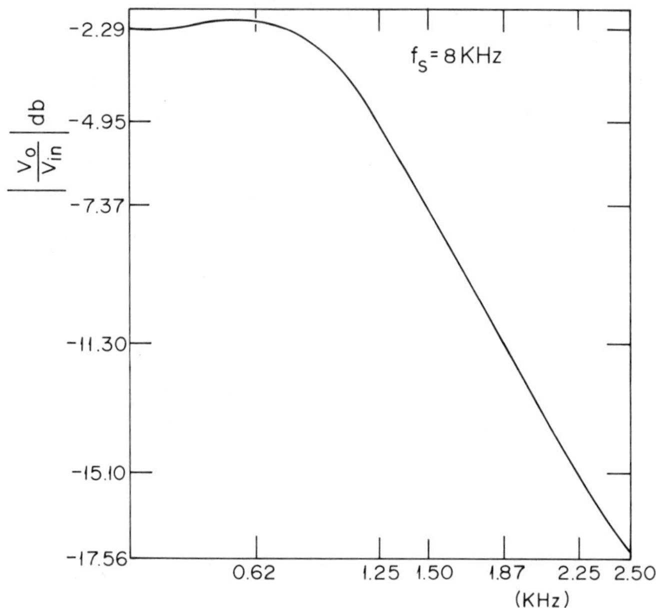

Fig. 3.4–6. Frequency response of low-pass SC active filter of Fig. 3.4–4, case b)

and

$$H(1) = -\frac{C_1}{C_2} \tag{24}$$

It should be mentioned that the spread of the capacitor ratio depends on the ratio f_0/f_c and the Q of the network; therefore some SC structures are prohibitive from an economical point of view. The important consideration is the total capacitance area required for the actual implementation. Typically capacitor ratios of less than 10 are very desirable.

The design technique discussed in this section does not depend on the filter topology and can be applied to any RC filter. In fact, the design technique can be extended to replace not only resistors, but a one-to-one replacement of branch impedances.[23] Next a design example based on a passive filter prototype is presented. See boxes 1, 2, 3, 6, 7, 9, and 14 of Fig. 3.1–16.

Example 3.4–2. *Elliptic FDNR filter.* Given the RLC prototype of Fig. 3.4–7(a), a third-degree elliptic filter, design an SC equivalent implementation.[24]

[23] J. A. Nossek and G. C. Temes, "Switched-Capacitor Filter Design Using Bilinear Element Modelling," *Proc. IEEE ISCAS 1980,* Houston, TX, April 1980, pp. 330–333.

[24] G. C. Temes et al., "Switched Capacitor Filter Design Using the Bilinear z-Transform," op. cit.

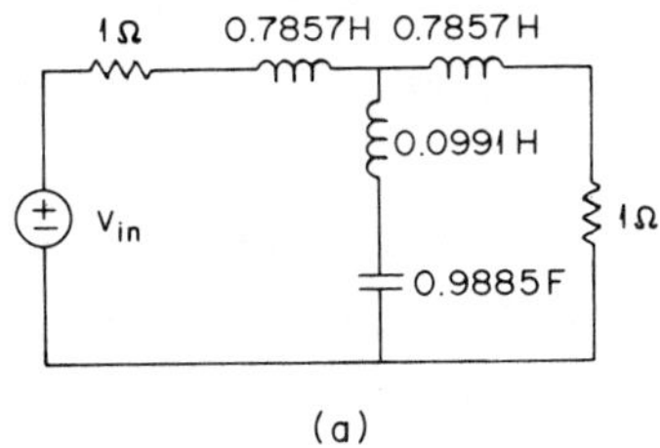

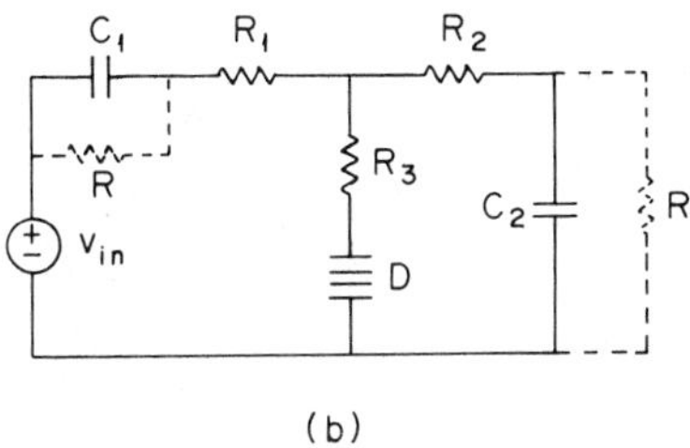

Fig. 3.4–7. Elliptic filter. (a) RLC prototype, (b) CRD prototype

To obtain an equivalent SC version from Fig. 3.4–7(a), several steps need to be carried out. First the RLC prototype is transformed into a CRD prototype,[25] where D is a *frequency-dependent negative resistor* FDNR. The transformed CRD prototype model is shown in Fig. 3.4–7(b). Next the active RC-FDNR realization[26] is obtained for a passband limit of 2.5 kHz, which is shown in Fig. 3.4–8. The added resistor R across the load and the source capacitors are required components to define the response at zero frequency. R is chosen as large as practical, say 0.5 MΩ. Finally, the SC-FDNR version is obtained by resistor substitution as given by Eq. (11). The impedance level can be conveniently chosen for MOS IC implementation or for a discrete breadboarding version. The SC-FDNR version is shown in Fig. 3.4–9.

In this section, a general design method for obtaining active SC filters from active RC prototypes has been presented. The main advantage of this approach is that it is based on the bilinear z-mapping. Consequently, it does not have stability problems for any sampling frequency or higher sensitivity to element variations due to imperfect emulation of the analog prototype. For instance, an active RC integrator, as shown in Fig. 2.5–1, can be realized

[25] L. P. Huelsman and P. E. Allen, op. cit., Chap. 6.

[26] G. C. Temes et al., "Switched Capacitor Filter Design Using the Bilinear z-Transform," op. cit.

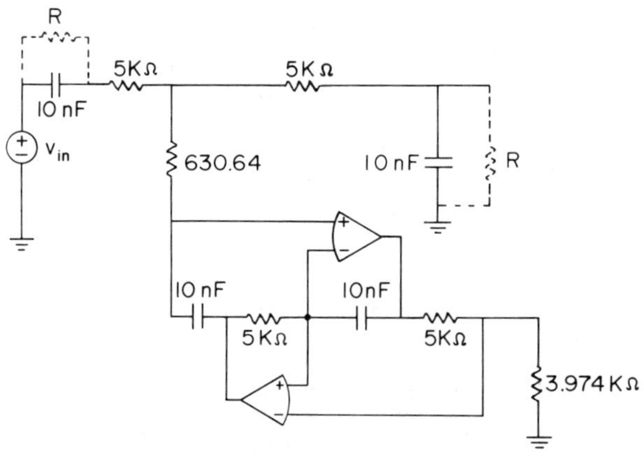

Fig. 3.4–8. RC-FDNR active realization.

by means of a stray-insensitive bilinear SC integrator[27] with sampled data input and output sampling at ϕ_1. This is illustrated in Fig. 3.4–10(a). The corresponding transfer function assuming an S/H input is

$$H^{oo}(z) = \frac{V_o^o}{V_{in}^o} = -\alpha \frac{1 + z^{-1}}{1 - z^{-1}} \tag{25}$$

Although this integrator consists of eight switches, two capacitors, and one op amp, and therefore has more components than the ones discussed in Section 2.5, it is an ideal approximation of the analog integrator of Fig. 2.5–1.

A noninverting integrator can be obtained if the switches S_2, S_4, S_6, and S_8 work in phase opposition to those shown in Fig. 3.4–10(a). Fig. 3.4–10(b) illustrates the z-domain flow diagram of Fig. 3.4–10(a). As mentioned previously, this design technique can be extended[28] to simulate inductors and lossless LC resonators. This design approach is illustrated in Fig. 3.1–16, via the blocks numbered 1, 2, 3, 6, 7, 9, and 14. Besides the design approach described here, another design approach directly in the z-domain is possible. That is, once r and θ are specified, the capacitor ratios as functions of r and θ can be determined, i.e., Eqs. (22) and (23). We have considered only a low-pass active RC and a passive RLC prototype to illustrate design using the bilinear z-transform. Many other prototypes can be used; see, for instance, Problems 3.18 to 3.21.

---

[27] A. Knob, "Novel Strays-Insensitive Switched-Capacitor Integrator Realising the Bilinear z-transform," *Electronics Letters,* Vol. 16, No. 5, February 1980, pp. 173–174.

[28] J. A. Nossek and G. C. Temes, op. cit.

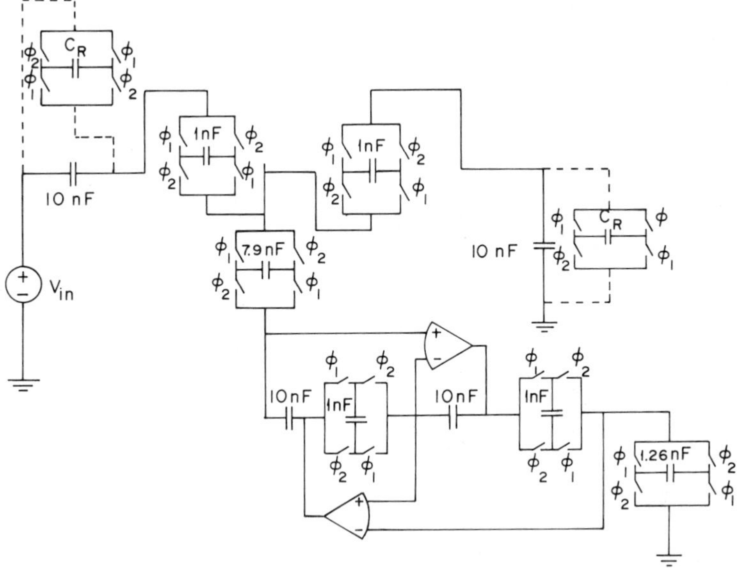

Fig. 3.4–9. SC realization of Fig. 3.4–8.

3.5 SC FILTERS USING THE BACKWARD Z-TRANSFORM COMPONENT SIMULATION

In the previous section a general technique of transforming an analog active RC network into an SC filter by replacing resistors with bilinear SC simulations was presented. In this section, we follow a similar procedure, but the resistors are replaced with the backward SC resistor simulation. The objective of this section is to obtain an alternative to the bilinear transformation with possible reduction, in some cases, of switches and parasitic capacitances associated with some of the bilinear SC topologies. The approach presented here follows that of boxes 1, 2, 3, 6, 8, 10, and 14 or 1, 2, 3, 6, 7, 9, and 14 of Fig. 3.1–16.

Let us consider the linear active RC filter of Fig. 3.4–1(a) of the previous section. If each branch resistor, R_m, is replaced by a two-terminal circuit containing a single capacitor, C_m, and two switches, as shown in Fig. 3.5–1, then it can be shown that the simulation of the resistor R_m by the SC C_m is equivalent to the backward mapping, i.e., $s \leftrightarrow (1 - z^{-1}/T)$. Notice that the resistors are simulated by capacitors with shunt switches, which are periodically closed at one particular clock period, say at ϕ_1. Thus, these simulated resistors operate only during the other clock period, say at ϕ_2. Thus, the RC circuits are designated even (ϕ_2) SC networks related to the

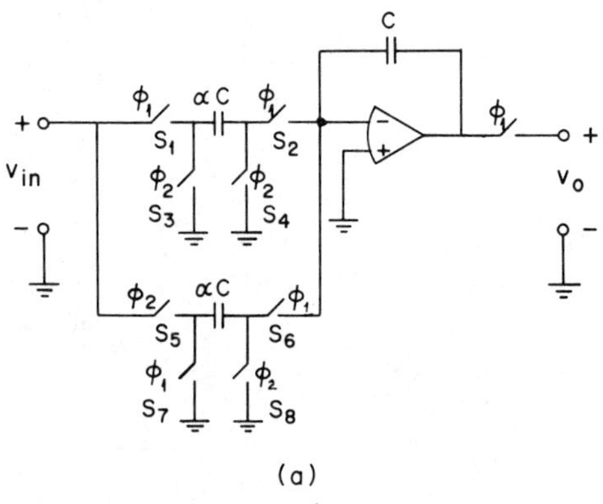

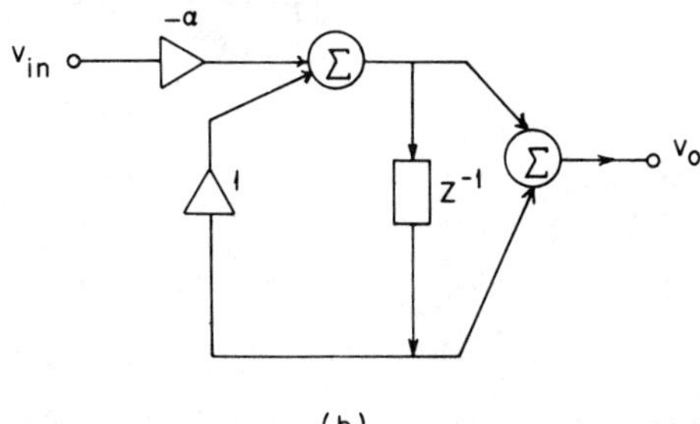

Fig. 3.4–10. (a) A stray-insensitive bilinear SC integrator. (b) Corresponding flow diagram.

incoming signals.[29] Assuming that all the switches in the circuit of Fig. 3.5–1 operate every T seconds, the charge conservation equation can be written for ϕ_2 as

$$q_m(n) - q_m(n-1) = C_m \, v_m(n) \tag{1}$$

Using the z-transform, Eq. (1) yields

$$Q_m(z) = C_m \frac{1}{1 - z^{-1}} V_m(z) \tag{2}$$

A comparison of Eq. (2) with Eq. (2) of the previous section shows that choosing $C_m = T/R_m$, the simulation of the resistor R_m by the SC C_m is equivalent to the backward mapping, discussed in Section 1.5 given as

[29] B. J. Hosticka and G. S. Moschytz, "Practical Design of Switched-Capacitor Networks for Integrated Circuit Implementation," *IEE, Electronic Circuits and Systems,* Vol. 3, March 1979, pp. 76–88.

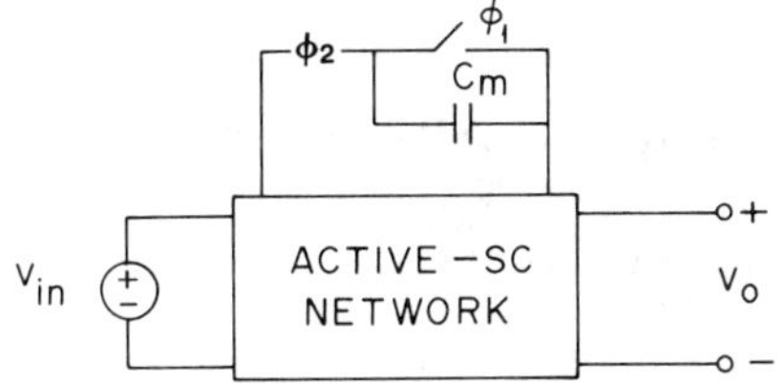

Fig. 3.5–1. Series SC filter simulation of the network of Fig. 3.4–1(a)

$$s \leftrightarrow \frac{1}{T}(1 - z^{-1}) \tag{3}$$

This mapping transforms a stable RC active filter into a stable SC active network, a property shared with the bilinear z-transformation of the previous section. The parallel SC resistor equivalent presented in Section 3.3 does not have this property regarding stability. If a sample-and-hold circuit is cascaded, the resulting transfer function designated as $H_{RE}(\omega)$ becomes

$$H_{RE}(\omega) = H(\omega)\frac{\sin \omega T}{\omega T} \tag{4}$$

where $H(\omega)$ is the backward transform of the RC characteristic, and $\sin \omega T/\omega T$ is the spectrum of a zero-order hold circuit.

Next, a set of design equations for first- and second-order filter sections, valid for both poles and zeros, is given. A first-order factor in the s-domain is

$$D(s) = s + \sigma \tag{5}$$

Based on a prewarping algorithm,[30] the corresponding z-domain first-order factor normalized to $T/2$ is

$$D(z) = 1 - az^{-1} \tag{6}$$

where

$$a = e^{-\sigma/f_c} = e^{-\sigma T} \tag{7a}$$

[30] U. W. Brugger and B. J. Hosticka, "A Prewarping Scheme for the Design of Switched-Capacitor Filters," *IEEE Proc. ISCAS,* April 1980, pp. 317–320.

and

$$\sigma \text{ (prewarped)} = \Sigma = e^{\sigma T} - 1 \tag{7b}$$

Now, consider a second-order factor

$$s^2 + \frac{\omega_i}{q_i} s + \omega_i^2 \tag{8}$$

The prewarped, normalized to $(T/2)$, parameters are described by

$$\Omega_i = \left[1 + e^{\omega_i T/q_i} - 2e^{\omega_i T/2q_i} \cos\left(\omega_i T\sqrt{1 - \frac{1}{4q_i^2}}\right)\right]^{1/2} \tag{9}$$

$$Q_i = \frac{\Omega_i}{e^{\omega_i T/q_i} - \Omega_i^2 - 1} \tag{10}$$

A simple design method for the design of an active SC filter from an active RC prototype based on an element-by-element resistor replacement via the backward transformation follows.

1. The critical (desired) frequencies ω_d's of the SC filter are not prewarped if $\omega_d T << 1$; otherwise the ω_d's and q's are prewarped to obtain the corresponding ω_a's (denormalized frequencies) and Q_a's of the active RC prototype filter, according to the relations (7), (9), and (10).
2. The active RC prototype is designed from the transform frequencies, ω_a's.
3. Each resistor R_m in the active RC circuit is replaced by an SC of value

$$C_m = \frac{T}{R_m} \tag{11}$$

4. A sample-and-hold circuit is added at the input. Typically a smoothing low-pass filter is connected at the output of the SC filter.

This backward difference design approach was introduced by Hosticka[31] for developing "even $-$ ϕ_2" SC circuits. Even (odd) circuits mean those whose inputs and outputs are sampled only at even (odd) time instants.

Given an original active RC prototype, all resistors, as mentioned before, are simulated by capacitors with shunt switches, which will be periodically

[31] B. J. Hosticka et al., op. cit.

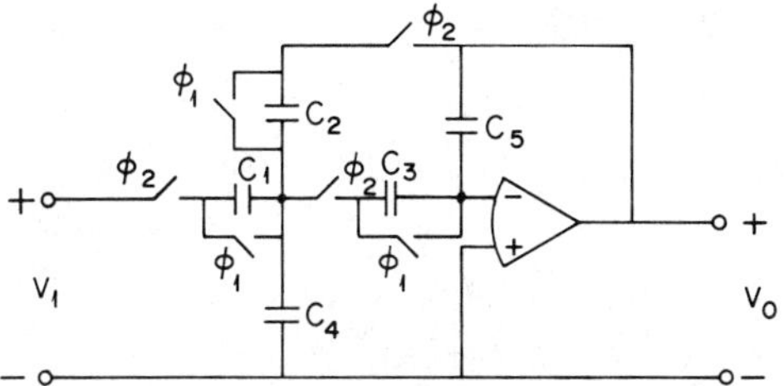

Fig. 3.5–2. Backward SC simulation of Fig. 3.4–3

closed at one clock period, i.e., ϕ_1. Those capacitors that correspond to true capacitors in the original RC prototype must then be decoupled during ϕ_1 times. That is, the charge in each real capacitor must remain unchanged during the ϕ_1 phase period.

Let us consider the low-pass filter of Fig. 3.3–1(a). The transfer function of this filter is

$$H(s) = \frac{-\dfrac{1}{R_1R_3C_4C_5}}{s^2 + \left(\dfrac{1}{R_1} + \dfrac{1}{R_2} + \dfrac{1}{R_3}\right)\dfrac{s}{C_4} + \dfrac{1}{R_2R_3C_4C_5}} \tag{12}$$

The equivalent backward SC version is shown in Fig. 3.5–2. Its corresponding transfer function is

$$H^{ee}(z) = \frac{-\dfrac{\alpha_1\alpha_3r^2}{A}}{1 - z^{-1}2r\cos\theta + r^2z^{-2}} \tag{13}$$

where the resistors have been replaced by their SC equivalents. That is, $R_1 = T/C_1 = T/\alpha_1C$, $R_2 = T/C_2 = T/(\alpha_2AC)$, $\mathrm{R}_3 = T/C_3 = T/\alpha_3C$, $C_4 = C$, and $C_5 = AC$. Therefore,

$$r^2 = \frac{1}{1 + \alpha_1 + \alpha_3 + \alpha_2A + \alpha_3\alpha_2} \tag{14}$$

$$2r\cos\theta = r^2(2 + \alpha_1 + \alpha_2A + \alpha_3) \tag{15}$$

The DC gain is

$$H^{ee}(1) = H_0 = \frac{-\alpha_1\alpha_3r^2}{A[1 - r^2(1 + \alpha_1 + \alpha_2A + \alpha_3)]} \tag{16}$$

A set of simple design equations can be obtained by setting $\alpha_1 = -\alpha_3 = \alpha_2 A$.

$$\alpha_1 = \frac{2}{3}\left(\frac{\cos\theta - r}{r}\right) \tag{17}$$

$$A = \frac{\alpha_1^2}{\dfrac{1}{r^2} - 3\alpha_1 - 1} \tag{18}$$

In general, it should be observed that the design of backward SC filters based on active RC prototypes can be accomplished by means of (1) using the design method, described before, based on an element-by-element resistor replacement via the backward transformation, or (2) direct design in the z-domain once the $H(z)$ has been determined. In the particular case of the filter of Fig. 3.5–2, Eqs. (17) and (18) were used to obtain approximations of r and θ as functions of ω_0 and Q through one of the s- to z-transformation discussed in Chapter 1, i.e., $r = e^{-\omega_0 T/2Q}$, and $\theta = (\omega_0 T/2)\sqrt{4 - 1/4Q^2}$. Clearly with either design approach there are limitations for each topology. For instance, the filter described by Eqs. (13) through (16) has the limitation that $\cos\theta > r$.

Let us now consider the RC bandpass filter of Fig. 3.5–3. Its transfer function is given by

$$H(s) = \frac{-\dfrac{s}{R_1C_4(1-1/k)}}{s^2 + s\left(\dfrac{1}{R_2C_3} + \dfrac{1}{R_2C_4} - \dfrac{1}{k-1}\dfrac{1}{R_1C_4}\right) + \dfrac{1}{R_1R_2C_3C_4}} \tag{19}$$

The equivalent backward SC version is shown in Fig. 3.5–4. Its equivalent z-domain transfer function is

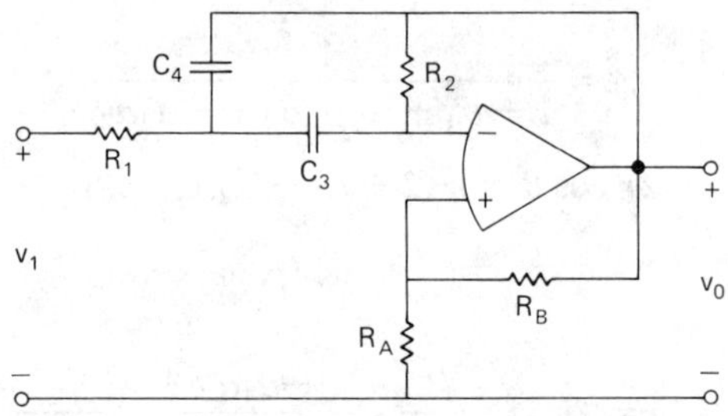

Fig. 3.5–3. Delyiannis RC-bandpass filter.

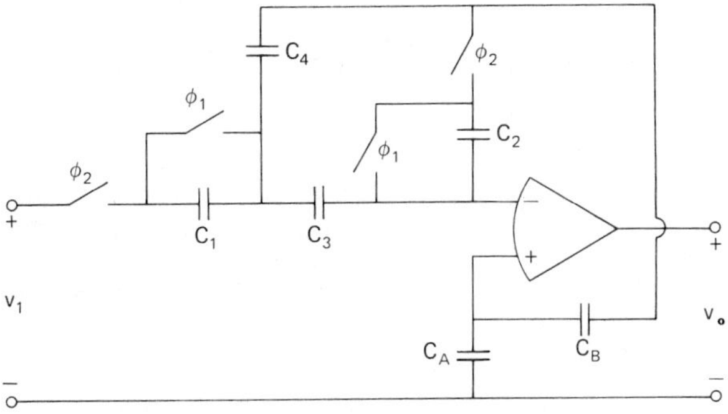

Fig. 3.5–4. Delyiannis SC-bandpass filter. $k = 1 + \frac{C_1}{C_B}$.

$$H^{ee}(z) = \frac{-\frac{C_1}{C_4}\frac{r^2}{1-1/k}(1-z^{-1})}{1-2r\cos\theta\, z^{-1}+r^2z^{-2}} \tag{20}$$

where

$$r^2 = \frac{1}{1+\frac{C_2}{C_3}+\frac{C_2}{C_4}+\frac{C_1C_2}{C_3C_4}-\frac{1}{k-1}} \tag{21}$$

$$2r\cos\theta = r^2\left(2+\frac{C_2}{C_3}+\frac{C_2}{C_4}-\frac{1}{k-1}\frac{C_1}{C_4}\right) \tag{22}$$

If $C_3 = C_4 = C$ and $C_1 = mC_2$, a set of design equations is given by

$$k = \frac{m}{2\left[2-(\cos\theta - r)\sqrt{\frac{m}{1-r^2}}\right]} \tag{23a}$$

$$\frac{C_2}{C} = \frac{2(\cos\theta - r)}{r\left(2-\frac{m}{k-1}\right)} \tag{23b}$$

where m is a constant.

Example 3.5–1. *First-order filter.* An RC active filter is shown in Fig. 3.5–5(a). It is desired to design an equivalent SC version with 3-dB cutoff frequency of f_0 = 1kHz. The transfer function of Fig. 3.5–5(a) is

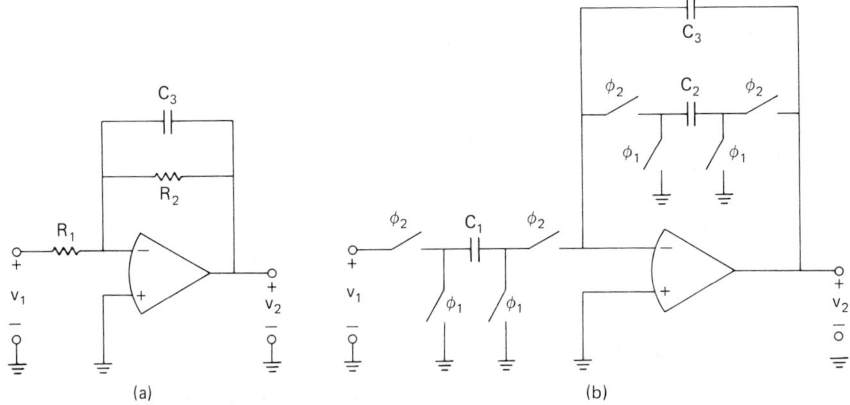

Fig. 3.5–5. (a) RC first-order prototype and (b) SC first-order filter.

$$H(s) = \frac{1}{R_1 C_3} \frac{1}{s + \dfrac{1}{R_2 C_3}}$$

The equivalent backward SC version is shown in Fig. 3.5–5(b). The corresponding $H(z)$ is

$$H(z) = -\frac{\dfrac{C_1}{C_3}}{1 + \dfrac{C_2}{C_3}} \frac{1}{1 - \dfrac{z^{-1}}{1 + \dfrac{C_2}{C_3}}}$$

Consider two cases; (1) $f_c = 8$ kHz and (2) $f_c = 64$ kHz. The predistorted normalized value of $\sigma = (1/R_2 C_3)$ from Eq. (7b) is

$$\Sigma = e^{\sigma T} - 1 \text{ and } f_{-3\ dB} = \Sigma f_c / 2\pi$$

The denormalized value $\Sigma' = f_c \Sigma$; thus the component values, assuming $R_1 = 1$ kohm, for the two cases are

(1)

$$\frac{C_1}{C_3} = 0.18992$$

$$\frac{C_2}{C_3} = 1.1933$$

(2)

$$\frac{C_1}{C_3} = .0164184$$

$$\frac{C_2}{C_3} = 0.10316$$

$f_{3\text{ dB}}$ (pred.) = 1.5193 kHz $\qquad$ $f_{3\text{ dB}}$ (pred.) = 1.1651 kHz

$f_c = 8$ kHz $\qquad$ $f_c = 64$ kHz

Example 3.5–2. *Single-amplifier infinite-gain low-pass filter.* It is desired to obtain the capacitor values of the filter of Fig. 3.5–2 for the design specifications of $\omega_0 = 2\pi \times 10^3$ rad/s, $Q = 0.7071$ and (a) $f_c = 50$ kHz, (b) $f_c = 8$ kHz. First, we design the active RC prototype with $\Omega = 0.131372$ and denormalized $\omega_0 = \Omega f_c = 2\pi \times 1.04543 \times 10^3$. The corresponding Q using Eq. (10) gives $Q = 0.74126$. The procedure given in Example 3.4–1 results in the following RC values

$$R_1 = R_2 = 1.3956 \text{ M}\Omega$$
$$R_3 = 1.44157 \text{ M}\Omega$$
$$C_4 = C = 240 \text{ pF and } C_5 = 48 \text{ pF}$$

For case (a) these values give

$$\alpha_1 = 0.0597$$
$$\alpha_3 = 0.0578$$
$$A = 0.2$$
$$\{\alpha_1 C,\ \alpha_2 AC,\ \alpha_3 C,\ C,\ AC\} \rightarrow \{1.033,\ 1.033,\ 1,\ 17.3,\ 3.46\}\ C_u$$

The total capacitance is approximately $24C_u$. We can design this filter directly using Eqs. (17) and (18), if we allow $\alpha_1 = \alpha_3 = \alpha_2 A$. Thus, $r = 0.9391$ and $\theta = 5.091°$, and $\alpha_1 = 0.04043$ and $A = 0.1297$.

For case (b)

$$R_1 = R_2 = 0.488 \text{ M}\Omega$$
$$R_3 = 11.766 \text{ M}\Omega$$
$$C_4 = C = 240 \text{ pF and } C_5 = 10.53 \text{ pF}$$

which results in

$$\alpha_1 = 1.066$$
$$\alpha_3 = 0.0443$$
$$A = 0.0439$$
$$\{\alpha_1 C,\ \alpha_2 AC,\ \alpha_3 C,\ C,\ AC\} \rightarrow \{24.3,\ 24.3,\ 1.32,\ 22.8,\ 1\}\ C_u$$

A total capacitance of $74C_u$ is required for this filter. Note that for certain specifications the design and component values are not unique.

In this section, a second general design method for obtaining active SC filters from active RC prototypes, has been discussed. The advantages of this approach are that it is also based on a stable s to z-transformation and that stray-insensitive SC resistor equivalents can be used. Due to the nature of this backward transformation for a given f_c, Q, and f_0, it may not be possible to obtain the capacitor ratios otherwise feasible through the bilinear z-transform component simulation.

3.6 ANALYSIS AND SIMULATION METHODS FOR SC NETWORKS

In this section we wish to examine the use of computer-aided analysis of SC networks. SC networks present a problem in this area, because most programs designed for the analysis of continuous analog networks are not appropriate for analog sampled data networks. Often it is not possible to breadboard the analog sampled data circuit before its fabrication as an integrated circuit. Some of the reasons are that appropriate devices are not available in breadboard form and that extra parasitics, which will not be present in the final form, will be introduced into the circuit. Consequently, it is very important to have some means of determining whether the analog sampled data circuit will meet the desired specifications. Computer simulation is a natural tool for this purpose. Frequently, the computer simulation is a small fraction of the manufacturing cost and can avoid circuit errors and optimize the performance.

There are various model levels at which the computer can simulate an analog sampled data network. The most elementary level is the simulation or evaluation of the transfer function $H(z)$ or the z-domain block diagram. At this level one can easily obtain the frequency response of an analog sampled data network. The next level in complexity simulates the switches, capacitors, and op amps directly. One can identify various blocks that are used often, such as the Type I DDI, and achieve the frequency response. This is the so-called block partitioning approach. The next level for methods of analyzing SC networks uses nodal analysis based on the charge conservation principles given previously. This level of modeling typically has a solution for each of the phases present. The most complex level in modeling simulates the individual MOS devices and capacitors. This level requires a very good understanding of the MOS model. We shall discuss examples of all four levels of simulation in this section.

We shall discuss each method of simulation in the order of increasing complexity. If $H(z)$ is known, then it is a simple matter to replace z by $\cos \omega T + j \sin \omega T$ and to evaluate the real and imaginary parts of $H(z)$ or the magnitude and argument of $H(z)$. An example illustrates the method.

Example 3.6–1. *Calculation of frequency response of* $H(z)$. Assume that $H(z)$ is given as

$$H(z) = K\left(\frac{1 + b_1 z^{-1} + b_2 z^{-2}}{1 + a_1 z^{-1} + a_2 z^{-2}}\right) \tag{1}$$

where

$$\begin{aligned} a_2 &= K_{3A},\ a_1 = K_{1A} + K_{2A}K_{3A} \\ b_2 &= K_{3B},\ b_1 = K_{1B} + K_{2B}K_{3B} \end{aligned} \tag{2}$$

and

$$K = \frac{1 + a_1 + a_2}{1 + b_1 + b_2} \tag{3}$$

Develop a program to plot the magnitude and phase of $H(z)$ from 0 to 500 Hz. Assume that $f_c = 10$ kHz and

$$\begin{aligned} K_{1A} &= K_{2A} = -0.9876883 \\ K_{3A} &= 0.8868803 \\ K_{1B} &= K_{2B} = -0.9876855 \\ K_{3B} &= 1.0 \end{aligned} \tag{4}$$

Table 3.6–1 shows the program developed for this purpose, and Fig. 3.6–1 illustrates the magnitude of the frequency response. The result is seen to be a 250-Hz notch.

If the SC network can be represented uniquely by a z-domain block diagram consisting of delays, summers, and multipliers, then digital filter simulators can be used for a means of analysis.[32,33] Several effects can be modeled analyti-

[32] T. N. Trick, J. Yau, and E. Sánchez-Sinencio, "Simulation of Fixed-Point Digital Filter Structures," *IEEE Proc. 1979 Int. Symposium on Circuits and Systems,* July 1979, pp. 370–371.

[33] DINAP—A Digital Network Analysis Program, Dept. of Electrical Engineering and Computer Sciences, Purdue University, Lafayette, IN.

Table 3.6–1 A FORTRAN Program for Evaluating the Frequency Response of a z-Domain Transfer Function

THIS PROGRAM CALCULATES THE MAGNITUDE AND PHASE OF A Z-DOMAIN TRANSFER FUNCTION. THE FUNCTION CALCULATED HERE IS A NOTCH BASED AT 250 HZ.

```
/*WATFIV
      COMPLEX*16 V,DCMPLX,Z,S,AG,G,OOPS*8
      REAL *8 WT,W,T,A1,A2,A3,B1,B2,B3,K1A,K2A,K3A,K1B,K2B,K3B,GAIN
     X, RE*4,IM*4,K,INC,DB
      PRINT 101
      PRINT 98
 98   FORMAT(1H—, 'MODIFIED NOTCH AT 250')
      PRINT 99
 99   FORMAT (1H0,7X, 'FREQUENCY',12X,'GAIN',15X,'PHASE',/,/)
      T = 1./10000.
      K1A= -.9876883*1.0
      K2A= -.9876883*1.0
      K3A=  .8868803*1.0
      K1B= -1.0 + (1.0 - .9894682)*1.0
      K2B= -1.0 + (1.0 - .9894682)*1.01
      K3B=  1.0
      B1= K1B + K2B*K3B
      B2= K3B
      A1= K1A + K2A*K3A
      A2= K3A
      K = 1.1028
      W = 0.
      INC = 6.28331853*1.
      DO 10 I = 1,500
      W = W + INC
      WT = W*T
      V = DCMPLX(DBLE(0.),WT)
      Z = CDEXP(V)
      G = K*(1. + B1/Z + B2/(Z*Z))/(1. + A1/Z + A2/(Z*Z))
      GAIN= CDABS(G)
      DB= 20.*DLOG10(GAIN)
      OOPS = G
      RE= REAL (OOPS)
      IM= AIMAG(OOPS)
      PHASE= 57.2958*ATAN2(IM,RE)
      FREQ= W/6.2831853
 10   PRINT 100 ,FREQ,DB,PHASE
100   FORMAT(5X,E13.5,10X,D14.7,5X,D14.7)
101   FORMAT (1H1)
      STOP
      END
//SDATA
/*END
```

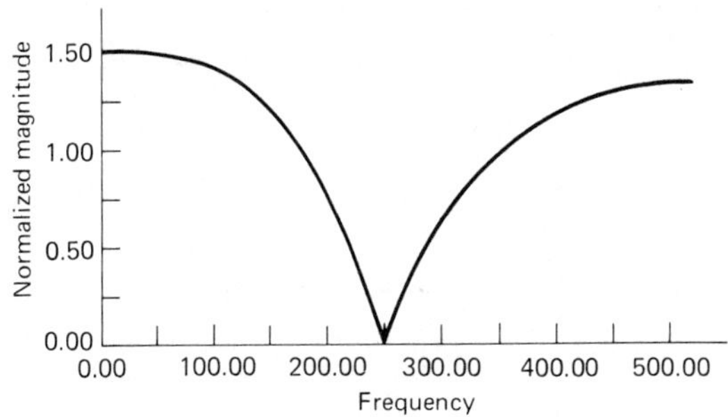

Fig. 3.6–1. Frequency response for Example 3.6–1.

cally in the z-domain, such as top plate parasitic capacitance, C_p. Another effect is the finite DC open loop voltage gain of the op amp, A_0. In practice the effective capacitor ratios are dependent on A_0. The effects of A_0 and C_p are described in this section from the computer simulation point of view. Further considerations and the derivation of these parameters will be presented in Chapter 8 and Section 5.5. Let us consider an example to illustrate the use of the digital filter simulators.

Example 3.6–2. *Integrator simulation.* Given the parallel SC integrator shown in Fig. 3.6–2, draw the digital filter equivalent circuit for the following two cases

1. Ideal op amp with top plate parasitic capacitance, C_p.
2. Finite DC open loop gain A_0 and no parasitic capacitance, C_p.

The equivalent digital filters are shown in Fig. 3.6–3. For Case (1), the parameters k_1 and k_2 are given by

$$k_1 = -\alpha - \frac{C_p}{C} \tag{5a}$$

$$k_2 = 1 \tag{5b}$$

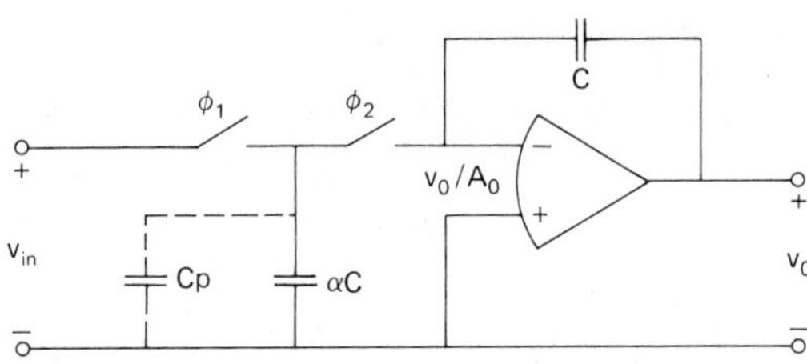

Fig. 3.6–2. Forward integrator with nonideal effects.

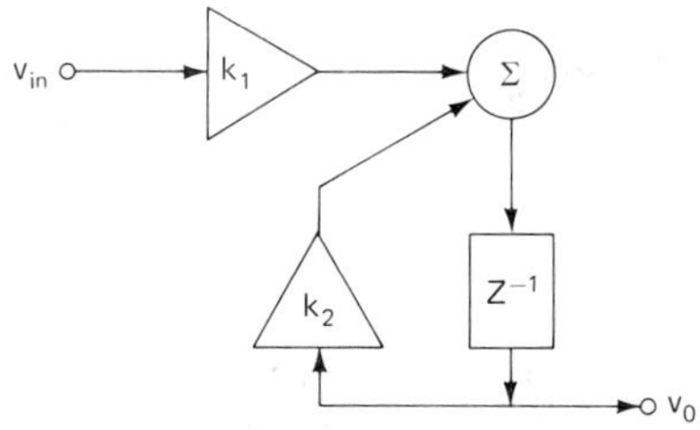

Fig. 3.6–3. Equivalent digital filter structure of Fig. 3.6–2.

For Case (2), the parameters k_1 and k_2 are

$$k_1 = \frac{-\alpha}{\frac{(1+\alpha)}{A_0} + 1} \tag{6}$$

$$k_2 = \frac{1}{\frac{\alpha}{1 + A_0} + 1} \tag{7}$$

The block diagrams can now be described to a digital filter simulator to obtain the frequency response.

In many SC structures,[34,35] it is possible to partition a network N into basic building blocks, N_b, that can be individually characterized mathematically. See, for example, the SC integrators considered in Section 2.5, which can be readily identified as the basic substructures of the networks, i.e., Figs. 3.3–5, 3.3–6, 3.3–7(b), 3.3–8, and the SC ladder filters to be considered in the next chapter, as well as the SC filters based on digital filter simulation. The partitioning approach to be presented identifies the interconnections of switches and capacitors that are clustered around each op amp in a network. Therefore, each op amp forms the focal point of a subnetwork N_b, for which the output voltage v_i is taken directly from the op amp output. A voltage v_j that acts as an input to N_b is the output voltage of the jth subnetwork. In this manner a vector of nodal voltages V of the circuit is defined as the nodal voltages at the nodes where the subnetwork interconnections occur. The order of the network equations is the same as the number of subnetwork modules in the network. These building blocks, shown in Fig. 3.6–4, include

[34] C. F. Lee, "An Investigation of Computer-Aided Analysis for Switched Capacitor Sampled-Data Filters, *Report T-16, U. of Illinois, CSL,* July 1981.

[35] C. F. Lee and W. K. Jenkins, "Computer Aided Analysis of Switched Capacitor Filters," *Thirteenth Annual Asilomar Conf. on Circuits, Syst., and Computers,* Monterey, CA, 1979, pp. 552–555.

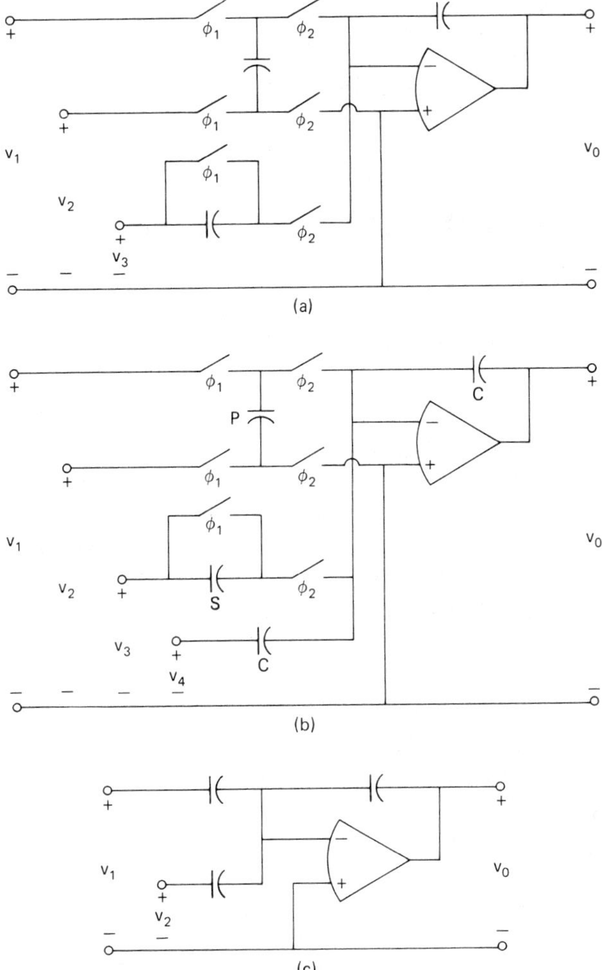

Fig. 3.6–4. Building blocks for switched capacitor networks. (a.) Integrator. (b.) Combined integrator-summer. (c.) Summer.

the series SC and parallel SC integrators, a combined integrator-summer, and a summer. The formulation of time domain analysis is presented here for a two-phase clock. The network model is illustrated in Fig. 2.3–5. Assigning ϕ_1 to correspond to the clock signal at time $(n - ½)T$ and ϕ_2 at time nT, and so on, the following general time equations can be written during ϕ_1 as

$$\begin{aligned}\mathbf{v}(n - ½) = \mathbf{F}\mathbf{v}(n-1) - \mathbf{A}_1\mathbf{v}_{p_1}(n-1) - \mathbf{D}_1\mathbf{v}(n - ½) \\ - \mathbf{E}_1\,\mathbf{v}_{\text{in}}(n - ½) + \mathbf{G}_1\,\mathbf{v}_{\text{in}}(n-1)\end{aligned} \tag{8}$$

and

$$\mathbf{v}_{p_2}(n-1) = \mathbf{B}_1\, \mathbf{v}(n-\tfrac{1}{2}) + \mathbf{C}_1\, \mathbf{v}_{\text{in}}(n-\tfrac{1}{2}) \qquad n = 1, 2 \ldots \tag{9}$$

During ϕ_2 we have

$$\begin{aligned} \mathbf{v}(n) = \mathbf{F}\, \mathbf{v}(n-\tfrac{1}{2}) - \mathbf{A}_2\, \mathbf{v}_{p_2}(n-\tfrac{1}{2}) - \mathbf{D}_2\, \mathbf{v}(n) \\ - \mathbf{E}_2\, \mathbf{v}_{\text{in}}(n) + \mathbf{G}_2\, \mathbf{v}_{\text{in}}(n-\tfrac{1}{2}) \end{aligned} \tag{10}$$

and

$$\mathbf{v}_{p_1}(n-\tfrac{1}{2}) = \mathbf{B}_2\, \mathbf{v}(n) + \mathbf{C}_2\, \mathbf{v}_{\text{in}}(n) \qquad n = 1, 2, \ldots \tag{11}$$

where $\mathbf{v}$ is the building block output node voltage vector, $\mathbf{v}_{p1}$ and $\mathbf{v}_{p2}$ are the ϕ_1 and ϕ_2 parallel SC branch voltage vectors, $\mathbf{v}_{\text{in}}$ is input node voltage vector, and $\mathbf{A}_1, \mathbf{A}_2, \ldots, \mathbf{G}_1, \mathbf{G}_2$ are all constant coefficient matrices. The dimension of the system is proportional to the number of operational amplifiers. The parallel SC voltages play the important role of establishing the appropriate initial conditions and hence have been specifically identified. The variables $\mathbf{v}_{p1}$ and $\mathbf{v}_{p2}$ can be eliminated in Eqs. (8) through (11), resulting, at ϕ_1, in

$$\begin{aligned} \mathbf{v}(n-\tfrac{1}{2}) = (\mathbf{I}+\mathbf{D}_1)^{-1}\,((\mathbf{F}-\mathbf{A}_1\mathbf{B}_2)\, \mathbf{v}(n-1) - \mathbf{E}_1\mathbf{v}_{\text{in}}(n-\tfrac{1}{2}) \\ + (\mathbf{G}_1-\mathbf{A}_1\mathbf{C}_2)\, \mathbf{v}_{\text{in}}(n-1)) \end{aligned} \tag{12}$$

During ϕ_2 we obtain

$$\begin{aligned} \mathbf{v}(n) = (\mathbf{I}+\mathbf{D}_2)^{-1}\,((\mathbf{F}-\mathbf{A}_2\mathbf{B}_1)\, \mathbf{v}(n-\tfrac{1}{2}) - \mathbf{E}_2\mathbf{v}_{\text{in}}(n) \\ + (\mathbf{G}_2-\mathbf{A}_2\mathbf{C}_1)\, \mathbf{v}_{\text{in}}(n-\tfrac{1}{2})) \end{aligned} \tag{13}$$

Because the equations are in general implicit, matrix inversion operations are required in the iteration process to solve for the point-by-point time domain solutions. All the constant coefficient matrices can be generated by inspection, and therefore a straightforward computer algorithm can be developed to construct these matrices from the user-specified parameter values and circuit topology. The **A, B,** and **C** matrices are responsible for parallel SC, whereas the **D, E, G,** and **F** matrices characterize both series SC and unswitched capacitors connected to the summing nodes of the op amps. Furthermore, the entries of the constant matrices are +1, −1, or a capacitor ratio. This is illustrated by the following four cases. (1) If a capacitor *j,* which is connected to the summing node of an op amp *i,* is a parallel SC transferring charge on the ϕ_2 phase, a capacitor ratio will enter location

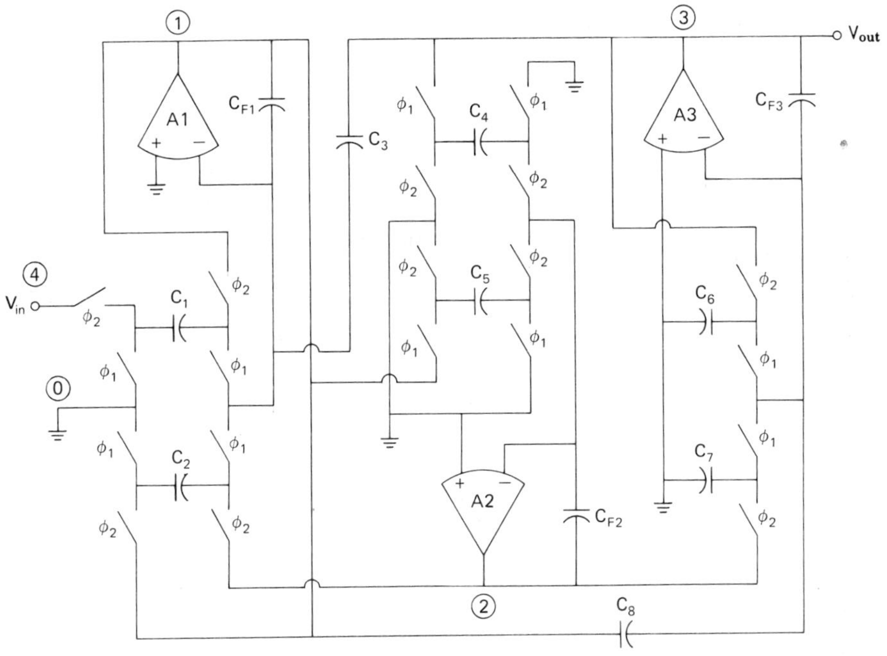

Fig. 3.6–5. Third-order elliptic filter for Example 3.6–3.

(i, j) of the $\mathbf{A}_1$ matrix. Also appropriate entries will enter the $\mathbf{B}_2$ and $\mathbf{C}_2$ matrices, depending on how the SC is connected during charging on the ϕ_1 phase. (2) If the capacitor is a series SC transferring charge to the op amp i from op amp k, then the capacitor ratio will enter location (i, k) of the $\mathbf{D}_1$ or $\mathbf{D}_2$ matrix, depending on the clock phase when the transfer takes place. (3) If the series SC is connected from $\mathbf{v}_{\text{in}}$, then $\mathbf{E}_1$ or $\mathbf{E}_2$ will be modified instead. Finally, (4) if an unswitched capacitor is transferring charge from another operational amplifier, then both $\mathbf{D}_1$ and $\mathbf{D}_2$, together with the $\mathbf{F}$ matrix, will be affected. Both $\mathbf{E}_1$ and $\mathbf{E}_2$ as well as $\mathbf{G}_1$ and $\mathbf{G}_2$ matrices will be modified if the transfer occurs from $\mathbf{v}_{\text{in}}$. Let us consider an example to illustrate the nature of the different matrices.

Example 3.6–3. *Third-order elliptic low-pass filter.* Obtain the matrices $\mathbf{A}_j$, $\mathbf{B}_j$, $\mathbf{D}_j$, $\mathbf{E}_j$, $\mathbf{G}_j$, $j = 1, 2$, and $\mathbf{F}$ for the filter shown in Fig. 3.6–5.

First, consider $\mathbf{A}_j$. The rows are the number of op amps and the columns are the parallel SC at ϕ_j. In this case

$$\mathbf{A}_1 = \begin{matrix} \\ 1 \\ 2 \\ 3 \end{matrix} \begin{matrix} C_1 & C_2 & C_6 & C_7 \\ \left[\begin{matrix} C_1/C_{F_1} \\ 0 \\ 0 \end{matrix}\right. & \begin{matrix} C_2/C_{F_1} \\ 0 \\ 0 \end{matrix} & \begin{matrix} 0 \\ 0 \\ C_6/C_{F_3} \end{matrix} & \left.\begin{matrix} 0 \\ 0 \\ C_7/C_{F_3} \end{matrix}\right] \end{matrix} \qquad (14)$$

$$\mathbf{A}_2 = \begin{array}{c} \\ 1 \\ 2 \\ 3 \end{array} \begin{array}{c} \begin{array}{cc} C_4 & C_5 \end{array} \\ \begin{bmatrix} 0 & 0 \\ C_4/C_{F_2} & C_5/C_{F_2} \\ 0 & 0 \end{bmatrix} \end{array} \tag{15}$$

Next label the polarities of all SC's and all the output of the op amps from 1 to N. Label the input node as $N + 1$, and the ground node as 0. The incidence matrices $\mathbf{B}_j$ ($j = 1, 2$) relate the parallel SC branch voltages with the output node voltage. The columns of $\mathbf{B}_j$ are the number of the op amps, and the rows of $\mathbf{B}_j$ are switched capacitors connected at the output of a particular op amp. The entry $\mathbf{B}_j$ is $+1$ (-1) if the polarity of the SC is $+1$ (-1). Thus in our example we obtain

$$\mathbf{B}_1 = \begin{array}{c} \\ C_4 \\ C_5 \end{array} \begin{array}{c} \begin{array}{ccc} 1 & 2 & 3 \end{array} \\ \begin{bmatrix} 0 & 0 & -1 \\ -1 & 0 & 0 \end{bmatrix} \end{array} \tag{16}$$

$$\mathbf{B}_2 = \begin{array}{c} \\ C_1 \\ C_2 \\ C_6 \\ C_7 \end{array} \begin{array}{c} \begin{array}{ccc} 1 & 2 & 3 \end{array} \\ \begin{bmatrix} 1 & 0 & 0 \\ -1 & 1 & 0 \\ 0 & 0 & 1 \\ 0 & 1 & 0 \end{bmatrix} \end{array} \tag{17}$$

The vectors $\mathbf{C}_j$ relate the parallel SC branch voltages with the input signal; note that the same number of rows is associated with $\mathbf{B}_j$ and $\mathbf{C}_j$.

$$\mathbf{C}_1 = \begin{array}{c} C_4 \\ C_5 \end{array} \begin{bmatrix} 0 \\ 0 \end{bmatrix} \tag{18}$$

and

$$\mathbf{C}_2 = \begin{array}{c} C_1 \\ C_2 \\ C_6 \\ C_7 \end{array} \begin{bmatrix} -1 \\ 0 \\ 0 \\ 0 \end{bmatrix} \tag{19}$$

The square $N \times N$ matrices $\mathbf{D}_j$ are constant coefficients, associated with the fixed (unswitched) capacitors and series SC's. Here the N rows indicate which node the component is connected to and the N columns which node the component is coming from

$$\mathbf{D}_1 = \mathbf{D}_2 = \begin{matrix} & 1 & 2 & 3 \\ 1 & 0 & 0 & C_3/C_{F_1} \\ 2 & 0 & 0 & 0 \\ 3 & C_8/C_{F_3} & 0 & 0 \end{matrix}$$

The matrices $\mathbf{E}_j$ and $\mathbf{G}_j$ relate the output nodes with the input signal; in our case

$$\mathbf{E}_1 = \mathbf{E}_2 = \mathbf{G}_1 = \mathbf{G}_2 = \begin{matrix} 1 \\ 2 \\ 3 \end{matrix} \begin{bmatrix} 0 \\ 0 \\ 0 \end{bmatrix}$$

If the z-domain equations to evaluate the frequency response are desired, Eqs. (8) through (11), after some matrix manipulation, can be expressed at ϕ_1 as

$$\begin{aligned}[\mathbf{P}_1\mathbf{U}_1(\mathbf{F} - \mathbf{A}_2\mathbf{B}_1) - z\mathbf{I}]\,\mathbf{V}(z) = [&\mathbf{P}_1\mathbf{U}_1(\mathbf{A}_2\mathbf{C}_1 - \mathbf{G}_2) \\ &+ \mathbf{P}_1(\mathbf{U}_1\mathbf{E}_2 + \mathbf{A}_1\mathbf{C}_2 - \mathbf{G}_1)z^{1/2} + \mathbf{P}_1\mathbf{E}_1\, z]\,\mathbf{V}_{\text{in}}(z)\end{aligned} \tag{20}$$

where

$$\begin{aligned}\mathbf{P}_1 &= (\mathbf{I} + \mathbf{D}_1)^{-1} \\ \mathbf{U}_1 &= (\mathbf{F} - \mathbf{A}_1\mathbf{B}_2)\mathbf{P}_2 \\ \mathbf{P}_2 &= (\mathbf{I} + \mathbf{D}_2)^{-1} \\ \mathbf{U}_2 &= (\mathbf{F} - \mathbf{A}_2\mathbf{B}_1)\mathbf{P}_1\end{aligned}$$

and at ϕ_2 as

$$\begin{aligned}[\mathbf{P}_2\mathbf{U}_2(\mathbf{F} - \mathbf{A}_1\mathbf{B}_2) - z\mathbf{I}]\,\mathbf{V}(z) = [&\mathbf{P}_2\mathbf{U}_2(\mathbf{A}_1\mathbf{C}_2 - \mathbf{G}_1) \\ &+ \mathbf{P}_2(\mathbf{U}_2\mathbf{E}_1 + \mathbf{A}_2\mathbf{C}_1 - \mathbf{G}_2)z^{1/2} + \mathbf{P}_2\mathbf{E}_2 z]\,\mathbf{V}_{\text{in}}(z)\end{aligned} \tag{21}$$

The frequency response is computed by setting $z = e^{j2\pi f/f_c}$. Figure 3.6–6 shows the simulated frequency response of the third-order elliptic filter of Example 3.6–3. Even with its limitations, this approach can be very useful in many practical SC networks. In fact the building blocks allowed in this approach include a large family of useful SC circuits. It should be mentioned that some SC filter structures require a sample-and-hold circuit at the input to avoid continuous signal feedthrough that causes frequency response distortion. The operation of the sample-and-hold circuit can be modeled by appropriate entries in the **E** and **G** matrices of the time domain equations or by modifying the powers of z in the frequency equations.

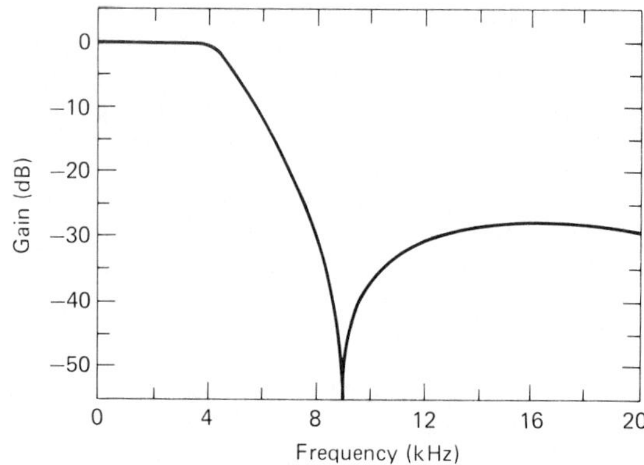

Fig. 3.6–6. Simulated frequency response for Fig. 3.6–5.

Most of the SC network analysis methods reported in the literature involve nodal analysis or modified nodal analysis.[36,37,38,39,40] The analysis is based on the nodal charge equation referred to as the *charge conservation* and expressed as Eqs. (15) and (16) of Section 2.3

$$q_L(t') = q_m(t) + q_c(t) \qquad t' > t \tag{22}$$

Equation (22) can be extended to include all the capacitor nodes in the network, that is

$$\mathbf{q}_L(t') = \mathbf{q}_m(t) + \mathbf{q}_c(t) \qquad t' > t \tag{23}$$

The matrix equation Eq. (23) corresponds to Kirchhoff's current law. Because an SC network has time-variant coefficients, the general matrix equation can be expressed as

$$\mathbf{q}_c(t') = [\mathbf{C}(t')]\, \mathbf{v}(t') - [\mathbf{C}(t)]\, \mathbf{v}(t) \tag{24}$$

[36] C. F. Kurth and G. S. Moschytz, "Nodal Analysis of Switched-Capacitor Networks," *IEEE Trans. Circuits and Systems,* Vol. CAS-26, No. 11, February 1979, pp. 93–105.

[37] M. L. Liou and Y. L. Kuo, "Exact Analysis of Switched Capacitor Circuits with Arbitrary Inputs," *IEEE Trans. Circuits and Systems,* Vol. CAS-26, No. 4, April 1979, pp. 213–223.

[38] Y. P. Tsividis, "Analysis of Switched Capacitive Networks," *IEEE Trans. Circuits and Systems,* Vol. CAS-26, No. 11, November 1979, pp. 935–947.

[39] F. Brglez, "SCOP, A Switched-Capacitor Optimization Program," *IEEE Proc. Int. Symp. Circuits and Systems,* April 1980, pp. 985–988.

[40] S. C. Fang and Y. P. Tsividis, "Modified Nodal Analysis with Improved Numerical Methods for Switched Capacitive Networks," *Proc. Int. Symp. Circuits and Systems,* April 1980, pp. 977–980.

During each phase period (i.e., ϕ_1 or ϕ_2), several subtopologies are generated. Each one may consist of either a set of nodes connected through closed switches or an isolated node. The concept of the switching matrix $\mathbf{S}_k$, is useful to keep track of which switches are closed or open.

$$\mathbf{S}_k\,\mathbf{q}_c(t') = \mathbf{S}_k\,\mathbf{C}\,\mathbf{v}(t') - \mathbf{S}_k\,\mathbf{C}\,\mathbf{v}(t) \tag{25}$$

where the entries $s_{k,ij}$ of the switching matrix S_k are

$$s_{k,ij} = \begin{cases} 1, & \text{if } i \text{ is the lowest numbered node of a separate part of the closed switch network, and node } j \text{ belongs to that separate part} \\ 0, & \text{otherwise} \end{cases} \tag{26}$$

Equation (26) is obtained by removing all the components in the original network, except the closed switches. The entries c_{ij} of $\mathbf{C}$, an $n \times n$ matrix, independent of switching, designated as the capacitance matrix, are given as follows

$$c_{ij} = \begin{cases} \text{Total capacitance permanently connected to node } i, & i = j \\ \text{Negative of total capacitance permanently connected between nodes } i \text{ and } j, & i \neq j \end{cases} \tag{27}$$

Furthermore, $\mathbf{q}_c(t')$ can be divided into $\mathbf{q}_D(t')$ and $\mathbf{q}_I(t')$, the total charge that has passed through the dependent voltage sources, and through the independent voltage sources, respectively. Then $\mathbf{q}_c(t')$ can be described as

$$\mathbf{q}_c(t') = [\mathbf{E}_I\ \mathbf{E}_D] \begin{bmatrix} \mathbf{q}_I(t') \\ \mathbf{q}_D(t') \end{bmatrix} \tag{28}$$

$\mathbf{E}_I$ and $\mathbf{E}_D$ are the node-to-branch incidence matrices associated with the independent and dependent voltage sources. Equation (25) yields

$$\mathbf{S}_k[\mathbf{E}_I\ \mathbf{E}_D] \begin{bmatrix} \mathbf{q}_I(t') \\ \mathbf{q}_D(t') \end{bmatrix} = \mathbf{S}_k\,\mathbf{C}\,\mathbf{v}(t') - \mathbf{S}_k\,\mathbf{C}\,\mathbf{v}(t) \tag{29}$$

It should be noticed that the closure of the switches follows Kirchhoff's voltage law (KVL), which can be characterized by

$$[\mathbf{S}_k^t - \mathbf{I}]\, \mathbf{v}(t') = 0 \tag{30}$$

where t denotes the transpose operation, and $\mathbf{I}$ is the identity matrix. $\mathbf{S}_k$ has zero rows for Eq. (29) exactly where Eq. (30) has nonzero rows and vice versa. Therefore Eqs. (29) and (30) can be combined as

$$[\mathbf{S}_k\, \mathbf{C} + \mathbf{S}_k^t - \mathbf{I}]\mathbf{v}(t') + \mathbf{S}_k[\mathbf{E}_I\ E_D]\begin{bmatrix}\mathbf{q}_I(t')\\ \mathbf{q}_D(t')\end{bmatrix} = \mathbf{S}_k\, \mathbf{C}\, \mathbf{v}(t) \tag{31}$$

Thus a general matrix equation can be written as

$$\begin{array}{c} p \\ r \\ N \\ \ \end{array}
\begin{array}{c}
\left[\begin{array}{c|c|c}
-\mathbf{E}_I^t & \multicolumn{2}{c}{\mathbf{O}} \\
\hline
\mathbf{D} + \mathbf{E}_D^t & \multicolumn{2}{c}{} \\
\hline
\mathbf{S}_k\mathbf{C} + \mathbf{S}_k^t - \mathbf{I} & \mathbf{S}_k\mathbf{E}_I & \mathbf{S}_k\mathbf{E}_D
\end{array}\right] \\
\quad N \qquad\qquad p \qquad r
\end{array}
\begin{array}{c}
\left[\begin{array}{c}
\mathbf{v}(t') \\ \hline \mathbf{q}_I(t') \\ \hline \mathbf{q}_D(t')
\end{array}\right] \\ 1
\end{array}
=
\begin{array}{c}
\left[\begin{array}{c}
\mathbf{v}_I(t') \\ \hline \mathbf{O} \\ \hline \mathbf{S}_k\mathbf{C}v(t)
\end{array}\right] \\ 1
\end{array}
\begin{array}{c} p \\ r \\ N \\ \ \end{array} \tag{32}$$

where N is the total number of nodes with p independent voltage sources and r linear dependent voltage sources. The voltages of the linear dependent sources can be expressed as a linear combination of the controlling node voltage

$$\mathbf{v}_D(t') = \mathbf{D}\, \mathbf{v}(t') \tag{33}$$

and $\mathbf{D}$ is the $r \times N$ matrix of the controlling voltage coefficients. Thus Eq. (32) for any time t in $(n - i,\ n - i + 1)$ can provide the node voltages and the total charge that passed through each source. In summary, a complete solution using the method described above can be obtained,[41] if the following data are specified:

1. Network topology (excluding switches) and element values;
2. Node pairs across which switches are connected;
3. The switching pattern with the "on" and "off" times for each switch.

---

[41] Y. P. Tsividis, op. cit.

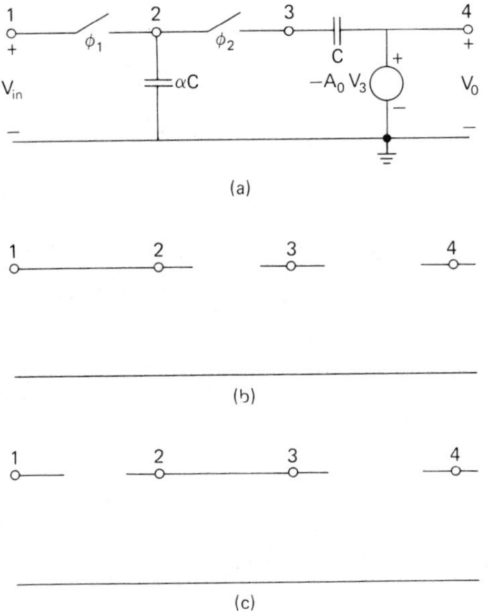

Fig. 3.6–7. (a) Equivalent circuit of Fig. 3.6–2 for $C_p = 0$. (b) Graph for ϕ_1, S_1; (c) Graph for ϕ_2, S_2.

Example 3.6–4. *Forward integrator.* Consider the circuit of Fig. 3.6–2 with $C_p = 0$ and a finite A_0. Using the analysis method discussed in this section, obtain the matrices C, S_1, and S_2 and the corresponding matrix equation (32).

A convenient equivalent circuit for the analysis is shown in Fig. 3.6–7(a). The capacitance matrix is

$$\mathbf{C} = \begin{bmatrix} 0 & 0 & 0 & 0 \\ 0 & \alpha C & 0 & 0 \\ 0 & 0 & C & -C \\ 0 & 0 & -C & C \end{bmatrix} \tag{34}$$

Notice that **C** is obtained by removing all the switches in the original circuit and then following Eq. (27). At ϕ_1

$$\mathbf{S}_1 = \begin{bmatrix} 1 & 1 & 0 & 0 \\ 0 & 0 & 0 & 0 \\ 0 & 0 & 1 & 0 \\ 0 & 0 & 0 & 1 \end{bmatrix} \tag{35}$$

At ϕ_2

$$\mathbf{S}_2 = \begin{bmatrix} 1 & 0 & 0 & 0 \\ 0 & 1 & 1 & 0 \\ 0 & 0 & 0 & 0 \\ 0 & 0 & 0 & 1 \end{bmatrix} \tag{36}$$

$\mathbf{S}_1$ and $\mathbf{S}_2$ are obtained following Eq. (26), and the corresponding graphs are shown in Figs. 3.6–7(b) and (c). The graphs are the closed switch networks obtained when all the elements in the original network, except the closed switches, are removed. Also the incidence matrices $\mathbf{E}_I$, $\mathbf{E}_D$, and $\mathbf{D}$ are given by

$$\mathbf{D} = [0 \quad 0 \quad -A_0 \quad 0] \tag{37}$$

$$\mathbf{E}_I = [-1 \quad 0 \quad 0 \quad 0]^t \tag{38}$$

$$\mathbf{E}_D = [0 \quad 0 \quad 0 \quad -1]^t \tag{39}$$

Then the general matrix equation, Eq. (32), for ϕ_1 is given by

$$\begin{bmatrix} 1 & 0 & 0 & 0 & 0 & 0 \\ 0 & 0 & A_0 & 1 & 0 & 0 \\ 0 & \alpha C & 0 & 0 & -1 & 0 \\ 1 & -1 & 0 & 0 & 0 & 0 \\ 0 & 0 & C & -C & 0 & 0 \\ 0 & 0 & -C & C & 0 & 0 \end{bmatrix} \begin{bmatrix} v_1 \\ v_2 \\ v_3 \\ v_4 \\ q_I \\ q_D \end{bmatrix} = \begin{bmatrix} v_{\text{in}}(t) \\ 0 \\ \alpha C\, v_2(t_1^-) \\ 0 \\ C\, v_3(t_1^-) - C\, v_4(t_1^-) \\ -C v_3(t_1^-) - C\, v_4(t^-) \end{bmatrix} \tag{40}$$

t_i^- indicates "the instant immediately before switching to the other phase period." Thus for ϕ_2 we obtain

$$\begin{bmatrix} 1 & 0 & 0 & 0 & 0 & 0 \\ 0 & 0 & A_0 & 1 & 0 & 0 \\ 0 & 0 & 0 & 0 & -1 & 0 \\ 0 & \alpha C & C & -C & 0 & 0 \\ 0 & 1 & -1 & 0 & 0 & 0 \\ 0 & 0 & -C & C & 0 & -1 \end{bmatrix} \begin{bmatrix} v_1 \\ v_2 \\ v_3 \\ v_4 \\ q_I \\ q_D \end{bmatrix} = \begin{bmatrix} v_{\text{in}}(t') \\ 0 \\ 0 \\ \alpha C\, v_2(t_2^-) + C\, v\,(t_2^-) - C\, v_4(t_2^-) \\ 0 \\ -C\, v_3(t_2^-) + C\, v_4(t_2^-) \end{bmatrix} \tag{41}$$

Next consider $\mathbf{I}_k$, which is defined as an $(N + p + r)$ by $(N + p + r)$ matrix premultiplying the unknown vector in Eq. (32). $\mathbf{A}_k$ represents the upper left $N \times p$ submatrix of $\mathbf{I}_k^{-1}$, and $\mathbf{B}_k = (\mathbf{I}_k^{-1})\, \mathbf{S}_k\, \mathbf{C}$, where $(\mathbf{I}_k^{-1})_R$ is the upper submatrix of $\mathbf{I}_k^{-1}$. Thus if Eq. (32) is solved for $\mathbf{v}(t)$ at any time t, we obtain

$$\mathbf{v}(t) = \mathbf{A}_k \, \mathbf{v}_I(t) + \mathbf{B}_k \, \mathbf{v}(t_k^-) \tag{42}$$

where t is a switching interval, i.e., $t = [(n-1)T, nT]$. Assuming the input is piecewise-constant, we can write Eq. (42) for $k = 1, 2$ as follows

$$\mathbf{v}_1(n-1) = \mathbf{A}_1 \, \mathbf{v}_I(n-1) + \mathbf{B}_1 \, \mathbf{v}_2(n-1) \tag{43}$$

and

$$\mathbf{v}_2(n) = \mathbf{A}_2 \, \mathbf{v}_I(n) + \mathbf{B}_2 \, \mathbf{v}_1(n-1) \tag{44}$$

If the frequency response is desired, we can take the z-transform of Eqs. (43) and (44) and obtain, after some manipulations

$$\mathbf{V}_1(z) = [\mathbf{I} - z^{-1}\mathbf{B}_1\mathbf{B}_2]^{-1} \, [\mathbf{A}_1 + z^{-1}\mathbf{B}_1\mathbf{A}_2] \, \mathbf{V}_I(z) \tag{45}$$

and

$$\mathbf{V}_2(z) = [\mathbf{I} - z^{-1}\mathbf{B}_2\mathbf{B}_1]^{-1} \, [\mathbf{A}_2 + z^{-1}\mathbf{B}_2\mathbf{A}_1]\mathbf{V}_I(z) \tag{46}$$

Also notice that typically a sample-and-hold circuit is used at the input and therefore its spectrum has to be considered. Thus, for a single sinusoidal input Eqs. (45) and (46) yield

$$H_i(\omega) = \frac{1}{T} H_i(e^{j\omega T}) \, P(\omega), \qquad i = 1, 2 \tag{47}$$

where $P(\omega)$ is the spectrum of the sample-and-hold, and $H_i(e^{j\omega T}) = V_i(z)/V_I(z)$ for $z = e^{j\omega T}$.

Only recently have computer methods been available for simulating SC networks at the MOS device level.[42] This level of simulation is probably the most useful to the designer who will implement his analog sampled data circuit design by integrated circuit technology. Many of the programs now used for continuous circuits can provide time domain analysis of switched capacitor circuits. These programs allow for full top-down design from principle to transistor level. In other words, one may be able to enter an equation, a flow diagram, an equivalent circuit model, circuit blocks, or the actual schematic diagram. The more difficult problem is that of analyzing the frequency response of such circuits. The technique used at present is to first

[42] H. F. DeMan, J. Rabaey, G. Arnout, and J. Vanderwalle, "Practical Implementation of a General Computer Aided Design Technique for Switched Capacitor Circuits," "*IEEE J. of Solid-State Circuits,* Vol. SC-15, No. 2, April 1980, pp. 190–200.

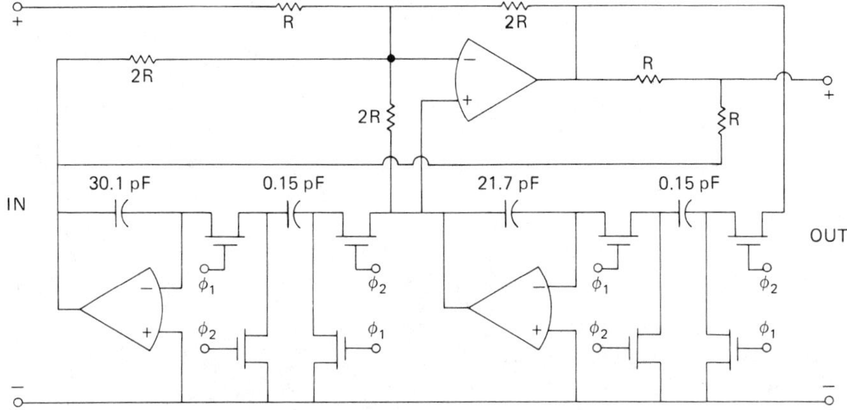

Fig. 3.6–8. 60 Hz notch filter for Example 3.6–5.

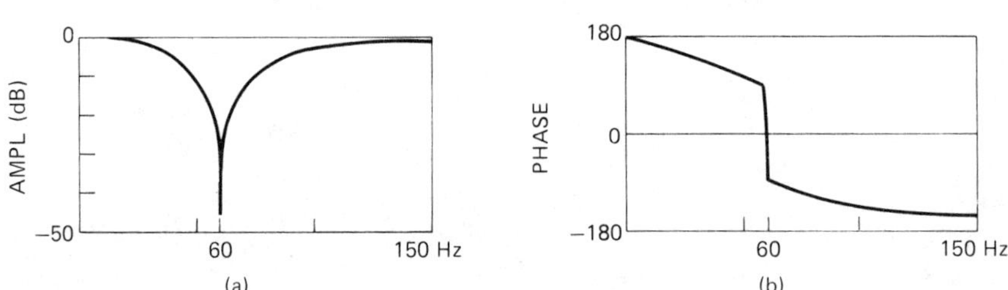

Fig. 3.6–9. (a.) Magnitude response and (b.) phase response for Fig. 3.6–8.

find the impulse response of the circuit. An algorithm can be developed that can handle both continuous and sampled input data and provide the frequency response by the use of fast Fourier transform (FFT) methods. The output is dependent upon the clocking sequence and whether or not a sample-and-hold circuit is used at the output. Two examples will be presented to demonstrate the capability of such a program.

Example 3.6–5. *Simulation of a 60-Hz notch for PCM filters.* Figure 3.6–8 shows the schematic of a 60-Hz notch filter that has a continuous input-output path consisting of resistors, as well as two noninverting Type I DDI's. The clock frequency is designated as 128 kHz. The DIANA program[43] was used to obtain the magnitude and phase plots shown in Fig. 3.6–9(a) and (b). The simulation of over 3000 time points, together with the FFT of over 16,000 points took 2 minutes of CPU time on an IBM 370/158.

This example shows that such a program has the ability to model the nonideal characteristics of the op amp. The next example given shows the influence of the clock sequence on the frequency response.

[43] Ibid.

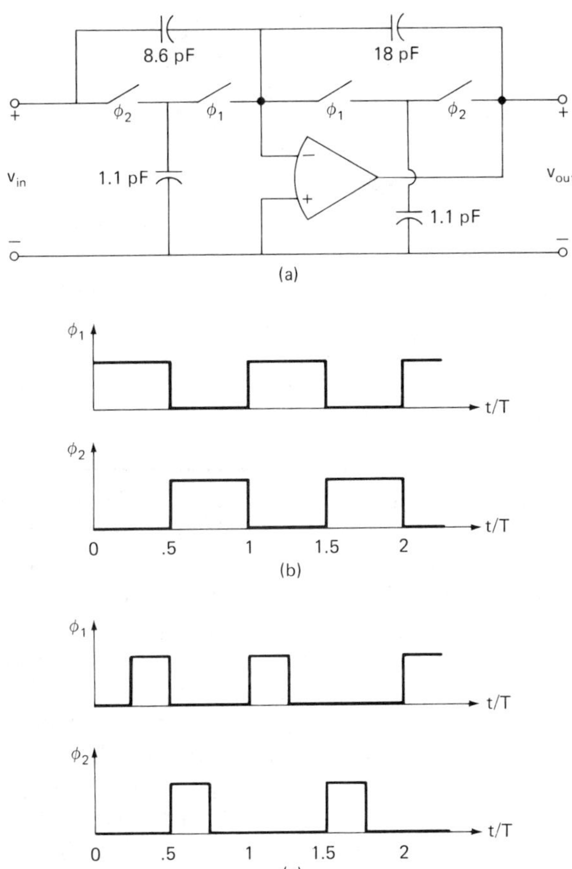

Fig. 3.6–10. (a.) The first order circuit used in Example 3.6–6. (b.) Clock sequence for 50% duty cycle. (c.) Clock sequence for 25% duty cycle.

Example 3.6–6. *Simulation of a pole-zero high-pass filter.* The circuit of Fig. 3.6–2 has been used to design a simple pole-zero filter for audio treble control shown in Fig. 3.6–10(a). The amplitude of the frequency response of this circuit is to be simulated using DIANA for the two clocking sequences shown in Fig. 3.6–10(b) and (c). The frequency response is shown in Fig. 3.6–11, with the frequency axis normalized to the sampling frequency. Curve 1 corresponds to the clock sequence in Fig. 3.6–10(b), and Curve 2 corresponds to the clock sequence in Fig. 3.6–10(c). Curve 3 is for the clock sequence of Fig. 3.6–10(c) with a sample-and-hold circuit on the input over the full period. The sin x/x effects due to the sample-and-hold are clearly evident. This example demonstrates the influence of duty cycle and continuous coupling and the usefulness of simulation in each of these cases.

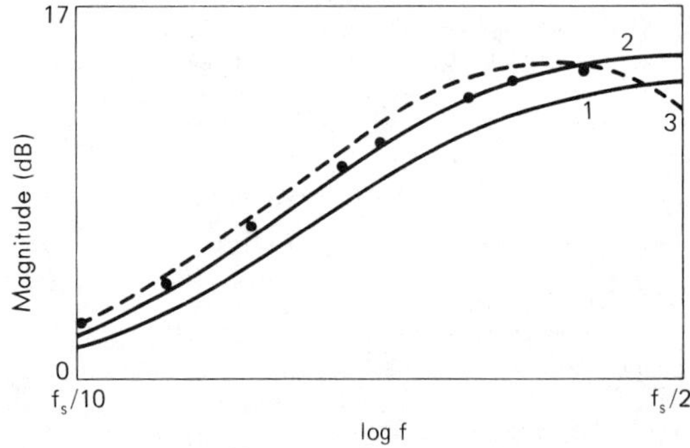

Fig. 3.6–11. Magnitude response of Fig. 3.6–10a. Curve 1 uses the clock sequence in Fig. 3.6–10b and Curve 2 uses the clock sequence of Fig. 3.6–10c. Curve 3 corresponds to a sample-and-hold on the input over the complete period and for the clock sequence of Fig. 3.6–10b.

Simulation methods for analog sampled data circuits is a very important part of the design procedure. Several levels of simulation have been presented in this section to help the designer determine which level of simulation is most appropriate for a given situation. The typical objective of most simulation programs is the frequency response of an SC circuit. Other analyses, such as noise and sensitivity, are also important and can be implemented by some of the simulation methods.[44] For the existing switched capacitor CAD programs, it is not possible to discuss and compare their numerical accuracies and computational efficiencies. Because actual running data are not available, the programs SCOP[45] and DIANA[46] are available only commercially. The programs developed in the University of Illinois[44] are available at little or no cost. SWITCAP, a program based on the Tsividis algorithm called is also available from Columbia University.

3.7 SUMMARY

This chapter has introduced the basic design techniques of filters that are implemented using analog sampled data methods. A review of continuous passive and active filter design was presented. It was seen that these methods are generally the starting point of the design techniques presented in this text for switched capacitor filters. Figure 3.1–16 was introduced to provide

---

[44] C. F. Lee, op. cit.

[45] F. Brglez, op. cit.

[46] H. F. Deman et al., op. cit.

a background for the classification of the various approaches that were presented in this chapter and will be presented in the next two chapters.

The left-hand path of Fig. 3.1–16, consisting of blocks 1, 2, 3, 6, 8, 10, and 14 or 1, 2, 3, 6, 7, 9, and 14, characterized the methods presented in this chapter. The performance characterization of SC filters permits a comparison of the various approaches and an evaluation of the performance. Three design approaches for SC filters were presented based upon the component simulation approach. The first method used the parallel SC circuit to replace resistors in an active RC realization. The parallel SC circuit was seen to be the equivalent of using the forward z-transform. The second method used the bilinear z-transform and the resulting SC circuits to replace resistors or branch impedances. The third method used the series SC circuit to replace resistors in active realizations. The series SC circuit is equivalent to using the backward z-transform. Examples of each of the methods were given to illustrate the design approach.

This chapter concluded with the very important topic of analysis and simulation methods for SC networks using the computer. Various levels of modeling the SC network and example algorithms were discussed. Many programs exist that provide the designer with the means to adequately model SC networks of reasonable complexity.

PROBLEMS

3.1 (Sec. 3.1). Figure P 3.1 shows a low-pass filter characteristic of those used in communications equipment. It has unity gain out to 3.4 kHz and falls below −32 dB by 4.6 kHz. The maximum ripple allowed in the passband is 0.1 dB. Determine the $H(s)$ that represents the filter characteristics.

3.2 (Sec. 3.1). Obtain an RLC circuit, and specify the component values for the following transfer function

$$H(s) = \frac{(s + j\omega_1)(s - j\omega_1)(s + j\omega_2)(s - j\omega_2)}{(s + a_0)(s + a_1 - jb_1)(s + a_1 + jb_1)(s + a_2 - jb_2)(s + a_2 + jb_2)}$$

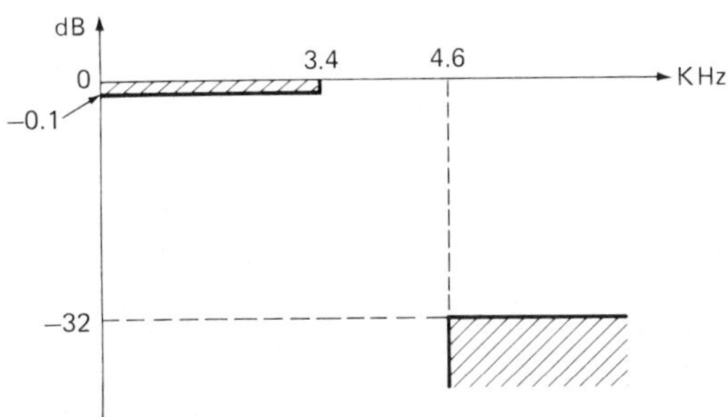

Fig. P 3.1.

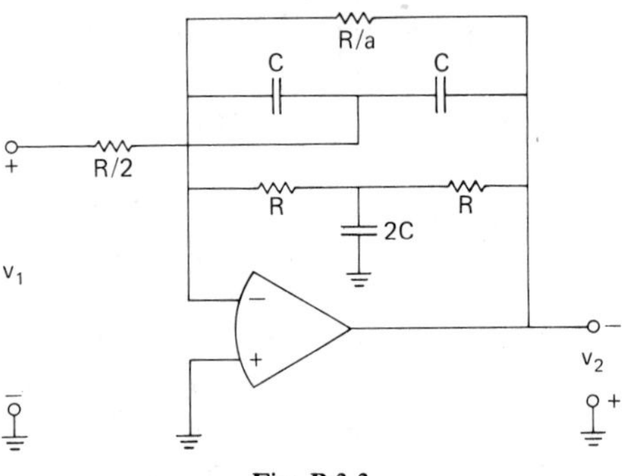

Fig. P 3.3.

where $\omega_1 = 40.792 \times 10^3$, $\omega_2 = 28.367 \times 10^3$, $a_0 = 15.474 \times 10^3$, $a_1 = 1.938 \times 10^3$, $b_1 = 22.604 \times 10^3$, $a_2 = 8.514 \times 10^3$, $b_2 = 17.831 \times 10^3$.

3.3 (Sec. 3.1). Obtain the transfer function of the circuit shown in Fig. P 3.3, and determine the design equations.

3.4 (Sec. 3.1). Substitute the resistors for the capacitors and vice versa in Fig. P. 3.3, and determine the transfer function.

3.5 (Sec. 3.2). Using the generalized chain rule of differentiation, show that

$$S_{\alpha_i}^{(f_0 T)} = S_r^{(f_0 T)} S_{\alpha_i}^r + S_\theta^{(f_0 T)} S_{\alpha_i}^\theta$$

$$S_{\alpha_i}^\theta = S_r^\theta S_{\alpha_i}^r + S_\theta^Q S_{\alpha_i}^\theta$$

where $S_r^Q = -\theta^2/(\omega_0 T)^2 \ln r$, $S_\theta^Q = \theta^2/(\omega_0 T)^2$, and from Eqs. (39) and (40) prove the validity of Eqs. (41) and (42).

3.6 (Sec. 3.2). Consider a third-order function described as $z^3 + B_2 z^2 + B_1 z + B_0$ and $(z + a)(z^2 - 2r \cos\theta\, z + r^2)$, where

$$B_2 = a - 2r\cos\theta$$

$$B_1 = r^2 - a\, 2r \cos\theta$$

$$B_0 = ar^2$$

Prove that the sensitivities $S_{\alpha_i}^\theta$, $S_{\alpha_i}^r$, and $S_{\alpha_i}^a$ are given by

$$S_{\alpha_i}^\theta = \frac{-ar(r - a\cos\theta - 2r\cos^2\theta)\, S_{\alpha_i}^{B_0} + B\,(r\cos\theta + a)\, S_{\alpha_i}^{B_1} + rB\,(r + a\cos\theta)\, S_i^{B_2}}{2r\theta \sin\theta\,(r^2 + a^2 + 2ar\cos\theta)}$$

$$S_{\alpha_i}^r = \frac{a(a + 2r\cos\theta)\, S_{\alpha_i}^{B_0} + B_1\, S_{\alpha_i}^{B_1} - B_2 a S_{\alpha_i}^{B_2}}{2(r^2 + a^2 + 2ar\cos\theta)}$$

$$S^{a}_{\alpha_i} = \frac{r^2 S^{B_0}_{\alpha_i} - B_1 S^{B_1}_{\alpha_i} + B_2 a S^{B_2}_{\alpha_i}}{r^2 + a^2 + 2ar \cos\theta}$$

3.7 (Sec. 3.3). Compute $S^{Q}_{\alpha_i}$, $S^{\theta}_{\alpha_i}$ of Fig. 3.3–1(c) when $\alpha_1 = \alpha_3 = \alpha_2 A$. Form a table for α_1, A, and LSC for the range of $Q = \{0.5, 1, 2, 4, 8, 16, 32, 64, 128\}$ and $f_0/f_c = \{0.02, 0.06, 0.10, 0.14, 0.18, 0.22, 0.26\}$.

3.8 (Sec. 3.3). Compute the sensitivities $S^{Q}_{\alpha_i}$, $S^{f_0}_{\alpha_i}$ of the circuit of Fig. 3.3–1(c) characterized by Eqs. (12)–(14).

3.9 (Sec. 3.3). Design an SC filter characterized by Eqs. (1) to (5b) with (a) $Q = 73$, $\theta \cong 13.18°$, and $f_0 = 3.75$ kHz. Compare results with Example 3.3–1. Repeat the first part if the SC filter is characterized by (b) $Q = 16$, $f_0 = 1633$ Hz, $f_c = 8$ kHz, and a peak gain of 10 dB at f_0.

3.10 (Sec. 3.3). Given the circuit of Fig. P 3.3, show that the transfer function is

$$H^{ee}(z) = \frac{-2\alpha_3 (1 - z^{-1})\, z^{-1}}{1 - (2 - 2\alpha_2\alpha_3 - 2\alpha_2 - \alpha_3^2) z^{-1} + (1 - 2\alpha_2) z^{-2}}$$

when the resistors R/a, R are replaced by parallel SC equivalent resistors $\alpha_2 C$, $\alpha_3 C$.

3.11 (Sec. 3.3). Design the bandpass filter of Fig. 3.3–2 with the specifications given in Problem 3.9. (b).

3.12 (Sec. 3.3). Obtain the capacitor ratio values of the filter of Fig. 3.3–5 for the following two cases (a) $Q = 73$, $\theta = 13.18°$, and (b) $Q = 16$, $\theta \cong 73.485°$, $f_c = 8$ kHz, and a peak gain of 10 dB at f_0. Make sure the outputs of the op amps are the same.

3.13 (Sec. 3.3). Repeat problem 3.12, but use the topology of Fig. 3.3–6.

3.14 (Sec. 3.3). Derive Eqs. (48) and (49) using charge conservation analysis.

3.15 (Sec. 3.3). Repeat problem 3.12, but use the structure of Fig. 3.3–9.

3.16 (Sec. 3.3). (a) Prove that the transfer functions of Fig. P 3.16 are given by

$$H^{ee}_{\text{LP}}(z) = \frac{\alpha_1\alpha_2 z^{-2}}{1 - (2 - \alpha_2 - \alpha_1\alpha_2) z^{-1} + (1 - \alpha_2) z^{-2}}$$

$$H^{ee}_{\text{BP}}(z) = \frac{-\alpha_1(1 - z^{-1}) z^{-1}}{1 - (2 - \alpha_2 - \alpha_1\alpha_2) z^{-1} + (1 - \alpha_2) z^{-2}}$$

(b) Design this filter with the specifications given in Problem 3.12.

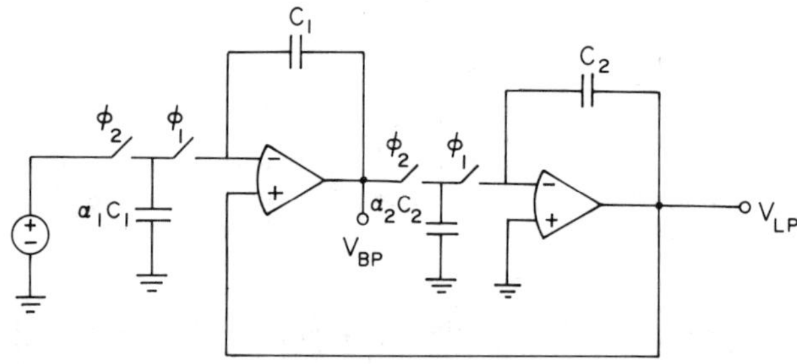

Fig. P 3.16.

3.17 (Sec. 3.3). Design a Tow-Thomas filter like the one shown in Fig. 3.3–7(b) that satisfies the following bandpass transfer function

$$H(z) = \frac{0.00164z^{-1}(1 - z^{-1})}{1 - 0.5455z^{-1} + 0.9229z^{-2}}$$

Assume that the clock frequency $f_c = 80$ kHz. Determine the total capacitance area for equal output levels V_{LP} and V_{BP}.

3.18 (Sec. 3.4). Given the RC prototype of Fig. P 3.3, design the bilinear component simulation SC filter version for the design specifications given in Problem 3.9.

3.19 (Sec. 3.4). Design a bandpass bilinear SC filter based on the Tow-Thomas RC biquad of Fig. 3.3–7(a) with the following specifications

1. $\theta = 7.3485°$, $H_0 = 1$, and $Q = 16$ and
2. $\theta = 73.49°$, $H_0 = 1$, and $Q = 16$.

3.20 (Sec. 3.4). Based on the prototype of Fig. P 3.20, whose transfer function is

$$H(s) = \frac{s^2}{s^2 + s\,(1/R_2C_1 + 1/R_2C_2) + 1/R_1R_2C_1C_2}$$

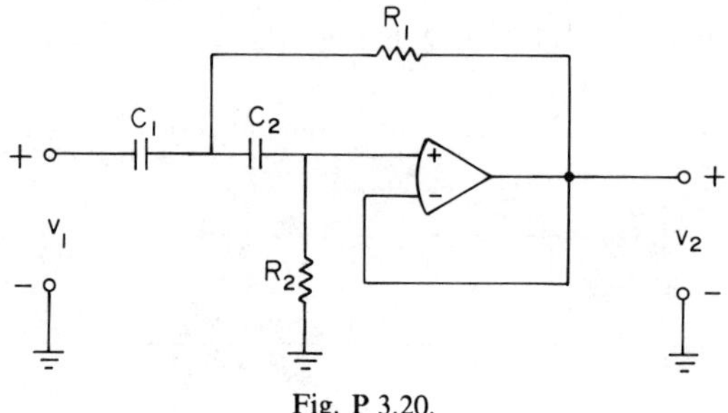

Fig. P 3.20.

design a high-pass bilinear component simulation SC filter for $f_h = 1000$ Hz, $Q = 0.7071$, and (a) $f_c = 8$ kHz and (b) $f_c = 64$ kHz.

3.21 (Sec. 3.4). Given the Delyiannis bandpass circuit of Fig. 3.5–3, design a bilinear component simulation SC filter with the same design specifications of Problem 3.19.

3.22 (Sec. 3.5). Obtain the capacitor ratio values of Fig. 3.5–5 for a cutoff frequency of 3.4 kHz, DC gain of 10 dB and (a) $f_c = 32$ kHz and (b) $f_c = 128$ kHz. Compare the total capacitance for both cases.

3.23 (Sec. 3.5). Obtain the values of A, α_1, $\alpha_2 A$, α_3, and C for the circuit of Fig. 3.5–2 when $f_0 = 4$ kHz, $f_c = 128$ kHz, and (a) $Q = 0.7071$ and (b) $Q = 16$. Refer to Eqs. (12) through (18).

3.24 (Sec. 3.5). For the SC bandpass filter of Fig. 3.5–4, obtain the capacitor ratios for the following specifications: $Q = 73$, $f_0 = 3.75$ kHz and (a) $f_c = 102.4$ kHz, and (b) 64 kHz, and (c) $Q = 16$, $f_0 = 1.633$ kHz, and $f_c = 8$ kHz.

3.25 (Sec. 3.5). Given the low-pass Tow-Thomas RC biquad of Fig. 3.3–7(a), obtain the backward z-transform (series) resistor equivalent for the following specifications: (a) $Q = 16$, $f_0 = 2041$ Hz, and $f_c = 100$ kHz and (b) $Q = 0.7071$, $f_0 = 1000$ Hz, and $f_c = 50$ kHz.

3.26 (Sec. 3.5). Based on the RC prototype of Fig. P 3.20, design a high-pass backward component simulation SC filter for $f_h = 1$ kHz, $Q = 0.7071$, and (a) $f_c = 8$ kHz and (b) $f_c = 64$ kHz.

3.27 (Sec. 3.5). Using the filter of Fig. P3.3, design an SC backward (series) equivalent for the following design specifications: $f_0 = 10^3$ Hz, $Q = 5$, and (a) $f_c = 8$ kHz, and (b) $f_c = 32$ kHz. Discuss the feasibility of both filters.

3.28 (Sec. 3.6). Obtain the matrices $\mathbf{F}$, $\mathbf{D}_1$, $\mathbf{D}_2$, $\mathbf{E}_1$, $\mathbf{E}_2$, $\mathbf{G}_1$, $\mathbf{G}_2$, $\mathbf{A}_1$, and $\mathbf{B}_1$ for the building block shown in Fig. P3.28.

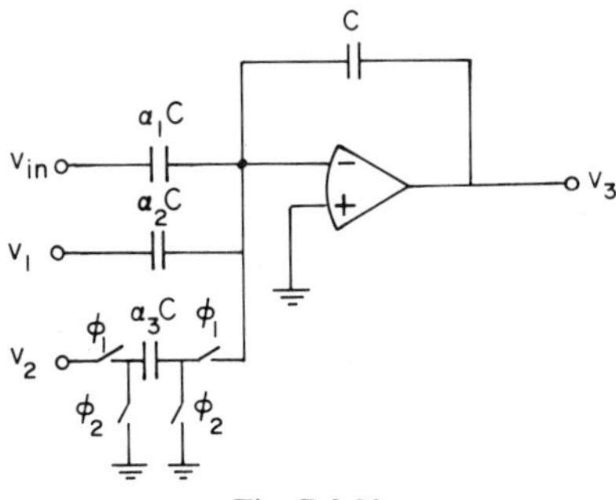

Fig. P 3.28.

4
Switched Capacitor LC Ladder Simulation

4.0 INTRODUCTION

This chapter introduces one of the most powerful and successful methods of synthesizing filters using switched capacitor networks. The method is based on the simulation of LC ladder networks. Negative feedback is used to simulate the voltage and current relations of a passive RLC ladder network. This method follows the path using blocks 1, 2, 3, 6, 7, 9, and 14 of Fig. 3.1-16. One form of this technique, which has become quite successful, is called the *leapfrog technique.*[1] It has resulted in useful and stable realizations and has the advantage of repeated use of nearly identical circuit components as building blocks of the realization. A further advantage of this method is that the sensitivity of the realization is the same as the original RLC network, which is typically very good.

The switched capacitor circuits developed in the previous two chapters will be used to achieve the passive ladder simulations of this chapter. The primary switched capacitor circuits that will be used include the integrator and the second-order realization. The filters considered in this chapter are constrained to the ladder topology, but include the low-pass, bandpass, high-pass, band-elimination, and general ladder networks. All these filters are capable of realizing $j\omega$ axis zeros. In addition to illustrating the method of realizing these filters with switched capacitor networks, methods of predistortion will be examined. This will permit the filter to more closely achieve its specifications, even though the clock frequency is not much larger than the filter passband frequencies.

4.1 GENERAL PRINCIPLES OF LC LADDER SYNTHESIS USING ACTIVE NETWORKS

The synthesis of LC ladder networks by active networks involves the use of active networks to simulate the branch voltage and current relationships

[1] F. E. J. Girling and E. F. Good, "Active Filters 12: The Leap-Frog or Active-Ladder Synthesis," *Wireless World,* Vol. 76, July 1970, pp. 341–345.

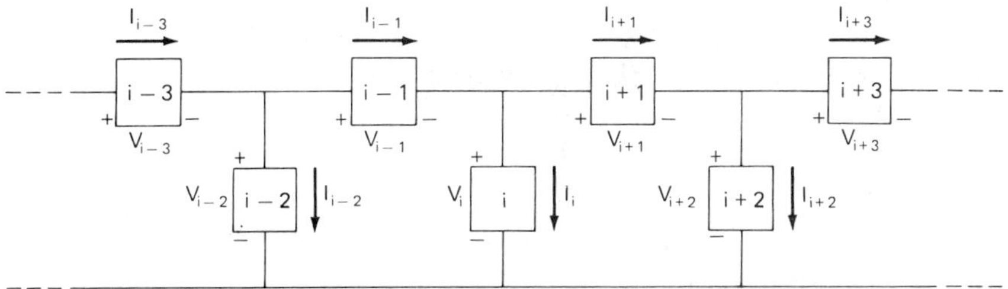

Fig. 4.1–1. The interior portion of a general LC ladder network.

of the LC ladder network. This method has been used very successfully in continuous active RC synthesis to generate low-sensitivity, high-order filter realizations. In this section we will develop the background necessary to use this synthesis approach with SC networks.

The interior portion of an LC ladder network is shown in Fig. 4.1–1. It is seen that the ladder network consists of series and shunt branches. In simple cases, the branch may consist of a single element, whereas, in the more complex cases, many elements may constitute the branch. One characteristic of the ladder RLC filter is that the interior branches consist of reactive elements only. Resistors are used only in the source branch or the load branch. It is necessary for the LC ladder to be terminated in a resistor, but it is not necessary for the LC ladder to have a finite source resistor. The methods to be developed can be used for either case.

In each branch of Fig. 4.1–1, the variables have been identified. Because the branch can be considered as a one-port or two-terminal network, the branch variables are voltage and current. The general method of ladder synthesis using active networks is to use the branch variables to generate equations that describe the ladder and then to use active networks to synthesize these equations. A problem exists, however, in that the number of variables required to characterize the LC ladder is less than the number of branch variables and in fact is typically half this number. It can be shown that the number of variables required to characterize an LC ladder network is equal to the number of series and shunt branches, providing there are no inductor cutsets or no capacitor loops.[2] This number of variables is called the *order of complexity* of a network.

The variables that characterize the network are called the *state variables.* The proper selection of the state variables from the branch variables of the

[2] N. Balabanian and T. A. Bickart, *Linear Network Theory—Analysis, Properties, Design and Synthesis,* Matrix Publishers, Inc., Beaverton, OR, 1981, p. 102.

LC ladder network is an important step in the methods presented in this chapter. In general, the state equations can be represented as

$$\frac{d}{dt}\mathbf{x} = f(\mathbf{x},\mathbf{u}) \tag{1}$$

where $\mathbf{x}$ is a vector of elements called the state variables, and $\mathbf{u}$ is a vector of sources. The advantage of the state variables as defined by Eq. (1) is that $\mathbf{x}$ can be found by integration of the matrix differential equation. This concept is particularly important to active network realizations of LC ladder networks, because the primary active network element is the integrator.

To be able to identify the appropriate state variables of the LC ladder network, let us consider a single branch of the LC ladder. The vector $\mathbf{x}$ is simply a single branch voltage or current variable, and the vector $\mathbf{u}$ is the other branch variable. If we assume the branch has only one reactive element, then the state equation for an inductor is

$$di/dt = (1/\mathrm{L})v \tag{2}$$

and for a capacitor is

$$dv/dt = (1/\mathrm{C})i \tag{3}$$

It is obvious that inductor currents and capacitor voltages are the proper state variables for the general form of Eq. (1). It is seen from Eqs. (2) and (3) that an inductor's current can be found by integrating its voltage and that a capacitor's voltage can be found by integrating its current.

If the branch contains multiple reactive elements, the same considerations hold. For example, consider a branch consisting of a series L and C. If we choose the voltage, v, across this branch as the state variable, then the state equation becomes

$$v = \mathrm{L}(di/dt) + (1/\mathrm{C})\int i\, dt \tag{4}$$

If instead, we choose the current, i, through the branch as the state variable, then the state equation becomes

$$d^2i/dt^2 = -(1/\mathrm{LC})i + (1/\mathrm{L})(dv/dt) \tag{5}$$

To solve for v of Eq. (4), it is necessary to both differentiate and integrate i. To solve for i of Eq. (5), it is necessary to integrate i twice and v once.

Therefore, to avoid differentiation, the state variable should be selected as the current of the series LC branch. It can be shown that the proper state variable for a branch consisting of an inductor in parallel with a capacitor is voltage (see Problem 4.3).

To demonstrate the method of simulating the LC ladder network, it is necessary to assume a set of state variables. For this purpose consider Fig. 4.1–2, where only the state variables are shown. If voltage is the state variable, then the branch is defined as an admittance branch, whereas, if the state variable is current, the branch is defined as an impedance branch. It can be shown that these definitions are consistent with the desired state equations.

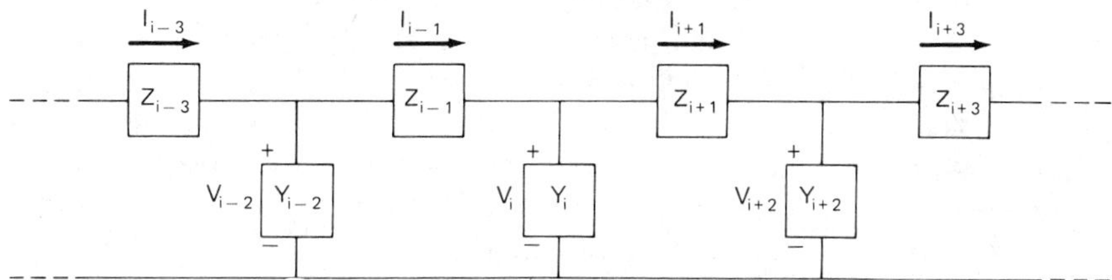

Fig. 4.1–2. The interior portion of an LC ladder network.

The following equations are written from Fig. 4.1–2 and express the state variable of one branch in terms of the state variables of its neighboring branches. These equations are

$$
\begin{aligned}
&\vdots \\
V_{i-2} &= (1/Y_{i-2})[I_{i-3} - I_{i-1}] \\
I_{i-1} &= (1/Z_{i-1})[V_{i-2} - V_i] \\
V_i &= (1/Y_i)[I_{i-1} - I_{i+1}] \\
I_{i+1} &= (1/Z_{i+1})[V_i - V_{i+2}] \\
V_{i+2} &= (1/Y_{i+2})[I_{i+1} - I_{i+3}] \\
&\vdots
\end{aligned}
\tag{6}
$$

It is useful at this point to introduce the definition of a voltage analog. This definition replaces the above currents by multiplying and dividing the above equations by an arbitrary resistance, R. When a current has been replaced by its voltage analog, the voltage analog will be designated with a prime. The advantage of using voltage analogs is that immittance functions become voltage transfer functions. The above equations can be rewritten as

$$
\begin{aligned}
&\vdots \\
V_{i-2} &= (1/RY_{i-2})[V'_{i-3} - V'_{i-1}] \\
V'_{i-1} &= (R/Z_{i-1})[V_{i-2} - V_i] \\
V_i &= (1/RY_i)[V'_{i-1} - V'_{i+1}] \\
V'_{i+1} &= (R/Z_{i+1})[V_i - V_{i+2}] \\
V_{i+2} &= (1/RY_{i+2})[V'_{i+1} - V'_{i+3}] \\
&\vdots
\end{aligned} \tag{7}
$$

where $V'_k = I_k R$.

In some of the literature regarding the synthesis of LC ladder networks by active networks,[3,4] writers have found it useful to introduce signal flow graphs to assist in the design of the filter. However, the concept and manipulations of the signal flow graphs can become a major effort and can mask the desired results. The approach taken here will be to write the appropriate state equations and to synthesize each equation directly using the switched capacitor circuits of the previous two chapters.

Next, the terminations of the LC ladder filter are considered. We shall assume that the filter will be voltage-driven, which leads to the two possible configurations shown in Fig. 4.1–3. In Fig. 4.1–3(a), the first complex element

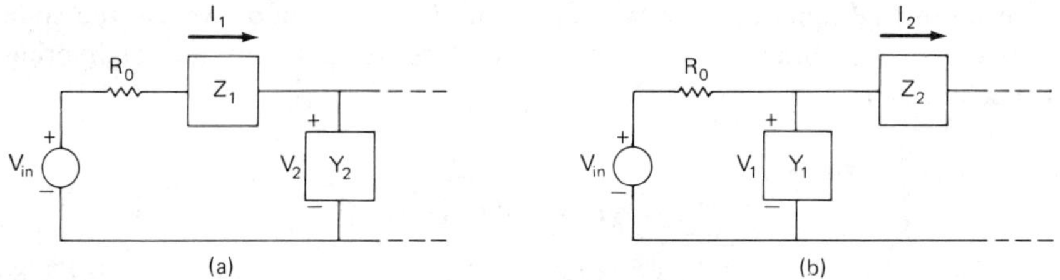

Fig. 4.1–3. Voltage driven source terminations of an LC ladder network. (a.) Left-most ladder element is a series impedance. (b.) Left-most ladder element is a shunt admittance.

of the LC ladder filter is considered to be a series impedance. In this case the state variable will be the current or the voltage analog of this current. The state equations that characterize the source part of this ladder network in Fig. 4.1–3(a) are

[3] A. S. Sedra and P. O Brackett, *Filter Theory and Design: Active and Passive,* Matrix Publishers, Inc., Portland, OR, 1978.

[4] L. P. Huelsman and P. E. Allen, *Introduction to the Theory and Design of Active Filters,* McGraw-Hill Book Co., New York, NY, 1980.

$$
\begin{aligned}
V_1' &= (R/Z_1)[V_{in} - (R_0/R)V_1' - V_2] \\
V_2 &= (1/RY_2)[V_1' - V_3'] \\
&\vdots
\end{aligned}
\tag{8}
$$

Similarily, the equivalent expressions that characterize the network of Fig. 4.1–3(b) are given as follows

$$
\begin{aligned}
V_1 &= (1/RY_1)[(R/R_0)(V_{in} - V_1) - V_2'] \\
V_2' &= (R/Z_2)[V_1 - V_3] \\
&\vdots
\end{aligned}
\tag{9}
$$

These expressions will be used to develop switched capacitor realizations of the source terminations of LC ladder networks when the Y's and Z's are specified.

Figure 4.1–4 shows two possible load terminations of an LC ladder network. The output voltage, V_{out}, of Fig. 4.1–4(a) could be expressed as

$$V_{out} = V_n = (R_n/R)V_{n-1}' \tag{10}$$

However, it is more economical, in terms of active devices, to treat Z_{n-1} and R_n as one element, which results in the following expression for V_{out}.

$$V_{out} = R_n[(V_{n-2} - V_{out})/Z_{n-1}] = (R/Z_{n-1})[(R_n/R)(V_{n-2} - V_{out})] \tag{11}$$

This expression is easily realized by solving for V_{out} that will typically lead to the expression for a damped integrator, which has been discussed previously.

The other possible load termination for the ladder is shown in Fig. 4.1–

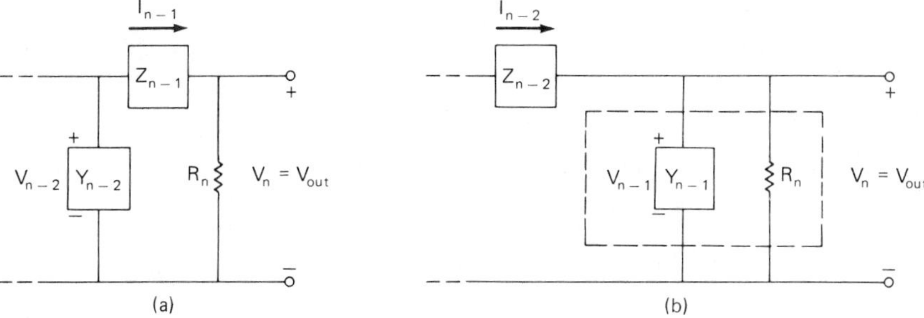

Fig. 4.1–4. Load terminations of an LC ladder network. (a.) Right-most ladder element is a series impedance. (b.) Right-most ladder element is a shunt admittance.

4(b). In this case, we shall consider R_n as part of the admittance, Y_{n-1}. Thus V_{out} can be expressed as

$$\begin{aligned} V_{\text{out}} &= (1/R_n Y_{n-1})[(R_n/R)V'_{n-2} - V_{\text{out}}] \\ &= (1/RY_{n-1})[V'_{n-2} - (R/R_n)V_{\text{out}}] \end{aligned} \tag{12}$$

This expression can be realized also by solving for V_{out} resulting in the equation of a damped integrator. In the high-pass and band-elimination filters, it will be necessary to modify these results to develop the desired state equations.

In LC ladder filters that realize finite zeros, it is found that inductor cutsets and capacitor loops can exist. Although it is not necessary that these conditions be removed, it permits the development of a general method of synthesizing LC ladder networks by active elements. Figure 4.1–5(a) shows the interior portion of an LC ladder network that has two loops of admittances. If these admittances were capacitors, as is the case in a low-pass elliptic filter, then a capacitor loop would be formed. We now show a method of removing

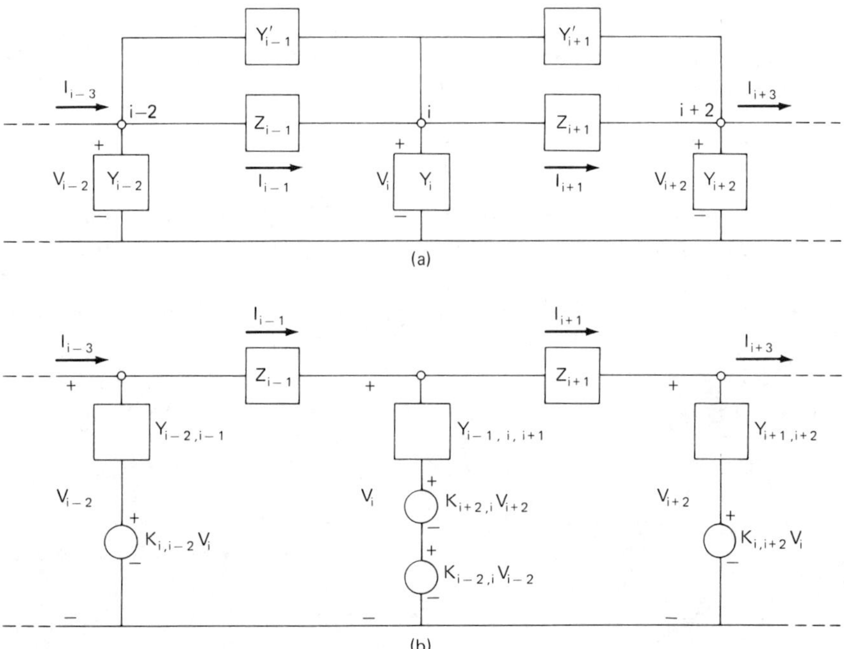

Fig. 4.1–5. (a.) Interior portion of an LC ladder network having shunt elements in the series branches. (b.) Equivalent circuit of (a.) using voltage-controlled, voltage-sources.

the bridging admittances, Y'_{i-1} and Y'_{i+1}, which will eliminate the admittance loop.

The method uses dependent sources to replace the bridging admittances in the LC ladder. The sum of the currents associated with node $i - 2$ can be written as

$$I_{i-3} - I_{i-1} = Y_{i-2}V_{i-2} + Y'_{i-1}[V_{i-2} - V_i] \tag{13}$$

This eqution can be solved for V_{i-2} to get

$$V_{i-2} = [Y'_{i-1}/(Y_{i-2} + Y'_{i-1})]V_i + (I_{i-3} - I_{i-1})/(Y_{i-2} + Y'_{i-1}) \tag{14}$$

or

$$V_{i-2} = K_{i,i-2}V_i + (I_{i-3} - I_{i-1})/Y_{i-2,i-1} \tag{15}$$

We see that the state variable of branch $i - 2$, V_{i-2}, is equal to the two terms of Eq. (15). The first term is proportional to the state variable in branch i, V_i, where the constant of proportionality is given as

$$K_{i,i-2} = Y'_{i-i}/(Y_{i-2} + Y'_{i-1}) \tag{16}$$

The second term is the relationship that is identical to that in Fig. 4.1–2, except that the shunt admittance, $Y_{i-2,i-1}$, is given as

$$Y_{i-2,i-1} = Y_{i-2} + Y'_{i-1} \tag{17}$$

The voltage analog form of Eq. (15) is given as

$$V_{i-2} = K_{i,i-2}V_i + (1/RY_{i-2,i-1})(V'_{i-3} - V'_{i-1}) \tag{18}$$

where the primed voltage variables are current analogs.

The sum of the currents associated with node i can also be written as

$$I_{i-1} - I_{i+1} = Y_iV_i - Y'_{i-1}(V_{i-2} - V_i) + Y'_{i+1}(V_i - V_{i+2}) \tag{19}$$

Solving for V_i gives

$$V_i = [Y'_{i-1}/(Y'_{i-1} + Y_i + Y'_{i+1})]V_{i-2} + [Y'_{i+1}/(Y'_{i-1} + Y_i + Y'_{i+1})]V_{i+2} + (I_{i+1} - I_{i-1})/(Y'_{i-1} + Y_i + Y'_{i+1}) \tag{20}$$

or

$$V_i = K_{i+2,i}V_{i-2} + K_{i+2,i}V_{i+2} + (I_{i-1} - I_{i+1})/(Y_{i-1,i,i+1}) \tag{21}$$

The voltage analog form of Eq. (21) is

$$V_i = K_{i-2,i}V_{i-2} + K_{i+2,i}V_{i+2} + (V'_{i-1} - V'_{i+1})/(RY_{i-1,i,i+1}) \tag{22}$$

The sum of the currents associated with the rightmost node, $i + 2$, can be written as

$$I_{i+1} - I_{i+3} = Y_{i+2}V_{i+2} - Y'_{i+1}(V_i - V_{i+2}) \tag{23}$$

Solving for V_{i+2} gives

$$V_{i+2} = [Y'_{i+1}/(Y'_{i+1} + Y_{i+2})]V_i + (I_{i+1} - I_{i+3})/(Y'_{i+1} + Y_{i+2}) \tag{24}$$

or

$$V_{i+2} = K_{i,i+2}V_i + (I_{i+1} - I_{i+3})/Y_{i+1,i+2} \tag{25}$$

The voltage analog form of Eq. (25) is

$$V_{i+2} = K_{i,i+2}V_i + (V'_{i+1} - V'_{i+3})/(RY_{i+1,i+2}) \tag{26}$$

The resulting circuit based on Eqs. (15), (21), and (25) is shown in Fig. 4.1–5(b). We see that the ladder of Fig. 4.1–5(a) has been reduced to the ladder of Fig. 4.1–2 with the addition of voltage-controlled voltage sources. Furthermore, each voltage source is controlled by a state variable.

Figure 4.1–6(a) shows the situation where an impedance cutset exists in the interior portion of an LC ladder network. If the impedances are inductors, then an inductance cutset exists. The impedances, Z'_{i-2}, Z'_i, and Z'_{i+2} can be eliminated in a manner similar to that of Fig. 4.1–5(a). The result is shown in Fig. 4.1–6(b), where current-controlled current sources are used (see Problem 4.5). Note that each current source is controlled by a state variable.

At this point we have developed the basic groundwork to synthesize passive LC ladder filters using active devices and specifically the switched capacitor circuits of the previous two chapters. The general approach to be used in the following four sections is summarized by the following steps.

1. Start from the desired LC low-pass prototype network that satisfies the filter specifications.

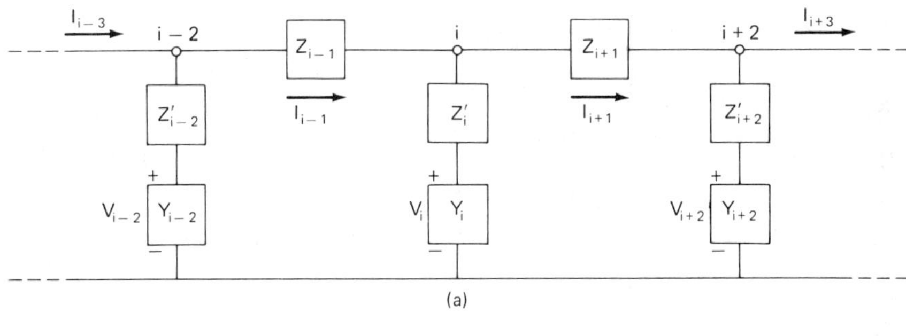

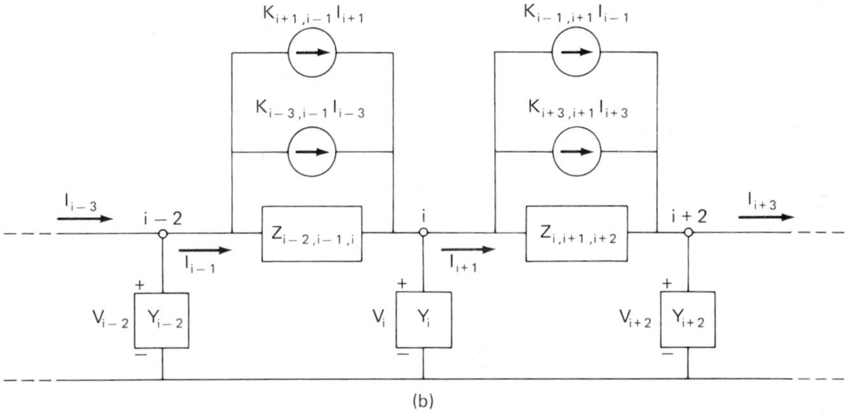

Fig. 4.1–6. (a.) Interior portion of an LC ladder network having series elements in the shunt branches. (b.) An equivalent circuit of (a.) using current-controlled, current-sources.

2. Eliminate all inductor cutsets or capacitor loops using the methods of Figs. 4.1–5 and 4.1–6.
3. Normalize and transform the network if the realization is not low-pass. Note that the voltage-controlled voltage sources and current-controlled current sources are unaffected by this step.
4. Select the state variables that will satisfy the state equation given in Eq. (1).
5. Solve for each state variable in terms of the other state variables and the output variable.
6. Synthesize each equation with the switched capacitor networks of the previous two chapters.

Table 4.1–1 illustrates these steps with respect to the type of filter that is to be realized. Each of these steps will be developed in more detail as they are discussed in the following sections. One can see that the results are very

Table 4.1–1 Illustration of the Application of the General Approach to Synthesizing LC Ladder Filters Using Switched Capacitor Circuits.

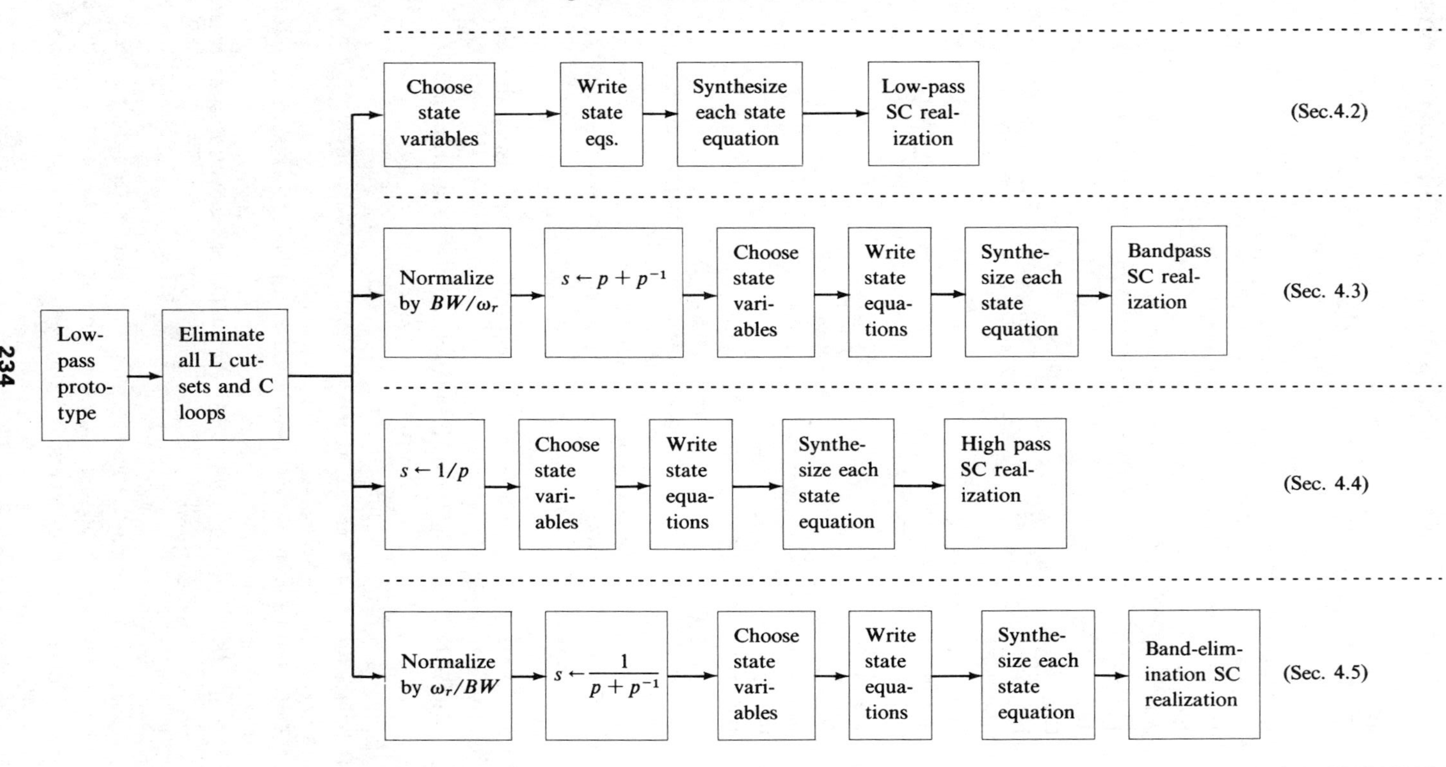

general and that the only limitation to the method is that it requires an RLC ladder realization as the starting point.

4.2 SWITCHED CAPACITOR LOW-PASS LADDER FILTERS

In this section we apply the ideas presented in the previous section to the realization of low-pass ladder filters using switched capacitor circuits. The type of switched capacitor networks that will be used in this section include amplifiers, integrators, and damped integrators. Two examples will be presented to illustrate the application of the principles of the previous section to the realization of switched capacitor filters. These examples are both fifth-order and include various types of terminations and filters having finite zeros on the imaginary frequency axis.

Figure 4.2–1 shows a fifth-order, doubly terminated, low-pass ladder filter. The components have been numbered sequentially from left to right. Note that the termination resistances are numbered distinctly from the closest reactance. This realization is called a *low-pass prototype realization* and can be obtained from the tables in Appendix A. The low-pass prototype will be assumed to be normalized to a cutoff frequency of 1 radian/second unless otherwise stated. The n subscript on the components indicates prototype values. The desired state variables are indicated on Fig. 4.2–1. They include I_1, the current through the series branch containing both R_{0n} and L_{1n}; V_2, the voltage across C_{2n}; I_3, the current through L_{3n}; V_4, the voltage across C_{4n}; and I_5, the current through L_{5n} (and R_{6n}). V_{in} is the input variable and is also indicated. Because the output variable of the filter is V_{out}, it is also indicated. If we know I_5, it is an easy matter to solve for V_{out}. In fact, we can use V_{out} in place of I_5 if we wish.

There being no inductor cutsets or capacitor loops, we proceed to write the state equations as follows. For the state variable, I_1, we may write

$$V_{\text{in}} - I_1(R_{0n} + sL_{1n}) - V_2 = 0 \tag{1}$$

Using current analogs and solving for V_1' ($V_1' = I_1 R$), we get

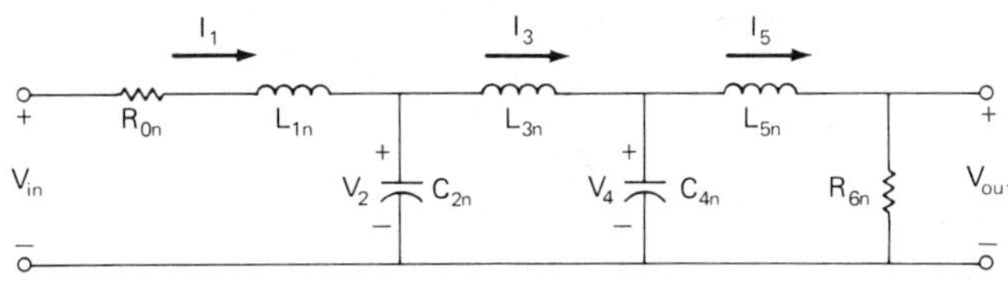

Fig. 4.2–1. A fifth-order, low-pass ladder filter.

$$V_1' = (R/sL_{1n})[V_{in} - V_2 - (R_{0n}/R)V_1'] \tag{2}$$

The fact that a state variable is in part equal to the integral of itself simply means that the integrator is damped, as we shall see. Moving to the state variable, V_2, we may write the equation that

$$I_1 - I_3 - sC_{2n}V_2 = 0 \tag{3}$$

or solving for V_2 and using voltage analogs, we get

$$V_2 = (1/sRC_{2n})[V_1' - V_3'] \tag{4}$$

The equation describing the state variable, I_3, can be written as

$$V_2 - sL_{3n}I_3 - V_4 = 0 \tag{5}$$

Using voltage analogs, we may solve for V_3' as

$$V_3' = (R/sL_{3n})[V_2 - V_4] \tag{6}$$

The equation describing the state variable, V_4, can be written as

$$I_3 - sC_{4n}V_4 - I_5 = 0 \tag{7}$$

Using voltage analogs, we may solve for V_4 as

$$V_4 = (1/sRC_{4n})[V_3' - V_5'] \tag{8}$$

Finally, the equation describing the state variable, I_5, can be written as

$$V_4 - sL_{5n}I_5 - I_5R_{6n} = 0 \tag{9}$$

which can be expressed in terms of voltage analogs as

$$V_5' = (R/sL_{5n})[V_4 - (R_{6n}/R)V_5'] \tag{10}$$

Because

$$V_{out} = I_5R_{6n} \tag{11}$$

we may replace the state variable, V_5', by

$$V_{out} = (R_{6n}/R)V'_5 \tag{12}$$

The method of generating these equations is obvious from the above relationships. Starting with Eq. (1), the equations were simply a loop equation followed by a node equation. The last equation, Eq. (11), was simply Ohm's law. The loop-node sequence can be used in practically all of the SC ladder realizations presented in this chapter.

The next step is the realization of five state variables, V'_1, V_2, V'_3, V_4, and V'_5 (or V_{out}). However, let us first consider the generalization of a procedure for realizing the state variables. As we examine the state variable equations of (2), (4), (6), (8), and (10), it is observed that the general form can be written as

$$V_j(s) = (1/s)[\pm A_i V_i(s) \pm A_j V_j(s) \pm A_k V_k(s)] \tag{13}$$

where $V_i(s)$, $V_j(s)$, and $V_k(s)$ are state variables, some of which may be voltage analogs of current that will be indicated by primes, and the As are constants that will depend on the components of the RLC ladder filter. From our study of SC integrators in Section 2.5, there are many choices we can make to realize Eq. (13). For the moment, let us consider the use of the Type I DDI integrators of Fig. 2.5–4(a) and Fig. 2.5–5. The choice between these two integrators will depend upon the sign of A.

A realization using the Type I DDI integrators of Eq. (13) is shown in Fig. 4.2–2, where we have arbitrarily assumed that A_i and A_k are positive and A_j is negative. It can be shown (see Problem 4.6) that the z-domain transfer function at ϕ_1 of Fig. 4.2–2 is

$$V_j(z) = [1/(z-1)][\alpha_{1j} V_i(z) - \alpha_{2j} V_j(z) + \alpha_{3j} V_k(z)] \tag{14}$$

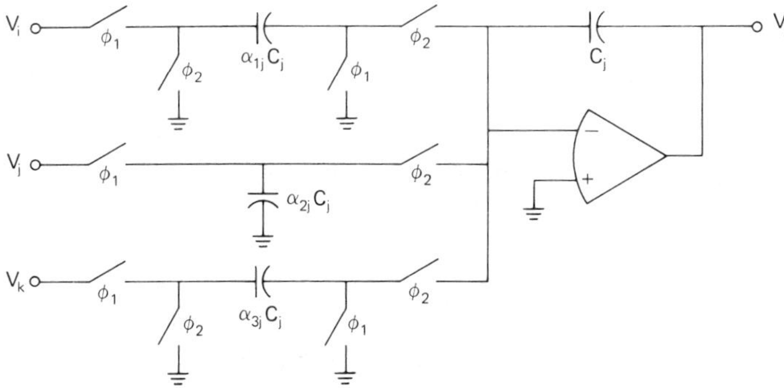

Fig. 4.2–2. General realization of state variable equation using Type I DDI switched capacitor integrators.

where the α_{lj} represents the ratio of the lth input capacitor to the feedback capacitor C_j, $l = 1,2,3$. Next, we make an important assumption that will be reconsidered in Sections 4.6 and 4.7. We assume that $\omega << \omega_c$, where ω is the highest frequency of interest in the filter passband, and ω_c is the clock frequency. With this assumption, we may replace z in Eq. (14) by $1 + sT$ to get

$$V_j(s) \cong (1/sT)[\alpha_{1j}V_i(s) - \alpha_{2j}V_j(s) + \alpha_{3j}V_k(s)] \tag{15}$$

Therefore, Fig. 4.2–2 becomes a realization of Eq. (13) with the assumed polarities of the A's.

The next step in the realization is to equate Eq. (13) with the assumed polarities to Eq. (15) in order to solve for the capacitor ratios, α. However, before this can be done it is necessary to normalize Eq. (15) because Eq. (13) is based on a frequency-normalized low-pass, prototype RLC ladder network. This normalization can be accomplished by replacing the clock period T by

$$T = T_n/\Omega_n \tag{16}$$

where Ω_n is defined as

$$\Omega_n = \frac{\omega_{co}}{\omega_n} = \frac{\text{actual cutoff frequency in rps}}{\text{normalized cutoff frequency in rps}} \tag{17}$$

Thus the normalized version of Eq. (15) can be written as

$$V_j(s) \cong (1/sT_n)[\alpha_{1j}V_i(s) - \alpha_{2j}V_j(s) + \alpha_{3j}V_k(s)] \tag{18}$$

where T_n is defined as

$$T_n = \Omega_n T \tag{19}$$

Now equating Eq. (13) to Eq. (18) results in equations of the form

$$\alpha_{lj} = A_l T_n = A_l T \Omega_n = A_l \Omega_n / f_c \tag{20}$$

where α_{lj} is the capacitor ratio of the lth input capacitor to the feedback capacitor C_j of the realization for the jth state variable. Equation (20) gives the general relationship that will be used in designing the capacitor ratios of the SC realizations of the ladder filters. In many cases, Eq. (18) will be simplified because of having fewer inputs or capacitor ratios that are equal.

Returning to the realization of the state variable equations of Fig. 4.2–1, we begin with Eq. (2). A realization of Eq. (2) using a Type I, DDI SC integrator is shown in Fig. 4.2–3(a). It is seen that, because the constant in Eq. (2) is the same for V_{in} and V_2, a single capacitor can be shared by

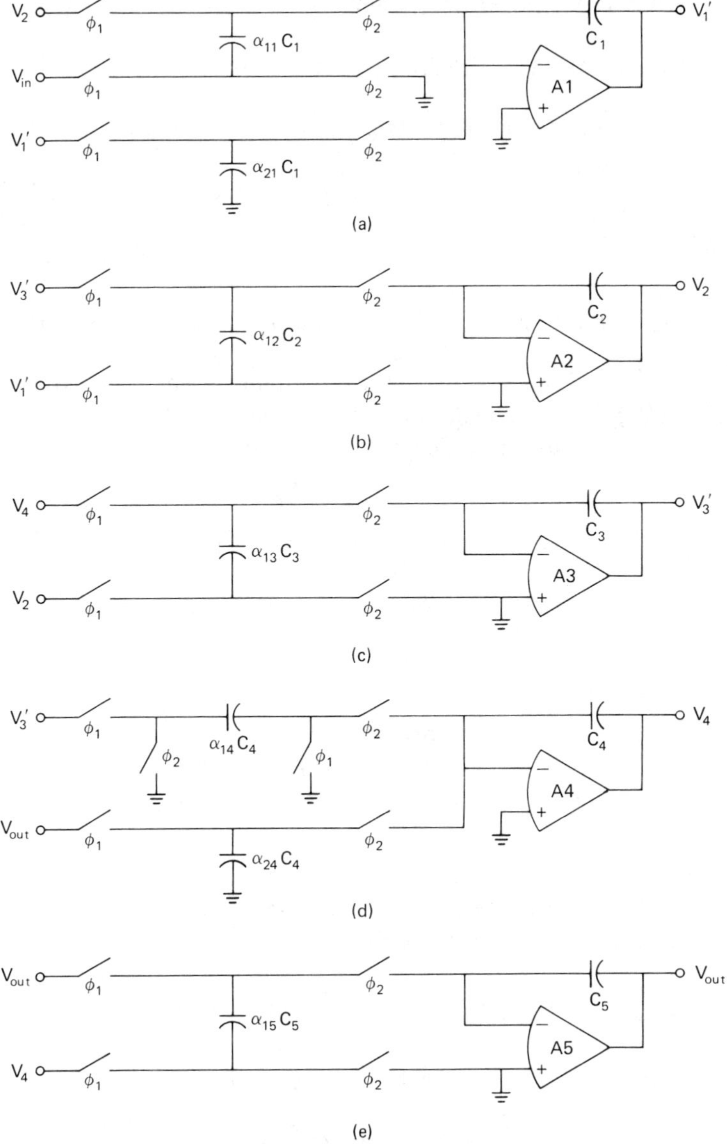

Fig. 4.2–3. Realizations for the state variables of Fig. 4.2–1.

these two inputs. It can be shown that the z-domain transfer function of Fig. 4.2–3(a) is

$$V_1'(z) = [1/(z-1)][\alpha_{11}(V_{in}(z) - V_2(z)) - \alpha_{21}V_1'(z)] \tag{21}$$

Assuming that $\omega << \omega_c$ and replacing z by $1 + sT_n$ gives

$$V_1'(s) \cong (1/sT_n)[\alpha_{11}(V_{in}(s) - V_2(s)) - \alpha_{21}V_1'(s)] \tag{22}$$

Equating Eq. (2) to Eq. (22) results in the capacitor ratios, which are

$$\alpha_{11} = (RT_n)/L_{1n} = (R\Omega_n T)/L_{1n} = (\mathrm{R}\Omega_n)/(f_c L_{1n}) \tag{23}$$

and

$$\alpha_{21} = (R_{on}T_n)/L_{1n} = (R_{on}\Omega_n T)/L_{1n} = (R_{on}\Omega_n)/(f_c L_{1n}) \tag{24}$$

Next, we consider that realization of the state variable, V_2, given by Eq. (4). Figure 4.2–3(b) is a realization of this state variable equation using a Type I, DDI SC integrator. The z-domain transfer function of this circuit can be found as

$$V_2(z) = [1/(z-1)][\alpha_{12}V_1'(z) - \alpha_{12}V_3'(z)] \tag{25}$$

Assuming that $\omega << \omega_c$ and replacing z by $1 + sT_n$ gives

$$V_2(s) \cong (1/sT_n)[\alpha_{12}V_1'(s) - \alpha_{12}V_3'(s)] \tag{26}$$

Equating Eq. (4) and Eq. (26) results in

$$\alpha_{12} = T_n/(RC_{2n}) = (\Omega_n T)/(RC_{2n}) = \Omega_n/(RC_{2n}f_c) \tag{27}$$

The next state variable equation is given in Eq. (6). Figure 4.2–3(c) can be shown to be a realization of this equation using a Type I, DDI SC integrator. The z-domain transfer function of the circuit can be found as

$$V_3'(z) = [\alpha_{13}/(z-1)][V_2(z) - V_4(z)] \tag{28}$$

Assuming that $\omega << \omega_c$ and replacing z by $1 + sT_n$ gives

$$V_3'(s) \cong (\alpha_{13}/sT_n)[V_2(s) - V_4(s)] \tag{29}$$

Equating Eq. (6) and Eq. (29) results in

$$\alpha_{13} = (RT_n)/L_{3n} = (R\Omega_n T)/L_{3n} = (R\Omega_n)/(f_c L_{3n}) \tag{30}$$

Although the state variable, V_4, has the same form as the state variable, V_2, it behooves us to treat it differently so that we can identify V_{out}. Substituting Eq. (12) into Eq. (8) results in

$$V_4 = (1/sRC_{4n})[V_3' - (R/R_{6n})V_{out}] \tag{31}$$

Using the variable V_{out} rather than V_5' will allow us to easily obtain V_{out} and save one op amp. Figure 4.2–3(d) is a realization of Eq. (13) using the Type I, DDI SC integrator. The z-domain transfer function of this circuit can be written as

$$V_4(z) = [1/(z-1)][\alpha_{14}V_3'(z) - \alpha_{24}V_{out}(z)] \tag{32}$$

Assuming $\omega << \omega_c$ and replacing z by $1 + sT_n$ gives

$$V_4(s) \cong (1/sT_n)[\alpha_{14}V_3'(s) - \alpha_{24}V_{out}(s)] \tag{33}$$

Equating Eqs. (8) and (33) results in

$$\alpha_{14} = T_n/(RC_{4n}) = (\Omega_n T/(RC_{4n}) = (\Omega_n T/(RC_{4n}f_c) \tag{34}$$

and

$$\alpha_{24} = T_n/(R_{6n}C_{4n}) = (\Omega_n T/R_{6n}C_{4n}) = \Omega_n/(R_{6n}C_{4n}f_c) \tag{35}$$

Eliminating the state variable, V_5', in Eq. (10) by the use of Eq. (12) results in

$$V_{out} = (R_{6n}/sL_{5n})[V_4 - V_{out}] \tag{36}$$

This realization is similar to the one for V_2 and is shown in Fig. 4.2–3(e). Using the above methods, we can find the capacitor ratio as

$$\alpha_{15} = R_{6n}T_n/L_{5n} = R_{6n}\Omega_n T/L_{5n} = R_{6n}\Omega_n/L_{5n}f_c \tag{37}$$

The SC realization of Fig. 4.2–1 is shown in Fig. 4.2–4 and simply connects all the blocks of Fig. 4.2–3. One notes that the result is an interconnection of integrator loops. In each loop there are two integrators. It is important

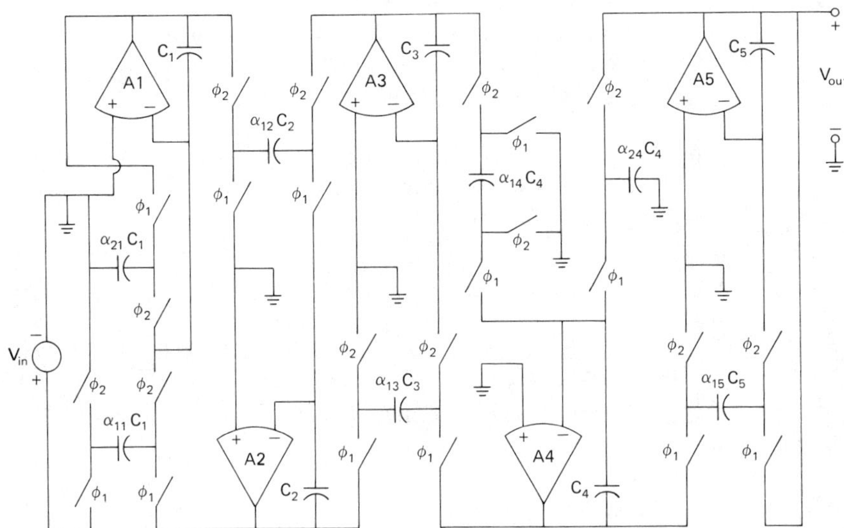

Fig. 4.2–4. Switched capacitor realization of Fig. 4.2–1 using Type I, DDI switched capacitor integrators.

that the switch phasing be designed for minimum phase around the loop. The output of each integrator should be sampled by the next integrator as soon as the new sample is available. Therefore, in one clock period the signal circulates around the internal feedback loop. This configuration is desirable in order to avoid a $T/2$ delay.[5] With the phasing as shown, the integrators become Type I LDI rather than Type I DDI. In this case, Table 2.5–2 shows that the phase shift of the Type I LDI integrator becomes ideal. This will become important when the highest frequency in the passband of the filter is not much less than the clock frequency. An example will illustrate the application of the above concepts.

Example 4.2–1. *An SC realization of a fifth-order, low-pass, Chebyshev Filter.* A fifth-order, low-pass, Chebyshev filter with a 1-dB ripple in the passband is to be designed for a cutoff frequency of 1000 Hz. The structure of Fig. 4.2–1 is to be used. Find the SC realization for the filter if the clock frequency is 100 kHz.

From Appendix A, we obtain the values of Fig. 4.2–1 as $R_{on} = 1$ ohm,

[5] G. M. Jacobs, D. J. Allstot, R. W. Brodersen, and P. R. Gray, "Design Techniques for MOS Switched Capacitor Ladder Filters," *IEEE Trans. on Circuits and Systems,* Vol. CAS-25, No. 12, December 1978, pp. 1014–1021. (This paper won the IEEE 1980 W. R. G. Baker Prize Award.)

$L_{1n} = 2.1349$ H, $C_{2n} = 1.0911$ F, $L_{3n} = 3.0009$ H, $C_{4n} = 1.0911$ F, $L_{5n} = 2.1349$ H, and $R_{6n} = 1$ ohm. Using Eqs. (23), (24), (27), (30), (34), (35), and (37) along with $\Omega_n = 2000\ \pi$, we get

$$\alpha_{11} = \alpha_{21} = 0.02943 \tag{38a}$$

$$\alpha_{12} = 0.05759 \tag{38b}$$

$$\alpha_{13} = 0.02094 \tag{38c}$$

$$\alpha_{14} = \alpha_{24} = 0.05759 \tag{38d}$$

$$\alpha_{15} = 0.02943 \tag{38e}$$

The resulting realization is shown in Fig. 4.2–5. The differences between Figs. 4.2–4 and 4.2–5 are due to the fact that $R_{0n} = R_{6n} = R$.

One item of concern in switched capacitor filters is the total amount of capacitance required in the realization. Large (or small) ratios of capacitors are particularily undesirable. In the above example we can find the total capacitance by the following method. If we assume that $\alpha_{lj}C_j$ is equal to a unit capacitance, C_{uj}, then C_j can be found by dividing C_{uj} by α_{lj}. In our case, we shall assume that $C_{u1} = C_{u2} = C_{u3} = C_{u4} = C_{u5} = C_u$. Therefore, the total capacitance in Example 4.2–1 is 155.596 C_u. If C_u is selected as

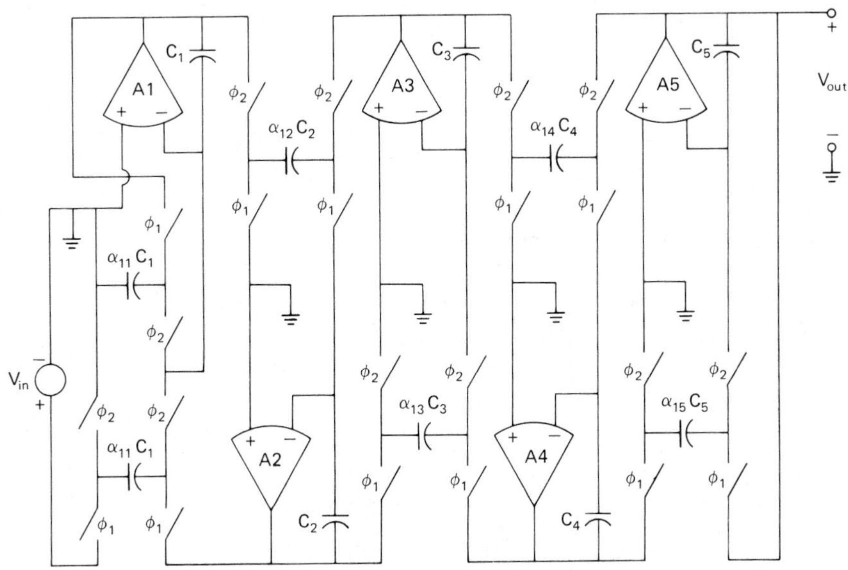

Fig. 4.2–5. Realization for Example 4.2–1.

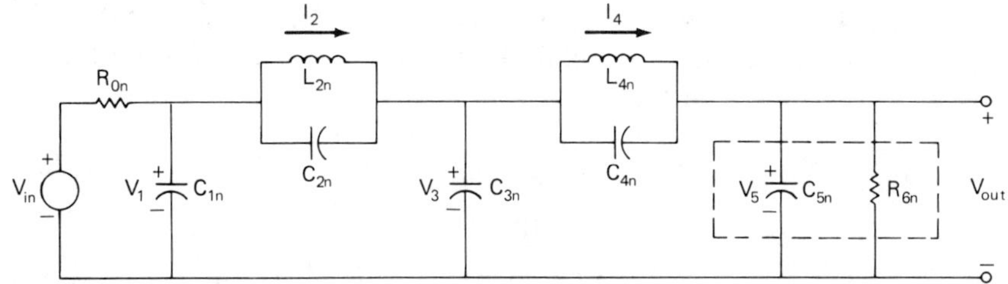

Fig. 4.2–6. A fifth-order, low-pass filter having $j\omega$ axis zeros.

1 pF, then a total of 155.596 pF of capacitance is required for the filter. It is important to keep this number as small as possible if the filter is to be integrated.

Let us consider next the realization of low-pass filters with finite $j\omega$ axis zeros. Figure 4.2–6 shows a fifth-order, low-pass filter that has two sets of complex conjugate zeros on the $j\omega$ axis. Applying the concepts of Section 4.1 to this filter results in the simplified equivalent filter of Fig. 4.2–7. The

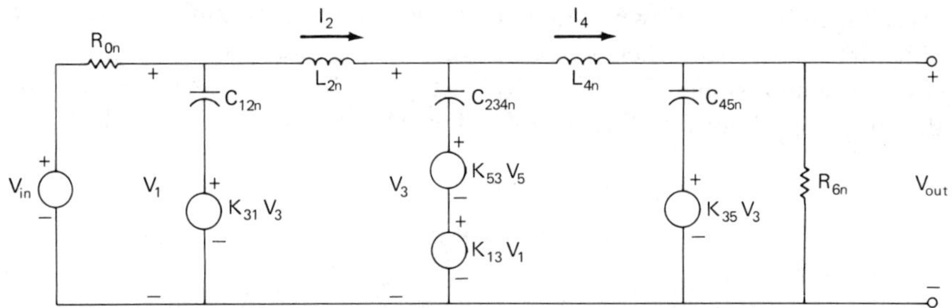

Fig. 4.2–7. An equivalent circuit for Fig. 4.2–6.

various elements of Fig. 4.2–7 can be related to Fig. 4.2–6 by the following expressions based on the results of Section 4.1.

$$C_{12n} = C_{1n} + C_{2n} \tag{39}$$

$$C_{234n} = C_{2n} + C_{3n} + C_{4n} \tag{40}$$

$$C_{45n} = C_{4n} + C_{5n} \tag{41}$$

$$K_{31} = C_{2n}/C_{12n} \tag{42}$$

$$K_{53} = C_{4n}/C_{234n} \tag{43}$$

$$K_{13} = C_{2n}/C_{234n} \tag{44}$$

and

$$K_{35} = C_{4n}/C_{45n} \tag{45}$$

Each of the state variables of Fig. 4.2–7 may be written as follows

$$V_1 = (1/sC_{12n})[(V_{in}/R_{0n}) - (V_1/R_{0n}) - (V_2'/R)] + K_{31}V_3 \tag{46}$$

$$V_2' = (R/sL_{2n})[V_1 - V_3] \tag{47}$$

$$V_3 = (1/sRC_{234n})[V_2' - V_4'] + K_{53}V_5 + K_{31}V_1 \tag{48}$$

$$V_4' = (R/sL_{4n})[V_3 - V_{out}] \tag{49}$$

and

$$V_{out} = V_5 = (1/sC_{45n})[(V_4'/R) - (V_{out}/R_{6n})] + K_{35}V_3 \tag{50}$$

In this case it is a simple matter to use the output voltage rather than V_5 as the fifth state variable. We note that the effect of the finite zeros has been to add an unintegrated term to the state equations.

For reasons that will shortly become evident, it is more useful to work with the negative of the output variable, V_3, designated as $\overline{V_3}$. The above equations are therefore rewritten as

$$V_1 = (1/sC_{12n})[(V_{in}/R_{0n}) - (V_1/R_{0n}) - (V_2'/R)] - K_{31}\overline{V_3} \tag{51}$$

$$V_2' = (R/sL_{2n})[V_1 + \overline{V_3}] \tag{52}$$

$$\overline{V_3} = (1/sRC_{234n})[V_4' - V_2'] - K_{53}V_5 - K_{31}V_1 \tag{53}$$

$$V_4' = (R/sL_{4n})[-\overline{V_3} - V_{out}] \tag{54}$$

and

$$V_{out} = V_5 = (1/sC_{45n})[(V_4'/R) - (V_{out}/R_{6n})] - K_{35}\overline{V_3} \tag{55}$$

where $\overline{V_3}$ is equal to $-V_3$.

The realization of the state equations of Eqs. (52) and (54) uses the methods previously developed, while the state equations of Eqs. (51), (53), and (55) will require the realization of nonintegrated terms. The realization of the nonintegrated terms can be easily accomplished by adding a summing capability with an unswitched capacitor. Because we have managed to get the signs of all the nonintegrated terms negative, this can be done using the SC realization of Fig. 4.2–8(a). The z-domain transfer function of Fig. 4.2–8(a) can be written as

$$V_1(z) = [1/(z-1)][-\alpha_{11}V_1(z) - \alpha_{21}V_2'(z) + \alpha_{31}V_{in}(z)] - \alpha_{41}\overline{V_3}(z) \tag{56}$$

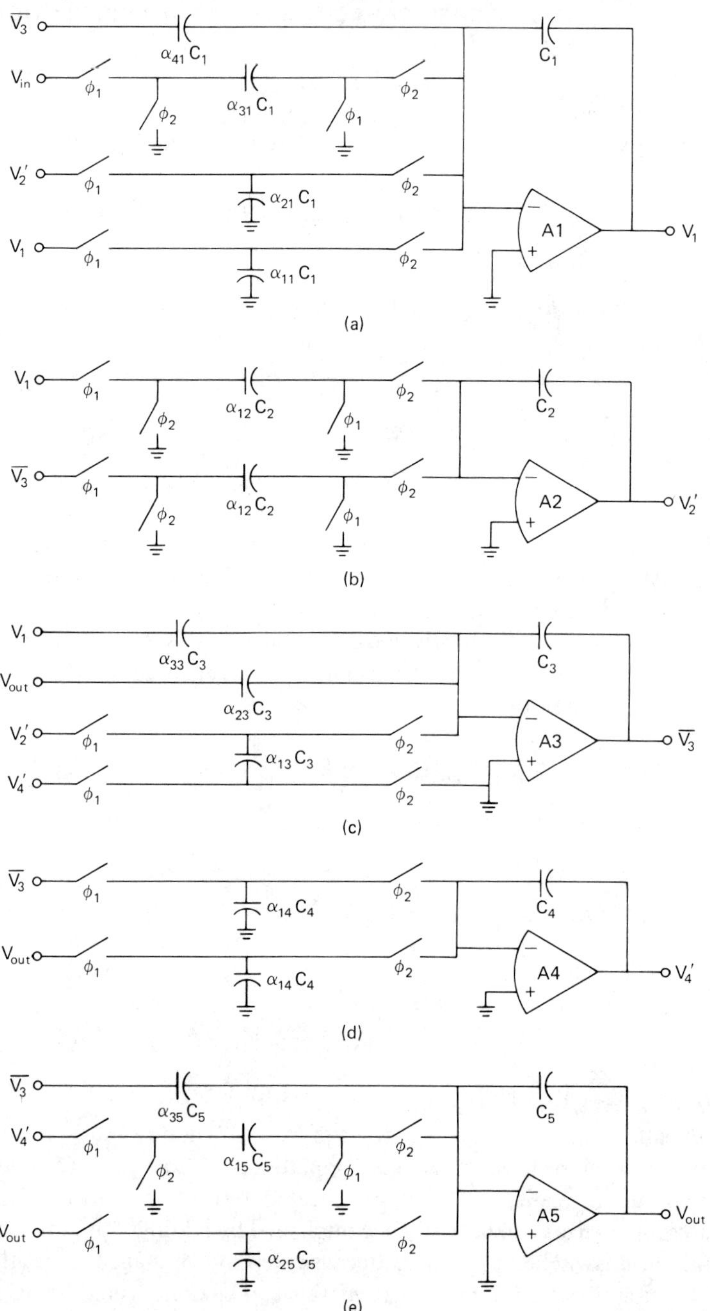

Fig. 4.2–8. Switched capacitor realizations of the filter of Fig. 4.2–7. (a.) Realization for V_1. (b.) Realization for V_2'. (c.) Realization for $\overline{V_3}$. (d.) Realization for V_4. (e.) Realization for V_{out}.

Assuming that $\omega << \omega_c$, so that z can be replaced by $1 + sT_n$, results in

$$V_1(s) \cong (1/sT_n)[-\alpha_{11}V_1(s) - \alpha_{21}V_2'(s) + \alpha_{31}V_{in}(s)] - \alpha_{41}\overline{V_3}(s) \quad (57)$$

Equating (51) to (57) gives the capacitor ratios

$$\alpha_{11} = \alpha_{31} = T_n/(C_{12n}R_{0n}) = (\Omega_n T)/(C_{12n}R_{0n}) = \Omega_n/(C_{12n}R_{0n}f_c) \quad (58)$$

$$\alpha_{21} = T_n/(C_{12n}R) = (\Omega_n T)/(C_{12n}R) = \Omega_n/(C_{12n}Rf_c) \quad (59)$$

and

$$\alpha_{41} = K_{31} \quad (60)$$

The next state variable is V_2' given by Eq. (52). Though this equation is similar to previous work, it is slightly different because the sign of V_3 has been changed. Figure 4.2–8(b) is a realization of this equation. The z-domain transfer function can be written as

$$V_2'(z) = [\alpha_{21}/(z-1)][V_1'(z) + \overline{V_3}(z)] \quad (61)$$

Assuming that $\omega << \omega_c$, so that z can be replaced by $1 + sT_n$, results in

$$V_2'(s) \cong (\alpha_{21}/sT_n)[V_1'(s) + \overline{V_3}(s)] \quad (62)$$

Equating (52) to (62) gives the capacitor ratio

$$\alpha_{21} = (T_nR/L_{2n}) = (\Omega_n TR/L_{2n}) = (\Omega_n R/L_{2n}f_c) \quad (63)$$

The third state variable is $\overline{V_3}$, expressed by Eq. (53). An SC realization for this equation is given in Fig. 4.2–8(c). The z-domain transfer function of this circuit is

$$\overline{V_3}(z) = [\alpha_{13}/(z-1)][V_4'(z) - V_2'(z)] - \alpha_{23}V_5(z) - \alpha_{33}V_1(z) \quad (64)$$

Assuming that $\omega << \omega_c$, so that z can be substituted by $1 + sT_n$, results in

$$\overline{V_3}(s) \cong (\alpha_{13}/sT_n)[V_4'(s) - V_2'(s)] - \alpha_{23}V_5(z) - \alpha_{33}V_1(z) \quad (65)$$

Equating (53) to (65) gives the following capacitor ratios

$$\alpha_{13} = T_n/(RC_{234n}) = (\Omega_n T/RC_{234n}) = \Omega_n/(RC_{234n} f_c) \tag{66}$$

$$\alpha_{23} = K_{53} \tag{67}$$

and

$$\alpha_{33} = K_{13} \tag{68}$$

Although the fourth state equation, Eq. (54), is similar to previous work, it is slightly different because of $\overline{V_3}$. Because both inputs to the integrator are negative, Fig. 4.2–8(d) is an SC realization of Eq. (54). The z-domain transfer function of this realization is

$$V_4'(z) = [-\alpha_{14}/(z-1)][\overline{V_3}(z) + V_{out}(z)] \tag{69}$$

Assuming once more that $\omega << \omega_c$, then z may be replaced by $1 + sT_n$ to get

$$V_4'(s) \cong (-\alpha_{14}/sT_n)[\overline{V_3}(s) + V_{out}(s)] \tag{70}$$

Equating (54) to (70) results in

$$\alpha_{14} = (RT_n/L_{4n}) = (R\Omega_n T/L_{4n}) = R\Omega_n/L_{4n} f_c) \tag{71}$$

The last state equation for Fig. 4.2–6 is given by Eq. (55). An SC realization for this equation is shown in Fig. 4.2–8(e). The z-domain transfer function of this circuit is

$$V_{out}(z) = [1/(z-1)][\alpha_{15} V_4'(z) - \alpha_{25} V_{out}(z)] - \alpha_{35} \overline{V_3}(z) \tag{72}$$

Assuming that $\omega << \omega_c$, then z may be replaced by $1 + sT_n$ to get

$$V_{out}(s) \cong (1/sT_n)[\alpha_{15} V_4'(s) - \alpha_{25} V_{out}(s)] - \alpha_{35} \overline{V_3}(s) \tag{73}$$

Equating (55) to (73) gives the following capacitor ratios

$$\alpha_{15} = T_n/(RC_{45}) = (\Omega_n T/RC_{45n}) = \Omega_n/(RC_{45n} f_c) \tag{74}$$

$$\alpha_{25} = T_n/(R_{6n} C_{45n}) = (\Omega_n T/R_{6n} C_{45n}) = \Omega_n/(R_{6n} C_{45n} f_c) \tag{75}$$

and

$$\alpha_{35} = K_{35} \tag{76}$$

Equations (58), (59), (60), (63), (66), (67), (68), (71), (74), (75), and (76) are the design equations for the filter of Fig. 4.2–6. Once each of the individual circuits has been realized, they can all be connected to form the SC realization of Fig. 4.2–6. Assuming that $R_{0n} = R_{6n} = R$ results in the realization of Fig. 4.2–9. The phasing of the switches in this realization has been selected so that minimum delay occurs around the integrator loops. An example will illustrate the application of the above concepts to the realization of a low-pass filter having finite $j\omega$ axis zeros.

Example 4.2–2. *SC realization of a fifth order, low-pass elliptic filter.* A low-pass elliptic filter is to be designed using the methods illustrated above. The realization must meet the specifications shown in Fig. 4.2–10(a), using the minimum number of op amps, and a clock frequency of 128 kHz.

Using the tables in Appendix A, we find that a fifth-order elliptic filter will satisfy the specification given in Fig.4.2–10(a). The resulting low-pass prototype filter is shown in Fig. 4.2–10(b). As this structure is identical to that in Fig. 4.2–6, we can use the above equations directly in this example. The first stage is designed using Eqs. (58) and (59), which give

$$\alpha_{11} = \alpha_{21} = \alpha_{31} = 0.1415 \tag{77}$$

Equation (60) gives

$$\alpha_{41} = 0.12834 \tag{78}$$

The second stage is designed from Eq. (63), which gives

$$\alpha_{12} = 0.13734 \tag{79}$$

The third stage is designed from Eqs. (66), (67), and (68), which give

$$\alpha_{13} = 0.07505 \tag{80}$$

$$\alpha_{23} = 0.19822 \tag{81}$$

and

$$\alpha_{33} = 0.06805 \tag{82}$$

The fourth stage is designed using Eq. (71), which gives

$$\alpha_{14} = 0.17845 \tag{83}$$

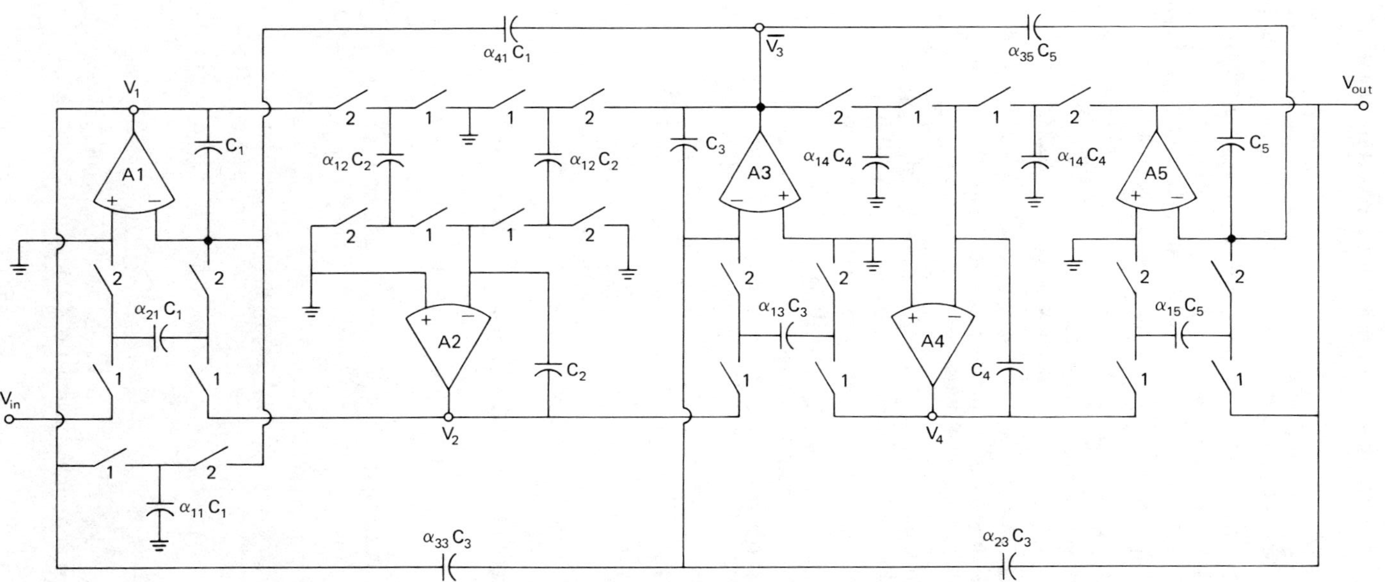

Fig. 4.2–9. Switched capacitor realization of Fig. 4.2–6 using Type I, DDI SC integrators. The switch phasing is indicated by 1 or 2 rather than by ϕ_1 or ϕ_2 for simplicity.

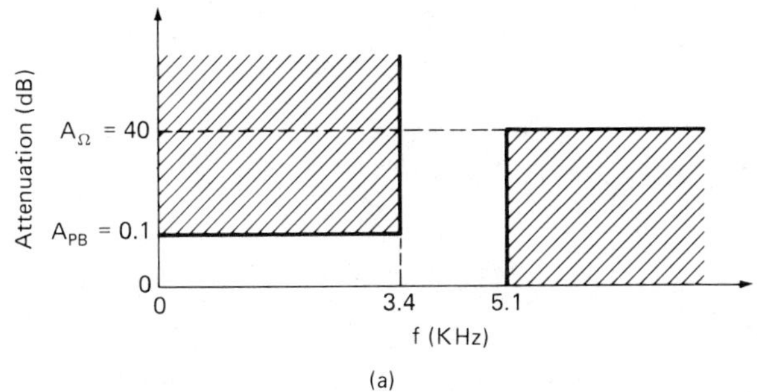

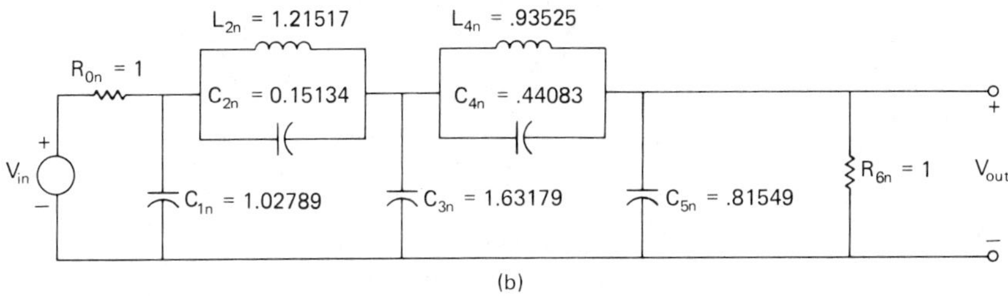

Fig. 4.2–10. (a.) Specifications for Ex. 4.2–2. (b.) Passive LC ladder realization of the specifications.

Finally the fifth stage is designed using (74), (75), and (76), which give

$$\alpha_{15} = \alpha_{25} = 0.13285 \tag{84}$$

and

$$\alpha_{35} = 0.35089 \tag{85}$$

Because $\alpha_{15} = \alpha_{25}$ and V_4' is positive and V_{out} is negative, then a single capacitor can be used to input both V_4' and V_{out}.

The resulting circuit is shown in Fig. 4.2–9. If we assume that $\alpha_{ej}C_j$ is equal to a unit capacitance, C_{uj}, then C_j can be found by dividing C_{uj} by α_{ej}. Using different values of C_u for each stage would provide the opportunity to scale the value of capacitors for each individual stage. For example, if all the capacitors of the ith stage were less than C_{ui}, it would be advantageous to scale C_{ui} so that these smaller capacitors would be equal to or greater than the smallest capacitor used in the filter realization. Figure 4.2–11 shows the magnitude of the response of the filter in this example using discrete components and with all $C_{uj} = 200$ pF.

This section has introduced the design of low-pass LC ladder filters using SC circuits. The result is a very practical structure with good performance

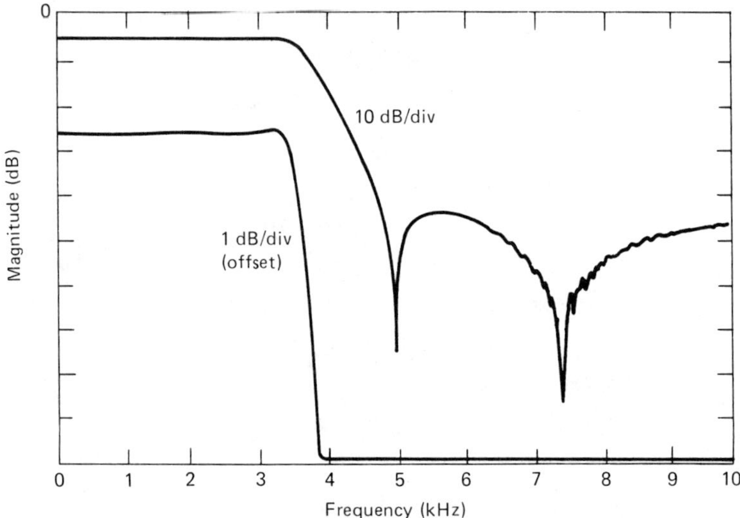

Fig. 4.2–11. Experimental response of Ex. 4.2–2.

and is suitable for fabrication by MOS technology. The capacitor ratios that determine the filter performance are among the smallest of those from any other filter method. It is to the designer's advantage to keep the capacitor ratios as small as possible to maintain accuracy (see Chapter 8) and to minimize the chip area. In Example 4.2–2, the largest-to-smallest capacitor ratio is $\alpha_{13} = 0.06805$, which is quite good for a filter of this complexity. The total capacitance is also of interest to the designer. Large capacitor ratios may be acceptable if the result is less total capacitance than other approaches. In the above example, the total capacitance in terms of a unit capacitance, C_u, is 57.4394 C_u, which again is very good for a fifth-order filter.

In considering the chip area, one must include the op amp. As a rule of thumb, an NMOS op amp is equivalent in area to approximately 50 to 60 pF of capacitance, and a CMOS op amp is equivalent to approximately 30 to 40 pF of capacitance. These values, of course, depend upon the various technologies. In the overall picture, one must consider not only the capacitance ratios and the area of the filter, but such other items as sensitivity, dynamic range, the effects of parasitics, clock feedthrough, and noise. These considerations will be examined in more detail in Chapter 8.

A very important consideration that will be postponed until later sections is the question of how good is the approximation that z can be replaced by $1 + sT_n$. As long as the filter passband frequency is much less than f_c, the approximation is good. The ratio of the clock frequency to the filter cutoff frequency was 100 in Example 4.2–1 and 37.6 in Example 4.2–2. If the clock frequencies are increased while everything else remains the same, the capacitor ratios must be increased. As a consequence, a compromise must be reached between the influence of the clock frequency and the capacitor ratios.

4.3 SWITCHED CAPACITOR BANDPASS LADDER FILTERS

In this section we extend the design method developed in the previous two sections to bandpass filters. Although the complexity will be twice that of a similar order low-pass filter, the ideas and concepts of design are identical. The basic building block becomes a differential input, second-order bandpass filter realization with infinite Q, rather than a differential integrator. The source and load terminations will lead to second-order bandpass realizations having a finite Q. The class of bandpass filters considered in this section has the property of geometrical symmetry about a given center frequency.

The methods of the previous two sections are easily extended through the use of the low-pass-to-bandpass transformation presented in Eq. (12) of Section 3.1. This transformation is given as

$$s = p + (\omega_r^2/p) \tag{1}$$

The influence of this transformation on the magnitude response and the normalized passive components was illustrated in Table 3.1–3. Normally, the cutoff frequency of the low-pass circuit, ω_{Tlp}, is 1 rps. If we define the bandwidth of the bandpass circuit as ω_{Tlp}, then we may write that

$$BW = \omega_{Tlp} = \omega_{T2} - \omega_{T1} \tag{2}$$

The center frequency of the bandpass realization is ω_r, and is related to ω_{T2} and ω_{T1} as

$$\omega_r = \sqrt{\omega_{T2}\omega_{T1}} \tag{3}$$

The transformation of Eq. (1) and the resulting relationship of Eq. (3) constrains the filters of this section to possess geometrical symmetry about ω_r.

The following procedure is used to achieve a specified BW and ω_r starting from the normalized low-pass prototype filter. First, frequency-normalize (see Table 3.1–2) the low-pass prototype from a filter with a cutoff frequency of 1 rps by a factor of

$$\Omega_b = \omega_r / BW \tag{4}$$

Using Eq. (4) in the normalization of $s = \Omega_b p$ causes the cutoff frequency of the low-pass, prototype to be less than 1 rps because ω_r is normally greater than BW. Next, apply the low-pass-to-bandpass transformation with $\omega_r = 1$ rps. The resulting filter will be called the bandpass prototype and will have a center frequency of 1 rps. The filter is then designed using the methods that will be described in this section. The final realization is achieved by simply frequency-denormalizing from the 1 rps center frequency to the desired center frequency. Tables 3.1–2 and 3.1–3 will be very useful in this section.

Let us demonstrate how to achieve a switched capacitor realization of a bandpass filter having all its zeros at the origin or infinity. Figure 4.3–1(a) shows a fourth-order, doubly terminated, low-pass prototype filter. Because there are no inductor cutsets or capacitor loops, we may proceed to the third step of the general approach outlined at the end of Section 4.1. In this step we first normalize all elements by Eq. (4), resulting in Fig. 4.3–1(b). In this step we use the specified values of ω_r and BW. Next, we apply Eq. (1) with $\omega_r = 1$ rps to each element to achieve the normalized, bandpass prototype of Fig. 4.3–1(c) (see Table 3.1–3). To simplify our notation we shall make the definition shown in Fig. 4.3–1(d), where the subscript B stands for the normalized bandpass prototype with $\omega_r = 1$ rps.

To determine the correct state variables for each branch, we examine the immittance functions of Fig. 4.3–2. The impedance of Fig. 4.3–2(a) is

$$Z_i(s) = \frac{V_i(s)}{I_i(s)} = \frac{s^2 + [1/(L_{Bi}C_{Bi})]}{s/L_{Bi}} \tag{5}$$

and

$$Y_j(s) = \frac{I_j(s)}{V_j(s)} = \frac{s^2 + [1/(L_{Bj}C_{Bj})]}{s/C_{Bj}} \tag{6}$$

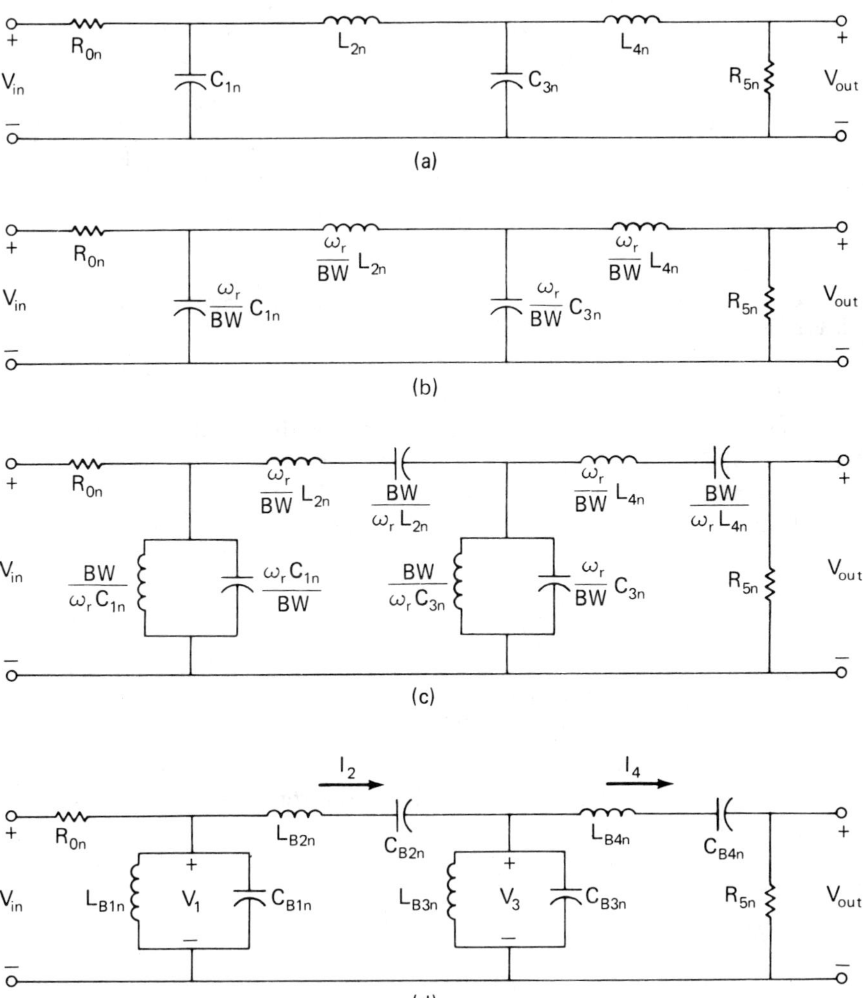

Fig. 4.3–1. (a.) Fourth-order, doubly terminated low-pass prototype filter. (b.) Frequency normalization of the low-pass prototype. (c.) Normalized, bandpass prototype filter. (d.) Simplification of notation of the bandpass prototype.

From Section 4.1, we saw that the proper state variable for the series LC branch is current, and for the parallel LC branch it is voltage. The desired state variables have been identified on Fig. 4.3–1(d).

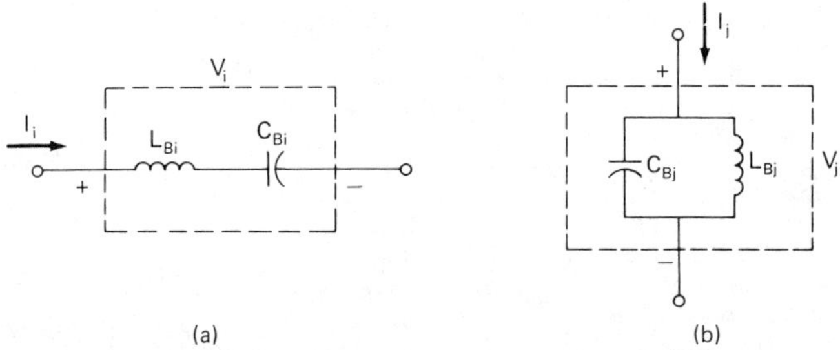

Fig. 4.3–2. (a.) An impedance branch of the bandpass prototype filter. (b.) An admittance branch of the bandpass prototype filter.

We may now proceed to write the expressions that give each of the state variables in Fig. 4.3–1(d) in terms of the others (or the input). Note that we anticipate using V_{out} as a state variable rather than I_4. From Fig. 4.3–1(d), Eq. (9) of Section 4.1, and Eq. (6) above, we write

$$V_1 = \frac{s/(RC_{B1n})}{s^2 + [1/(L_{B1n}C_{B1n})]}[(R/R_{0n})(V_{\text{in}} - V_1) - V_2'] \tag{7}$$

$$= \frac{s/(RC_{B1n})}{s^2+1}[(R/R_{0n})(V_{\text{in}} - V_1) - V_2']$$

As all LC products are equal to unity, the interior state variables are expressed as

$$V_2' = \frac{sR/L_{B2n}}{s^2+1}[V_1 - V_3] \tag{8}$$

and

$$V_3 = \frac{s/(RC_{B3n})}{s^2+1}[V_2' - V_4'] = \frac{s/(RC_{B3n})}{s^2+1}[V_2' - (R/R_{5n})V_{\text{out}}] \tag{9}$$

From Fig. 4.1–3(a), Eq. (11) of Section 4.1, and Eq. (5) above, we can write the output state variable V_{out} as

$$V_{\text{out}} = \frac{sR_{5n}/L_{B4n}}{s^2+1}[V_3 - V_{\text{out}}] \tag{10}$$

In Eqs. (7) and (10), the state variables are functions of themselves. This is due to the source and load terminations and implies that the Q of the second-order bandpass filter will be finite.

The last step in the synthesis procedure of Section 4.1 is to realize Eqs. (7) through (10) with switched capacitor realizations. It can be seen that a second-order, bandpass realization is required. In addition, the realization must have the capability of differential input and infinite Q. Such realizations were discussed in the previous chapter. Let us consider Fig. 4.3–3(a), which is a differential input version of Fig. 3.3–7(b). We have arbitrarily interchanged the position of the switched capacitors in the loop, so that the integrator

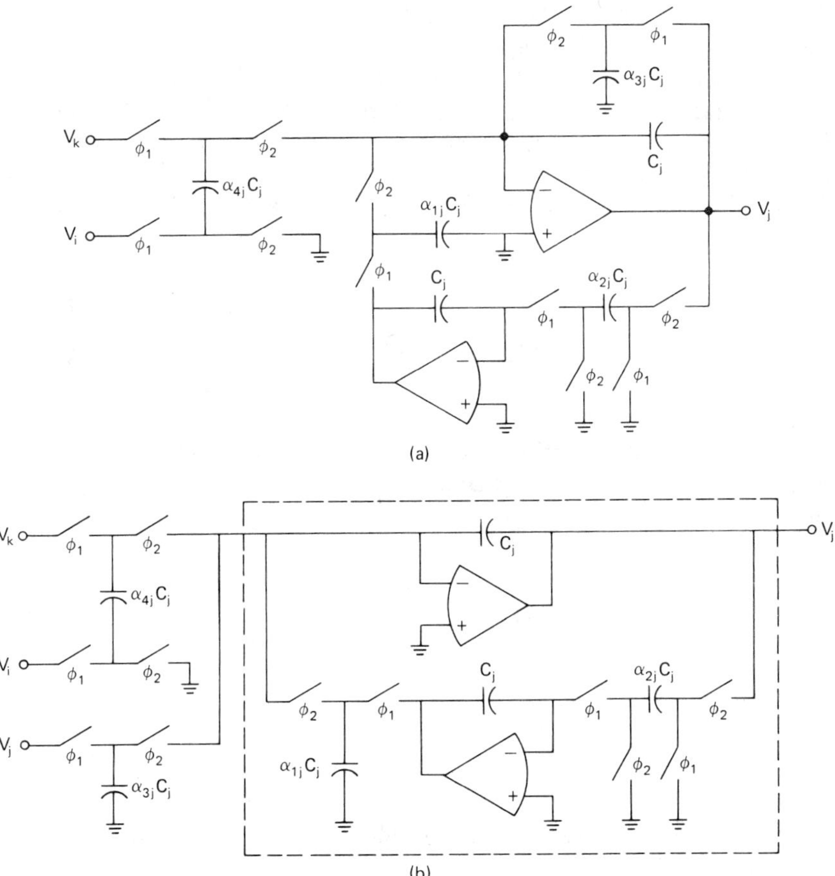

Fig. 4.3–3. (a.) A second-order, differential-input, switched capacitor bandpass realization. (b.) Redrawing of (a.) to emphasize the block format.

between V_{BP} and V_{LP} of Fig. 3.3–7(b) now becomes noninverting instead of inverting. Either configuration may be used with no difference in the following. The z-domain transfer function of Fig. 4.3–3(a) was given in Eq. (48) of Section 3.3. If we assume that the filter frequency, f_r, is much less than the clock frequency, f_c, then z^{-1} can be approximated by $1 - sT$. Problem 4.16 shows that Eq. (48) of Section 3.3 simplifies to

$$V_j(s) \cong \frac{(\alpha_{4j}/T)s}{s^2 + (\alpha_{3j}/T)s + (\alpha_{1j}\alpha_{2j}/T^2)}[V_i(s) - V_k(s)] \tag{11}$$

where α_{ij} is the ratio of the ith capacitor of the jth stage to the integrating capacitors, C_j, of the jth stage.

The synthesis of the state variables can be easily accomplished by redrawing Fig. 4.3–3(a) as shown in Fig. 4.3–3(b). Here we have disconnected the damping switched capacitor around the first integrator and are treating it as another input. Again, assuming that z^{-1} can be approximated by $1 - sT$, we find the approximate s-domain transfer function for Fig. 4.3–3(b) as

$$V_j(s) \cong \frac{s}{s^2 + (\alpha_{1j}\alpha_{2j}/T^2)}\left[-\frac{\alpha_{3j}}{T}V_j(s) + \frac{\alpha_{4j}}{T}V_i(s) - \frac{\alpha_{4j}}{T}V_k(s)\right] \tag{12}$$

This result is easily obtained from Eq. (11) or from Eq. (46) of Section 3.3.

Before Eq. (12) can be equated to the output variable expressions, the clock period, T, must be normalized. Using the normalization of Eq. (16) in Section 4.2 results in

$$V_j(s) \cong \frac{s}{s^2 + (\alpha_{1j}\alpha_{2j}/T_n^2)}\left[-\frac{\alpha_{3j}}{T_n}V_j(s) + \frac{\alpha_{4j}}{T_n}V_i(s) - \frac{\alpha_{4j}}{T_n}V_k(s)\right] \tag{13}$$

where T_n is given by

$$T_n = \Omega_n T \tag{14}$$

In this case we redefine Ω_n slightly different from Eq. (17) of Section 4.2 as

$$\Omega_n = \frac{\omega_r}{1 \text{ rps}} = \frac{\text{actual center frequency in rps}}{1 \text{ rps}} \tag{15}$$

We also note that if $R \neq R_{0n}$ in Eq. (7) that the V_k and V_i inputs of Fig. 4.3–3 cannot be applied differentially to the same capacitor, but must be applied to individual capacitors.

Proceeding as before, the state variables of Eqs. (7) through (10) are realized next. If $R = R_{0n}$ (see Problem 4.17 for the case where $R_{0n} \neq R$), then equating Eq. (7) to Eq. (13) gives

$$\alpha_{11} = \alpha_{21} = T_n = \Omega_n T = \omega_r / f_c \tag{16a}$$

$$\alpha_{31} = (BW\,T)/RC_{1n} \tag{16b}$$

and

$$\alpha_{41} = (BW\Omega_n T)/\omega_r R_{0n} C_{1n} = (BW\,T)/(R_{0n} C_{1n}) \tag{16c}$$

The second stage is designed by equating Eq. (8) to Eq. (13), resulting in

$$\alpha_{12} = \alpha_{22} = \Omega_n T \tag{17a}$$

$$\alpha_{32} = 0 \tag{17b}$$

and

$$\alpha_{42} = (R\,BW\,T)/L_{2n} \tag{17c}$$

Similarly, the third stage is designed by equating Eq. (9) to Eq. (13) to get

$$\alpha_{13} = \alpha_{23} = \Omega_n T \tag{18a}$$

$$\alpha_{33} = (BWT)/(R_{5n} C_{3n}) \tag{18b}$$

$$\alpha_{43} = (BW\,T)/(RC_{3n}) \tag{18c}$$

Lastly, the fourth stage is designed by equating Eq. (10) to Eq. (13) giving

$$\alpha_{14} = \alpha_{24} = \Omega_n T \tag{19a}$$

$$\alpha_{34} = 0 \tag{19b}$$

and

$$\alpha_{44} = (BW\,R_{5n}\,T)/L_{4n} \tag{19c}$$

The realization of Fig. 4.3–1(a) is shown in Fig. 4.3–4. Note that if $R = R_{5n} = R_{0n}$ that considerable simplification results in the first and last stages.

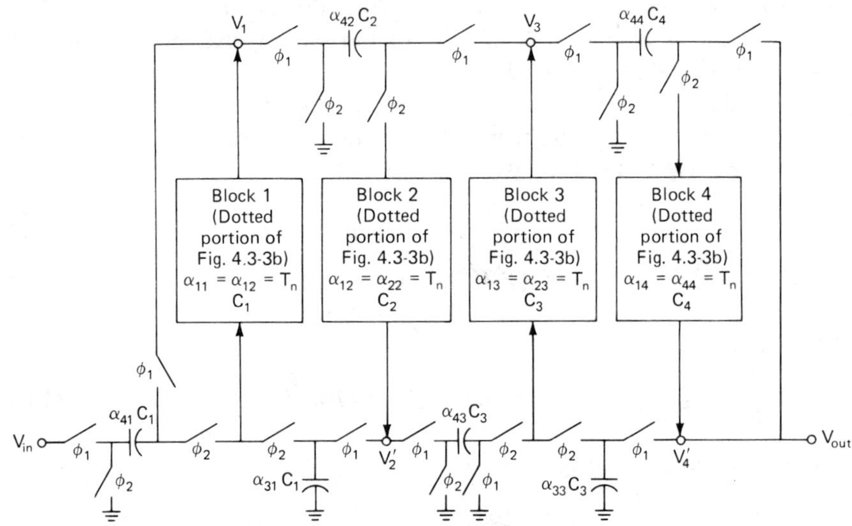

Fig. 4.3–4. A SC realization of the filter of Fig. 4.3–1(d).

Each of the blocks is identical and equivalent to the dotted box in Fig. 4.3–3(b), where Eqs. (16a), (17a), (18a), and (19a) give the values.

Example 4.3–1. *Design of a fourth-order bandpass filter.* Use the realization of Fig. 4.3–4 to obtain an SC realization of a fourth-order (eight-pole), Butterworth bandpass filter having a center frequency of 3000 Hz, a bandwidth of 600 Hz, and a clock frequency of 128 kHz. The low-pass prototype obtained from Table 3.1–3 and Fig. A 4(a) is shown in Fig. 4.3–5(a), with the element number corresponding to the presentation in this chapter rather than that used in Fig. A4(a). Choosing $R = 1$ ohm gives, from Eqs. (16), (17), (18), and (19), the following values

Stage 1:

$$\alpha_{11} = \alpha_{21} = 0.1743$$
$$\alpha_{31} = \alpha_{41} = 0.03848$$

Stage 2:

$$\alpha_{12} = \alpha_{22} = 0.1743$$
$$\alpha_{32} = 0$$
$$\alpha_{42} = 0.01594$$

Stage 3:

$$\alpha_{13} = \alpha_{23} = 0.1743$$
$$\alpha_{33} = 0$$
$$\alpha_{43} = 0.01594$$

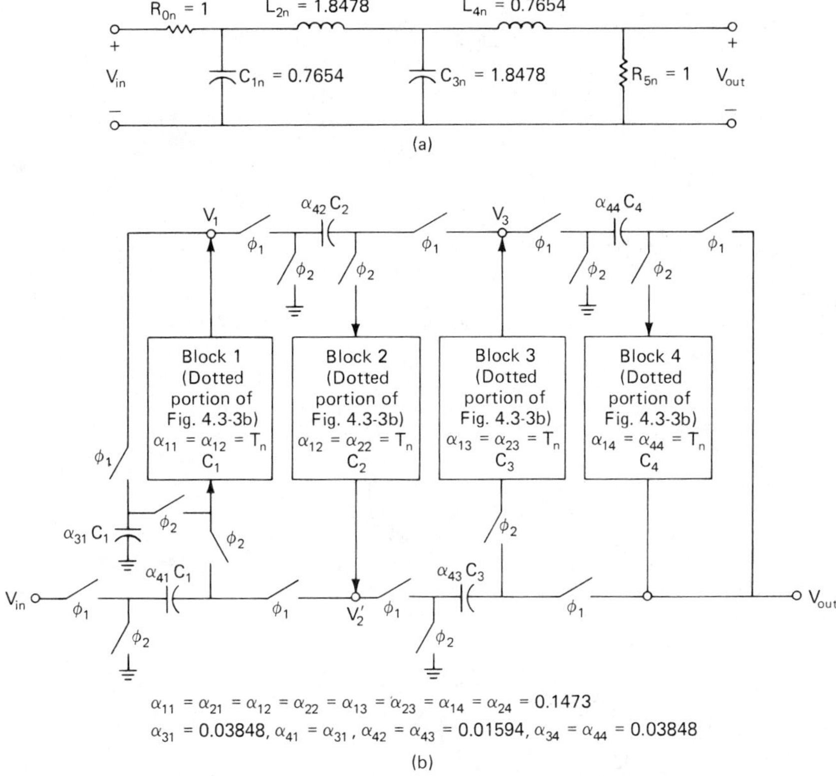

Fig. 4.3–5. (a.) Low-pass prototype for Ex. 4.3–1. (b.) Realization of Ex. 4.3–1.

Stage 4:

$$\alpha_{14} = \alpha_{24} = 0.1743$$
$$\alpha_{34} = \alpha_{44} = 0.03848$$

The largest capacitor ratio is seen to be 0.01594 or 62.7353, which is large in part due to the fact that f_r is 3000 Hz, and f_c is 128 kHz. The realization of Fig. 4.3–5(a) is shown in Fig. 4.3–5(b).

Let us next consider the realization of bandpass filters that have finite $j\omega$ axis zeros. Figure 4.2–6 of the previous section illustrated a fifth-order, low-pass filter that had four finite, $j\omega$ axis zeros. The next step in the design procedure was to eliminate the capacitor loops resulting in Fig. 4.2–7. To obtain the normalized, prototype bandpass circuit, we normalize Fig. 4.2–7 by Eq. (4) to get Fig. 4.3–6(a). The next step is to use the low-pass-to-bandpass

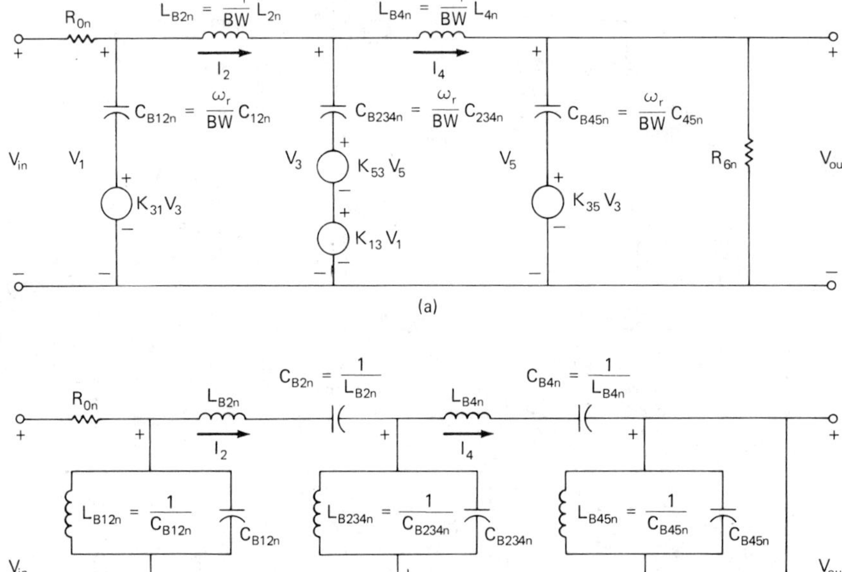

Fig. 4.3–6. (a.) Normalization of Fig. 4.2–9 by ω_r/BW. (b.) Bandpass circuit resulting from the low-pass to bandpass transformation applied to Fig. 4.3–6(a.).

transformation of Eq. (1) with ω_r equal to unity. The resulting circuit is shown in Fig. 4.3–6(b). It is interesting to note that the voltage-controlled voltage sources are independent of the low-pass-to-bandpass transformation. It can be demonstrated (see Problem 4.23) that Fig. 4.3–6(b) results if one first applies the low-pass-to-bandpass transformation to Fig. 4.2–6 and then removes the branches in parallel with the series branches, similar to what was done in going from Fig. 4.1–5(a) to Fig. 4.1–5(b).

Each of the state variables of Fig. 4.3–6(b) can be written as

$$V_1 = \frac{s/(RC_{B12n})}{s^2+1}\left[(R/R_{0n})(V_{\text{in}} - V_1) - V_2'\right] + K_{31}V_3 \tag{20a}$$

$$V_2' = \frac{sR/(L_{B2n})}{s^2+1}\left[V_1 - V_3\right] \tag{20b}$$

$$V_2 = \frac{s/(RC_{B\,234n})}{s^2+1}[V_2' - V_4'] + K_{53}V_5 + K_{13}V_1 \tag{20c}$$

$$V_4' = \frac{sR/(L_{B\,4n})}{s^2+1}[V_3 - V_{out}] \tag{20d}$$

and

$$V_{out} = V_5 = \frac{s/(RC_{B\,45n})}{s^2+1}[V_4' - (R/R_{6n})V_{out}] + K_{35}V_3 \tag{20e}$$

As in Section 4.2, we choose the negative value of V_3 to simplify the realization. Thus, Eq. (20) becomes

$$V_1 = \frac{s/(RC_{B\,12n})}{s^2+1}[(R/R_{0n})V_{in} - (R/R_{0n})V_1 - V_2'] - K_{31}\overline{V_3} \tag{21a}$$

$$V_2' = \frac{sR/(L_{B\,2n})}{s^2+1}[V_1 + \overline{V_3}] \tag{21b}$$

$$\overline{V_3} = \frac{s/(RC_{B\,234n})}{s^2+1}[V_4' - V_2'] - K_{53}V_5 - K_{13}V_1 \tag{21c}$$

$$V_4' = \frac{sR/(L_{B\,4n})}{s^2+1}[-\overline{V_3} - V_{out}] \tag{21d}$$

and

$$V_{out} = V_5 = \frac{s/(RC_{B\,45n})}{s^2+1}[V_4' - (R/R_{6n})V_{out}] - K_{35}\overline{V_3} \tag{21e}$$

We note that three of the state equations are of the form $ks/(s^2+1)$ plus a constant. Unfortunately, the second-order structure of Fig. 4.3–3 cannot realize this form. One possible solution is shown in Fig. 4.3–7(a). The approximate s-domain transfer function of Fig. 4.3–7(a) can be found as (see Prob. 4.24)

$$V_{out}(s) \cong \frac{\alpha_{41}sT_n}{(sT_n)^2 + \alpha_{11}\alpha_{21}}V_1(s) - \alpha_{31}\left[\frac{(sT_n)^2 + (\alpha_{11}\alpha_{51}/\alpha_{31})}{(sT_n)^2 + \alpha_{11}\alpha_{21}}\right]V_2(s) \tag{22}$$

If we let

$$\alpha_{51} = \alpha_{21}\alpha_{31} \tag{23}$$

then eq. (22) simplifies to

$$V_{out}(s) = \frac{s(\alpha_{41}/T_n)}{s^2 + (\alpha_{11}\alpha_{21}/T_n^2)} V_1(s) - \alpha_{31} V_2(s) \tag{24}$$

which provides the desired realization of the bandpass, second-order term plus a constant. Fig. 4.3–7(b) will be used to represent the dotted portion of Fig. 4.3–7(a) and is also equivalent to the dotted portion of Fig. 4.3–3(b) with the addition of the second input indicated by the circled 2.

We now proceed to realize each of the above expressions using Fig. 4.3–7 as the basic block. Figure 4.3–8(a) shows a realization for Eq. (21a). The approximate s-domain transfer function of Fig. 4.3–8(a) can be found as

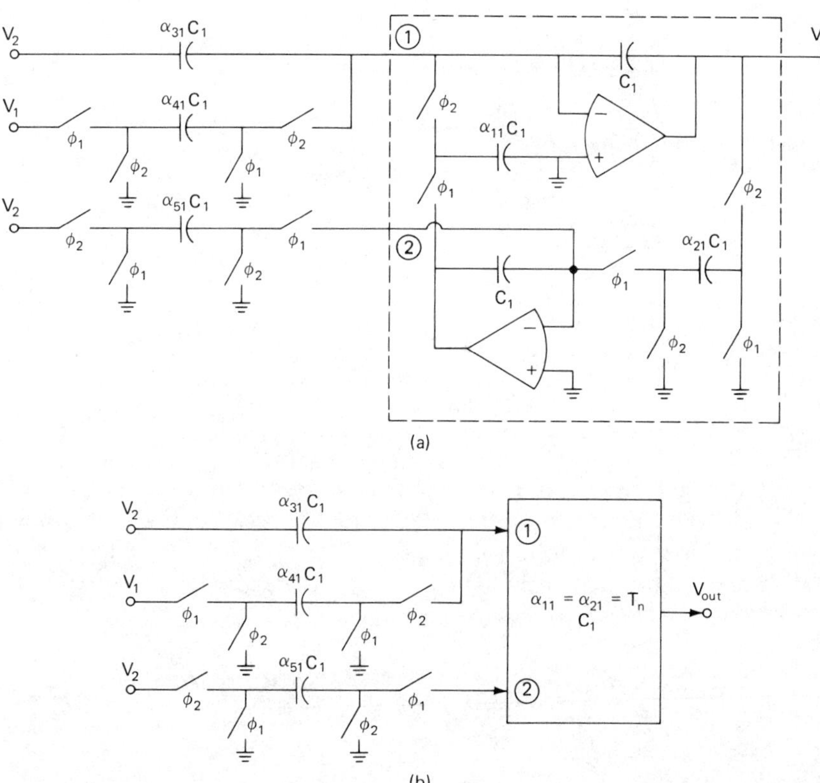

Fig. 4.3–7. (a.) Realization of a bandpass transfer function plus a constant. (b.) Simplified equivalent circuit of (a.).

$$V_1(s) \cong \frac{s}{s^2 + (\alpha_{11}\alpha_{21}/T_n^2)} \left[-\frac{\alpha_{31}}{T_n} V_1 - \frac{\alpha_{41}}{T_n} V_2' + \frac{\alpha_{51}}{T_n} V_{in} \right] - \alpha_{61} \overline{V_3} \tag{25}$$

Equating Eq. (21a) with Eq. (25) gives

$$\alpha_{11} = \alpha_{21} = T_n = \Omega_n T \tag{26a}$$

$$\alpha_{31} = \frac{\Omega_n T}{R_{0n} C_{B\,12n}} = \frac{BW\,T}{R_{0n} C_{12n}} = \frac{BW\,T}{R_{0n}(C_{1n} + C_{2n})} \tag{26b}$$

$$\alpha_{41} = \frac{\Omega_n T}{R C_{B\,12n}} = \frac{BW\,T}{R C_{12n}} = \frac{BW\,T}{R(C_{1n} + C_{2n})} \tag{26c}$$

and

$$\frac{\alpha_{61}}{\alpha_{21}} = K_{31} = C_{2n}/(C_{1n} + C_{2n}) = \alpha_{51} \tag{26d}$$

Figure 4.3–8(b) is a realization for Eq. (21b). The approximate s-domain transfer function can be found as

$$V_2'(s) \cong \frac{s}{s^2 + (\alpha_{12}\alpha_{22}/T_n^2)} \left[\frac{\alpha_{32}}{T_n} V_1 + \frac{\alpha_{42}}{T_n} \overline{V_3} \right] \tag{27}$$

Equating Eq. (21b) and Eq. (27) results in

$$\alpha_{12} = \alpha_{22} = T_n = \Omega_n T \tag{28a}$$

and

$$\alpha_{32} = \alpha_{42} = (R\,BW\,T)/(L_{2n}) \tag{28b}$$

Figure 4.3–8(c) is a realization of Eq. (21c). The approximate s-domain transfer function can be found as

$$\overline{V_3(s)} \cong \frac{s}{s^2 + (\alpha_{13}\alpha_{23}/T_n^2)} \left[\frac{\alpha_{33}}{T_n} (V_4' - V_2') \right] - \alpha_{43} V_5 - \alpha_{53} V_1 \tag{29}$$

Equating Eq. (21c) to Eq. (29) gives

$$\alpha_{13} = \alpha_{23} = T_n = \Omega_n T \tag{30a}$$

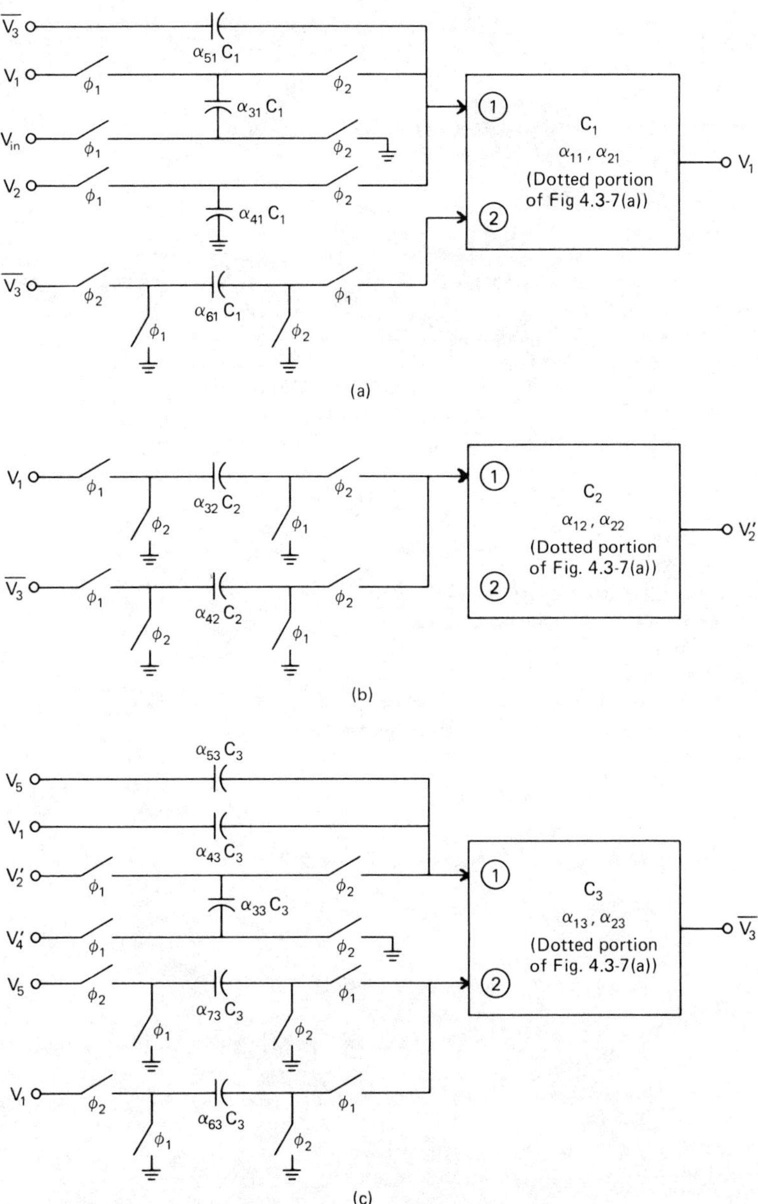

Fig. 4.3–8. SC realizations for the individual stages of Fig. 4.3–6(b). (a.) through (e.) are for stages 1 through 5 respectively.

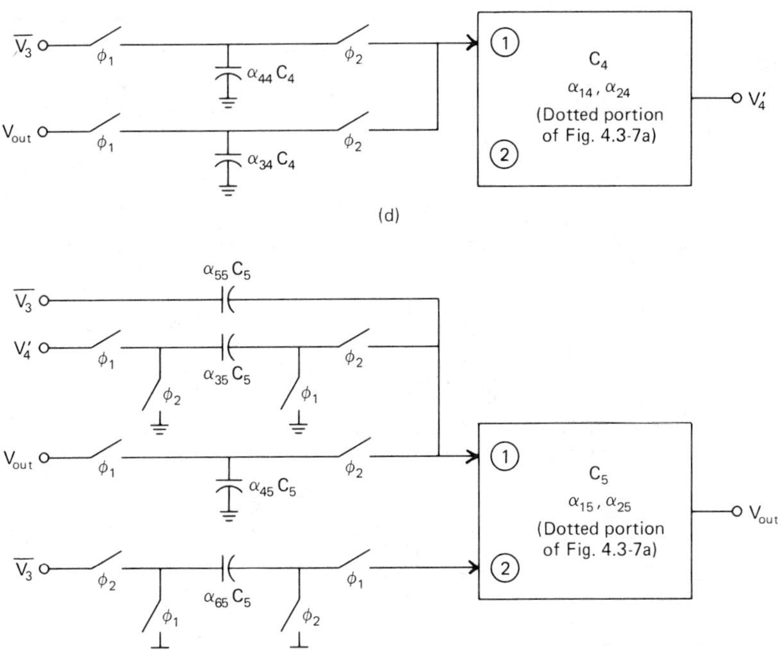

Fig. 4.3–8. (*Continued*)

$$\alpha_{33} = \frac{\Omega_n T}{RC_{B234n}} = \frac{BW\,T}{RC_{234n}} = \frac{BW\,T}{R(C_{2n} + C_{3n} + C_{4n})} \tag{30b}$$

$$\alpha_{43} = K_{13} = C_{2n}/(C_{2n} + C_{3n} + C_{4n}) \tag{30c}$$

$$\alpha_{53} = K_{53} = C_{4n}/(C_{2n} + C_{3n} + C_{4n}) \tag{30d}$$

$$\alpha_{63} = \alpha_{23}\alpha_{43} = K_{13}\Omega_n T \tag{30e}$$

and

$$\alpha_{73} = \alpha_{23}\alpha_{53} = K_{53}\Omega_n T \tag{30f}$$

Figure 4.3–8(d) is a realization of Eq. (21d). The approximate s-domain transfer function is

$$V_4'(s) = \frac{s}{s^2 + (\alpha_{14}\alpha_{24}/T_n^2)}\left[-\alpha_{44}\overline{V_3} - \alpha_{34}V_{out}\right] \tag{31}$$

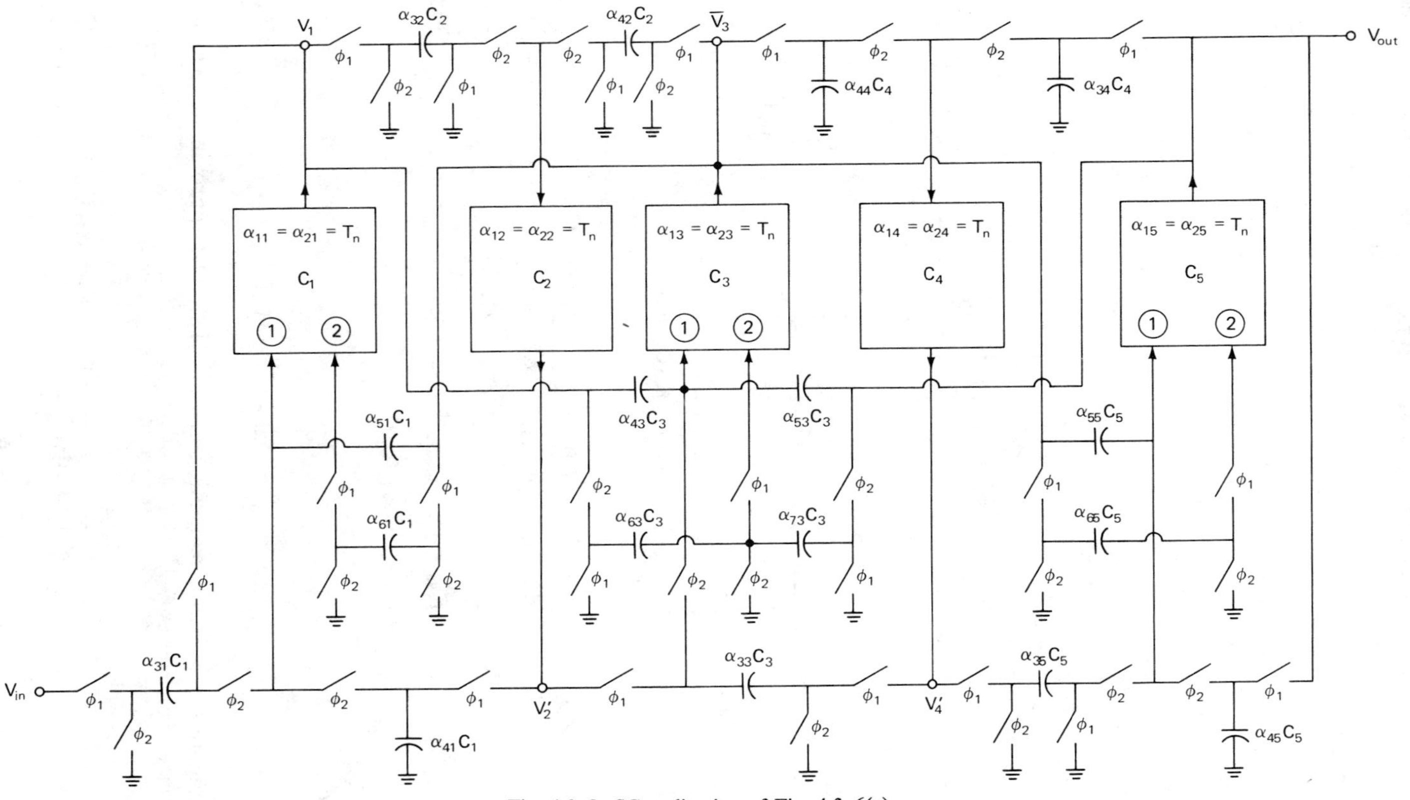

Fig. 4.3–9. SC realization of Fig. 4.3–6(a).

Equating Eq. (21d) to Eq. (31) gives

$$\alpha_{14} = \alpha_{24} = T_n = \Omega_n T \tag{32a}$$

and

$$\alpha_{34} = \alpha_{44} = (R\ BW\ T)/L_{4n} \tag{32b}$$

Finally, Fig. 4.3–8(e) is a realization for Eq. (21e). The approximate s-domain transfer function of this circuit is

$$V_{\text{out}} = V_5 \cong \frac{s}{s^2 + (\alpha_{15}\alpha_{25}/T_n^2)} \left[\frac{\alpha_{35}}{T_n} V_4' - \frac{\alpha_{45}}{T_n} V_{\text{out}} \right] - \alpha_{55} \overline{V_3} \tag{33}$$

Equating Eq. (21e) to Eq. (33) results in

$$\alpha_{15} = \alpha_{25} = T_n = \Omega_n T \tag{34a}$$

$$\alpha_{35} = \frac{\Omega_n T}{RC_{B\,45n}} = \frac{BW\,T}{RC_{45n}} = \frac{BW\,T}{R(C_{4n} + C_{5n})} \tag{34b}$$

$$\alpha_{45} = \frac{\Omega_n T}{R_{6n} C_{B\,45n}} = \frac{BW\,T}{R_{6n} C_{45n}} = \frac{BW\,T}{R_{6n}(C_{4n} + C_{5n})} \tag{34c}$$

$$\alpha_{55} = K_{35} = C_{4n}/(C_{4n} + C_{5n}) \tag{34d}$$

and

$$\alpha_{25}\alpha_{55} = K_{35}\Omega_n T \tag{34e}$$

The similarities between the low-pass filter with $j\omega$ axis zeros in Section 4.2 and the above bandpass filter with $j\omega$ axis zeros should be noted. In fact, if the dotted box of each of the circuits in Fig. 4.3–8 were replaced by an integrator the circuits in Fig. 4.2–8 would result. Figure 4.3–9 shows the realization of Fig. 4.3–6(a) assuming $R_{0n} = R_{6n} = R$. The phasing of the switches has been selected so that minimum loop delay results.

Example 4.3–2. *SC realization of a fifth-order, bandpass, elliptic filter.* A band-pass elliptic filter is to be designed using the methods illustrated above. The realization must meet the specifications shown in Fig. 4.3–10, using a minimum number of op amps. The clock frequency is 128 kHz.

We see from the specifications that Example 4.2–2 provides the initial steps in the design procedure. Using Fig. 4.2–10(b) and the fact that $\Omega_b =$

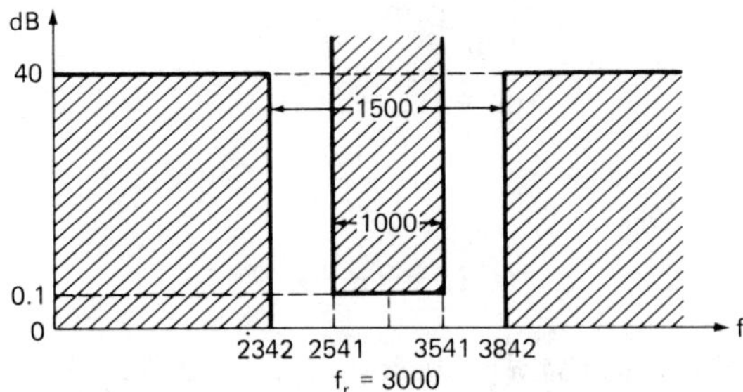

Fig. 4.3–10. Specifications for Ex. 4.3–2.

⅓ permits the design of each stage using the above results. The ratios in Fig. 4.3–8(a) are found from Eq. (26) as

$$\alpha_{11} = \alpha_{12} = 0.14726$$
$$\alpha_{31} = \alpha_{41} = 0.04163$$
$$\alpha_{51} = 0.12834$$

and

$$\alpha_{61} = 0.01890$$

As $\alpha_{41} = \alpha_{51}$, a single capacitor $\alpha_{41}C_1$ can be charged differentially. The ratios of Fig. 4.3–8(b) are found from Eq. (28) and are

$$\alpha_{12} = \alpha_{22} = 0.14726$$

and

$$\alpha_{32} = \alpha_{42} = 0.04040$$

The ratios of Fig. 4.3–8(c) are determined from Eq. (30) as

$$\alpha_{13} = \alpha_{23} = 0.14726$$
$$\alpha_{33} = 0.02207$$
$$\alpha_{43} = 0.06805$$
$$\alpha_{53} = 0.19822$$

and

$$\alpha_{63} = 0.01002$$

The ratios of Fig. 4.3–8(d) are found from Eq. (32) as

$$\alpha_{14} = \alpha_{24} = 0.14726$$

and

$$\alpha_{34} = \alpha_{44} = 0.05249$$

Last, the capacitor ratios of the terminating stage of Fig. 4.3–8(e) are found from Eq. (34) as

$$\alpha_{15} = \alpha_{25} = 0.14726$$
$$\alpha_{35} = \alpha_{45} = 0.03907$$
$$\alpha_{55} = 0.35089$$

and

$$\alpha_{65} = 0.05167$$

Because α_{35} and α_{45} are identical, a single capacitance $\alpha_{45} C_5$ can be charged differentially. The resulting realization is the same as Figure 4.3–9.

In this section we have extended the ideas of SC ladder filter synthesis to symmetrical bandpass filters. The basic building block is a second-order bandpass realization with the capability of summing both positively and negatively. The extension was based on the well-known low-pass-to-band-pass transformation. Two items not covered in this section are nonsymmetrical bandpass filters and the influence of the clock frequency upon the SC realization. The results of this section assume that z can be approximated by $1 + sT$. Although valid in many cases where larger capacitor ratios can be tolerated, this approximation may not be a valid approximation in general. Both items will be considered in later sections of this chapter.

4.4 SWITCHED CAPACITOR HIGH-PASS LADDER FILTERS

In this section, we continue to build on the general development given in Section 4.1 by applying the methods to the realization of high-pass filters. It will be seen that, because of the low-pass-to-high-pass transformation, the more natural building block for the high-pass filter is the differentiator rather than the integrator. A little cleverness in expressing the relationships will permit the use of the integrator with a slight modification. One important problem in SC high-pass filter realizations is that the capacitor ratios become large because the cutoff frequency is typically much less than the clock frequency. To obtain high-pass filters with a reasonable passband, it will be necessary to have large capacitor ratios. Several alternatives to this problem will be considered.

The design procedure begins by using the low-pass-to-high-pass transformation given in Eq. (10) of Section 3.1. We shall assume that the low-pass network has been normalized to a cutoff frequency of 1 rps ($\omega_0 = 1$) so that the low-pass-to-high-pass transformation is

$$s = 1/p \tag{1}$$

It is seen that the resulting high-pass filter obtained by Eq. (1) will also have a cutoff frequency of 1 rps. The low-pass-to-high-pass transformation applied to a low-pass structure, such as Fig. 4.2–1, causes the series elements of the ladder to become capacitors and the shunt elements of the ladder to become inductors. Selecting the capacitor voltages and the inductor currents as state variables implies that the desired state variable for the series branch would be voltage and for the shunt branch would be current. The expressions for the state variables become slightly more complicated than those for the low-pass case. However, we shall show in the following how to generate expressions for each state variable that are compatible for realization by SC integrators with a slight modification.

Let us consider the SC realization of a fourth-order, singly terminated, high-pass filter. A low-pass prototype of a fourth-order, singly terminated filter is shown in Fig. 4.4–1(a). Applying the low-pass-to-high-pass transformation of Eq. (1) results in Fig. 4.4.–1(b). The subscript H is used to indicate that the circuit is the high-pass prototype filter. The state variables are V_1, I_2, V_3, I_4, and V_{out} and are identified in Fig. 4.4–1(b). Actually, V_{out} and I_4 are redundant, and I_4 will be eliminated later in the design procedure. Let us begin by writing a nodal equation at node A, which is

$$sC_{H1n}V_1 = I_2 + I_4 + (V_{out}/R_{5n}) \tag{2}$$

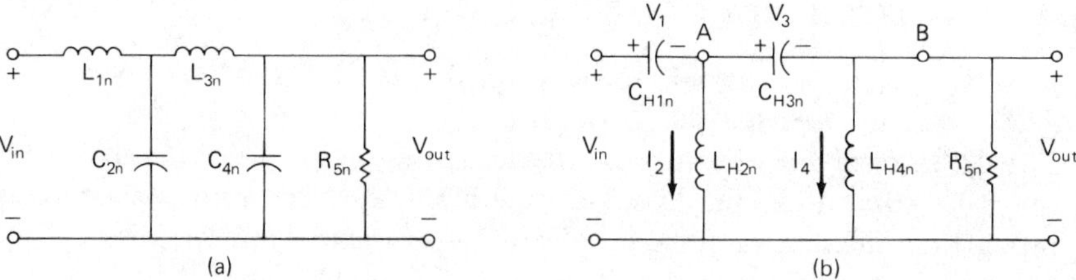

Fig. 4.4–1. (a.) Low-pass prototype fourth-order filter. (b.) High-pass prototype of (a.).

or

$$V_1 = \frac{1}{sC_{H1n}}\left(I_2 + I_4 + \frac{V_{out}}{R_{5n}}\right) = \frac{1}{sC_{H1n}R}\left(V_2' + V_4' + \frac{R}{R_{5n}}V_{out}\right) \quad (3)$$

Next, we derive an expression for the state variable I_2 by writing a loop equation that includes V_{in}, C_{H1n}, and L_{H2n}. This equation is

$$V_{in} = V_1 + sL_{H2n}I_2 = V_1 + \left(\frac{sL_{H2n}}{R}\right)V_2' \quad (4)$$

or

$$V_2' = \frac{R}{sL_{H2n}}(V_{in} - V_1) \quad (5)$$

Writing a nodal equation at node B results in

$$sC_{H3n}V_3 = I_4 + (V_{out}/R_{5n}) = (V_4'/R) + (V_{out}/R_{5n}) \quad (6)$$

or

$$V_3 = \frac{1}{sC_{H3n}}\left(\frac{V_4'}{R} + \frac{V_{out}}{R_{5n}}\right) \quad (7)$$

An expression for the state variable I_4 can be found by writing a loop equation that includes L_{H2n}, C_{H3n}, and L_{H4n}. This equation is

$$sL_{H4n}I_4 = sL_{H2n}I_2 - V_3 \quad (8)$$

or

$$V_4' = \frac{R}{sL_{H4n}}\left(\frac{sL_{H2n}}{R}V_2' - V_3\right) = \frac{L_{H2n}}{L_{H4n}}V_2' - \frac{R}{sL_{H4n}}V_3 \quad (9)$$

Though there are only four state variables in this problem, a fifth relationship can be written that will allow the output voltage, V_{out}, to replace one of the above four state variables. This expression is

$$V_{out} = V_{in} - V_1 - V_3 \quad (10)$$

and is obtained by writing a loop equation around the input and output and only the series branches.

Equations (3), (5), (7), (9), and (10) constitute a nonunique set of equations that describe the circuit of Fig. 4.4–1(b). We see that the realization of Eqs. (3), (5) and (7) require differential integrators, and that Eq. (9) requires a multiple-input integrator with the ability to have nonintegrated inputs summed with the integrated inputs. The realization of Eq. (10) requires a summer only. We observed in Fig. 4.2–8 a realization for an integrator that is capable of summing an inverted, nonintegrated input with the integrated inputs. Unfortunately, the expressions for the general high-pass filter require that the nonintegrated input be summed in a noninverting manner, as is the case for the above. This problem will be solved in the same manner as was done for the low-pass filter in Section 4.2, which had $j\omega$ zeros. The appropriate state variables will be defined as negative.

The number of op amps required in the realization may be reduced by combining the five state equations into four and still permit each equation to be realized by an integrator. The simplest approach is to use Eq. (10) and replace either V_1 or V_3 in Eqs. (3), (5), (7), and (9). In this manner, we guarantee that we can still use integrators and yet save one op amp. Let us choose to eliminate V_3. Thus Eq. (10) is written as

$$V_3 = V_{in} - V_1 - V_{out} \tag{11}$$

Because V_3 occurs only in Eqs. (7) and (9), we substitute Eq. (11) into those equations to get

$$V_{out} = V_{in} - V_1 - \frac{1}{sC_{H3n}} \left(\frac{V_4'}{R} + \frac{V_{out}}{R_{5n}} \right) \tag{12}$$

and

$$V_4' = \frac{L_{H2n}}{L_{H4n}} V_2' - \frac{R}{sL_{H4n}} (V_{in} - V_1 - V_{out}) \tag{13}$$

The desired set of equations to describe Fig. 4.4–1(b) is given by Eqs. (3), (5), (12), and (13).

The next step is to realize these equations. A method is shown in Fig. 4.4–2(a) that is very useful for high-pass filters because generally they are not as straightforward as the low-pass or bandpass filters. In Fig. 4.4–2(a), we use symbolic blocks to represent the integrator and the summing junction. The sign of the integration is given in the left part of the integrator symbol, and the integrator time constant is shown in the triangle. The circles at the input of the integrator symbol represent the various weighting functions that are applied to each input to obtain the desired integrating time constant for that particular input. In some cases, which we will see later, the nonintegrated inputs are also weighted. The state variables plus V_{in} are all available by simply connecting each of the vertical lines to the correct output or input signal. It can be seen that Fig. 4.4–2(a) synthesizes Eq. (3), (5), (12), and (13) and as a result is a functional realization of Fig. 4.4–1(b).

To complete the design procedure, we need to redefine the polarities of some or all of the state variables in order for all nonintegrated state variables to be inverting. We see that the realizations for V'_4 and V_{out} all have noninverting inputs that are not integrated. Through a process of trial and error, we select various combinations of redefined state variables to attempt to achieve the situation where all nonintegrated variables are subtractive. If it is not possible to do this, then rearrangement of Eqs. (3), (5), (12), and (13) may permit the desired result. Figure 4.4–2(b) shows that if V_1, V'_2, and V_{out} are defined as $\overline{V_1}$, $\overline{V'_2}$, and $\overline{V_{out}}$, where $\overline{V_i} = -V_i$, then Eqs. (3), (5), (12), and (13) can be written as

$$\overline{V_1} = \frac{1}{sC_{H1n}}\left(\frac{\overline{V'_2}}{R} - \frac{V'_4}{R} + \frac{\overline{V_{out}}}{R_{5n}}\right) \tag{14}$$

$$\overline{V'_2} = -\frac{R}{sL_{H2n}}\,(V_{in} + \overline{V_1}) \tag{15}$$

$$\overline{V_{out}} = -V_{in} - \overline{V_1} + \frac{1}{sC_{H3n}}\left(\frac{V'_4}{R} - \frac{\overline{V_{out}}}{R_{5n}}\right) \tag{16}$$

and

$$V'_4 = -\frac{L_{H2n}}{L_{H4n}}\overline{V'_2} - \frac{R}{sL_{H4n}}\,(V_{in} + \overline{V_1} + \overline{V_{out}}) \tag{17}$$

The manipulations to achieve a realization that can be implemented by switched capacitor circuits can be done either on the functional block diagram of Fig. 4.4–2(a), resulting in Fig. 4.4–2(b), or on Eqs. (3), (5), (12), and (13) resulting in Eqs. (14), (15), (16), and (17).

A switched capacitor realization of Fig. 4.4–1(b) is shown in Fig. 4.4–3. This realization was developed by directly synthesizing Fig. 4.4–2(b) using stray insensitive integrators of Figs. 2.5–5 and 2.5–8(b). The nonintegrated variables are realized by coupling the input variable through a capacitor to

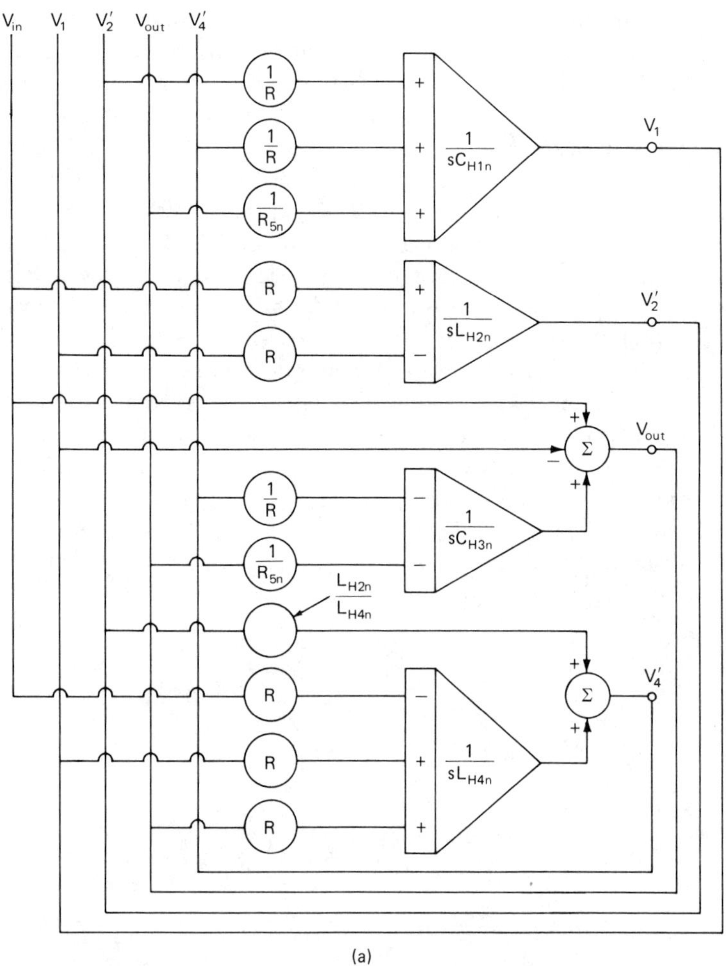

(a)

Fig. 4.4–2. (a) Functional block diagram of Fig. 4.4–1(b.).

the summing junction of the op amp. Though other configurations of the switched capacitor integrator could be used, this form has the most generality and the best performance, although the number of capacitors can be reduced when R_{5n} is equal to R by charging a single capacitance differentially.

Example 4.4–1. *SC realization of a fourth-order, singly terminated, high-pass filter.* A high-pass filter is to be designed using the methods above. The filter is to be a 1-dB Cheybshev, fourth-order, singly terminated, high-pass filter. The cutoff frequency is to be 100 Hz and the clock frequency 100 kHz.

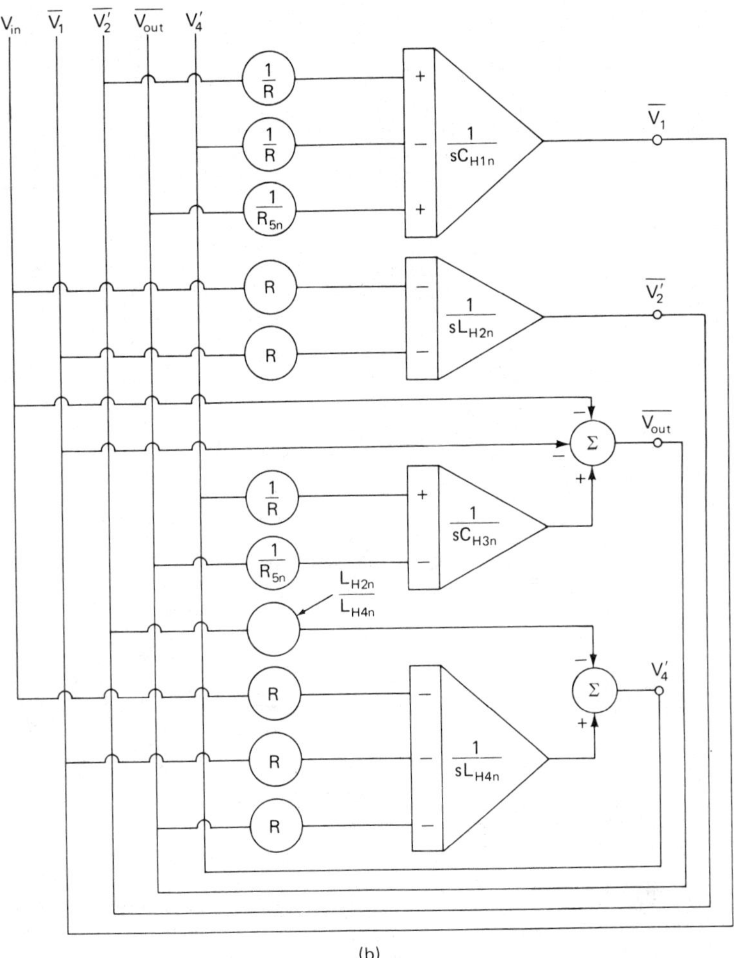

Fig. 4.4–2. (b) Functional block diagram equivalent of Fig. 4.4–2(a) with all nonintegrated inputs having a minus sign.

The values of Fig. 4.4–1(a) that satisfy the specifications are $L_{1n} = 1.2817$ H, $C_{2n} = 1.9093$ F, $L_{3n} = 1.4126$ H, $C_{4n} = 1.0495$ F, and $R_{5n} = 1$ ohm. Applying the low-pass-to-high-pass transformation gives the component values of Fig. 4.4–1(b) as $C_{H1n} = 0.7802$ F, $L_{H2n} = 0.5238$ H, $C_{H3n} = 0.7079$ F, $L_{H4n} = 0.9528$, and $R_{5n} = 1$ ohm.

If we assume that $\omega << \omega_c$, then, from Fig. 4.4–3, we can write that

$$\overline{V_1} = \frac{\alpha_{11}}{sT_n}\overline{V_2'} - \frac{\alpha_{21}}{sT_n}V_4' - \frac{\alpha_{31}}{sT_n}\overline{V_{out}} \tag{18}$$

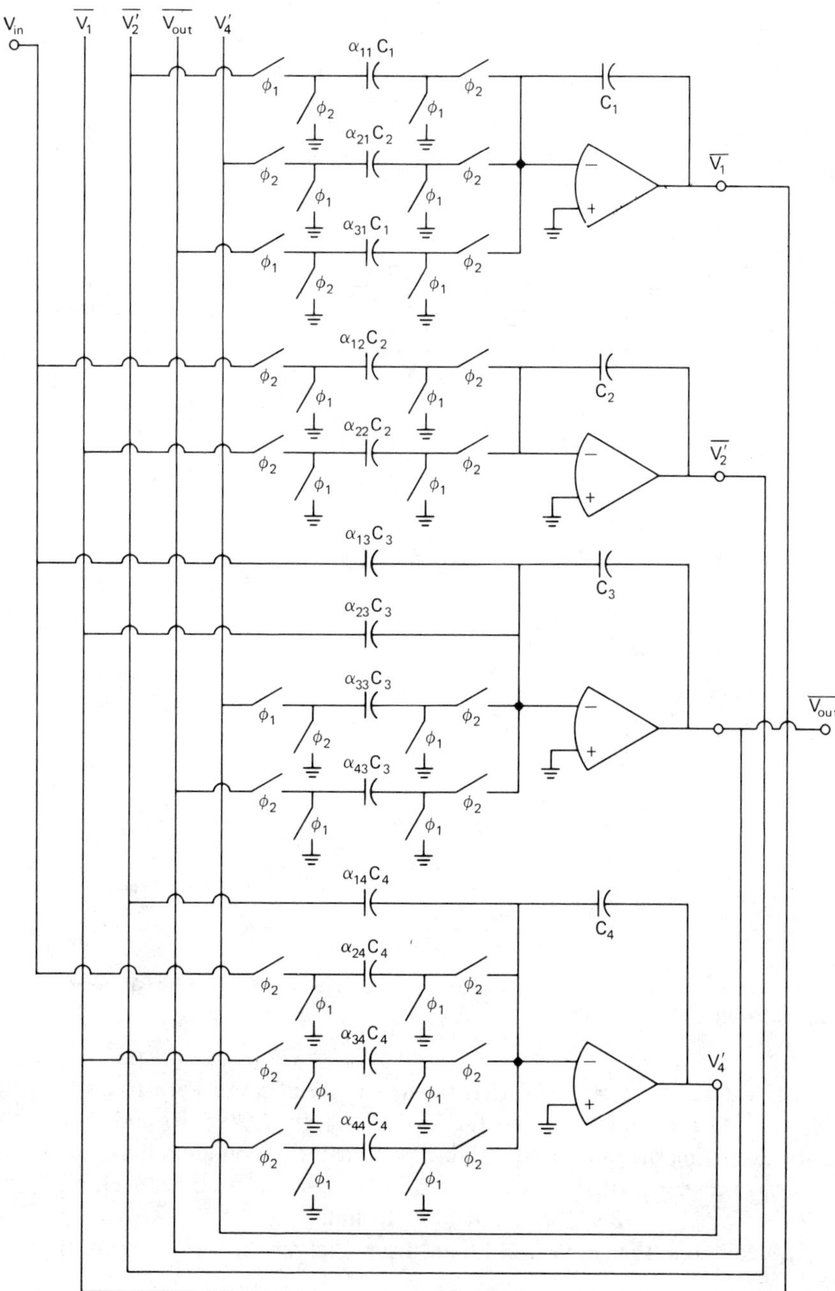

Fig. 4.4–3. SC realization of Fig. 4.4–2(b).

$$\overline{V_2'} = -\frac{\alpha_{12}}{sT_n} V_{in} - \frac{\alpha_{22}}{sT_n} \overline{V_1} \tag{19}$$

$$\overline{V_{out}} = -\alpha_{13} V_{in} - \alpha_{23} \overline{V_1} + \frac{\alpha_{33}}{sT_n} V_4' - \frac{\alpha_{43}}{sT_n} \overline{V_{out}} \tag{20}$$

and

$$V_4' = -\alpha_{14} \overline{V_2'} - \frac{\alpha_{24}}{sT_n} V_{in} - \frac{\alpha_{34}}{sT_n} \overline{V_1} - \frac{\alpha_{44}}{sT_n} \overline{V_{out}} \tag{21}$$

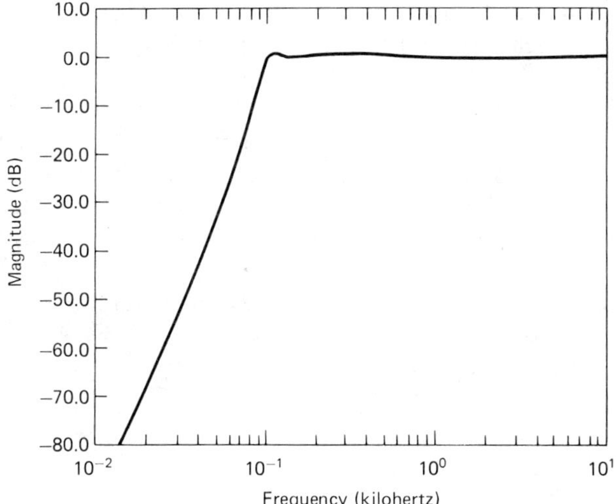

Fig. 4.4–4. Ideal response for Example 4.4–1.

Equating Eqs. (18) through (21) to Eqs. (14) through (17) results in

$$\alpha_{11} = \alpha_{21} = \alpha_{31} = 0.008533$$
$$\alpha_{12} = \alpha_{22} = 0.011996$$
$$\alpha_{13} = \alpha_{23} = 1, \ \alpha_{33} = \alpha_{43} = 0.008876$$

and

$$\alpha_{14} = 0.5497, \ \alpha_{24} = \alpha_{34} = \alpha_{44} = 0.0.006594$$

Figure 4.4–4 shows the ideal frequency response for the filter of this example.

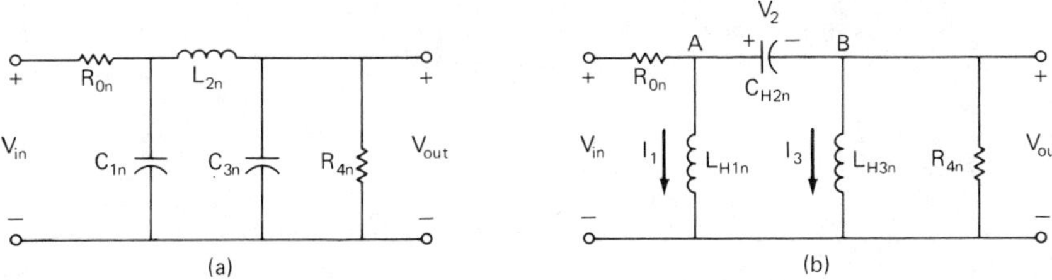

Fig. 4.4–5. (a.) Doubly-terminated, third-order, low-pass filter. (b.) High-pass equivalent of (a.).

We note with some concern that the ratios of capacitors in Example 4.4–1 are as large as 150 to 1. These ratios are so large because the clock frequency is so much greater than the filter cutoff frequency (100 kHz compared with 100 Hz).

To further illustrate the high-pass filter design method, let us consider the doubly-terminated third-order filter shown in Fig. 4.4–5(a). Transforming this low-pass prototype to a high-pass prototype results in each high-pass element having the reciprocal value of the low-pass element value. The state variables are I_1, V_2, I_3, and V_{out} and are identified on the high-pass prototype shown in Fig. 4.4–5(b). Writing a loop equation consisting of the elements V_{in}, R_{On}, and L_{H1n} using voltage analogs of current results in

$$V_{in} = R_{On}\left(\frac{V_1'}{R} + sC_{H2n}V_2\right) + \frac{sL_{H1n}}{R}V_1' \tag{22}$$

or

$$V_1' = \frac{R}{sL_{H1n}}V_{in} - \frac{R_{On}}{sL_{H1n}}V_1' - \frac{RR_{On}C_{H2n}}{L_{Hn1}}V_2 \tag{23}$$

Writing a nodal equation at node B using voltage analogs of currents results in

$$sC_{H2n}V_2 = (V_3'/R) + (V_{out}/R_{4n}) \tag{24}$$

or

$$V_2 = \frac{1}{sC_{H2n}}\left(\frac{V_3'}{R} + \frac{V_{out}}{R_{4n}}\right) \tag{25}$$

Writing a loop equation consisting of the elements L_{H1n}, C_{H2n}, and L_{H3n} using voltage analogs of current gives

$$\frac{V_1' \, sL_{H1n}}{R} = V_2 + \frac{sL_{H3n}}{R} V_3' \tag{26}$$

or

$$V_3' = \frac{R}{sL_{H3n}} \left(\frac{sL_{H1n}}{R} V_1' - V_2 \right) \tag{27}$$

However, we note that

$$V_3' = \frac{R}{sL_{H3n}} V_{\text{out}} \tag{28}$$

A fourth equation, which is independent of frequency, can be generated as follows

$$V_{\text{in}} = R_{0n} \left(I_1 + I_3 + \frac{V_{\text{out}}}{R_{4n}} \right) + V_2 + V_{\text{out}} \tag{29}$$

or

$$V_{\text{out}} = \frac{1}{1 + (R_{0n}/R_{4n})} \left[V_{\text{in}} - \left(\frac{R_{0n}}{R} \right) (V_1' + V_3') - V_2 \right] \tag{30}$$

Equations (23), (25), (29), and (30) constitute one set of equations expressing each state variable in terms of the others that can be realized by integrators and summers.

As the filter is third-order, only three op amps should be required. To reduce the number of op amps (equations) to three, let us eliminate one of the state variables. Substituting Eq. (25) into Eqs. (23) and (30) gives

$$V_1' = \frac{R}{sL_{H1n}} V_{\text{in}} - \frac{R_{0n}}{sL_{H1n}} V_1' - \frac{R_{0n}}{sL_{H1n}} V_3' - \frac{R_{0n} R}{sR_{4n} L_{H1n}} V_{\text{out}} \tag{31}$$

and

$$V_{\text{out}} = \left(\frac{1}{1 + (R_{0n}/R_{4n})} \right) \left[V_{\text{in}} - \left(\frac{R_{0n}}{R} \right) (V_1' + V_3') - \frac{V_3'}{sRC_{H2n}} - \frac{V_{\text{out}}}{sR_{4n} C_{H2n}} \right] \tag{32}$$

Equations (31) and (32), along with Eq. (28), represent three expressions of the state variables V_1', V_3', and V_{out} that describe the filter of Fig. 4.4–5(b).

Figure 4.4–6(a) shows a functional block diagram realization of the filter. A quick check of the signs of the nonintegrated variables shows that the sign of V_{in} of Eq. (32) must be changed. In Fig. 4.4–6(b) we have defined $\overline{V}_{in}$ as $-V_{in}$, which changes the sign of all inputs from V_{in}. Thus, Fig. 4.4–6(b) is in a form by which it can be directly realized by SC networks. Problem 4.31 shows an alternative method of redefining the state variables to satisfy the realization. Figure 4.4–7 shows an SC realization of Fig. 4.4–6(b).

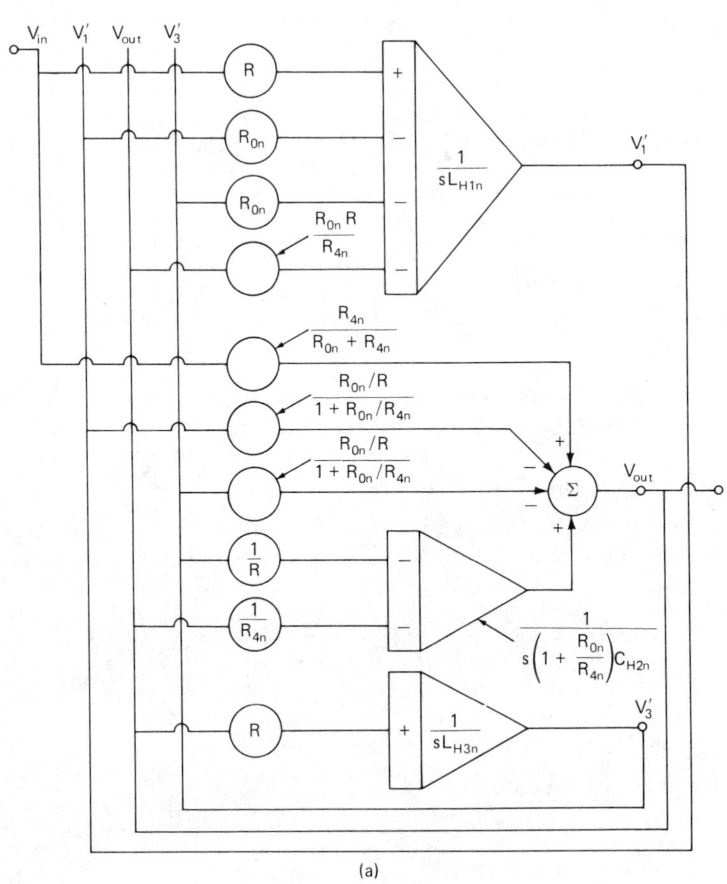

Fig. 4.4–6. (a) Functional block diagram realization of Fig. 4.4–5(b).

Example 4.4–2. *An SC realization of a third-order, doubly terminated high-pass filter.* A high-pass, third-order, Chebyshev filter with a 1-dB passband ripple is to be designed using the above method. The filter is doubly terminated and has a cutoff frequency of 100 Hz and a clock frequency of 100 kHz.

The element values of Fig. 4.4–5(a) are $R_{0n} = 1$ ohm, $C_{1n} = 2.0236$ F, $L_{2n} = 0.9941$ H, $C_{3n} = 2.0236$ F, and $R_{4n} = 1$ ohm. The corresponding values of Fig. 4.4–5(b) are $L_{H1n} = 0.4942$ H, $C_{H2n} = 1.0059$ F, and $L_{H3n} = 0.4942$ H. If we assume that $\omega << \omega_c$, then, from Fig. 4.4–7, we may express the state variables as

$$V_1' = \frac{-\alpha_{11}}{sT_n}\overline{V_{\text{in}}} - \frac{\alpha_{21}}{sT_n}V_1' - \frac{\alpha_{31}}{sT_n}V_3' - \frac{\alpha_{41}}{sT_n}V_{\text{out}} \tag{33}$$

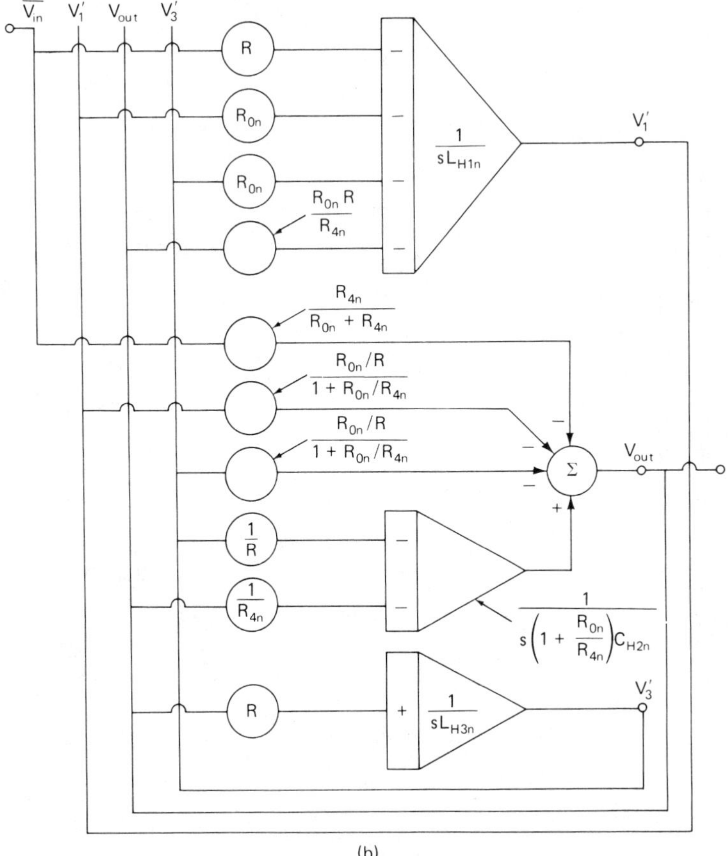

Fig. 4.4–6. (b) Modification of Fig. 4.4–6(a) to change all non-integrated variables of inverting.

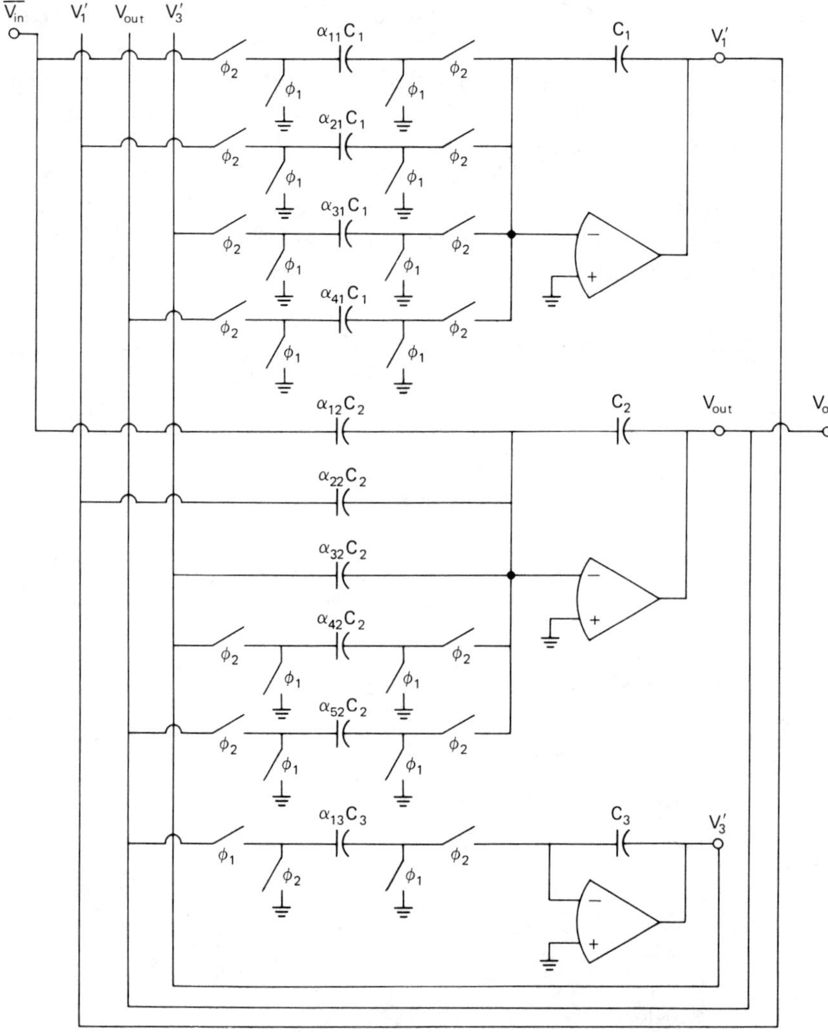

Fig. 4.4–7. SC realization of Fig. 4.4–6(b).

$$V_{out} = -\alpha_{12}\,\overline{V_{in}} - \alpha_{22}\,V_1' - \alpha_{32}\,V_3' - \frac{\alpha_{42}}{sT_n}\,V_3' - \frac{\alpha_{52}}{sT_n}\,V_{out} \tag{34}$$

and

$$V_3' = (\alpha_{13}/sT_n)\,V_{out} \tag{35}$$

Equating Eqs. (33), (34), and (35) to Eqs. (31), (32), and (28), respectively, results in the design of the capacitor ratios as given below

$$\alpha_{11}=\alpha_{21}=\alpha_{31}=\alpha_{41}=0.01271$$
$$\alpha_{12}=\alpha_{22}=\alpha_{32}=0.5000,\ \alpha_{42}=\alpha_{52}=0.003123$$

and

$$\alpha_{13}=0.01271$$

Figure 4.4–8 shows the ideal performance of this filter.

Finally, let us consider the realization of a high-pass filter having finite $j\omega$ axis zeros. Figure 4.4–9(a) is the low-pass prototype of a third-order having one pair of complex conjugate zeros on the $j\omega$ axis. The first step in the design procedure is to eliminate the C_{2n} capacitor resulting in the equivalent circuit of Fig. 4.4–9(b). The next step is to apply the low-pass-to-high-pass transformation, resulting in Fig. 4.4–9(c). It can be shown that Fig. 4.4–9(c) results if the low-pass-to-high-pass transformation is applied first to Fig. 4.4–9(a) and then the bridging element is removed. The reason the approaches are identical is that the voltage-controlled voltage-sources are independent of the transformation. The state variables of Fig. 4.4–9(c) are I_1, V_2, I_3, and V_{out}. V_1 and V_3 are identified only because they control the two dependent voltage sources.

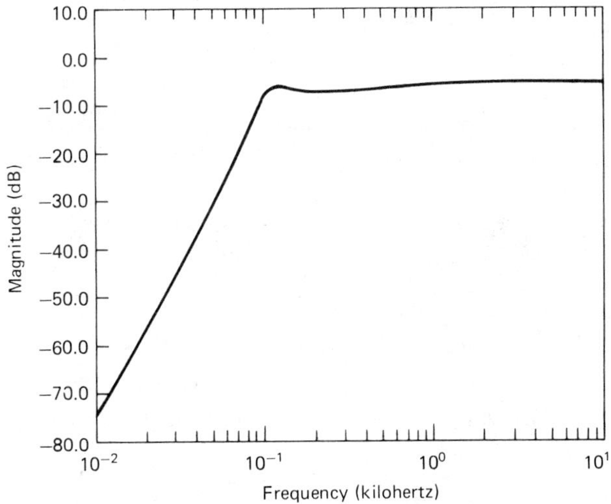

Fig. 4.4–8. Ideal response of Example 4.4–2.

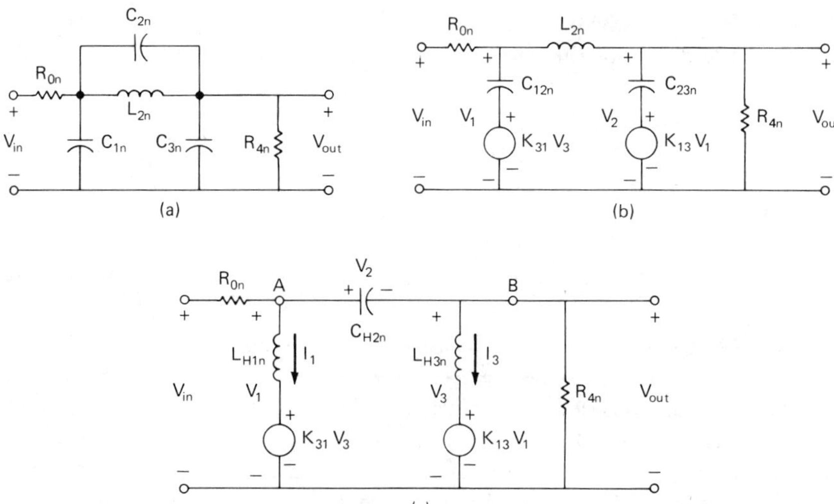

Fig. 4.4–9. (a.) Low-pass prototype, third-order, elliptic filter. (b.) Elimination of the capacitor pi. (c.) High-pass prototype of Fig. 4.4–9(a.).

The expressions for the state variables are generated in the same manner as before by writing an appropriate set of node-loop equations. We begin by writing a loop equation around the loop containing V_{in}, R_{0n}, L_{H1n}, and $K_{31}V_3$. This equation is

$$V_{\text{in}} = R_{0n}(I_1 + sC_{H2n}V_2) + sL_{H1n}I_1 + K_{31}V_{\text{out}} \tag{36}$$

Solving for V_1' using voltage analogs of currents gives

$$V_1' = \frac{R}{sL_{H1n}}V_{\text{in}} - \frac{R_{0n}}{sL_{H1n}}V_1' - \frac{R_{0n}RC_{H2n}}{L_{H1n}}V_2 - \frac{K_{31}R}{sL_{H1n}}V_{\text{out}} \tag{37}$$

In developing this equation, we have used the fact that $V_3 = V_{\text{out}}$. An expression for the state variable V_2 can be found by writing a nodal equation at node B in Fig. 4.4–9(c). This nodal equation is

$$sC_{H2n}V_2 = I_3 + (V_{\text{out}}/R_{4n}) \tag{38}$$

Solving for V_2 using voltage analogs of currents gives

$$V_2 = \frac{1}{sC_{H2n}}\left[\frac{V_3'}{R} + \frac{V_{\text{out}}}{R_{4n}}\right] \tag{39}$$

An expression for the state variable I_3 (V_3') can be found by writing a loop equation about the loop consisting of L_{H3n}, R_{4n}, and $K_{13}V_1$. This equation is

$$sL_{H3n}I_3 = V_{\text{out}} - K_{13}[V_{\text{in}} - R_{0n}(I_1 + sC_{H2n}V_2)] \tag{40}$$

Solving for V_3' using voltage analogs of currents results in

$$V_3' = \frac{R}{sL_{H3n}}\left[V_{\text{out}} - K_{13}V_{\text{in}} + K_{13}R_{0n}\left(\frac{V_1'}{R} + sC_{H2n}V_2\right)\right] \tag{41}$$

A fourth equation can be written that will permit the simplification of the realization. This equation is

$$V_{\text{in}} = R_{0n}[I_1 + I_3 + (V_{\text{out}}/R_{4n})] + V_2 + V_{\text{out}} \tag{42}$$

or

$$V_{\text{out}} = \frac{1}{1 + (R_{0n}/R_{4n})}\left[V_{\text{in}} - \frac{R_{0n}}{R}(V_1' + V_3') - V_2\right] \tag{43}$$

Equations (37), (39), (41), and (43) constitute a satisfactory set of expressions for the state variables of Fig. 4.4–9(c). As before, this set can be reduced to a minimum number of op amps by eliminating one of the output variables. Let us choose to substitute Eqs. (39) into Eqs. (37), (41), and (43) to get

$$V_1' = \frac{R}{sL_{H1n}}V_{\text{in}} - \frac{R_{0n}}{sL_{H1n}}V_1' - \frac{R_{0n}}{sL_{H1n}}V_3' - \frac{R}{sL_{H1n}}\left(\frac{R_{0n}}{R_{4n}} + K_{31}\right)V_{\text{out}} \tag{44}$$

$$V_{\text{out}} = \frac{1}{1 + (R_{0n}/R_{4n})}\left[V_{\text{in}} - \frac{R_{0n}}{R}V_1' - \frac{R_{0n}}{R}V_3' - \frac{V_3'}{sRC_{H2n}} - \frac{V_{\text{out}}}{sR_{4n}C_{H2n}}\right] \tag{45}$$

and

$$V_3' = \frac{R}{sL_{H3n}}V_{\text{out}} - \frac{RK_{13}}{sL_{H3n}}V_{\text{in}} + \frac{R_{0n}K_{13}}{sL_{H3n}}V_1' + \frac{R_{0n}K_{13}}{sL_{H3n}}V_3' + \frac{R_{0n}RK_{13}}{sR_{4n}L_{H3n}}V_{\text{out}} \tag{46}$$

Equations (44), (45), and (46) represent the equations that will be used to realize Fig. 4.4–9(c).

Figure 4.4–10(a) shows a functional block diagram of Fig. 4.4–9(c) based on Eqs. (44), (45), and (46). To make all nonintegrated variables negative, $\overline{V_{in}} = -V_{in}$ has been defined. Figure 4.4–10(b) illustrates the final realization in functional block diagram form. Figure 4.4–11 gives an SC realization of

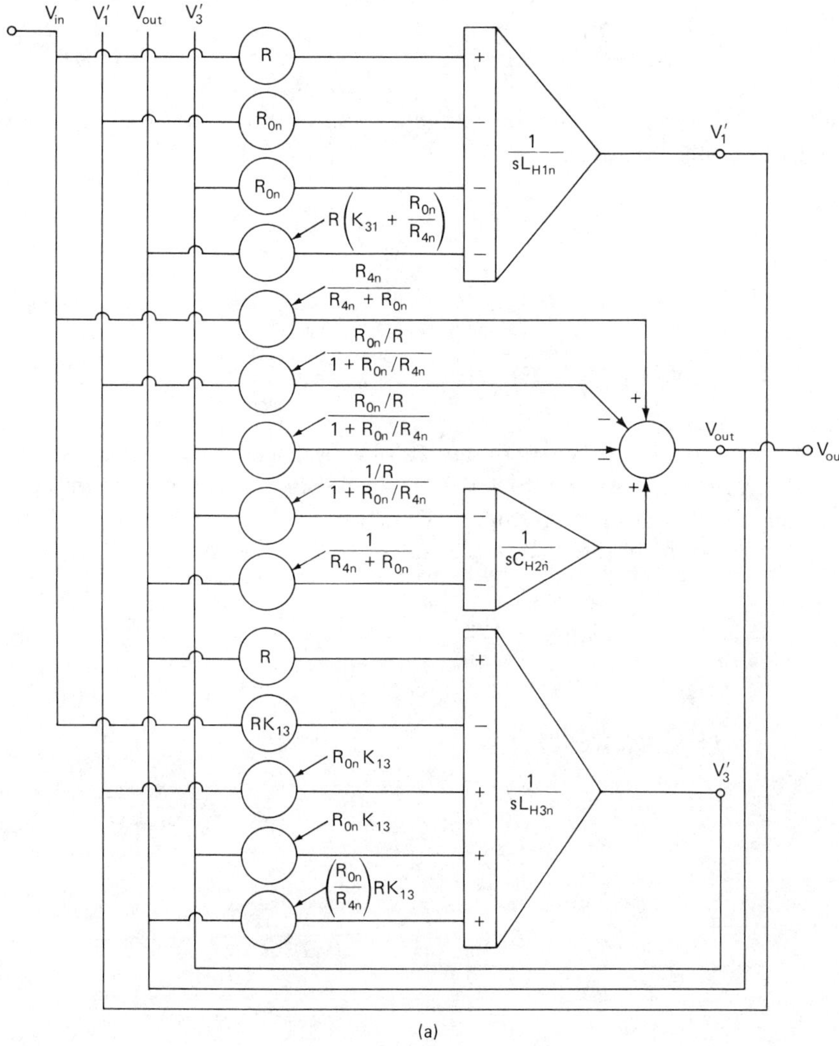

Fig. 4.4–10. (a) Functional block diagram realization of Fig. 4.4–9(b).

the third-order, high-pass filter having finite $j\omega$ axis zeros. An example follows to illustrate the application of this method.

Example 4.4–3. An SC realization of a third-order, doubly terminated, elliptic high-pass filter. A high-pass, third-order, elliptic filter with 1-dB passband ripple is to be designed. The filter is to be doubly terminated and have a cutoff frequency of 100 Hz and a clock frequency of 100 kHz.

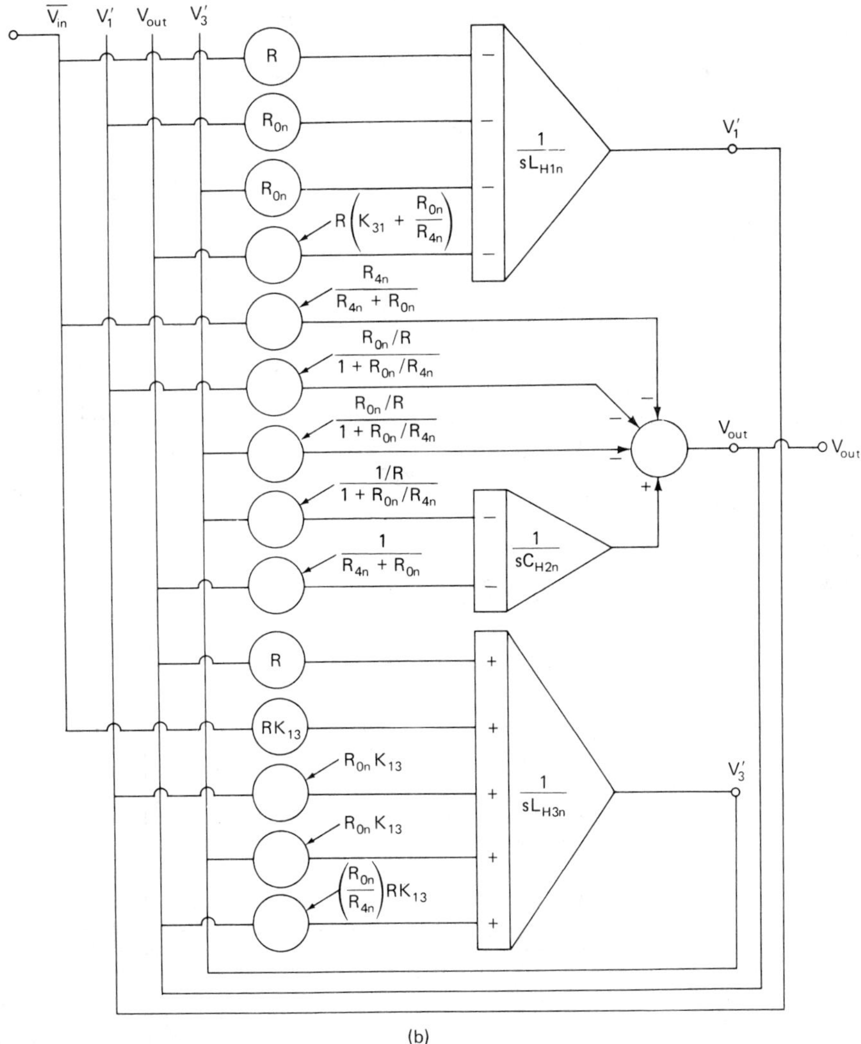

Fig. 4.4–10. (b) Modification of Fig. 4.4–10(a) to allow all nonintegrated variables to be inverting.

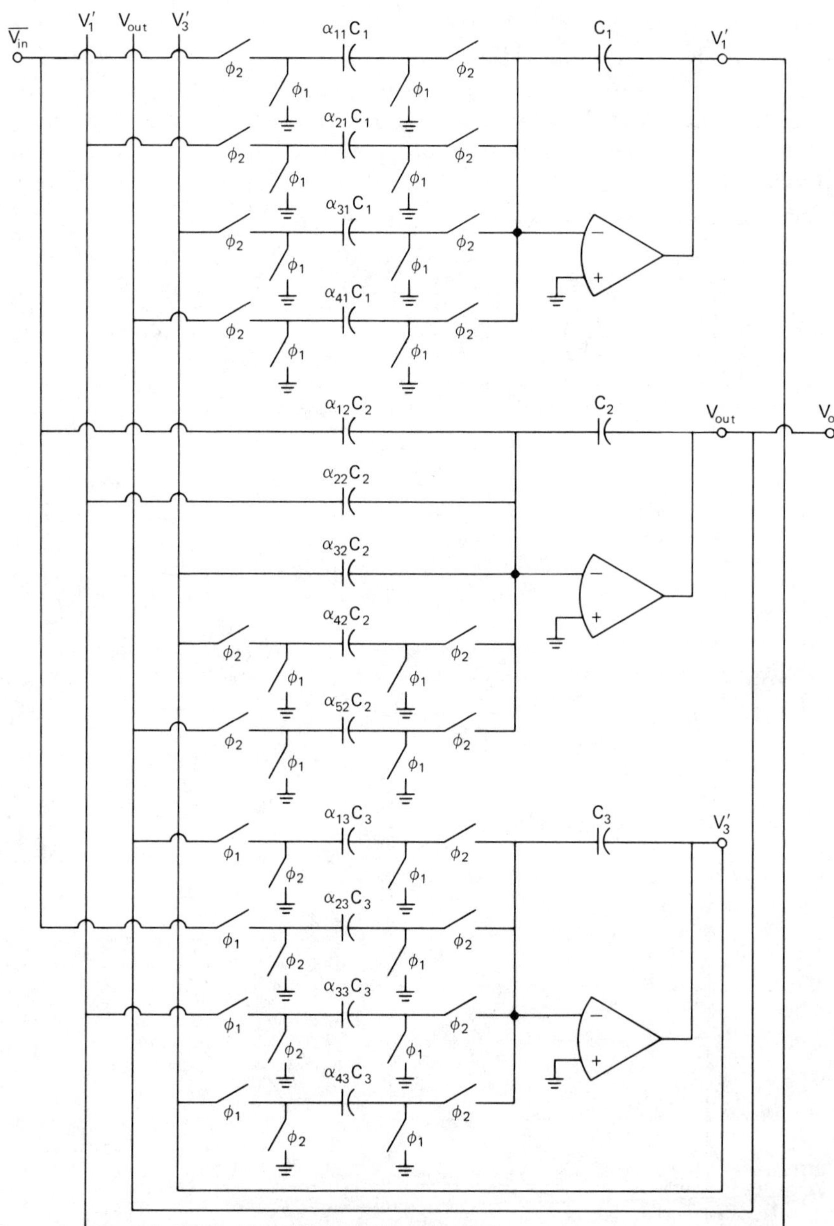

Fig. 4.4–11. SC realization of Fig. 4.4–10(b).

The element values for Fig. 4.4–9(a) are $R_{0n} = 1$ ohm, $C_{1n} = 1.85199$ F, $L_{2n} = 0.85903$ H, $C_{2n} = 0.22590$ F, $C_{3n} = 1.85199$ F, and $R_{4n} = 1$ ohm. The values of Fig. 4.4–9(b) are $C_{12n} = 2.07789$ F $= C_{23n}$ and $K_{13} = K_{31} = 0.10872$. The high-pass values of Fig. 4.4–9(c) are $L_{H1n} = 0.48126$ H, $C_{H2n} = 1.16410$ F, and $L_{H3n} = 0.48126$ H. Assuming that $\omega << \omega_c$, we can write, from Fig. 4.4–11, that

$$V_1' = -\frac{\alpha_{11}}{sT_n}\overline{V_{\text{in}}} - \frac{\alpha_{21}}{sT_n}V_1' - \frac{\alpha_{31}}{sT_n}V_3 - \frac{\alpha_{41}}{sT_n}V_{\text{out}} \tag{47}$$

$$V_{\text{out}} = -\alpha_{12}\overline{V_{\text{in}}} - \alpha_{22}V_1' - \alpha_{32}V_3' - \frac{\alpha_{42}}{sT_n}V_3' - \frac{\alpha_{52}}{sT_n}V_{\text{out}} \tag{48}$$

and

$$V_3' = \frac{\alpha_{13}}{sT_n}V_{\text{out}} + \frac{\alpha_{23}}{sT_n}\overline{V_{\text{in}}} + \frac{\alpha_{33}}{sT_n}V_1' + \frac{\alpha_{43}}{sT_n}V_3' \tag{49}$$

Equating (47), (48), and (49) to (44), (45), and (46), respectively, yields

$$\alpha_{11} = \alpha_{21} = \alpha_{31} = 0.013056,\ \alpha_{41} = 0.014475$$
$$\alpha_{12} = \alpha_{22} = \alpha_{32} = 0.5000,\ \alpha_{42} = \alpha_{52} = 0.0053975$$

and

$$\alpha_{13} = 0.013056,\ \alpha_{23} = \alpha_{33} = \alpha_{43} = 0.0014194$$

Figure 4.4–12 shows the ideal performance of Fig. 4.4–11.

In this section we have presented a method of designing SC high-pass filters, assuming that the sampling frequency is large. One of the results has been that the capacitor ratios have become quite large, because the cutoff frequency of the high-pass filter is much less than the sampling frequency. There are several possible solutions to this problem. One is to reduce the sampling frequency, which will cause the passband of the high-pass filter to be reduced. A second method is to use a capacitive divider arrangement.[6] Figure 4.4–13(a) shows an SC integrator with a capacitor ratio of 100. Figure 4.4–13(b) shows how a capacitance divider can be used to reduce the capacitor

[6] T. Hui and D. J. Allstot, "MOS Switched Capacitor Highpass/Notch Ladder Filters," *Proc. of International Symp. on Circuits and Systems,* Houston, TX, April 1980, pp. 309–312.

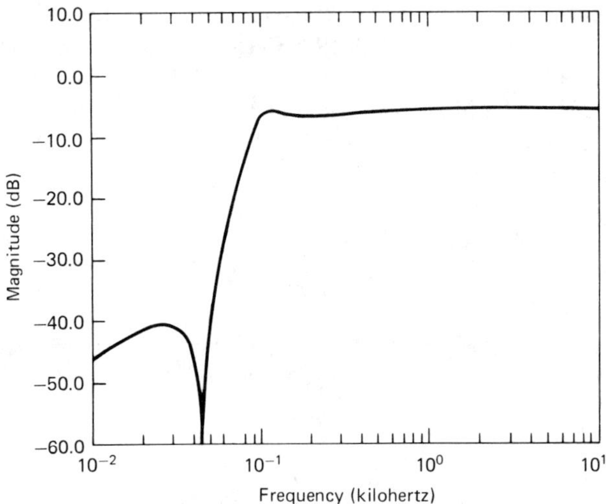

Fig. 4.4–12. Ideal frequency response of Example 4.4–3.

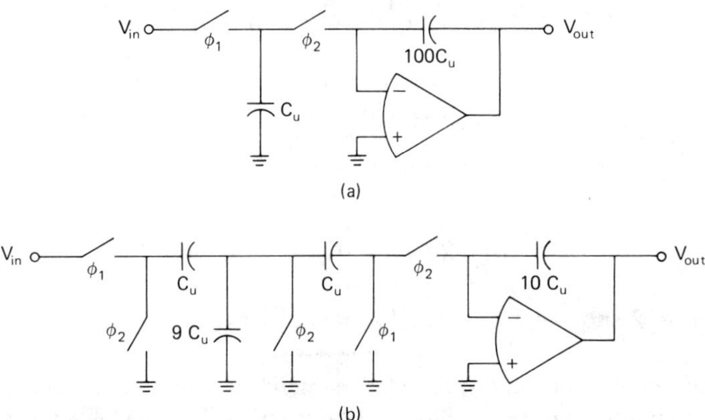

Fig. 4.4–13. (a.) SC integrator with a capacitance ratio of 100 : 1. (b.) Equivalent SC integrator using a capacitance divider to reduce the capacitor ratios.

ratio. The remaining sections of this chapter present other methods of designing high-pass filters that may lead to reduced capacitor ratios.

4.5 SWITCHED CAPACITOR BAND-ELIMINATION AND GENERAL RLC FILTERS

In this section we complete the development of the synthesis of RLC ladder filters. The technique will be extended to symmetrical band-elimination filters by the use of low-pass-to-band-elimination transformation. It will also be shown that the technique can be applied to ladder filters that are not low-pass, bandpass, high-pass, or band-elimination. Any filter that can be expressed in terms of an RLC ladder network can be realized by the methods developed so far. These filters will be called *general RLC filters* and often are the result of a design program using optimization techniques. Their primary distinction from the filters presented so far is that they do not use the transformations from one type of filter to another. Examples are included in this section to illustrate the concepts and methods.

We begin by substituting the low-pass-to-bandpass transformation of Eq. (12) in Section 3.1 into the low-pass-to-high-pass transformation of Eq. (10) in Section 3.1 to get

$$s = \frac{1}{p + (\omega_r^2/p)} \tag{1}$$

We have assumed that ω_0 in Eq. (10) of Section 3.1 is unity. Table 3.1–3 illustrates the influence of the low-pass-to-band-elimination transformation on the magnitude response and upon the low-pass prototype passive components. If we define ω_{Thp} as the bandwidth of the high-pass filter, then the bandwidth, BW, of the band-elimination filter can be expressed as

$$\mathrm{BW} = \omega_{Thp} = \omega_{T2} - \omega_{T1} \tag{2}$$

This bandwidth is related to the distance from the upper edge of the ripple bandwidth of the lower passband to the lower edge of the ripple bandwidth of the upper passband. The geometric center frequency of the band-elimination filter is ω_r and is related to ω_{T1} and ω_{T2} as

$$\omega_r = \sqrt{\omega_{T2}\omega_{T1}} \tag{3}$$

We see that the band-elimination filter possesses the same geometrical symmetry of the bandpass filters of Section 4.3.

The design procedure closely follows that of bandpass filters given in Section 4.3. The objective of the following procedure is to achieve an SC realization of a band-elimination filter given the desired BW and ω_r. We begin with a normalized, low-pass prototype filter that will meet the desired specifications of the band-elimination filter transformed to a low-pass format. To have the center frequency of the band-elimination prototype filter at 1 rps, we normalize the low-pass prototype filter by a factor of

$$\Omega_s = \frac{\mathrm{BW}}{\omega_r} \tag{4}$$

Using Eq. (4) in the normalization of $s = \Omega_{s}p$ causes the cutoff frequency of the low-pass prototype to be greater than 1 rps because ω_r is normally greater than BW. The next step is to apply the low-pass-to-band-elimination transformation with $\omega_r = 1$ rps. The resulting filter will be a band-elimination filter with a center frequency of 1 rps. The state variables are identified, and an expression for each variable is developed in terms of the others and the input. After these expressions are simplified, SC realizations are developed. The SC realization for the band-elimination filter will be a modification of the SC bandpass realization used in Section 4.3. Frequency-denormalizing the SC realization results in the desired SC band-elimination filter.

Let us demonstrate how to use the above method to achieve an SC band-elimination realization starting from the third-order, low-pass, doubly terminated prototype of Fig. 4.5–1(a). Frequency-normalizing this filter by Eq.

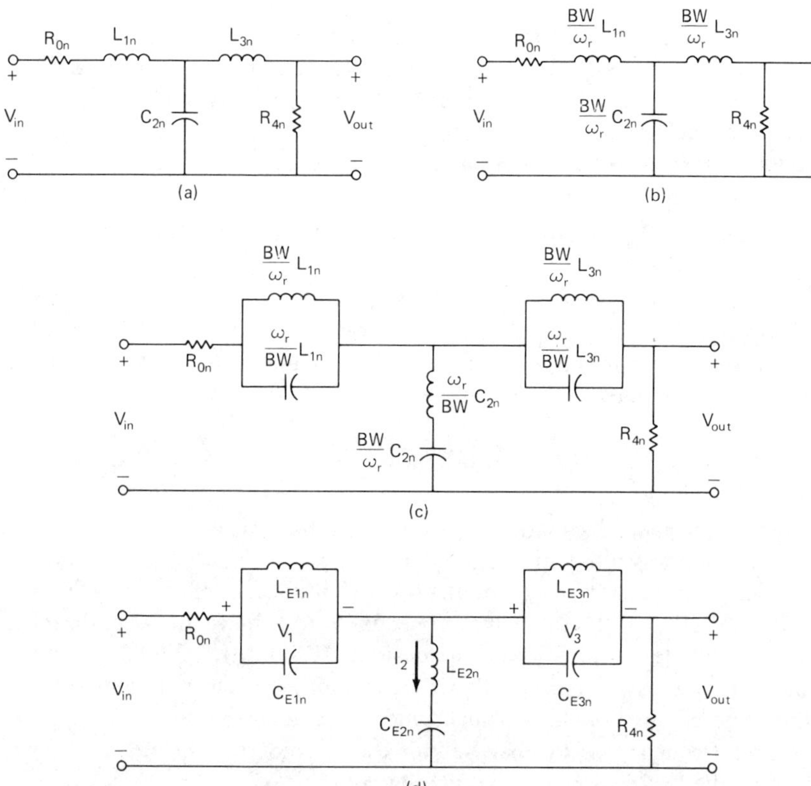

Fig. 4.5–1. (a) Low-pass prototype. (b.) Frequency normalized low-pass prototype. (c.) Band-elimination prototype filter. (d.) Band-elimination filter with redefined components.

(4) results in Fig. 4.5–1(b). Next, we apply Eq. (1) with $\omega_r = 1$ rps to each element of Fig. 4.5–1(b) to obtain the normalized, band-elimination prototype shown in Fig. 4.5–1(c). To simplify our notation, we shall make the definition shown in Fig. 4.5–1(d), where the subscript E stands for the normalized band-elimination prototype with $\omega_r = 1$ rps.

We have noted before that, if the branches of the ladder consist of a parallel inductor and capacitor, choosing the state variable as voltage will always result in realizable expressions using integrator or bandpass structures. Similarly, choosing current as the state variable for a series inductor and capacitor branch is desirable. We have defined the state variables of Fig. 4.5–1(d) as V_1, I_2, V_3, and V_{out}. The expressions for each state variable are easily generated by expressing the state variable of a branch in terms of the other branch variable, which is in turn expressed in terms of the other state variables. For example, the state variable V_1 can be expressed as

$$V_1 = Z_1 I_1 = \left(\frac{s/C_{E1n}}{s^2+1}\right) I_1 = \left(\frac{s/C_{E1n}}{s^2+1}\right)\left[\frac{V_{\text{in}} - V_1 - (I_2/Y_2)}{R_{0n}}\right] \tag{5}$$

The state variable I_2 can be expressed as

$$I_2 = Y_2 V_2 = Y_2(V_3 + V_{\text{out}}) = \left(\frac{s/L_{E2n}}{s^2+1}\right)(V_3 + V_{\text{out}}) \tag{6}$$

The expression for the state variable V_3 is

$$V_3 = Z_3 I_3 = \left(\frac{s/C_{E3n}}{s^2+1}\right)\frac{V_{\text{out}}}{R_{4n}} \tag{7}$$

The last state variable, V_{out}, is written by summing voltage drops around the input, output, and series branches and is

$$V_{\text{out}} = V_{\text{in}} - R_{0n}[I_2 + (V_{\text{out}}/R_{4n})] - V_1 - V_3 \tag{8}$$

or

$$V_{\text{out}} = \left(\frac{1}{1 + (R_{0n}/R_{4n})}\right)[V_{\text{in}} - R_{0n}I_2 - V_1 - V_3] \tag{9}$$

Because the filter is third-order, Eqs. (5), (6), (7), and (9) can be reduced to three equations. Using Eq. (6) to eliminate I_2 gives

$$V_1 = \left(\frac{s/C_{E1n}}{s^2+1}\right)\left[\frac{V_{\text{in}} - V_1 - V_3 - V_{\text{out}}}{R_{0n}}\right] \tag{10}$$

$$V_3 = \left(\frac{s/C_{E3n}}{s^2+1}\right)\left[\frac{V_{\text{out}}}{R_{4n}}\right] \tag{11}$$

and

$$V_{\text{out}} = \left(\frac{1}{1+(R_{0n}/R_{4n})}\right)\left[V_{\text{in}} - V_1 - V_3 - \frac{sR_{0n}/L_{E2n}}{s^2+1}(V_3 + V_{\text{out}})\right] \tag{12}$$

The next step is to propose SC realizations for Eqs. (10), (11), and (12). Equations (10) and (11) can be realized using the bandpass circuit of Fig. 4.3–3. A method of achieving a realization of Eq. (12) using the basic structure of Fig. 4.3–3 was shown in Fig. 4.3–7(a). Here we see that if the variable, V_2, is to be both summed and multiplied by the bandpass function and summed, a realization results if V_2 is applied to the inverting inputs of both op amps of Fig. 4.3–3. It can be shown (see Problem 4.38) that the z-domain transfer function of Fig. 4.3–7(a) is

$$V_{\text{out}}(z) = -\alpha_{31}\, V_2(z) + \frac{\alpha_{41}(z-1)}{(z-1)^2 + \alpha_{11}\alpha_{21}}\, V_1(z) \tag{13}$$

where $\alpha_{51} = \alpha_{21}\alpha_{31}$. If the sampling frequency is assumed to be higher than f_r, then V_{out} can be expressed as

$$V_{\text{out}}(s) \cong -\alpha_{31}\, V_2(s) + \frac{sT_n\alpha_{41}}{(sT_n)^2 + \alpha_{11}\alpha_{21}}\, V_1(s) \tag{14}$$

As before we shall adopt the realization of Fig. 4.3–7(a) for the cases where we have state variables that are functions of other variables that are not multiplied by the bandpass transfer function. The number 1 or 2 at the input of the block in Fig. 4.3–7(b) that represents the two-integrator loop in Fig. 4.3–7(a) will be used to identify which op amp input the input-switched capacitors are connected to.

To use Fig. 4.3–7(a), it is necessary to change the sign of the summed variables to negative. This can be accomplished by changing the sign of V_{in} in Eq. (12) using the same notation as before. The final form of the expressions for the state variables describing Fig. 4.5–1(d) are given as

$$V_1 = \left(\frac{s/(C_{E1n}R_{0n})}{s^2+1}\right)[-\overline{V_{\text{in}}} - V_1 - V_3 - V_{\text{out}}] \tag{15}$$

$$V_3 = \left(\frac{s/(C_{E3n}R_{4n})}{s^2+1}\right) V_{out} \tag{16}$$

and

$$V_{out} = \left(\frac{1}{1+(R_{0n}/R_{4n})}\right) \left[-\overline{V_{in}} - V_1 - V_3 - \left(\frac{sR_{0n}/L_{E2n}}{s^2+1}\right)(V_3 + V_{out})\right] \tag{17}$$

Using the basic scheme of Fig. 4.3–7(a), we propose Fig. 4.5–2(a) as a realization for Eq. (15). If we assume that z^{-1} can be approximated by $1 - sT_n$, we find the approximate s-domain transfer function for Fig. 4.5–2(a) as

$$V_1 \cong \frac{s}{s^2 + (\alpha_{11}\alpha_{21}/T_n^2)} \left[-\frac{\alpha_{31}}{T_n} V_{in} - \frac{\alpha_{41}}{T_n} V_1 - \frac{\alpha_{51}}{T_n} V_3 - \frac{\alpha_{61}}{T_n} V_{out}\right] \tag{18}$$

where we have assumed that the clock period, T, has been normalized by Eq. (15) of Section 4.3. Equating Eqs. (15) and (18) gives

$$\alpha_{11} = \alpha_{21} = T_n = \Omega_n T = \omega_r/f_c \tag{19a}$$

and

$$\alpha_{31} = \alpha_{41} = \alpha_{51} = \alpha_{61} = T_n/(C_{E1n}R_{0n}) = (\text{BW } T L_{1n})/R_{0n} \tag{19b}$$

Equation (16) can be realized by Fig. 4.5–2(b). Assuming that z^{-1} can be approximated by $1 - sT_n$ results in the approximate s-domain transfer function of

$$V_3 \cong \frac{-s}{s^2 + (\alpha_{12}\alpha_{22}/T_n^2)} \left(\frac{\alpha_{32}}{T_n}\right) V_{out} \tag{20}$$

Equating Eqs. (16) and (20) gives

$$\alpha_{12} = \alpha_{22} = T_n = \Omega_n T = \omega_r/f_c \tag{21a}$$

and

$$\alpha_{32} = T_n/(C_{E3n}R_{4n}) = (\text{BW } T L_{3n})/R_{4n} \tag{21b}$$

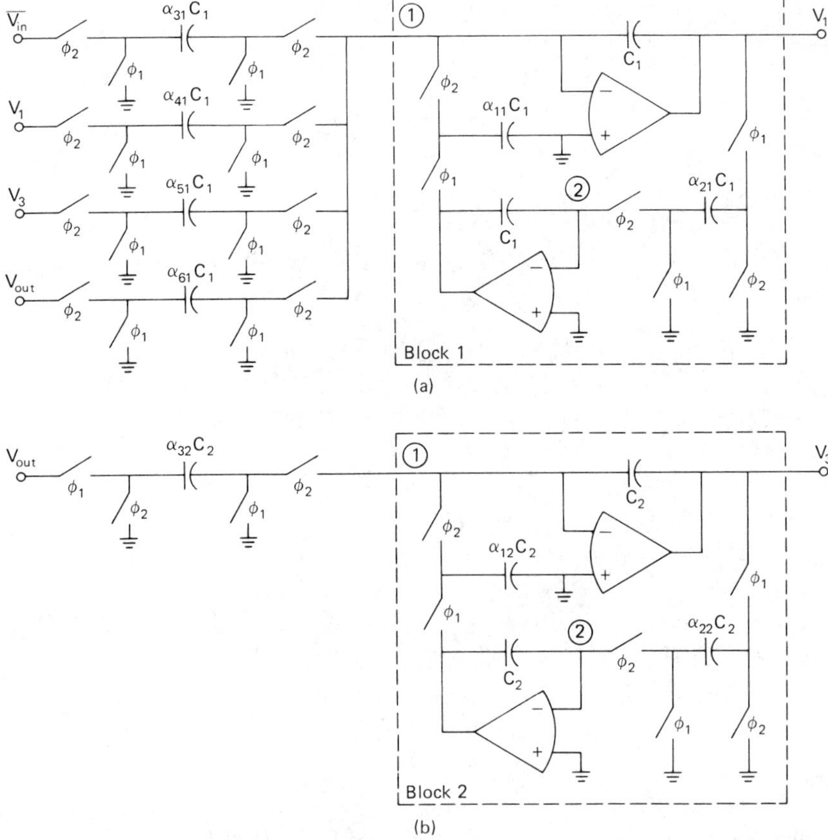

Fig. 4.5–2. (a.) Realization of Eq. (15). (b.) Realization of Eq. (16). (c.) Realization of Eq. (17).

Equation (17) is realized by the configuration in Fig. 4.5–2(c). A tradeoff may be made between the number of capacitors and the number of op amps by breaking Eq. (12) into two equations requiring one op amp each. Assuming that z^{-1} can be approximated by $1 - sT_n$ permits one to express the approximate s-domain transfer function of Fig. 4.5–2(c) as

$$V_{\text{out}} \cong -\alpha_{33}\overline{V_{\text{in}}} - \alpha_{43}V_1 - \alpha_{53}V_3 - \frac{s}{s^2 + (\alpha_{13}\alpha_{23}/T_n^2)}\left[\frac{\alpha_{93}}{T_n}V_3 + \frac{\alpha_{103}}{T_n}V_{\text{out}}\right] \quad \text{(22a)}$$

where

$$\alpha_{63} = \alpha_{23}\alpha_{33}, \qquad \alpha_{73} = \alpha_{23}\alpha_{43}, \qquad \alpha_{83} = \alpha_{23}\alpha_{53} \quad \text{(22b)}$$

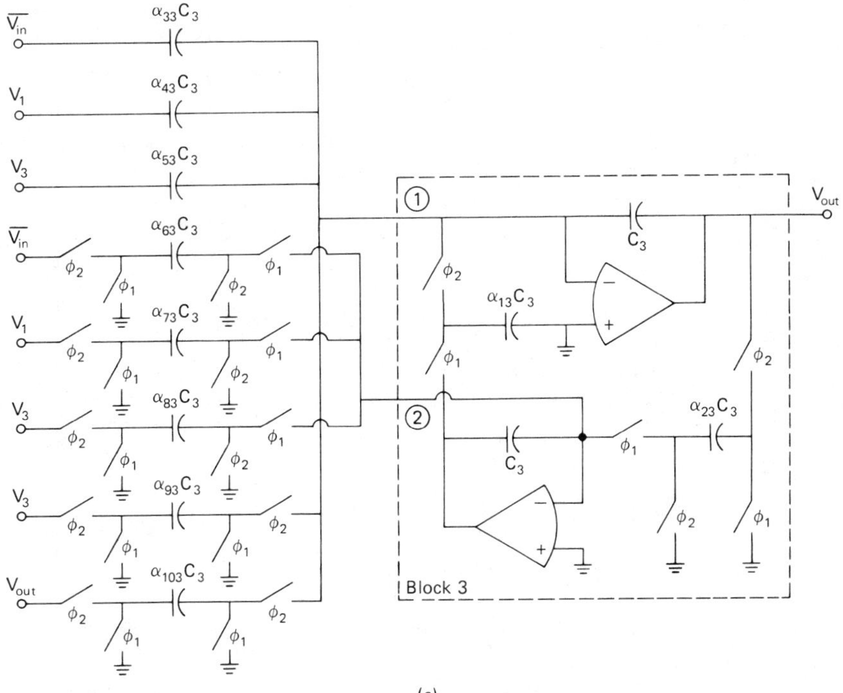

(c)

Fig. 4.5–2. (*Continued*)

Equating Eqs. (17) and (22a) gives

$$\alpha_{13} = \alpha_{23} = T_n = \Omega_n T = \omega_r / f_c \tag{23a}$$

$$\alpha_{33} = \alpha_{43} = \alpha_{53} = R_{4n}/(R_{4n} + R_{on}) \tag{23b}$$

$$\alpha_{93} = \alpha_{103} = \frac{T_n R_{on} R_{4n}}{L_{E2n}(R_{4n} + R_{on})} = \frac{\text{BW } T R_{on} R_{4n} C_{2n}}{(R_{4n} + R_{on})} \tag{23c}$$

and

$$\alpha_{63} = \alpha_{73} = \alpha_{83} = \alpha_{23} R_{4n}/(R_{4n} + R_{on}) \tag{23d}$$

The realization of Fig. 4.5–1(d) is shown in Fig. 4.5–3. Note that if $R = R_{4n} = R_{on}$, considerable simplification results in the first and last stages. Each of the blocks is identical and equivalent to the dotted box in Fig. 4.3–7(a).

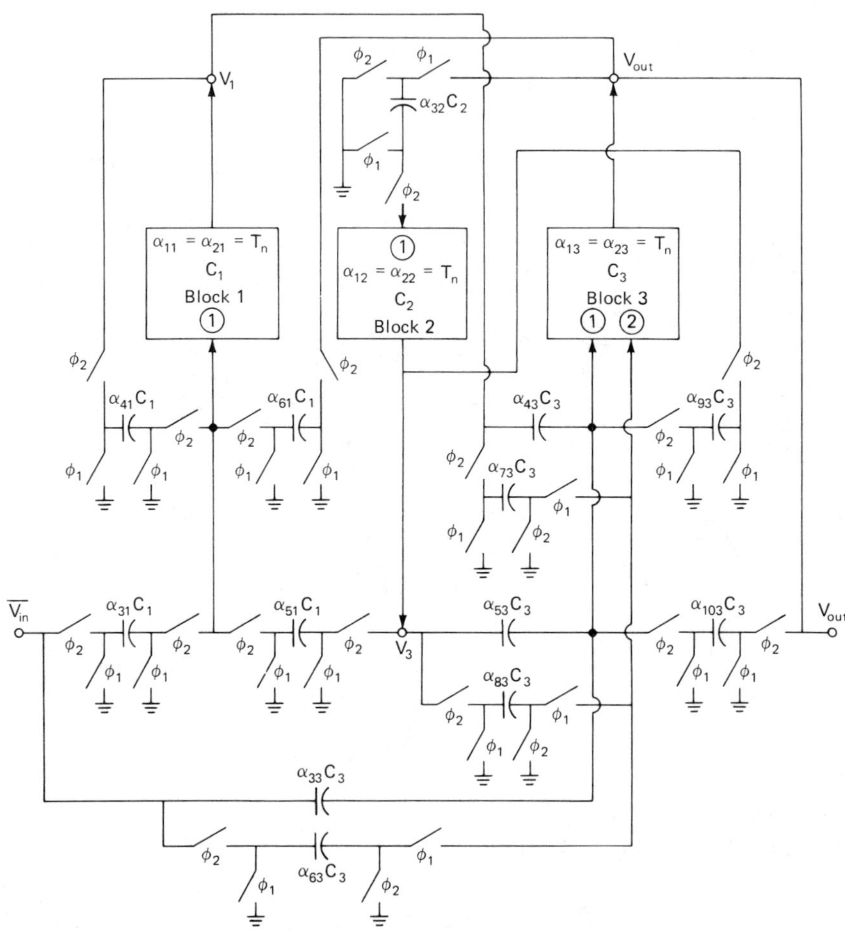

Fig. 4.5–3. SC realization of Fig. 4.5–1(d).

Example 4.5–1. *Design of a third-order, band-elimination filter.* Use the design method presented above to realize a third-order, Chebyshev, band-elimination filter having a center frequency of 1 kHz, a bandwidth of 500 Hz, and a clock frequency of 100 kHz. The filter is to have a 1-dB ripple in the passband regions.

If we select Fig. 4.5–1(a) as the low-pass prototype for this realization, then we can use the above formulas to design the desired filter. The values of the components of Fig. 4.5–1(a) are $R_{0n} = 1$ ohm, $L_{1n} = 2.0236$ H, $C_{2n} = 0.9941$ F, $L_{3n} = 2.0236$ H, and $R_{4n} = 1$ ohm. Choosing $R = 1$ ohm gives the following capacitor ratios

Stage 1:

$\alpha_{11} = \alpha_{21} = 0.06283$

$\alpha_{31} = \alpha_{41} = \alpha_{51} = \alpha_{61} = 0.25429$

Stage 2:

$\alpha_{12} = \alpha_{22} = 0.06283$

$\alpha_{32} = 0.25429$

Stage 3:

$\alpha_{13} = \alpha_{23} = 0.06283$

$\alpha_{13} = \alpha_{43} = \alpha_{53} = 0.500$

$\alpha_{93} = \alpha_{103} = 0.01562$

$\alpha_{63} = \alpha_{73} = \alpha_{83} = 0.03142$

The largest capacitor ratio is seen to be 0.01562 or 63.662, which is primarily because the bandwidth is one-half the center frequency and the clock frequency is 100 times the center frequency. Figure 4.5–4 shows the ideal frequency response of the filter designed in this example.

Let us next consider the realization of band-elimination filters that have finite $j\omega$ axis zeros. Figure 4.5–5(a) is the low-pass prototype of a third-order filter having finite $j\omega$ axis zeros. We follow the same procedure that was used in the previous sections for filters with finite $j\omega$ axis zeros. Figure 4.5–5(b) shows the results after eliminating the capacitor loop consisting of C_{1n}, C_{2n}, and C_{3n}. The next step is to normalize this modified low-pass prototype by Eq. (4), resulting in Fig. 4.5–5(c). This step is followed by the application of the low-pass-to-band-elimination transformation of Eq. (1) with ω_r equal to 1 rps. The result is shown in Fig. 4.5–5(d). We again

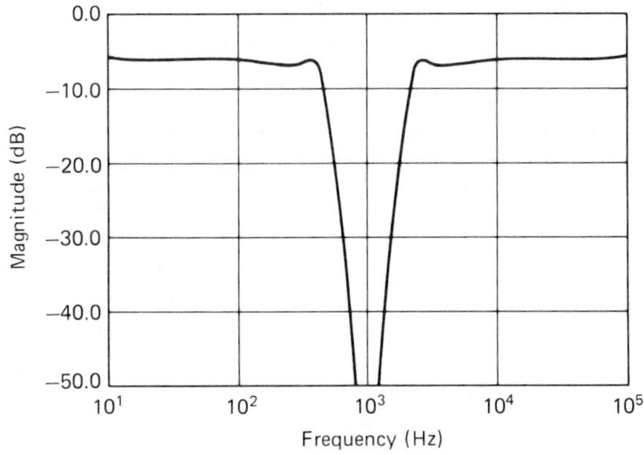

Fig. 4.5–4. Ideal frequency response for Example 4.5–1.

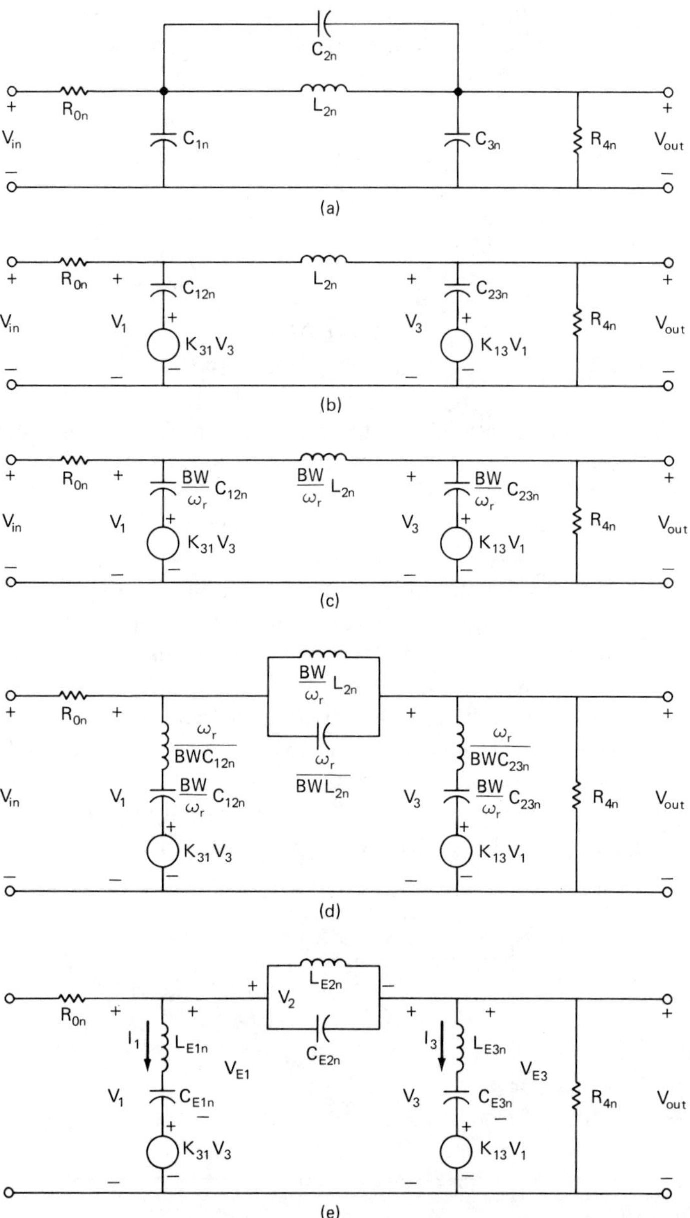

Fig. 4.5–5. (a.) Low-pass prototype. (b.) Removal of the capacitor loop. (c.) Frequency normalization of (b.). (d.) Band-elimination prototype. (e.) Simplification in notation for (d.).

note that the voltage-controlled voltage-sources are not affected by the transformation. Figure 4.5–5(e) gives the band-elimination prototype filter with the indicated simplifications in the notation. The state variables have been selected as I_1, V_2, I_3, and V_{out}.

Each of the state variables can be expressed using the method illustrated for Fig. 4.5–1(d). The state variables of Fig. 4.5–5(e) can be expressed as

$$V_1' = Y_1 V_{E1} = \frac{sR}{(s^2+1)L_{E1n}}(V_1 - K_{31}V_3)$$

$$= \frac{sR}{(s^2+1)L_{E1n}}[V_2 + (1-K_{31})V_{\text{out}}] \tag{24}$$

$$V_2 = Z_2 I_2 = \frac{s}{(s^2+1)RC_{E2n}}\left(V_3' + \frac{R}{R_{4n}}V_{\text{out}}\right) \tag{25}$$

$$V_3' = Y_3 V_{E3} = \frac{sR}{(s^2+1)L_{E3n}}(V_{\text{out}} - K_{13}V_1)$$

$$= \frac{sR}{(s^2+1)L_{E3n}}[V_{\text{out}}(1-K_{13}) - K_{13}V_2] \tag{26}$$

and

$$V_{\text{out}} = V_{\text{in}} - V_2 - (R_{0n}/R)[V_1' + V_3'] - (R_{0n}/R_{4n})V_{\text{out}} \tag{27}$$

or

$$V_{\text{out}} = \frac{R_{4n}}{R_{4n}+R_{0n}}\left[V_{\text{in}} - V_2 - \frac{R_{0n}}{R}V_1' - \frac{R_{0n}}{R}V_3'\right] \tag{28}$$

Equations (24), (25), (26), and (28) can be realized directly (see Problem 4.44), but an op amp summer is required in addition to the three circuits similar to Fig. 4.3–3. In the approach we take here, we shall reduce the number of state equations from four to three by substituting Eq. (24) into Eq. (28) to obtain

$$V_{\text{out}} = \frac{R_{4n}R_{0n}}{L_{E1n}(R_{4n}+R_{0n})}\frac{s}{s^2+1}[-(1-K_{31})V_{\text{out}} - V_2]$$
$$+ \frac{R_{4n}}{R_{4n}+R_{0n}}\left[-\overline{V_{\text{in}}} - V_2 - \frac{R_{0n}}{R}V_3'\right] \tag{29}$$

Equations (25), (26), and (29) are suitable for realization using the SC circuits of Fig. 4.3–7. Figure 4.5–6(a) is a realization for Eq. (25). If we assume that z^{-1} can be approximated by $1 - sT_n$, then the s-domain transfer function for Fig. 4.5–6(a) can be written as

$$V_2 \cong \frac{s}{s^2 + (\alpha_{11}\alpha_{21}/T_n^2)}\left(\frac{\alpha_{31}}{T_n}V_3' + \frac{\alpha_{41}}{T_n}V_{\text{out}}\right) \tag{30}$$

Equating (25) to Eq. (30) gives

$$\alpha_{11} = \alpha_{21} = T_n = \Omega_n T = \omega_r/f_c \tag{31a}$$

$$\alpha_{31} = T_n/(RC_{E2n}) = (\text{BWL}_{2n}/f_c R) \tag{31b}$$

and

$$\alpha_{41} = T_n/(R_{4n}C_{E2n}) = (\text{BWL}_{2n}/f_c R_{4n}) \tag{31c}$$

A realization for Eq. (26) is shown in Fig. 4.5–6(b). Assuming that z^{-1} can be approximated by $1 - sT_n$ allows the s-domain transfer function to be approximated as

$$V_3' \cong \frac{s}{s^2 + (\alpha_{12}\alpha_{22}/T_n^2)}\left(\frac{\alpha_{32}}{T_n}V_{\text{out}} - \frac{\alpha_{42}}{T_n}V_2\right) \tag{32}$$

Equating Eq. (26) to Eq. (32) gives

$$\alpha_{12} = \alpha_{22} = T_n = \Omega_n T = \omega_r/f_c \tag{33a}$$

$$\alpha_{32} = T_n R(1 - K_{13})/L_{E3n} = \text{BWC}_{23n}R(1 - K_{13})/f_c \tag{33b}$$

and

$$\alpha_{42} = T_n R K_{13}/L_{E3n} = \text{BWC}_{23n}RK_{13}/f_c \tag{33c}$$

Using both inputs of Fig. 4.3–7 permits the realization of Eq. (29) by Fig. 4.5–6(c). Using the high-frequency sampling assumption yields the following approximate s-domain transfer function

$$V_{\text{out}} \cong -\alpha_{33}V_{\text{in}} - \alpha_{43}V_2 - \alpha_{53}V_3' - \frac{s}{s^2 + (\alpha_{13}\alpha_{23}/T_n^2)}\left(\frac{\alpha_{93}}{T_n}V_{\text{out}} + \frac{\alpha_{103}}{T_n}V_2\right) \tag{34}$$

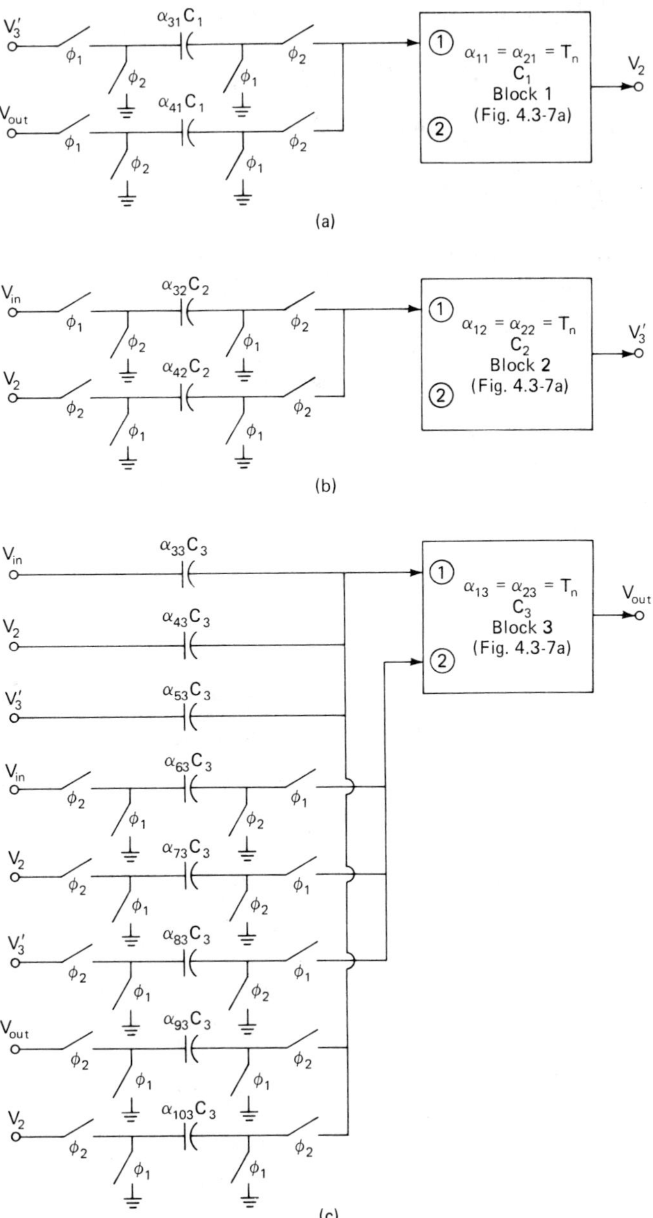

Fig. 4.5–6. (a.) SC realization of Eq. (25). (b.) SC realization of Eq. (26). (c.) SC realization of Eq. (29).

Equating Eq. (29) to Eq. (34) gives

$$\alpha_{13} = \alpha_{23} = T_n = \Omega_n T = \omega_r / f_c \tag{35a}$$

$$\alpha_{33} = \alpha_{43} = R_{4n}/(R_{4n} + R_{0n}) \tag{35b}$$

$$\alpha_{53} = \frac{R_{0n} R_{4n}}{R(R_{4n} + R_{0n})} \tag{35c}$$

$$\alpha_{63} = \alpha_{73} = \alpha_{23} R_{4n}/(R_{4n} + R_{0n}) \tag{35d}$$

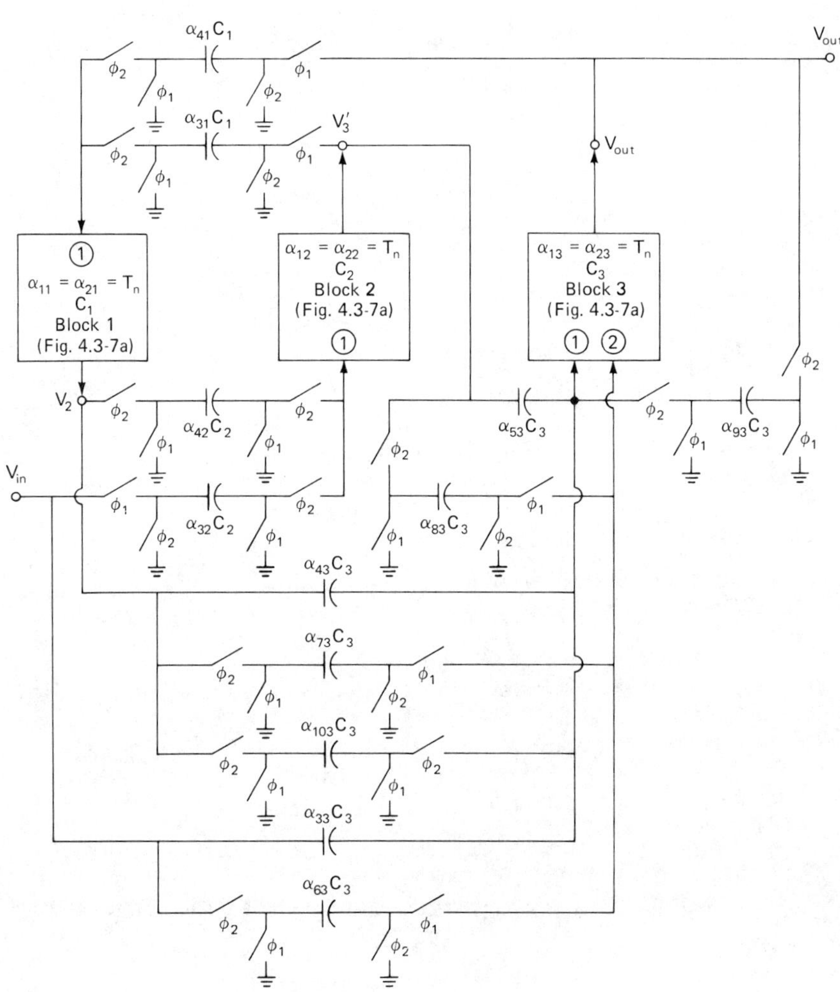

Fig. 4.5–7. SC realization of Fig. 4.5–6(e).

$$\alpha_{83} = \frac{R_{0n} R_{4n} \alpha_{23}}{R(R_{4n} + R_{0n})} \tag{35e}$$

$$\alpha_{93} = \frac{R_{4n} R_{0n}(1 - K_{31}) T_n}{(R_{4n} + R_{0n}) L_{E1n}} = \frac{R_{4n} R_{0n} \mathrm{BWC}_{12n}(1 - K_{31})}{(R_{4n} + R_{0n}) f_c} \tag{35f}$$

and

$$\alpha_{103} = \frac{R_{4n} R_{0n} T_n}{(R_{4n} + R_{0n}) L_{E1n}} = \frac{R_{4n} R_{0n} \mathrm{BWC}_{12n}}{(R_{4n} + R_{0n}) f_c} \tag{35g}$$

Figure 4.5–7 shows a realization of Fig. 4.5–5(e) in terms of the second-order, bandpass blocks. Each of these blocks is identical and equivalent to the dotted box in Fig. 4.3–7(a).

Example 4.5–2. *Design of a third-order, band-elimination having finite $j\omega$ axis zeros.* Use the above method to design a third-order, elliptic, band-elimination filter having a bandwidth of 500 Hz, a center frequency of 1000 Hz, a clock frequency of 100 kHz, and a 1-dB ripple in the passbands. The attenuation in the stopband must be greater than 30 dB. In Appendix A, we select a design having 34 dB of attenuation in the stopband and a ratio of ω_{SB} to ω_{PB} (see Fig. 3.1–2) of 2. This ratio implies that the attenuation of 34 dB will exist over a bandwidth of 250 Hz geometrically centered at 1000 Hz. The values of the low-pass prototype of Fig. 4.5–5(a) are R_{0n} = 1 ohm, C_{1n} = 1.85199 F, L_{2n} = 1.30964 H, C_{2n} = 0.3972 F, C_{3n} = 1.85199 F, and R_{4n} = 1 ohm. Choosing R = 1 ohm gives the following capacitor ratios

Stage 1:

$$\alpha_{11} = \alpha_{21} = 0.06283$$
$$\alpha_{31} = \alpha_{41} = 0.04387$$

Stage 2:

$$\alpha_{12} = \alpha_{22} = 0.06283$$
$$\alpha_{32} = 0.05818$$
$$\alpha_{42} = 0.01248$$

Stage 3:

$$\alpha_{13} = \alpha_{23} = 0.06283$$
$$\alpha_{33} = \alpha_{43} = \alpha_{53} = 0.5000$$
$$\alpha_{63} = \alpha_{73} = \alpha_{83} = 0.5\alpha_{23} = 0.031416$$
$$\alpha_{93} = 0.02909$$
$$\alpha_{103} = 0.03533$$

The synthesis method developed for ladder filters of this and the previous sections is also applicable to general RLC ladder filters. The primary difference between general RLC ladders and RLC ladders generated by transformation from the low-pass prototype is that a mixture of state equations may be used. For example, some of the state equations may be realized by a differencing integrator, and others may require a bandpass realization. An example of a general RLC ladder filter is shown in Fig. 4.5–8. This filter cannot be generated from a low-pass prototype by applying any of the transformations considered so far. We shall assume that the component values of Fig. 4.5–8 have been frequency-normalized, although this may not be the case. If the filter has not been normalized, then we use T rather than T_n as the clock period. Keeping with the ideas concerning the desired state variables, we select I_1, I_2, and V_{out} as state variables. These state variables can be expressed in terms of the remaining state variables and the input, V_{in}, as shown below

$$I_1 = (1/sL_{1n})[V_{in} - R_{0n}I_1 - V_{out}] \tag{36}$$

$$I_2 = \frac{s/L_{2n}}{s^2 + (1/L_{2n}C_{2n})} V_{out} \tag{37}$$

and

$$V_{out} = (I_1 - I_2)R_{3n} \tag{38}$$

Equation (37) may be eliminated by substituting it into Eq. (38) to get

$$V_{out} = -\frac{R_{3n}}{R}\overline{V_1'} - \frac{sR_{3n}/L_{2n}}{s^2 + (1/L_{2n}C_{2n})} V_{out} \tag{39}$$

Equation (36) can be rewritten using voltage analogs of current as

$$\overline{V_1'} = \frac{R}{sL_{1n}}\left[-V_{in} - \frac{R_{0n}}{R}\overline{V_1'} + V_{out}\right] \tag{40}$$

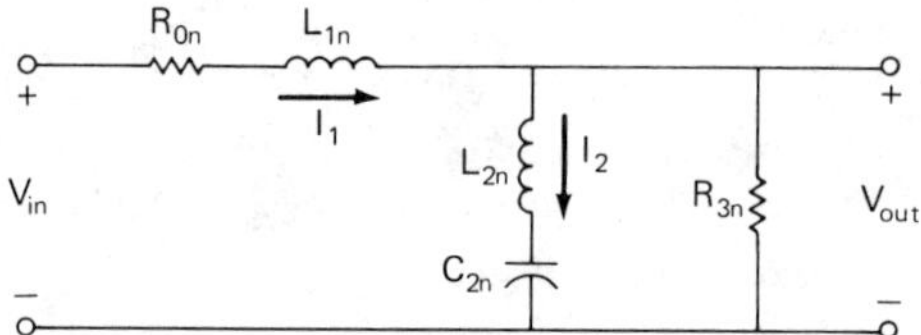

Fig. 4.5–8. Third-order, general RLC filter.

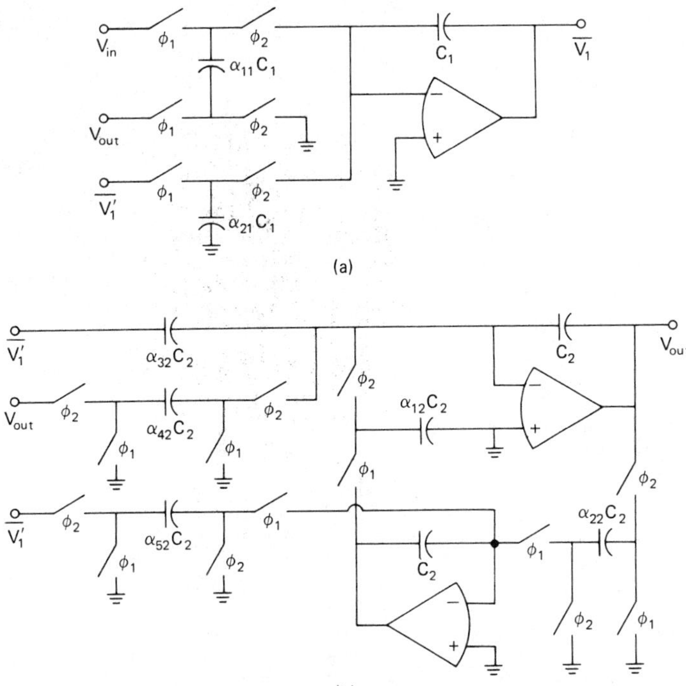

Fig. 4.5–9. (a.) SC realization of Eq. (40). (b.) SC realization of Eq. (39).

Equations (39) and (40) are the reduced set of state-variable equations that will be used to synthesize Fig. 4.5–8.

Fig. 4.5–9(a) is a realization of Eq. (40). It can be shown that the approximate s-domain transfer function is

$$\overline{V_1'} \cong \frac{\alpha_{11}}{sT_n}(V_{\text{out}} - V_{\text{in}}) - \frac{\alpha_{12}}{sT_n}\overline{V_1'} \tag{41}$$

Equating Eqs. (40) and (41) gives

$$\alpha_{11} = \frac{RT_n}{L_{1n}} = \frac{R\Omega_n}{L_{1n}f_c} \tag{42a}$$

$$\alpha_{12} = \frac{R_{\text{on}}T_n}{L_{1n}} = \frac{R_{\text{on}}\Omega_n}{L_{1n}f_c} \tag{42b}$$

Figure 4.5–9(b) is a realization of Eq. (39) using the scheme proposed in Fig. 4.3–7. The approximate s-domain transfer function is written as

$$V_{\text{out}} \cong -\alpha_{32}\,\overline{V_1'} - \frac{s\alpha_{42}/T_n}{s^2 + (\alpha_{12}\alpha_{22}/T_n^2)}\,V_{\text{out}} \tag{43}$$

Equating Eqs. (39) and (43) results in

$$\alpha_{12} = \alpha_{22} = T_n/(L_{2n}C_{2n})^{1/2} = \Omega_n/[(L_{2n}C_{2n})^{1/2}f_c] \tag{44a}$$

$$\alpha_{32} = R_{3n}/R \tag{44b}$$

and

$$\alpha_{42} = R_{3n}T_n/L_{2n} = R_{3n}\Omega_n/L_{2n}f_c \tag{44c}$$

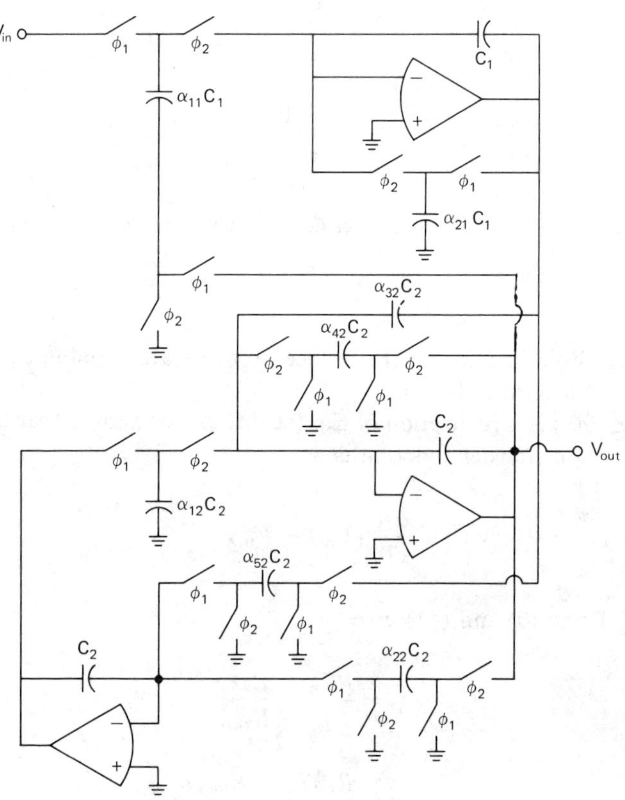

Fig. 4.5–10. SC realization of Fig. 4.5–8.

Figure 4.5–10 shows an SC realization of Fig. 4.5–8 using the realizations of Fig. 4.5–9.

The above example illustrates how the method is applied to general RLC ladder filters. Sections 4.1 through 4.5 constitute a very general and flexible method of realizing SC filters. Ladder filters exhibit very good sensitivity properties and typically require capacitance ratios that are more closer to unity than all other forms of filter realizations. The next two sections will examine the influence of the clock frequency upon the design and performance of the filter. All the previous work in this chapter has assumed that the ratio of the clock frequency to the filter cutoff frequency or center frequency has been large enough so that the z-domain variable z can be replaced by $1 + sT$. The methods developed are still important, as in many cases the high sampling frequency assumed is valid or the filters designed using this assumption are the starting point for methods that incorporate the effect of the sampling frequency.

4.6 NONIDEAL PERFORMANCE OF SC LADDER FILTERS

In the previous sections of this chapter we have made several assumptions that will be examined in this section. It has been assumed that a high clock frequency was used and that the influence of stray capacitance could be ignored. The objective of this section is to review the results of the previous sections from the viewpoint that these assumptions are no longer valid.

It is important to consider the effects of stray capacitance upon the realization. The best realizations for this purpose are the noninverting Type I LDI and inverting Type II LDI shown in Fig. 4.6–1. In the previous section we did not make a special effort to use stray-insensitive configurations. For example, in Fig. 4.2–5, which pertains to Example 4.2–1, only inverting Type I LDI's are used. This circuit can be converted into a stray-insensitive realization by the use of the integrators in Fig. 4.6–1. The result is shown in Fig. 4.6–2. An interesting fact can be observed from the circuit in Fig. 4.6–2. We note that the capacitor used to sample V_{in} is equal to $2\alpha_{11}C_1$. If this

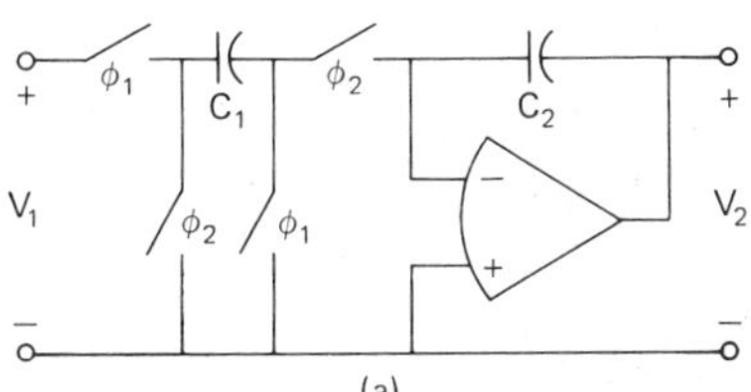

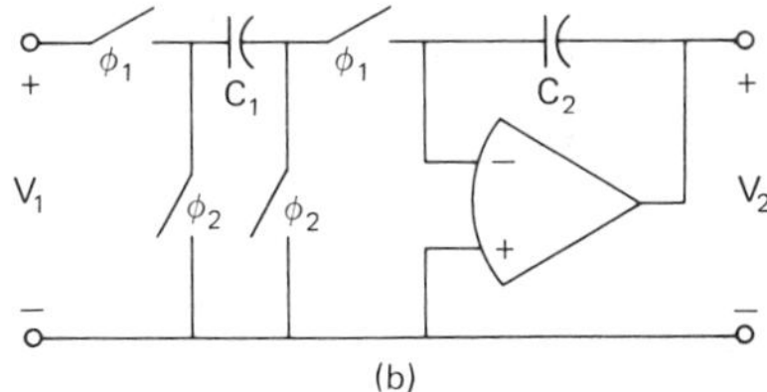

Fig. 4.6–1. Stray insensitive integrators. (a.) Noninverting Type I LDI. (b.) Inverting Type II LDI.

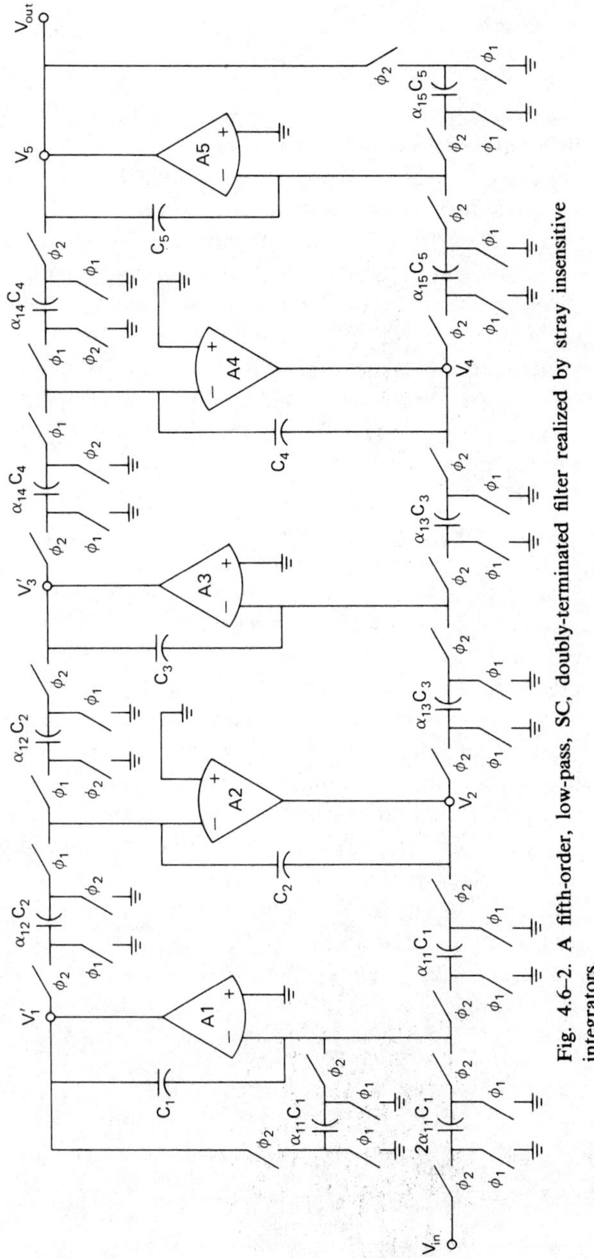

Fig. 4.6–2. A fifth-order, low-pass, SC, doubly-terminated filter realized by stray insensitive integrators.

capacitor only samples V_{in}, then it may be used as shown to provide a gain adjustment for the entire filter. The value of $2\alpha_{11}C_1$ causes a 6-dB gain, which is used to compensate for the 6-dB loss of doubly terminated RLC passive ladder filters. We observe that the capacitor in Fig. 4.2–5, which samples V_{in}, also samples the output of $A2$. Therefore this scaling capability is not available. This is just another example of the additional flexibility of SC over continuous methods of circuit realization.

Next let us consider what happens to the SC filter when the sampling frequency, ω_c, is not much greater than the filter frequency ω. Because we are using SC integrators to simulate the RLC passive ladders, Table 2.5–2 becomes important in this consideration. Regardless of the integrator used, Table 2.5–2 shows that there will always be a magnitude error that increases as ω approaches ω_c. Figure 2.5–6 shows how this error manifests itself in

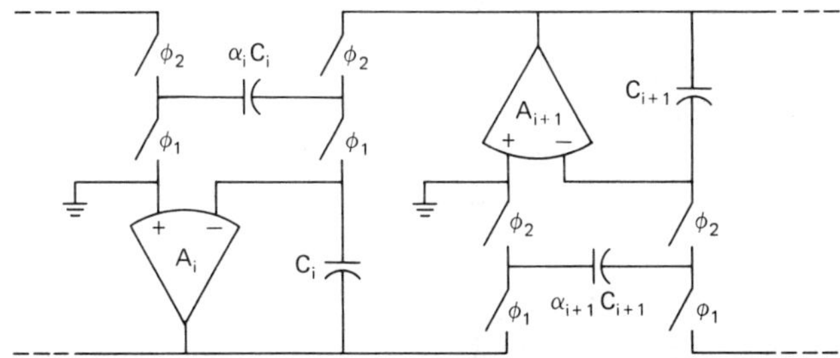

Fig. 4.6–3. Portion of the interior realization of a passive ladder network.

the frequency response. Figure 4.6–3 represents the i and $i + 1$ integrators of a low-pass ladder simulation using Type I LDIs. The feedback path through the two integrators can be shown to have a phase shift of exactly zero degrees. This result is a consequence of the integrators having an ideal Q factor.[7,8] In general we may express the transfer function of an SC integrator as

$$H_I(e^{j\omega T}) = \pm \frac{\omega_0}{j\omega}(1 + \epsilon)e^{j\theta} \tag{1}$$

where ϵ is the fractional magnitude error, and θ is the phase error. The second and third columns of Table 2.5–1 give ϵ in percentage and θ in

[7] L. T. Bruton, "Low-Sensitivity Digital Filters," *IEEE Trans. on Circuits and Systems,* Vol. CAS-22, No. 3, March 1975, pp. 168–176.

[8] K. Martin, "Improved Circuits for the Realization of Switched Capacitor Filters," *IEEE Trans. on Circuits and Systems,* Vol. CAS-27, No. 4, April 1980, pp. 237–244.

degrees. If θ is much less than 360° (or 2π) and ϵ is much less than unity, then Eq. (1) can be approximated as

$$H_I(e^{j\omega T}) \approx \pm \frac{\omega_0}{j\omega}(1+\epsilon)(1+j\theta) \approx \pm \left[\frac{\omega_0}{j\omega}(1+\epsilon) + \frac{\omega_0\theta}{\omega}\right] \tag{2}$$

However, the Q factor of an integrator whose transfer function is

$$H_I(j\omega) = \pm \left[\frac{1}{R(\omega) + jX(\omega)}\right] \tag{3}$$

is given as

$$Q = \frac{X(\omega)}{R(\omega)} \tag{4}$$

Comparing Eqs. (2) and (3) results in the Q factor in terms of ϵ and θ as

$$Q \cong \frac{1+\epsilon}{\theta} \approx \frac{1}{\theta} \tag{5}$$

We see that because $\theta = 0$ for the LDI, their Q factor is infinity if ϵ can be ignored. We shall show later in this section how to accomplish a predistortion that will minimize the effects of ϵ.

Unfortunately, the LDI realization of Fig. 4.2–5 is not without several disadvantages. One of these disadvantages is the stray-sensitive realizations that were solved in the realization of Fig. 4.6–2. A second problem involves an additional delay due to the source and load terminations. Consider the load termination and the previous stage of Fig. 4.2–5. The first thing to be noted is that the output stage is not necessarily a noninverting Type I LDI because of the presence of the feedback path around $A5$. This feedback path has an extra half-period of delay. This can be clearly seen by observing that, during the ϕ_2 clock phase, $\alpha_{14}C_4$ of $A4$ is sampling the output of $A5$. At the next phase period, ϕ_1, $\alpha_{15}C_5$ of $A5$ samples both the output of $A4$ and the *previous output of A5,* which was sampled by $A4$ a half-period earlier. This extra delay will cause an error in the simulated termination resistance. For practical high-order filters, the error in the response due to this incorrect termination is often negligible until ω begins to approach ω_c. The terminations for the stray-insensitive version also suffer an error in the simulated termination.

Let us examine more closely the problem presented by the terminations

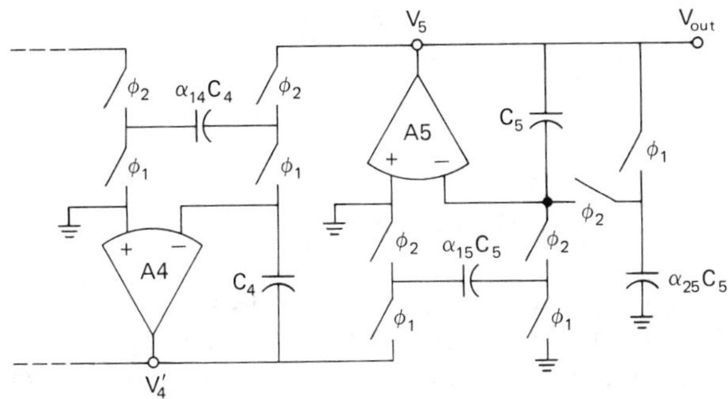

Fig. 4.6–4. An equivalent circuit for the last two stages of Fig. 4.2–5(e).

of the passive RLC ladder filters. Figure 4.6–4 shows the load and previous stage of the fifth-order, low-pass realization of Fig. 4.2–5. We have redrawn the output stage to simplify the analysis and to make comparisons with earlier work. $\alpha_{15}C_5$ of the fifth integrator has been separated into two capacitors, $\alpha_{15}C_5$ and $\alpha_{25}C_5$, which are normally equal and can be lumped into a single capacitor. The transfer function of Fig. 4.6–4 from V_4' to V_5 is sampled during the ϕ_2 phase period and is found from Eq. (21) of Section 2.6 as

$$H_{54}(z) = \frac{V_5(z)}{V_4'(z)} = \frac{\alpha_{15}z^{-1}}{1 - z^{-1} + \alpha_{25}\, z^{-1}} \tag{6}$$

where C_1, C_2, and C_4 of Fig. 2.6–8 are $\alpha_{15}C_5$, $\alpha_{25}C_5$, and C_5 of Fig. 4.6–4, respectively. The frequency response of Eq. (6) can be expressed as

$$H_{54}(j\omega) = \frac{\alpha_{15}}{\cos \omega T_n - 1 + \alpha_{25} + j \sin \omega T_n} \tag{7}$$

If the load termination of Fig. 4.2–5 is to be realized by RC active means, then the desired frequency response of the last integrator can be found from Eq. (36) of Section 4.2 and is given as

$$H_{54}(j\omega) = \frac{V_{out}(j\omega)}{V_4(j\omega)} = \frac{1}{j\omega(L_{5n}/R_{6n}) + 1} \tag{8}$$

A method of predistortion to account for the errors in the simulated termination resistance amounts to equating Eqs. (7) and (8) and solving for the capacitor ratios α_{15} and α_{25}. This approach results in the following relationships

$$\alpha_{15} = \frac{T_n R_{6n}}{L_{5n}} \frac{\sin \omega T_n}{\omega T_n} \tag{9}$$

and

$$\alpha_{25} = \alpha_{15} + 1 - \cos \omega T_n \tag{10}$$

As ωT_n approaches zero, these equations reduce to Eq. (37) in Section 4.2.

To complete the predistortion, we must select a value of ω at which to complete our design. The two expressions given in Eqs. (7) and (8) will be equated at this frequency. A reasonable value to pick might be the cutoff frequency of the filter, ω_0, or the reciprocal time constant of the continuous terminating integrator stage, $\omega_5 = R_{6n}/L_{5n}$. In many cases, these two frequencies are very close. If we select ω_5, then Eqs. (9) and (10) become

$$\alpha_{15} = \frac{T_n R_{6n}}{L_{5n}} \frac{\sin \omega_5 T_n}{\omega_5 T_n} = \sin \omega_5 T_n \tag{11}$$

and

$$\alpha_{25} = \sin \omega_5 T_n + 1 - \cos \omega_5 T_n \tag{12}$$

An example will illustrate the use of these relationships.

Example 4.6–1. *Predistorting for termination errors.* In Example 4.2–1, a fifth-order, low-pass ladder filter was designed, resulting in the realization of Fig. 4.2–5. Find new values for the load termination stage if the filter cutoff frequency is (a) 1000 Hz and (b) 10,000 Hz, if the clock frequency remains constant at 100 kHz.

a. For $f_0 = 1000$ Hz, we get

$$\alpha_{15} = 0.02943 \tag{13a}$$

and

$$\alpha_{25} = 0.02986 \tag{13b}$$

where normally α_{15} is equal to α_{25}.

b. For $f_0 = 10{,}000$ Hz, we get

$$\alpha_{15} = 0.2901 \tag{14a}$$

and

$$\alpha_{25} = 0.3331 \tag{14b}$$

Again, α_{15} is normally equal to α_{25}. These values should be compared with the values of Eq. (38e) in Section 4.2.

Similar procedures may be used to obtain predistorted values for the load termination with different realizations, such as that shown in Fig. 4.6–2, which uses the stray-insensitive configuration. If the filter is doubly terminated, then a half-delay error can also occur in the source termination. A scheme that uses a complex conjugate approach with one termination having a half-period delay and the other termination having a half-period advance has been used with good results.[9]

Fortunately, the LDI integrators of Figs. 4.2–5 and 4.6–2 have only a magnitude error and no phase error. However, if ω begins to approach ω_c, it is necessary to compensate for the effects of the magnitude errors. We shall discuss several schemes that are suitable for compensating the magnitude errors of LDIs. The problem is much more serious with DDI's, for there is both a magnitude and a phase error. We shall not consider the DDI compensation techniques here.[10]

The first scheme is the simplest to use. Here predistortion is applied to the integrators themselves. The transfer function of an ideal SC simulation of an analog integrator can be expressed as

$$H_I(j\omega) = \frac{-1}{j\omega \dfrac{C_2 T_I}{C_1}} \tag{15}$$

which was given originally in Eq. (14) of Section 2.5. The actual transfer function of SC simulation of an analog integrator can be expressed as

$$H_A(j\omega) = H^{oe}(e^{j\omega T}) = -\frac{C_1}{j\omega T C_2} \frac{\dfrac{\omega T}{2}}{\sin \dfrac{\omega T}{2}} \tag{16}$$

---

[9] D. J. Allstot, "MOS Switched Capacitor Ladder Filters," Ph.D. thesis, EE Dept., University of California, Berkeley, California 94720.

[10] T. C. Choi and R. W. Brodersen, "Considerations for High-Frequency Switched-Capacitor Ladder Filters," *IEEE Trans. on Circuits and Systems,* Vol. CAS-27, No. 6, June 1980, pp. 545–552.

which was derived in Eq. (18) of Section 2.5. If we define ω as the discrete time frequency in radians/second and Ω as the continuous time frequency in radians/second, then equating Eqs. (15) to (16) and solving for T_I gives

$$T_I = \frac{2}{\Omega} \sin\left(\frac{\omega T}{2}\right) \tag{17}$$

If we replace the clock period T in the equations involving integrator simulation of reactive elements with T_I, then a predistortion for the effects of the magnitude will be accomplished. We must select a value for ω and Ω in Eq. (17). Typically the cutoff frequency of the filter, Ω_0, is used for both.

Example 4.6–2. *Predistortion of the integrators of Example 4.2–1.* Use the predistortion technique explained above to predistort the interior stages of the fifth-order filter of Example 4.2–1. Solving for T_I gives 9.9984 μs. This gives an f_{CI} of 100,016 Hz. Therefore all values of integrating capacitors should be multiplied by 1.00016. We note that, because $f_0 = 0.01\ f_c$, the magnitude error is negligible. If $f_c = 10$ kHz rather than 100 kHz, this ratio would be 1.0166.

A second scheme applies predistortion to the roots of the filter. One may use the LDI transformation given as

$$s = \frac{1}{T}(z^{-1/2} - z^{1/2}) = \frac{1}{T} \sinh\left(\frac{s'}{2f_c}\right) \tag{18}$$

to accomplish this predistortion. If we define $s = \alpha_{di} + j\omega_{di}$ as the predistorted continuous roots and $s' = \alpha_i' + j\omega_i'$ as the discrete time roots, then we may solve for α_{di} and ω_{di} as

$$\alpha_{di} = 2f_c \cos\left(\frac{\omega_i'}{2f_c}\right) \sinh\left(\frac{\alpha_i'}{2f_c}\right) \tag{19}$$

and

$$\omega_{di} = 2f_c \sin\left(\frac{\omega_i'}{2f_c}\right) \cosh\left(\frac{\alpha_i'}{2f_c}\right) \tag{20}$$

These equations can be used to calculate the predistorted root positions. Figure 4.6–5 illustrates the location of the predistorted roots with respect to the ideal Chebyshev roots, which are on the ellipse. Unfortunately it is necessary to resynthesize the passive RLC structure before the SC design of the ladder filter can begin.

The methods presented above can be applied to other types of filters besides the low-pass. As one succeeds in operating closer to the sampling frequency, the requirements for the antialiasing filter will become more severe. Another

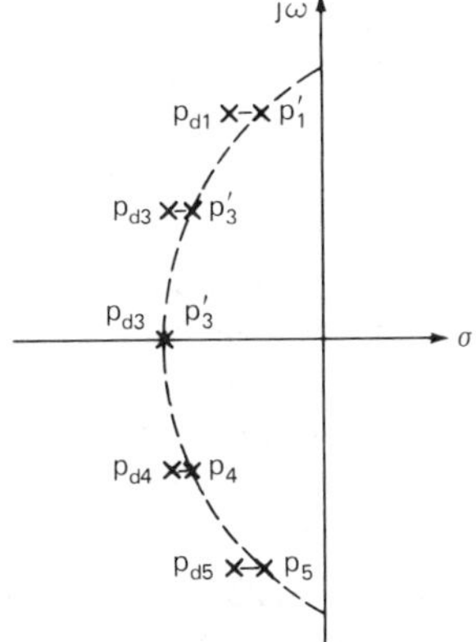

Fig. 4.6–5. Predistortion of the initial pole positions. p_{di} are the predistorted poles and p_i are the discrete-time poles placed at the desired locations.

consideration is important if a sample-and-hold circuit is used. We have seen that the frequency response of a sample-and-hold circuit is a sin x/x form. The filter to be realized may have to be predistorted independent of the above considerations to result in the desired response.

Another method of compensation transforms the s-domain specifications into the z-domain. A realization is proposed in the z-domain, and the design is completed by solving a set of simultaneous equations. This method has resulted in realizations that are termed *exact SC ladder filters.* This category of filters is discussed in the next section, which completes the considerations given in this section toward achieving SC filters not requiring high sampling frequencies.

It was seen in this section that phase errors due to the filter termination loops caused deviations in the passband magnitude response for LDI-transformed, RLC, low-pass, ladder filters. These phase errors increase as the ratio of the clock frequency to the filter cutoff frequency decreases. Recently, it was shown how these phase errors can be eliminated by using bilinear terminations.[11,12] As a result, the accuracy of the passband magnitude can be greatly improved.

4.7 EXACT SC LADDER FILTER DESIGN

The requirement for a high sampling rate ($\omega << \omega_c$) in SC filters has led to the development of design methods that result in SC filters meeting their specifications at all frequencies up to and including half of the sampling

---

[11] R. B. Datar and A. S. Sedra, "Exact Design of Strays-Insensitive Switched-Capacitor Ladder Filters," *Proc. of 1983 IEEE Inter. Sym. on Circuits and Systems,* May 1983, pp. 590–593, Newport Beach, CA.

[12] R. D. Davis, "The Use of Bilinear Terminations in Switched Capacitor Ladder Filters," *Proc. of 1983 IEEE Inter. Sym. on Circuits and Systems,* May 1983, pp. 785–788, Newport Beach, CA.

frequency. The approach is based on the fact that SC filters are essentially discrete-time networks and should therefore be designed in the discrete-time domain, rather than the continuous-time domain. This method, which is called the *exact method,* uses one of the s- to z-domain transformations to convert the continuous-time specifications into discrete-time specifications. Next the z-domain performance of a discrete-time realization is related to the specifications in the z-domain to complete the design. The method is illustrated in Table 4.7–1 on a general basis. The difference in the methods presented here are due to different s- to z-domain transformations or different types of discrete-time realizations. The primary disadvantage of the exact method is the requirement to solve a set of nonlinear equations.

Let us first consider an exact method based on the Type I LDI of Section 2.5, Eq. (12).[13,14] The transfer function of the Type I LDI is given as

$$H^{oe}(z) = \pm \frac{C_1/C_2\, z^{-1/2}}{1 - z^{-1}} = \frac{\pm C_1/C_2}{z^{1/2} - z^{-1/2}} \tag{1}$$

Equating Eq. (1) to Eq. (14) of Section 2.5 results in what is called the *LDI transform,* given as

$$s = \frac{1}{T}(z^{1/2} - z^{-1/2}) \tag{2}$$

For the discrete-time filter to retain the essential properties of the analog filter, the imaginary axis in the s-plane should map onto the unit circle in the z-domain. To demonstrate that Eq. (2) partially satisfies this requirement, let $s = \sigma + j\Omega$ and $z = e^{j\omega T}$, where Ω and ω are the continuous-time and discrete-time frequencies, respectively. Equation (2) is now given as

$$\sigma + j\Omega = j\frac{2}{T}\sin\left(\frac{\omega T}{2}\right) \tag{3}$$

Equating the real and imaginary parts of Eq. (3) gives

$$\sigma = 0 \tag{4}$$

and

$$\Omega = \frac{2}{T}\sin\left(\frac{\omega T}{2}\right) \tag{5}$$

[13] R. D. Davis and T. N. Trick, "Optimization Design of Low-Pass Switched-Capacitor Ladder Filters," *IEEE Trans. on Circuits and Systems,* Vol. CAS-27, No. 6, June 1980, pp. 522–527.

[14] T. C. Choi and R. W. Brodersen, "Considerations for High-Frequency Switched-Capacitor Ladder Filters," *IEEE Trans. on Circuits and Systems,* Vol. CAS-27, No. 6, June 1980, pp. 545–552.

Table 4.7–1 General Method for Exact SC Filter Design.

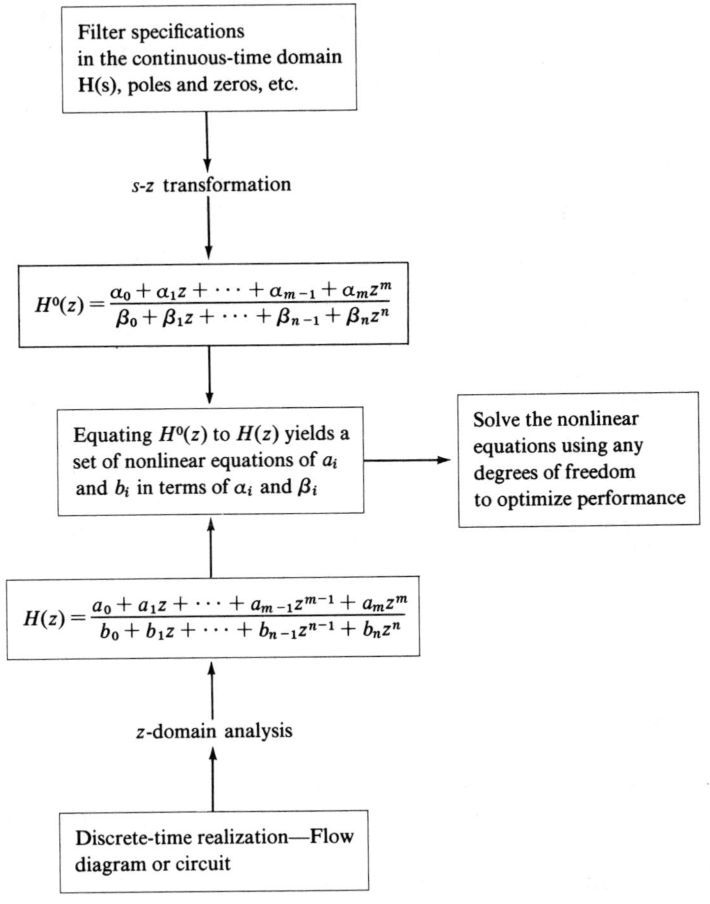

Equation (5) is illustrated in Fig. 4.7–1, where Ω is plotted as a function of ω. We see that the imaginary axis of the s-plane in the range of

$$-\frac{1}{\pi T}<\frac{\Omega}{2\pi}<\frac{1}{\pi T} \tag{6}$$

is mapped onto the unit circle over the range of

$$-\frac{1}{2T}<\frac{\omega}{2\pi}<\frac{1}{2T} \tag{7}$$

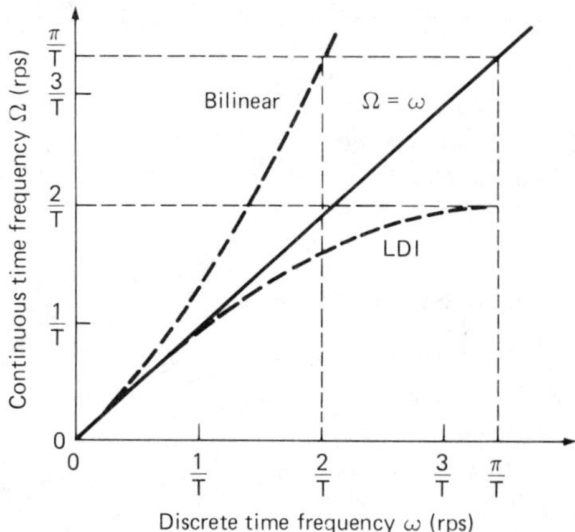

Fig. 4.7–1. Warping of the analog (continuous time) frequency by the bilinear and the LDI transformations.

Only over the above ranges will the transformation of Eq. (2) have the necessary property of a one-to-one mapping. The bilinear transformation, which is given as

$$\Omega = \frac{2}{T} \tan \frac{\omega T}{2} \tag{8}$$

and will be used later in this section is also illustrated in Fig. 4.7–1 and has the desirable one-to-one mapping property on the entire unit circle.

Because Ω and ω are not linearly related, the critical frequencies of the analog filter must be "prewarped" according to Eq. (5). To illustrate this, consider the design specifications for the discrete-time low-pass filter shown in Fig. 4.7–2, where ω_p and ω_s are the passband and stopband frequencies, respectively. In order that the discrete-time filter meet the desired specifications, Ω_p and Ω_s of the continuous-time filter must be calculated from Eq. (5).

The above approach will result in $H^0(z)$ of Table 4.7–1. It now becomes necessary to propose a discrete-time realization that can be used to synthesize $H^0(z)$. Though many approaches could be used, we shall illustrate the concepts through the use of the third-order, doubly terminated filter in Fig. 4.7–3. Using the methods of Section 4.2, we can develop the realization of Fig. 4.7–4 using Type I LDIs. As we have seen in the last section, the load and source terminations contribute a half-delay error and prevent the straightforward use of Eq. (5) in the design. Even though the terminations will cause errors, we will find the transfer function in the z-domain and

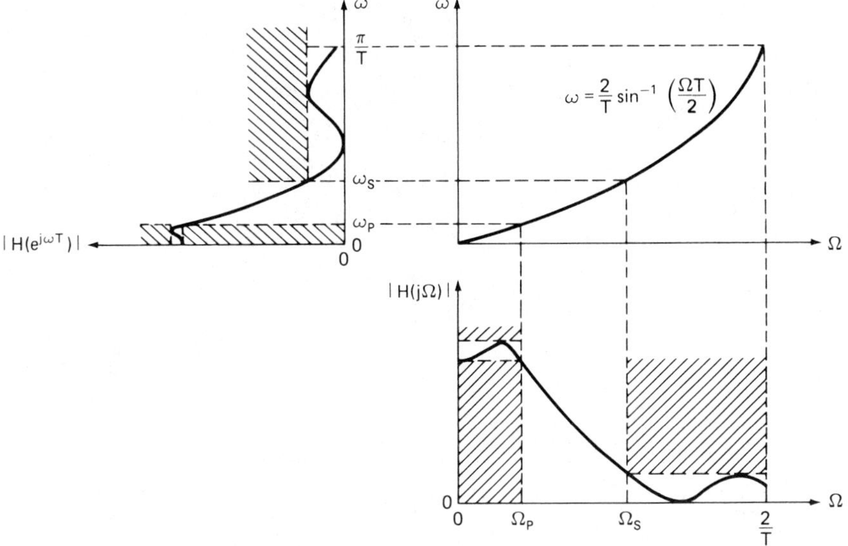

Fig. 4.7–2. Application of eq. (5) to prewarp the analog filter specifications.

complete the design by solving the set of nonlinear equations. This will result in an exact realization that avoids the termination error problem. Figure 4.7–5 shows a z-domain block diagram circuit equivalent of Fig. 4.7–4. Each of the integrator gain constants is designated by a_i for the ith integrator. The transfer function of this block diagram is

$$\frac{V_{out}(z)}{V_{in}(z)} = \frac{a_1\, a_2\, a_3\, z^{3/2}}{z^3 f_1(a_1,a_2,a_3) - z^2 f_2(a_1,a_2,a_3) + z f_3(a_1,a_2,a_3) - 1} \tag{9}$$

where

$$f_1(a_1,a_2,a_3) = 1 + a_1 + a_3 + a_1 a_3 \tag{10}$$

$$f_2(a_1,a_2,a_3) = (a_1 + a_3)(a_2 - 2) + a_1 a_3(2a_2 - 1) - 3 \tag{11}$$

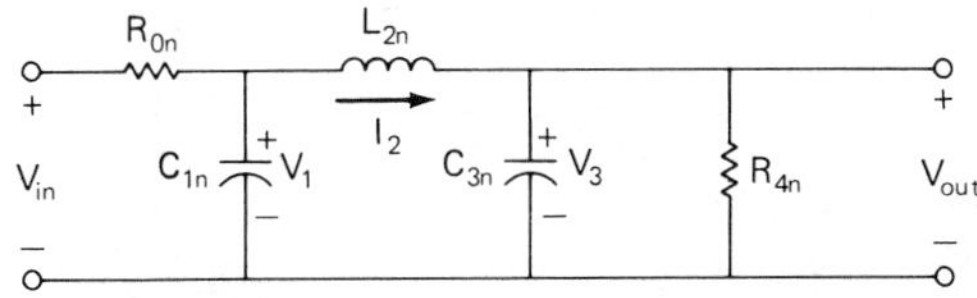

Fig. 4.7–3. Circuit diagram of a third-order, low-pass filter.

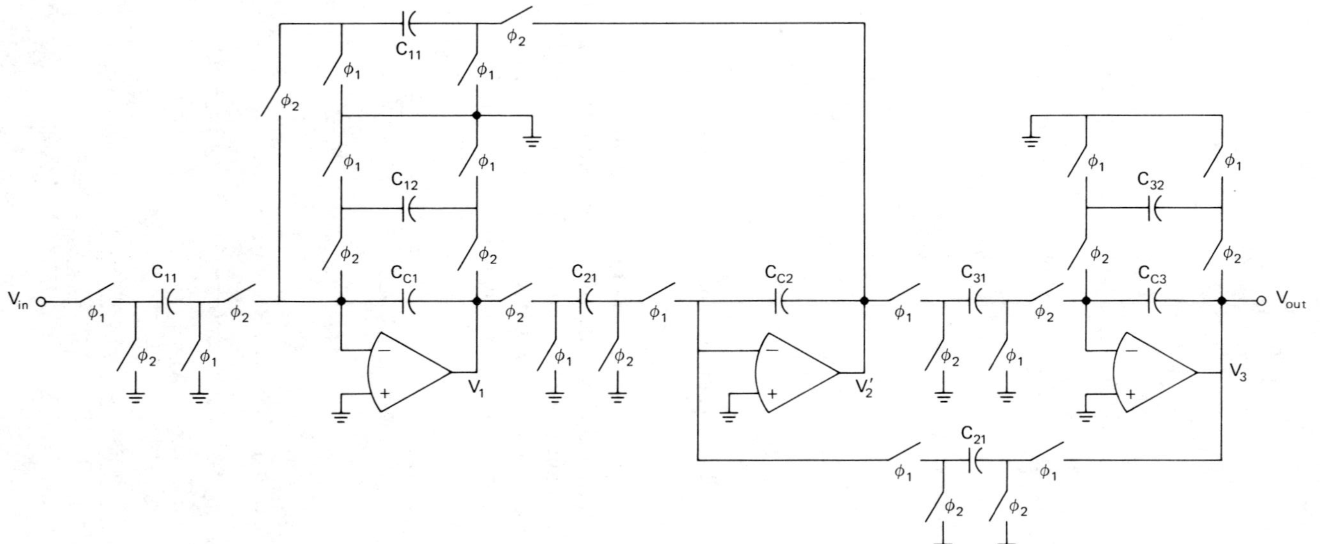

Fig. 4.7–4. SC realization of Fig. 4.7–3.

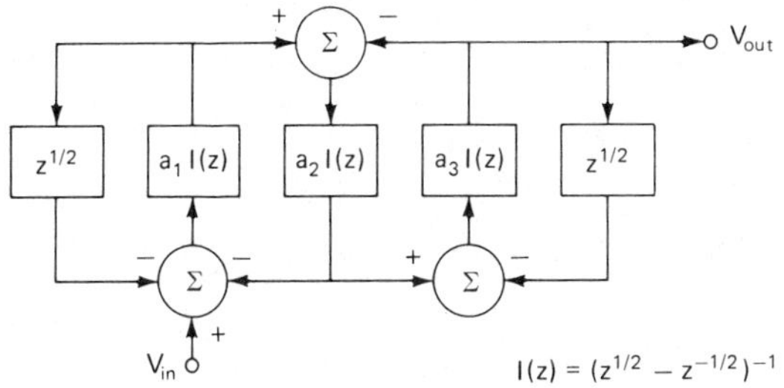

Fig. 4.7–5. Block diagram of Fig. 4.7–4.

and

$$f_3(a_1,a_2,a_3) = (a_1 + a_3)(1 - a_2) + 3 \tag{12}$$

The design is completed by transforming the s-domain specification into the z-domain using Eq. (5) and solving the resulting nonlinear equations. The following example demonstrates the method.

Example 4.7–1. *Exact design of a third-order SC filter.* Consider the design of a third-order, doubly terminated, Chebyshev filter with a ripple of 0.1 dB, a ripple bandwidth of 11 kHz, and a sampling frequency of 100 kHz. The values of C_{1n}, L_{2n}, and C_{3n} are 1.49249 F, 1.66011 H, and 1.49249 F, respectively.[15] First let us consider the design of the filter without using Eq. (5). The normalized values of a_1, a_2, and a_3 are 0.67002, 0.60237, and 0.67002, respectively. Substituting these values into Eq. (9) and plotting the magnitude response gives Fig. 4.7–6(a).

To obtain the correct values of the a_i's the ripple bandwidth is prewarped according to Eq. (5) to yield a value of

$$f_{\text{ripple}} = 10.782 \text{ kHz}$$

With this value of ripple bandwidth, the three poles of the analog prewarped filter are

$$p_1 = -65.676 \text{ rps}$$

$$p_2,\, p_3 = -32{,}838 \pm j\ 81{,}716 \text{ rps}$$

[15] R. D. Davis and T. N. Trick, op. cit.

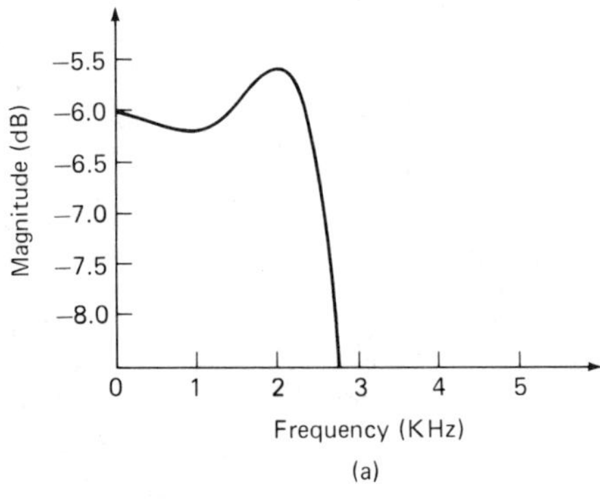

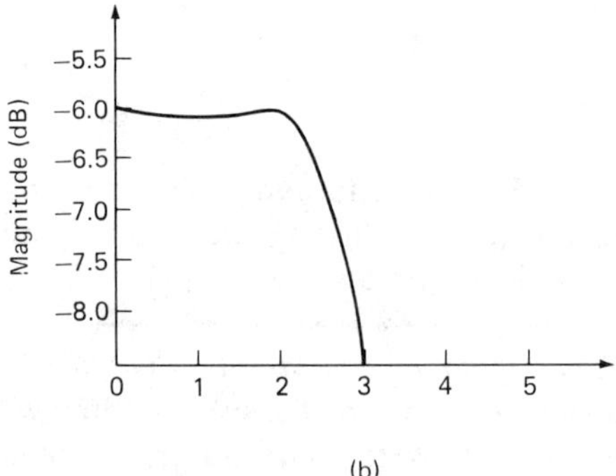

Fig. 4.7–6. (a.) Magnitude response of the third-order filter of Fig. 4.7–5 with the incorrect integrator gain constants. (b.) Magnitude response of the third-order filter of Fig. 4.7–5 with the optimized integrator gain constants.

Substituting these roots into Eq. (2) with $T = 10^{-5}$ and taking only the roots inside the unit circle gives the z-domain poles as

$$z_1 = 0.5244$$

$$z_2,\ z_3 = 0.4732 \pm j\ 0.5157$$

The z-domain transfer function resulting from these poles is

$$\frac{V_{\text{out}}(z)}{V_{\text{in}}(z)} = \frac{K_2\, z^{3/2}}{3.8929\, z^3 - 5.7254\, z^2 + 3.8388\, z - 1}$$

Setting these coefficients equal to Eqs. (10) through (12) and using the Newton-Raphson method of solving for the values of a_i, using the incorrect ones as the starting points, gives a_1, a_2, and a_3 as 0.5911, 0.5884, and 1.4466, respectively. The value of K_2 is found to be 0.5031, which gives the −6-dB passband gain at DC. Figure 4.7–6(b) shows the correct response of the filter.

The second category of exact design methods is based upon the use of the bilinear transformation. The primary difference between the approaches that use the bilinear transformation is how the discrete-time realization is developed. The bilinear z-transform alleviates the high sampling requirement and has the advantage of mapping the entire $j\omega$ axis in the continuous-time domain onto the entire unit circle in the discrete-time domain, as noted in Fig. 4.7–1. After the s-domain filter has been prewarped by the bilinear z-transform, then one can either replace the integrators of the prewarped s-domain realization by bilinear integrators (see Fig. 2.5–14(a)) or transform the prewarped filter specifications to the z-domain and solve for the parameters of the z-domain through a set of nonlinear equations.

The approach now to be described uses the bilinear z-transform and uses only the stray-insensitive configurations of the integrators. This method applies to any filter function that can be realized using cascade or coupled biquad structures. These include: even-order cascade filters,[16] bandpass filters derived from leapfrog low-pass prototypes as done above, bandpass filters using the follow-the-leader or PRB structure,[17] and general parameter filters designed using either the nodal voltage simulation approach[18] or Szentirmai's multiple-loop feedback synthesis method.[19] Although the method proposed here can be applied with varying degrees of success to other types of filters, we shall consider only the bandpass filter.

The realization of the biquad transfer function is the key to this approach. An s-domain biquadratic transfer function can be expressed as

$$H(s) = \frac{a_2 s^2 + a_1 s + a_0}{s^2 + b_1 s + b_0} \tag{13}$$

A normalized version of the bilinear transformation given in Eq. (15) of Section 1.5 can be written as

---

[16] K. Martin and A. S. Sedra, "Stray-Insensitive Switched-Capacitor Filters based on the bilinear z-transform," *Electronics Letters,* Vol. 15, June 1979, pp. 365–366.

[17] A. S. Sedra and P. O. Brackett, *Filter Theory and Design: Active and Passive,* Matrix Publishers, Portland, OR, 1978.

[18] M. Yoshihiro, A. Nishihara, and T. Yanagisawa, "Low-Sensitivity Active and Digital Filters Based on the Node-Voltage Simulation of LC Ladder Structures," *1977 IEEE Int. Symposium on Circuits and Systems,* Phoenix, AZ, April 1977.

[19] G. Szentirmai, "Synthesis of Multiple Feedback Active Filters," *Bell System Tech. J.,* Vol. 52, April 1973, pp. 527–555.

$$s = \frac{z-1}{z+1} \tag{14}$$

Substituting this transformation into Eq. (13) gives

$$H(z) = \frac{m(a_0 + a_1 + a_2)z^2 + 2m(a_0 - a_2)z + m(a_0 - a_1 + a_2)}{z^2 + 2m(b_0 - 1)z + m(1 + b_0 - b_1)} \tag{15}$$

where

$$m = 1/(1 + b_0 + b_1) \tag{16}$$

Figure 4.7–7 shows a realization of Eq. (15).[20] Assuming that the output of the circuit in Fig. 4.7–7 is sampled at the end of clock phase ϕ_2, then

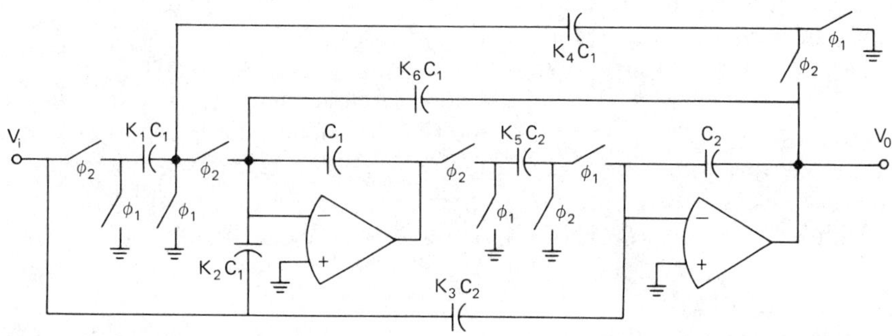

Fig. 4.7–7. A biquad circuit capable of realizing the bilinear z-transform of any second-order transfer function, $H(s)$.

its transfer function can be found by writing difference equations (for the charge transfer) and then taking z-transforms. The result is

$$\frac{V_0(z)}{V_i(z)} = \frac{K_3 z^2 + (-2K_3 + K_1K_5 + K_2K_5)z + (-K_2K_5 + K_3)}{z^2 + (-2 + K_4K_5 + K_5K_6)z + (1 - K_5K_6)} \tag{17}$$

The pole design equations can be obtained by equating the coefficients of the like powers of z in the denominators of Eqs. (15) and (17). The result is

[20] K. Martin and A. S. Sedra, "Exact Design of Switched-Capacitor Bandpass Filters Using Coupled-Biquad Structure," *IEEE Trans. on Circuits and Systems,* Vol. CAS-27, No. 6, June 1980, pp. 469–475.

$$K_4K_5 = 4\ mb_0 \tag{18}$$

$$K_5K_6 = 2\ mb_1 \tag{19}$$

These are two equations with three unknowns K_4, K_5, and K_6. The one available degree of freedom may be used to maximize the dynamic range of the biquad. A near optimum solution is obtained by choosing equal time constants for the two integrators

$$K_4 = K_5 \tag{20}$$

The zero design equations can be obtained by equating the coefficients of the corresponding z terms in the numerators of Eqs. (15) and (17). The results for important special cases are

a. *Low-Pass* $(a_1 = a_2 = 0)$

$$K_2 = 0,\ K_1K_5 = 4\ ma_0,\ \text{and}\ K_3 = ma_0 \tag{21}$$

b. *Bandpass* $(a_0 = a_2 = 0)$

$$K_1 = 0,\ K_2K_5 = 2\ ma_1,\ \text{and}\ K_3 = ma_1 \tag{22}$$

c. *High-Pass* $(a_0 = a_1 = 0)$

$$K_1 = K_2 = 0,\ \text{and}\ K_3 = ma_2 \tag{23}$$

d. *Notch* $(a_1 = 0)$

$$K_2 = 0,\ K_1K_5 = 4\ ma_0,\ \text{and}\ K_3 = m(a_0 + a_2) \tag{24}$$

Although the circuit of Fig. 4.7–7 realizes a negative bandpass function, it will be seen that a positive bandpass is also needed. A circuit that realizes a positive bandpass function is given in Fig. 4.7–8.[21] Design equations, obtained in a manner identical to that used for the circuit of Fig. 4.7–7 are given by

$$K_4K_5 = 4\ mb_0 \tag{25}$$

$$K_4K_6 = 2\ mb_1 \tag{26}$$

$$(K_4 = K_5,\ \text{for near maximum dynamic range}) \tag{27}$$

$$K_2K_5 = 2\ ma_1 \tag{28}$$

$$K_3 = ma_1\left(1 + \frac{b_1}{b_o}\right) \tag{29}$$

[21] Ibid.

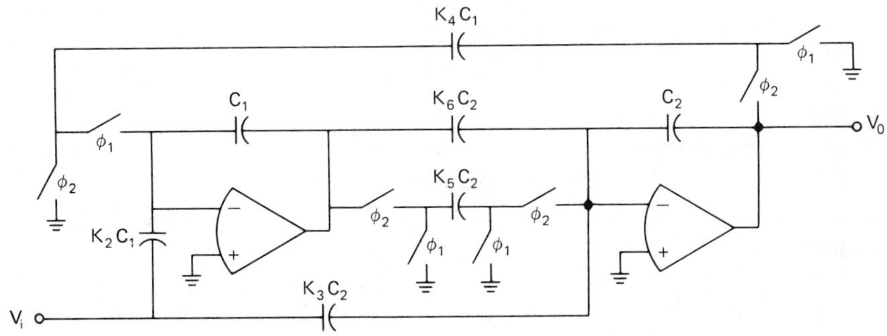

Fig. 4.7–8. A circuit for realizing the bilinear z-transform of a noninverting second-order band-pass transfer function.

where

$$m = 1/(1 + b_0 + b_1) \tag{30}$$

The biquad circuits of Figs. 4.7–7 and 4.7–8 will be used as building blocks in the coupled structures to be discussed. Coupling can be achieved by using additional feed-in components. To minimize chip areas switches may be shared among the different inputs, as will be illustrated in the following.

From the active RC filter literature,[22] it is known that geometrically symmetric band-pass filter functions can be realized using a variety of low-sensitivity coupled-biquad structures, including leapfrog and follow-the-leader feedback (FLF). In the following we shall present a systematic procedure for the design of such filters in switched capacitor form. We will use the leapfrog case as an example.

Given attenuation specifications versus frequency f, the first step in the design is to prewarp the specifications versus the continuous-time frequency variable Ω using the relationship

$$\Omega = \frac{2}{T} \tan\left(\frac{\omega T}{2}\right) = \frac{2}{T} \tan\left(\frac{\pi f}{f_c}\right) \tag{31}$$

Next, the prewarped specifications are made geometrically symmetric, such that

$$\Omega_{P1}\,\Omega_{p2} = \Omega_{S1}\,\Omega_{S2} = \Omega_0^2 \tag{32}$$

[22] A. S. Sedra and P. O. Brackett, *Filter Theory and Design: Active and Passive,* Matrix Publishers, Portland, OR, 1978.

where Ω_{P1} and Ω_{P2} are the edges of the passband, Ω_{P0} is the geometric center of the passband, and Ω_{S1} and Ω_{S2} are the edges of the stopbands. The specifications of the corresponding normalized low-pass prototype (LPP) can then be obtained.

The next task is to find a doubly terminated LC ladder realization for the LPP. This can be either an all-pole one (e.g., Butterworth or Chebyshev) or one with finite transmission zeros (e.g., inverse Chebyshev or elliptic). In the following we shall treat the two cases separately.

If an all-pole realization is required, filter tables or closed-form expressions can be used to obtain the element values of the LC ladder. As an example assume that the LPP specifications can be realized by a third-order Chebyshev filter whose LC ladder realization is shown in Fig. 4.7–9(a). Figure 4.7–9(b) shows a block diagram for the leapfrog realization of this LC ladder network. The summing coefficients (shown between parentheses) are for the purpose of scaling for maximum dynamic range. These factors are obtained from an analysis of the LC ladder network assuming a constant amplitude, say 1 volt, swept-frequency input signal. Then the maxima of the frequency spectra of capacitor voltages and inductor currents are evaluated. Thus, $\hat{V}_1$ denotes the maximum of the spectrum of the voltage across C_1. Including these scaling factors will enable the active circuit to saturate at the same input signal level. The factor k in the input summing coefficient determines the filter gain, as it is the ratio of the desired maximum gain of the bandpass filter to the maximum gain of the LC network realization of the LPP.

Applying the low-pass to bandpass transformation

$$s \longrightarrow \frac{s^2 + \Omega_0^2}{sB} \tag{33}$$

where $B = \Omega_{P1} - \Omega_{P1}$, to the leapfrog block diagram of Fig. 4.7–9(b) results in the coupled-biquad structure of Fig. 4.7–9(c). Note that this structure uses bandpass circuits of both polarities. These bandpass sections can now be realized using the switched capacitor circuits of Figs. 4.7–7 and 4.7–8. The resulting switched capacitor filter will exactly meet its specifications, irrespective of how low the clock frequency is. Furthermore the circuit will have good sensitivity and dynamic range properties[23] and is completely insensitive to strays.

---

[23] Although the scaling employed equalizes the peak value of the voltages at the output of the various biquads, it does not ensure that the output of the other op amp in the biquad will be scaled correctly. However, choosing the two time constants of each biquad equal approximately equalizes the peak output voltages of each op amp in the biquad.

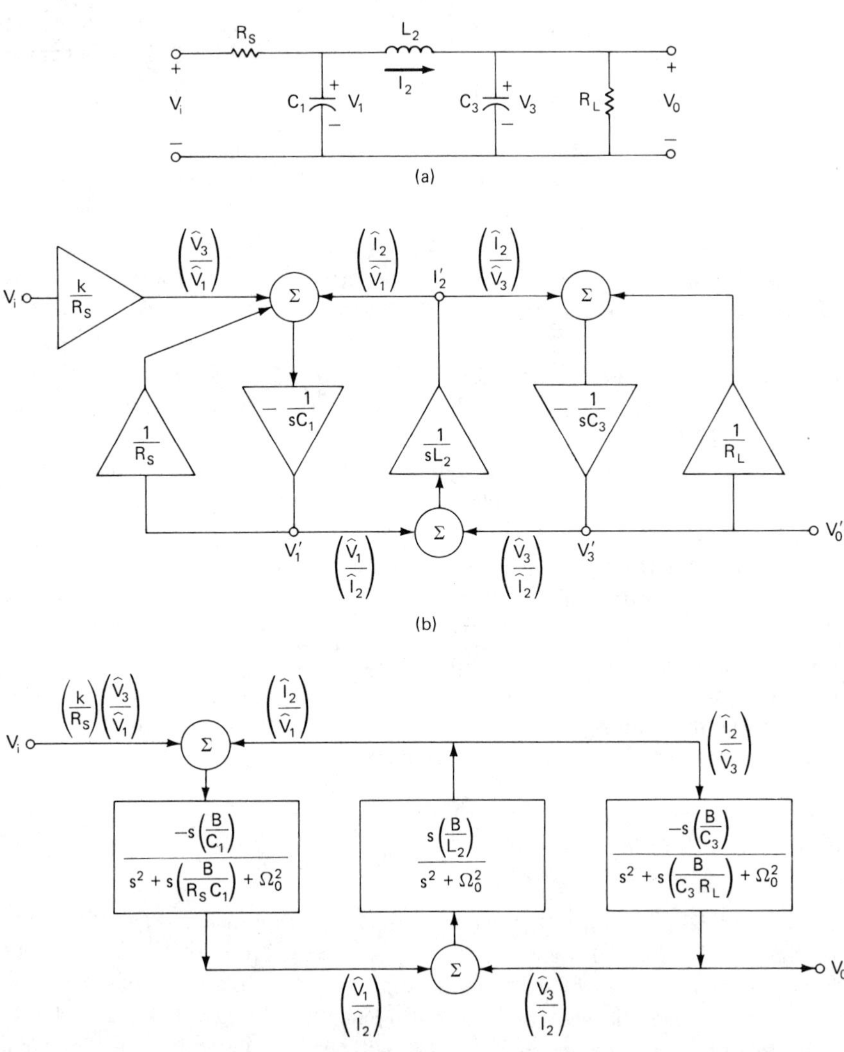

Fig. 4.7–9. Illustration of the design of an all-pole bandpass filter using the leap frog structure. (a.) LC realization of the low-pass prototype. (b.) Active simulation of the operation of the LC network (the summing factors are for the purpose of maximizing the dynamic range). (c.) The bandpass filter in signal flowgraph form.

If a realization with finite transmission zeros is required (as is normally the case), the design proceeds in much the same manner using the technique illustrated in Section 4.1. Consider the third-order elliptic, LPP of Fig. 4.7–10(a). Using the technique of Fig. 4.1–5 results in the modified circuit shown

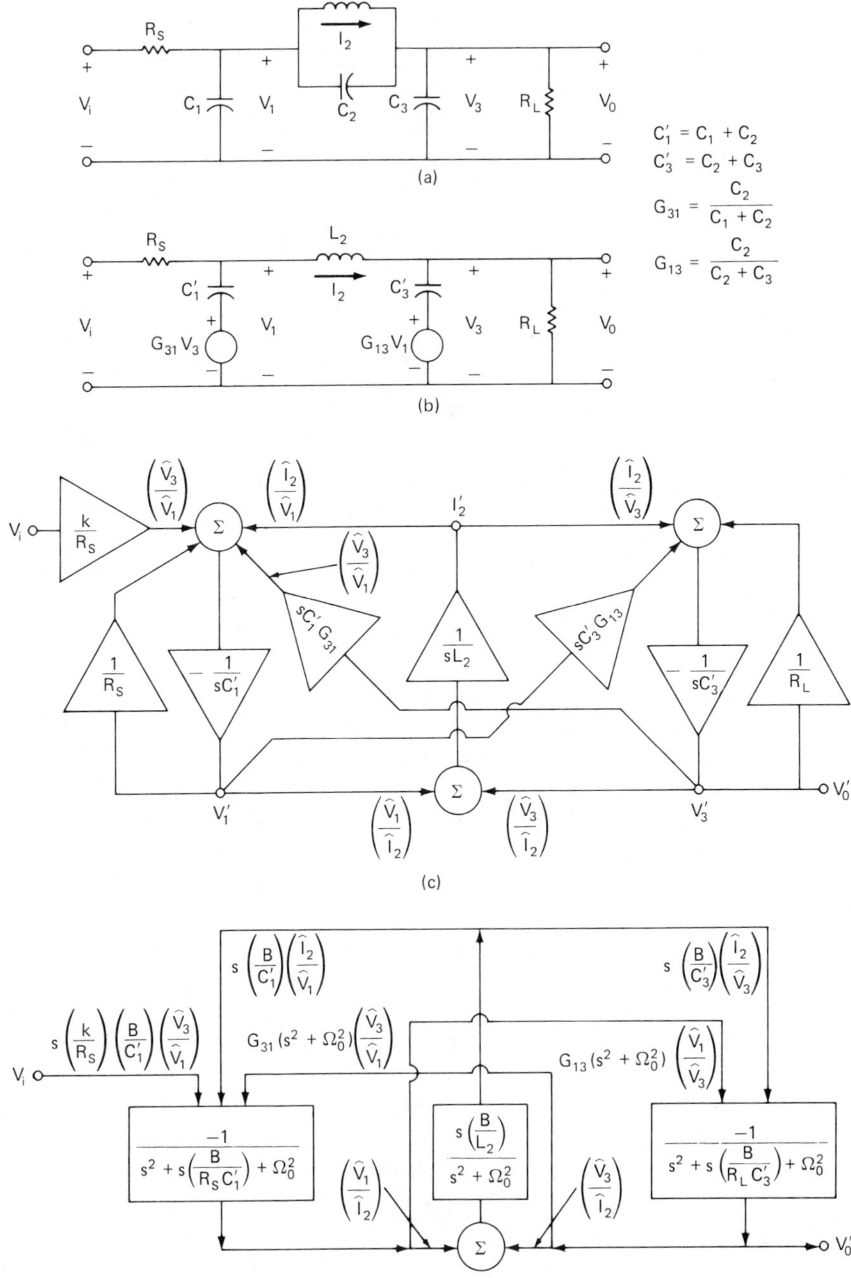

Fig. 4.7–10. (a.) LC realization of the low-pass prototype. (b.) The LC realization after the elimination of C_2. (c.) Active simulation of (b.). (d.) The bandpass filter realization obtained by applying the low-pass to bandpass transformation to (c.).

in Fig. 4.7–10(b). A block diagram realization of the operational simulation of this latter circuit is shown in Fig. 4.7–10(c). Again the summing coefficients, shown between parentheses, are included for optimizing the filter dynamic range.

Application of the low-pass to bandpass transformation of Eq. (33) results in the block diagram realization of Fig. 4.7–10(d). It should be noted that some sections have more than one input with different transmission zeros. For instance, the first block has three inputs, two with bandpass functions and one with a notch function. Because all three functions have the same poles, they can be realized by one biquad with three different sets of feed-in components. Such blocks can be easily realized with the switched capacitor biquads, as will be illustrated in the following.

General-parameter bandpass filters, that is, those not derived from a low-pass prototype, can also be realized in a coupled, switched capacitor biquad structure using the bilinear z-transform. First the specifications have to be prewarped. Then, a suitable LC ladder realization of the prewarped specifications is obtained using LC synthesis programs.[24] Subsequently, the method proposed by Yoshihiro et al. can be used to obtain a coupled-biquad realization.[25] It should be noted, however, that this method does not result in a full simulation of the ladder operation. Rather, the interaction of the nodal voltages is simulated. Thus, voltages in the active circuit correspond to the nodal voltages of the LC ladder. Nevertheless, good sensitivity behavior has been observed in realizations obtained in this manner.

Once a coupled-biquad structure has been obtained, the circuits Figs. 4.7–7 and 4.7–8 can be used to realize the different biquads. In the following the method will be illustrated using an eighth-order example.

Example 4.7–2. *Design of a general-parameter bandpass filter.* Let it be required to design a switched capacitor bandpass filter whose attenuation function meets the following specifications:

Passband:	0.9 kHz to 1 kHz, equiripple with 0.25-dB maximum deviation
Upper Stopband:	$A \geq 50$ dB for $f \geq 1.05$ kHz
Lower Stopband:	$A \leq 39$ dB for $f \leq 0.95$ kHz
Clock Frequency:	$f_c = 8$ kHz

---

[24] W. M. Snelgrove and A. S. Sedra, "FILTOR 2: A Computer-Aided Filter Design Program, Dept. of Electrical Engineering, University of Toronto, 1979.

[25] Yoshihiro, et al., op. cit.

After prewarping the specifications and using a filter approximation program and an LC synthesis program,[26] the doubly terminated LC ladder of Fig. 4.7–11 was obtained. The element values of this ladder are given in Table 4.7–2. The ladder was analyzed using a variable-frequency input sinusoid with 1-volt amplitude. The analysis provides the peaks of the spectra of all node voltages, given in Table 4.7–2. These values will be used in scaling the active realization to maximize its dynamic range. To obtain a coupled-

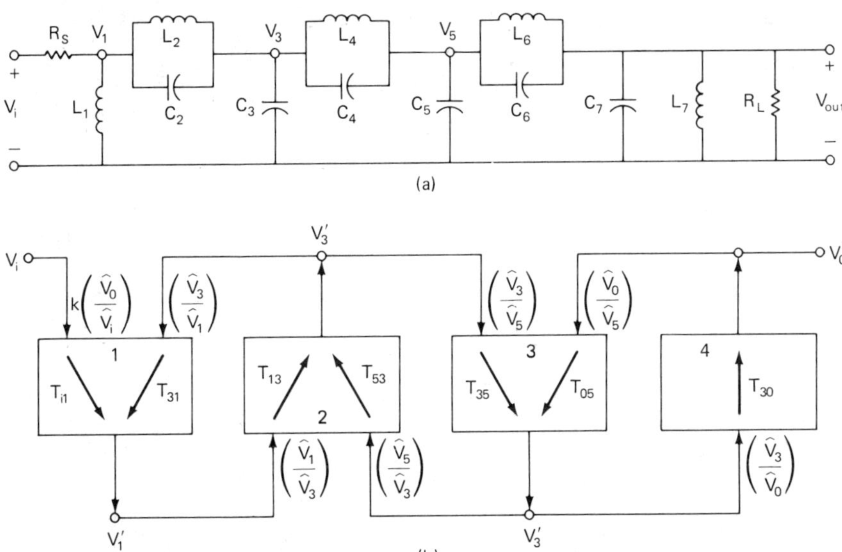

Fig. 4.7–11. Illustration of the design of a general parameter bandpass filter using the nodal voltage simulation method. (a.) LC ladder realization of an eighth-order bandpass filter. (b.) Block diagram of the active simulation. The various transfer functions are given in the text and the component values in Table 4.7–2.

biquad simulation of the LC ladder in Fig. 4.7–11, Yoshihiro's method was used as follows. Equations are written to express the voltage at each node in terms of the voltages at the preceding and succeeding nodes

$$V_1 = T_{i1}\, V_i + T_{31}\, V_3 \tag{34}$$

$$V_3 = T_{13}\, V_1 + T_{53}\, V_5 \tag{35}$$

$$V_5 = T_{35}\, V_3 + T_{05}\, V_0 \tag{36}$$

$$V_0 = T_{50}\, V_5 \tag{37}$$

[26] Snelgrove et al., op. cit.

Table 4.7–2 Design Data for the Bandpass Filter Example of Figs. 4.7–11 and 4.7–12.

Component Values for the LC Ladder of Fig. 4.7–11

$R_s = 1\Omega$	$L_1 = 1.0412H$	$L_2 = 0.23533H$	$L_4 = 0.028068H$
$L_6 = 0.004084H$	$L_7 = 0.00030626H$	$C_2 = 36.239F$	
$C_3 = 40.240F$	$C_4 = 184.03F$	$C_5 = 461.70F$	
$C_6 = 1076.5F$	$C_7 = 21068F$	$R_L = 0.0014375\Omega$	

Analysis of the LC Ladder with $1V$ Input Sinusoidal Yields the Following peak values

$\hat{V}_1 = 0.817V$	$\hat{V}_3 = 0.376V$	$\hat{V}_5 = 0.125V$	$\hat{V}_0 = 0.019V$

Capacitor Ratios for the Circuit in Fig. 4.7–12

$K_{0,1} = .08232$	$K_1 = .2222$	$K_{2,1} = .7007$
$K_{0,2} = .02884$	$K_{1,2} = .7007$	$K_{4,2} = .4390$
$K_{2,3} = .1687$	$K_{4,3} = .7287$	$K_{6,3} = .2164$
$K_{2,4} = .9644$	$K_{3,4} = .7287$	$K_{6,4} = .2431$
$K_{4,5} = .2860$	$K_{6,5} = .7485$	$K_{8,5} = .09932$
$K_{4,6} = .3299$	$K_{5,6} = .7485$	$K_{8,6} = .1003$
$K_{6,7} = .3350$	$K_7 = .0724$	$K_{8,7} = .7300$
$K_{6,8} = .3299$	$K_{7,8} = .7300$	$K_{4,1} = .2630$

where

$$T_{i1} = \frac{V_1}{V_i}\bigg|_{V_3=0}, \qquad T_{31} = \frac{V_1}{V_3}\bigg|_{V_i=0} \tag{38}$$

$$T_{13} = \frac{V_3}{V_1}\bigg|_{V_5=0}, \qquad T_{53} = \frac{V_3}{V_5}\bigg|_{V_1=0} \tag{39}$$

$$T_{35} = \frac{V_5}{V_3}\bigg|_{V_0=0}, \qquad T_{05} = \frac{V_5}{V_0}\bigg|_{V_3=0} \tag{40}$$

$$T_{50} = \frac{V_0}{V_5} \tag{41}$$

Each pair of these transfer functions (i.e., T_{i1} and T_{31}) has the same poles but different zeros. This enables the realization of each pair of transfer functions in one biquad circuit. The result is that, for the eighth-order filter under discussion, four biquads are needed for a total of eight op amps. It should be noted, however, that care has to be exercised in choosing a particular LC realization (out of the many possible ones) so that the different transfer functions involved are of second-order. This is usually possible for bandpass filters whose transmission zeros are restricted to the $j\omega$-axis. No special attention has to be given to minimizing the spread in the values of the LC network components, for the active simulation will be scaled for optimum dynamic range.

Continuing with our example, the various functions involved are given by

Biquad #1:

$$T_{i1} = \frac{s\left(\dfrac{1}{C_2 R_2}\right)}{s^2 + s\left(\dfrac{1}{C_2 R_s}\right) + \dfrac{1}{C_2(L_1 \parallel L_2)}} \tag{42}$$

$$T_{31} = \frac{s + \left(\dfrac{1}{L_2 C_2}\right)}{s^2 + s\left(\dfrac{1}{C_2 R_2}\right) + \dfrac{1}{C_2(L_1 \parallel L_2)}} \tag{43}$$

Biquad #2:

$$T_{13} = \frac{C_2}{C_2 + C_3 + C_4} \frac{s + \dfrac{1}{L_2 C_2}}{s^2 + \dfrac{1}{(L_2 \parallel L_4)(C_2 + C_3 + C_4)}} \tag{44}$$

$$T_{53} = \frac{C_4}{C_2 + C_3 + C_4} \frac{s^2 + \dfrac{1}{L_4 C_4}}{s^2 + \dfrac{1}{(L_2 \parallel L_4)(C_2 + C_3 + C_3)}} \tag{45}$$

Biquad #3:

$$T_{35} = \frac{C_4}{C_4 + C_5 + C_6} \frac{s^2 + \dfrac{1}{L_4 C_4}}{s^2 + \dfrac{1}{(L_4 \parallel L_6)(C_4 + C_5 + C_6)}} \tag{46}$$

$$T_{05} = \frac{C_6}{C_4 + C_5 + C_6} \frac{s^2 + \dfrac{1}{L_6 C_6}}{s^2 + \dfrac{1}{(L_4 \parallel L_6)(C_4 + C_5 + C_6)}} \tag{47}$$

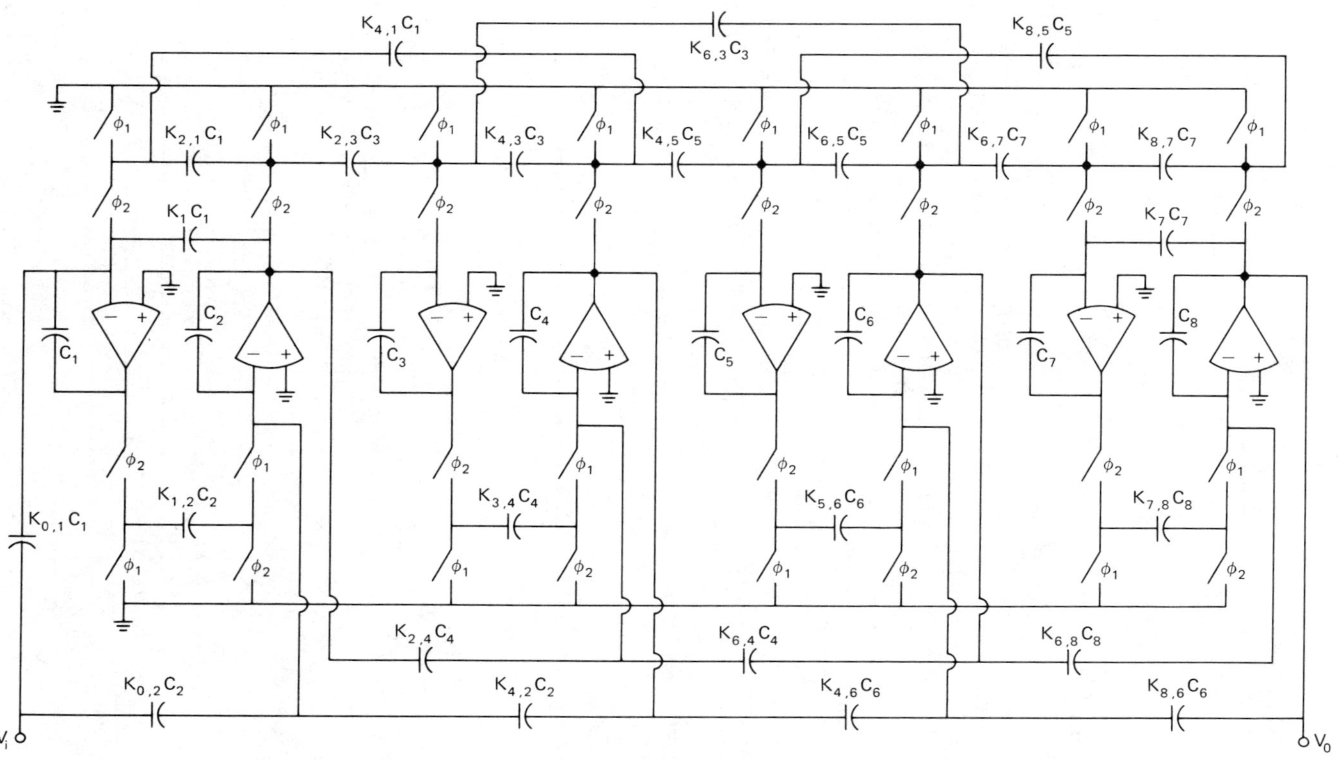

Fig. 4.7–12. Switched capacitor realization of Example 4.7–2.

Biquad #4:

$$T_{50} = \frac{C_6}{C_6 + C_7} \frac{s^2 + \dfrac{1}{L_6 C_6}}{s^2 + s\dfrac{1}{(C_6 + C_7)R_L} + \dfrac{1}{(L_6 \| L_8)(C_6 + C_7)}} \tag{48}$$

A block diagram of the realization is given in Fig. 4.7–11(b). To optimize the dynamic range, the various transfer functions have been multiplied by appropriate scaling coefficients whose values were obtained from the ladder analysis. This scaling will ensure that the output op amps of the four biquads will saturate at the same input signal level. By choosing the two time constants of each biquad equal, the other amplifier in the biquad will saturate at nearly the same signal level.

The complete circuit of the switched capacitor filter is given in Fig. 4.7–12, with the value of capacitor ratios given in Table 4.7–2. It is interesting to note that scaling the switched capacitor filter reduced the spread in element values from (6.9×10^7) in the LC ladder to 35 in the switched capacitor circuit.

A discrete prototype of the circuit in Fig. 4.7–12 was built using capacitors selected to 0.3% typical accuracy. The capacitor values ranged from 95 pF to 3.3 nF. The op amps used were LM 355, and the switches were MC 14016 CMOS transmission gates. The filter magnitude response was measured using an HP 3580A spectrum analyzer (100-Hz bandwidth setting) and plotted

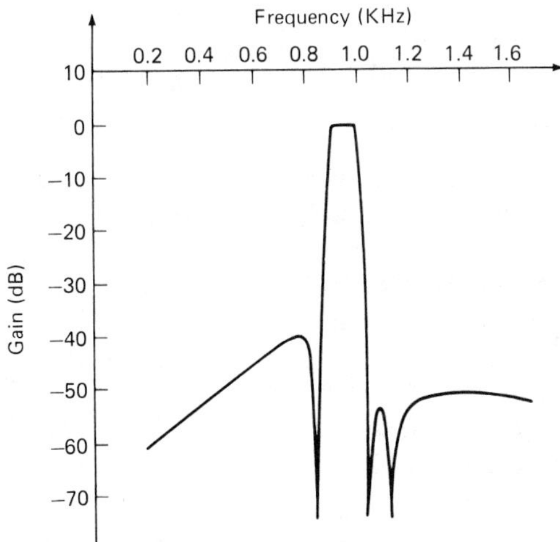

Fig. 4.7–13. Complete magnitude response of the circuit in Fig. 4.7–12.

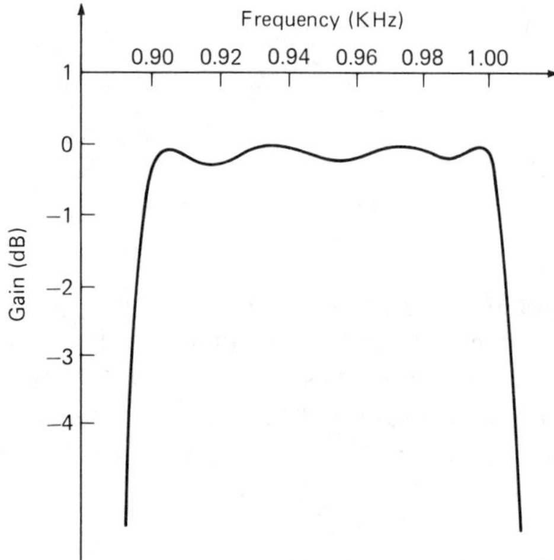

Fig. 4.7–14. Expanded response of the magnitude of Fig. 4.7–12 in the passband.

using an *XY* plotter. The complete response is shown in Fig. 4.7–13, and an expanded view of the passband is shown in Fig. 4.7–14.

This method enables the exact design of switched capacitor bandpass filters using low-sensitivity, coupled-biquad structures. Because the method is based on the bilinear z-transform, the resulting filters meet their specifications, irrespective of how close the clocking frequency is relative to the signal frequencies. The resulting circuits are completely insensitive to stray capacitances, allowing the designer to use smaller capacitors and to reduce the IC chip area.

4.8 SUMMARY

This chapter has presented methods that can be used to design SC ladder filters. The methods are based on the identification of the proper state variables and the synthesis of the resulting state equations. The methods were applicable to any type of RLC ladder filter and were illustrated for the low-pass, bandpass, high-pass, and band-elimination filters. These methods resulted in a straightforward design of SC filters that exhibit good performance and sensitivity properties.

Unfortunately, space does not permit the inclusion of another method of designing SC ladder filters, which uses voltage inverter switches.[27] This

[27] A. Fettweis, D. Herbst, B. Hoefflinger, J. Pandel, and R. Schweer, "MOS Switched-Capacitor Filters Using Voltage Inverter Switches," *IEEE Trans. on Circuits and Systems,* Vol. CAS-27, No. 6, June 1980, pp. 527–538.

method follows blocks 1, 2, 3, 6, 7, 11, and 14 of Fig. 3.1–16. The voltage inverter switch method is based on the principle of the resonant-transfer filter and can be built using capacitors, switches, and an element called a *voltage inverting switch.* The voltage inverting switch is implemented from capacitors, switches, and an op amp. This method is based on the use of the bilinear z-transform and thus does not imply any sampling rate restriction other than that resulting from the Nyquist condition. The method allows the design of filters for which there exists a one-to-one correspondence with doubly terminated, lossless, prototype filters. The voltage inverting switch method in its primitive form has floating capacitor nodes, which make it very sensitive to parasitic capacitances. Several schemes have been proposed to eliminate this problem and have attained some degree of success.[28] The voltage inverting switch approach should be considered as an alternative or complementary approach to the design of ladder filters using SC techniques.

In the first five sections, the design approach assumed that the sampling rate was high. In the last two sections, the effects of the clock frequency upon the filter performance were considered. A method of accounting for some of the nonideal effects of switched capacitor ladder filters was given. These nonideal effects included stray capacitances and the sampling frequency. The nonideal effects due to operational amplifiers will be considered in the next chapter. A method of designing exact SC filter simulations was also presented. This method eliminated the high sampling rate requirement, but requires the solution of a set of nonlinear equations.

PROBLEMS

4.1 (Sec. 4.1). What is the number of variables or the complexity of (a) Fig. 3.1–15 or (b) Fig. 3.4–7a?

4.2 (Sec. 4.1). What is the number of variables or the complexity of Fig. P 4–2?

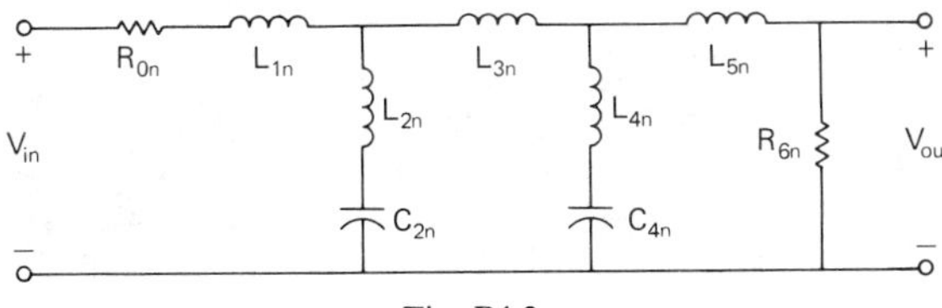

Fig. P4.2.

[28] G. C. Temes and R. Gregorian, "Compensation for Parasitic Capacitances in Switched-Capacitance Filters," *Electronics Letters,* Vol. 15, June 21, 1979, pp. 377–379.

4.3 (Sec. 4.1). Show that for a branch consisting of a parallel capacitor C and inductor L the proper state variable is voltage if the state equation is to be solved using only integration.

4.4 (Sec. 4.1). Find the equivalent expressions for the state equations of Eqs. (8) and (9) if the Thevenin sources of Fig. 4.1–3(a) and 4.1–3(b) are replaced by a Norton source. The value of the Norton current source is I_{in} and the shunt resistor is R_0.

4.5 (Sec. 4.1). Develop the equivalent relationships for Fig. 4.1–6 that are given in Eqs. (18), (22), and (26) for Fig. 4.1–5.

4.6 (Sec. 4.2). (a) Derive the expression for the z-domain transfer function of Fig. 4.2–2 given in Eq. (14) when the output is sampled at ϕ_1. (b) Use the high sampling assumption and derive Eq. (15) from Eq. (14).

4.7 (Sec. 4.2). Repeat Example 4.2–1 for a similar filter but having a 0.5-dB ripple in the passband and a clock frequency of 200 kHz. Give the total capacitance in terms of a common unit capacitance, C_u.

4.8 (Sec. 4.2). Repeat Example 4.2–1 using the RLC ladder structure shown in Fig. P 4–8.

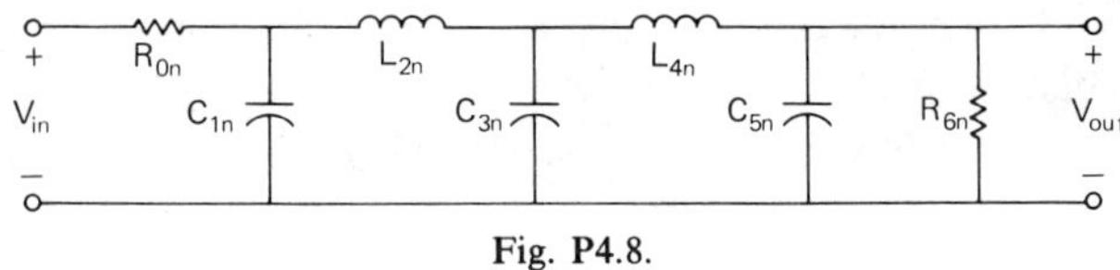

Fig. P4.8.

4.9 (Sec. 4.2). Repeat Example 4.2–1 using a singly terminated RLC ladder structure. What is the total capacitance in terms of a common unit capacitance, C_u?

4.10 (Sec. 4.2). Develop a third-order, doubly terminated, Butterworth SC realization using the RLC ladder structure of Fig. P 4–10.

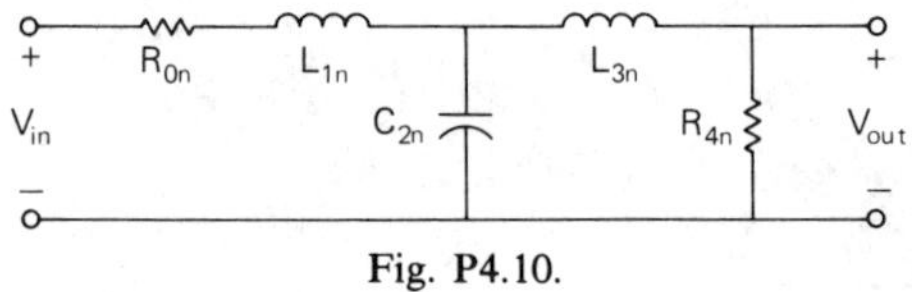

Fig. P4.10.

4.11 (Sec. 4.2). Repeat problem 4.10 assuming that the source resistor is twice the value of the load resistor.

4.12 (Sec. 4.2). Repeat Problem 4.8 assuming that the source resistor is half the value of the load resistor.

4.13 (Sec. 4.2). Repeat Example 4.2–2 using the RLC ladder structure of Fig. P 4–2. What is the total capacitance in terms of a common unit capacitance, C_u?

4.14 (Sec. 4.2). Use the RLC ladder structure of Fig. P 4–14 to achieve an SC realization of a third-order elliptic filter having a passband ripple of 0.1 dB, a cutoff frequency of 1 kHz, an attenuation of at least 24 dB for frequencies greater than 2 kHz, and a clock frequency of 100 kHz. Give the total capacitance of this filter in terms of C_u, a common unit capacitance.

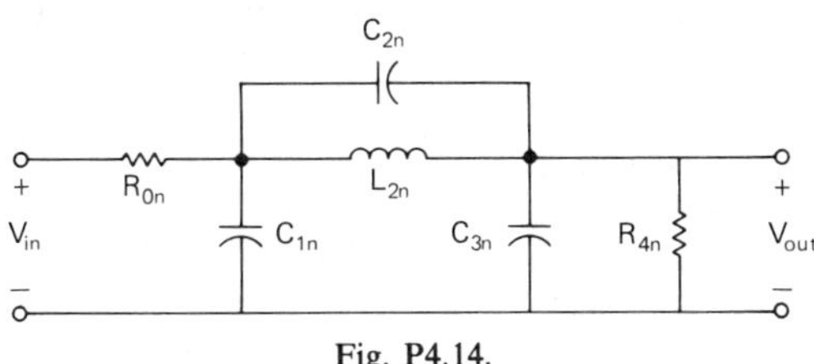

Fig. P4.14.

4.15 (Sec. 4.2). Find an SC realization for the third-order, elliptic filter shown in Fig. P 4–15. Assume that the cutoff frequency is 1 kHz and the clock frequency is 100 kHz. Use a suitable computer program (see Section 3.6) to obtain the passive and SC realization frequency responses from 10 Hz to 100 kHz.

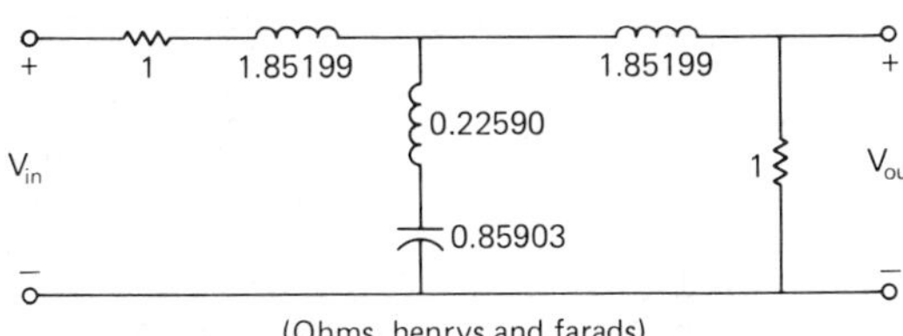

Fig. P4.15.

4.16 (Sec. 4.3). Show that the z-domain transfer function given in Eq. (48) of Section 3.3 can be used to derive Eq. (11), assuming that f_r is much less than f_c.

4.17 (Sec. 4.3). Redevelop Eqs. (16) through (19) for the case where R_{0n} is not equal to R.

4.18 (Sec. 4.3). Find the total capacitance of Example 4.3–1 in terms of C_u, a common capacitance.

4.19 (Sec. 4.3). Repeat Example 4.3–1 for a Chebychev bandpass filter having a 1-dB passband ripple. Give the total capacitance of this filter in terms of C_u, a common unit capacitance.

4.20 (Sec. 4.3). Develop a switched capacitor realization for a third-order (six-pole), Chebyshev, bandpass filter, using the low-pass prototype filter shown in Fig. P 4–10. The bandpass filter is to have a center frequency of 1 kHz, a bandwidth of 500 Hz, a 1-dB ripple in the passband, and a clock frequency of 100 kHz. (Optional) Plot the passive and SC realization frequency responses from 1 Hz to 100 kHz.

4.21 (Sec. 4.3). Develop a switched capacitor realization for a third-order (six-pole), Butterworth, bandpass filter, using the low-pass prototype filter shown in Fig. P 4–21. The bandpass filter is to have a center frequency of 1000 Hz, a bandwidth of 500 Hz, and a clock frequency of 100 kHz. (Optional) Plot the passive and SC realization frequency responses from 1 Hz to 100 kHz.

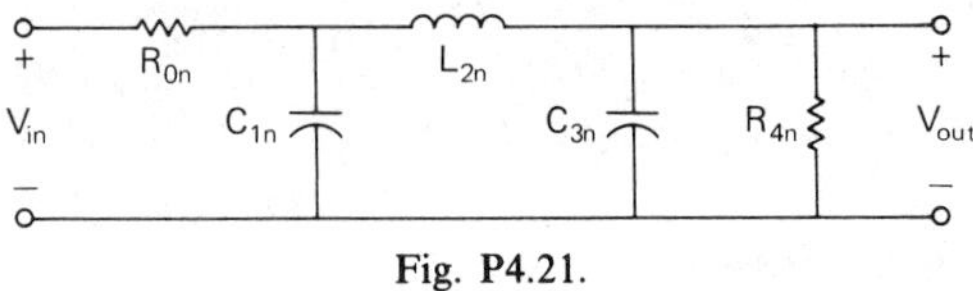

Fig. P4.21.

4.22 (Sec. 4.3). Repeat Problem 4.21 starting with the low-pass prototype filter given in Fig. P 4–22.

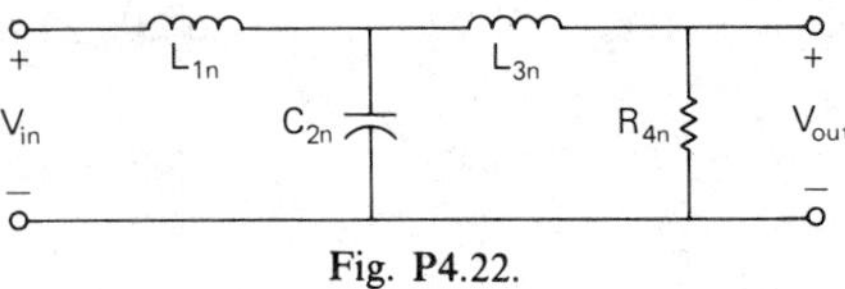

Fig. P4.22.

4.23 (Sec. 4.3). Show that Fig. 4.3–6(b) results if one first applies the low-pass-to-bandpass transformation to Fig. 4.2–6 and then removes the branches in parallel with the series branches, similar to what was done in going from Fig. 4.1–5(a) to Fig. 4.1–5(b).

4.24 (Sec. 4.3). Use the high sampling frequency assumption to show that the approximate s-domain transfer function of Fig. 4.3–7(a) is given by Eq. (22).

4.25 (Sec. 4.3). Find the total capacitance of Example 4.3–2 in terms of C_u, a common unit capacitance.

4.26 (Sec. 4.3). Repeat Example 4.3–2 using the low-pass prototype of Fig. P 4–2.

4.27 (Sec. 4.3). (a) Develop the design equations for the capacitor ratios of Fig. 4.3–7 for a switched capacitor realization of a third-order (six-pole), bandpass filter, starting with the low-pass prototype of Fig. P 4–14. (b) Design a third-order bandpass filter using the results of (a) that has a center frequency of 1000 Hz, a 1-dB ripple bandwidth of 500 Hz, and a maximum attenuation outside of a bandwidth of 1000 Hz of greater than 34 dB.

4.28 (Sec. 4.3). Repeat problem 4.27 using the low-pass prototype shown in Fig. P 4–15.

4.29 (Sec. 4.4). (a) Find the total capacitance of Example 4.4–1 in terms of C_u, a common unit capacitance. (b) Repeat part (a) if the clock frequency is 10 kHz.

4.30 (Sec. 4.4). Develop a switched capacitor realization of a third-order, Butterworth, high-pass filter using the low-pass prototype of Fig. P 4–22. The filter is to have a cutoff frequency of 100 Hz and a clock frequency of 50 kHz.

4.31 (Sec. 4.4). Develop a different set of state equations for Fig. 4.4–5(b) that yields a different functional realization for this filter compared with that of Fig. 4.4–6(a).

4.32 (Sec. 4.4). Calculate the total capacitance of Example 4.4–3 in terms of C_u, the common unit capacitance.

4.33 (Sec. 4.4). Develop a switched capacitor realization for a third-order, high-pass, elliptic filter using the prototype given in Fig. P 4–15. Assume that the cutoff frequency is 1 kHz and the clock frequency is 100 kHz. (Optional) Plot the passive and SC realization frequency responses from 10 Hz to 100 kHz.

4.34 (Sec. 4.4). Repeat Problem 4.14 except let the filter be high-pass.

4.35 (Sec. 4.4). Design a switched capacitor high-pass filter using the low-pass prototype shown in Fig. P 4–35. The filter is to have a cutoff frequency of 1000 Hz, a passband ripple of 1 dB, and an attentuation greater than 35 dB for all frequencies below 500 Hz. The clock frequency is 100 kHz.

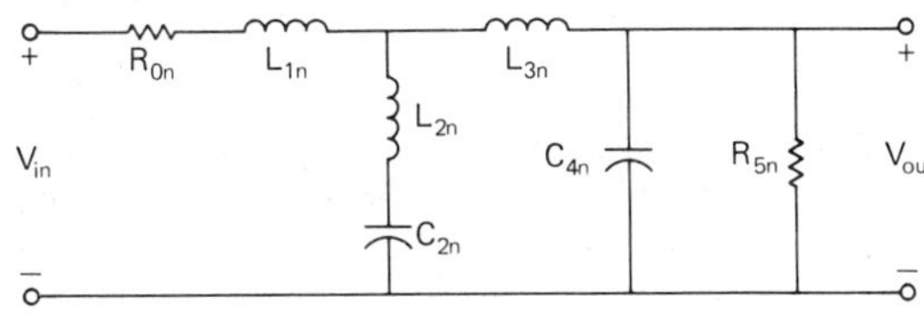

Fig. P4.35.

4.36 (Sec. 4.4). Find the s-domain transfer function of Fig. 4.4–13(b) and compare it with that of Fig. 4.4–13(a).

4.37 (Sec. 4.5). Give a functional realization of Eqs. (5), (6), (7), and (9) that does not require variables to be summed with variables that are passed through the band-pass circuit.

4.38 (Sec. 4.5). Show that the z-domain transfer function of Fig. 4.3–7(a) is given by Eq. (13).

4.39 (Sec. 4.5). Find the total units of capacitance of Example 4.5–1 in terms of the common unit capacitance, C_u.

4.40 (Sec. 4.5). Design a switched capacitor realization for a third-order (six-pole), Chebyshev, band-elimination filter having a center frequency of 1 kHz, a bandwidth of 500 Hz, and a clock frequency of 100 kHz. The filter is to have a 1-dB ripple in the passband regions. Use the low-pass prototype of Fig. P 4–10. (Optional) Plot the passive and SC realization frequency responses from 1 Hz to 100 kHz.

4.41 (Sec. 4.5). Repeat Problem 4.40 using the low-pass prototype of Fig. P 4–22.

4.42 (Sec. 4.5). Repeat problem 4.40 using the low-pass prototype of Fig. P 4–21.

4.43 (Sec. 4.5). Show that Fig. 4.5–5(d) results if the low-pass-to-band-elimination transformation is first applied to Fig. 4.5–5(a) and then the capacitor loops are eliminated.

4.44 (Sec. 4.5). Give a functional realization of Fig. 4.5–5(e), based on Eqs. (24), (25), (26), and (28), that uses Fig. 4.3–7(a).

4.45 (Sec. 4.5). Find the total capacitance of Example 4.5–2 in terms of the common unit capacitance, C_u.

4.46 (Sec. 4.5). Design a band-elimination switched capacitor filter starting with the low-pass prototype of Fig. P 4–14. The filter should have a center frequency of 1000 Hz, a bandwidth of 1000 Hz, a 1-dB ripple in the passbands, and a minimum attenuation of 34 dB over the bandwidth of 500 Hz.

4.47 (Sec. 4.5). Repeat Problem 4.45 for the low-pass prototype shown in Fig. P 4–15.

4.48 (Sec. 4.5). Develop the design equations that give the capacitor ratios as a function of the component values of Fig. P 4–48 and the clock frequency similar to the relationships in Eqs. (42) and (44).

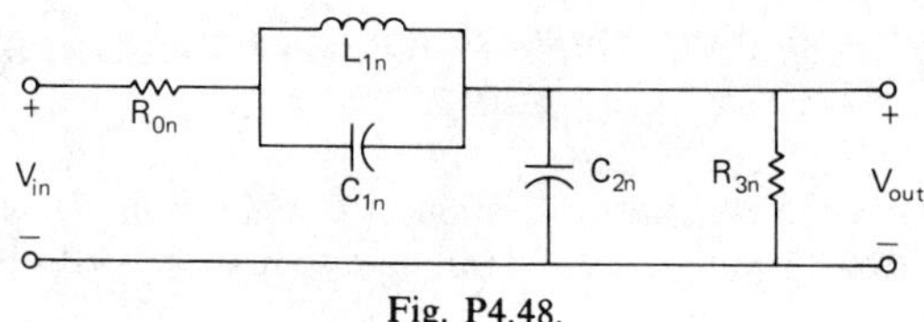

Fig. P4.48.

4.49 (Sec. 4.6). Develop from Fig. 4.2–9 a stray-insensitive realization of the fifth-order, elliptic filter of Example 4.2–2.

4.50 (Sec. 4.6). Repeat Example 4.6–1 using ω_0 rather than ω_5.

4.51 (Sec. 4.6). Repeat Example 4.6–1 for the filter of Example 4.2–2.

4.52 (Sec. 4.6). Repeat Example 4.6–2 for the filter of Example 4.2–2.

5
Switched Capacitor *Z*-domain Filters

5.0 INTRODUCTION

The common factor in the design approaches for realizing SC filters discussed in the previous chapters is the replacement of passive elements. Resistors of RC active filters were replaced by SC simulations, and the elements of RLC ladders were replaced by SC integrators.

In this chapter we take a different approach. This approach consists in transforming an analog filter transfer function $H(s)$ into a discrete-time filter transfer function $H(z)$. In Appendix B the details of transforming $H(s)$ into an equivalent $H(z)$, through the bilinear transformation, are presented. In fact, as shown in Fig. 3.1–16, $H(z)$, the transfer function in the z-domain, might be obtained directly given the filter specifications (box 1), and by means of approximation (box 2), we can define a set of pole and zero locations in the z-plane. $H(z)$ is then manipulated to form structures that have the same overall transfer function. The structures generated could be new structures or ones already implemented with SC building blocks.[1,2] This design approach is carried out by means of combining building blocks to realize a certain $H(z)$.[3,4] This transfer function can be of second or higher order; these topics are discussed in Sections 5.2 and 5.3. $H(z)$ can also be implemented by emulating known digital filter topologies.[5,6] This approach is presented in Section 5.1. Finally, in this chapter, several nonidealities of SC filters are

---

[1] W. K. Jenkins, T. N. Trick, and E. El-Masry, "New Realizations for Switched-Capacitor Filters," *Twelfth Annual Asilomar Conference on Circuits, Systems, and Computers,* Pacific Grove, CA, November 1978, pp. 694–698.

[2] G. Szentirmai and G. C. Temes, "Switched Capacitor Building Blocks," *Thirteenth Annual Asilomar Conference on Circuits, Systems and Computers,* November 1979, pp. 542–545.

[3] W. K. Jenkins, T. N. Trick, and E. El-Masry, ibid.

[4] G. Szentirmai and G. C. Temes, ibid.

[5] W. Kuntz, "A New Sample-and-Hold Device and Its Application to the Realization of Digital Filters," *Proc. IEEE* (Lett.), Vol. 56, November 1968, pp. 2092–2093.

[6] S. Nishimura, K. Hirano, and R. N. Pal, "A New Class of Very Low Sensitivity and Low Roundoff Noise Recursive Digital Filter Structures," *IEEE Trans. Circuits and Systems,* Vol. CAS-28, December 1981, pp. 1152–1158.

considered in detail. An example comparing the gain-bandwidth nonideal considerations versus the ideal case is presented to conclude this chapter.

5.1 SC FILTER DESIGN BY DIGITAL FILTER EMULATION

The basic operations involved in the implementation of digital filters are addition, subtraction, multiplication of a signal by a constant, and the delay of a signal. In Section 2.4 we discussed the implementations of these basic operations. Therefore we can realize digital filter emulation by means of SC operations.[7,8,9] In this section we consider the basic, simple, second-order structures, but the approach can be applied to higher order topologies.[10] Boxes 1, 2, 3, 4, 5, 12, and 14 of Fig. 3.1–16 characterize the method presented in this section.

Let us consider two common second-order digital filter structures, the direct and the coupled forms, illustrated in Fig. 5.1–1. A possible implementa-

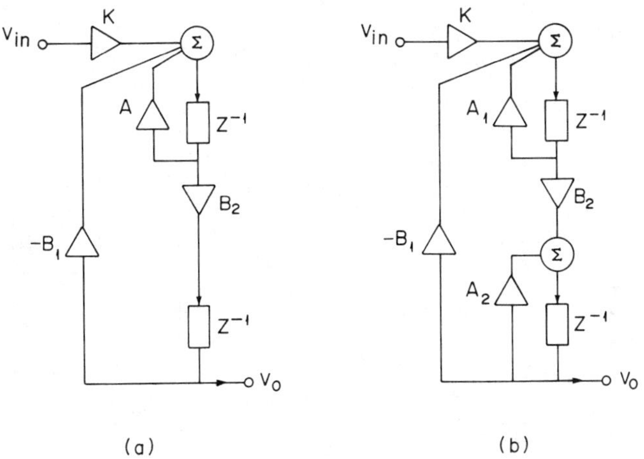

Fig. 5.1–1. Common digital filter structures (a) Direct form and (b) Coupled form.

---

[7] E. Sánchez-Sinencio, P. E. Allen, J. I. Arreola, and J. L. Gómez, "A Direct Design Approach for Analog Sampled Data Filters," *Proc. Thirteenth Annual Asilomar Conference on Circuits, Systems, and Computers,* Pacific Grove, CA. November 1979, pp. 538–541.

[8] I. A. Young and D. A. Hodges, "MOS Switched-Capacitor Analog Sampled Data Direct-Form Recursive Filters," *IEEE J. Solid-State Circuits,* Vol. SC-14, December 1979, pp. 1029–1033.

[9] J. J. Mulawka, "Switched Capacitor Analogue Delays Comprising Unity Gain," *Electronics Letters,* Vol. 17, April 1981, pp. 276–277.

[10] A. V. Oppenheim and R. W. Schafer, *Digital Signal Processing,* Prentice-Hall, Inc., Englewood Cliffs, NJ, 1975.

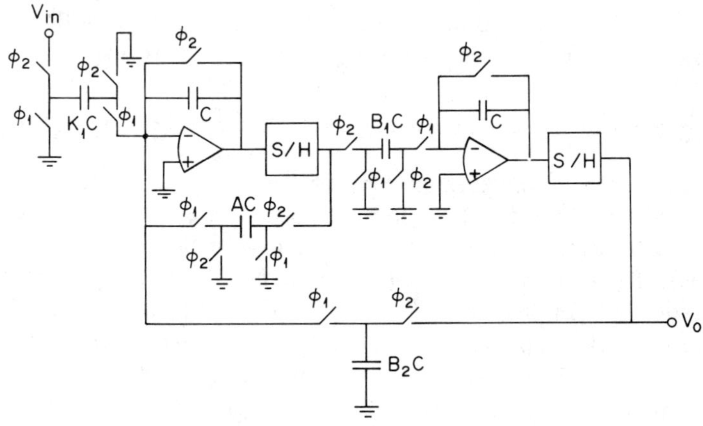

Fig. 5.1–2 (a). Analog implementation of the Direct form of Fig. 5.1–1 (a).

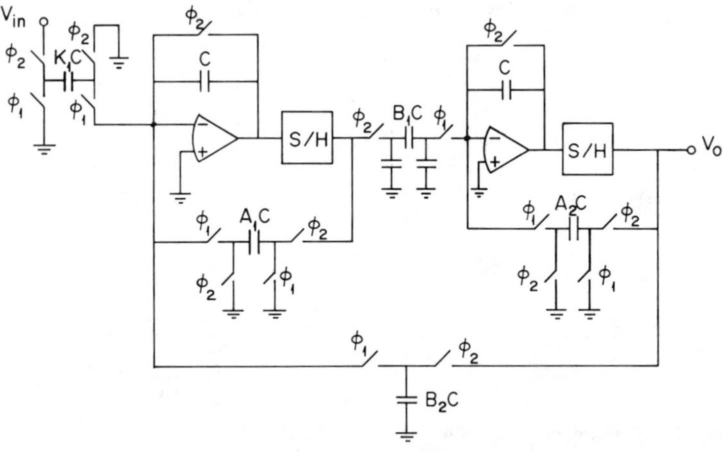

Fig. 5.1–2 (b). Analog implementation of the Coupled form of Fig. 5.1–1 (b).

tion for each of the structures is shown in Fig. 5.1–2. Other implementations are possible for different z^{-1} implementations (see Problem 5.1).[9]

Typically, the design of the analog SC filters based on digital filter emulation is carried out in the z-domain. Considering first just poles, assume that a complex pair of poles is required at $z_{1,2} = r\,e^{\pm j\theta}$. The transfer function of an all-pole structure is of the form

$$H(z) = \frac{k}{(z - r\,e^{j\theta})(z - r\,e^{-j\theta})} \tag{1}$$

or

$$H(z) = \frac{kz^{-2}}{1 - 2r\cos\theta\, z^{-1} + r^2 z^{-2}} \tag{2}$$

The transfer functions for the direct $H_D(z)$ and the coupled form $H_C(z)$ are

$$H_D(z) = \frac{-KB_2 z^{-2}}{1 - Az^{-1} + B_1 B_2 z^{-2}} \tag{3}$$

and

$$H_C(z) = \frac{-KB_2 z^{-2}}{1 - (A_1 + A_2)z^{-1} + (A_1 A_2 + B_1 B_2)z^{-2}} \tag{4}$$

One criterion for comparison of structures, as discussed in Section 3.2, is the sensitivity of r, θ, and H_m with respect to the coefficients, i.e., the capacitor ratios designated α. We define

$$S_\alpha^r = \frac{\alpha}{r}\frac{\partial r}{\partial \alpha} \tag{5}$$

$$S_\alpha^\theta = \frac{\alpha}{\theta}\frac{\partial \theta}{\partial \alpha} \tag{6}$$

and

$$S_\alpha^{H_{\max}} = \frac{2r^2}{1 - r^2} S_\alpha^r - \frac{\theta}{\tan\theta} S_\alpha^\theta \tag{7}$$

where $H_{\max}$ is given by Eq. (15) of Section 3.2. If Q is defined as $\theta/(-2\ln r)$, the S_α^Q is sometimes required and is expressed as

$$S_\alpha^Q = S_\alpha^\theta - \frac{1}{\ln r} S_\alpha^r \tag{8}$$

Table 5.1–1 Sensitivity of Direct and Coupled Forms in the z-Domain

STRUCTURE AND PARAMETERS	$S^{\theta}_{x_1}$	$S^{\theta}_{x_2}$	$S^{H_m}_{x_2}$	$S^{H_m}_{x}$
Direct: $x_1 = A = 2r\cos\theta$, $x_2 = B_1 = B_2 = r$	$\frac{-1}{\theta \tan \theta}$	$\frac{1}{\theta \tan \theta}$	$\frac{1}{\tan^2 \theta}$	$\frac{2r^2}{1-r^2} - \frac{1}{\tan^2 \theta}$
Coupled: $x_1 = A_1 = A_2 = r\cos\theta$, $x_2 = B_1 = B_2 = r\sin\theta$	$-\frac{\sin 2\theta}{2\theta}$	$\frac{\sin 2\theta}{2\theta}$	$\frac{2r^2}{1-r^2}\cos^2\theta + \frac{\sin 2\theta}{2 \tan \theta}$	$\frac{2r^2\sin^2\theta}{1-r^2} - \frac{\sin 2\theta}{2 \tan \theta}$

Table 5.1–1 summarizes the results of sensitivity for the direct and coupled forms. The coupled form is superior to the direct form regarding sensitivity, although it requires an extra coefficient.

It has been recognized that digital filters derived with accumulator type elements,[11,12,13] rather than with unit delay elements, have better sensitivity properties. This is especially critical for very narrow band filters, i.e., poles near $z = 1$. The basic idea of the approach using the accumulator type elements is to translate the original pole location from a high sensitivity region into a low-sensitivity region.[11,12] In practice this can be accomplished by using the following transformation

$$\hat{z} = e(z - d) \tag{9}$$

where $\hat{z}$ is the new z-plane. Thus, to realize filters in the z-plane, we require the implementation of z^{-1}, that is

$$\hat{z}^{-1} = \frac{e^{-1} z^{-1}}{1 - dz^{-1}} \tag{10}$$

A $\hat{z}^{-1}$ block analog implementation is shown in Fig. 5.1–3.[14] Neglecting the effect of the parasitic capacitors, there are several particular cases of

---

[11] J. Szczupak and S. K. Mitra, "On Digital Filter Structures with Low Coefficient Sensitivities," *Proc. IEEE,* Vol. 66, September 1978, pp. 1082–1083.

[12] R. C. Agarwal and C. S. Burrus, "New Recursive Digital Filter Structures Having Very Low Sensitivity and Round-off Noise," *IEEE Trans. on Circuits and Systems,* Vol. CAS-22, December 1975, pp. 921–927.

[13] S. Nishimura et al., op. cit.

[14] This structure has been discussed in some detail in Section 2.6.

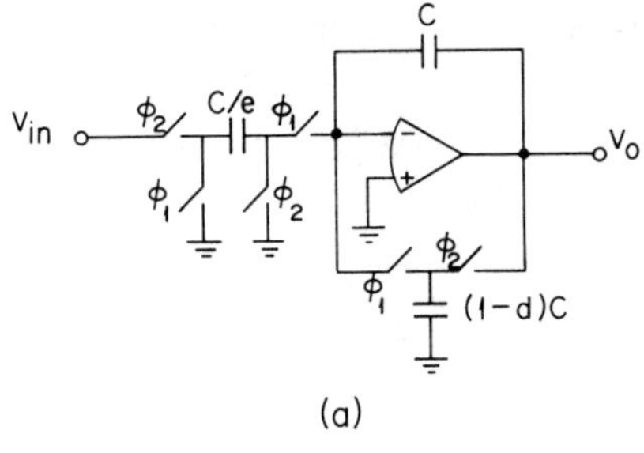

(a)

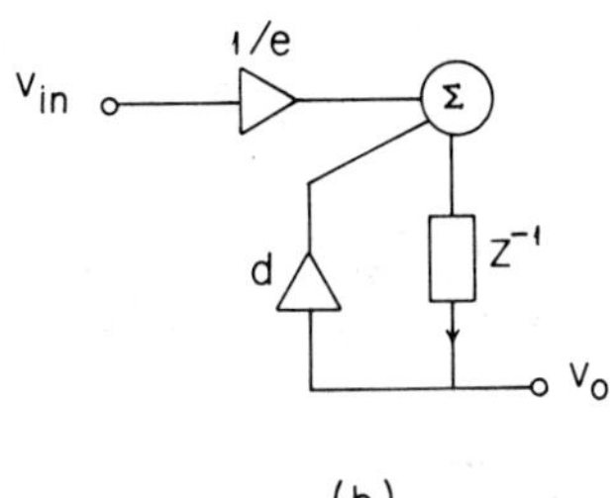

(b)

Fig. 5.1–3. Analog general block $\hat{z}^{-1}$. (a) Implementation, (b) Block diagram.

Fig. 5.1–3 that can be derived: (1) if $d = 1$ the forward integrator is obtained; (2) if $d = 0$, a multiply-and-delay element z^{-1}/e can be obtained.

We can use the circuit of Fig. 5.1–3 as the general basic block to derive low-sensitivity structures.[15] However, the simplicity of the forward integrator is very attractive. This accumulator (forward integrator) contains only a capacitor in the feedback path. Therefore, we will use this simpler version of Fig. 5.1–3.

Next, we will show with a simple first-order filter the sensitivity improvement when an accumulator is employed rather than a delay element and the pole is near $z = 1$. Consider a low-pass filter characterized by

$$H(z) = \frac{k}{z - a} = \frac{kz^{-1}}{1 - az^{-1}}, \qquad a \leq 1 \tag{11}$$

The corresponding S_a^H yields

$$S_a^H = \frac{a}{H}\frac{\partial H}{\partial a} = \frac{a}{z - a} \tag{12}$$

[15] J. Szczupak and S. K. Mitra, op. cit.

Now consider the case when an accumulator is employed, that is

$$\hat{z} = z - 1 \tag{13}$$

or

$$z = \hat{z} + 1 \tag{14}$$

Thus

$$H(\hat{z}) = \frac{k}{\hat{z} + (1 - a)} = \frac{k}{\hat{z} + \hat{a}} \tag{15}$$

where $\hat{a} = 1 - a$. Note that both $H(z)$ and $H(\hat{z})$ represent the same transfer function. The sensitivity of $H(\hat{z})$ with respect to $\hat{a}$ is given by

$$S_{\hat{a}}^{H} = \frac{-\hat{a}}{\hat{z} + \hat{a}} \tag{16}$$

or

$$S_{\hat{a}}^{H} = \frac{a - 1}{z - a} \tag{17}$$

Therefore, we can conclude from Eqs. (12) and (17) that

$$|S_{\hat{a}}^{H}| < |S_{a}^{H}|, \qquad \text{for } a \text{ close to } 1 \tag{18}$$

i.e., $\omega = 0$, $a = 0.99$, $S_{\hat{a}}^{H} = -1$, and $S_{a}^{H} = 99$.

Next, we consider the case of the direct and coupled forms using accumulators. The transfer function for a general second-order filter is given by

$$H(z) = \frac{b_2 z^{-2} + b_1 z^{-1} + b_0}{1 - a_1 z^{-1} + a_2 z^{-2}} \tag{19}$$

The corresponding $H(z)$ using accumulators of the form $\hat{z}^{-1}$ is given by

$$H(\hat{z}) = \frac{\beta_2 \hat{z}^{-2} + \beta_1 \hat{z}^{-1} + \beta_0}{1 + \alpha_1 \hat{z}^{-1} + \alpha_2 \hat{z}^{-2}} \tag{20}$$

where

$$\beta_0 = b_0,\ \beta_1 = 2b_0 + b_1, \qquad \beta_2 = b_1 + b_2 + b_0$$
$$\alpha_1 = 2 - a_1, \qquad \alpha_2 = 1 + a_2 - a_1 \tag{21}$$

The direct and coupled forms using $\hat{z}^{-1}$ are shown in Fig. 5.1–4. The all-pole transfer functions are given by

$$H_D(\hat{z}) = \frac{-k\hat{z}^{-2}B_2}{1 + A\hat{z}^{-1} + B_1B_2\hat{z}^{-2}} \tag{22}$$

and

$$H_C(\hat{z}) = \frac{-k\hat{z}^{-2}B_2}{1 + (A_1 + A_2)\hat{z}^{-1} + (A_1A_2 + B_1B_2)\hat{z}^{-2}} \tag{23}$$

The corresponding relations of the parameters are
Direct form $\hat{z}^{-1}$:

$$A = 2 - 2r\cos\theta$$
$$B_1B_2 = 1 + r^2 - 2r\cos\theta \tag{24}$$

Coupled form $\hat{z}^{-1}$:

$$A_1 + A_2 = 2 - 2r\cos\theta$$
$$A_1A_2 + B_1B_2 = 1 + r^2 - 2r\cos\theta \tag{25}$$

Table 5.1–2 shows the sensitivity results for these two structures.

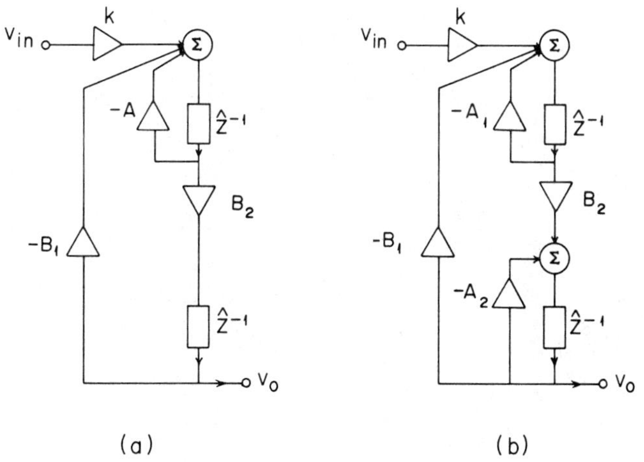

Fig. 5.1–4. Digital filter structures using accumulators. (a) Direct form. (b) Coupled form.

Table 5.1–2 Sensitivity of Direct and Coupled Forms Using Forward Accumulators

STRUCTURE AND PARAMETERS	$S_{x_1}^{\theta}$	$S_{x_2}^{\theta}$	$S_{x_1}^{H_m}$	$S_{x_2}^{H_m}$
Direct: $x_1 = A = 2(1 - r\cos\theta)$ $x_2 = B_1 = -B_2$ $= (1 + r^2 - 2r\cos\theta)^{1/2}$	$\frac{(1 - r\cos\theta)(r - \cos\theta)}{\theta\, r^2 \sin\theta}$	$\frac{1 - 2r\cos\theta + r^2}{4r^2\theta\tan\theta}$	$\frac{-2(1 - r\cos\theta)}{1 - r^2}$ $-\frac{(r - \cos\theta)(1 - r\cos\theta)}{r^2\sin\theta\tan\theta}$	$\frac{1 + r^2 - 2r\cos\theta}{2\, r^2(1 - r^2)}$ $-\frac{1 + r^2 - 2r\cos\theta}{4\, r^2 \tan^2\theta}$
Coupled: $x_1 = A_1 = A_2 = 1 - r\cos\theta$ $x_2 = B_1 = B_2 = r\sin\theta$	$\frac{(1 - r\cos\theta)(1 - \cos 2\theta)}{2\theta r}$	$\frac{\sin 2\theta}{2\theta}$	$-\frac{2r\cos\theta(1 - r\cos\theta)}{1 - r^2}$ $\frac{-(1 - \cos 2\theta)(1 - r\cos\theta)}{-2r\tan\theta}$	$\frac{2r\sin^2\theta}{1 - r^2}$ $-\frac{\sin 2\theta}{2\tan\theta}$

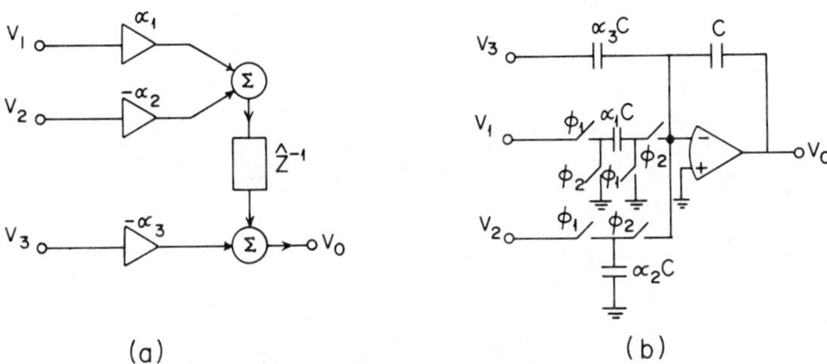

Fig. 5.1–5. Basic accumulator-multiplier/adder. (a) Block diagram and (b) Implementation.

The implementation of the coupled and direct form of Fig. 5.1–4 using the forward integrators ($\hat{z}^{-1}$) can be carried out using the basic block shown in Fig. 5.1–5. Thus the implementations based on Fig. 5.1–5 of the direct and coupled form are shown in Fig. 5.1–6.

Example 5.1–1. *Sensitivity comparison.* Let us consider the four structures discussed before. The poles in the z-domain to be implemented by structures of Fig. 5.1–2 and 5.1–6 are located at $r = 0.99$ and $\theta = 6.2°$. The information from Tables 5.1–1 and 5.1–2 for this numerical example is given in Table 5.1–3, where LSC stands for the *largest to smallest capacitor ratio.* It can be seen that for large θ, the LSC decreases.

Consider another pole location, $\theta = 19.414°$ and $r = 0.98076$; the corresponding table of sensitivities is shown in Table 5.1–4.

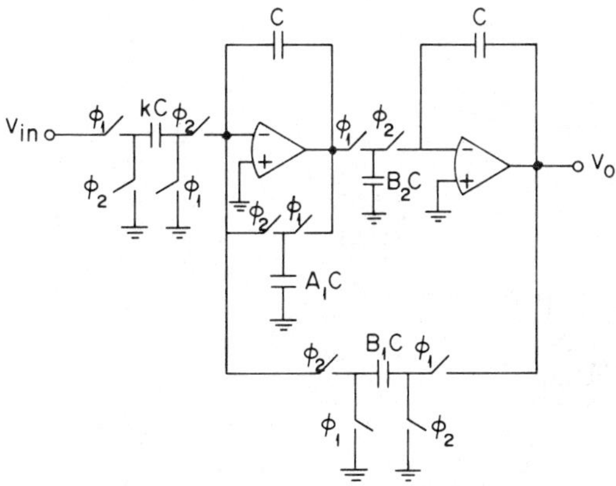

Fig. 5.1–6. (a) Direct form with $\hat{z}^{-1}$.

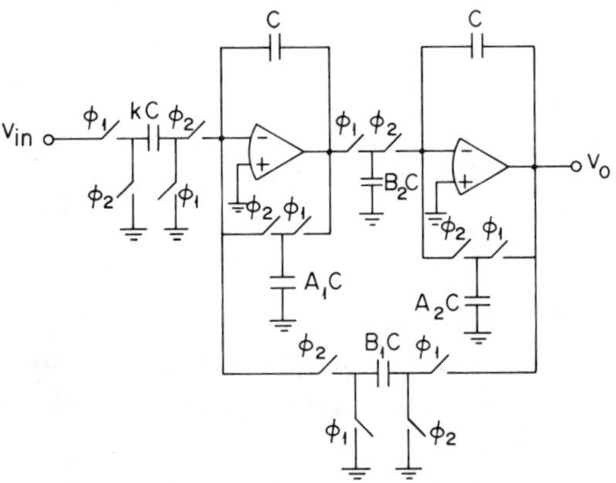

Fig. 5.1–6. (b) Coupled form with $\hat{z}^{-1}$.

Table 5.1–3 Sensitivities for $\theta = 6.2°$ and $r = 0.99$

STRUCTURE AND PARAMETERS	$S_{x_1}^{\theta}$	$S_{x_2}^{\theta}$	$S_{x_1}^{H_m}$	$S_{x_2}^{H_m}$
Direct with Delays $x_1 = 1.9684$ $x_2 = 0.99$ LSC = 1.98	−85.07	85.07	84.73	13.77
Coupled with Delays $x_1 = 0.984209$ $x_2 = 0.106919$ $LSC = 9.352$	−0.992	0.992	98.34	0.16
Direct with Accumulators $x_1 = 0.31581$ $x_2 = .10808$ $LSC = 31.6$	−0.0057	0.2534	−1.581	0.04698
Coupled with Accumulators $x_1 = 0.01579$ $x_2 = 0.1069$ $LSC = 63.3$	.0017	0.992	−1.563	0.1722

Table 5.1–4 Sensitivities for $\theta = 19.414°$ and $r = 0.98076$

STRUCTURE AND PARAMETERS	$S^{\theta}_{x_1}$	$S^{\theta}_{x_2}$	$S^{H_m}_{x_1}$	$S^{H_m}_{x_2}$
Direct with Delays $x_1 = 1.84$ $x_2 = 0.9807$ $LSC = 1.9$	−8.374	8.374	8.05	42.43
Coupled with Delays $x_1 = 0.925$ $x_2 = 0.3239$ $LSC = 3.0$	−0.925	0.925	45.79	4.68
Direct with Accumulators $x_1 = 0.15$ $x_2 = 0.111$ $LSC = 9$	0.026	0.2435	−3.96	1.292
Coupled with Accumulators $x_1 = 0.075$ $x_2 = 0.3259$ LSC = 13.3	0.025	0.925	−3.66	4.68

From the previous sensitivity results, we can make the following observations:

1. The sensitivities of the structures using accumulators, as expected, are lower than the ones using delay elements.
2. The smaller the sensitivities the larger the LSC.
3. An increase of θ usually results in a decrease in the LSC.
4. The direct form structure with accumulators performs better than the other structures, regarding sensitivity.
5. There is a clear tradeoff between the LSC and sensitivity.

Next, we will discuss the implementation of zeros. Let us consider the implementation of the zeros with the direct or coupled form using accumulators. One possible implementation of zeros is shown in Fig. 5.1–7. The SC C_R in Fig. 5.1–7(b) is a very small capacitor used to provide a DC path for the output summer. The corresponding transfer function is given by

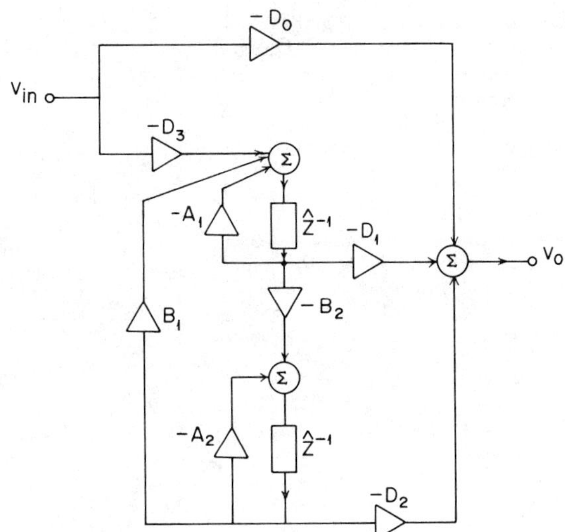

Fig. 5.1–7. (a) Block diagram of a biquad with a summer.

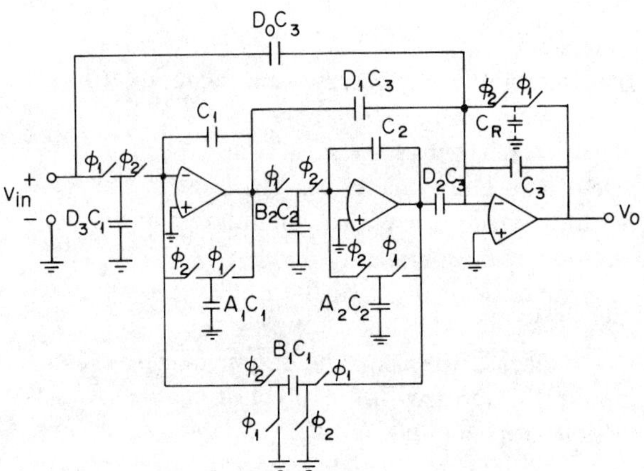

Fig. 5.1–7. (b) Implementation of a biquad with a summer.

$$H(\hat{z}) = \frac{-D_0 - [D_0(A_1 + A_2) - D_1 D_3]\hat{z}^{-1} - [D_0(A_1 A_2 + B_1 B_2) + D_3 B_2 D_2 - D_3 D_1 A_2]\hat{z}^{-2}}{1 + (A_1 + A_2)\hat{z}^{-1} + (A_1 A_2 + B_1 B_2)\hat{z}^{-2}} \tag{26}$$

Observe that the sign of D_3 can be easily changed by using an inverting SC topology, i.e., Fig. 2.6–5. Thus the circuit has additional flexibility.

If the direct form is used to implement the poles, say $A_1 = 0$ and the values of A_2, B_1, and B_2 are modified accordingly, the transfer function becomes

$$H(\hat{z}) = \frac{-D_0 - [D_0 A_2 - D_1 D_3]\hat{z}^{-1} - [D_0 B_1 B_2 + D_3 B_2 D_2 - D_3 D_1 A_2]\hat{z}^{-2}}{1 + A_2\hat{z}^{-1} + (B_1 B_2)\hat{z}^{-2}} \tag{27}$$

The choice of the structure will depend on the particular location of the desired poles and zeros and their corresponding coefficient values of $\hat{z}^0$, $\hat{z}^{-1}$, and $\hat{z}^{-2}$.

Another alternative to implement a biquad is shown in Fig. 5.1–8, where only two op amps are involved. The transfer function is given by

$$H(\hat{z}) = \frac{-D_0 - (D_0 A_1 + D_1 - D_3 B_2)\hat{z}^{-1} - (-D_2 B_2 + D_1 A_1)\hat{z}^{-2}}{1 + (A_1 + A_2)\hat{z}^{-1} + (A_1 A_2 + B_1 B_2)\hat{z}^{-2}} \tag{28}$$

It should be noticed that the opposite sign values of D_1 and D_2, if required, can be obtained.

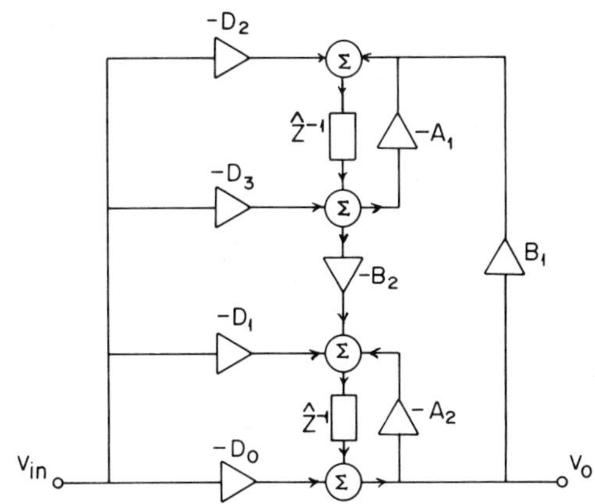

Fig. 5.1–8. (a) Block diagram of a two op amp biquad.

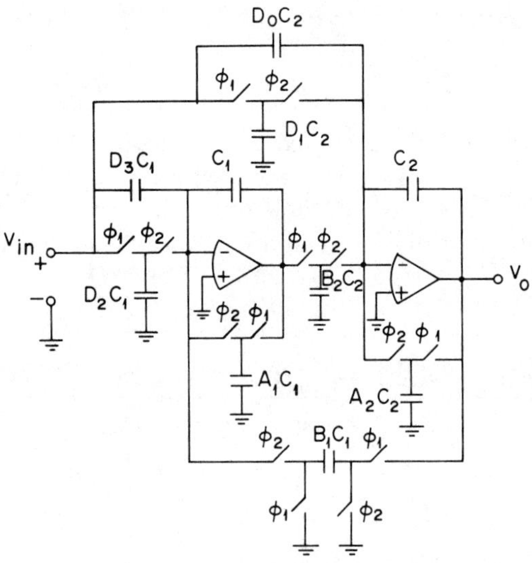

Fig. 5.1–8. (b) Implementation of a two op amp biquad.

Example 5.1–2. *Bandpass structure.* Consider the implementation of a bandpass structure with zeros at ± 1 and poles at $0.95\ e^{\pm 6.2^\circ}$. Using the structure of Fig. 5.1–8(b), we obtain

$$A_1 = A_2 = 1 - r\cos\theta = 0.0555$$

$$B_1 = B_2 = 0.1026$$

$$D_0 = 1$$

$$D_0 A_1 + D_1 - D_3 B_2 = 2$$

$$-D_2 B_2 + D_1 A_1 = 0$$

then

$$D_1 = 2.13431$$

$$D_2 = 1.1545$$

$$D_3 = 1.85$$

$$LSC = 38.42$$

The computation of the total capacitance is left as a problem (Problem 5.4).

5.2 BUILDING-BLOCK APPROACH TO SECOND-ORDER SC FILTERS

A possible systematic design approach of sampled data filters is shown in Fig. 3.1–16, through boxes 1, 2, 3, 4, 5, 12, and 14. The approach presented in this section starts with $H(z)$ or $H(\hat{z})$ and, by means of building blocks, several structures are generated that have the same overall transfer functions $H(\hat{z})$. The steps involved from mapping $H(s)$ to $H(z)$ are discussed in Appendix B, in particular when bilinear mapping is employed.

There are several basic building blocks available to obtain different SC topologies that can implement a specific $H(z)$ or $H(\hat{z})$. One building block proposed by Szentirmai and Temes[16] is shown in Fig. 5.2–1.

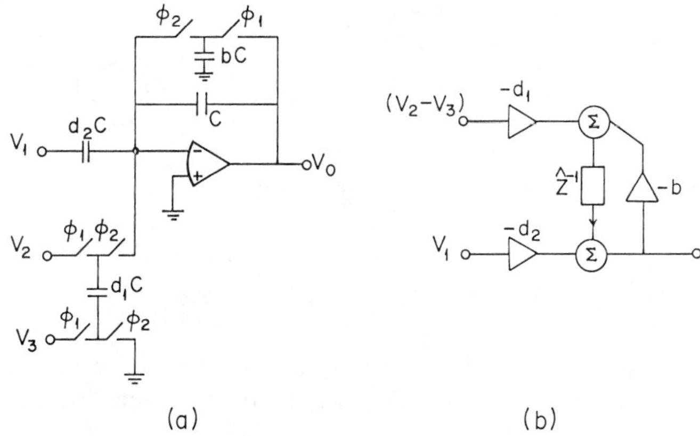

Fig. 5.2–1. A basic building block (a) Implementation and (b) Equivalent flow diagram.

The z-transform of the output voltage is given by

$$V_0 = -\frac{d_2(1 - z^{-1})V_1 + d_1 z^{-1}(V_2 - V_3)}{1 + (1 - b)z^{-1}} \tag{1}$$

and if

$$z = \hat{z} + 1 \tag{2}$$

then

[16] G. Szentirmai and G. C. Temes, op. cit.

$$V_0 = \frac{-d_1(V_2 - V_3)\hat{z}^{-1}}{1 + b\hat{z}^{-1}} - \frac{d_2 V_1}{1 + b\hat{z}^{-1}} \tag{3}$$

Now our first objective is to find SC implementations that realize a general second-order transfer function

$$H(z) = -K \frac{\beta_0 + \beta_1 \hat{z}^{-1} + \beta_2 \hat{z}^{-2}}{1 + \alpha_1 \hat{z}^{-1} + \alpha_2 \hat{z}^{-2}} \tag{4}$$

where the only restriction is that this $H(\hat{z})$ represents a stable system, i.e., $r < 1$. One possible implementation is shown in Fig. 5.2–2; this biquad only uses two op amps. The transfer function, when the output is V_2, is given by

$$H_2(\hat{z}) = \frac{b_6 b_2 + (b_1 b_6 + b_2 b_5 - a_1)\hat{z}^{-1} + (b_1 b_5 - a_1 b_4)\hat{z}^{-2}}{1 + (b_3 + b_4 + b_2 b_5)\hat{z}^{-1} + (b_3 b_4 + b_1 b_5)\hat{z}^{-2}} \tag{5}$$

and when the output is taken at V_1, we obtain

$$H_1(\hat{z}) = -\frac{b_6 + (b_5 + b_3 b_6)\hat{z}^{-1} + (b_3 b_5 + a_1 b_5)\hat{z}^{-2}}{1 + (b_3 + b_4 + b_2 b_5)\hat{z}^{-1} + (b_3 b_4 + b_1 b_5)\hat{z}^{-2}} \tag{6}$$

$H_1(\hat{z})$ is not as general as $H_2(\hat{z})$, i.e., $H_1(\hat{z})$ cannot realize a high-pass filter. However, for certain cases $H_1(\hat{z})$ may yield more convenient coefficient values

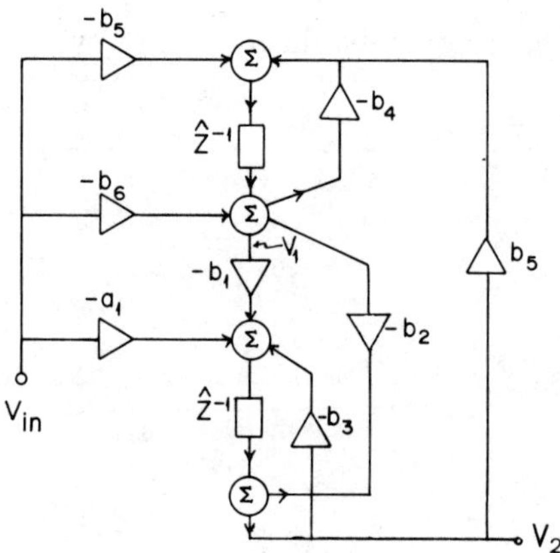

Fig. 5.2–2. (a) Flow diagram of a two op amp biquad section.

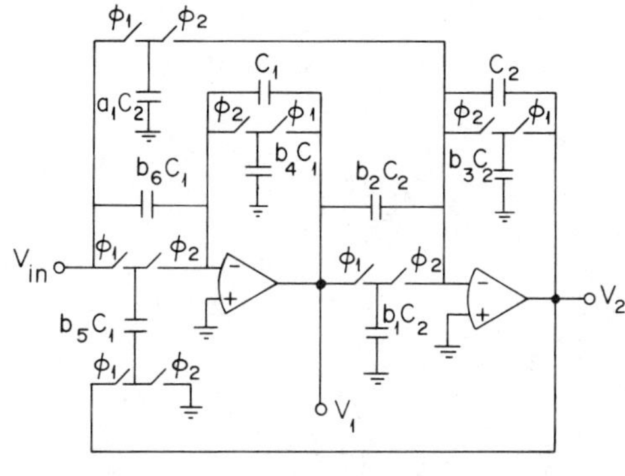

(b)

Fig. 5.2–2. (b) SC implementation of a two op amp biquad section.

than $H_2(\hat{z})$. Table 5.2–1 gives the forms of $H(\hat{z})$ for different types of filters in the $\hat{z}$ plane.

For the situation of a high-sampling rate, as discussed in Section 1.6, we can make the following approximations between the *s*- and *z*-domain. For high Q values

$$Q \cong \frac{\theta}{2(1-r)} \tag{7}$$

which gives

$$r \cong 1 - \frac{\theta}{2Q} \tag{8}$$

For small θ, we can make the following approximations

$$\theta \cong 2\pi \frac{f_0}{f_s} \tag{9}$$

$$\cos\theta \cong 1 - \frac{\theta^2}{2} \tag{10}$$

Therefore

$$\alpha_1 = 2 - 2r\cos\theta \cong \frac{\theta}{Q} + \theta^2 \tag{11}$$

Table 5.2–1. Generic Biquadratic Transfer Functions in the $\hat{z}$-domain, $\alpha = 2r\cos\theta$ and $\beta = r^2$

GENERIC FORM	NUMERATOR $N(z)$	NUMERATOR $N(\hat{z})$
LP 20 (bilinear transform)	$K(1 + z^{-1})^2$	$K(1 + 4\hat{z}^{-1} + 4\hat{z}^{-2})$
LP 11	$Kz^{-1}(1 + z^{-1})$	$K(2\hat{z}^{-1} + 1)\hat{z}^{-1}$
LP 10	$K(1 + z^{-1})$	$K(2\hat{z}^{-2} + 3\hat{z}^{-1} + 1)$
LP 02 (forward transform)	Kz^{-2}	$K\hat{z}^{-2}$
LP 01	Kz^{-1}	$K(1 + \hat{z}^{-1})\hat{z}^{-1}$
LP 00 (backward transform)	K	$K(1 + \hat{z}^{-1})^2$
BP 10 (bilinear transform)	$K(1 - z^{-1})(1 + z^{-1})$	$K(1 + 2\hat{z}^{-1})$
BP 01 (forward)	$Kz^{-1}(1 - z^{-1})$	$K\hat{z}^{-1}$
BP 00 (backward)	$K(1 - z^{-1})$	$K(1 + \hat{z}^{-1})$
HP	$K(1 - z^{-1})^2$	K
LPN	$K(1 + \epsilon z^{-1} + z^{-2})$, $\epsilon > \alpha/\sqrt{\beta}, \beta > 0$	$K[1 + \hat{z}^{-1}(2 + \epsilon) + (2 + \epsilon)\hat{z}^{-2}]$
HPN	$K(1 + \epsilon z^{-1} + z^{-2})$, $\epsilon < \alpha/\sqrt{\beta}$, $\beta > 0$	$K[1 + \hat{z}^{-1}(2 + \epsilon) + (2 + \epsilon)\hat{z}^{-2}]$
AP	$K(\beta + \alpha z^{-1} + z^{-2})$	See Eq. (21), Sect. 5.1

and

$$\alpha_2 = 1 + r^2 - 2r\cos\theta \tag{12}$$

$$\alpha_2 \cong \theta^2\left(1 + \frac{1}{4Q^2}\right) \cong \theta^2 \tag{13}$$

Now let us consider the case of a bandpass filter. Assume that $H_1(\hat{z})$ is employed with $b_4 = b_2 = 0$, and a_1 is inverted; a_1 becomes $-a_1$. Then Eq. (6) becomes

$$H_1(\hat{z}) = \frac{b_6 + (b_5 + b_3 b_6)\hat{z}^{-1} + (b_3 b_5 - a_1 b_5)\hat{z}^{-2}}{1 + (b_3)\hat{z}^{-1} + (b_1 b_5)\hat{z}^{-2}} \tag{14}$$

We can then relate Eq. (14) to Eq. (4) to obtain the design equations for a bandpass filter, which are

$$b_6 = K \tag{15}$$

$$b_5 + b_3 b_6 = 2K \tag{16}$$

$$b_3 b_5 - a_1 b_5 = 0 \tag{17}$$

$$b_3 = \frac{\theta}{Q} + \theta^2 \tag{18}$$

$$b_1 b_5 = \theta^2 \tag{19}$$

or

$$Q = \frac{\sqrt{b_1 b_5}}{b_3 - b_1 b_5} \tag{20}$$

$$r = 1 - \frac{\theta}{2Q} \tag{21}$$

It should be emphasized that the above equations are valid for $\theta << 1$ and $Q >> 1$. Let us consider a numerical example where the above equations do not exactly hold.

Example 5.2–1. *Second-order bandpass.* Design a bandpass filter using $H_1(\hat{z})$ of Eq. (14) with $b_4 = b_2 = 0$, and $a_1 = -a_1$ with the following characteristics.[17]

$K = 0.0605$, $r = 0.9803$, and $\theta = 19.3377°$. The resulting exact capacitor ratios are given by

$$b_3 = 0.150 \qquad b_6 = 0.0605$$
$$b_5 = 0.112 \qquad a_1 = 0.150$$
$$b_1 = 0.992$$

The approximated values using the above equations, Eqs. (15) through (21), result in

$$b_3 \cong 0.153 \qquad b_6 = 0.0605$$
$$b_5 \cong 0.1117 \qquad a_1 \cong 0.153$$
$$b_1 = 1.01456$$

In this example the LSC is 16.5. The total capacitance, *TC,* is obtained by dividing all the capacitors by the smallest one, that is

$$b_3 = 2.479 \qquad b_6 = 1$$
$$b_5 = 1.851 \qquad a_1 = 2.479$$
$$b_1 = 16.3967$$

Thus, including the two feedback capacitors C_1 and C_2, $TC = 24.33 + 2(16.528) \cong 58\ C_u$, where C_u is the unit capacitance. It should be observed that our capacitor ratios can be divided into two groups $\{b_3C_2, b_1C_2, a_1C_2, C_2\}$ and $\{b_5C_1, b_6C_1, C_1\}$; in this way each group requires 15.28 C_u and 19.38 C_u, respectively, and the total capacitance $TC \cong 35\ C_u$. Finally, for a near-optimal dynamic range, the output signals of both op amps should be in the same range. This can be accomplished by scaling C_1, b_2C_2, and b_1C_2 in Fig. 5.2–2(b).

There are many possible ways to generate the particular structures of Fig. 5.2–2. These structures will have different LSC capacitance areas and sensitivities. It appears that large (small) LSC are related to small (large) sensitivities. We will illustrate this effect by means of examples.

Next, we consider another basic building block, the most important characteristic of which is immunity to parasitic capacitances. The inconvenience

[17] G. Szentirmai et al., op. cit.

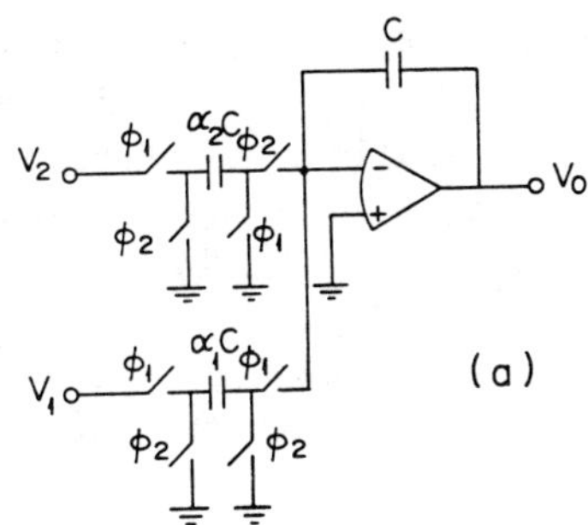

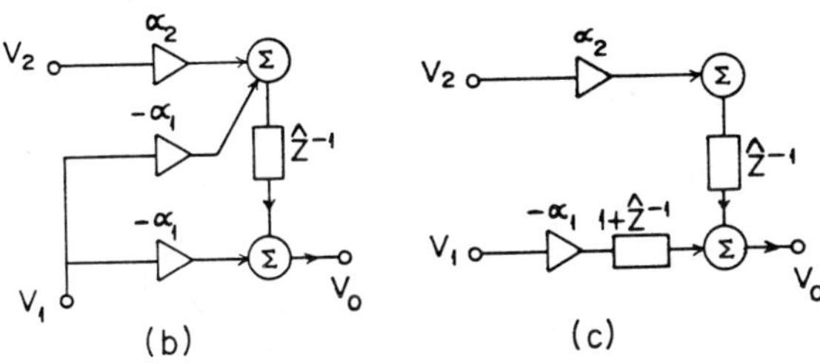

Fig. 5.2–3. Stray insensitive basic building block. (a) SC Implementation, (b) Flow diagram, (c) Alternative flow diagram.

is that we have only two types of integrators when we sample both input and output during the same phase clock, a positive forward and a negative backward integrator. This building block is shown in Fig. 5.2–3. The output V_0 can be written as

$$V_0 = \frac{\alpha_2 z^{-1}}{1 - z^{-1}} V_2 - \frac{\alpha_1}{1 - z^{-1}} V_1 \tag{22}$$

or

$$V_0 = \alpha_2 \hat{z}^{-1} V_2 - \alpha_1 (1 + \hat{z}^{-1}) V_1 \tag{23}$$

where $\hat{z}$ is defined as before: $\hat{z} = z - 1$. Several important particular cases are described next. Assume $V_2 = V_1 = V_{in}$, then

$$V_0 = (\alpha_2 \hat{z}^{-1} - \alpha_1 - \alpha_1 \hat{z}^{-1}) V_{in} \tag{24}$$

or

$$V_0 = (-\alpha_1 + (\alpha_2 - \alpha_1)\hat{z}^{-1}) V_{in} \tag{25}$$

1. If $\alpha_2 > \alpha_1$, the building block of Fig. 5.2–3 is reduced to the one shown in Fig. 5.2–4(a), that is, $\alpha_2 C$ is replaced by another capacitor of value $(\alpha_2 - \alpha_1)C$, and $\alpha_1 C$ is replaced by an unswitched capacitor $\alpha_1 C$.
2. If $\alpha_2 < \alpha_1$, then $\alpha_1 C$ in Fig. 5.2–3 is replaced by another capacitor of value $(\alpha_1 - \alpha_2)C$, and $\alpha_2 C$ is replaced by an unswitched capacitor of value $\alpha_2 C$. This is illustrated in Fig. 5.2–4(b).
3. If $\alpha_1 = \alpha_2 = \alpha$, it can be seen that the two SCs, $\alpha_1 C$ and $\alpha_2 C$, can be replaced by just one unswitched capacitor of value αC. This is shown in Fig. 5.2–4(c).

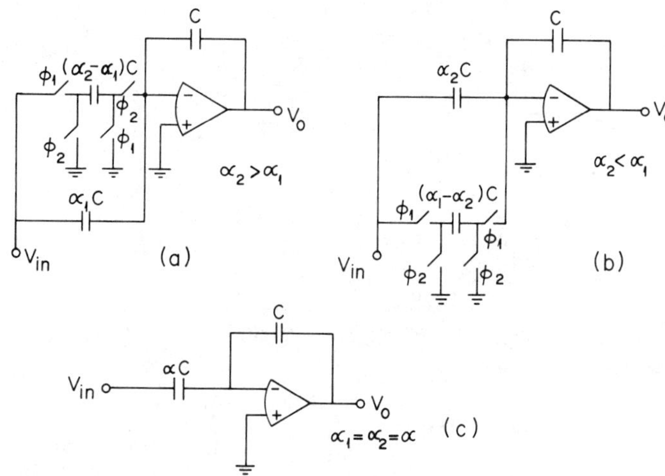

Fig. 5.2–4. Simplifications of Fig. 5.2–3, for the particular cases of (a) $\alpha_2 > \alpha_1$, (b) $\alpha_2 < \alpha_1$, and (c) $\alpha_1 = \alpha_2 = \alpha$.

Next two SC biquad topologies, based on the basic building block of Fig. 5.2–3, are discussed in detail.

Type 1. This structure,[18] shown in Fig. 5.2–5(a), requires 2 op amps, 8 capacitor ratios (coefficients), and a total of 10 capacitors. Figs. 5.2–5(b) and 5.2–5(c) show the implementation with the maximum and minimum number of switches, respectively. Because we have two op amp outputs, we can obtain two transfer functions with identical poles, but with significantly different zeros. Thus, for simplicity of analysis, we can assume the feedback capacitors D and B to be unity

[18] P. E. Fleisher and K. R. Laker, "A Family of Active Switched Capacitor Biquad Building Blocks," *The Bell System Technical Journal,* Vol. 58, December 1979, pp. 2235–2269.

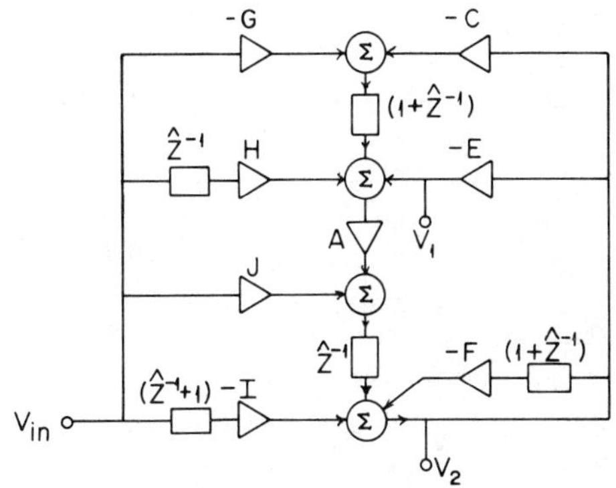

Fig. 5.2–5. (a) A general SC biquad flow diagram type 1.

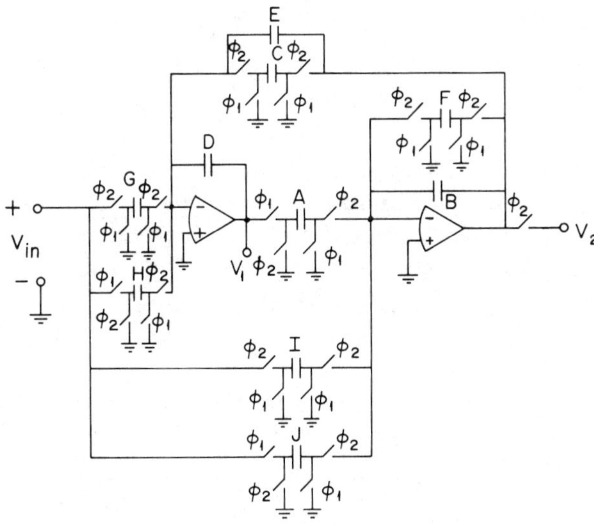

Fig. 5.2–5. (b) SC implementation with maximum number of switches.

$$H_1(\hat{z}) = \frac{V_1}{V_{in}}$$

$$= \frac{(IC + IE - FG - G) + (FH + H - G - JC - JE + IE + 2IC - 2FG)\hat{z}^{-1} + (HF + IC - JC - GF)\hat{z}^{-2}}{1 + F + (F + AE + AC)\hat{z}^{-1} + AC\hat{z}^{-2}} \quad (26)$$

and

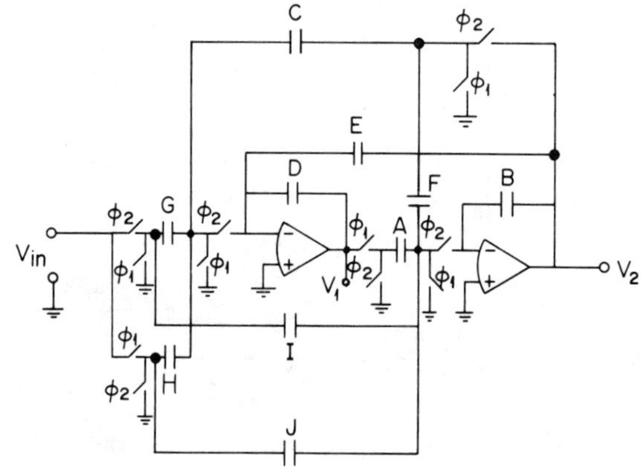

Fig. 5.2–5. (c) SC implementation minimum switch configuration.

$$H_2(\hat{z}) = \frac{V_2}{V_{in}} = -\frac{I + (I + AG - J)\hat{z}^{-1} + (AG - AH)\hat{z}^{-2}}{1 + F + (F + AE + AC)\hat{z}^{-1} + AC\hat{z}^{-2}} \tag{27}$$

There are several possible simplifications of the general structure of Fig. 5.2–5. Two particular cases are discussed.[19] We will call $F = 0$ *Type 1E* and $E = 0$ *Type 1F.* Again each structure is characterized by two transfer functions; when A is arbitrarily chosen as unity, we obtain

$$H_{E1}(\hat{z}) = \frac{(IC + IE - G) + (H - G - JC - JE + IE + 2IC)\hat{z}^{-1} + (IC - JC)\hat{z}^{-2}}{1 + (E + C)\hat{z}^{-1} + C\hat{z}^{-2}} \tag{28}$$

$$H_{E2}(\hat{z}) = -\frac{I + (I + G - J)\hat{z}^{-1} + (G - H)\hat{z}^{-2}}{1 + (E + C)\hat{z}^{-1} + C\hat{z}^{-2}} \tag{29}$$

and

$$H_{F1}(\hat{z}) = \frac{(IC - FG - G) + (FH + H - G - JC + 2IC - 2FG)\hat{z}^{-1} + (HF + IC - JC - GF)\hat{z}^{-2}}{1 + F + (F + C)\hat{z}^{-1} + C\hat{z}^{-2}} \tag{30}$$

---

[19] P. E. Fleisher and K. R. Laker, op. cit.

$$H_{F2}(\hat{z}) = -\frac{I + (I + G - J)\hat{z}^{-1} + (G - H)\hat{z}^2}{1 + F + (F + C)\hat{z}^{-1} + C\hat{z}^{-2}} \tag{31}$$

As we did before for a high sampling rate, we can obtain the relations of Q, ω_0, and the coefficient parameters of the structures. For Type 1E, we can write for any A

$$\alpha_1 = \frac{\theta}{Q} + \theta^2 = AE + CA \tag{32}$$

$$\alpha_2 \cong \theta^2 = AC \tag{33}$$

or

$$\theta = \omega_0 T \cong (AC)^{1/2} \tag{34}$$

and

$$Q \cong \frac{1}{E}\left(\frac{C}{A}\right)^{1/2} \tag{35}$$

For Type 1F, similarly

$$\theta = \omega_0 T \cong \left(\frac{AC}{1 + DF}\right)^{1/2} \tag{36}$$

and

$$Q = \left[AC\left(1 + \frac{B}{F}\right)\right]^{1/2} \tag{37}$$

We can observe that for both circuits, ω_0 and Q are determined by ratios of four or five capacitors. In general, the sensitivities are $|S_p^{\omega_0}| \leq 1/2$ and $|S_p^Q| \leq 1$, where p denotes any capacitor in the F or E circuits.

It should be observed that the different types of filters in the $\hat{z}$-domain discussed in Table 5.2–1, can be accomplished by the E circuit, the F circuit, or both. There is no basic procedure to determine whether circuit E or F, output V_1, or output V_2 will render minimum capacitance area for a particular set of design specifications. Therefore, each design problem should be considered separately. The final design must also include dynamic range considerations. These considerations have been discussed in Section 3.2.

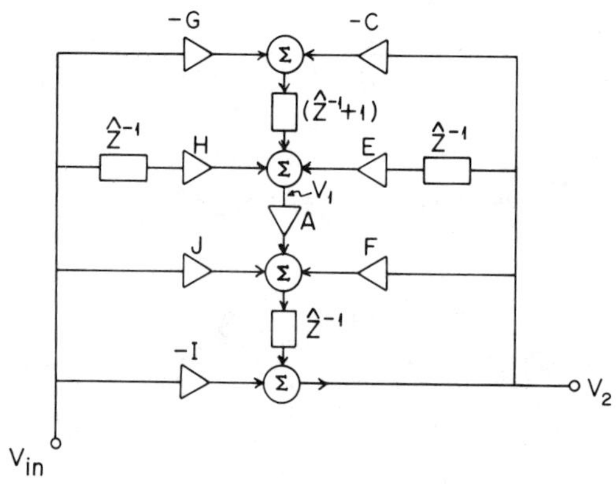

Fig. 5.2–6. (a) A general SC biquad flow diagram, Type 2.

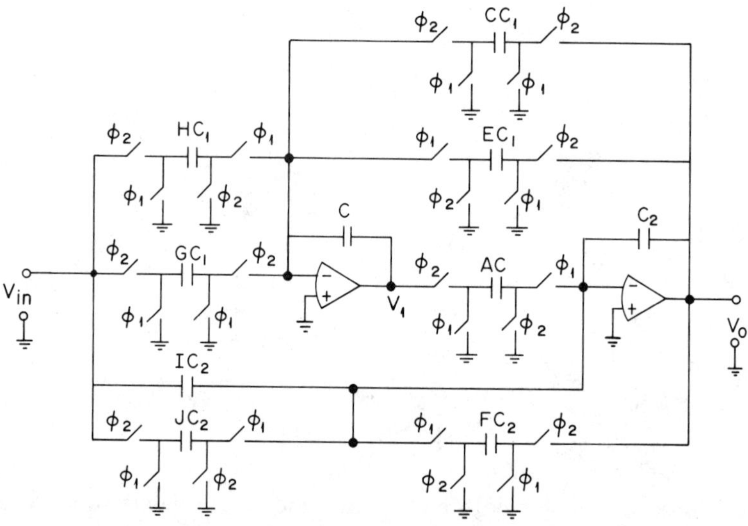

Fig. 5.2–6. (b) SC implementation of Fig. 5.2–6(a).

Type 2. This structure, shown in Fig. 5.2–6(a), also requires two op amps. The basic difference from the topology of Type 1 is that it involves positive and negative feedback. Also zeros are implemented in a different form. Figures 5.2–6(b) and 5.2–6(c) show the implementation of the structure. Assuming $B = D = 1$, the following transfer functions can be derived:

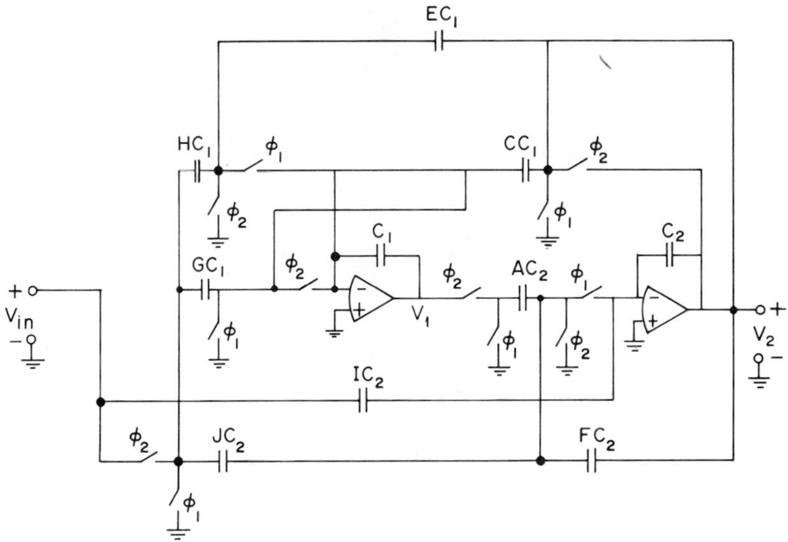

Fig. 5.2–6. (c) SC implementation of Fig. 5.2–6(a) with a minimum number of switches.

$$H_1(\hat{z}) = \frac{(IC - G) + (G + GF + H - JC - IE)\hat{z}^{-1} + (GF + JE - JC - HF)\hat{z}^{-2}}{1 + (CA - F)\hat{z}^{-1} + (CA - AE)\hat{z}^{-2}} \tag{38}$$

and

$$H_2(\hat{z}) = -\frac{I + (GA - J)\hat{z}^{-1} + (GA - HA)\hat{z}^{-2}}{1 + (CA - F)\hat{z}^{-1} + (CA - AE)\hat{z}^{-2}} \tag{39}$$

Next, we derive the relations between ω_0, Q, and the parameters (coefficients) of the structure

$$\frac{\theta}{Q} + \theta^2 = AC - F \tag{40}$$

$$\theta^2 \cong AC - AE \tag{41}$$

then

$$\omega_0 T = \sqrt{A(C - E)} \tag{42}$$

and

$$Q = \frac{\sqrt{A(C-E)}}{AE-F} \tag{43}$$

In this case the sensitivities, due to the positive feedback, might be greater than 1 for some particular set of capacitor ratios. The sensitivity expressions are

$$S_A^Q = -\frac{AE+F}{2(AE-F)} \tag{44}$$

$$S_E^Q = \frac{E(AE+F-2AC)}{2(C-E)(AE-F)} \tag{45}$$

$$S_C^Q = \frac{C}{2(C-E)} \tag{46}$$

$$S_F^Q = \frac{F}{AE-F} \tag{47}$$

$$S_A^{\omega_0} = \frac{1}{2} \tag{48}$$

$$S_C^{\omega_0} = \frac{C}{2(C-E)} \tag{49}$$

$$S_E^{\omega_0} = -\frac{E}{2(C-E)} \tag{50}$$

$$S_F^{\omega_0} = 0 \tag{51}$$

There is a tradeoff between sensitivity and total capacitance area. Type 2 can be designed to give moderate sensitivity values with a relatively reduced area. Again it cannot be claimed in general that this structure is better than Type 1 or vice versa. No systematic design procedure is given regarding minimum area and small sensitivities. Each design problem has to be considered as a particular case for one to be able to choose the most appropriate structure. Type 2 can give parameter values with small areas at the expense of relatively high sensitivity values. However, a compromise between area and sensitivity can be met. One possible set of design equations for the poles follows. Let $CA = k$. Thus we can solve for E and F

$$F = k - \alpha_1 \tag{52}$$

$$AE = k - \alpha_2 \tag{53}$$

choose A, E, and C to compromise area, sensitivity, and dynamic range.

Finally, notice in this structure that if $C > E$ and $G > H$, we can simplify the SC topology as shown in Fig. 5.2–4. Furthermore, for some particular design considerations, it might be possible to make $F = 0$, resulting in a reduction in the sensitivity values, i.e.

$$S_A^Q = -\frac{1}{2} \tag{54}$$

$$S_F^Q = 0 \tag{55}$$

Also note that $H_1(\hat{z})$ or $H_2(\hat{z})$ for Type 2 can implement the different types of filter in the $\hat{z}$-domain presented in Table 5.2–1.

The capacitor scaling considerations discussed in Section 3.2 also apply to these structures. Consider the following scaling for Fig. 5.2–5(b).

$$H_1(\hat{z}) \rightarrow x\, H_1(\hat{z}) \tag{56}$$

which implies the scaling

$$(D, A) \rightarrow \left(\frac{1}{x} A, \frac{1}{x} D\right) \tag{57}$$

For $H_2(z)$, we obtain

$$H_2(\hat{z}) \rightarrow y\, H_2(\hat{z}) \tag{58}$$

which implies the scaling

$$(B, C, E, F) \rightarrow \left(\frac{1}{y} B, \frac{1}{y} C, \frac{1}{y} E, \frac{1}{y} F\right) \tag{59}$$

It is possible to generate other second-order SC topologies different from Types 1 and 2. Fig. 5.2–7 shows the flow diagram of a very general biquad.[20] Type 1 and Type 3 can be obtained as particular cases of Fig. 5.2–7.[21] The Type 3 shown in Fig. 5.2–8 has a transfer function given by

$$H(z) = -\frac{I + \hat{z}^{-1}(G + H)A + GA\hat{z}^{-2}}{1 + \hat{z}^{-1}(C + E)A + CA\hat{z}^{-2}} \tag{60}$$

[20] E. I. El-Masry, "Stray-Insensitive Active Switched Capacitor Biquad," *Electronics Letters,* Vol. 16, June 1980, pp. 480–481.

[21] K. Martin and A. S. Sedra, "Exact Design of Switched-Capacitor Bandpass Filters using Coupled-Biquad Structures," *IEEE* Trans. *Circuits and Systems,* Vol. 27, June 1980, pp. 469–474.

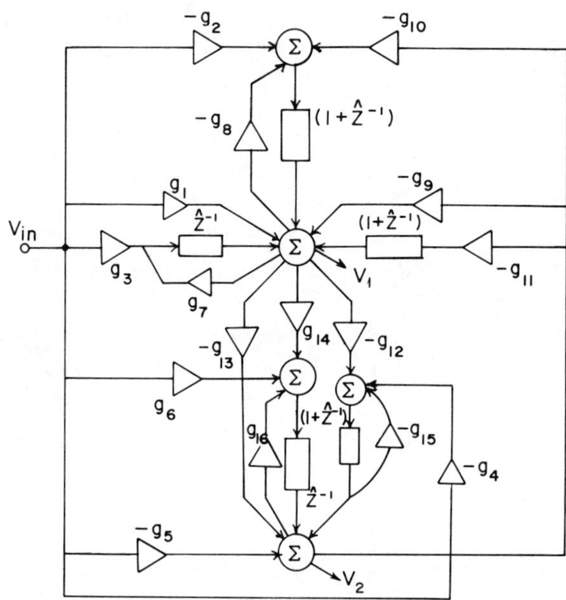

Fig. 5.2–7. A general biquad flow-diagram.

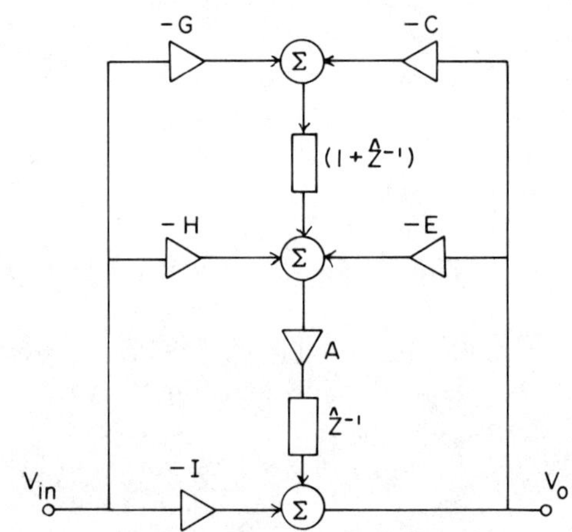

Fig. 5.2–8. General SC biquad diagram, Type 3.

A useful noninverting bandpass topology is shown in Fig. 5.2–9. Its transfer function is

$$H(z) = \frac{H(A+E) - I + HA\hat{z}^{-1}}{1 + \hat{z}^{-1}(A+E)C + AC\hat{z}^{-2}} \tag{61}$$

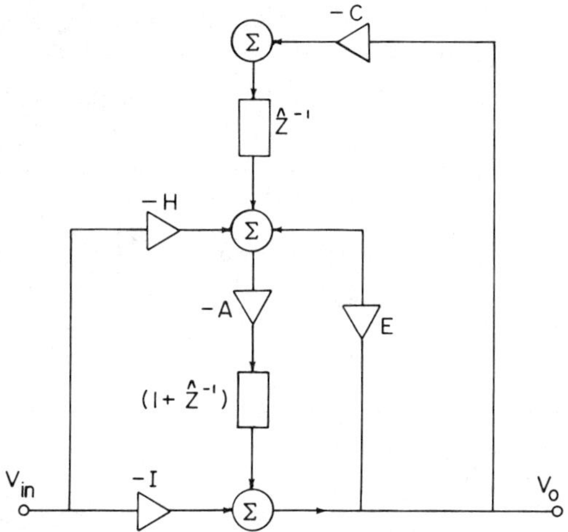

Fig. 5.2–9. A noninverting SC bandpass flow-diagram.

The SC implementations of flow diagrams of Figs. 5.2–7 to 5.2–9 are left as problems.

Next several design examples are considered.

Example 5.2–2. *Low-pass notch circuit.* The transfer function of a low-pass notch filter in the s-domain is

$$H(s) = \frac{0.891975\, s^2 + (1.140926 \times 10^8)}{s^2 + 356.0475s + (1.140926 \times 10^8)}$$

Note that this $H(s)$ has a notch frequency of $f_z = 1800$ Hz, $Q_p = 30$ at $f_p = 1700$ Hz and 0-dB DC gain. The sampling period is $T = 1/128$ kHz.

Design an SC filter of Type 1 of Fig. 5.2–5. Using the bilinear transformation described in Appendix B, we obtain

$$H(z) = 0.89093 \frac{1 - 1.99220z^{-1} + z^{-2}}{1 - 1.990292z^{-1} + 0.997232z^{-2}}$$

or

$$H(\hat{z}) = 0.89093 \frac{1 + 0.0078\hat{z}^{-1} + 0.0078\hat{z}^{-2}}{1 + 0.00971\hat{z}^{-1} + 0.00694\hat{z}^{-2}}$$

The design values for Type 1, $H_{E2}(\hat{z})$ and $H_{F2}(\hat{z})$ are given in Table 5.2–2. To obtain the maximum gain at the output of each op amp for each configuration, the capacitor values A and D were scaled as described before, i.e., $x = 12.0365$ for the $H_{E2}(\hat{z})$ and $x = 11.9124$ for $H_{F2}(\hat{z})$. All the results are described in Table 5.2–2, where the minimum capacitance is rescaled to 1 pF, and the other capacitors are scaled accordingly.

Table 5.2–2 Low-pass notch realization Type 1

CAPACITOR (PF)	E-CIRCUIT			F-CIRCUIT		
	INITIAL	DYNAMIC RANGE ADJUSTED	FINAL	INITIAL	DYNAMIC RANGE ADJUSTED	FINAL
A	1.0000	0.08308	1.0000	1.0000	0.08395	30.1895
B	1.0000	1.0000	12.0365	1.0000	1.0000	359.629
C	0.00694	0.00694	2.5035	0.00696	0.00696	1.0000
D	1.0000	0.08308	29.9613	1.0000	0.08395	12.0591
E	0.00277	0.00277	1.0000	—	—	—
F	—	—	—	0.00278	0.00278	1.0000
G	0.00694	0.00694	2.5035	0.00696	0.00696	1.0000
H	—	—	—	—	—	—
I	—	—	—	—	—	—
J	—	—	—	—	—	—
K(I = J)	0.89093	0.89093	10.7238	0.89340	0.89340	321.293
ΣC (pF)	—	—	59.7	—	—	726.1

Example 5.2–3. *Low-pass notch circuit 2.* Design an SC filter of Type 2 of Fig. 5.2–6 for the same transfer function of Example 5.2–1. From Eq. (39) we have

$$CA - F = 0.00971$$

$$CA - AE = 0.00694$$

$$GA - J = 0.00694$$

$$GA - HA = 0.00694$$

$$I = 0.8903$$

Let $F = 0$. Solving the above simultaneous equations, we obtain the capacitor ratio values given in Table 5.2–3. The final values are adjusted for 10 dB maximum output for each op amp. The minimum capacitance is rescaled to the unit capacitance C_u and the other capacitors, accordingly.

Table 5.2-3 Low-pass notch realization Type 2

CAPACITOR	INITIAL	DYNAMIC RANGE ADJUSTED	FINAL
C_1	1.0	1.6536	31.437
C_2	1.0	1.0	19.0114
AC_2	0.0526	0.08698	1.6536
CC_1	0.1845	0.1845	3.5076
EC_1	0.0526	0.0526	1.0
FC_2	—	—	—
GC_1	0.1320	0.1320	2.5095
HC_1	—	—	—
IC_2	0.8903	0.8903	16.9258
JC_2	—	—	—
$\Sigma C(Cu)$	62.965	61.187	61.187

5.3 BUILDING-BLOCK APPROACH TO HIGHER-ORDER SC FILTERS

There are several distinct approaches to the design of SC active filters with an order greater than 2, similar to the case of analog active filters.[22] One approach is based on the fact that passive LC ladder filters have good sensitivity properties, as was shown in Chapter 4. There also exists an extensive background pertaining to the design of good quality LC ladder filters. This topic has been covered in detail in Chapter 4.

Another important approach to the design of SC filters starts directly in the sampled data (z-variable) domain[23] from the required transfer function $H(z)$. We can obtain $H(z)$ by using the approximation methods and computer programs or by using a mapping from the s-domain to the z-domain as shown in Appendix B. At this point it is possible to use a structure of coupled biquadratic sections for even-order filters or to combine a number of simple (first-, second-, or third-order) switched capacitor circuits to realize $H(z)$ of even or odd order. For the filters of even order, our basic building blocks might be the biquads developed in Section 5.2. The easiest way of coupling the biquads is by just cascading them. The biquads are typically designed to have very low impedance; therefore no interaction between the different

[22] Adel S. Sedra and P. O. Brackett, "*Filter Theory and Design Active and Passive,* Matrix Publishers, Inc. Champaign, IL 1978.

[23] R. Gregorian, "Switched-Capacitor Filter Design Using Cascade Sections," *IEEE Trans. on Circuits and Systems,* Vol. CAS-27, June 1980, pp. 515–521.

sections in the cascade design exists. However, there is one major disadvantage of the cascade design method, namely, poor sensitivity to component variations, especially for filters with rigorous requirements. For the latter case other low-sensitivity structures should be employed, i.e., coupling biquadratic sections through negative feedback, which is essentially an attempt to simulate the internal interactions involved in LC ladder filters discussed in Chapter 4.

Next, we present a number of first- and third-order building blocks that will complement the second-order discussed in the previous section. The common factor among the circuits to be presented is that they are stray-insensitive. Most of them are based on the stray-insensitive building block of Fig. 5.2–3.

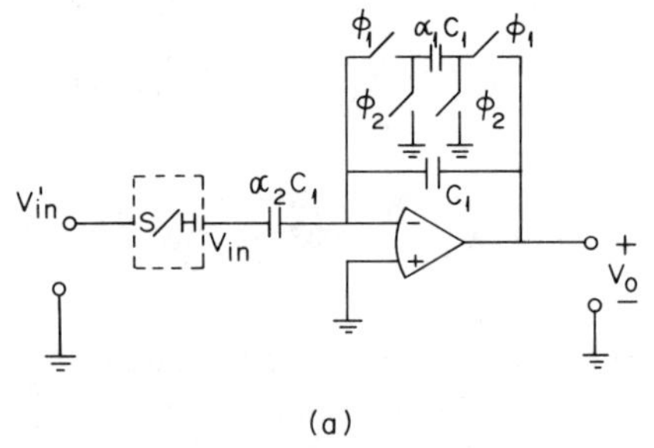

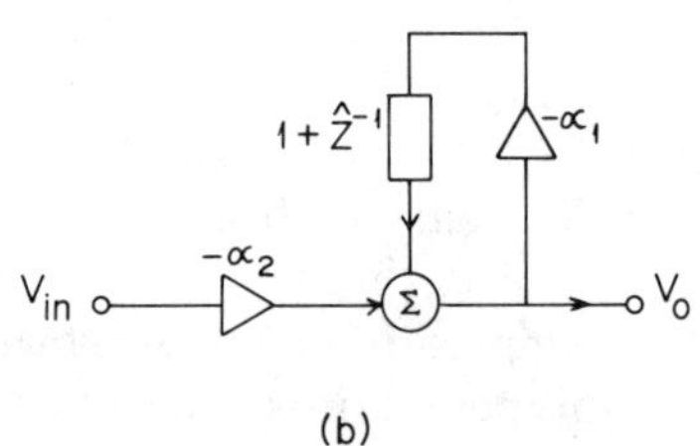

Fig. 5.3–1. A first-order SC high-pass (inverting) filter. (a) Implementation and (b) Flow diagram.

Figure 5.3–1 shows a first-order, stray-insensitive high-pass filter stage that can be assumed to be associated with a separate sample-and-hold stage that is used at the input.[24] Typically, however, the first-order circuit forms part of an output section of a higher-order filter and V_{in} is, in general, a sampled-and-held signal. The charge conservation equation for Fig. 5.3–1 can be written as

---

[24] R. Gregorian, op. cit.

$$(1+\alpha)C_1\left[v^o_{\text{out}}(n)-\frac{v^o_{\text{out}}(n-1)}{1+\alpha_1}\right]=-\alpha_2 C_1\left[v^o_{\text{in}}(n)-v^o_{\text{in}}(n-1)\right] \quad (1)$$

Note that $v^e_0(n - ½) = v^o_0(n - 1)$. Taking the z-transform, the transfer function is

$$H^{oo}(z)=\frac{V^o_{\text{out}}(z)}{V^o_{\text{in}}(z)}=-\frac{\alpha_2}{1+\alpha_1}\frac{1-z^{-1}}{1-\dfrac{z^{-1}}{1+\alpha_1}} \quad (2)$$

or using $\hat{z}$ notation

$$H^{oo}(\hat{z})=-\frac{\alpha_2}{1+\alpha_1}\frac{1}{1+\dfrac{\alpha_1}{1+\alpha_1}\hat{z}^{-1}} \quad (3)$$

Now, a first-order, low-pass section[25] is shown in Fig. 5.3–2. This section can be operated on an antialiased input during both clock phases ϕ_1 and ϕ_2. In fact, this operation is equivalent to interpolation and effectively doubles the inherent sampling frequency. It should be noticed that doubling the input sampling rate makes the requirements less stringent on the antialiasing section used in higher-order filters. The transfer function of this low-pass section is given by

$$H(z)=\frac{V^o_0(z)}{V^o_{\text{in}}(z)+V^e_{\text{in}}(z)}=\frac{\dfrac{\alpha_2+\alpha_1 z^{-1}}{1+\alpha_3}}{\left[1-\left(1-\dfrac{\alpha_3}{1+\alpha_3}\right)z^{-1}\right]\left(1+z^{1/2}\right)} \quad (4)$$

or using $\hat{z}$ notation

$$H(\hat{z})=\frac{\dfrac{\alpha_2+\alpha_i\hat{z}^{-1}}{1+\alpha_3}}{\left[1+\dfrac{\alpha_3}{1+\alpha_3}\hat{z}^{-1}\right]\left[1+(1+\hat{z})^{1/2}\right]} \quad (5)$$

[25] P. E. Fleischer et al., "An NMOS Building Block for Telecommunication Applications," *IEEE Trans. on Circuits and Systems,* Vol. CAS-24, June 1980, pp. 552–559.

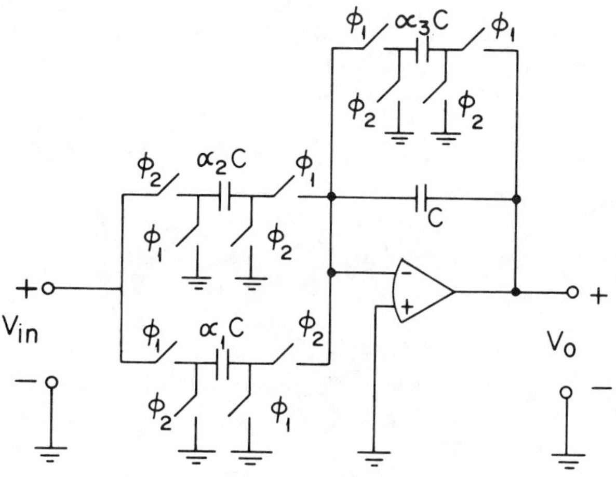

Fig. 5.3–2. A first-order SC low-pass filter.

where

$$\alpha_i = \alpha_1 + \alpha_2 \tag{6}$$

Another high-pass section is shown in Fig. 5.3–3(a). This HP section can be used as a final section of a higher-order filter, for it serves to cancel the DC offsets accumulated through the preceding sections. This filter is buffered with a unity-gain amplifier, included in Fig. 5.3–3(a). Using the charge conservation method discussed in Section 2.3, we can write for ϕ_2

$$(1 + \alpha_1 + \alpha_2)\, v_O(n) = (\alpha_1 + \alpha_2)\, v_O(n-1) + \alpha_1\,(v_{in}(n) - v_{in}(n-1)) \tag{7}$$

Thus the transfer function $H(z)$ for ϕ_2 can be written as

$$H(z) = \frac{V_O(z)}{V_{in}(z)} = \frac{\alpha_1}{1+\alpha_1+\alpha_2}\,\frac{1 - z^{-1}}{1 - \dfrac{\alpha_1+\alpha_2}{1+\alpha_1+\alpha_2} z^{-1}} \tag{8}$$

The reader should observe from Fig. 5.3–3(a) that the switches for ϕ_1 and ϕ_2, together with the capacitor C are equivalent to an SC backward resistor,

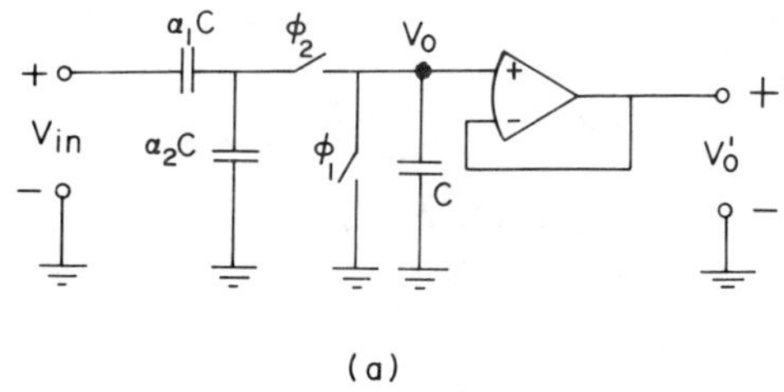

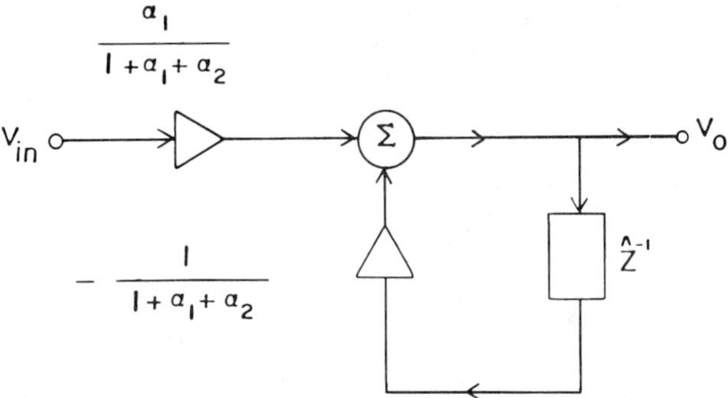

Fig. 5.3–3. A first-order SC high-pass (noninverting) filter with a buffer. (a) Implementation and (b) Flow diagram.

or

$$H(\hat{z}) = \frac{\alpha_1}{1+\alpha_1+\alpha_2} \frac{1}{1+\dfrac{1}{1+\alpha_1+\alpha_2}\hat{z}^{-1}} \tag{9}$$

as shown in Section 3.5. From this viewpoint the RC prototype of Fig. 5.3–3(a) consists of two capacitors, $\alpha_1 C$ and $\alpha_2 C$, the latter connected in parallel with a resistor of value T/C. The flow diagram representing Eq. (9) is shown in Fig. 5.3–3(b). We see that Figs. 5.3–1 (inverting) and 5.3–3 (noninverting) are both first-order, high-pass filters and are characterized by Eqs. (2) and (8).

Let us now consider several topologies for third-order SC filters. Figure 5.3–4(a) shows a third-order, low-pass (elliptic high-pass) filter for an input V_{inLP} (V_{inHP}). For the input $V_{\text{inLP}} \neq 0$ and $V_{\text{inHP}} = 0$, the circuit needs no S/H at its input, because the capacitor $\alpha_3 C_1$ performs the sampling function. However, for the input $V_{\text{inHP}} \neq 0$ and $V_{\text{inLP}} = 0$, a sampled-and-held signal

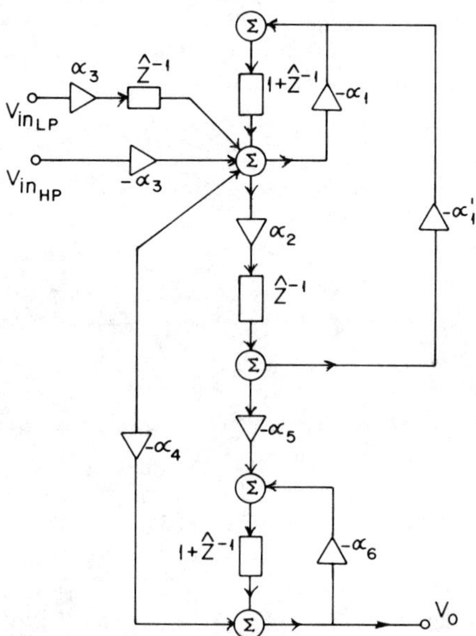

Fig. 5.3–4. (a) Flow diagram for a third-order filter; low-pass for input V_{inLP} and high-pass for input V_{inHP}.

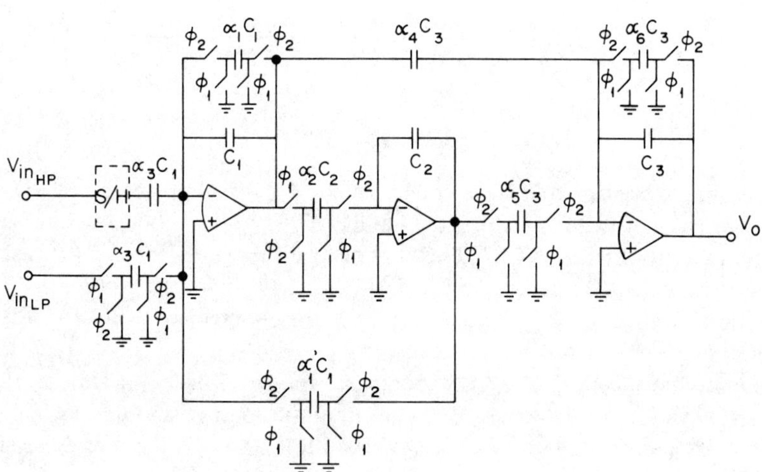

Fig. 5.3–4. (b) Implementation of Fig. 5.3–4(a).

is required. The transfer functions for the low-pass and high-pass are given by

$$H_{LP}(\hat{z}) = \frac{-\alpha_3\,\alpha_4}{(1+\alpha_1)(1+\alpha_6)} \cdot \frac{\left(1+\dfrac{\alpha_2\alpha_5}{\alpha_4}\hat{z}^{-1}+\dfrac{\alpha_2\alpha_5}{\alpha_4}\hat{z}^{-2}\right)\hat{z}^{-1}}{\left(1+\dfrac{\alpha_6\hat{z}^{-1}}{1+\alpha_6}\right)\left(1+\dfrac{\alpha_1+\alpha_1'\alpha_2}{1+\alpha_1}\hat{z}^{-1}+\dfrac{\alpha_1'\alpha_2}{1+\alpha_1}\hat{z}^{-2}\right)} \tag{10}$$

or

$$H_{LP}(z) = \frac{-\alpha_3\alpha_4}{(1+\alpha_1)(1+\alpha_6)} \cdot \frac{\left[1-\left(2-\dfrac{\alpha_2\alpha_5}{\alpha_4}\right)z^{-1}+z^{-2}\right]z^{-1}}{\left(1-\dfrac{z^{-1}}{1+\alpha_6}\right)\left(1-\dfrac{2+\alpha_1-\alpha_1'\alpha_2}{1+\alpha_1}z^{-1}+\dfrac{z^{-2}}{1+\alpha_1}\right)} \tag{11}$$

and for the elliptic, high-pass filter, we obtain

$$H_{HP}(\hat{z}) = \frac{-\alpha_3\alpha_4}{(1+\alpha_1)(1+\alpha_6)} \frac{1+\dfrac{\alpha_2\alpha_5}{\alpha_4}\hat{z}^{-1}+\dfrac{\alpha_2\alpha_5}{\alpha_4}\hat{z}^{-2}}{\left(1+\dfrac{\alpha_6\hat{z}^{-1}}{1+\alpha_6}\right)\left(1+\dfrac{\alpha_1+\alpha_1'\alpha_2}{1+\alpha_1}\hat{z}^{-1}+\dfrac{\alpha_1'\alpha_2}{1+\alpha_1}\hat{z}^{-2}\right)} \tag{12}$$

or

$$H_{HP}(z) = \frac{-\alpha_3\alpha_4}{(1+\alpha_1)(1+\alpha_6)} \frac{(1-z^{-1})\left[1-\left(2-\dfrac{\alpha_2\alpha_5}{\alpha_4}\right)z^{-1}+z^{-2}\right]}{\left(1-\dfrac{z^{-1}}{1+\alpha_6}\right)\left(1-\dfrac{2+\alpha_1-\alpha_1'\,\alpha_2}{1+\alpha_1}z^{-1}+\dfrac{z^{-2}}{1+\alpha_1}\right)} \tag{13}$$

It should be observed that expression (12) does not represent exactly an elliptic transfer function, because there is no transmission zero at $\hat{z} = -2$, corresponding to $z = -1$ in the z-plane and to $s \rightarrow \infty$ in the s-plane. Fortunately, if the sampling frequency $f_s >> f_p$, (i.e., $\theta << 1$) where f_p is the passband frequency, the effect of the missing zero on the passband response will be negligible. Regarding the high-pass notch transfer function, it contains one zero at DC (i.e., $\hat{z} = 0$ or $z = 1$), two conjugate complex poles, and one real pole. Thus as seen in Table 5.2–1, Eq. (13) is appropriate for a third-order, elliptic, high-pass.

One of the main disadvantages of the structure of Fig. 5.3–4 is that if the pole Q is very high or the pole angle is very small, the natural modes will be very close to the unit circle. This typically results in a large spread of capacitor values in the circuit. This can be easily observed from the second-order denominator of Eq. (13), where

$$2r \cos \theta = \frac{2 + \alpha_1 - \alpha_1' \alpha_2}{1 + \alpha_1} \tag{14}$$

and

$$r^2 = \frac{1}{1 + \alpha_1} \tag{15}$$

From Eq. (15), α_1 is given by

$$\alpha_1 = \frac{1 - r^2}{r^2} \tag{16}$$

Clearly if $r \rightarrow 1$ then $\alpha_1 \rightarrow 0$, and LSC considerations make the circuit economically prohibitive.

Another third-order, elliptic, high-pass structure, shown in Fig. 5.3–5, with better capacitance-value spread properties is discussed next. The transfer function is given by

$$H(\hat{z}) = \frac{V_1(\hat{z})}{V_{\text{in}}(\hat{z})}$$

$$= \frac{-\alpha_3(1 + \alpha_4\alpha_5\, \hat{z}^{-1} + \alpha_4\alpha_5\, \hat{z}^{-2})/(1 + \alpha_1)}{\left[1 + \hat{z}^{-1}\left(1 + \alpha_4\alpha_5 - \dfrac{1 - \alpha_1'\alpha_2}{1 + \alpha_1}\right) - \hat{z}^{-2}\left(2\alpha_4\alpha_5 - \dfrac{\alpha_1'\alpha_2 - \alpha_4\alpha_5 + \alpha_2\alpha_4\alpha_6}{1 + \alpha_1}\right) + \hat{z}^{-3}\left(\alpha_4\alpha_5 + \dfrac{1 - \alpha_4\,(\alpha_5 + \alpha_2\alpha_6)}{1 + \alpha_1}\right)\right]} \tag{17}$$

or

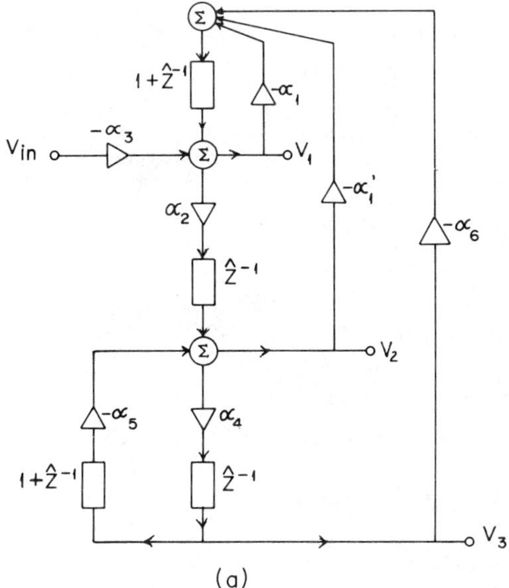

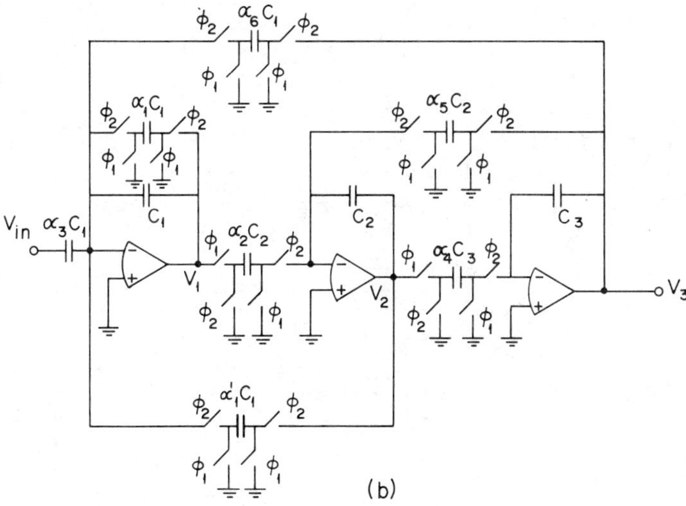

Fig. 5.3–5. Third-order, elliptic, high-pass SC filter with good capacitance-value spread properties. (a) Flow diagram and (b) Implementation.

$$H(z) = \frac{V_1(z)}{V_{\text{in}}(z)}$$

$$= \frac{-\alpha_3(1-z^{-1})(1-(2-\alpha_4\alpha_5)z^{-1}+z^{-2})/(1+\alpha_1)}{\left[1 - z^{-1}\left(2-\alpha_4\alpha_5+\dfrac{\alpha_1'\alpha_2}{1+\alpha_1}\right)+z^{-2}\left(1+\dfrac{2-\alpha_4\alpha_5+\alpha_2\alpha_4\alpha_6-\alpha_1'\alpha_2}{1+\alpha_1}\right) - \dfrac{z^{-3}}{1+\alpha_1}\right]} \tag{18}$$

Let us compare the denominator of $H(z)$ of Eq. (18) with the third-order, natural mode polynomial

$$\begin{aligned}(1-r_1z^{-1})&(1-2r\cos\theta z^{-1}+r^2z^{-2})\\ &= 1-(2r\cos\theta+r_1)z^{-1}+(2rr_1\cos\theta+r^2)z^{-2}-r_1r^2z^{-3}\end{aligned} \tag{19}$$

Equating the coefficient of z^{-3} yields

$$r_1r^2 = \frac{1}{1+\alpha_1} \tag{20}$$

Therefore α_1 becomes

$$\alpha_1 = \frac{1}{r_1r^2} - 1 \tag{21}$$

Usually, $r_1 < r$, and thus, even when $r \cong 1$, α_1 will not be as small as in Eq. (16). Hence, this structure has a smaller capacitance-value spread than the previous one shown in Fig. 5.3–4.

Next a presentation of a general third-order SC filter topology called the *triquad* is given. This topology is based on the biquad Type 2, discussed in the previous section, and is shown in Fig. 5.3–6. The transfer function is given by

$$H(\hat{z}) = -\frac{\alpha_0+\alpha_1\hat{z}^{-1}+\alpha_2\hat{z}^{-2}+\alpha_3\hat{z}^{-3}}{1+\beta_1\hat{z}^{-1}+\beta_2\hat{z}^{-2}+\beta_3\hat{z}^{-3}} \tag{22}$$

where

$$\begin{aligned}\alpha_0 &= A_5\\ \alpha_1 &= A_2B_6\\ \alpha_2 &= B_6\,(A_0B_5-A_3+A_2)\\ \alpha_3 &= B_5B_6A_0\end{aligned} \tag{23}$$

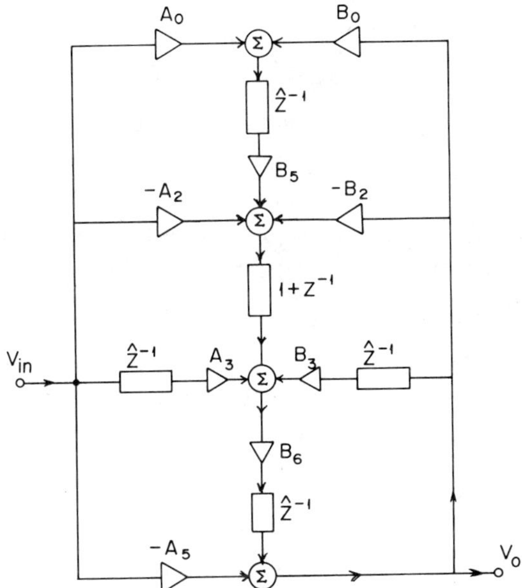

Fig. 5.3–6. (a) Flow diagram of a general third-order SC filter "Triquad."

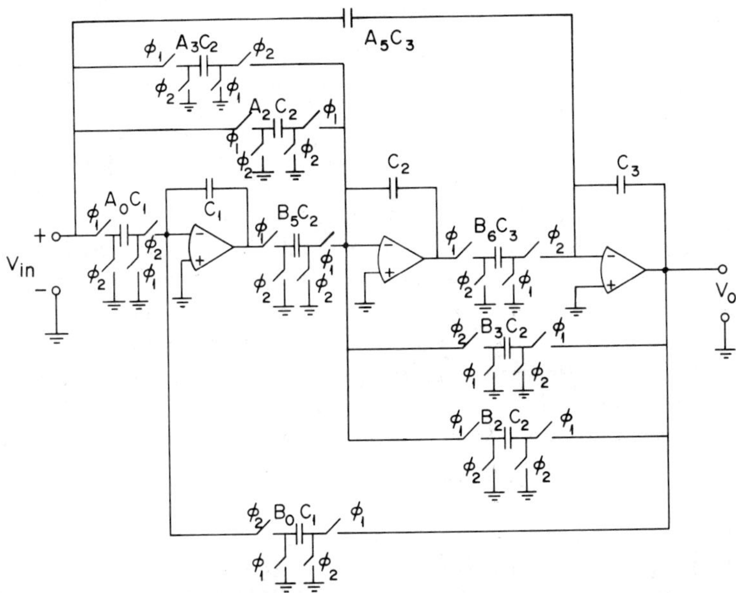

Fig. 5.3–6. (b) SC realization of a general third-order SC filter "Triquad."

and

$$\beta_1 = B_2 B_6$$
$$\beta_2 = B_6 (B_0 B_5 - B_3 + B_2) \tag{24}$$
$$\beta_3 = B_0 B_5 B_6$$

It should be pointed out that $\alpha_0 = A_5$ may become very small for certain design specifications, thereby causing a large capacitance-value spread. Fortunately, neglecting A_5 is equivalent to factoring out a zero at $z = -1$ (or $\hat{z} = -2$), which corresponds to $s \to \infty$ i.e.

$$\alpha_0 \hat{z}^3 + \alpha_1 \hat{z}^2 + \alpha_2 \hat{z} + \alpha_3 = (\hat{z} + 2)(\alpha_1' \hat{z}^2 + \alpha_2' \hat{z} + \alpha_3') \tag{25}$$

where

$$\alpha_1' = \alpha_0 = A_2 B_6$$
$$\alpha_2' = \alpha_1 - 2\alpha_0 \tag{26}$$
$$\alpha_3' = \alpha_2 - 2\alpha_1 + 4\alpha_0$$

For small ratios of f_p/f_s, the difference between $H'(\hat{z})$, where

$$H'(\hat{z}) = \frac{\alpha_1' \hat{z}^{-1} + \alpha_2' \hat{z}^{-2} + \alpha_3 \hat{z}^{-3}}{1 + \beta_1 \hat{z}^{-1} + \beta_2 \hat{z}^{-2} + \beta_3 \hat{z}^{-3}} \tag{27}$$

and $H(\hat{z})$ of Eq. (22) is essentially only a DC shift.

The LSC can be very small at the expense of high coefficient sensitivity. This is because positive and negative feedback are used. In practice moderate LSC's (i.e., 1:6) result in reasonable coefficient sensitivity values. Table 5.3–1 shows the generality of the structure for the basic type of filter LP, BP, HP, and notch.

The filter blocks described earlier in this section, in addition to the biquads of the previous chapter, provide the basic building blocks to design SC filters in a cascade form. Of course other building blocks can be used to design in a cascade form.[26] For instance, the one-op amp structures presented in Sections 3.3 to 3.5 could be used.

---

[26] U. W. Brugger, D. C. von Grünigen, and G. S. Moschytz, "A Comprehensive Procedure for the Design of Cascade Switched-Capacitor Filters," *IEEE Trans. Circuits and Systems,* Vol. CAS-28, August 1981, pp. 803–810.

Table 5.3–1. Coefficient Values of $H(\hat{z})$ of Eq. (22) for conventional types of filters

FILTER TYPE	ZEROS IN S-PLANE	ZEROS IN Z-PLANE	COEFFICIENT OF $H(\hat{z})$				CORRESPONDING CAPACITOR RATIOS			
			α_0	α_1	α_2	α_3	A_0	A_2	A_3	A_5
Low pass	∞, ∞	−1 1	0	0	0	1	$B_0^{\dagger}$	0	A_0B_5	0
Band pass	$0, \infty$	−1 1	0	0	1	0	0	0	A_3^*	0
High pass	0, 0	1	1	0	0	0	0	0	0	A_5^*
Notch	$\pm j\omega_0$	θ_0	0	1	$2(1-\cos\theta_0)$	$2(1-\cos\theta_0)$	$\frac{\alpha_2}{B_5B_6}$	$\frac{1}{B_6}$	$\frac{1^{\dagger *}}{B_6}$	0

† $A_0 = B_0$ to obtain a 0 dB dC gain.
* A_3 and A_5 to be chosen to give the required center (cutoff) frequency gain.
†* A_0, A_2 and A_3 can be scaled to obtain the desired DC gain.

The sampling between building blocks connected in cascade determines the appropriate overall transfer function describing the filter. If the input and output signals are synchronously sampled at the same clock period (say even, ϕ_2), the total "even" transfer function is obtained by multiplying the single "even" transfer functions. For the special case that the input is coupled into the circuit only during the even phase (i.e., $H^{oe} = H^{oo} = 0$) and the output is sampled during both phases, a total transfer function H_t can be defined as

$$H_t = \frac{V_o^e + V_o^o}{V_{\text{in}}^e} = H^{ee} + H^{eo} \tag{28}$$

This total transfer function is of special interest if the output is held over a full clock period, i.e., if $H^{eo} = z^{-1/2} H^{ee}$, then

$$H_t = (1 + z^{-1/2}) H^{ee}$$

Note that when the basic sections are cascaded to build higher-order filters, the switches between two sections can be omitted if the transfer functions H_i^{oe} are zero. Next we present an example of a cascade design approach.

Example 5.3–1. *Design of a voiceband filter.* Design a low-pass filter with the following specifications:

Passband	$0 < f < 3.4$ kHz	Equiripple with 0.125-dB maximum deviation
Stopband	$4.6\text{ kHz} < f < \infty$	Minimum stopband loss $= -32$ dB

Two blocks are connected in cascade to obtain the filter that satisfies the design specifications. A biquad Type 2 and the triquad of Fig. 5.3–6 are used.

A design approach, illustrated in Fig. 3.1–16 (boxes 1, 2, 3, 4, 5, 13, and 14), starts with the design specifications and standard filter data tables, which yields the corresponding $H(s)$. Then $H(s)$ is transformed through the bilinear mapping into the corresponding $H(z)$, and mapped into $H(\hat{z})$. To obtain $H(\hat{z})$ given $H(s)$, we can use the program S2ZHAT described in Appendix B. The final fifth-order transfer function is given by

$$H(\hat{z}) = \frac{e_0 + e_1\hat{z}^{-1} + e_2\hat{z}^{-2}}{1 + f_1\hat{z}^{-1} + f_2\hat{z}^{-2}} \cdot \frac{e_3 + e_4\hat{z}^{-1} + e_5\hat{z}^{-2} + e_6\hat{z}^{-3}}{1 + f_3\hat{z}^{-1} + f_4\hat{z}^{-2} + f_5\hat{z}^{-3}}$$

where

$$
\begin{array}{ll}
e_0 = 0.4333600 & \\
e_1 = 0.0219690 & f_1 = 0.16008 \\
e_2 = 0.0219690 & f_2 = 0.0219690 \\
e_3 = 0.0163720 & f_3 = 0.1858600 \\
e_4 = 0.0345150 & f_4 = 0.0370950 \\
e_5 = 0.0053147 & f_5 = 0.0035431 \\
e_6 = 0.0035431 &
\end{array}
$$

It has been pointed out that the zeros at infinity in the s-domain are transformed into zeros at $z = -1$ ($\hat{z} = -2$), and they can be factored out.

The capacitor values of the biquad as functions of C_u are

$$
\begin{array}{ll}
G = A = C_u & \mathrm{TC} = 27.36\ C_u \\
E = 6.2866\ C_u & \mathrm{LSC} = 9.13 \\
C_1 = 4.9856\ C_u & \\
I = 3.9566\ C_u & \\
C_2 = 9.1301\ C_u &
\end{array}
$$

and the corresponding values for the triquad are

$$
\begin{array}{ll}
B_6 = B_0 = A_0 = A_2 = A_3 = C_u & C_2 = 3.054\ C_u \\
C_1 = 10\ C_u & C_3 = 10\ C_u \\
B_5 = 1.082\ C_u & B_3 = 4.652\ C_u \\
 & B_2 = 3.649\ C_u
\end{array}
$$

According to the simplifications of Fig. 5.2–4, in this triquad, because $A_3 = A_2$, we can use only one unswitched capacitor (say A_2). Besides, if $B_2 > B_3$, then B_3 becomes an unswitched capacitor, and B_3 is modified as $(B_2 - B_3)$ and results in a series SC type; thus TC = 33.8 C_u, and LSC = 10. The experimental results are shown in Fig. 5.3–7.

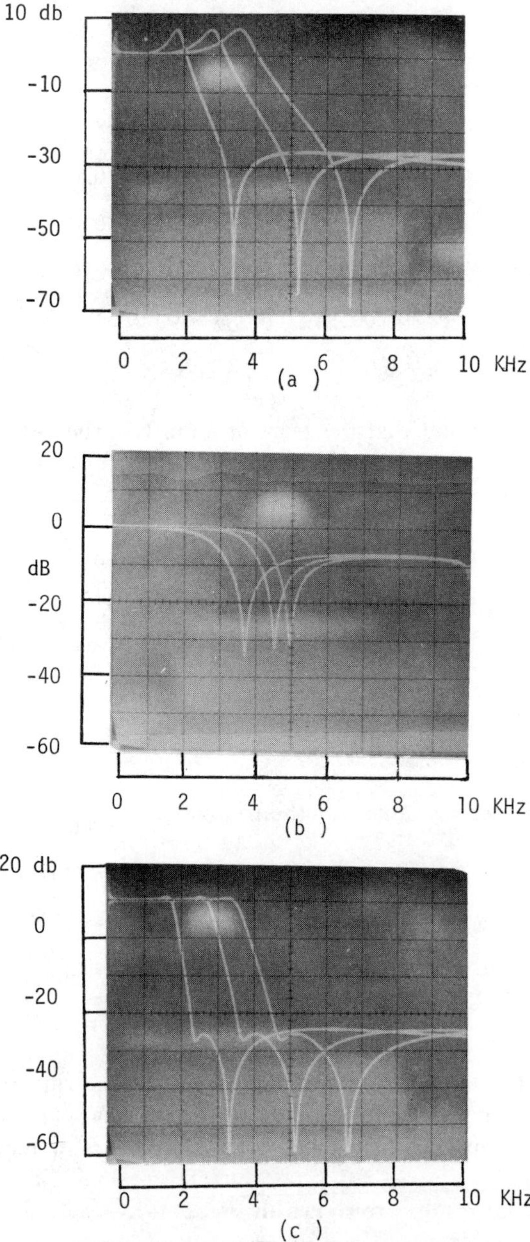

Fig. 5.3–7. Experimental results of Example 5.3–1. Magnitude response with clock frequencies of 64 KHz, 100 KHz, and 128 KHz. (a). Triquad. (b) Biquad. (c) Fifth-order filter.

5.4 NONIDEAL CONSIDERATIONS OF SC FILTERS

There are parasitic properties and nonideal effects of SC components that have to be considered in actual IC fabrications. In this section we discuss some of the most common deviations from ideal behavior.

Most critical for capacitors is the parasitic capacitance from the lower metallization or polysilicon to the substrate. In fact, this parasitic capacitance value may range from 5 to 20% of the desired capacitance. Fortunately, the desired capacitance can be constructed in such a manner that at least one end of the capacitor is at virtual ground point or is grounded or connected to a low impedance source, i.e., an op amp output.[27, 28] In a careful design most of these parasitics can be reduced or eliminated. The value of the parasitic capacitance of the top plate varies from 0.1 to 1% of of the MOS capacitor itself, depending on layout and technology. This top plate capacitance in some SC configurations may have an effect, i.e., parallel SC configuration. Consider the parasitics to the substrate in Fig. 5.4–1(a): C_{p1} and C_{p3} have no effect, but C_{p2} has an effect that can be taken into account; its value can be estimated if the desired switched capacitor C is itself grounded. There are situations where the bottom plate of C is also switched, as in Fig. 5.4–1(b), where all parasitic effects have been eliminated. Another situation where the parasitic capacitance has to be considered is shown in Fig. 5.4–1(c), a differential integrator, where $C_{p\text{II}}$ does not have any effect; it is alternatingly charged to voltages V_2 and V_3, but does not influence the charge entering the integrating capacitance C. The influence of the top parasitic, $C_{p\text{I}}$, however, must be considered. The output voltage V_0, assuming $V_3 = 0$ for simplicity, is given as

$$V_0 = -\frac{\alpha z^{-1}(V_1 - V_2) + (C_{p\text{I}}/C)z^{-1}V_1}{1 - z^{-1}} \tag{1}$$

We can observe from Eq. (1) that if $C_{p\text{I}}/C$ is not negligible compared to α, then we must use Eq. (1) for design considerations. In general, it is advisable whenever possible to use the capacitors with the stray-insensitive SC structures.[29]

---

[27] G. Szentirmai and G. C. Temes, "Switched Capacitor Building Blocks," op. cit.

[28] B. J. Hosticka and G. S. Moschytz, "Practical Design of Switched Capacitor Networks for Integrated Circuit Implementation," *IEE Electronic Circuits and Systems,* Vol. 3, March 1979, pp. 76–88.

[29] P. E. Fleischer et al., "Parasitic Compensated Switched-Capacitor Circuits," *Electronics Letters,* Vol. 17, November 1981, pp. 929–931.

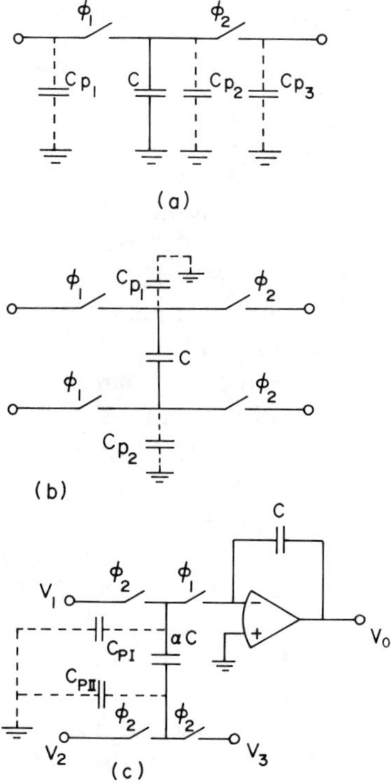

Fig. 5.4–1. Parasitic capacitances; (a) Parallel SC, (b) Parasitic free, and (c) Differential integrator.

Next we consider the influence of MOS switches and the finite gain of op amps on SC circuits. The off-resistance of the MOSFET switches is usually high enough so that they can be considered to be open circuits. R_{on} depends on the actual dimension of the switch (see Section 8.3). Commercial CMOS analog switches have values for R_{on} between 40 and 70 ohms. A reasonable value of an on-resistance for a MOSFET in an IC is of the order of from 2 to 10 kohms, depending on the size of the MOSFET. The analysis of the effects of the on-resistance R_{on} of the switches is directly involved with those of the op amp. The op amp can be characterized by

$$A(s) = -\frac{GB}{s + \omega_1} \tag{2}$$

where $GB = A_0\omega_1$ is the gain-bandwidth product, A_0 is the finite DC gain, and ω_1 is the lower dominant pole. We also will consider the op amp output

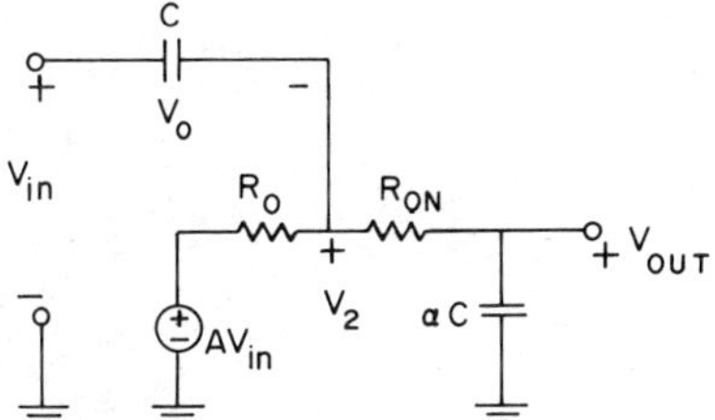

Fig. 5.4–2. Equivalent op amp circuit for a charging half-period.

resistance, R_0. To numerically evaluate the effects of these parameters on the SC filter performance,[30] let us assume the following pessimistic values

$$A_0 = 500 \qquad \omega_1 = 2\pi \times 10^3$$
$$R_{on} = 4 \text{ kohms} \qquad R_0 = 3 \text{ kohms}$$

First, we consider that the amplifier is charging one or several capacitors. Figure 5.4–2 shows the equivalent circuit; the relations among the node voltages are

$$V_{in} = \left[\frac{s(R_0 + R_{on})\,\alpha\, C + 1}{(s\alpha CR_{on} + 1)(1 - A) + s\alpha CR_0)} \right] \frac{V_0}{s} \tag{3}$$

and

$$V_{out} = \frac{AV_0/s}{(s\alpha CR_{on} + 1)(1 - A) + s\alpha CR_0} \tag{4}$$

Now substituting Eq. (2) into Eq. (4), we obtain

$$V_{out} = \frac{\dfrac{-A_0\omega_1}{\alpha C\,(R_{on} + R_0)}\,(V_0/s)}{s^2 + \dfrac{\alpha R_{on} C\,(1 + A_0)\,\omega_1 + 1 + \omega_1}{\alpha C\,(R_{on} + R_0)}\,s + \dfrac{\omega_1(1 + A_0)}{\alpha C\,(R_{on} + R_0)}} \tag{5}$$

The exact waveform can be obtained, but the important steps are to determine the final value of V_{out} and to measure how fast V_{out} will reach this final value. The final value is given by

[30] G. Szentirmai et al., op. cit.

$$V_{out}\Big|_{t\to\infty} = -\frac{A_0}{1+A_0} V_0 \tag{6}$$

The denominator of Eq. (5) determines the poles. Let us consider the pole locations for two cases

1. $C = 50$ pF and $\alpha = 1$.

The poles are located at

$$s_{1,2} \cong -2.33 \times 10^6 \pm j\, 1.89 \times 10^6$$

2. $C = 50$ pF and $\alpha = 0.1$.

$$s_1 \cong -3.33 \times 10^6$$

$$s_2 \cong -27.0 \times 10^6$$

The dominant term is the real part of the complex root. If we assume that five time constants are necessary to complete the charging of C, then it can be shown that the sampling frequency must be less than 230 kHz.

Next, the analysis of the effect of R_{on} and A while discharging αC through the following amplifier stage is considered. Let us consider an integrator stage, shown in Fig. 5.4–3, where R_{on} is the on-resistance of one or possibly two switches and both capacitors αC and C have initial voltages designated by V_1 (0) and V_2 (0), respectively. The output voltage is given by

$$V_{out} = \frac{(R_0 + A/sC)\, V_1(0) + (A(R_{on} + 1/s\alpha C) - R_0)\, V_2(0)/s}{(1 - A)(R_{on} + 1/s\alpha C) + (R_0 + 1/sC)} \tag{7}$$

Again we are mainly interested in determining the final value of V_{out} and a measure of how fast V_{out} reaches its final value. The latter is given by

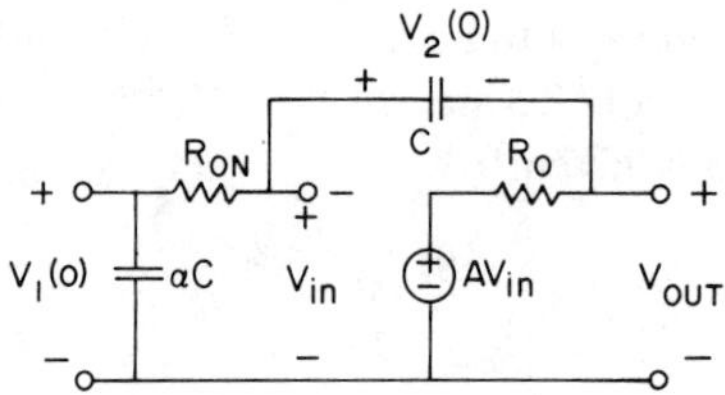

Fig. 5.4–3. Equivalent op amp circuit for a discharging half-period.

$$V_{out}\bigg|_{t\to\infty} = -\frac{\alpha V_1(0) + V_2(0)}{1 + (1+\alpha)/A_0} \tag{8}$$

The second term $(1+\alpha)/A_0$ in the denominator of Eq. (8) represents the error. Compensation for this error is possible if a reasonable estimate or measurement of A_0 can be obtained. The poles of Eq. (7) can be found by substituting Eq. (2) into Eq. (7); we obtain the characteristic polynomial

$$P(s) = \frac{\alpha C(R_0 + R_{on})}{\omega_1} s^2 + \left(\frac{1+\alpha}{\omega_1} + A_0 \alpha C R_{on} + \alpha C(R_{on} + R_0)\right) s + (1 + \alpha + A_0) \tag{9}$$

Considering practical values, we can simplify Eq. (9) to

$$P(s) = \frac{\alpha C(R_0 + R_{on})}{\omega_1} s^2 + \left(\frac{1+\alpha}{\omega_1} + A_0 \alpha C R_{on}\right) s + A_0 \tag{10}$$

As in the charging case, we consider that $\alpha = 1$. The dominant (controlling) terms are

$$s_1 = -1.49 \times 10^6$$

$$s_2 = -6.02 \times 10^6$$

Considering that it takes five time constants for the transients to settle, s_1 indicates an upper limit of about 130 kHz for the sampling frequency. From the above numerical examples, it can be observed that the lower value of the roots of s determines the highest usable sampling frequency. This root is generally proportional to *GB*. Thus high-frequency applications will require op amps with larger *GB*'s or smaller capacitances and small R_{on} (larger switch geometries).

In the case of integrators, which are basic building blocks, if A_0 is finite the effect is to produce a lossy integrator by moving the pole away from $z = 1$. Let us consider the case of a stray-insensitive, backward (series) integrator (see Fig. 5.2–3(a)) for $\alpha = \alpha_1$ and V_1.[31] The output voltage during clock phase ϕ_1 is given by

$$v_0^o(n) = v_c^o(n) - \frac{v_0^o(n)}{A_0} \tag{11}$$

[31] K. Martin, "Switched-Capacitor Networks," Ph.D. thesis, University of Toronto, 1980.

where $v_c(n)$ is the voltage across the feedback capacitor C during ϕ_1. The charge conservation equation at the negative terminal of the op amp gives

$$v_c^o(n) - v_c^e\left(n - \frac{1}{2}\right) + \left[\frac{1}{A_0} v_0^o(n) + v_1^o(n)\right]\alpha_1 = 0 \tag{12}$$

But $v_c^e(n - ½)$ is simply given by

$$v_c^e\left(n - \frac{1}{2}\right) = v_0^e\left(n - \frac{1}{2}\right) + \frac{v_0^e\left(n - \frac{1}{2}\right)}{A_0} = v_0^o(n-1) + \frac{v_0^o(n-1)}{A_0} \tag{13}$$

Substituting Eqs. (12) and (13) into Eq. (11) and taking the z-transform of both sides yields

$$H^{oo}(z) = \frac{\dfrac{-\alpha}{\left[1 + \dfrac{1}{A_0}(1+\alpha)\right]}}{1 - z^{-1}\dfrac{1 + \dfrac{1}{A_0}}{1 + \dfrac{1}{A_0}(1+\alpha)}} \tag{14}$$

Equation (14) can be written as

$$H^{oo}(\omega) = \frac{-\alpha\, e^{j\omega T/2}}{j\, 2 \sin(\omega T/2)} \frac{1}{1 + \dfrac{1}{A_0}\left(1 + \dfrac{\alpha}{2}\right) - j\dfrac{\alpha}{2\, A_0 \tan(\omega T/2)}} \tag{15}$$

Using a similar analysis (see Problem 5.14), the transfer function of the noninverting (forward) integrator of Fig. 5.2–3(a) for V_2 can be obtained

$$\frac{V_0(\omega)}{V_2(\omega)} = \frac{\alpha\, e^{-j\omega T/2}}{j\, 2 \sin(\omega T/2)} \frac{1}{1 + \dfrac{1}{A_0}\left(1 + \dfrac{\alpha}{2}\right) - j\dfrac{\alpha}{2\, A_0 \tan(\omega T/2)}} \tag{16}$$

Figure 5.4–4 shows the simulated effect that A_0 has on the frequency response of a fifth-order Chebyshev[32] SC low-pass ladder filter. In the lower trace,

[32] D. J. Allstot, "MOS Switched Capacitor Ladder Filters," Ph.D. thesis, U. of California Berkeley, 1980.

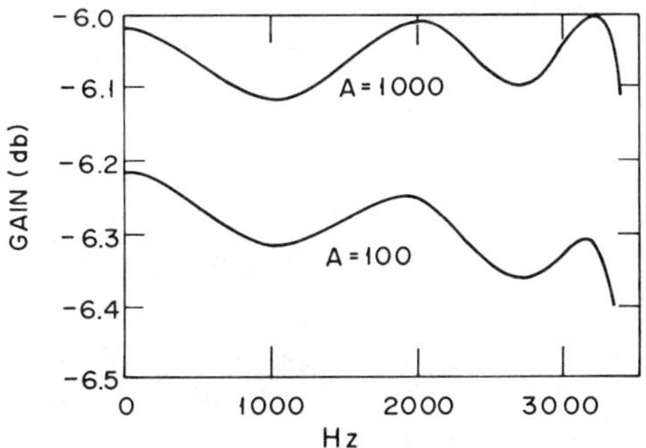

Fig. 5.4-4. Effect of open-loop DC again on response of a fifth-order Chebyshev.

all five op amps have an open loop gain of 10^2. In this case, the filter has a DC gain error of 0.22 dB with a passband droop of 0.09 dB due to the poles being pushed away from the imaginary axis. In the upper trace, the gain A_0 is 10^3, and the filter response shows a DC gain error of only 0.02 dB. Hence, it can be concluded that, for this type of filter, op amps with gains of 10^3 can be used to realize filters with very accurate responses when the pole Qs are smaller than 10. For high-Q realizations, A_0 is required to be proportionately higher. The actual Q_A can be expressed in terms of the desired Q by

$$Q_A = \left[\frac{Q}{A_0 + 2Q}\right]\left[1 + A_0^2 + \frac{A_0}{Q}\right]^{1/2} \approx \left(1 - \frac{2Q}{A_0}\right)Q \tag{17}$$

The center frequency also depends on A_0, i.e.

$$\omega_A = \omega_0\left[\frac{1}{A_0 + 1}\right]\left[A_0^2 + \frac{A_0}{Q} + 1\right]^{1/2} \approx \omega_0\frac{A_0}{A_0 + 1} \tag{18}$$

For $A = 10^3$, the difference between ω_0 and ω_A is less than 0.1% and can therefore be neglected for most cases.

It should be observed that independently of the A_0 effects the actual peak frequency is given by

$$\omega_{\text{peak}} = \omega_0\left[1 - \frac{1}{2Q^2}\right]^{1/2} \tag{19}$$

and the actual Q value is

$$Q_A = Q\left[\frac{1}{1-\dfrac{1}{4Q^2}}\right]^{1/2}, \; Q > \frac{1}{2} \tag{20}$$

For example if $Q = 1$, then $\omega_{\text{peak}} = 0.707\ \omega_0$, and $Q_A = 1.1547$. In general, the discrepancy in the approximate expression increases as Q decreases.

Next we consider the analysis of SC networks with op amps modeled by one dominant pole. A systematic analysis approach for SC networks when the op amps (OA's) are characterized with one dominant pole is described as follows. The OA is characterized by

$$\frac{dv_0(t)}{dt} + GB\ v_a(t) = 0 \tag{21}$$

where GB is the OA gain-bandwidth product, $v_a(t)$ is the differential input voltage of the OA, and $v_0(t)$ is the output voltage of the OA.

The analysis considerations are:

1. The clock phase periods are arbitrarily defined as

$$(n+i) < \frac{t}{T} \leq \left(n + \frac{1}{2} + i\right) \qquad \text{for } \phi_1\text{, odd phase}$$

$$\left(n + \frac{1}{2} + i\right) < \frac{t}{T} \leq (n+i+1) \qquad \text{for } \phi_2\text{, even phase}$$

$$i = \cdots -2, -1, 0, 1, 2, \cdots$$

2. The input voltage $v_i(t)$ is in general assumed invariant during the clock phases, i.e.

$$v_i(t) = v_i^o\left(n - \frac{1}{2}\right)T = v_i^e\,(n-1)T, \forall\,(n-1)\,T \leq t \leq nT \tag{22}$$

3. $v_0(t)$ is continuous for all time, which implies that no jumps occur at switch transitions, from t_0 to t_1, that is

$$v_0(t)\Big|_{t=t_1} = v_0(t_0) \tag{23}$$

In general when $v_a(t)$ is determined for one particular topology and one phase, Eq. (21) results in

$$\frac{dv_0(t)}{dt} + a\, v_0(t) = B(t) + b \tag{24}$$

where a is a function of GBk, k is a topology-dependent voltage divider, b depends on initial voltage conditions on the capacitors and on step-input signals, as well as on the topology, and $B(t)$ is a function of topology, as well as of the time-dependent input signals. A general solution of Eq. (24) can be expressed as

$$v_0(t) = e^{-at} \int e^{at} \left[B(t) + b \right] dt + ce^{-at} \tag{25}$$

Assuming that $B(t)$ is an exponential type function, i.e. e^{ht}, c can be determined by considering the following initial condition

$$v_0(t) \Big|_{t = t_1} = v_0(t_1) \tag{26}$$

where t_1 is the transition time evaluated at $(n + ½ + i)T$. c is given as

$$c = \left[v_0(t_1) - \frac{e^{ht_1}}{a + h} - \frac{b}{a} \right] e^{at_1} \tag{27}$$

Thus we obtain

$$v_0(t) = v_0(t_1)\, e^{-(t-t_1)a} + \frac{b}{a} \left[1 - e^{-(t-t_1)a} \right] + \frac{1}{a + h} \left[e^{ht} - e^{-at + t_1(a+h)} \right] \tag{28}$$

The particular case of $e^{ht} = 0$ gives

$$v_0(t) = v_0(t_1) e^{-(t-t_1)} a + \frac{b}{a} \left(1 - e^{-a(t-t_1)}\right) \tag{29}$$

In order to simplify the analysis, Table 5.4–1 is provided for different cases. The use of Table 5.4–1 is illustrated through examples. Note the polarity of the initial conditions on the capacitors. Case 8 is useful in determining the updated initial conditions at t_2 due to previous initial conditions at t_1. Furthermore, observe that nonfeedback capacitors connected directly at op amp outputs will not affect (due to analysis consideration 3) the values of a and b/a.

Table 5.4–1. Capacitive Circuit Solutions

CIRCUIT	a	b/a
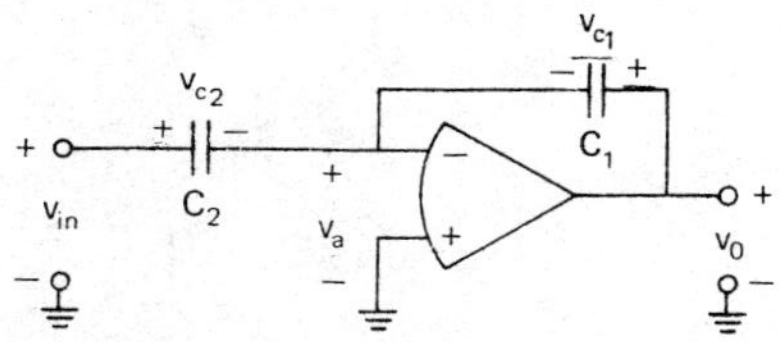 Case 1	$\dfrac{GBC_1}{C_1 + C_2}$	$v_{c_1}(t_1) - \dfrac{C_2}{C_1}[v_{in}(t) - v_{c_2}(t_1)]$
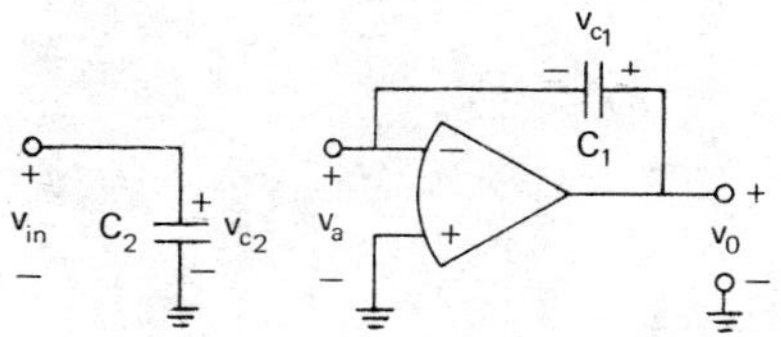 Case 2	GB	$v_{c_1}(t_1)$
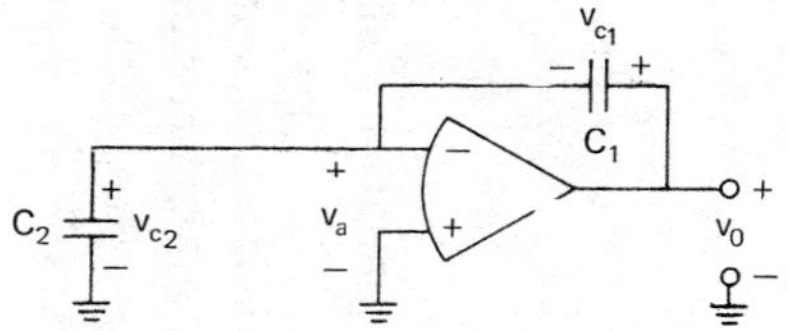 Case 3	$\dfrac{GBC_1}{C_1 + C_2}$	$v_{c_1}(t_1) - \dfrac{C_2}{C_1} v_{c_2}(t_1)$

Circuit		Output
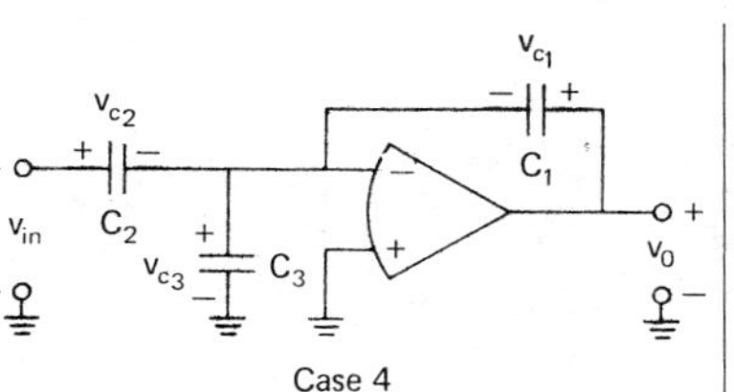 Case 4	$\dfrac{GBC_1}{C_1 + C_2 + C_3}$	$v_{c_1}(t_1) - \dfrac{C_2}{C_1}[v_{in}(t) - v_{c_2}(t_1)] - \dfrac{C_3}{C_1} v_{c_3}(t_1)$
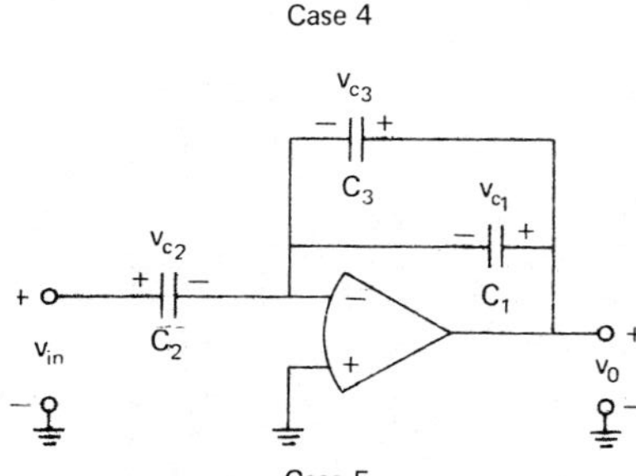 Case 5	$\dfrac{GB(C_1 + C_3)}{C_1 + C_2 + C_3}$	$\dfrac{C_1 v_{c_1}(t_1)}{C_1 + C_3} - \dfrac{C_2}{C_1 + C_3}[v_{in}(t) - v_{c_2}(t_1)] + \dfrac{C_3}{C_1 + C_3} v_{c_3}(t_1)$
Case 6	$\dfrac{GBC_1}{C_1 + C_2 + C_3}$	$v_{c_1}(t_1) - \dfrac{C_2}{C_1}[v_{in_1}(t) - v_{c_2}(t_1)] - \dfrac{C_3}{C_1}[v_{in_2}(t) - v_{c_3}(t_1)]$

Table 5.4–1 Capacitive Circuit Solutions (*continues*)

Circuit		Solution
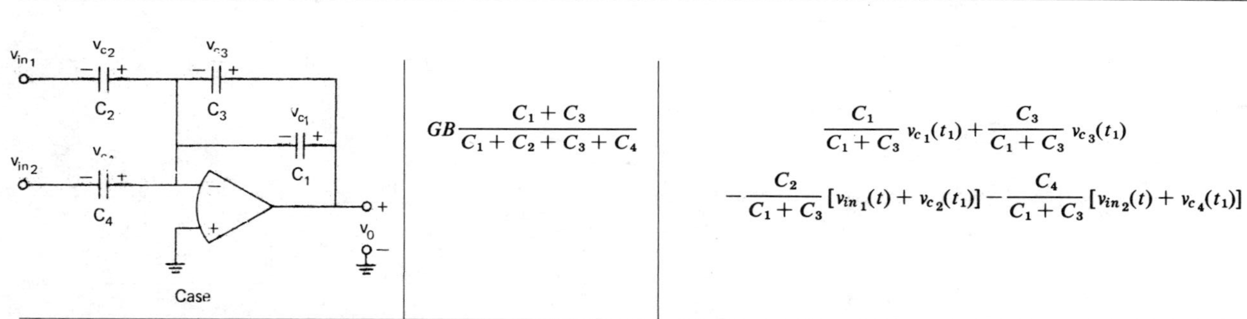	$GB\dfrac{C_1+C_3}{C_1+C_2+C_3+C_4}$	$\dfrac{C_1}{C_1+C_3}v_{c_1}(t_1)+\dfrac{C_3}{C_1+C_3}v_{c_3}(t_1)-\dfrac{C_2}{C_1+C_3}[v_{in_1}(t)+v_{c_2}(t_1)]-\dfrac{C_4}{C_1+C_3}[v_{in_2}(t)+v_{c_4}(t_1)]$
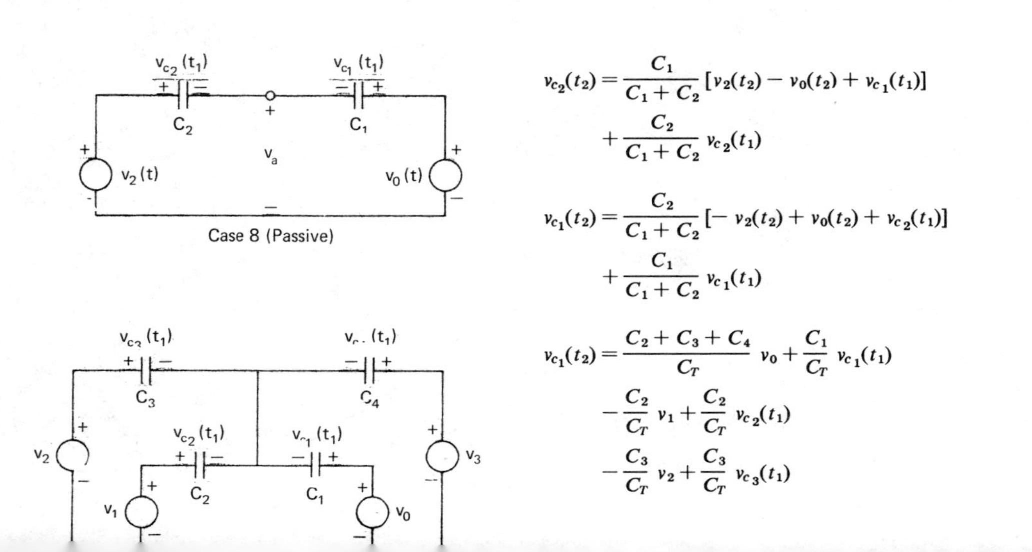 Case 8 (Passive)		$v_{c_2}(t_2)=\dfrac{C_1}{C_1+C_2}[v_2(t_2)-v_0(t_2)+v_{c_1}(t_1)]+\dfrac{C_2}{C_1+C_2}v_{c_2}(t_1)$ $v_{c_1}(t_2)=\dfrac{C_2}{C_1+C_2}[-v_2(t_2)+v_0(t_2)+v_{c_2}(t_1)]+\dfrac{C_1}{C_1+C_2}v_{c_1}(t_1)$
		$v_{c_1}(t_2)=\dfrac{C_2+C_3+C_4}{C_T}v_0+\dfrac{C_1}{C_T}v_{c_1}(t_1)-\dfrac{C_2}{C_T}v_1+\dfrac{C_2}{C_T}v_{c_2}(t_1)-\dfrac{C_3}{C_T}v_2+\dfrac{C_3}{C_T}v_{c_3}(t_1)$

$$-\frac{C_4}{C_T} v_3 + \frac{C_4}{C_T} v_{c_4}(t_1)$$

$$\begin{aligned} v_{c_2}(t_2) = &-\frac{C_1}{C_T} v_0 + \frac{C_1}{C_T} v_{c_1}(t_1) \\ &+ \frac{C_1 + C_3 + C_4}{C_T} v_1 + \frac{C_2}{C_T} v_{c_2}(t_1) \\ &- \frac{C_3}{C_T} v_2 + \frac{C_3}{C_T} v_{c_3}(t_1) \\ &- \frac{C_4}{C_T} v_3 + \frac{C_4}{C_T} v_{c_4}(t_1) \end{aligned}$$

$$\begin{aligned} v_{c_3}(t_2) = &-\frac{C_1}{C_T} v_0 + \frac{C_1}{C_T} v_{c_1}(t_1) \\ &- \frac{C_2}{C_T} v_1 + \frac{C_2}{C_T} v_{c_2}(t_1) \\ &+ \frac{C_1 + C_2 + C_4}{C_T} v_2 + \frac{C_3}{C_T} v_{c_3}(t_1) \\ &- \frac{C_4}{C_T} v_3 + \frac{C_4}{C_T} v_{c_4}(t_1) \end{aligned}$$

$$\begin{aligned} v_{c_4}(t_2) = &-\frac{C_1}{C_T} v_0 + \frac{C_1}{C_T} v_{c_1}(t_1) \\ &- \frac{C_2}{C_T} v_1 + \frac{C_2}{C_T} v_{c_2}(t_1) \\ &- \frac{C_3}{C_T} v_2 + \frac{C_3}{C_T} v_{c_3}(t_1) \\ &+ \frac{C_1 + C_2 + C_3}{C_T} v_3 + \frac{C_4}{C_T} v_{c_4}(t_1) \end{aligned}$$

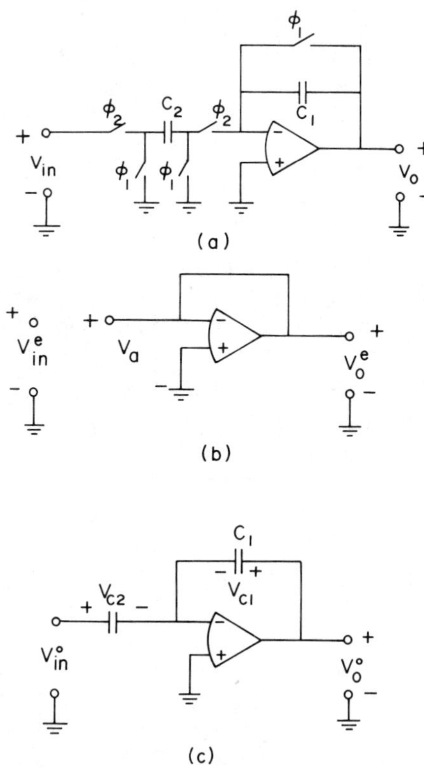

Fig. 5.4–5. (a) SC amplifier, (b) Equivalent circuit at ϕ_1, and (c) Equivalent circuit at ϕ_2.

Example 5.4–1. *SC amplifier.*[33] Compute the transfer function of the amplifier shown in Fig. 5.4–5(a) when the op amp is characterized by Eq. (21). Let us consider the circuit at ϕ_1 as shown in Fig. 5.4–5(b). The solution given by Eq. (29) and a and b/a from case 2 with $v_{C_1} = 0$ results in

$$v_0^o\left(n - \frac{1}{2}\right) = v_0^e(n - 1)e^{-a_1 T/2} \tag{30}$$

where $a_1 = GB$. Next, the solution during ϕ_2 is obtained; see Fig. 5.4–5(c). Again using Eq. (29) and Table 5.4–1, case 1, yields

[33] E. Sánchez-Sinencio, J. Silva-Martínez, and R. Alba-Flores, "Effects of Finite Operational Amplifier Gain-Bandwidth Product on a Switched Capacitor Amplifier," *Electronics Letters,* Vol. 17, July 1980, pp. 509–510.

$$v_0^e(n) = v_0^o\left(n - \frac{1}{2}\right) e^{-a_2 T/2} - \frac{C_2}{C_1} v_{\text{in}}^e(n)[1 - e^{-a_2 T/2}] \tag{31}$$

where $a_2 = GBC_1/(C_1 + C_2)$. Substituting $v_0^o(n - \frac{1}{2})$ from Eq. (30) into Eq. (31) and taking the z-transform, we obtain

$$H^{ee}(z) = \frac{-\dfrac{C_2}{C_1}(1 - e^{-a_2 T/2})z}{z - e^{-(a_1 + a_2)T/2}} \tag{32}$$

The ideal transfer function becomes

$$H^{ee}(z)\bigg|_{GB \to \infty} = -\frac{C_2}{C_1} \tag{33}$$

If the op amp is characterized by

$$A(s) = \frac{GB}{s + \omega_1} \tag{34}$$

The effect of ω_1 is generally small; see Problem 5.15.

Example 5.4–2. *Inverting forward integrator.*[34] Obtain the transfer function of the integrator shown in Fig. 5.4–6(a). At ϕ_1, Fig. 5.4–6(b), we identify case 2 of Table 5.4–1. Thus, we have

$$v_{C_2}^o\left(n - \frac{1}{2}\right) = v_{\text{in}}^o\left(n - \frac{1}{2}\right) \tag{35}$$

and

$$v_0^o\left(n - \frac{1}{2}\right) = v_0^e(n-1)e^{-a_1 T/2} + v_{C_1}^e(n-1)(1 - e^{-a_1 T/2}) \tag{36}$$

where $a_1 = GB$. Now at ϕ_2 and Fig. 5.4–6(c), we obtain for case 3 that

[34] R. L. Geiger and E. Sánchez-Sinencio, "A Comparison of Switched Capacitor Integrators with Respect to the Operational Amplifier Gain-Bandwidth Product," *Proc. 23rd Midwest Symposium on Circuits and Systems,* August 1980, pp. 444–448.

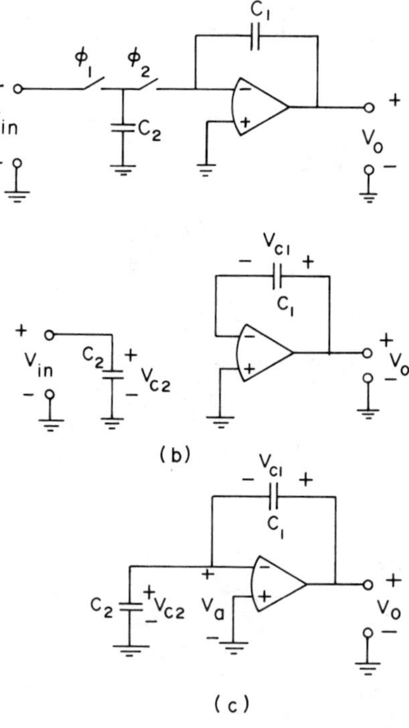

Fig. 5.4-6. (a) Noninverting SC integrator, (b) Equivalent circuit at ϕ_1, and (c) Equivalent circuit at ϕ_2.

$$v_0^e(n) = v_0^o\left(n - \frac{1}{2}\right) e^{-a_2 T/2} + \left[v_{C1}^o\left(n - \frac{1}{2}\right) - \frac{C_2}{C_1} v_{C2}^o\left(n - \frac{1}{2}\right)\right](1 - e^{-a_2 T/2}) \quad (37)$$

where $a_2 = GBC_1/(C_1 + C_2)$; then substituting Eqs. (35) and (36) into Eq. (37) yields

$$v_0^e(n) = [v_0^e(n-1)e^{-a_1 T/2} + v_{C1}^e(n-1)(1 - e^{-a_1 T/2})]e^{-a_2 T/2} + (1 - e^{-a_2 T/2})\left[v_{C1}^e(n-1) - \frac{C_2}{C_1} v_{in}^o\left(n - \frac{1}{2}\right)\right] \quad (38)$$

$v_{C1}^e(n - 1)$ is required to obtain an equation with only $v_0^e(n)$ and $v_{in}^o(n - ½)$. This can be accomplished using case 8 of Table 5.4-1. Thus

$$v_{C_1}^{e}(n-1) = \frac{C_2}{C_1+C_2}\left[v_0^{e}(n-1) - v_{\text{in}}^{o}\left(n-\frac{3}{2}\right)\right] + \frac{C_1}{C_1+C_2} v_{C_1}^{e}(n-2) \quad (39)$$

Note that $v_{C_1}^{o}\left(n-\frac{3}{2}\right) = v_{C_1}^{e}(n-2)$. Now, we can take the z-transform of Eqs. (38) and (39) and obtain

$$H^{eo}(z) = -z^{-1/2}\frac{\frac{C_2}{C_1}(1-e^{a_2T/2}) + z^{-1}\frac{C_2}{C_1+C_2}e^{-a_2T/2}(1-e^{-a_1T/2})}{(1-z^{-1})\left(1-\frac{C_1}{C_1+C_2}e^{-(a_1+a_2)T/2}z^{-1}\right)} \quad (40)$$

To determine $H^{oo}(z)$ we take the z-transform of Eqs. (36) and (39) and obtain $V_0^{o}(z)/V_{\text{in}}^{o}(z)$

$$H^{oo}(z) = \frac{[C_2 + C_1e^{-a_1T/2}(1-z^{-1})]z^{-1/2}H^{eo}(z) - C_2z^{-1}(1-e^{-a_1T/2})}{C_1+C_2-C_1z^{-1}} \quad (41)$$

Observe that

$$H^{eo}(z)\Bigg|_{a_1,a_2\to\infty} = \frac{-\frac{C_2}{C_1}z^{-1/2}}{1-z^{-1}} \quad (42)$$

and

$$H^{oo}(z)\Bigg|_{a_1,a_2\to\infty} = -\frac{C_2}{C_1}\frac{z^{-1}}{1-z^{-1}} \quad (43)$$

The transfer functions of the stray-insensitive basic building block of Fig. 5.2–3, when the op amp is characterized by Eq. (21), can be derived in a similar way (see Problems 5.16 and 5.17). The transfer functions of the noninverting integrator are

$$H^{eo}(z) = z^{-1/2}\frac{\alpha_2(1-e^{-a_2T/2}) + \frac{z^{-1}e^{-a_2T/2}(1-e^{-a_1T/2})}{1+\alpha_2}}{(1-z^{-1})\left(1-z^{-1}\frac{e^{-(a_1+a_2)T/2}}{1+\alpha_2}\right)} \quad (44)$$

This transfer function, Eq. (44), is also referred to as the *noninverting* LDI. The $H(z)$ when the output and input are sampled at the same clock phase is given by

$$H^{oo}(z) = \frac{\alpha_2 z^{-1}(1 - e^{-(a_1+a_2)T/2})}{(1 - z^{-1})\left(1 - \dfrac{z^{-1}\, e^{-(a_1+a_2)T/2}}{1+\alpha_2}\right)} \tag{45}$$

where $a_1 = GB$, and $a_2 = GB/(1 + \alpha_2)$. Note that Eqs. (44) and (45) reduce to the ideal LDI cases and noninverting forward integrators when $a_1, a_2 \longrightarrow \infty$. The transfer functions of the inverting integrator of Fig. 5.2–3 are

$$H^{oo}(z) = -\frac{\alpha_1(1 - e^{-a_2 T/2}) + \dfrac{z^{-1}\, e^{-a_2 T/2}(1 - e^{-a_1 T/2})}{1+\alpha_1}}{(1 - z^{-1})\left(1 - \dfrac{z^{-1}\, e^{-(a_1+a_2)T/2)}}{1+\alpha_1}\right)} \tag{46}$$

and

$$H^{eo}(z) = -\,\alpha_1 z^{-1/2}\frac{1 - e^{-(a_1+a_2)T/2}}{(1 - z^{-1})\left(1 - \dfrac{z^{-1}\, e^{-(a_1+a_2)T/2}}{1+\alpha_1}\right)} \tag{47}$$

where $a_1 = GB$, and $a_2 = GB/(1 + \alpha_1)$. The ideal cases, inverting LDI, and backward integrators are obtained when $a_1, a_2 \longrightarrow \infty$.

5.5 GAIN-BANDWIDTH EFFECTS ON THE PERFORMANCE OF A PAIR OF SC BIQUADS

In this section we present the analysis of the pair of stray-insensitive, second-order bandpass filters shown in Fig. 5.5–1.[35] It is shown that, although both filters ideally have the same transfer function when the op amps are modeled with one dominant pole, their frequency response differs significantly.

The biquad 2 of Fig. 5.5–1(b) consists of a noninverting integrator connected in a feedback loop with an inverting integrator. In biquad 2, the operational

[35] J. Silva-Martínez, E. Sánchez-Sinencio, and A. S. Sedra, "Effects on Performance of a Pair of SC Biquads due to the Op Amp Gain-Bandwidth Product," *Proc. ISCAS IEEE*, Rome, May 1982, pp. 373–376.

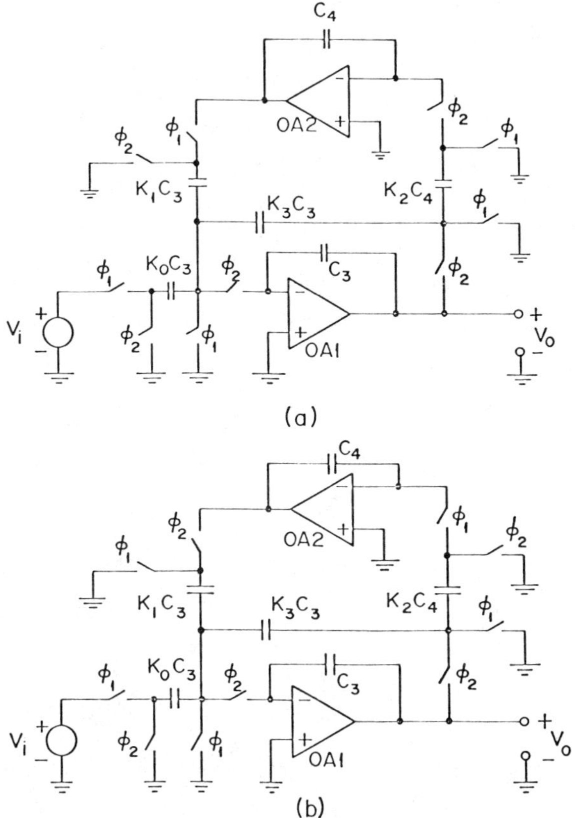

Fig. 5.5–1. SC bandpass filters; (a) Biquad 1, coupled form and (b) Biquad 2, decoupled.

amplifier OA1 receives a step input at the beginning of ϕ_2 and OA2 receives a step input at the beginning of ϕ_1. The analysis of the biquad 2, although algebraically tedious, is straightforward and uses the approach described before in Section 5.4. The ideal transfer function is

$$H_{I_2}(z) = \frac{[K_0/(1+K_3)]\,(z-1)}{z^2 - z(2+K_3-K_1K_2)/(1+K_3) + 1/(1+K_3)} \tag{1}$$

The analysis of the decoupled biquad 2 involves the solution of six equations. For ϕ_1, OA1 involves case 2 and OA2 case 1. For ϕ_2, OA1 can be considered case 7 and OA2 case 1. Also for each phase clock, case 8 or 9 is employed. These six equations are derived in Problem 5.19.

In the biquad 1 of Fig. 5.5–1(a), OA1 receives a step input at the beginning of ϕ_2, and OA2 also receives its input during ϕ_2. However, because OA1

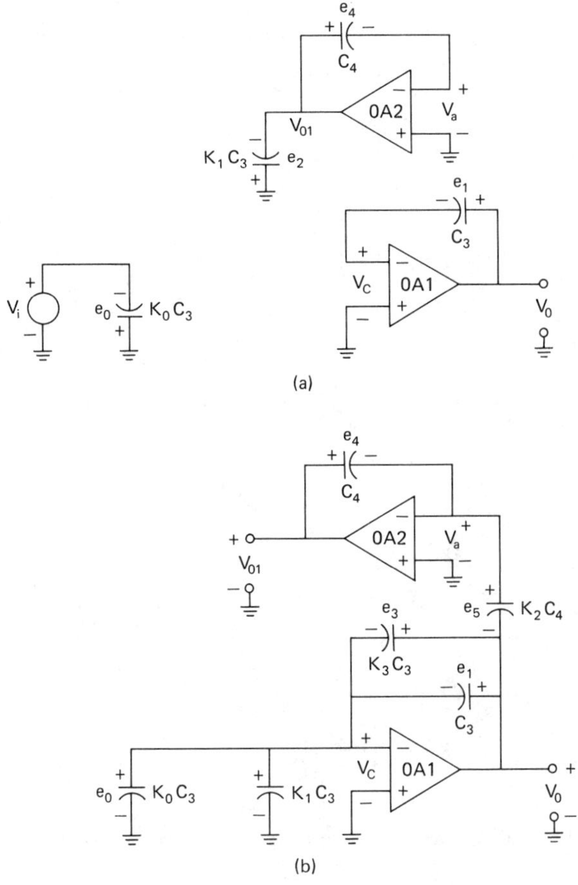

Fig. 5.5–2. Biquad 1; (a) During ϕ_1 and (b) During ϕ_2.

is characterized by Eq. (1) and is not ideal, its output cannot change immediately, and therefore the input to OA2 is not a step but rather an exponential ramp. Thus, the analysis approach described in Section 5.4 is applicable. In this case, we use Eq. (24) of the previous section containing both $B(t)$ and b. The solution $v_0(t)$ is given by Eq. (28) rather than by Eq. (29). Figure 5.5–2 illustrates the equivalent circuits for both clock phases. It can be seen from Fig. 5.5–2(a) during ϕ_1 that for both op amps case 1 applies. Figure 5.5–2(b) is valid during ϕ_2. OA1 results in case 9 with $V_{in1} = V_{in2} = 0$. For the OA2, case 1 results, although $v_{in}(t)$ is not a step but an exponential ramp, and the equation characterizing OA2 is given by Eq. (24) of Section 5.4, and its solution is given by Eq. (28). Rather than showing the algebraically complicated transfer functions for these two filters, the results are shown

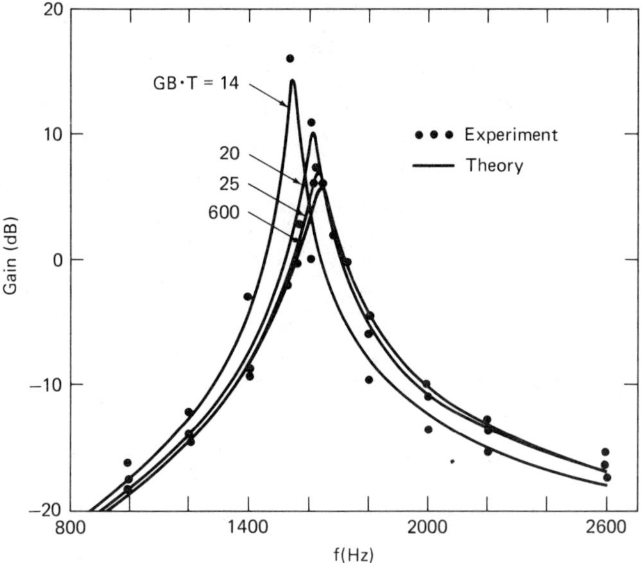

Fig. 5.5–3. (a) Frequency response of the decoupled form.

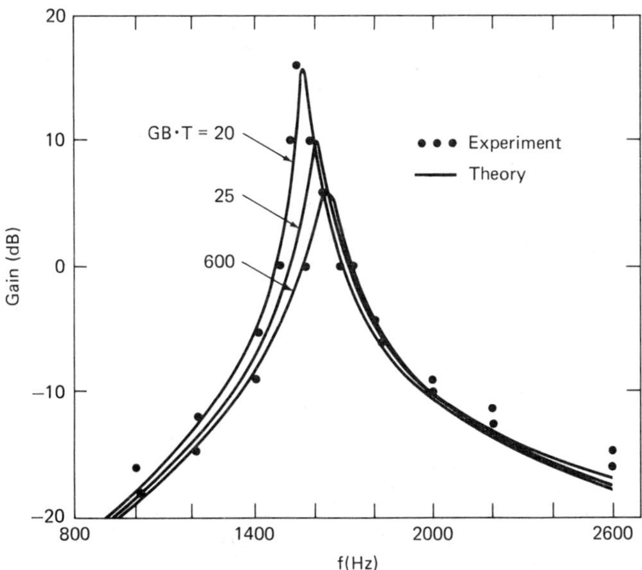

Fig. 5.5–3. (b) Frequency response of the coupled form.

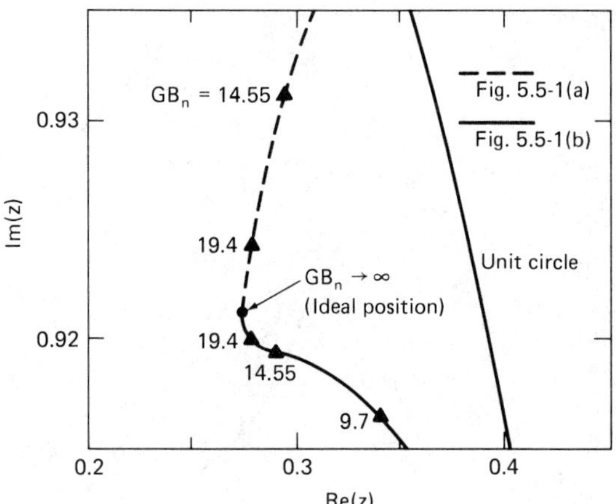

Fig. 5.5–4. Root-locus of both biquads.

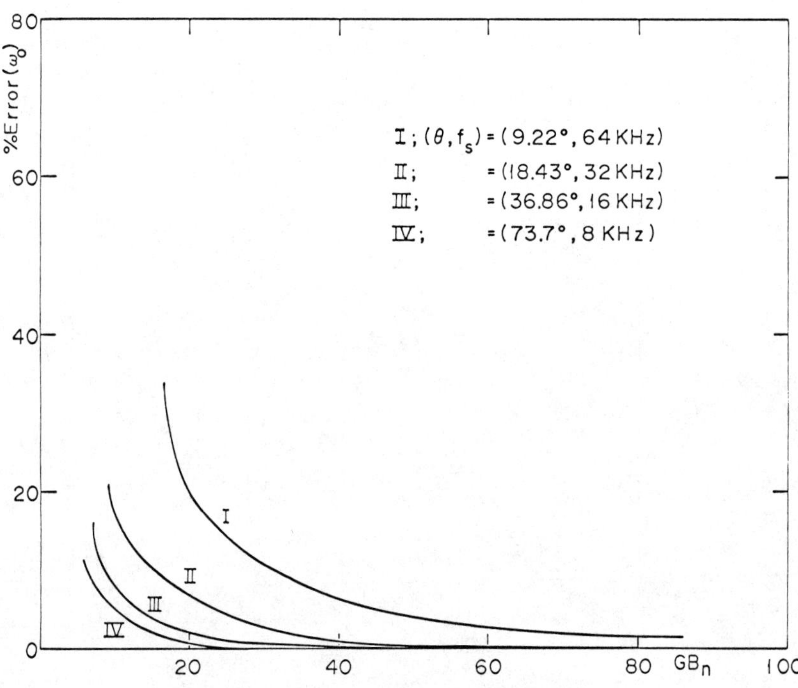

Fig. 5.5–5. (a) Deviations of ω_0 vs. GB_n (decoupled form).

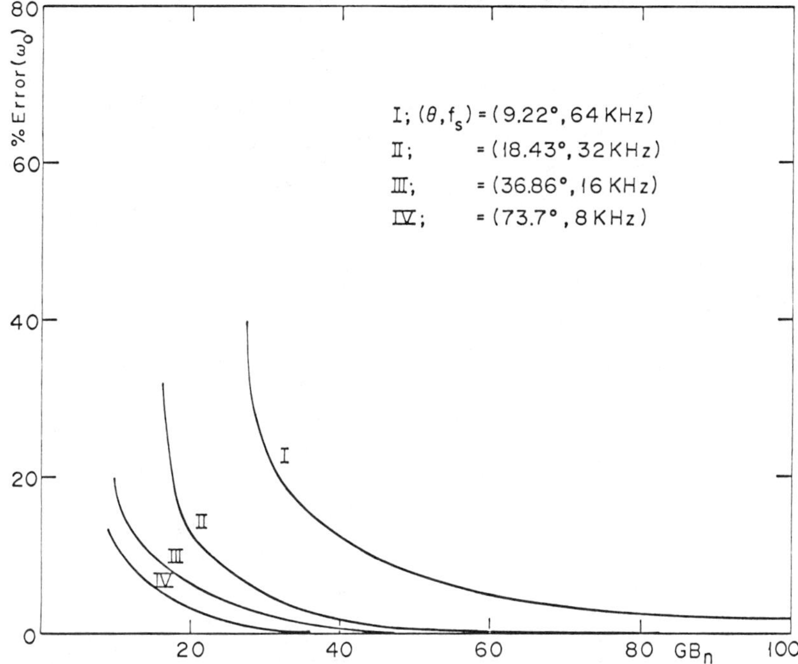

Fig. 5.5–5. (b) Deviations of ω_0 vs. GB_n (coupled form).

in plots. It can be shown that ideal transfer functions for both filters (see Problem 5.21) are given by Eq. (1).

To compare both biquads, an example is given with the following specifications. Assume, for the SC bandpass filter of Fig. 5.5–1a, that the center frequency is $f_0 = 1633$ Hz, ($f_0 = \omega_0/2\pi$), that the quality factor is $Q = 16$, and that the sampling frequency is $f_s = 8$ kHz. The frequency responses of the biquads 2 and 1 are shown in Figs. 5.5–3(a) and 5.5–3(b), respectively, where $GB_n = GB/\omega_0$. It can be observed that, for $GB/f_0 = GBT = 14$, the coupled biquad 1 becomes unstable and therefore is not plotted in Fig. 5.5–3(b). We note that the Q enhancement and ω_0 shift is more severe than the decoupled biquad 2. This can be fully appreciated by inspecting the root-locus plots for both structures shown in Fig. 5.5–4. The deviations of ω_0 and Q for both structures are shown in Figs. 5.5–5 and 5.5–6, where $\theta = \omega_0 T$.

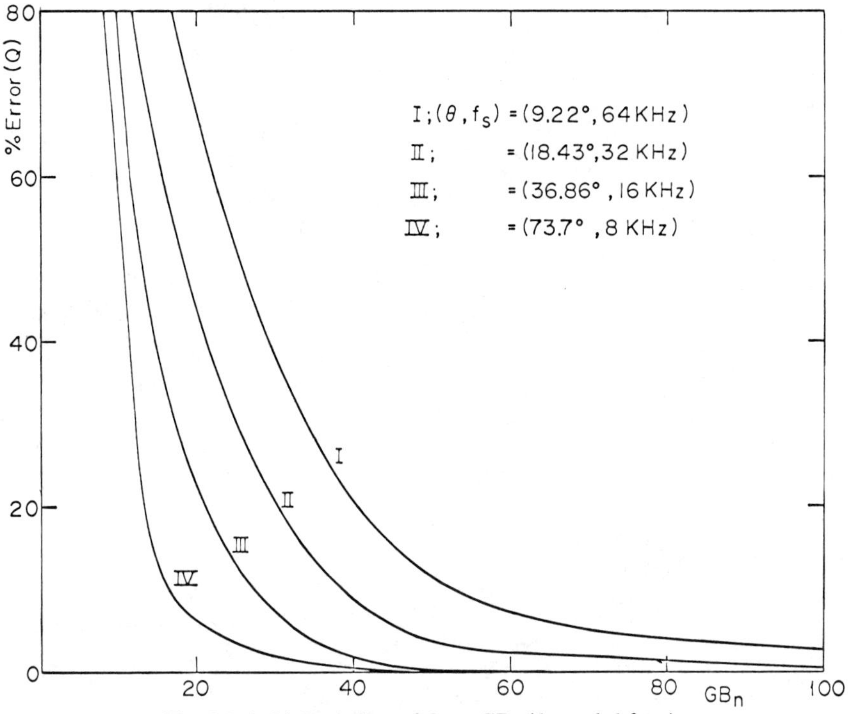

Fig. 5.5–6. (a) Deviations of Q vs. GB_n (decoupled form).

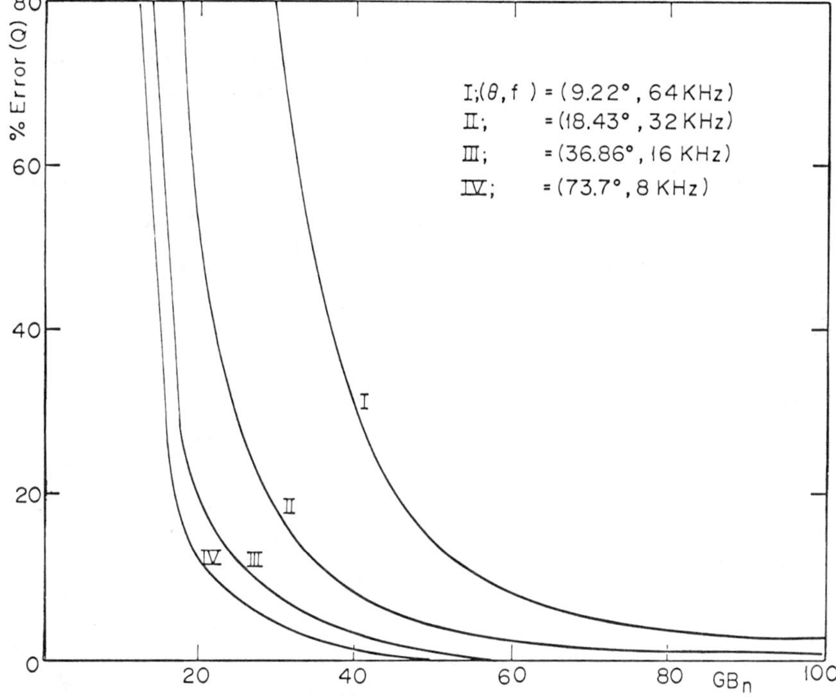

Fig. 5.5–6. (b) Deviations of Q vs. GB_n (coupled form).

5.6 SUMMARY

This chapter has introduced a design approach for SC filters that consists in transforming the analog filter transfer function into the discrete-time filter transfer function and realizing the discrete-time transfer function. This method has taken advantage of the existing information on digital filters. Techniques to synthesize higher-order filters were presented. These techniques include the straightforward cascade approach and the use of higher-order digital filter structures. The nonideal performance of SC filters was also examined in this chapter. It was seen that SC filters are limited by essentially the time required to completely transfer the charge from one capacitor to another.

This chapter concludes the study on switched capacitor techniques applied to filters. Many interesting existing and new ideas have not been included because of a lack of space. The focus has been on the basic ideas that are reasonably accepted in this rapidly growing new field. A good understanding of these basic concepts should allow a quick grasp of the new developments that are sure to come.

The next two chapters will examine nonfilter applications of switched capacitor circuits. These chapters will point out the versatility of the switched capacitor technique and its applicability to many different types of analog signal processing functions, of which filters are only a part. The last chapter will provide a more detailed examination of the technology that is suitable for fabricating the circuits considered so far. Many of the choices, assumptions, and constraints are based on information contained in this chapter. The reader interested only in filter applications should read Chapter 8 next to achieve a general picture of switched capacitor filters.

PROBLEMS

5.1 (Sec. 5.1). Prove that the circuit of Fig. P 5–1 has the following transfer function

$$H_i^{ee}(z) = \frac{V_0^e(z)}{V_i^e(z)} = \frac{C_1}{C_0} \frac{z^{-1}}{\left(1 + \frac{1}{A_0} + \frac{C_1}{A_0 C_0}\right)\left(1 + \frac{1}{A_0}\right)}$$

where A_0 is the open loop gain, and that

$$H_1^{ee}(z) \bigg|_{A_0 \to \infty} = \frac{C_1}{C_0} z^{-1} \quad \text{(noninverting)}$$

$$H_2^{ee}(z) \bigg|_{A_0 \to \infty} = -\frac{C_2}{C_0} z^{-1} \quad \text{(inverting)}$$

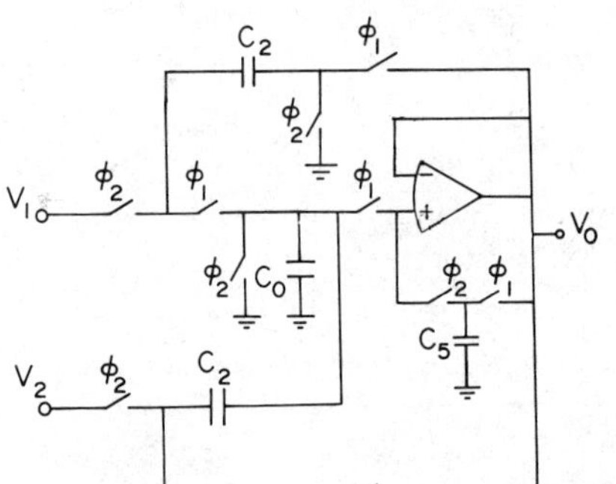

Fig. P5.1

5.2 (Sec. 5.1). Obtain the analog implementations of Fig. 5.1–1 using the z^{-1} implementations of Fig. P 5–1. Design the filters for $r = 0.99$ and $\theta = 6.2°$; compare with Example 5.1–1.

5.3 (Sec. 5.1). Design a low-pass SC filter using topologies of Figs. 5.1–2(a) and (b) for the following two cases: (1) $r = 0.960677$ and $\theta = 73.54°$, (2) $\theta = 9.186°$ and $r = 0.99003$. Compare sensitivities using Table 5.1–1 when the output voltage level of each op amp is the same.

5.4 (Sec. 5.1). Compute the total capacitance of Example 5.1–2 for the case of identical op amp output voltage levels. Also compute the sensitivities.

5.5 (Sec. 5.1). Using the filter of Fig. 5.1–8(a) for the case where $A_1 = 0$, $B_1 = -B_1$, $B_2 = -B_2$ and the $\hat{z}^{-1}$ accumulator associated with A_2 becomes $(1 + \hat{z}^{-1})$, obtain the transfer function and design a filter with the specifications of Problem 5.3. Give the SC implementation of this filter.

5.6 (Sec. 5.2). Design a bandpass filter using the circuit of Fig. 5.2–2(b) with the maximum output signal of both op amps near 10 dB and $r = 0.960677$ and $\theta = 73.54°$. Obtain the total capacitance required.

5.7 (Sec. 5.2). Obtain the SC implementation of the flow diagram of Fig. 5.2–8 for the transfer function

$$H(z) = \frac{0.1953\ (1 - z^{-1})}{1 - 0.5455z^{-1} + 0.9229z^{-2}}$$

5.8 (Sec. 5.2). Same as Problem 5.7, but with the flow diagram of Fig. 5.2–9. Compute the total capacitance when the op amps have the same output level.

5.9 (Sec. 5.2). Obtain the SC implementation of the flow diagram of Fig. 5.2–7. Implement the transfer function given in Problem 5.7 for the particular case of $g_2 = g_4 = g_7 = g_8 = g_{10} = g_{12} = g_{13} = g_{15} = g_{16} = 0$.

5.10 (Sec. 5.2). Design an SC filter using biquad Type 2 for the design specifications of Problem 5.3. Find a compromise between total capacitance and sensitivities.

5.11 (Sec. 5.3). Using one first-order section and two second-order sections of Fig. 5.2–5 (biquad Type 1), design the voiceband filter specified in Example 5.3–1.

5.12 (Sec. 5.3). Same as Problem 5.11, except that biquads are Type 2.

5.13 (Sec. 5.3). Using the third-order low-pass filter of Fig. 5.3–4 and the biquad Type 3 of Fig. 5.2–8, design a cascade filter for the specifications given in Example 5.3–1.

5.14 (Sec. 5.4). Assuming a finite A_0, obtain $H(z)$ for the stray-insensitive noninverting integrator.

5.15 (Sec. 5.4). Show that if the op amp is characterized by

$$A(s) = \frac{GB}{s + \omega_1}$$

the transfer function of Example 5.4–1 becomes

$$H^{ee}(z) = \frac{\dfrac{-GBC_2}{GBC_1 + \omega_1(C_1 + C_2)}(1 - e^{-a_2 T/2})}{1 - e^{-(a_1 + a_2 + \omega_1)T/2}\, z^{-1}}$$

5.16 (Sec. 5.4). Show that the transfer function of a noninverting forward integrator is given by Eqs. (44) and (45). Note from Table 5.4–1 that cases 2, 3, and 8 are employed to obtain Eqs. (44) and (45).

5.17 (Sec. 5.4). Show that the transfer function of an inverting (series) backward integrator is given by Eqs. (46) and (47). The solution of this problem involves cases 2, 1, and 8 of Table 5.4–1.

5.18 (Sec. 5.4). Plot the magnitude and phase errors in percentages versus (GBT/α) for Eqs. (45) and (46) for $\omega_0 T = \alpha = \alpha_1 = \alpha_2 = 0.1\ \pi$.

5.19 (Sec. 5.5). Write the z-transform equations characterizing the biquad 2 of Fig. 5.5–1(b) for both clock phases.

5.20 (Sec. 5.5). Write the z-transform equations characterizing the biquad 1 of Fig. 5.5-1(a) during ϕ_1 and ϕ_2.

5.21 (Sec. 5.5). Obtain the ideal transfer functions of biquads of Fig. 5.5–1. If the open loop gain of the op amps are finite, determine the corresponding transfer functions.

6
Applications of Switched Capacitor Circuits

6.0 INTRODUCTION

Up to this point, switched capacitor circuits have been primarily applied to analog filters. It is the objective of this and the next chapter to present nonfilter applications of switched capacitor circuits. Such applications are very important from the viewpoint that switched capacitor circuits can be easily implemented by MOS technology. This provides the opportunity for the designer to consider either an analog or a digital approach to processing signals. As a result, the performance and area of an integrated signal-processing system can be optimized for a given application.

This chapter first considers the general aspects of signal-processing systems. The influence of the applications and technology upon the signal-processor are considered. In nonfilter applications, the comparator becomes an important component. The various types of comparators suitable for MOS technology are introduced. Next these comparators are used to investigate switched capacitor circuits that realize rectifiers and other nonlinear circuits. Another useful area of switched capacitor circuits is that of modulators and multipliers. These circuits permit the realization of many useful signal-processing functions. Switched capacitor circuits have also found applications in oscillators and phase-locked loops, which are presented. The chapter closes with a consideration of programmable SC circuits. These circuits are very useful in adaptive applications.

6.1 ANALOG SIGNAL-PROCESSING[1]

Figure 6.1–1 shows a simple block diagram of a typical signal-processing system. As IC technology begins to exploit VLSI techniques, we will find many systems with this format being built on a single chip. A forerunner

[1] This section is based on lectures and notes by R. W. Brodersen given at the NATO Advanced Study Institute on Design Methodologies for VLSI in Louvain-le-neuve, Belgium, July 1980.

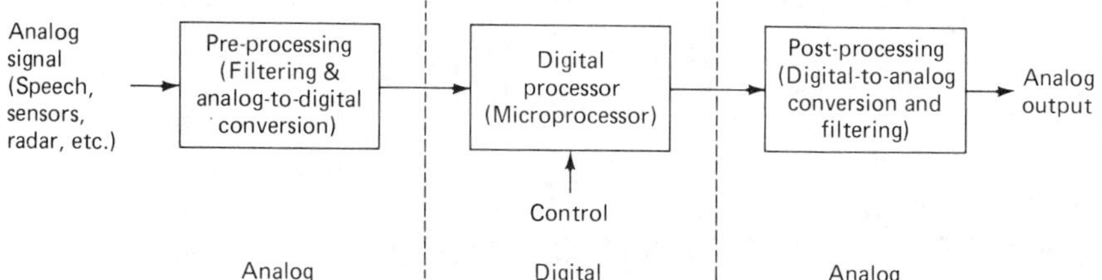

Fig. 6.1–1. A block diagram of a typical signal processing system.

of this approach is the analog signal-processor.[2] The advent of the analog sampled data techniques and MOS technology has made the design of a general signal-processor an interesting and challenging task. One must examine the specifications and decide what part of the system should be analog and what part should be digital. In most cases the input signal is analog. It could be a speech signal, a sensor output, a radar return, etc. The first block of Fig. 6.1–1 is a preprocessing block. Typically this block will consist of filters and an analog-to-digital converter. Often there are very strict speed and accuracy requirements on the components in this block. The next block of the analog signal-processor is essentially a microprocessor. The obvious advantage is that the function of the processor can be controlled and changed very easily. Finally, it may be necessary to achieve an analog output. In this case a postprocessing block is necessary, and it will typically contain a digital-to-analog converter and some filtering. The interesting decision on the system designer's part is deciding where to place the interfaces indicated by the dotted lines.

For signal-processing probably the most important system consideration is the bandwidth of the signal to be processed. A graph of the bandwidths of a variety of signals is given in Fig. 6.1–2. The bandwidths in this figure cover the enormous range of 10 orders of magnitude in frequency. At the low end are the seismic signals that do not extend much below 1 Hz because of the absorption characteristics of the earth. At the other extreme are the microwave signals that are not used much above 30 GHz because of the difficulties in performing even the simplest forms of signal-processing at higher frequencies.

To perform signal-processing over this range of frequencies, a variety of techniques have been developed that are almost exclusively analog above 10 MHz and digital below 100 Hz, as shown in Fig. 6.1–3. In the overlap

[2] M. Townsend, M. Hoff, Jr., and R. Holm, "An NMOS Microprocessor for Analog Signal Processing," *IEEE J. of Solid-State Circuits,* Vol. SC-15, No. 1, February 1980, pp. 33–38.

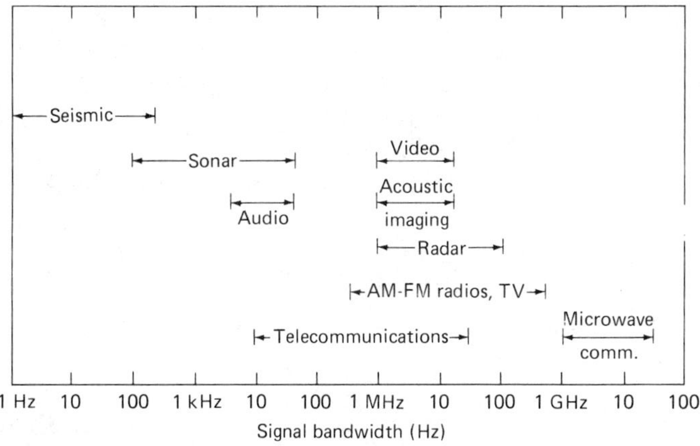

Fig. 6.1–2. Bandwidths of signals used in signal processing applications.

region a tradeoff must be made between the accuracy and flexibility of a digital approach and the low cost, power, and size of analog techniques.

By using the fact that VLSI is simply the continuing development of MOS technology, it is possible to extrapolate the trends of the past 10 years to obtain predictions about the capability of MOS technology in the next 10 years. Some of the implications of these predictions have been considered. It can be shown that even at the limits of scaling, the processing bandwidth of MOS technology will have improved only to the point where it is equivalent to today's advanced bipolar technologies. Because of the increased capital

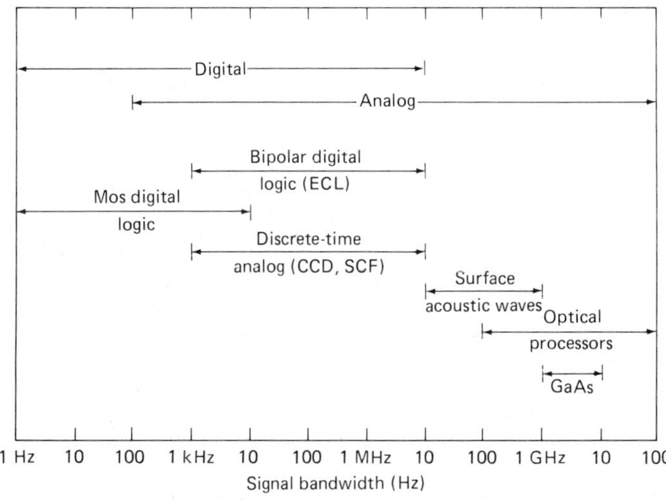

Fig. 6.1–3. The signal bandwidths which can be processed by present day technologies.

costs of implementing VLSI technology, the access to the technology will become increasingly limited. The result of this trend is that the development of VLSI signal-processing systems will be by large vertically integrated system/IC manufacturers. Because the VLSI signal-processing chips will be able to implement significant system functions, it will be necessary for the system designer to be involved in the chip design. This will be made possible by standardized design methods, such as gate arrays, standard cells, and PLAs, which will allow system designers to design the VLSI chips themselves. This high level of system integration will allow widespread use of sophisticated signal-processing algorithms because of the reductions in cost, size, and power of the signal-processing systems.

The fastest moving boundary in Fig. 6.1–3 is the upper limit of the MOS-LSI digital logic, which, as the technology progresses to VLSI, should be able to process signals by the year 1990 at bandwidths of 1–10 MHz. This will be accomplished by using greatly increased density along with moderately increased device speeds. However, as can be seen in Fig. 6.1–3, bipolar digital techniques are already able to process at these rates. Therefore VLSI will not make possible increased processing rates over what can now be achieved with processors built with bipolar technology, but rather will offer the primary advantages of reduced cost, size, and power requirements of the MOS-VLSI signal processors. This will make it possible for signal-processing techniques to make an impact in such cost-sensitive areas as consumer products that until now have been unable to afford the costs of using sophisticated signal-processing techniques.

There are a number of circuits that indicate that MOS technology has the capability at this time (1980) to process signals with bandwidths up to 10 kHz, as is indicated in Fig. 6.1–2.[3–8] The most impressive of these chips are the signal processors for telecommunication applications developed by Bell Laboratories[3] and Nippon Electric Co.,[4] both of which have the capability

---

[3] J. R. Boddie et al., "A Digital Signal Processor for Telecommunications Applications," *Int. Solid State Circuits Conf. Tech. Dig.,* San Francisco, February 1980, pp. 44–45.

[4] Y. Kawakami et al., "A Single Chip Signal Processor for Voiceband Applications," Ibid., pp. 40–41.

[5] Y. Chen and D. L. Duttweiler, "A 35,000 Transistor VLSI Echo Canceller," Ibid., pp. 42–43.

[6] M. Townsend et al., "An NMOS Microprocessor for Analog Signal Processing," *IEEE J. of Solid-State Circuits,* February 1980, pp. 33–38.

[7] R. W. Blasco, "VMOS Chip Joins Microprocessor to Handle Signals in Real Time," *Electronics,* Vol. 52, No. 18, August 30, 1979.

[8] C. A. T. Salama, "VLSI Technology for Telecommunication IC's," *IEEE J. of Solid-State Circuits,* Vol. SC-14, No. 6, August 1981, pp. 253–260.

to implement over 40 second-order filters at a sampling rate of 8 kHz, with a digital word length that yields a dynamic range on the order of 80 dB.

6.2 VOLTAGE COMPARATORS

In many nonfilter applications of switched capacitor networks, the ability to detect a voltage threshold is very important. Comparators are devices that have two stable output states. The states indicate whether an input voltage has crossed a threshold imposed by one or more other voltages, which are either fixed or variable. Figure 6.2–1(a) indicates the functional definition of a voltage comparator. The comparator output can be defined as

$$V_{\text{out}} = \begin{cases} V_{OH}, & V_A \geq V_B \\ V_{OL}, & V_A < V_B \end{cases} \tag{1}$$

where V_{OH} is the upper limit and V_{OL} is the lower limit of output voltage of the comparator. Figure 6.2–1(b) illustrates the transfer function of a voltage comparator. V_B can be considered as the reference voltage that defines the value of V_A at which the output changes state.

The comparator of Fig. 6.2–1(a) is called *noninverting,* because the output goes from the low state to the high state when the voltage V_A becomes larger than V_B. An inverting comparator is achieved when V_A is used as the reference and V_B is the input. One problem that can occur with the comparator of Fig. 6.2–1(a) happens when V_A gets closer to V_B. If V_A is a

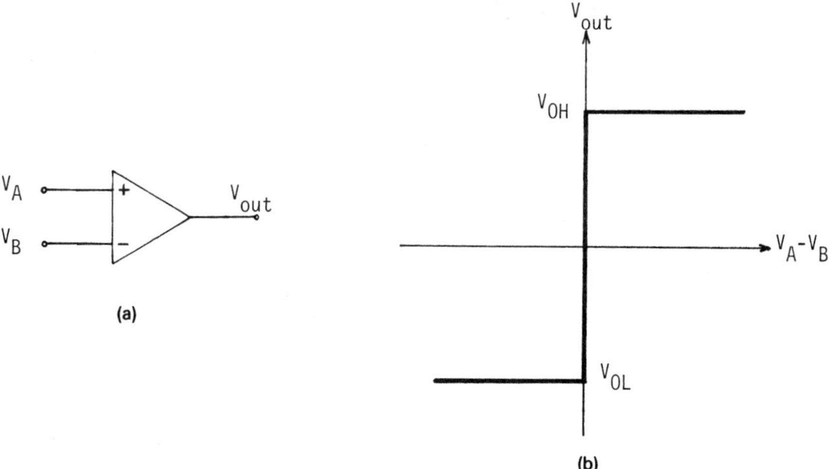

Fig. 6.2–1. (a) Voltage comparator. (b) Transfer characteristics of the voltage comparator.

noisy waveform, at some instants of time it will exceed V_B, and at others it will be less than V_B. Under these conditions the output of the comparator "chatters" between its two states. This behavior can be eliminated by causing the comparator to have hysteresis through the use of positive feedback. This could result in the characteristic shown in Fig. 6.2–2.

A comparator can be built using three different approaches, which are all compatible with MOS technology. These approaches are (1) high-gain amplifier, (2) regenerative, and (3) charge-differencing. The high-gain amplifier approach is widely used because it is essentially an op amp in that it usually has a high-gain, low-drift, differential-input circuit. The regenerative approach is best represented by the Schmitt trigger and the flip-flop. The charge-differencing method uses switches and capacitors, along with a low-gain open loop type of comparator, to detect voltage thresholds. Each of these methods will be further developed below.

The open loop comparator is typically an op amp without internal compensations (see Chapter 8 for further details). In a real comparator, the slope of the vertical transition in Fig. 6.2–1(a) is not infinite. One requirement for the comparator is that the gain must be large enough to ensure a satisfactory threshold-detection capability. If the comparator must be able to distinguish between a value of V_B and $V_B + \epsilon$, then it is easy to see that the gain of the comparator must be at least

$$\text{Gain} = \frac{V_{OH} - V_{OL}}{\epsilon} \tag{2}$$

Typical gains of MOS op amps vary from 1000 for NMOS to 10,000 for CMOS. Assuming a voltage swing of 10 V shows that the resolving capability

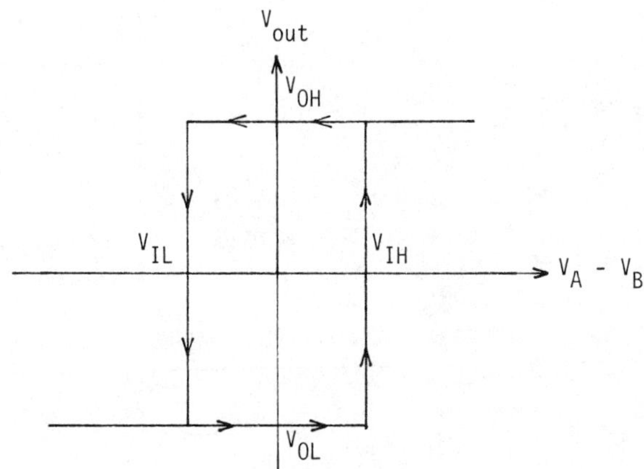

Fig. 6.2–2. Voltage comparator with hysteresis.

of an MOS comparator will vary from 1 mV to 10 mV. This assumes that the offset of the op amp itself is negligible, which is generally not the case. Another requirement of the open loop comparator is that the common mode-rejection ratio must be large, so that the comparator has the same resolving capability when the reference voltage (V_B in Fig. 6.2–1(a)) is at zero or near V_{OH} or V_{OL}.

The offset voltage of the op amp can be reduced by the use of a cancellation technique using switches and a capacitor. One possible circuit is shown in Fig. 6.2–3(a). V_{OS} represents the input offset voltage of the op amp (see Chapter 8). During the ϕ_2 phase period, the capacitor C is charged to V_{OS} with the polarity shown. During the ϕ_1 phase period, C provides a voltage between V_A and the minus terminal of the op amp that cancels V_{OS}. Though

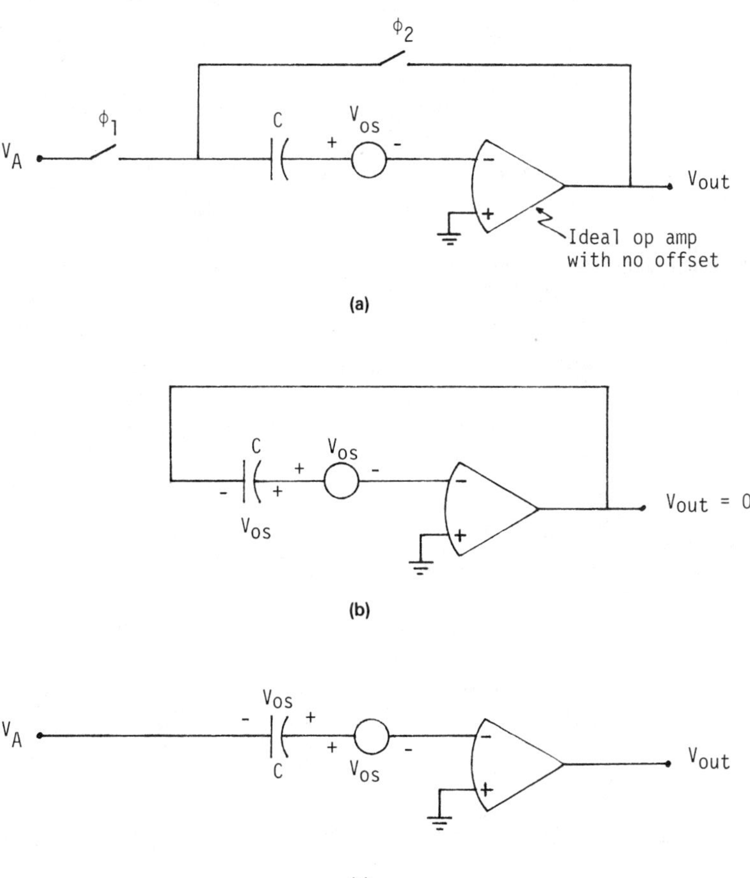

Fig. 6.2–3. (a) Circuit to cancel the effects of V_{os}. (b) Phase ϕ_2. (c) Phase ϕ_1.

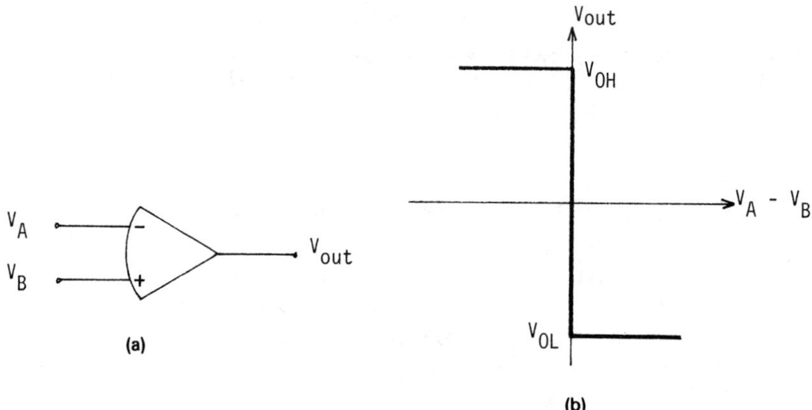

Fig. 6.2–4. (a) Inverting open-loop voltage comparator and (b) its transfer characteristics.

this method greatly reduces the offset of the op amp, second-order effects prevent perfect cancellation.

An inverting comparator using an op amp is shown in Fig. 6.2–4(a) and is obtained by reversing the terminals of Fig. 6.2–1(a). The transfer characteristic of this comparator is shown in Fig. 6.2–4(b). An inverting comparator with hysteresis is shown in Fig. 6.2–5(a). To see how this comparator works, let us assume that V_A is less than V_{IH} and that V_{out} is V_{OH}. Therefore V_B is equal to $V_{IH} = V_{OH}\ R_1/(R_1 + R_2)$. When V_A exceeds this value, the output goes to V_{OL}. With $V_{out} = V_{OL}$, V_B is now $V_{IL} = V_{OL}\ R_1/(R_1 + R_2)$. If V_{OL} is less than V_{OH}, then V_{IL} is less than V_{IH}. When V_A goes below V_{IL}, then the output switches back to V_{OH}. We see that R_1 and R_2 can be

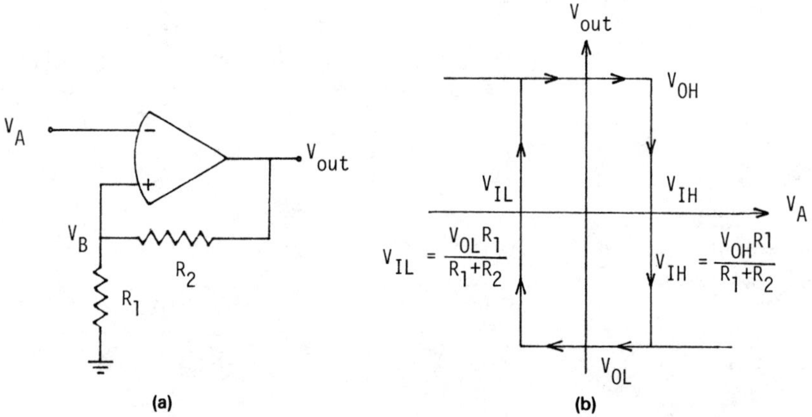

Fig. 6.2–5. (a) Inverting comparator with hysteresis. (b) Transfer characteristic.

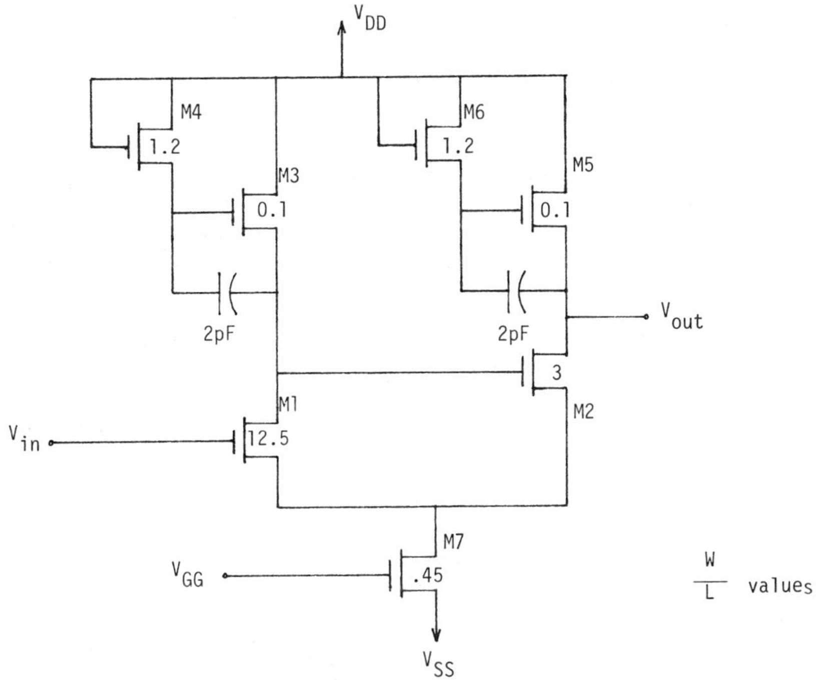

Fig. 6.2–6. NMOS Schmitt trigger.

used to determine the hysteresis width. It is a good practice to limit the output swings of the amp if possible. The threshold of Fig. 6.2–5(a) can be shifted by applying the grounded end of R_1 to a reference voltage (see Problem 6.5).

Other methods of realizing the open loop comparator have been used in MOS technology. They include the use of a chopper amplifier, which reduces the offset problem, and a dynamic, offset-nulled comparator.[9,10]

The second class of comparators consists of regenerative circuits. Regenerative circuits have the advantage that large gain is not required. Instead, the change of states occurs through the use of positive feedback. Figure 6.2–6 shows a Schmitt trigger that uses enhancement NMOS transistors. The devices M3 and M4 and M5 and M6 serve as loads for M1 and M2, respectively. The 2pF capacitor is a means of providing extra gate drive for M3 or M5,

---

[9] A. B. Hamade, "A Single Chip ALL-MOS 8-Bit A/D Converter," *IEEE J. of Solid-State Circuits,* Vol. SC-13, No. 6, December 1978, pp. 785–791.

[10] B. Fotouhi and D. A. Hodges, "High-Resolution A/D Conversion in MOS/LSI," *IEEE J. of Solid-State Circuits,* Vol. SC-14, No. 6, December 1979, pp. 920–926.

permitting it to bring the drain of M1 or M2 almost up to V_{DD}. The operation of the circuit can be described as follows. Assume that V_{in} is very large and negative. Therefore M1 is off and M3 will be on. The output voltage will be determined by the on resistances of M2, M5, and M7. Let us designate this level as V_{OL}. As V_{in} is increased, there will be some voltage V_{IH} where M1 begins to turn on. As M1 turns on, M2 turns off, and V_{out} is pulled to V_{DD}. This transition is not regenerative. However, once M1 is on, then as V_{in} is decreased it will reach a point where M1 begins to turn off. As M1 turns off, M2 turns on regeneratively through the source-coupled feedback path of M1 and M2. The transfer characteristics of Fig. 6.2–6 are shown in Fig. 6.2–7 and are for the device sizes and voltages indicated. Although there are many variations of the Schmitt trigger, the principle of operation is similar to that of Fig. 6.2–6.

Another type of regenerative comparator is based on the method used to sense the state of dynamic memories. This method is called a *sense amplifier* and consists of a strobed flip-flop. Figure 6.2–8 shows a flip-flop with storage capacitors to ground at each output of the flip-flop. The circuit works as follows. Assume that the strobe is low, so that M5 is off. The flip-flop consisting of M1 through M4 is deactivated, since no current can flow through any of the devices. This mode is called the *memory mode,* because the capacitors C_1 and C_2 will "remember" the states of V_1 and V_2. During this time C_1 and C_2 can also be changed to new values of V_1 and V_2, respectively. When the strobe turns on the flip-flop, then it will go to the state corresponding to the values of V_1 and V_2. For example, if V_1 is greater than V_2, then when the strobe (clock) pulse is applied to M5, the voltage at the gate of

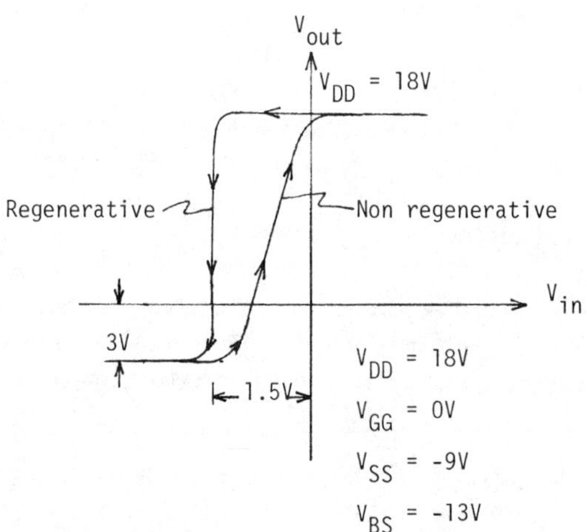

Fig. 6.2–7. Voltage transfer characteristics of Fig. 6.2–6.

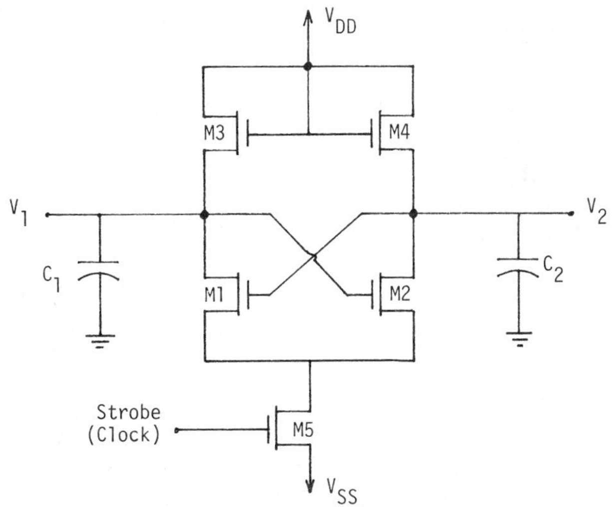

Fig. 6.2–8. Strobed flip-flop.

M2 will be higher than the voltage at the gate of M1. If the devices are matched, then the flip-flop will regeneratively switch to the state with V_1 high and V_2 low. This circuit can detect differences in V_1 and V_2 to within 5–10 mV. A circuit showing how the clocks are applied and how the outputs are buffered is illustrated in Fig. 6.2–9. During the sample mode, ϕ_1 is high and ϕ_2 is low. The voltages V_1 and V_2 are applied to C_1 and C_2. During the compare mode, ϕ_1 is low and ϕ_2 is high. If V_2 is greater than V_1 then Q is true. If V_1 is greater than V_2, then $\overline{Q}$ is true.

A slightly different approach to the strobed comparator of Fig. 6.2–8 is shown in Fig. 6.2–10.[11] This circuit uses a differential amplifier to provide gain to the sensed voltages and to provide common mode rejection. M6 and M7 are long devices that apply the output of the differential amplifier, consisting of M1, M2, M3, M4, and M5, to the flip-flop and also serve as loads. The state of the flip-flop is amplified by M10 through M13 when the comparator is strobed. This comparator uses 1 mW of power, requires approximately 90 mils^2 of area, and can detect thresholds lower than 1 mV. Unfortunately it is sensitive to offset voltages.

A third method of realizing comparators compatible with MOS technology is called the *charge-differencing, or charge-balancing, method.*[12] To illustrate

---

[11] A. Yukawo, "A Highly Sensitive Strobed Comparator," *IEEE J. of Solid-State Circuits,* Vol SC-16, No. 2, April 1981, pp. 109–113.

[12] Y. S. Yee, L. M. Terman, and L. G. Heller, "A 1 mV MOS Comparator," *IEEE J. of Solid-State Circuits,* Vol SC-13, No. 3, June 1978, pp. 294–298.

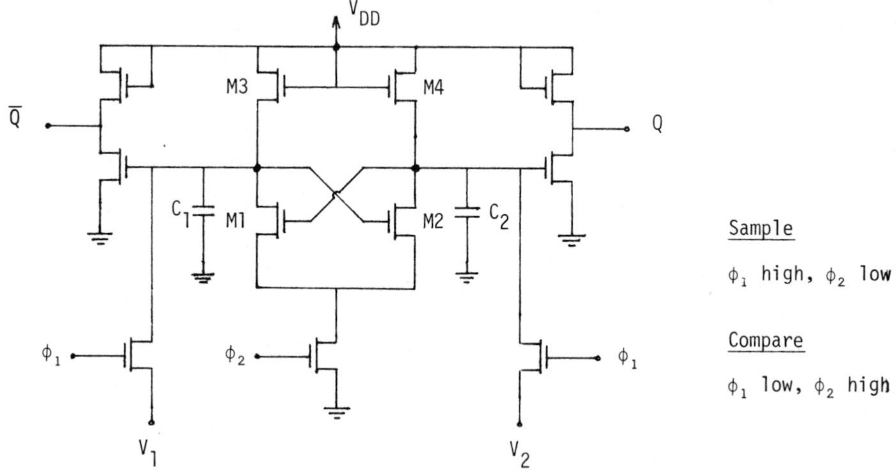

Fig. 6.2–9. Sense amplifier used as a comparator.

the principle of operation, consider Fig. 6.2–11. V_A and V_B are the two input signals applied to the comparator. C_1 and C_2 are the sampling capacitors, and C_p is a parasitic capacitor. With the clocking sequence shown, at the end of ϕ_2, V_0 is given as

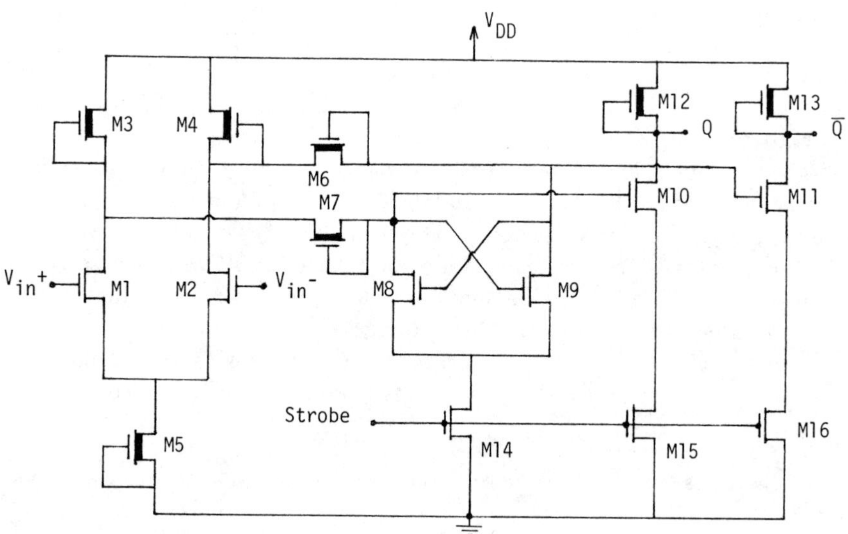

Fig. 6.2–10. Strobed comparator.

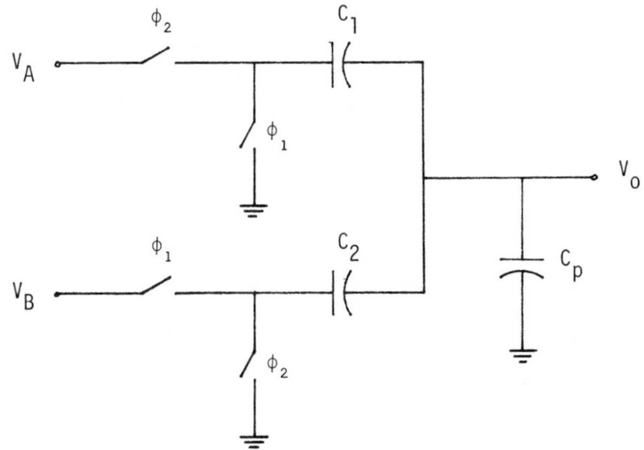

Fig. 6.2–11. Charge balancing comparator.

$$V_0 = \frac{C_1}{C_1 + C_2 + C_p} V_A - \frac{C_2}{C_1 + C_2 + C_p} V_B = \frac{(V_A - V_B)}{2 + \dfrac{C_p}{C}} \tag{3}$$

if $C_1 = C_2 = C$. To minimize the effects of C_p, C must be much greater than C_p. Because of the small value of V_0, it is typically applied to an amplifier to enhance the threshold-detection capability. The charge-balancing method is compatible with switched capacitor circuits, but suffers from a poor common-mode-rejection ratio and is sensitive to power supply changes.

The design and function of a comparator can become very important in many MOS analog circuits. The comparator is an important component in the following section and in the following chapter on analog-to-digital and digital-to-analog converters. Not only must the comparator have a very small threshold-detection capability, but this capability must be maintained over large common-mode voltage ranges. In the work that follows, the comparator will be given the symbol of Fig. 6.2–1, although it could be implemented by any of the methods discussed in this section.

6.3 RECTIFIERS AND NONLINEAR CIRCUITS

In Chapter 2 we developed many circuits that are useful as building blocks in analog sampled data circuit design. These circuits included amplifiers, integrators, differentiators, and general first-order networks. However, these circuits were all linear, and there are many applications that require the

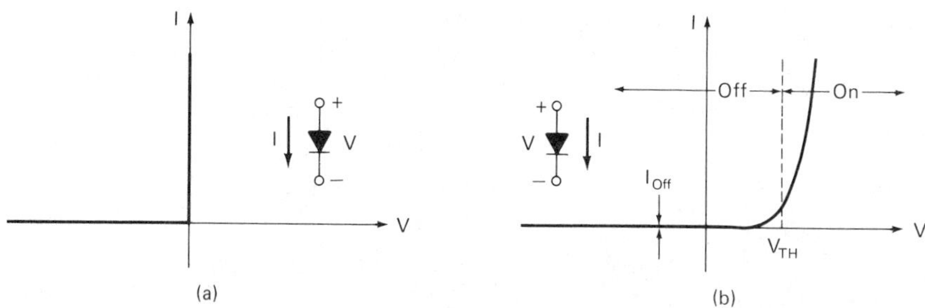

Fig. 6.3–1. (a) Ideal diode characteristics. (b) PN silicon diode characteristics.

ability to rectify.[13,14] It is the purpose of this section to develop MOS analog sampled data circuits that can rectify and to use these circuits to do nonlinear signal-processing.

Figure 6.3–1(a) shows the voltage-current characteristics of an ideal diode. When V is greater than zero the diode is a short circuit, and when V is less than zero the diode is an open circuit. An actual silicon pn diode is shown in Fig. 6.3–1(b). We see that the actual diode has an arbitrary voltage V_{TH} at which for V greater than V_{TH}, we assume the pn diode is on, and for V less than V_{TH}, we assume the pn diode is off. In the on region we observe that there is a nonzero resistance that is nonlinear. In the off region, the resistance is very large, but is in parallel with a small current, I_{OFF}.

These imperfections of the pn diode have been removed by the use of the op amp and negative feedback. Figure 6.3–2(a) shows a continuous realization of what is called a *precision diode circuit.* In reality it is a circuit that realizes the voltage transfer function shown in Fig. 6.3–2(b) containing a breakpoint. The circuit works in the following manner. When V_{in} is less than zero, a negative voltage is applied to the inverting terminal of the op amp, causing the output voltage, V_A, to be very large and positive. Therefore, D_1 is on and D_2 is off. As a result, the circuit has negative feedback from V_A through the diode D_1 and R_2 back to the inverting input. If the actual output is taken at V_{out}, then the diode D_1 is inside the feedback loop, and its turn-on voltage V_{TH} is divided by the loop gain. As a result, the breakpoint is very precise. The transfer function of the circuit when $V_{in} < 0$ is given as

$$V_{out} = -(R_2/R_1)\ V_{in} \tag{1}$$

---

[13] J. Wait, L. Huelsman, and G. Korn, *Introduction to Operational Amplifier Theory and Applications,* Chap. 3, McGraw-Hill, New York, 1975.

[14] D. Sheingold, *Nonlinear Circuits Handbook,* Analog Devices, Inc., Norwood, MA 02062, 1974.

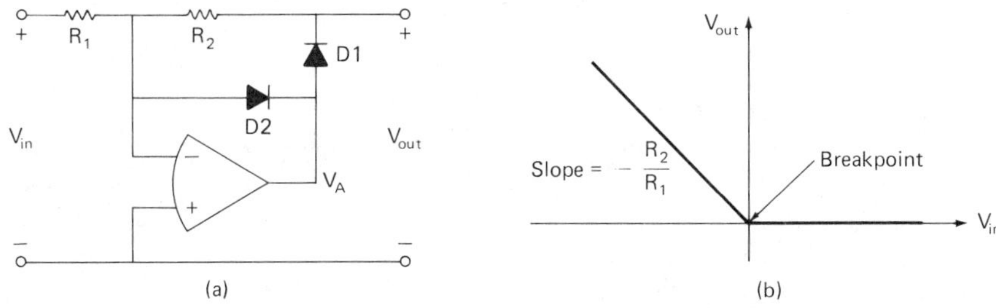

Fig. 6.3–2. (a) Precision diode circuit. (b) Transfer function characteristics.

When V_{in} becomes positive, then D_1 turns off and D_2 turns on. The output voltage of the op amp is at approximately V_{TH} of D_2, but the output voltage of the circuit is zero because no current is flowing in R_2 (assuming the load requires no current) and the left-hand side of R_2 is connected to a virtual ground. Diode D_2 can be eliminated, but serves a very useful purpose in keeping the op amp from being in an open loop state during the time when V_{in} is greater than zero.

Figure 6.3–3 shows how the precision diode may be reversed by reversing the diode directions. The breakpoint may be shifted to the left or the right by summing in a negative or positive voltage, respectively, to the inverting input of the operational amplifier, as shown in Fig. 6.3–4. The output voltage in Fig. 6.3–4 is given as

$$V_{out} = \begin{cases} -\dfrac{R_2}{R_1} V_{in} + \dfrac{E_R R_2}{R_3}, & V_{in} > -E_R \dfrac{R_1}{R_3} \\ 0 \quad , & V_{in} < -E_R \dfrac{R_1}{R_3} \end{cases} \tag{2}$$

where E_R is a reference voltage, and V_{in} is the input signal to the comparator. The slope of the transfer function may be changed by cascading any of the above circuits with an inverter.

What we seek now is an analog sampled data equivalent of the precision

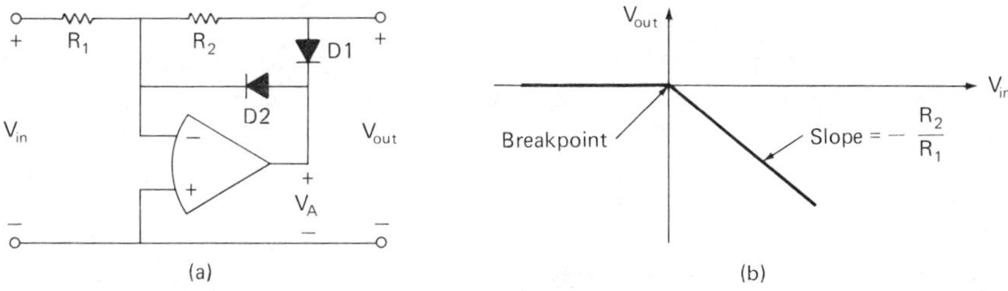

Fig. 6.3–3. (a) Precision diode which is inverting and on in the 4th quadrant. (b) Transfer characteristics of (a).

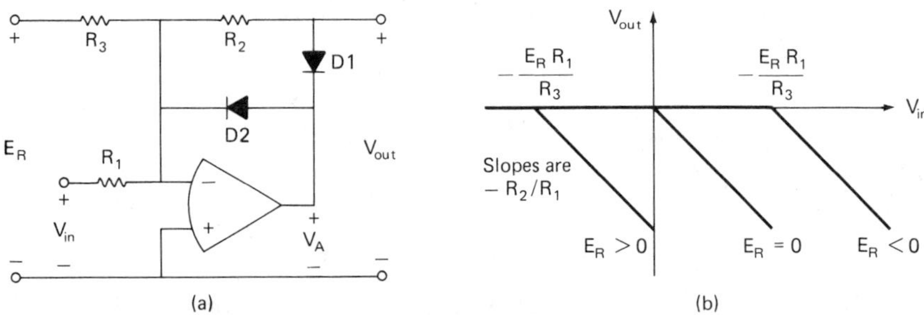

Fig. 6.3–4. (a) Method of shifting the breakpoint. (b) Transfer characteristics of (a).

diode. Such a realization is shown in Fig. 6.3–5(a).[15] This circuit requires both an op amp and a comparator. The method of operation is very simple, as will now be shown. We first note that C_1 is essentially a parallel switched resistor realization, except that C_1 is connected to ground through the NMOS transistor M1. If M1 were on, the transfer function of this circuit would be

$$V_{out} = -(C_1/C_2)\, V_{in}\, z^{-1/2} \tag{3}$$

When the voltage V_i is greater than zero, the comparator output is positive and will be at V_{OH}. When the voltage V_i is less than zero, the output voltage of the comparator will be negative and at V_{OL}. Therefore, when V_{in} is positive, M1 will be off, and the effective value of C_1 is zero. Thus from Eq. (3) we get $V_{out} = 0$. However, when V_{in} is negative, M1 is on, and the effective value of C_1 is C_1, and the output voltage is given by Eq. (3). As a result, we obtain the transfer function of Fig. 6.3–5(b), which is expressed as

$$V_{out} = \begin{cases} 0 \quad , & V_{in} > 0 \\ -\dfrac{C_1}{C_2} V_{in} z^{-1/2}, & V_{in} < 0 \end{cases} \tag{4}$$

The quadrant in which Eq. (3) is valid can be reversed by reversing the polarity of the comparator, as shown in Fig. 6.3–6. The sharpness of the breakpoint depends upon the gain of the comparator, but will typically be very sharp.

The SC precision diode circuit can have the breakpoint shifted to the

[15] B. J. Hosticka and G. S. Moschytz, "Practical Design of Switched-Capacitor Networks for Integrated Circuit Implementation," IEE *Electronic Circuits and Systems,* Vol. 3, No. 2, March 1979, pp. 76–88.

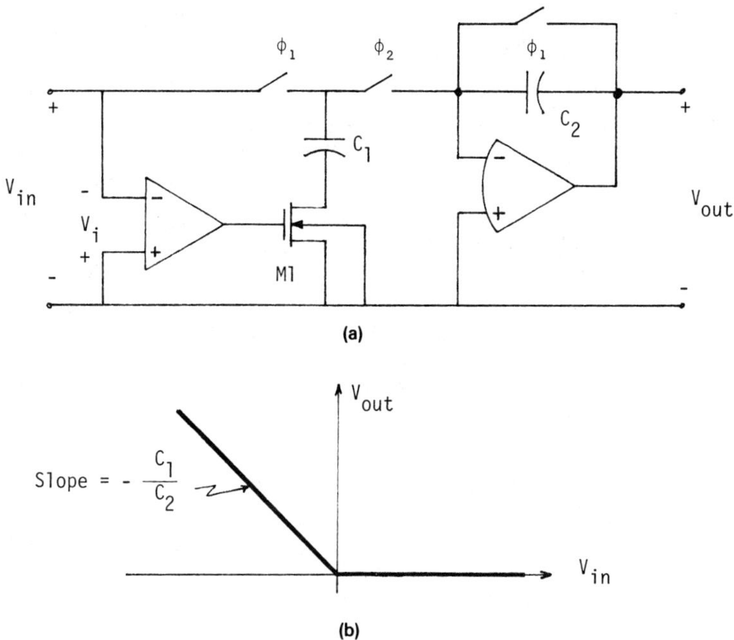

Fig. 6.3–5. (a) SC realization of Fig. 6.3–2a. (b) Transfer characteristics of (a).

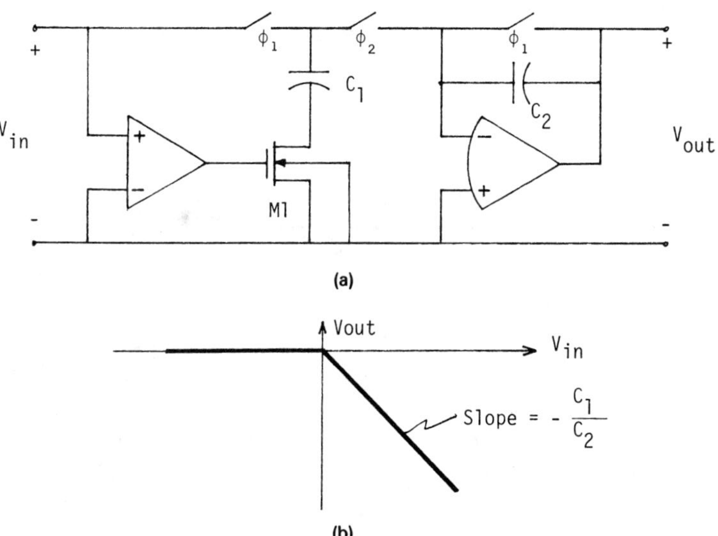

Fig. 6.3–6. (a) SC realization of Fig. 6.3–3a. (b) Transfer characteristics of (a).

left or right by using the circuit shown in Fig. 6.3–7(a), where a voltage of E_R has been applied to the noninverting terminal of the comparator. We note when E_R is not zero that the output will jump when the comparator switches. Therefore, it is necessary to add the additional circuitry shown to cancel this jump and to achieve the characteristics of Fig. 6.3–7(b). Although one could cascade the realization of Figs. 6.3–5(a), 6.3–6(a), or 6.3–7(a) with an inverter to change the quadrant in which Eq. (3) holds, a different method exists, which we shall present. Figure 6.3–8(a) shows a noninverting amplifier with M1 in series. When V_{in} is greater than zero, M1 is on, and the transfer function of this circuit is

$$V_{out} = (C_1/C_2)\ V_{in}z^{-1/2} \tag{5}$$

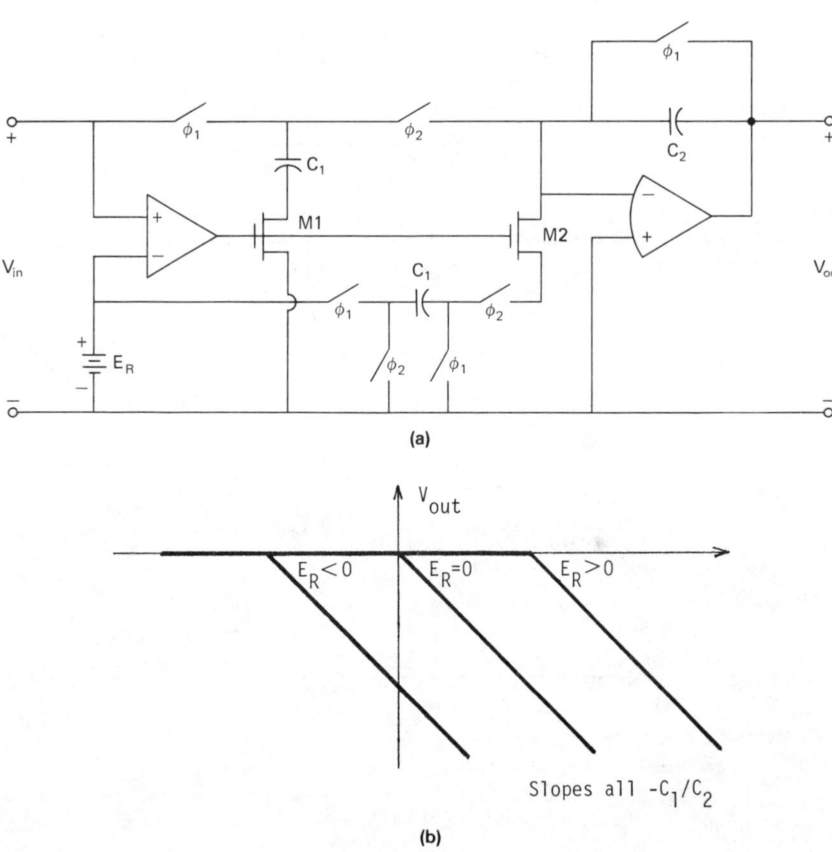

Fig. 6.3–7. (a) SC precision diode realization with a method of shifting the break point. (b) Transfer characteristics of (a) for various values of E_R.

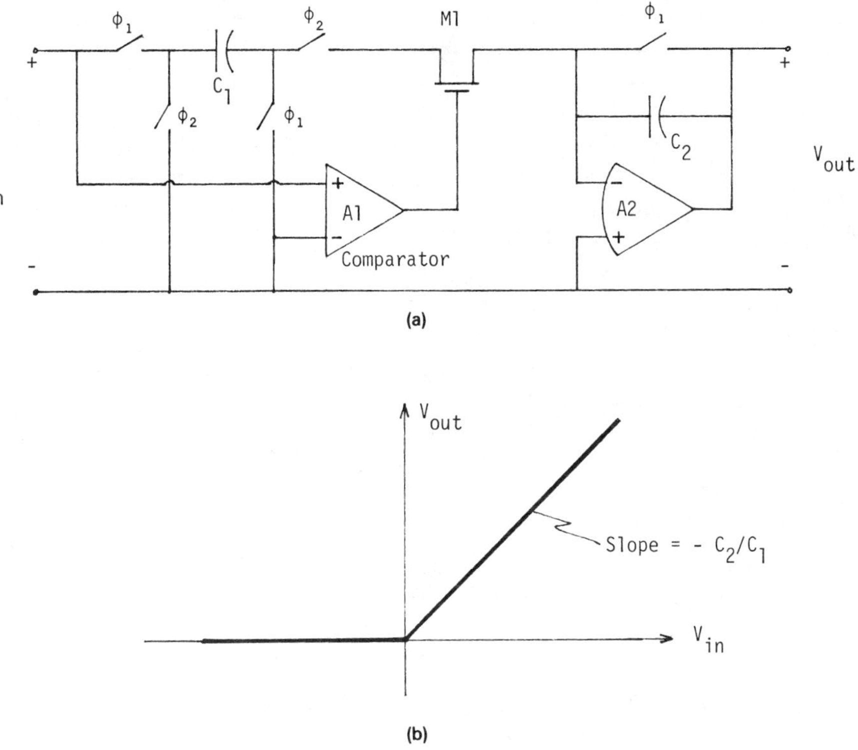

Fig. 6.3–8. (a) Non-inverting SC precision diode. (b) Transfer characteristics of (a).

When V_{in} is less than zero, V_{out} is zero because M1 is off. As a result, we obtain the transfer function of Fig. 6.3–8(a) as

$$V_{out} = \begin{cases} (C_1/C_2)\, V_{in} z^{-1/2}, & V_{in} > 0 \\ 0 \quad , & V_{in} < 0 \end{cases} \tag{6}$$

If the polarity of the comparator is reversed, then Eq. (5) is valid in the third quadrant.

Typically we do not like to use a series switch because the terminal potentials are never well defined. In Fig. 6.3–8(a) however, if the source is connected to the inverting terminal of the op amp, then it is equivalent to the grounded source realization. Thus the comparator output voltages can control M1 on a well-defined basis. The grounded terminal of the comparator can be applied to a potential for purposes of shifting the breakpoint, as we have seen previously in Fig. 6.3–7(a).

The diode circuits developed here give us a very flexible set of tools with which to do waveshaping. Many applications require only a single breakpoint, as will be illustrated by the following example.

Example 6.3–1. *SC absolute-value detector.* The methods of this section are to be used to realize the transfer function of Fig. 6.3–9(a). To develop the realization, let us decompose Fig. 6.3–9(a) into that shown in Fig. 6.3–9(b). Here we can see that the realization consists of a linear noninverting amplifier of unity gain summed with a precision diode of the type of Fig. 6.3–5(a) with a gain of −2. The resulting realization is shown in Fig. 6.3–9(c). When $V_{in} > 0$, then M1 is off and the gain is +1. When $V_{in} < 0$, M1 is on and the gain is the sum of + 1 and −2, giving an overall gain of −1. If the comparator polarity is reversed, then Fig. 6.3–9(c) is flipped about the V_{in} axis, and V_{out} is always less than zero.

In some applications, multiple breakpoints are desired. One can increase the number of breakpoints in the manner shown in Fig. 6.3–10(a), where each breakpoint not at $V_{in} = 0$ requires the additional circuitry shown to offset the jump caused by a nonzero value of V_{in}. The ith breakpoint can be adjusted by E_i, and the slope of the line for $-E_{i-1} < V_{in} < E_i$ is given as

$$\begin{array}{c}\text{Slope of}\\ i\text{th segment}\end{array} = -\frac{\sum_{k=1}^{i} C_k}{C_o}, \quad -E_{i-1} < V_{in} < -E_i \tag{7}$$

Reversing the polarity of all the comparators and changing the polarities of the reference voltages will move the characteristic in the second quadrant of Fig. 6.3–10(b) to the fourth quadrant, as shown by the dotted curve in Fig. 6.3–10(b). An inverter can be used to achieve operation in the first or third quadrants.

All the above multiple segment realizations have a monotonically increasing slope. A monotonically decreasing slope realization is given in Fig. 6.3–11(a). The decreasing slope is achieved by opposing the charge transfer of C_i by smaller, oppositely phased capacitors C_1, C_2, · · · . Again the ith breakpoint can be adjusted by E_i, and the slope of the ith line segment is given as

$$\begin{array}{c}\text{Slope of}\\ i\text{th segment}\end{array} = \left[\frac{\sum_{k=0}^{i} C_k}{C_i}\right]^{-1}, \quad -E_{i-1} < V_{in} < E_i \tag{8}$$

Figure 6.3–11(b) shows the transfer function of Figure 6.3–11(a). It is seen that C_0 causes the curve to go through the origin with a nonzero slope. This characteristic could be achieved for Fig. 6.3–10(b) by adding a capacitor

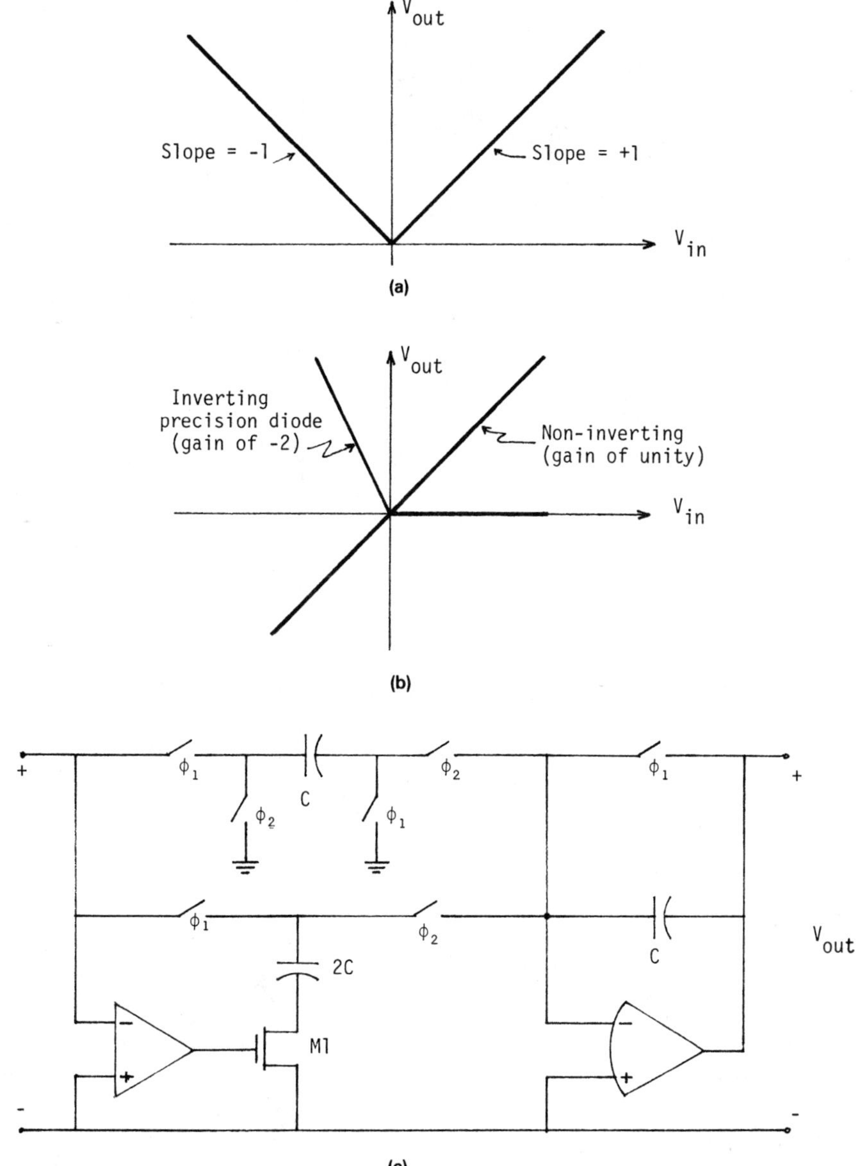

Fig. 6.3–9. (a) Transfer function for Ex. 6.3–1. (b) Decomposition of (a). (c) Realization of (a).

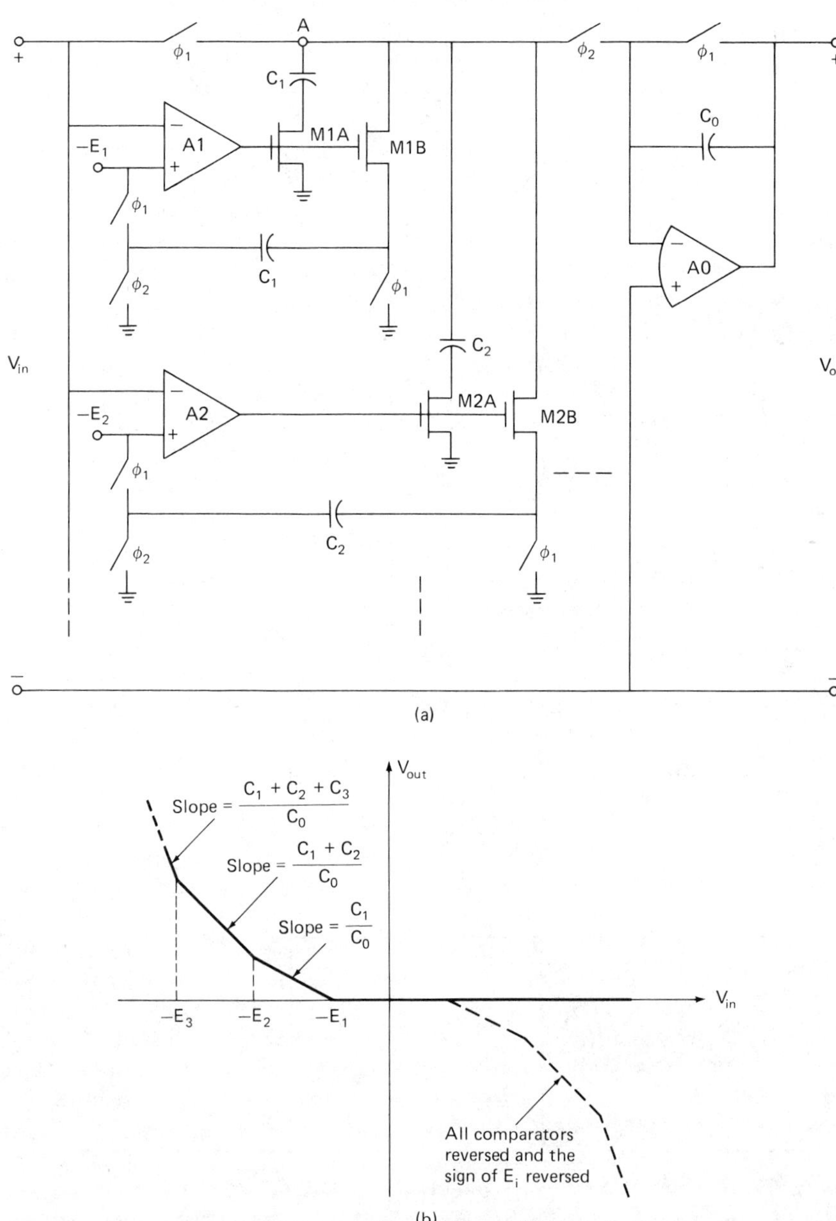

Fig. 6.3–10. (a) Monotonically increasing slope realization. (b) Transfer characteristics of (a).

from point A to ground in Fig. 6.3–10(a). Changing all comparator polarities and using an inverter will permit one to achieve a monotonically decreasing slope in any quadrant. Combinations of the previous techniques can be used to realize any general-shaped transfer function.

Example 6.3–2. *A triangle-to-sine converter.* An 18-volt peak-to-peak triangle waveform is to be converted to a 20-volt peak-to-peak sinusoid using Fig.

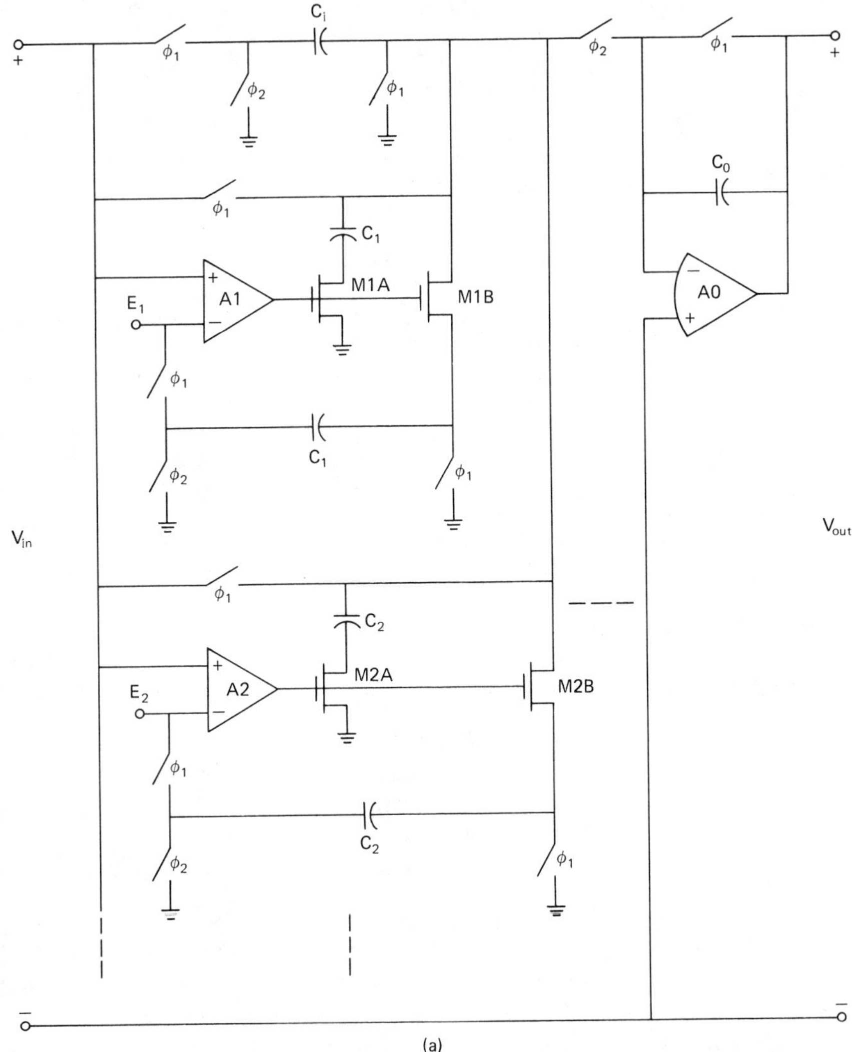

Fig. 6.3–11. (a) Realization of a multi-segment monotonically decreasing transfer curve.

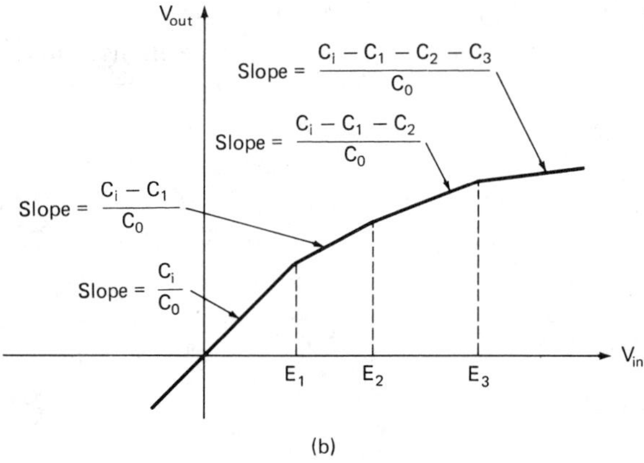

Fig. 6.3–11. (b) Transfer characteristics of (a).

6.3–11(a) with a seven-segment approximation. From 0 to $T/4$, the input waveform is

$$V_{in}(t) = \frac{36}{T}t = \frac{18}{\pi}\omega t, \quad 0 < t < \frac{T}{4} \tag{9}$$

At $t = T/4$, V_{out} must be 10. Therefore we can write V_{out} as

$$V_{out}(t) = 10 \sin\left[\frac{\pi}{18} V_{in}(t)\right] = 10 \sin \omega t \tag{10}$$

Plotting V_{out} (t) versus V_{in} (t) for values of t from 0 $T/4$ results in Fig. 6.3–12(a). Symmetry will be used to simplify our considerations. One possible set of breakpoints and slopes is given below.

Segment	*Breakpoint*	*Slope*
0	–	1.745
1	2.58 volts	1.274
2	5.73 volts	.645
3	7.62 volts	.166

The 0 segment goes through the origin and therefore does not have a breakpoint. The other three segments have the same slope as above, but the breakpoints are negative. The seven-segment approximation is plotted in Fig. 6.3–12(a). A realization of the seven-segment approximation is given in Fig. 6.3–12(b). At 1 kHz, the total harmonic distortion was 2.6%.

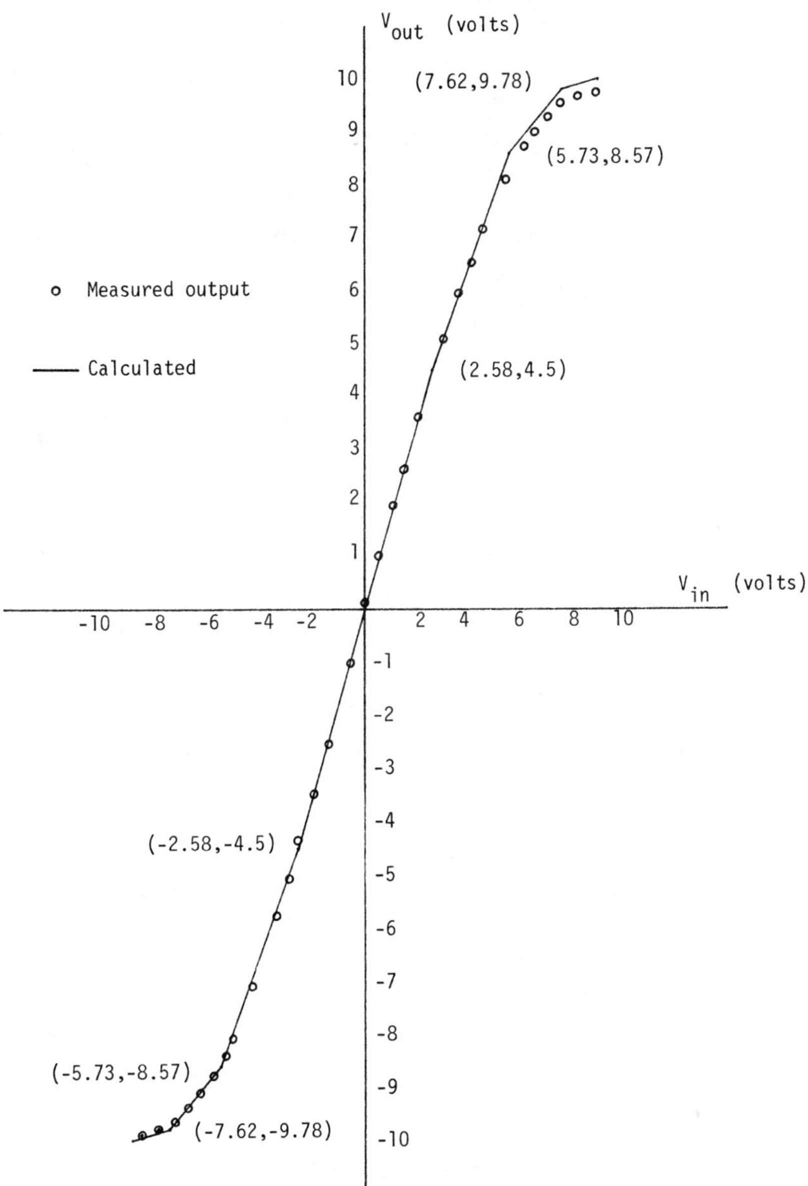

Fig. 6.3–12a. Seven segment approximation to $V_{out} = 10 \sin (\pi/18\ V_{in})$.

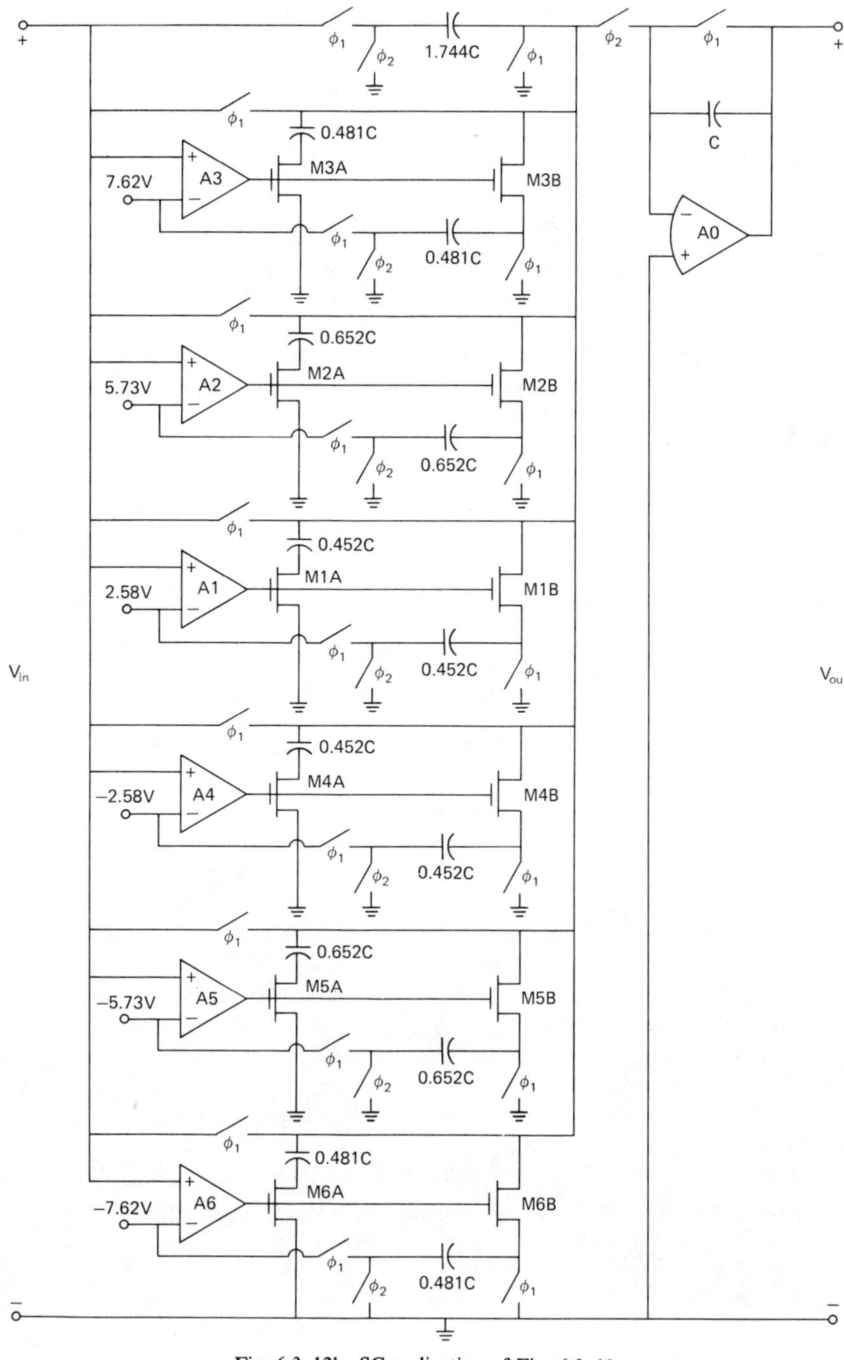

Fig. 6.3–12b. SC realization of Fig. 6.3–12a.

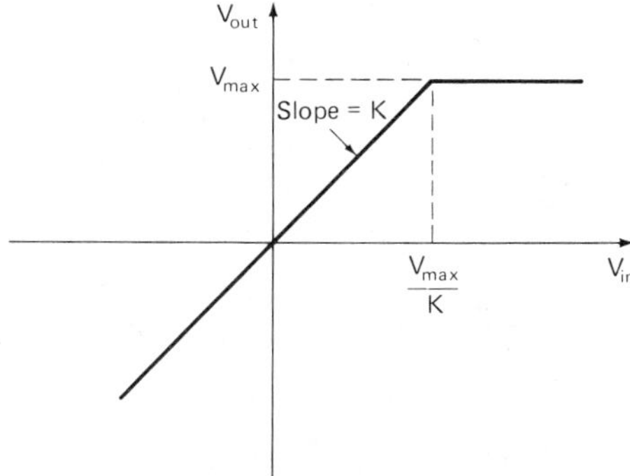

Fig. 6.3–13. Illustration of output signal limiting.

In many cases it is desired to limit the signal as illustrated in Fig. 6.3–13. Unfortunately, this requirement does not work well with SC circuits. The reason is that V_{out} determines when the gain is to be reduced to zero. It may turn out that V_{max} is reached during a charge transfer that has not been completed. It is very difficult to halt the charge transfer so that $V_{out} = V_{max}$ can be maintained. One method that can be used to limit the output is to modify Fig. 6.3–11(a). If C_i/C_o is equal to K, then set $E_1 = V_{max}/K$, and pick C_1 very large. Under these conditions the curve of Fig. 6.3–13 will be approximated. Unfortunately, to maintain $V_{out} = V_{max}$ for values of V_{in} greater than V_{max}/K requires a large capacitor and therefore large area. In many cases hard limiting is not necessary, so that this is a satisfactory approach.

The ability to approximate a transfer function by piecewise linear methods is a very important capability. Unfortunately, SC methods of accomplishing this require two op amps (assuming the comparator is roughly the equivalent of an op amp), whereas continuous methods require only one. However, the full capability of using SC methods for realizing nonlinear transfer functions has not yet been explored. For example, it may be possible to multiplex one op amp to do the job of both in Fig. 6.3–5(a).

6.4 MODULATORS AND MULTIPLIERS

The analog modulator or multiplier has found many areas of applications in recent years.[16] A multiplier is defined as a circuit that can provide the

[16] D. H. Sheingold, *Nonlinear Circuits Handbook,* Analog Devices, Inc., Norwood, MA, 02062, 1974.

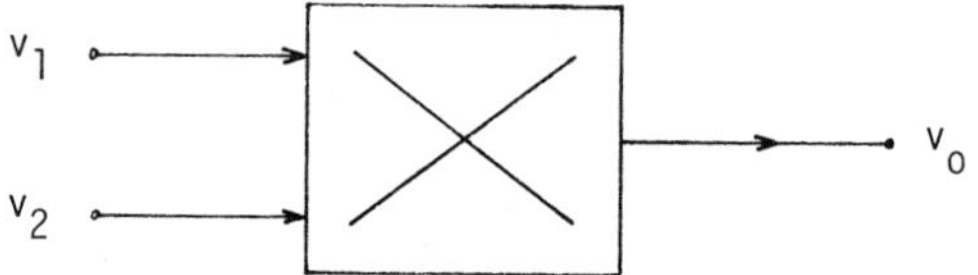

Fig. 6.4–1. Schematic for a multiplier.

product of two signals. If Fig. 6.4–1 represents the symbol for a multiplier, then we may characterize a multiplier as

$$v_o = K_m\, v_1 v_2 \tag{1}$$

where K_m is a multiplying constant, and V_1 and V_2 are the two input voltage signals that are to be multiplied. If the input signals have a range of $-V_{max}$ to $+V_{max}$, then K_m is typically equal to

$$K_m = \frac{1}{V_{max}} \tag{2}$$

so that the output is never larger than V_{max}. As the inputs can be positive or negative, it is necessary to conserve the polarity relationships between the two inputs, such that each of the inputs can be positive or negative and the output will have the polarity determined by Eq. (1). Such a multiplier is called a *four-quadrant multiplier.*

A *modulator* is a special case of a multiplier where the four-quadrant capability of linear multiplication is not necessary. In a modulator, linear response is required of only one input. This input is often called the *modulating input.* The other input is often called the *carrier input.* The modulator is very suitable for SC implementation, which results in many interesting applications of SC circuits.

A simple SC modulator using the ideas of the previous section is shown in Fig. 6.4–2. V_1 is the carrier input, and V_2 is the modulating input. A1 is an op amp, and A2 is a comparator. When V_2 is greater than zero, M1 is off, and the transfer function of the circuit is

$$V_o(z) = z^{-1/2}\, V_1(z) \tag{3}$$

This expression reduces to a gain of unity, if the carrier frequency is much less than the clock frequency. If V_2 is less than zero, then M1 is on, and the transfer function of Fig. 6.4–2 becomes

$$V_o(z) = -z^{-1/2}\, V_1(z) \tag{4}$$

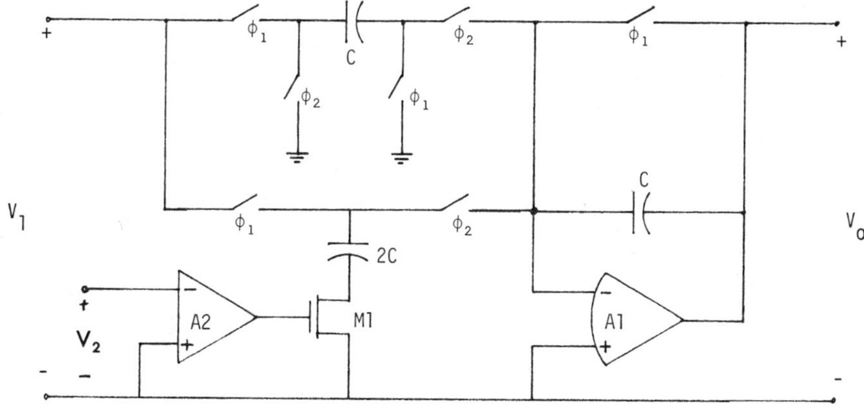

Fig. 6.4–2. A SC realization of a modulator.

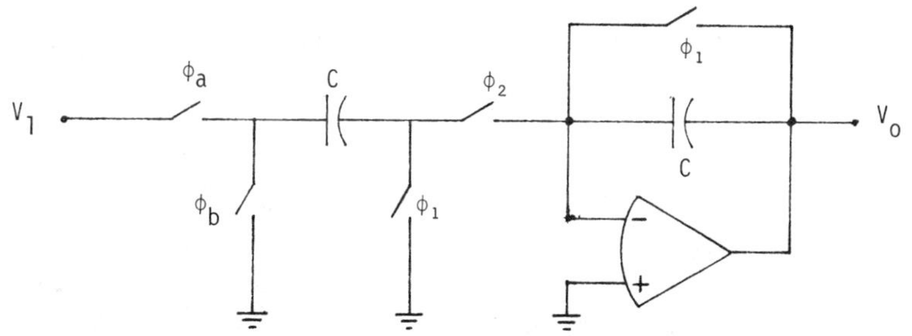

Fig. 6.4–3. SC realization of a modulator using different switch sequences.

The polarity of V_2 causes the carrier to be multiplied by a plus or minus unity value.

Another modulator is shown in Fig. 6.4–3.[17,18] In this modulator if $\phi_a = \phi_1$ and $\phi_b = \phi_2$, Eq. (3) results. If $\phi_a = \phi_2$ and $\phi_b = \phi_1$ Eq. (4) results. Typically a comparator will also be needed to implement the correct switch phasing, as determined by the modulation input, V_2. One possible implementation of Fig. 6.4–3, which uses the threshold of an MOS inverter, is shown in Fig. 6.4–4(a). A simple logic system is used to steer the ϕ_1 and ϕ_2 clocks to the correct switch. Fig. 6.4–4(b) shows a photomicrograph of Fig. 6.4–4(a) implemented by MOS technology.

Figure 6.4–5 shows the waveforms that result from the above circuits. It

[17] J. Bingham, "SC Modulators," Switched-Capacitor Workshop, February 10–12, 1980, Palo Alto, CA.

[18] K. Martin and A. S. Sedra, "Switched-Capacitor Building Blocks for Adaptive Systems," *IEEE Trans. on Circuits and Systems,* Vol. CAS-28, No. 6, June 1981, pp. 576–584.

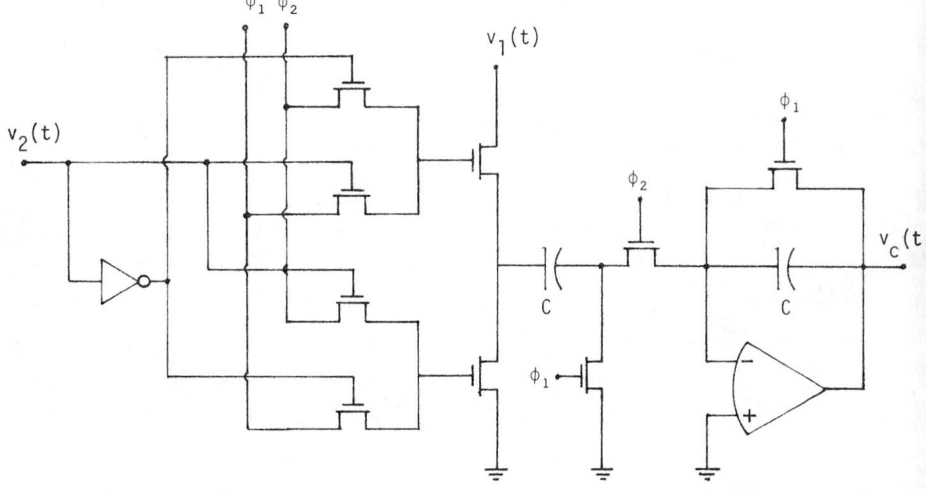

Fig. 6.4–4a. Realization of Fig. 6.4–3 which is compatible with MOS technology.

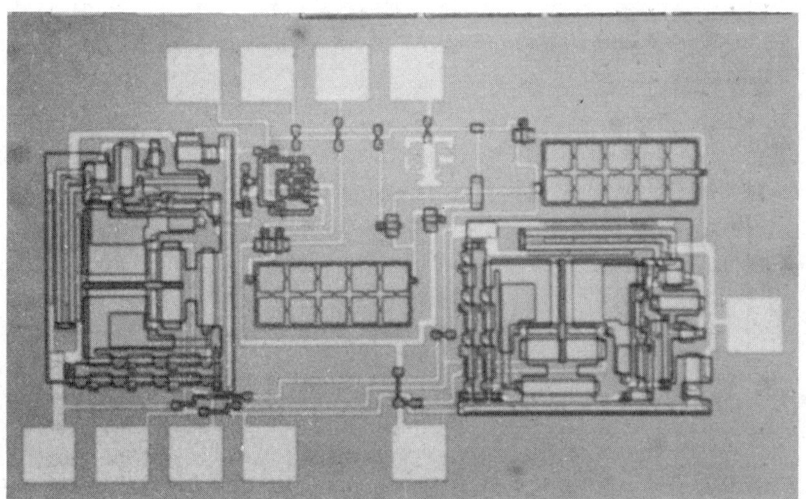

Fig. 6.4–4b. Photomicrograph of Fig. 6.4–4(a) implemented by MOS technology.

is seen that the modulation signal multiplies the carrier by ± 1, depending upon its value with respect to a threshold (usually zero). An important application of the modulator is that of a synchronous demodulator. A block diagram of a synchronous demodulator is shown in Fig. 6.4–6. We shall assume that

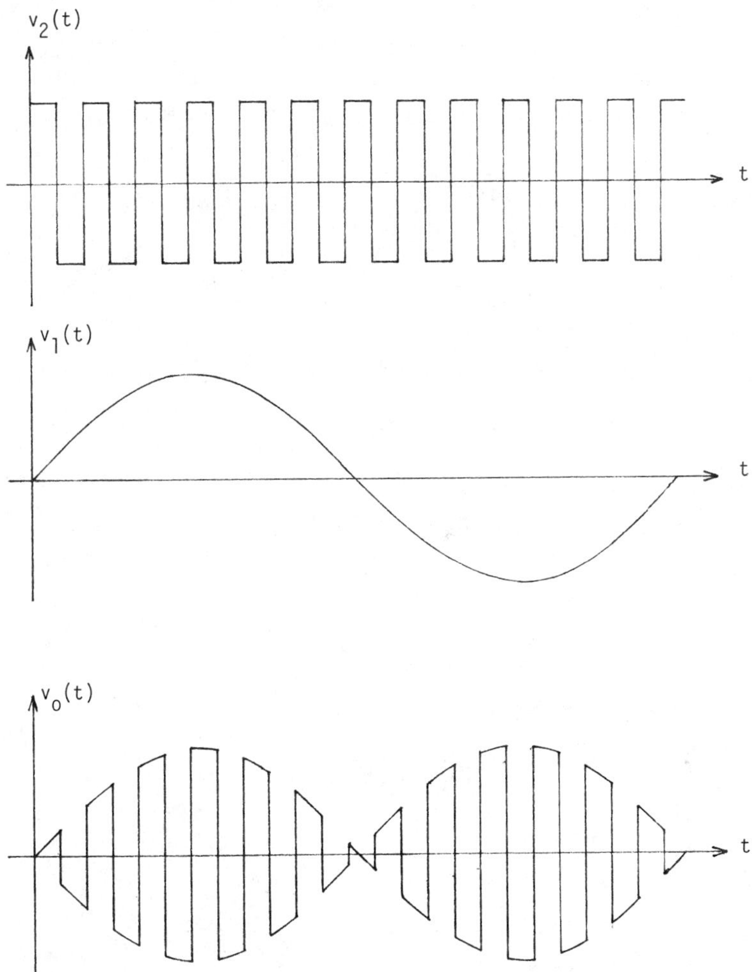

Fig. 6.4–5. Illustration of the waveforms for SC modulators.

the modulator input is a sinusoid, although it makes no difference whether it is a square wave or sine wave. If the carrier is

$$v_c(t) = E_c \cos \omega_c t \tag{5}$$

and the modulation is

$$v_m(t) = E_m \cos \omega_m t \tag{6}$$

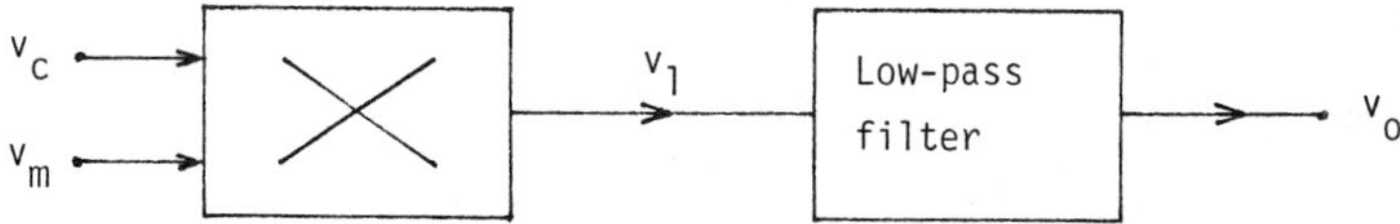

Fig. 6.4–6. Synchronous demodulator.

then $v_1(t)$ can be found to be

$$v_1(t) = K_1 E_m \cos \omega_m t \sum_{n=1}^{\infty} A_n \cos n\omega_c t \tag{7}$$

where K_1 is the modulator gain constant and

$$A_n = E_c \frac{\sin (n\pi/2)}{n\pi/2} \tag{8}$$

If the passband of the low-pass filter is chosen to be approximately equal to ω_c, the resulting output of the system is of the form

$$v_o(t) = \frac{2K_1E_1}{\pi} (\cos \omega_m t)(\cos \omega_c t) \tag{9}$$

where

$$E_1 = E_m E_c \tag{10}$$

Using trigonometric identities, we can express the output waveform of Eq. (9) as

$$v_o(t) = \frac{K_1E_1}{\pi} [\cos (\omega_c + \omega_m)t + \cos (\omega_c - \omega_m)t] \tag{11}$$

Equation (11) implies that, in this type of modulator, the output does not contain any energy at the carrier frequency. Instead, the total energy of the output is concentrated at two discrete frequencies, $(\omega_c + \omega_m)$ and $(\omega_c - \omega_m)$, which are known as the *upper* and *lower sidebands.* Because there is no component of the output at the carrier frequency, this type of modulation is known as *suppressed carrier modulation,* and it comes about from the basic null-suppression property of a modulator.

If the modulating signal contains a DC component, i.e.

$$v_m(t) = E_m(1 + m \cos \omega_m t) \tag{12}$$

then the output is of the form

$$v_o(t) = E_0[\cos \omega_c t + (m/2)\cos(\omega_c + \omega_m)t + (m/2)\cos(\omega_c - \omega_m)t] \quad (13)$$

which corresponds to conventional amplitude modulation (AM). This signal is made up of a carrier component at frequency ω_c, as well as the two symmetrical sideband components at frequencies $(\omega_c + \omega_m)$. The coefficient m is known as the *modulation index* and is restricted to $0 < m < 1$. Figure 6.4–7 shows a realization of Fig. 6.4–6 using SC networks.

In addition to amplitude modulation, balanced modulator circuits can be used for a number of other related applications. Figure 6.4–8 shows a phase-detection circuit. If two input signals at the same frequency ω_c, but having a relative phase difference ϕ, are applied to the inputs, as shown in Fig. 6.4–8, the output of the modulator would be of the form

$$v_o(t) = \frac{K_1 E_1 E_2}{2}[\cos(2\omega_c t + \phi) + \cos \phi] \quad (14)$$

The AC component of Eq. (14) is at twice the carrier frequency and can be readily filtered out by a low-pass filter. The remaining DC component of the output is proportional to the cosine of the phase angle

$$v_o' = V_o(0) = \frac{K_1 E_1 E_2}{2}\cos \phi \quad (15)$$

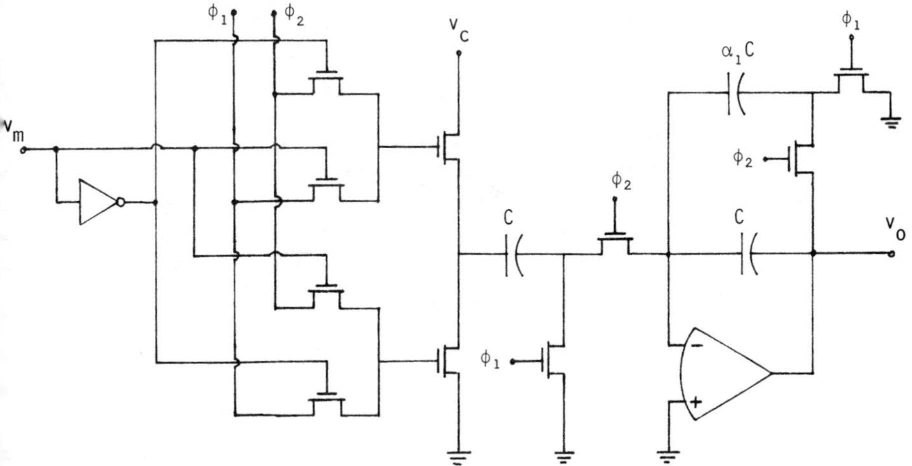

Fig. 6.4–7. SC realization of Fig. 6.4–6.

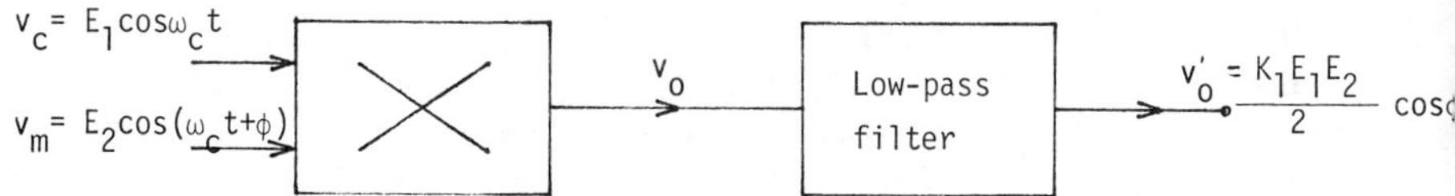

Fig. 6.4–8. Phase detection circuit.

Conversely, if the modulator output is AC-coupled, then the DC term in Eq. (14) is filtered out, resulting in an AC signal at twice the input frequency, and the circuit operates as a frequency-doubler.

Figure 6.4–9 shows the use of a modulator for synchronous AM detection. If an amplitude modulated signal is applied to the linear input of a balanced modulator and an unmodulated carrier is applied to the switching input, neglecting the higher-order harmonics of the switching waveform, one obtains an output of the form

$$v_o(t) = K_1[V_m(t) \cos \omega_c t] [\cos \omega_c t] \tag{16}$$

The output waveform of Eq. (14) can also be written as

$$v_o(t) = \frac{K_1 V_m(t)}{2} [1 + \cos 2\omega_c t] \tag{17}$$

If the high-frequency component of the output at twice the carrier frequency is filtered out by a low-pass filter, the remaining portion of $v_o(t)$ is proportional to the demodulated output

$$v_o' = v_o(t) \Big|_{\substack{\text{LP} \\ \text{filtered}}} = (K_1/2)\, V_m(t) \tag{18}$$

In the case of conventional double-sideband AM signals, the unmodulated carrier reference necessary for synchronous detection can be obtained by passing the AM signal through a "limiter" circuit, which removes the modulation and generates a switching waveform at the carrier frequency. By multiplying this switching waveform with the modulated input, as shown in Fig. 6.4–9, one can extract the demodulated output.

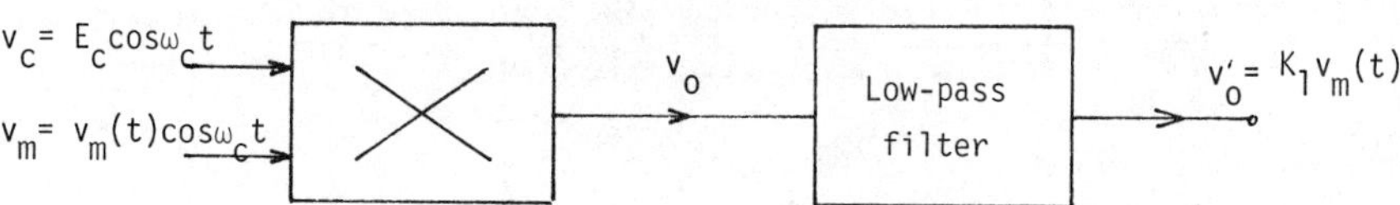

Fig. 6.4–9. Synchronous AM detector.

The phase-detection property of the demodulator circuit given in Eq. (15) can also be used for frequency discriminator applications or for demodulating frequency modulated (FM) signals. For this application the modulated circuits are interconnected with a broadband limiter and a bandpass phase-shift network, as shown in Fig. 6.4–10. The limiter circuit removes all the amplitude modulation from the input FM signal and generates a high-level switching signal $S_1(t)$ at the same frequency as the input. The bandpass phase-shift network is tuned to the carrier frequency ω_c of the input FM signal. It produces a second signal $S_2(t)$ to drive the balanced modulator. If Q is the selectivity of the bandpass network, then the phase shift ϕ introduced by this network for small frequency deviations $\Delta\omega$ in the vicinity of ω_c can be expressed as

$$\phi \approx -\pi/2 \pm 2\,Q\frac{\Delta\omega}{\omega_c} \tag{19}$$

for $(2Q\Delta\omega) << \omega_c$. The low-pass filtered output of the modulator is related to the frequency deviations of the input signal as

$$v_o(t) = \Bigg|_{\substack{\text{LP}\\ \text{filtered}}} \approx K\frac{\Delta\omega}{\omega_c} = K\frac{\Delta f}{f_c} \tag{20}$$

where K is the combined gain of the filter and the detector sections, and $f = (\omega/2\pi)$. Since, in an FM signal, the frequency deviation of the carrier, $\Delta\omega$, represents the information, the output given by Eq. (20) represents the demodulated signal.

In many applications a four-quadrant multiplier having linear multiplication is necessary. Such multipliers have been successfully built as bipolar

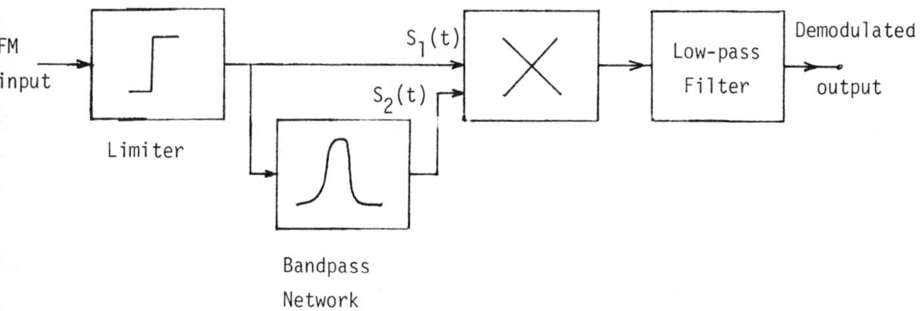

Fig. 6.4–10. Use of a modulator as an FM detector and discriminator.

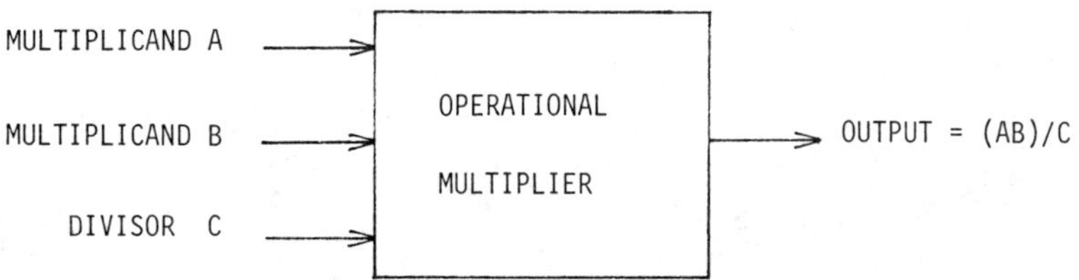

Fig. 6.4–11. Block diagram of an operational multiplier.

integrator circuits and have been widely used.[19,20] Unfortunately, the transconductance of the MOS transistor is too low to permit an MOS version of this very useful circuit, although a CMOS version may be realistic. A better approach is to examine ways to implement an SC multiplier that are compatible with both MOS technology and SC techniques. One such approach is called the *operational multiplier.*[21] The principle of the operational multiplier can be described using the block diagram of Fig. 6.4–11. This principle is simply to simultaneously operate on one of the multiplicands and the divisor in such a manner that the divisor is equal to the remaining multiplicand times a constant. The operated value of the multiplicand will be equal to the product of the multiplicands divided by the divisor. For example, suppose that the operated value of the divisor C is equal to the multiplicand B and is expressed as

$$\text{Operated value of } C = \mathrm{O}(C) = KC = B \tag{21}$$

If the remaining multiplicand, A, is operated on in the same manner as the divisor C, then we can write

$$\text{Operated value of } A = \mathrm{O}(A) = KA \tag{22}$$

From Eq. (19) we see that $K = B/C$. If the output is equal to the operated value of A, then the output can be expressed as

$$\text{Output} = \mathrm{O}(A) = \frac{AB}{C} \tag{23}$$

Thus multiplication of A and B and division by C has been achieved.

[19] B. Gilbert, "A Precise Four-Quadrant Multiplier with Subnanosecond Response," *IEEE J. of Solid-State Circuits,* Vol. SC-3(4), 1968, pp. 365–373.

[20] B. Gilbert, "A New High Performance Monolithic Multiplier Using Active Feedback," *IEEE J. of Solid-State Circuits,* Vol. SC-9, December 1974, pp. 364–373.

[21] P. E. Allen and W. H. Cantrell, "Considerations for an Analog Four Quadrant SC Multiplier," *Proc. of 2nd Caltech Conf. on VLSI,* Jan. 1981, pp. 313–327.

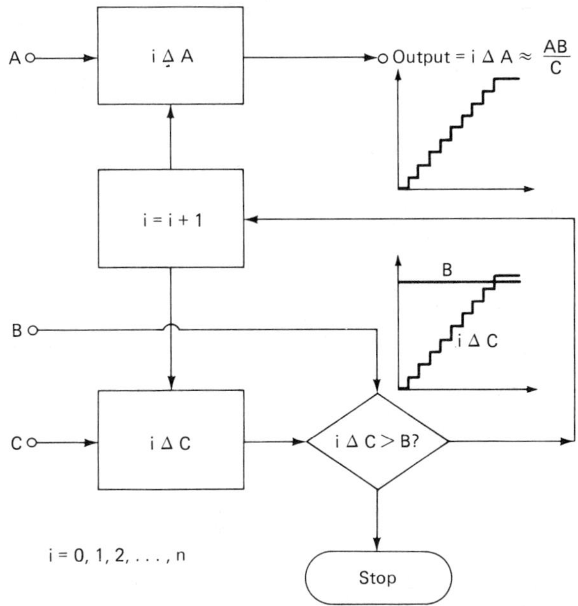

Fig. 6.4–12. Counter implementation of the operational multiplier.

One possible implementation of the operational multiplier is shown in Fig. 6.4–12. This approach is called the *counter approach* and simply adds a small portion of C designated as ΔC until $n\Delta C$ is equal to or greater than B. Simultaneously, ΔA has been accumulated n times using the same value of Δ. The constant K in Eqs. (21) and (22) becomes $n\Delta$. To achieve high accuracy, the incremental constant Δ must be very small. Unfortunately this means that n must be large, and the time required to complete the multiplication becomes too large.

A better approach uses a successive approximation technique and is shown in Fig. 6.4–13. After n steps the master accumulator approximates the value of V_B as

$$V_B \approx \sum_{i=0}^{n} \frac{b_i}{2^i} V_C \tag{24}$$

where b_i is $+1(-1)$ at sample i if the output of the master accumulator at $i-1$ was less than V_B (more than V_B). Since the slave accumulator tracks the sign decisions of the master accumulator, then

$$V_{\text{out}} = \sum_{i=0}^{n} \frac{b_i}{2^i} V_A \tag{25}$$

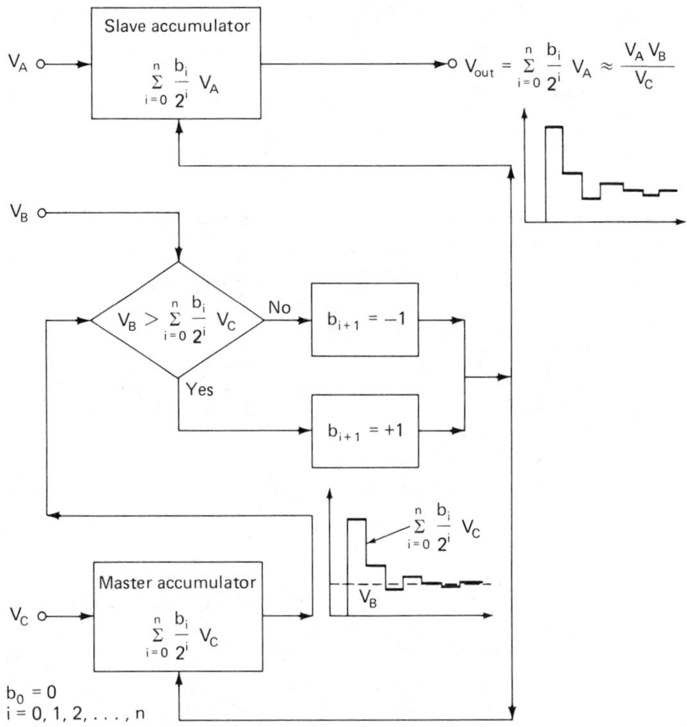

Fig. 6.4–13. Successive approximation implementation of the operational multiplier.

Solving Eqs. (24) and (25) gives

$$V_{out} \approx \frac{V_A V_B}{V_C} \tag{26}$$

to within $\pm\ V_C/2^{n+1}$.

One possible realization of the operational multiplier is shown in Fig. 6.4–14. The multiplication sequence begins by sampling V_A and V_C on the left-most capacitors. All other capacitors are discharged. At $i = 1$, the discharged capacitors sample ½ of V_C and V_A and add it to the master and slave accumulators, respectively. At $i = 2$, the discharged transfer capacitors sample ¼ of V_C and V_A and, depending upon the magnitude of $V_C/2$ and V_B, add or subtract from the accumulator. This process repeats for n times, which is determined by the desired accuracy and multiplication time. At $i = n$, the output of the slave accumulator is sampled, and the circuit is reset to begin another multiplication. Fig. 6.4–14(b) is a microphotograph of an NMOS implementation of Fig. 6.4–14(a).

In order to avoid Nyquist effects, the operational multiplier must sample the input waveforms at a frequency higher than the frequency of the input waveforms. If the clock frequency (ϕ_1,ϕ_2) is 1 MHz and $n = 10$, then the operational multiplier should provide good performance through audio fre-

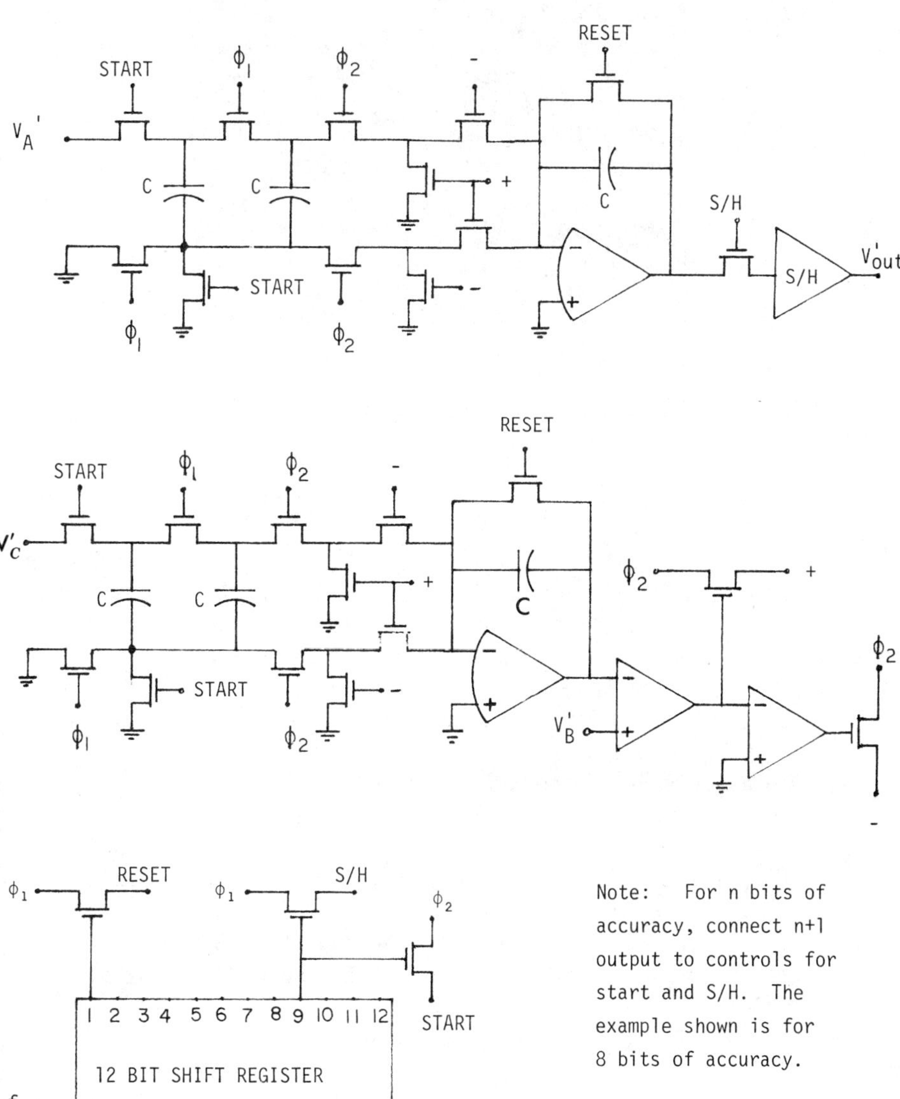

Fig. 6.4–14a. Accumulator realization for 4-quadrant SC multiplier of Fig. 6.4–13.

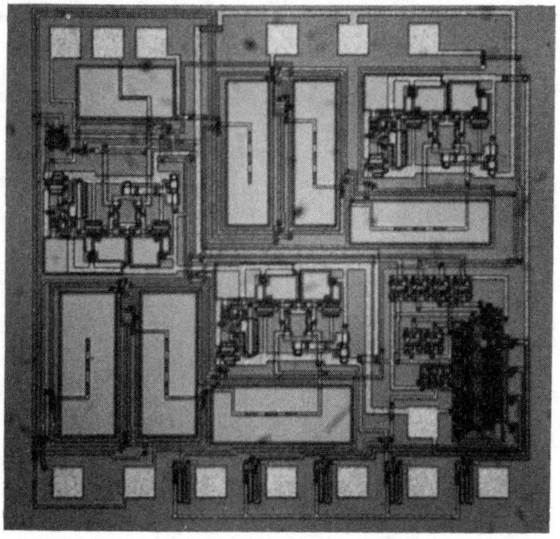

Fig. 6.4–14b. Microphotograph of an MOS implementation of Fig. 6.4–14a.

quencies. The multiplier of Fig. 6.4–14 has the advantage over conventional bipolar transconductance multipliers that the multiplication is independent of the input frequency. The limits of n are determined by both the comparator resolution and the feedthrough of the clock (see Chapter 8).

Another approach to implementing the four-quadrant multiplier is the quarter-square technique[22] used in the early analog computers. Fig. 6.4–15 shows how this might be accomplished. The most difficult part is the piecewise approximations to the second-order functions.

The availability of a linear four-quadrant multiplier permits the realization of many signal-processing functions. Such functions include division, AGC circuits, approximation to trigonometric and transcendental functions, approximation of exponents, and function-fitting. Whether or not one desires to use these techniques will depend upon whether there are any significant advantages over digital means of implementing the desired signal-processing circuit.

---

[22] G. A. Korn and T. M. Korn, *Electronic Analog and Hybrid Computers,* 2 ed., McGraw-Hill, New York, 1972.

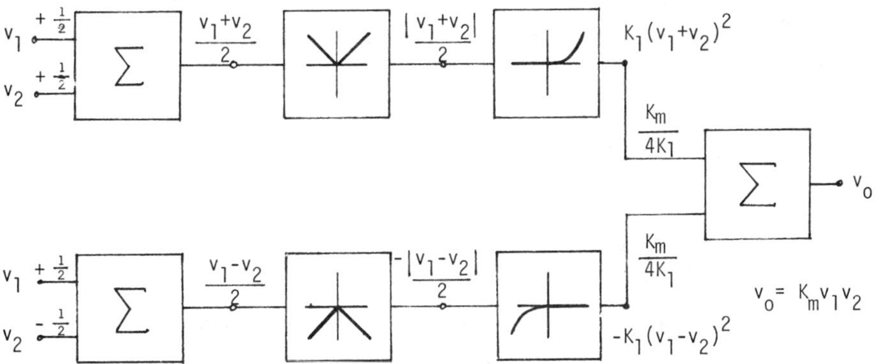

Fig. 6.4–15. Block diagram of a quarter-square multiplier circuit.

6.5 OSCILLATORS

In this section we will examine SC realizations of tuned and untuned oscillators and phase-shifters. These circuits represent another interesting class of applications for which SC methods are well suited. Because all SC circuits require a clock that is an oscillator, one may wonder about circuits that use something to create the same thing. However, there are several good reasons for examining SC oscillators. One of these reasons is that the clock frequency is a square wave whose frequency is generally very high. We shall see that the SC oscillator is in effect a frequency divider resulting in a lower frequency sinusoid. It will be shown that the frequency division ratio is not restricted to integers. Another reason to examine SC oscillators is to achieve quadrature oscillation, which leads to phase-shifting networks. For many oscillator applications, the frequency of the oscillator should be variable. The clock oscillator provides a convenient means of tuning the oscillation frequency of the SC oscillator. Finally, SC oscillators can provide precise waveforms from an imprecise clock.

The first category of oscillators considered is called the *tuned oscillator.* The RC oscillator is a subcategory of this class of oscillators. The tuned oscillator is almost a pure sinusoid and typically has very good frequency stability. It requires tuned circuits to maintain the frequency. The amplitude is determined by some form of soft-limiting, which can require a complex circuit. One of the disadvantages of the tuned oscillator is that it is difficult to vary the frequency. LC oscillators and quadrature oscillators also belong in the tuned oscillator class. The tuned oscillator works on the principle of having the open loop gain of a feedback loop become exactly $1\underline{/0^\circ}$ at the desired frequency of oscillation. The amplitude of the oscillator is determined by the limiting network. Figure 6.5–1 shows the analog form of a Wien

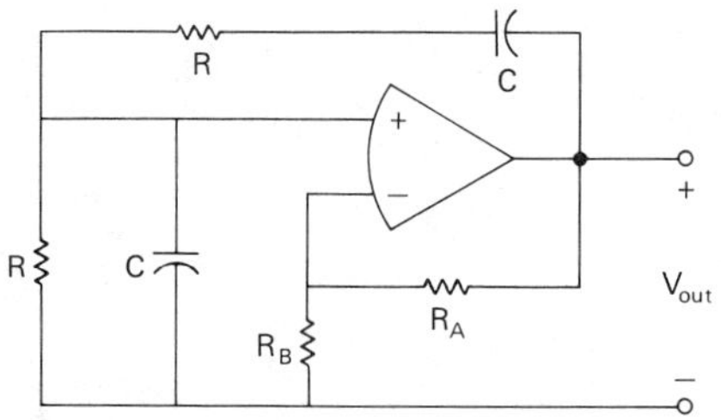

Fig. 6.5–1. Analog form of a Wien bridge RC oscillator.

bridge oscillator. Our approach is simply to replace the resistors with an appropriate SC simulation. One possible circuit is shown in Fig. 6.5–2. The frequency of oscillation of the Wien bridge oscillator of Fig. 6.5–1 is given as

$$\omega_0 = 1/RC \tag{1}$$

and the gain necessary for oscillation is

$$K = 1 + \frac{R_A}{R_B} \tag{2}$$

Replacing the two resistors designated as R by parallel SC resistors results in Fig. 6.5–2. For the moment, we shall ignore resistors R_A and R_B; we shall show how they can be realized shortly. Analysis of the transfer function of the SC feedback network, assuming that $f_c >> f_0$ results in (see Problem 6.21)

$$\frac{V_i(s)}{V_{out}(s)} \approx \frac{sT\dfrac{C}{C_R}}{\left(\dfrac{s\,TC}{C_R}\right)^2 + \dfrac{3s\,TC}{C_R} + 1} = \frac{\dfrac{s\,C_R}{TC}}{s^2 + \dfrac{3s\,C_R}{TC} + \left(\dfrac{C_R}{TC}\right)^2} \tag{3}$$

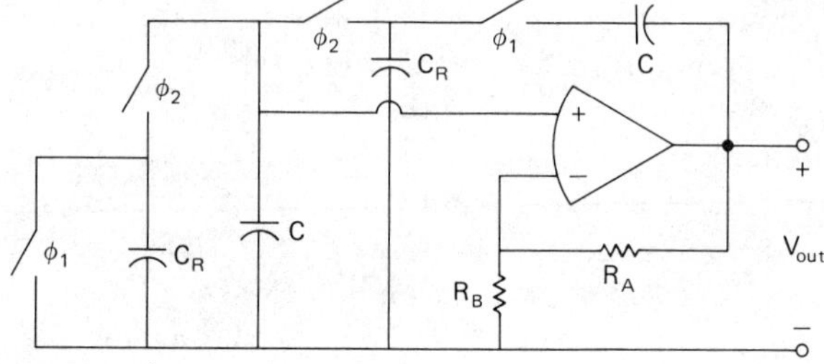

Fig. 6.5–2. SC realization of Fig. 6.5–1.

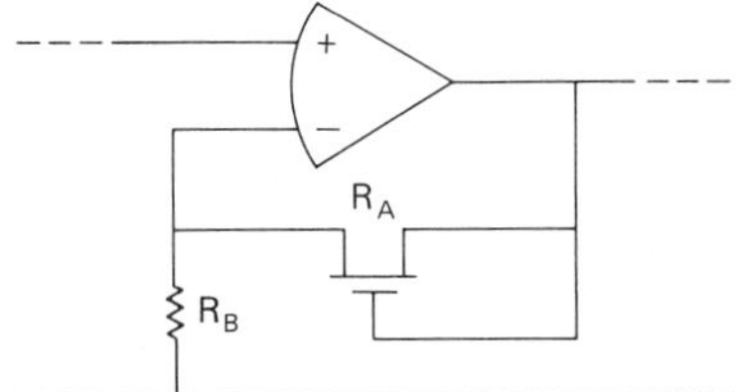

Fig. 6.5–3. A possible limiter for the amplifier of Fig. 6.5–2.

Therefore the frequency of the oscillator is

$$\omega_0 \approx \frac{C_R}{CT} = \frac{C_R}{C} f_c \tag{4}$$

We can now see how the clock frequency can be used to control the oscillator frequency. The result is essentially a frequency division of f_c by the factor $2\pi C/C_R$. An important component that is missing from both oscillators (Figs. 6.5–1 and 6.5–2) is the amplitude-limiting circuit. One method of getting a possible limiting circuit without too much extra effort would be to try to take advantage of the nonlinearity of the enhancement resistor (see Chapter 8). As the voltage increases across this element (Fig. 8.4–1), the resistance decreases. If we replace R_B and R_A of Fig. 6.5–2 by a diffused or polysilicon resistor and an enhancement resistor, respectively, we obtain Fig. 6.5–3. The amplifier gain is given by Eq. (2), and, as the amplitude increases, the voltage across R_A increases. However, because R_A decreases in value, the open loop gain of the oscillator will also decrease. If R_B is made from the source-drain diffusion, then the temperature and voltage coefficients of the two resistors should be pretty well matched, resulting in a stable amplitude control.

Another method of limiting the amplitude is shown in Fig. 6.5–4.[23] This method rectifies and integrates the output to get a DC voltage that is proportional to the amplitude of the sinusoidal out. This DC voltage is applied to the gate of an MOS device (M1) which is in series with a resistor to form R_B. The gain of the integrator can be adjusted to vary the amplitude of the oscillator.

An oscillator similar to Fig. 6.5–1 is shown in Fig. 6.5–5(a). An SC realization of this oscillator is shown in Fig. 6.5–5(b). If $C_1 = C_4 = C_R$ and $C_2 = C_3 = C_1$, then it can be shown that (see Problem 6.22)

$$f_{osc} \approx \frac{1}{2\pi}\frac{C_R}{C} f_c \tag{5}$$

[23] T. R. Viswanathan, K. Singhal, and G. Metzker, "Applications of Switched-Capacitor Resistors in RC Oscillators," *Electronics Letters,* Vol. 14, No. 20, September 28, 1978, pp. 659–660.

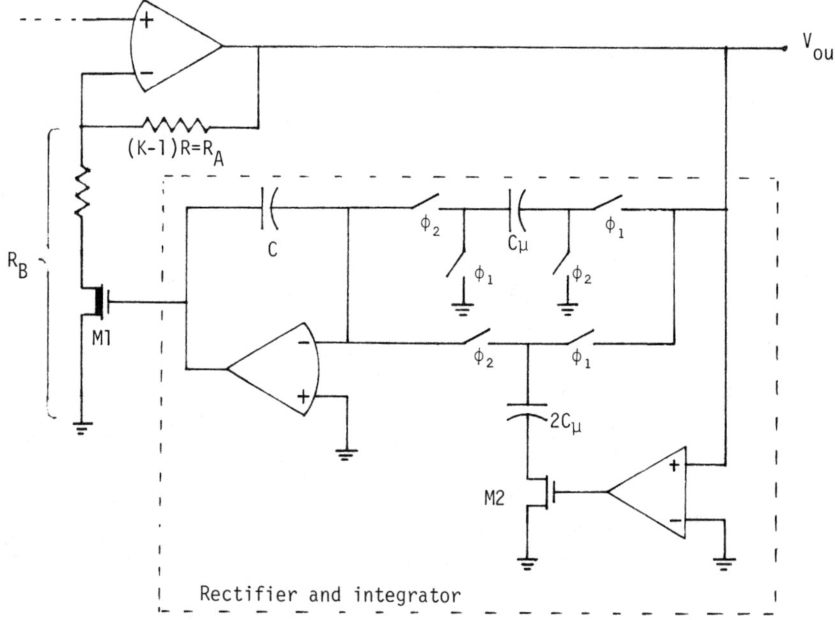

Fig. 6.5–4. SC amplitude limiting circuit for oscillators.

A familiar single-amplifier RC oscillator is the RC phase-shift oscillator shown in Fig. 6.5–6(a). This form is but one of many possible versions of the RC phase-shift oscillator. The MOS device serves as a transconductance amplifier, and its output resistance plus any load resistance is lumped into R_L. The frequency of oscillation can be found as (see Problem 6.23)

$$f_{osc} = \frac{\sqrt{3}}{2\pi} \frac{\sqrt{1 + (R/R_L)}}{RC} \tag{6}$$

and the gain necessary for oscillation is

$$A = -g_m R_L = -\left(3\frac{R}{R_L} + 14 + 12\frac{R_L}{R}\right) \tag{7}$$

For the SC version of Fig. 6.5–6(b)[24], the frequency of oscillation is given as

[24] E. A. Vittoz, "Micropower Switched-Capacitor Oscillator," *IEEE J. of Solid-State Circuits*, Vol. SC-14, No. 3, June 1979, pp. 622–624.

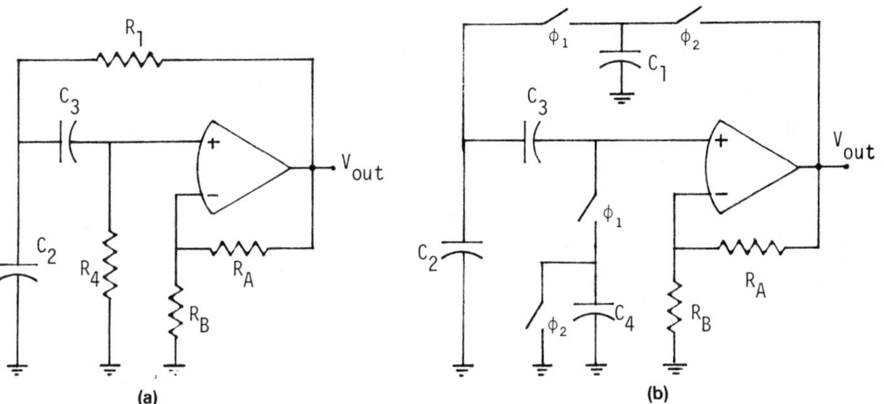

Fig. 6.5–5. (a) A RC oscillator. (b) A SC realization of (a).

$$f_{osc} \approx \frac{\sqrt{3}\sqrt{1+(C_L/C_R)}}{2\pi} \frac{C_R}{C} f_c \tag{8}$$

and the gain is

$$A \approx -\frac{3C_L}{C_R f_c} - 14 - 12\frac{C_R f_c}{C_L} \tag{9}$$

The amplitude can be stabilized by designing the DC bias current through the MOS device so that it is inversely proportional to the amplitude of the oscillators. As the output grows, the DC bias current will decrease. Because g_m is proportional to the square of the bias current, the gain A will decrease if the oscillation grows, resulting in amplitude stabilization.

Oscillators can also be constructed by cascading integrators. The advantage of this approach is that, if one is consistent, the exact discrete-time domain oscillation frequency can be easily found. Consider the two-integrator loop of Fig. 6.5–7(a), which uses an inverting and a noninverting Type I LDI integrator. From Eq. (19) of Section 2.5, we may write the loop gain of Fig. 6.5–7(a) as

$$\text{Loop gain} = \left(\frac{\omega_0}{\omega}\right)^2 \left(\frac{\omega T/2}{\sin\dfrac{\omega T}{2}}\right)^2 \tag{10}$$

where

$$\omega_0 = C_1/TC_2 = \frac{C_1}{C_2} f_c \tag{11}$$

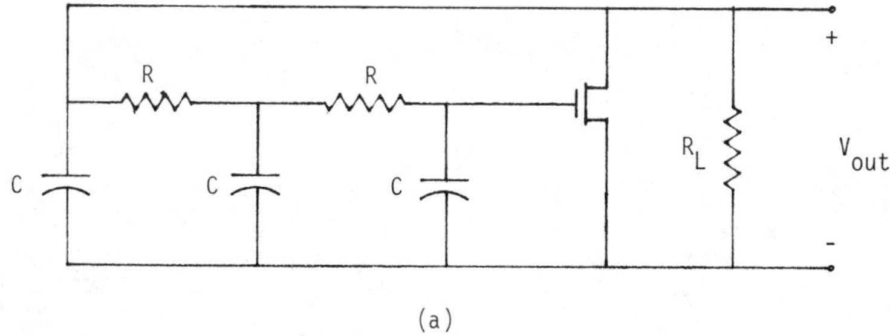

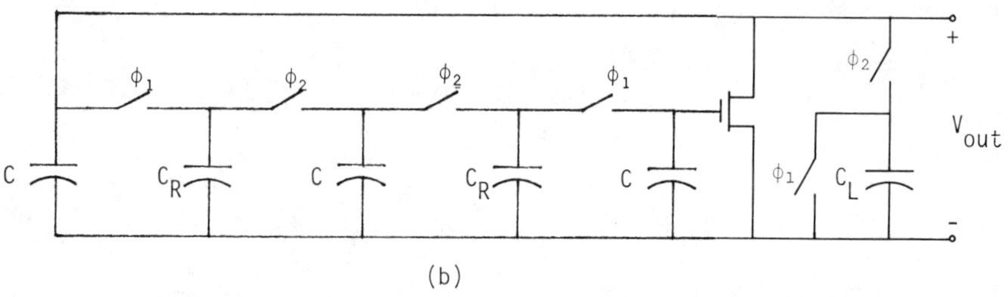

Fig. 6.5–6. (a) RC phase shift oscillator. (b) SC realization of (a).

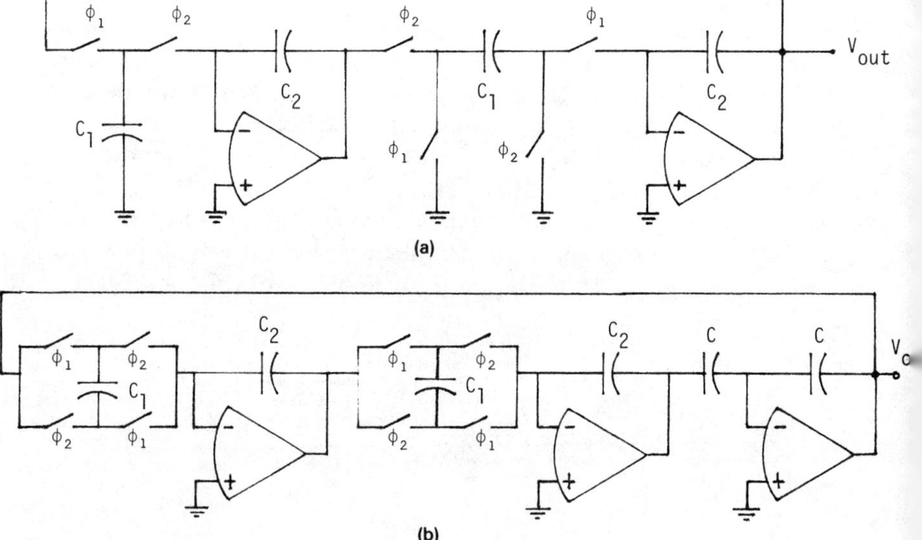

Fig. 6.5–7. Switched capacitor oscillator using (a) LDI integrators and (b) bilinear integrators.

Setting the loop gain equal to unity gives

$$f_{osc} = \frac{f_c}{\pi} \sin^{-1}\left(\frac{C_1}{2C_2}\right) \tag{13}$$

Similar relationships may be found for Fig. 6.5–7(b), (see Prob. 6.24). Some form of amplitude-limiting will be necessary for the oscillators of Fig. 6.5–7 to produce a distortion-free sinusoid.

The second class of oscillator is the *untuned oscillators.* This class of oscillators has been widely integrated using bipolar technology. The basic operation of untuned oscillator is shown in the block diagram of Fig. 6.5–8. We see that all the components of this block diagram are compatible with the techniques we have studied in switched capacitor circuits. The untuned oscillator provides a square wave and a triangular wave. If a sinusoid is desired, then a circuit that converts the triangle waveform into a sinusoid is required. One technique is to use the nonlinear waveshaping circuits discussed in Section 6.3. Another possibility is to use the large signal characteristics of an MOS differential amplifier to approximate the triangle-to-sine conversion. The untuned oscillator has the advantage of not requiring a tuned circuit, so that the frequency can be easily controlled or varied. The amplitude is also well controlled and does not require a limiting circuit. The only disadvantage is that a sine wave is difficult to obtain.

Figure 6.5–9 shows a block diagram of the untuned oscillator and its principles of operation. We note that the basic ingredients of the untuned oscillator are an integrator and a bistable circuit. The operation of the untuned oscillator can be described as follows. During the time interval from 0 to T_1, the integrator is integrating the voltage L_- provided by the bistable. If K is the constant of integration, we find that the value of V_T at $t = T_1$ is given as

$$v_T(T_1) = S_+ = KL_+T_1 + S_- \tag{14}$$

where all these terms are defined in Fig. 6.5–9. From Eq. (14) we get

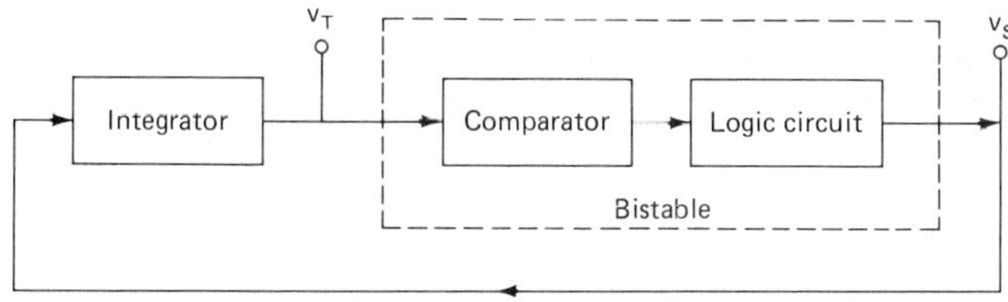

Fig. 6.5–8. Basic block diagram of an untuned oscillator.

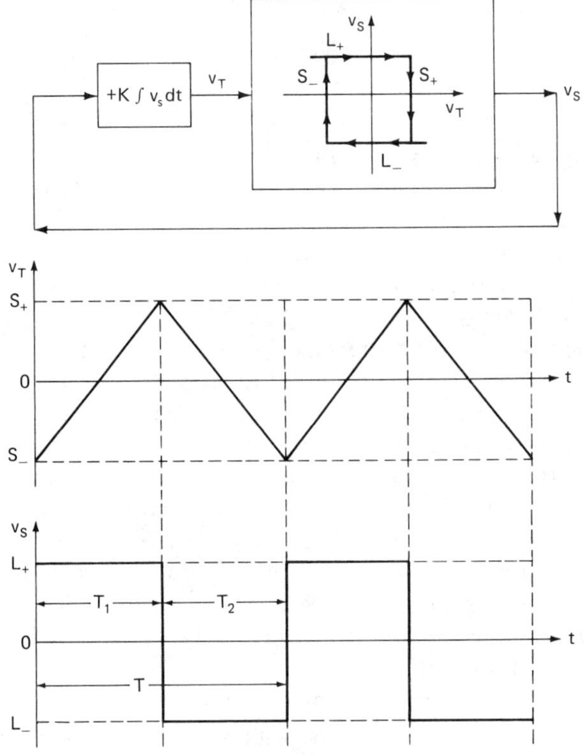

Fig. 6.5–9. Block diagram of the untuned oscillator and waveforms.

$$T_1 = \frac{S_+ - S_-}{KL_+} \tag{15}$$

We note that the integrator is a noninverting integrator. At $t = T_1$, the bistable switches from L_+ at the output to L_-. Therefore during the time interval from T_1 to T, the period of the oscillator, the integrator integrates L_-. Assume that $t = T_1 = 0$ for the following analysis. At $t = T_2$, the value of v_T is given as

$$v_T(T_2) = S_- = KL_-T_2 + S_+ \tag{16}$$

From Eq. (16), we can solve for T_2 as

$$T_2 = \frac{S_- - S_+}{KL_-} = -\frac{S_+ - S_-}{KL_-} \tag{17}$$

Therefore, the period of the oscillator T can be written as

$$T = T_1 + T_2 = \frac{1}{K}\left[\left(\frac{S_+ - S_-}{L_+}\right) - \left(\frac{S_+ - S_-}{L_-}\right)\right] \tag{18}$$

If $S_+ = -S_- = S$ and $L_+ = -L_- = L$, then Eq. (18) reduces to

$$T = 4S/(KL) \tag{19}$$

or the frequency of the untuned oscillator is

$$f_0 = KL/(4S) \tag{20}$$

These results are general and may be used for all oscillators that fit the block diagram of Fig. 6.5–9.

Figure 6.5–10 shows an analog untuned oscillator. If the op amps are powered from a positive voltage whose magnitude is V_{PS} and a negative voltage whose magnitude is V_{NS}, then it can be shown that $L_+ = V_{PS}$, $L_- = -V_{NS}$, $S_+ = V_{PS}[R_3/(R_2 + R_3)]$, and $S_- = -V_{NS}[R_3/(R_2 + R_3)]$. The constant of integration is $K = 1/R_1C$. Assuming that the magnitude of V_{PS} is equal to the magnitude of V_{NS}, we find that the frequency of oscillation is

$$f_0 = \frac{1}{4R_1C}\left(\frac{R_3}{R_2 + R_3}\right) \tag{21}$$

An SC realization of Fig. 6.5–10 is shown in Fig. 6.5–11. The resistors R_2 and R_3 have been replaced by capacitors. This should cause no real problems, for this op amp is at either its positive limit or its negative limit.

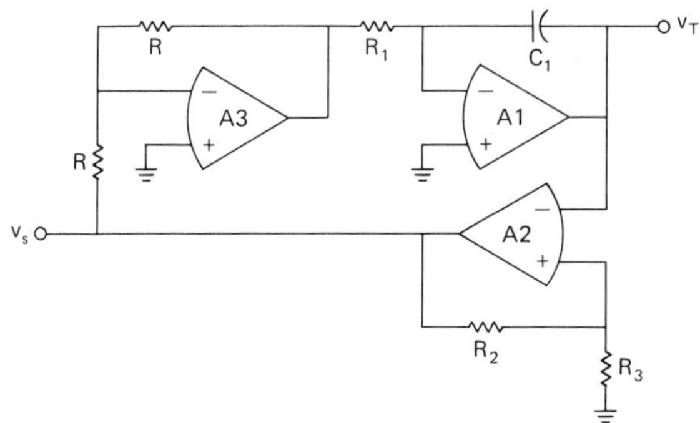

Fig. 6.5–10. An analog untuned oscillator.

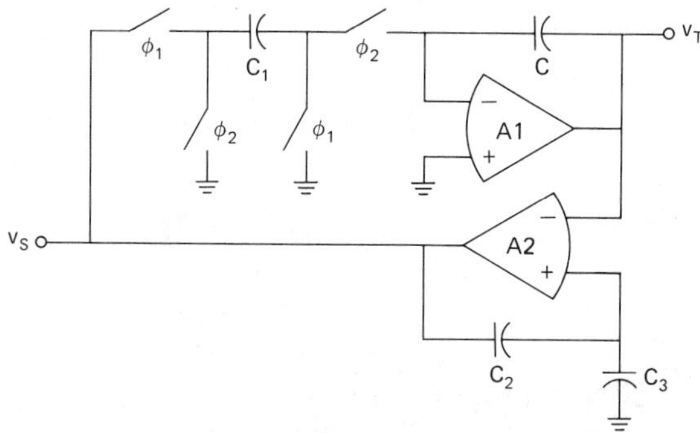

Fig. 6.5–11. SC realization of Fig. 6.5–10.

Thus only one SC simulated resistor is needed. The frequency of oscillation for this circuit is

$$f_0 = \frac{C_1 f_c}{4C}\left(\frac{C_2}{C_2 + C_3}\right) \tag{22}$$

A simpler version of this circuit is shown in Fig. 6.5–12. It can be shown that the frequency of oscillation is given by

$$f_0 = \frac{C_1 f_c}{2C}\left[\ln\left(\frac{1+\alpha}{1-\alpha}\right)\right] \tag{23}$$

where

$$\alpha = C_2/(C_2 + C_3) \tag{24}$$

The above results assume that f_c is much greater than f_0.

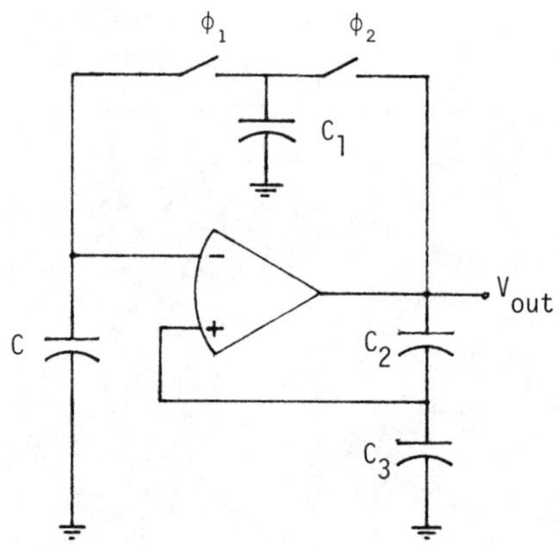

Fig. 6.5–12. A simple one-amplifier oscillator.

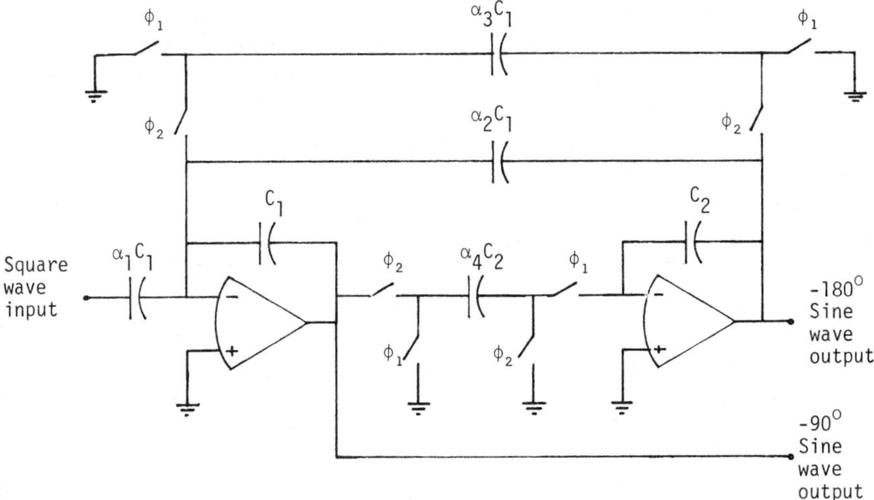

Fig. 6.5–13. A second-order bandpass filter used to convert a square wave into two sinusoids with 90° phase shift between them.

A third method of generating a sinusoid is to divide down the clock frequency and to apply this square wave to a high-Q band-pass network.[25] Figure 6.5–13 shows a second-order bandpass filter that is used to convert a square wave into two sinusoids with a phase shift of 90° between them. A Q of 33 gives a sinusoid with less than 0.5% harmonic distortion for the −180° sine output and less than 1% for the −90° sine output. Assuming that the square wave has a frequency of ω_0 and a zero-to-peak amplitude of V_{SQ}, it can be shown that

$$\alpha_1 = \frac{\pi}{4} \frac{V_{\text{sin}}}{V_{SQ}} \frac{1}{Q} \tag{25}$$

$$\alpha_2 = \frac{1}{Q} \tag{26}$$

and

$$\alpha_3 = \alpha_4 = 2 \sin\left(\frac{\omega_0 T}{2}\right) \tag{27}$$

[25] K. Martin and A. S. Sedra, "Switched-Capacitor Building Blocks for Adaptive Systems," *IEEE Trans. in Circuits and Systems,* Vol. CAS-28, No. 6, June 1981, pp. 576–584.

where V_{sin} is the zero-to-peak amplitude of the sine wave outputs. An interesting application of this circuit is to add weighted versions of the two outputs to get a low distortion sinusoid having any desired phase shift with respect to the square wave input.

This section has shown the method of developing oscillator realizations using SC techniques. Examples of the two basic categories of oscillators have been presented. Many other circuits and techniques are possible and may become important in a given application.

6.6 PHASE-SHIFTERS AND EQUALIZERS

An application that has many similarities to oscillators is that of *phase shifting.* In many analog signal-processing applications the need to phase-shift a sinusoid occurs. In this section we shall consider several methods of accomplishing phase-shifting using SC networks.

A simple passive, first-order, continuous-time, phase-shifting network is shown in Fig. 6.6–1(a). It is easy to show that the phase shift from V_{in} to V_{out} is

$$\theta(j\omega) = \text{Arg}\left[\frac{V_{\text{out}}(j\omega)}{V_{\text{in}}(j\omega)}\right] = \pi - 2\tan^{-1}(\omega R_2 C_1) \tag{1}$$

It can be shown that the magnitude is frequency-independent. A switched capacitor realization of Fig. 6.6–1(a) is shown in Fig. 6.6–1(b). Under high sampling conditions, the phase shift of this network is

$$\theta(j\omega) \approx \pi - 2\tan^{-1}\left(\frac{2\pi C_1}{C_2}\frac{f}{f_c}\right) \tag{2}$$

Another first-order phase-shifter is shown in Fig. 6.6–2(a). The phase shift of this network is given by Eq. (1). As in Fig. 6.6–1(a), the magnitude is frequency-independent. One possible SC realization is shown in Fig. 6.6–2(b). If sufficient negative DC feedback exists around the phase-shifter, then the two resistors designated R could be replaced by capacitors. Under high sampling frequency conditions, the phase shift of Fig. 6.6–2(b) is given by Eq. (2).

The ability to simulate a negative resistance leads to a third realization of a first-order phase-shifter. The continuous-time realization is shown in Fig. 6.6–3(a). The phase shift of this network is

$$\theta(j\omega) = \text{Arg}\left[\frac{V_{\text{out}}(j\omega)}{V_{\text{in}}(j\omega)}\right] = -2\tan^{-1}(\omega RC) \tag{3}$$

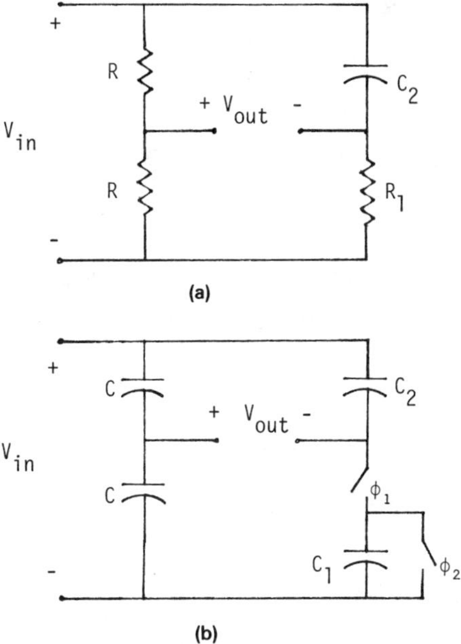

Fig. 6.6–1. (a) Continuous-time, first-order, passive phase shifter. (b) Discrete-time version of (a).

One possible SC realization (see Problem 6.29) is shown in Fig. 6.6–3(b). Under the assumption of a high sampling rate, it may be shown that the phase shift of the discrete-time realization is

$$\theta(j\omega) \approx -2\tan^{-1}\left(2\pi\frac{f}{f_c}\right) \tag{4}$$

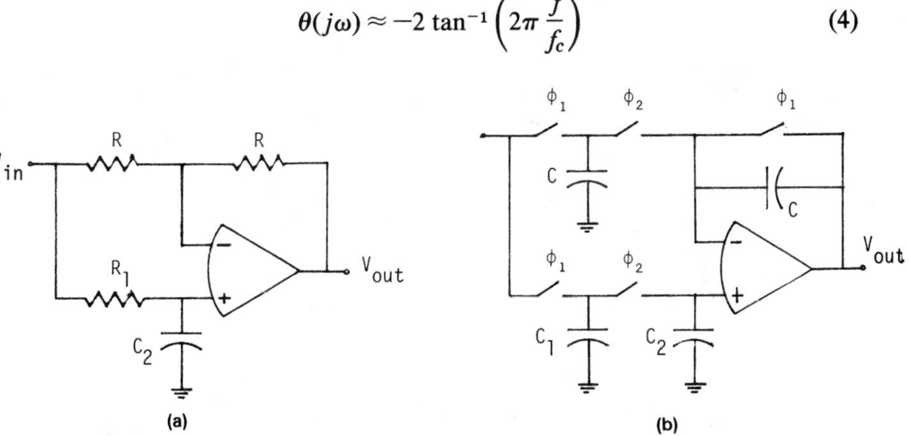

Fig. 6.6–2. (a) Continuous-time, first-order, active phase shifter. (b) Discrete-time version of (a).

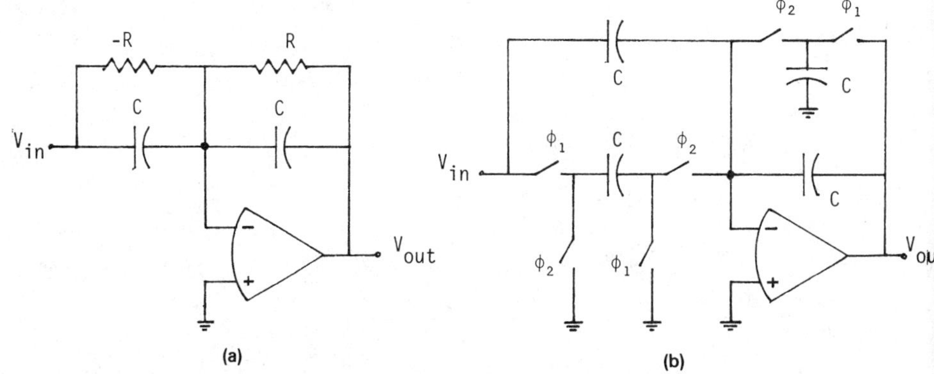

Fig. 6.6–3. A practical SC phase shifter. (a) Continous-time realization. (b) Discrete-time realization.

Second-order phase-shifters realize the all-pass transfer function

$$H(s) = \pm K \frac{s^2 - \dfrac{\omega_0}{Q}s + \omega_0^2}{s^2 + \dfrac{\omega_0}{Q}s + \omega_0^2} \tag{5}$$

in order to achieve a phase shift of

$$\theta(j\omega) = -2 \tan^{-1}\left[\frac{\dfrac{\omega}{\omega_0}\dfrac{1}{Q}}{1 - \left(\dfrac{\omega}{\omega_0}\right)^2}\right] \tag{6}$$

We see that a second-order phase-shifter can provide a phase shift between 0 and −360°. One possible continuous-time realization is shown in Fig. 6.6–4(a). The transfer function of this circuit is

$$\frac{V_{out}(s)}{V_{in}(s)} = -\left[\frac{s^2 - \omega_1 s + \omega_1\omega_2}{s^2 + \omega_1 s + \omega_1\omega_2}\right] \tag{7}$$

where $\omega_1 = 1/(R_1C_1)$, and $\omega_2 = 1/(R_2C_2)$. From Eqs. (5), (6), and (7), we see that the phase shift of Fig. 6.6–4(a) is given as

$$\theta(j\omega) = -2 \tan^{-1}\left[\frac{\omega\,\omega_1}{\omega_1\,\omega_2 - \omega^2}\right] = -2 \tan^{-1}\left[\frac{\dfrac{\omega_1}{\omega}}{\dfrac{\omega_1\omega_2}{\omega^2} - 1}\right] \tag{8}$$

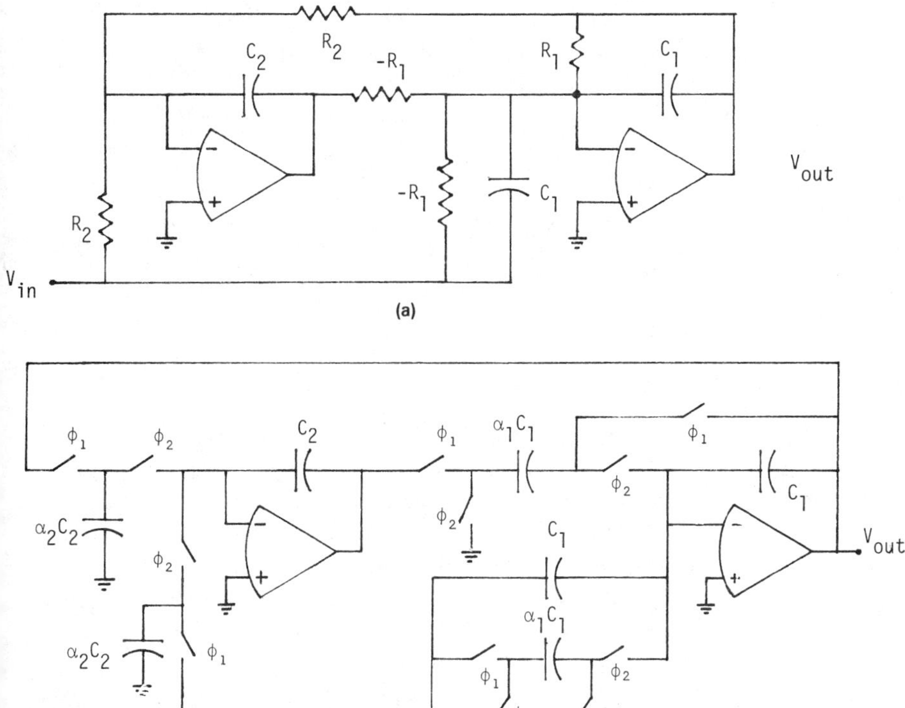

Fig. 6.6–4. (a) Second-order, continuous-time phase shifter. (b) Second-order, discrete-time phase shifter.

Figure 6.6–4(b) shows a discrete-time realization of Fig. 6.6–4(a). The phase of this circuit under the high sampling assumption is given as

$$\theta(j\omega) \approx -2\tan^{-1}\left[\frac{\dfrac{2\pi\alpha_1 f}{f_c}}{\dfrac{4\pi^2\alpha_1\alpha_2 f^2}{f_c^2} - 1}\right] \tag{9}$$

This circuit is an example of one possible second-order phase-shifter. Similar ideas may be used to generate other second-order phase-shifters.

6.7 PHASE-LOCKED LOOPS

Another area of application of SC networks is the phase-lock loop. The objective of the phase-lock loop is to provide a DC voltage that is proportional

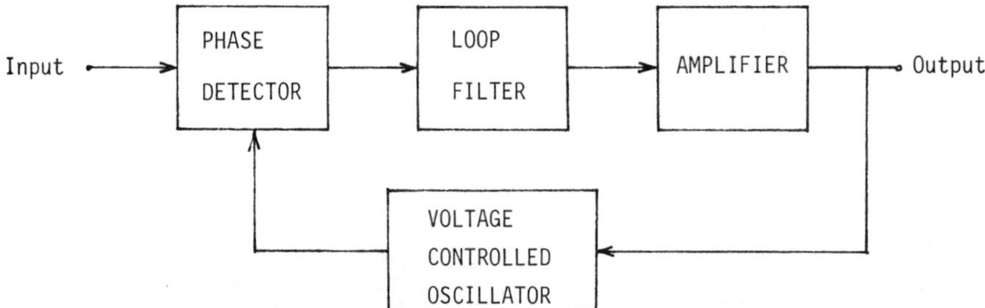

Fig. 6.7–1. Block diagram of a phase-lock loop.

to the phase of an applied sinusoid. A block diagram of a phase-lock loop is shown in Fig. 6.7–1. It is seen that a phase-lock loop consists of a negative feedback loop in which the output of a phase-detector is rectified and applied to a voltage-controlled oscillator. The oscillator applies a periodic waveform to the phase-detector, which is compared with the phase of the input signal. The phase-lock loop has many applications, such as FM demodulation, frequency synchronization, signal conditioning, frequency multiplication and division, frequency translation, and AM detection.[26]

A more detailed block diagram of the phase-lock loop is shown in Fig. 6.7–2. In the following analysis, the phase-lock loop is assumed to be locked. The input signal is assumed to be a sinusoid given as

$$v_i(t) = V_p \sin(\omega t + \theta_i) \tag{1}$$

If the phase shift of the voltage-controlled oscillator is θ_{osc}, then the output of the phase comparator is

$$V_b = K_D(\theta_i - \theta_{osc}) \tag{2}$$

The phase of the voltage-controlled oscillator as a function of time is actually equal to the integral of the frequency of the voltage-controlled oscillator, as shown by

$$\omega_{osc}(t) = \frac{d\theta_{osc}(t)}{dt} \tag{3}$$

---

[26] A. B. Grebene, *Analog Integrated Circuit Design,* Van Nostrand Reinhold, New York, 1972, pp. 306–310.

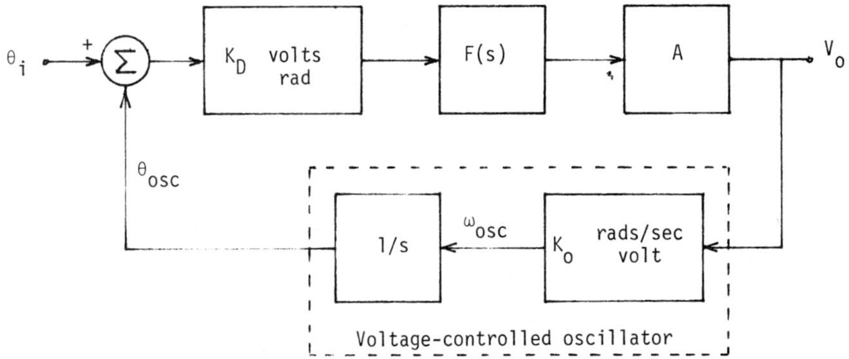

Fig. 6.7–2. More detailed block diagram of Fig. 6.7–1.

Thus in the block diagram of Fig. 6.7–2, this integration is represented in the voltage-controlled oscillator block as $1/s$. If we assume that the oscillator frequency is given as

$$\omega_{\text{osc}} = \omega_{\text{o}} + K_{\text{o}} V_o \tag{4}$$

where ω_{o} is the free-running frequency of the oscillator when $V_o = 0$, then the transfer function of the phase-lock loop becomes

$$\frac{V_o(s)}{\theta_i(s)} = \frac{sK_D F(s) A}{s + K_D K_0 A F(s)} \tag{5}$$

or

$$\frac{V_o(s)}{\omega_i(s)} = \frac{1}{s}\frac{V_o(s)}{\theta_i(s)} = \frac{K_D\, F(s)\, A}{s + K_D K_0 A F(s)} \tag{6}$$

If $F(s) = 1$, then the loop inherently has a first-order, low-pass transfer response. The loop bandwidth is defined as $K_v = K_0 K_D A$. Various types of filters have been used to achieve different dynamic performance of the phase-lock loop, and details can be found elsewhere.[27]

The loop lock range is defined as the range of frequencies about ω_{o} for which the phase-lock maintains the relationship that

$$\omega_i = \omega_{\text{osc}} \tag{7}$$

[27] P. R. Gray and R. G. Meyer, *Analysis and Design of Analog Integrated Circuits,* John Wiley and Sons, New York, 1977, pp. 575–593.

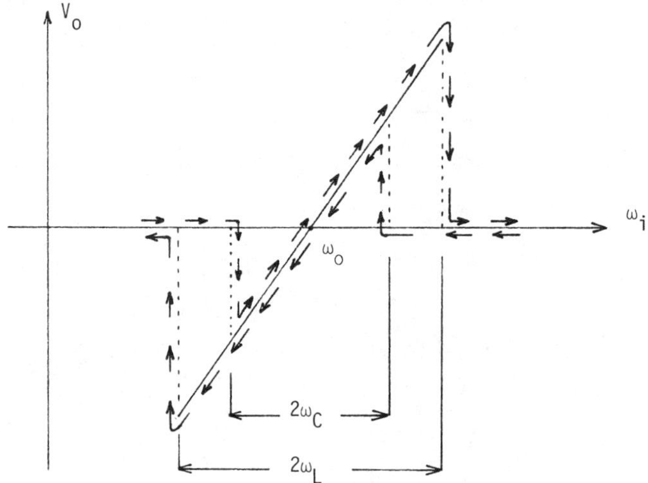

Fig. 6.7–3. Illustration of loop and capture range for a phase-lock loop.

If the phase detector can determine the phase difference between θ_i and θ_{osc} over a $\pm\,\pi/2$ range, then the loop lock range is expressed as

$$\omega_L = \pm\, \Delta\omega_{osc} = K_D A\, K_0\left(\pm\frac{\pi}{2}\right) = \pm\, K_v \frac{\pi}{2} \tag{8}$$

The capture range is the range of input frequencies for which an initially unlocked loop will lock on an input signal. If $F(s) = 1$, then the capture range is equal to the lock range. If $F(s) = (1 + s/\omega_1)^{-1}$, then the capture range is less than the lock range. If the capture range is designated as $2\omega_C$, then Fig. 6.7–3 illustrates the relationships between the loop and capture ranges of a phase-lock loop with $F(s) = (1 + s/\omega_1)^{-1}$. The loop and capture ranges are very dependent upon the loop bandwidth, K_v. If K_v decreases, the capture time increases, the capture range decreases, and the interference rejection properties of the phase-lock loop improve.

A simple switched-capacitor phase-lock loop suitable for sine wave or triangular wave inputs is shown in Fig. 6.7–4.[28] To understand how this phase-lock loop operates, let us consider the case when the loop is in lock and the input is a sine wave. Figure 6.7–5 shows the input waveform together with that of the VCO (ϕ_2) and the inverted form of the latter (ϕ_1). As shown, the frequency of the VCO is equal to that of the input, but a static

[28] K. Martin and A. S. Sedra, "Switched-Capacitor Building Blocks for Adaptive Systems," *IEEE Trans. on Circuits and Systems,* Vol. CAS-28, No. 6, June 1981, pp. 576–584.

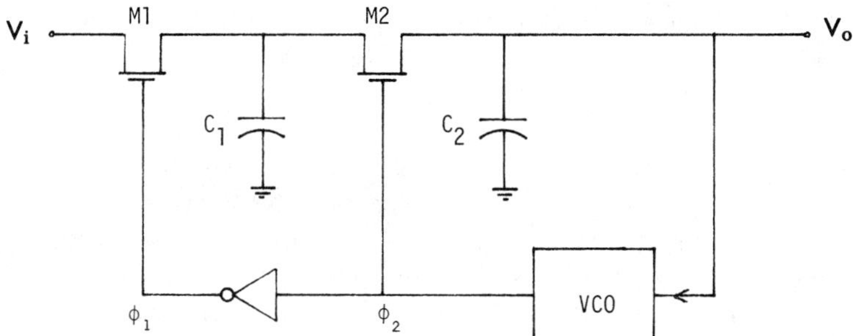

Fig. 6.7–4. A simple SC phase-lock loop.

phase error exists. The phase error is essential for it produces, at the output of the low-pass filter, the DC voltage required to shift the VCO frequency from its free-running value to a value equal to that of the input frequency.

As can be seen from Fig. 6.7–5, the switching waveform ϕ_1 causes samples of the input signal to appear across capacitor C_1. If the input signal is sampled at time t_1, then C_1 will be charged to a voltage V_L given by

$$V_L = V_p \sin \theta_1 \tag{9}$$

where V_p is the peak voltage of the input sine wave, and θ_1 is the phase error. Switching waveform ϕ_2 connects capacitors C_1 and C_2, which tends

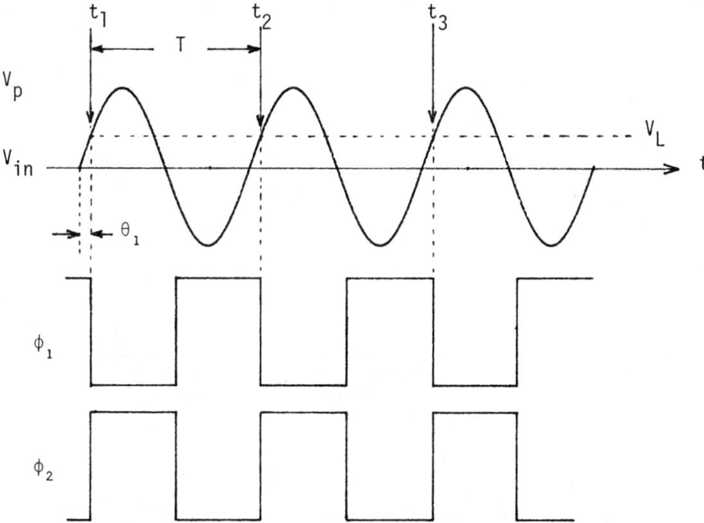

Fig. 6.7–5. Waveforms of SC phase-lock loop when in lock.

to bring the voltage across C_2 closer to V_L. In the lock condition, the input signal will always be sampled at times when the voltage is equal to V_L. This DC voltage will be the value required to cause the VCO frequency to equal the input frequency.

Let us now assume that the input frequency is slowly increased (that is, the period T is decreased). It can be seen that the samples across C_1 will increase in magnitude, which will tend to increase V_o, thus increasing the frequency of the VCO, which keeps the loop in lock. In the new lock condition, though, the input signal samples will be higher in magnitude than V_L, and the DC voltage V_o will be correspondingly larger. The static phase error θ_1 will also be higher. The loop will remain in lock until the phase error θ_1 reaches 90°, which corresponds to the input signal being sampled at its peak value of V_p. Therefore the lock range is given as

$$\omega_L = \pm K_0 K_p \tag{10}$$

From the previous considerations we see that the output of the phase detector in Fig. 6.7–4 is equivalent to the voltage sampled by C_1. This voltage is given as

$$v_b(t) = V_p \sin\left[\theta_i(t) - \theta_{osc}(t)\right] \tag{11}$$

If the phase difference is small, we can approximate $v_b(t)$ as

$$v_b(t) \approx V_p \left[\theta_i(t) - \theta_{osc}(t)\right] \tag{12}$$

Assuming the VCO is described by Eq. (4) allows us to write the transfer function of Fig. 6.7–4 as

$$\frac{V_o(s)}{\theta_i(s)} = \frac{s\, V_p\, F(s)}{s + K_0 V_p F(s)} \tag{13}$$

For the low-pass SC filter incorporated into Fig. 6.7–4, we can see that

$$F(j\omega) = \frac{1}{j\left(1 + \dfrac{2C_2}{C_1}\right)\sin\left(\dfrac{\omega T}{2}\right) + \cos\left(\dfrac{\omega T}{2}\right)} \tag{14}$$

which leads to a second-order loop transfer function.

One of the disadvantages of this phase-lock loop is that the lock range and the frequency response are dependent upon the magnitude of the input

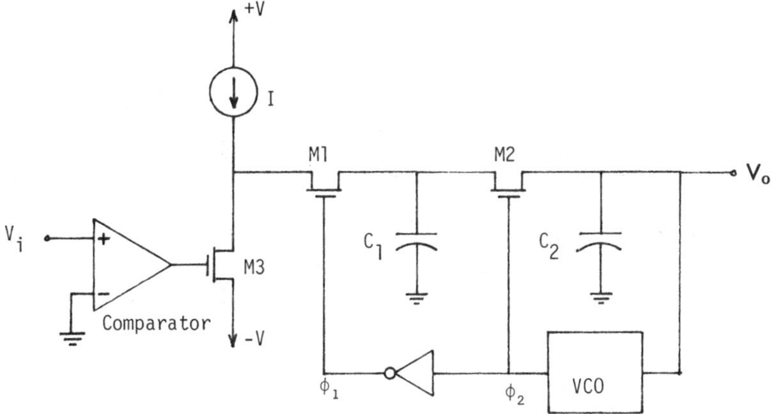

Fig. 6.7–6. Modification of Fig. 6.7–4 so that a square wave input may be applied.

voltage. Also, this phase-lock loop will not work for square wave inputs. A method to eliminate this restriction would be to use the square wave to generate a voltage ramp across C_1 using the scheme of Fig. 6.7–6. Here the square wave input is used to switch M3 on and off, causing C_1 to be charged by I. When M3 is turned off, the voltage across C_1 will be the steady-state voltage V_L required to keep the VCO frequency equal to the input frequency.

For proper operation of the phase-lock loop, ϕ_1 and ϕ_2 must be nonoverlapping. This can be achieved by the circuit of Fig. 6.7–7, where feedback around the NOR gates ensures that the one phase can go high only if the other phase is low. Longer nonoverlap times may be required and can be achieved by adding some time delay to the NOR-feedback paths.

A suitable VCO can be implemented using MOS technology, as shown in Fig. 6.7–8. This circuit is a source-coupled astable multivibrator whose frequency is controlled by V_o. Let us assume that the circuit is completely symmetrical about a vertical centerline. The voltage V_o applied to the gates of M5 and M6 will generate a constant current I in both drains. If M1 is on, then M2 is off. Therefore $I_1 = 2I$, $I_2 = 0$, and $I_c = I$. As C changes

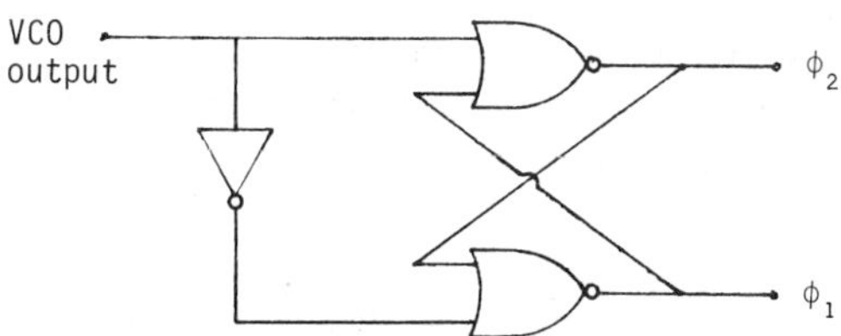

Fig. 6.7–7. A circuit for generating non-overlapping clocks from a single clock.

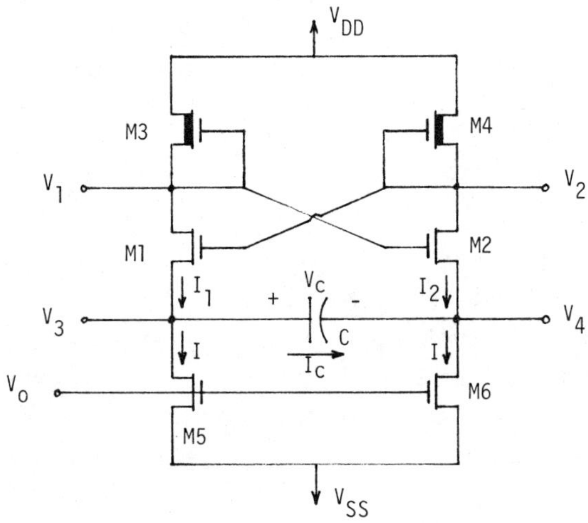

Fig. 6.7–8. Source-coupled astable multivibrator.

in this manner, V_3 increases and will eventually turn M1 off. As M1 turns off, M2 turns on, due to the regenerative coupling. Now $I_1 = 0$, $I_2 = 2I$, and $I_c = -I$. Therefore, V_4 begins to increase as C is charged in the reverse direction. Eventually M2 turns off because of the increasing V_4, and the cycle repeats. Ideal waveforms are shown in Fig. 6.7–9. If I is expressed as (see Chapter 8)

$$I = \frac{\beta}{2}(V_T - V_o)^2 \tag{15}$$

then the frequency of the VCO is given as

$$\omega_{osc} = \frac{\beta\,(V_T - V_o)^2 \pi}{4\,CV_T} \tag{16}$$

Although the nonlinear relationship is satisfactory, the squared relationship can be eliminated by using resistors (R_1) in series with the source of M5 and M6 to obtain an oscillation frequency of

$$\omega_{osc} \approx \frac{\pi\,V_o}{2\,V_T R_1 C} \tag{17}$$

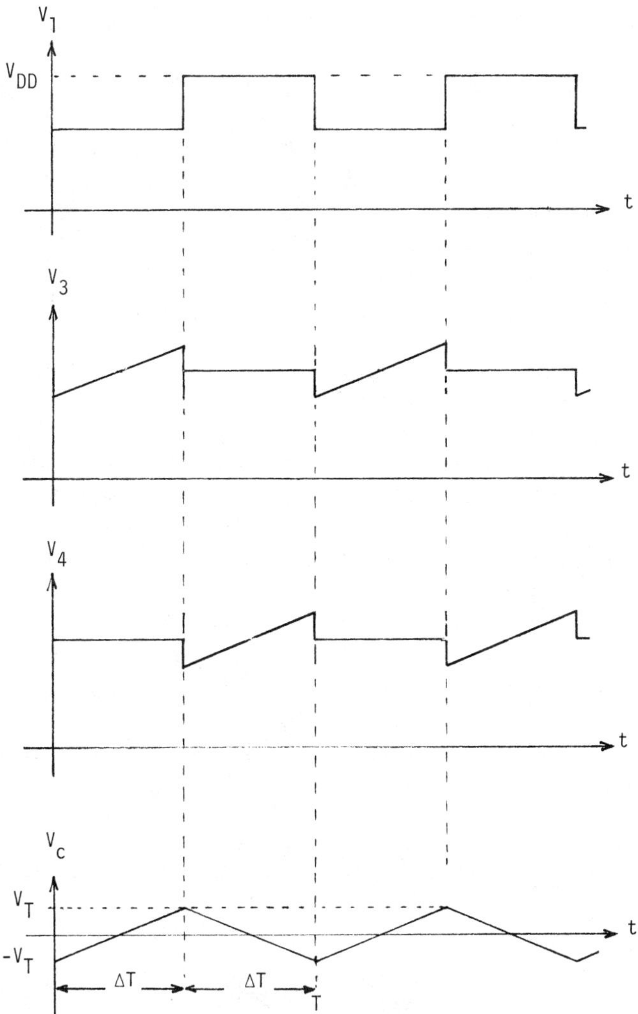

Fig. 6.7–9. Waveforms of Fig. 6.7–8.

Other VCOs that use a voltage-controlled current source to charge and discharge a capacitor would also be applicable for the phase-lock loop application.

The phase detector of Fig. 6.4–8 could be used in place of the phase detector in Fig. 6.7–4. This circuit has the advantage of being insensitive to the type of input waveform. The phase-lock loop is a very versatile building block and greatly extends the application areas for SC circuits.

6.8 PROGRAMMABLE SC CIRCUITS

In many signal-processing applications it is desirable to change the characteristics of a circuit. This change is called *programming* and can be done on a real-time basis. Circuits that are capable of programming on a real-time basis are called *dynamically programmable.* All other forms of programmable circuits are called *statically programmable.* There are many reasons to have programmable circuits. Dynamically programmable circuits can be used in adaptive circuits, such as vibrational analysis, adaptive filters, music synthesizers, and format speech synthesizers. Statically programmable circuits are generally used as a means to redesign the function of a circuit without starting from the beginning. Some statically programmable circuits are capable of only one-time programming, whereas others can be reprogrammed many times, but not on a real-time basis.

In switched capacitor circuits, the only elements suitable for programming are the capacitors. The capacitors can be varied by a number of means, which include fuses, switches, lasers, and electron beams. Of these, only the switch offers convenient reprogramming capability. Inasmuch as all programming schemes change the capacitors, let us consider how a capacitor array can be programmed. The first consideration is area. The silicon area required for the capacitor arrays should be as small as possible while having a large enough unit capacitor size to provide good relative accuracy. If the programming is to be done on a real-time basis, the arrays must be designed so that capacitor switching during the dynamic programming will not distort the response or create large displacement currents.

An eight-bit programmable capacitor is shown in Fig. 6.8–1: d_o through d_7 are binary inputs that have the value of 0 or 1. If the bit d_i is high (has a value of 1), then the capacitor with the value of 2^iC_u is connected between x and x'. If d_i is low (has the value of 0), then the capacitor of value 2^iC_u is connected between x and ground. If x' is at ground potential (such as across the null port of an op amp), then the voltage across the capacitors in the array is the same whether they are in or out of the array. If x is at the output of an op amp, then the capacitors from x to ground are driven as a capacitive load.

Another programmable capacitor array is shown in Fig. 6.8–2. In this array all switches are turned on except one. Some of the capacitors are connected in parallel between x and x', and the remaining capacitors are connected in parallel between x and ground. Fig. 6.8–3 shows how the capacitor arrays of Figs. 6.8–1 and 6.8–2 can be used with an op amp.

Although there are many different areas of application for programmable SC circuits, we shall illustrate a second-order filter and a second-order equal-

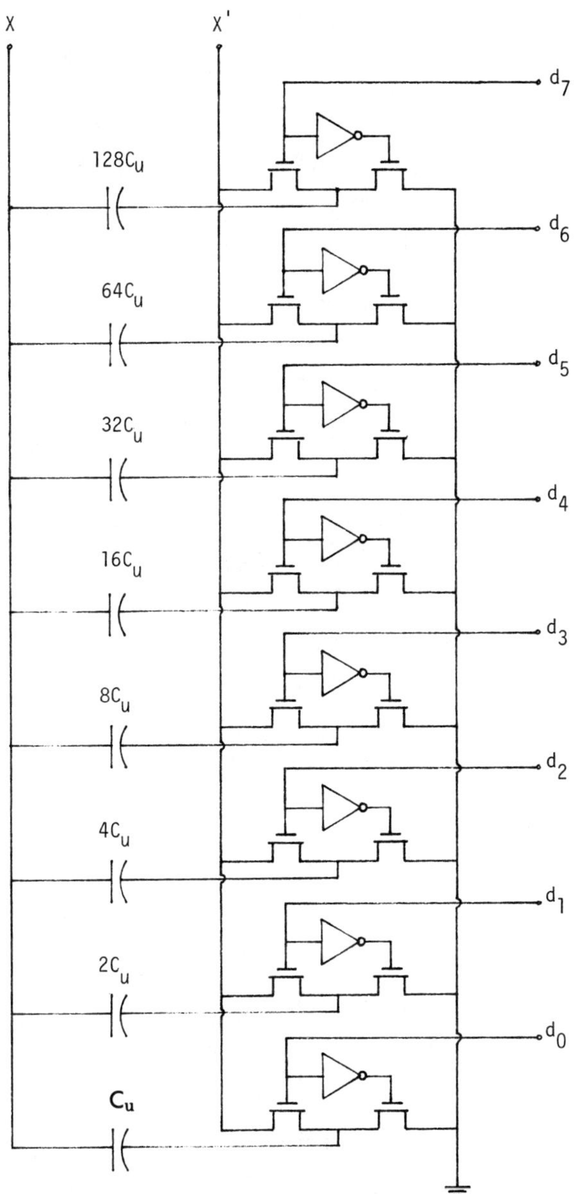

Fig. 6.8–1. A 8-bit, binary weighted capacitor array.

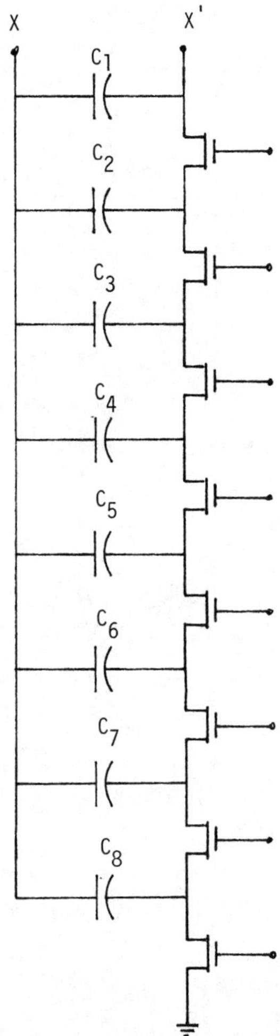

Fig. 6.8–2. An 8 segment capacitor array.

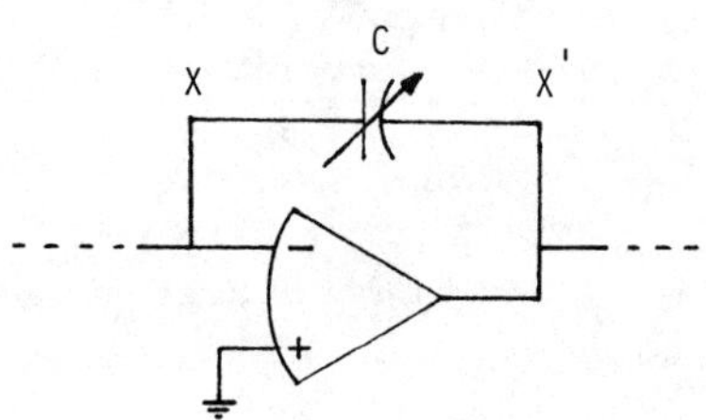

Fig. 6.8–3. Application of a programmable capacitor array in SC circuits.

izer. Figure 6.8–4 shows a second-order switched capacitor filter.[29,30] Assuming a high sampling frequency allows one to derive the relationships for the Q and the center frequency f_o as

$$C_Q = QC_u \tag{1}$$

and

$$C_{f_o} = \frac{f_c}{2\pi f_o} C_u \tag{2}$$

If the input is applied to V_{BPi}, then the output is a second-order bandpass filter with a gain of unity at f_o. If the input is applied to V_{LPi}, the output is a low-pass second-order filter with a DC gain of unity. The programmable capacitors of Figs. 6.8–1 and 6.8–2 can be connected across each op amp to achieve an electrically programmable SC filter. It should be observed that the pole frequency, f_o, of the filter is proportional to the clock frequency. Thus, the clock frequency can be used to adjust the value of f_o.

The above concepts have been used to produce a commercially available monolithic programmable filter array that consists of four second-order circuits like Fig. 6.8–4.[31] Each section is independently accessed through a digital 10-bit control word that determines the Q and f_o of a pole pair. Thirty-two values of Q are possible for each pole pair using the approach shown in Fig. 6.8–1. The f_o control used the configuration of Fig. 6.8–2 where 32 logarithmically spaced frequencies over an octave frequency range were available. One of seven possible frequency ranges could be selected by varying the clock frequency. A structural block diagram of the filter is shown in Fig. 6.8–5.[32] It is possible to permanently program the filter through the use of fusible lines. Figure 6.8–6 shows how this could be accomplished. A photomicrograph of the quad programmable filter is shown in Fig. 6.8–7.

As a second example of programmable SC circuits, consider the second-order equalizer of Fig. 6.8–8.[33] The function of the equalizer is to provide a more uniform delay response. The indicated capacitors are made program-

---

[29] D. J. Allstot, R. W. Brodersen, and P. R. Gray, "An Electrically-Programmable Switched Capacitor Filter," *IEEE J. of Solid-State Circuits,* Vol. SC-14, December 1979, pp. 1034–1041.

[30] D. B. Cox and L. T. Lin, "A Realtime Programmable Switched Capacitor Filter, *Digest of Technical Papers, 1980 IEEE International Conf. on Solid-State Circuits,* February 1980, pp. 94–95.

[31] R5610 Quad Programmable Filter Array, EG&G Reticon, 345 Potrero Avenue, Sunnyvale, CA 94086.

[32] D. B. Cox and L. T. Lin, Ibid.

[33] K. Martin and A. S. Sedra, "Switched-Capacitor Building Blocks for Adaptive Systems," *IEEE Trans. on Circuits and Systems,* Vol. CAS-28, No. 6, June 1981, pp. 526–584.

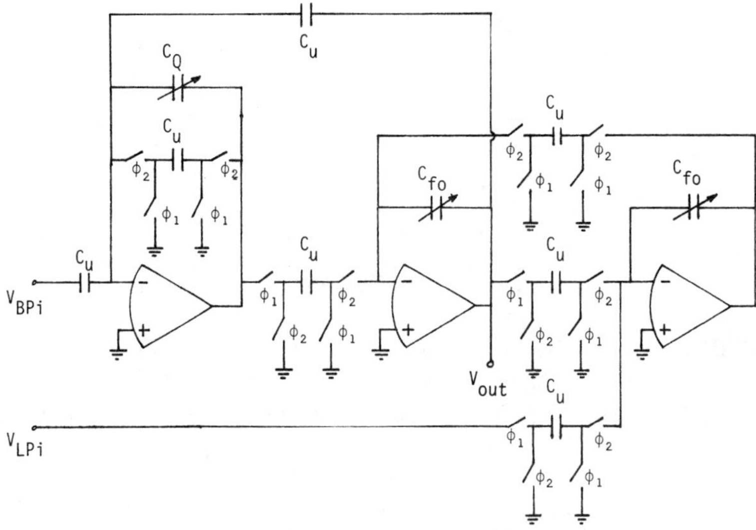

Fig. 6.8–4. A SC second-order programmable filter.

mable by using the technique of Fig. 6.8–1. The exact transfer function of Fig. 6.8–8 is given as

$$\frac{V_o(e^{j\omega T})}{V_i(e^{j\omega T})}=\left(\frac{\alpha_G\alpha_0}{\alpha_1}\,e^{-j\omega T/2}\right)\left[\frac{1-\left(\dfrac{2\alpha_5+\alpha_3}{\alpha_0\alpha_2}\right)2\sin^2\left(\dfrac{\omega T}{2}\right)+j\dfrac{\alpha_3}{\alpha_0\alpha_2}\sin\omega T}{1-\left(\dfrac{2+\alpha_4}{\alpha_1\alpha_2}\right)2\sin^2\left(\dfrac{\omega T}{2}\right)+j\dfrac{\alpha_4}{\alpha_1\alpha_2}\sin\omega T}\right] \quad (3)$$

Suitable design equations to achieve a denominator with pole frequency ω_0 and Q_0, a numerator with zero frequency ω_z and Q_z, and a low frequency gain G are

$$\alpha_1\alpha_2=\frac{2}{\left[\dfrac{1}{2\sin^2\left(\dfrac{\omega_0 T}{Q}\right)}-\dfrac{1}{Q_0\sin\omega_0 T}\right]} \quad (4)$$

$$\alpha_4=\frac{\alpha_1\alpha_2}{Q_0\sin\omega_0 T} \quad (5)$$

$$\alpha_0=\alpha_1 \quad (6)$$

$$\alpha_3 = \frac{\alpha_0\alpha_2}{2}\left[\frac{1}{2\sin^2\left(\frac{\omega_z T}{2}\right)} - \frac{1}{Q_z \sin \omega_z T}\right] \tag{7}$$

$$\alpha_5 = \frac{\alpha_0\alpha_2}{Q_z \sin \omega_z T} \tag{8}$$

and

$$\alpha_G = G \tag{9}$$

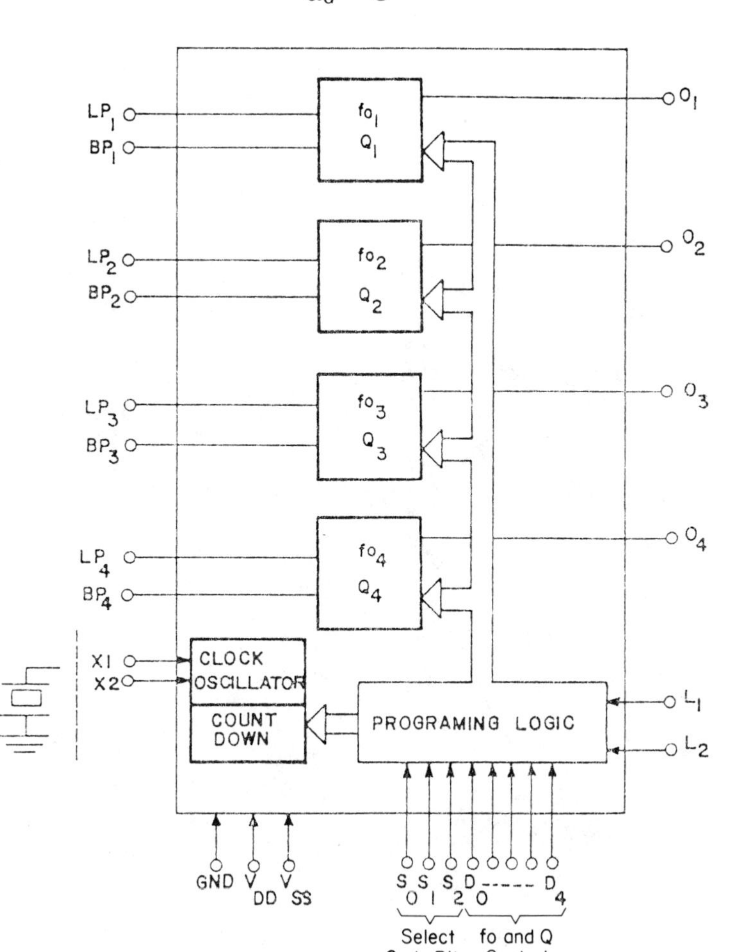

Fig. 6.8–5. Structural block diagram of a quad second-order programmable filter. (© 1980 IEEE)

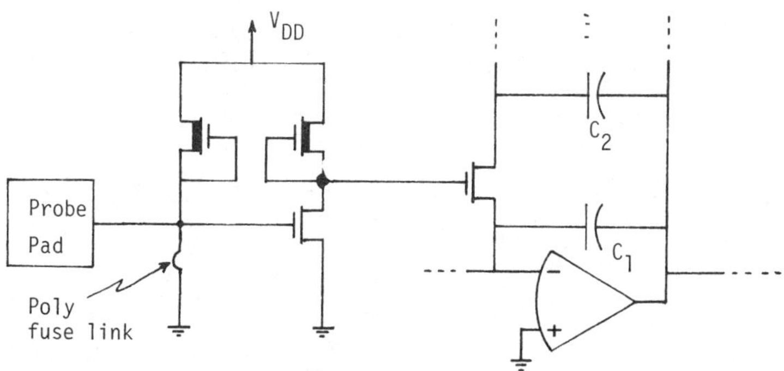

Fig. 6.8–6. A method of permanently programming a capacitor array.

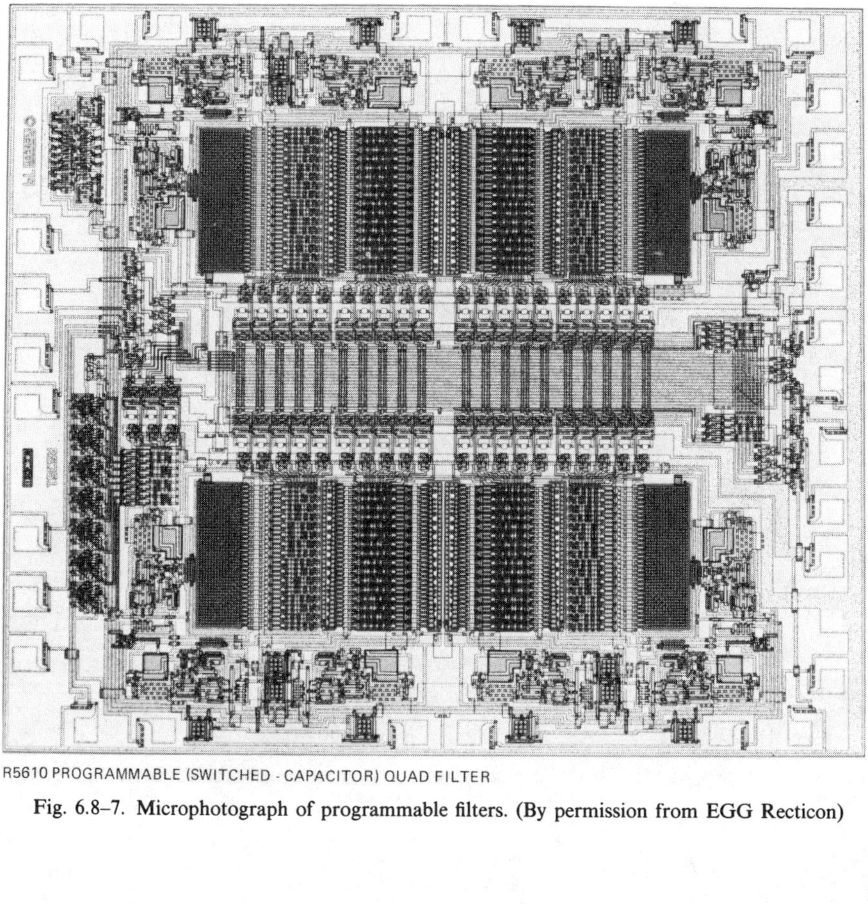

R5610 PROGRAMMABLE (SWITCHED - CAPACITOR) QUAD FILTER

Fig. 6.8–7. Microphotograph of programmable filters. (By permission from EGG Recticon)

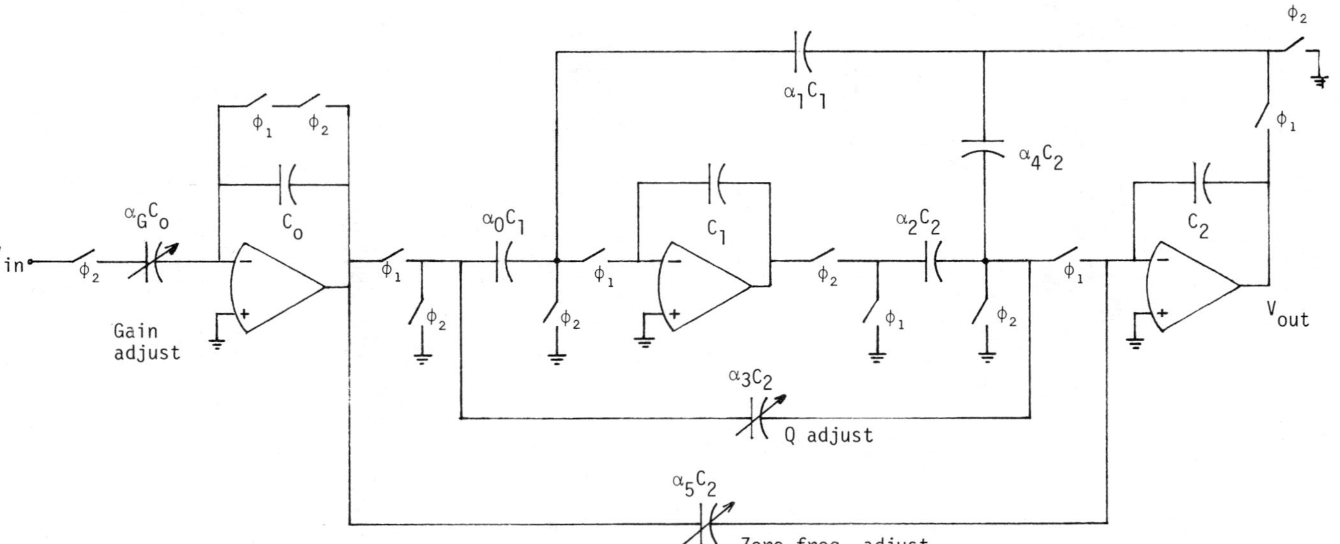

Fig. 6.8–8. A second-order equalizer with adjustable gain, zero, Q and zero frequency.

The ratio α_1/α_2 should be chosen to optimize the dynamic range. This choice will depend upon the frequency of the pole of the channel being equalized.

The equalizer of Fig. 6.8–8 is shown as an adaptive equalizer in Fig. 6.8–9. It is assumed that the equalizer is placed in cascade with the transmission channel. During an initial setup time the channel is taken out of general use, and two sine waves with different frequencies are transmitted through it. After passing through the channel and equalizer, the resulting output signals are compared with "perfect" versions of the inputs. Any detected errors are used to cause the equalizer to adapt until the total system has an essentially flat response and a 0-dB gain. Applications where such a system could be useful include equalizing the public address system of an auditorium and improving the echo cancellation of a two-to-four wire converter (in a telecommunications system).

The transmission channel is assumed to be characterized by an all-pole low-pass function of second-order or less. The complex-conjugate poles of the channel are assumed to be between 2 kHz and infinity. Furthermore, the low-frequency gain of the channel is assumed to be between 1/20 and 2.

In the adaptive equalization system of Fig. 6.8–9, a 0.5-kHz sine wave is used to adapt the system gain to 0 dB. This is done by using Fig. 6.8–10(a) to find the real part of the difference between the channel input and

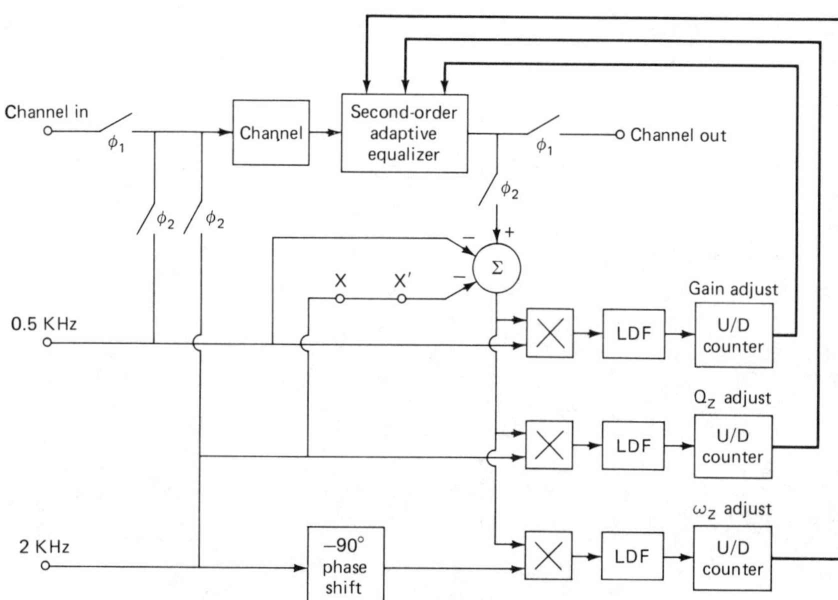

Fig. 6.8–9. A second-order adaptive equalizer.

the channel output at 0.5 kHz. It is seen that Fig. 6.8–10(a) uses the blocks developed in Sections 6.4 and 6.6. The real part of the transfer function at 0.5 kHz is used as an input to an up/down converter. The digital output of the up/down is used to control the gain adjust capacitor, $\alpha_G C_u$. This negative feedback loop will cause the up/down converter to oscillate about zero. This will occur when the real part of the system input is equal to the real part of the system output. Because the channel and the equalizer are assumed to have negligible phase shift at 0.5 kHz, the system will adapt to a 0-dB system gain.

A 2-kHz signal is used to adjust the zeros of the equalizer. If the equalizer has adapted "perfectly" to exactly a 0-dB gain and a flat response, then the system input should equal the system output at 2 kHz. The real and imaginary parts of any difference signal between the input and output are obtained using the techniques in Fig. 6.8–10. These orthogonal error signals are then used to control up/down counters that adjust the Q-factor and the frequency of the equalizer's zeros.

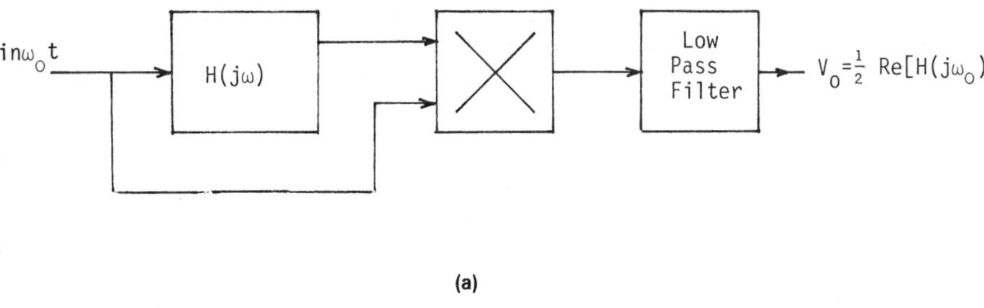

(a)

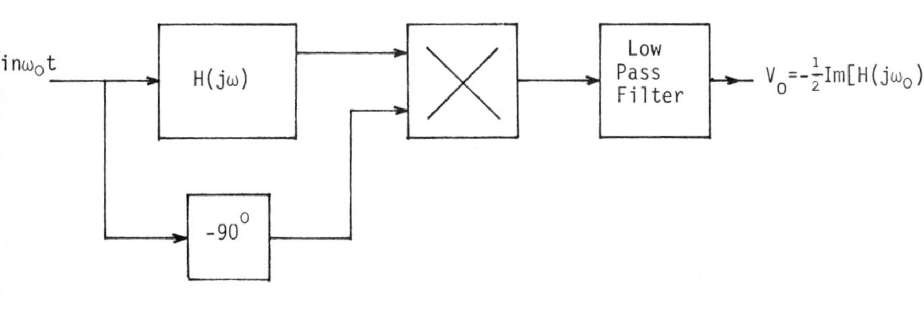

(b)

Fig. 6.8–10. A means of (a) finding the real part or (b) imaginary part of a transfer function.

This adaptation algorithm was chosen because it can be shown that for system phase errors of less than $\pm 90°$:

1. The real part of the difference between the system output and the system input is a monotonically decreasing function of the Q-factor of the equalizer's zeros.
2. The imaginary part of the difference between the system output and the system input is a monotonically decreasing function of the resonant frequency of the equalizer's zeros.

Furthermore, it was experimentally found that convergence was obtained even for initial phase errors larger than $\pm 90°$.

The integrating action of the up/down counters can be used to do much of the low-pass filtering required in measuring the real and imaginary error components. This allows for a smaller capacitor spread in the damped integrators. To ensure that errors due to aliasing do not occur, the counters must be clocked at a frequency equal to or larger than the signal frequencies.

In the preceding explanation it was assumed that after perfect adaptation the system would have 0-dB gain and 0° phase shift at 2 kHz. This ignores the effects of the equalizer's poles (the pole frequency was taken as 8 kHz and the pole Q-factor was $1/\sqrt{2}$), which were included for stability purposes. At 2 kHz the overall system attenuation due to these poles is negligible, but the phase shift is quite noticeable. If the equalizer were designed to adapt until there was no phase shift at 2 kHz, the equalizer's zeros would be placed at a frequency lower than that of the channel's poles to cancel the extra phase shift (of the equalizer's poles). This would result in a nonflat system frequency response. To eliminate this error the 2-kHz input signal was phase-shifted before it was subtracted from the equalizer's output (a phase-shifter was introduced between points x and x' of Fig. 6.8–9). Thus during adaptation, the equalizer's zeros are placed so that at 2 kHz there is a finite overall phase-shift equal to that caused by the equalizer's pole only. This gives flat response up to 8 kHz. The phase-shifted version of the 2 kHz input signal was obtained by adding part of the in-phase component and part of the quadrature component.

The up/down counters must be of the saturating type. This means that if the state of all 0's or all 1's is reached, the counters are inhibited until the up/down control signal changes polarity. Furthermore, to guarantee accurate adaptation, the threshold level of the up/down control must be exactly 0 V. Otherwise offset error will occur. This can be achieved by including a comparator before the up/down counter (or possibly a simple differential pair on its input). Finally, after convergence is achieved, the up/down counters must be inhibited to preserve the equalizer shape during actual system opera-

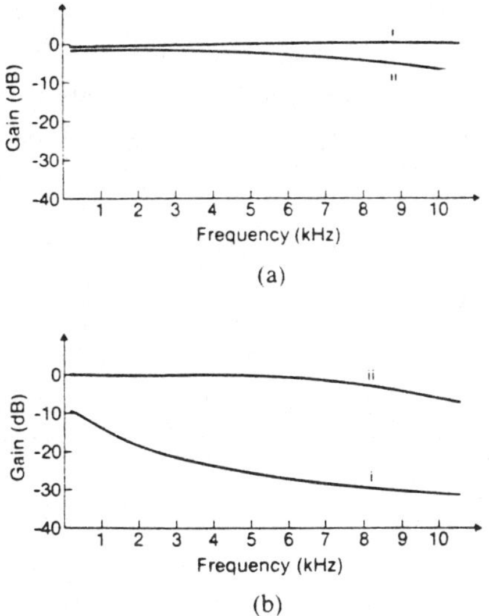

Fig. 6.8–11. Measured channel response (trace i) and channel and equalizer response (trace ii) for (a) a perfect channel and (b) a channel with a first-order pole at 0.5 KHz.

tion. The channel equalizer of Fig. 6.8–9 has been built and tested using discrete circuitry.[34] Figure 6.8–11(a) shows the measured channel response and the equalizer response for a perfect channel. Figure 6.8–11(b) shows the same waveforms when the channel has a first-order pole at 0.5 kHz. It is seen that the equalizer does a reasonable job of maintaining a flat magnitude.

Two examples of programmable SC circuits have been presented in this section. Their purpose was to illustrate the type of programmable circuit that can be implemented. Many other programmable circuits are possible using the concepts introduced here. The advantage of programmable circuits is their flexibility, and the disadvantage is the additional complexity and area that they require.

6.9 SUMMARY

It has been demonstrated in this chapter that SC circuit techniques find applications beyond that of filters. The forte of SC circuits has been to provide a means where practical analog circuits can be built using MOS integrated circuit technology. The key factor has been that the circuit performance can be related to capacitor ratios, which can be as acccurate as 0.3%. Though the most difficult circuits to build are filters, it has been relatively straight-

---

[34] K. Martin and A. S. Sedra, Ibid.

forward to extend SC techniques to nonfilter applications. Although the SC circuits in this chapter require considerable area, the functions they perform would typically require as much area if done by digital techniques. SC circuits have truly created a viable choice between analog and digital techniques for designing signal-processing circuits.

PROBLEMS

6.1 (Sec. 6.1). For the region in Fig. 6.1–3 where analog and digital bandwidths overlap, list considerations that may be important in choosing one approach versus the other.

6.2 (Sec. 6.2). Fig. P 6–2 shows another scheme for canceling the voltage offset of an op amp used as a comparator. Analyze this circuit to show how cancellation of the input offset voltage occurs.

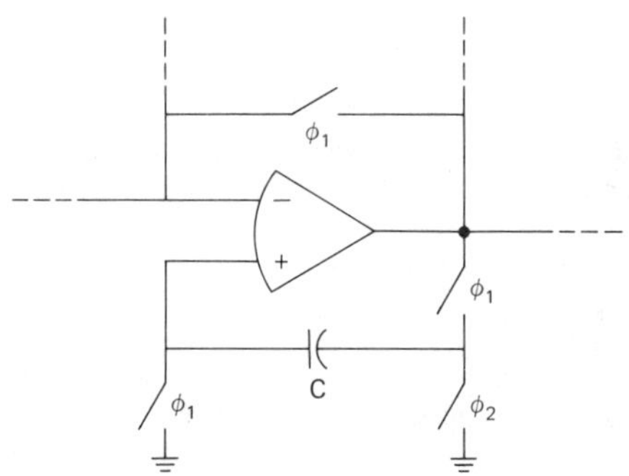

Fig. P6.2.

6.3 (Sec. 6.2). If for the comparator of Fig. 6.2–5, $V_{OH} = 10$ volts and $V_{OL} = -10$ volts, what is the relationship between R_1 and R_2 that gives $V_{IL} = -5$ volts and $V_{IH} = 5$ volts?

6.4 (Sec. 6.2). Assume that V_{OH} and V_{OL} of Fig. 6.2–5 are equal to the power supplies of the op amp and are available for biasing purposes. Show how to modify Fig. 6.2–5 so that if $V_{OH} = 10$ volts and $V_{OL} = -10$ volts $V_{IL} = 4$ volts and $V_{IH} = 6$ volts.

6.5 (Sec. 6.2). Assume an arbitrary voltage is available to apply between R_1 of Fig. 6.2–5 and ground. What value of this arbitrary voltage will meet the specifications of Problem 6.4?

6.6 (Sec. 6.2). Show how to convert Fig. 6.2–5 into a noninverting comparator with hysteresis, and give expressions for V_{IH} and V_{IL} in terms of V_{OH}, V_{OL}, and the components of the realization.

6.7 (Sec. 6.2). Assume that V_{in} in Fig. 6.2–6 is sufficiently negative so that M1 is off, M2, M5, and M7 are on and can be represented as resistors that are inversely proportional to the W/L ratios given on the schematic. Use this assumption and the values of V_{DD} and V_{SS} given in Fig. 6.2–7 to calculate V_{OL}, and compare to Fig. 6.2–7.

6.8 (Sec. 6.3). By using methods similar to that in Fig. 6.3–7, show how to shift the breakpoint of Fig. 6.3–5 to the left or right.

6.9 (Sec. 6.3). Sketch the transfer function of V_{out} versus V_{in} for Fig. 6.3–8 if the comparator input polarities are reversed.

6.10 (Sec. 6.3). Repeat Example 6.3–1 if (a) the output signal is to be the negative magnitude of the input signal or (b) if the output signal is to be 10 times the magnitude of the input signal.

6.11 (Sec. 6.3). Design a switched capacitor realization of the transfer function shown in Fig. P 6–11.

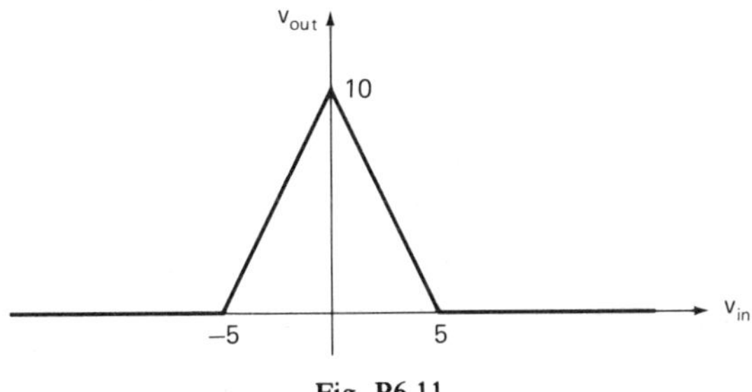

Fig. P6.11.

6.12 (Sec. 6.3). Repeat Problem 6.11 if the transfer function of v_{out} versus v_{in} is rotated 180° about the v_{in} axis.

6.13 (Sec. 6.3). Design a switched capacitor realization of the transfer function shown in Fig. P 6–13.

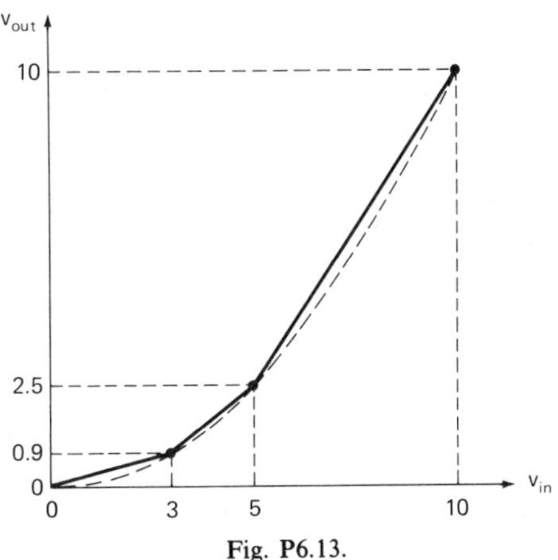

Fig. P6.13.

6.14 (Sec. 6.3). Repeat Problem 6.13 if the transfer function of v_{out} versus v_{in} is rotated 180° about the v_{out} axis.

6.15 (Sec. 6.3). Show by the addition of switches only how to use a single op amp to realize Fig. 6.3–5(a).

6.16 (Sec. 6.4). With the inputs of Eqs. (3) and (4) applied to Fig. 6.4–6, derive the output given in Eq. (5).

6.17 (Sec. 6.4). Repeat Problem 6.16 if $v_m(t)$ is given by Eq. (10) rather than Eq. (4).

6.18 (Sec. 6.4). Show that the output of the modulator of Fig. 6.4–8 is given by Eq. (12).

6.19 (Sec. 6.4). Show that the output of the modulator of Fig. 6.4–9 is given by Eq. (14) or Eq. (15).

6.20 (Sec. 6.4). What is the range of uncertainty in the operational multiplier scheme proposed in Fig. 6.4–13?

6.21 (Sec. 6.5). Show that the approximate s-domain transfer function of the passive simulation of Fig. 6.5–2 is given by Eq. (3).

6.22 (Sec. 6.5). (a) Find the frequency of oscillation for the RC oscillator of Fig. 6.5–5(a). (b) Find the frequency of oscillation of the SC oscillator of Fig. 6.5–5(b) and compare with part (a).

6.23 (Sec. 6.5). (a) Find the frequency of oscillation for the RC oscillator of Fig. 6.5–6(a). Assume that the MOS device is modeled by the transconductance g_m and the channel resistance is much greater than R_L. (b) Find the frequency of oscillation of the SC oscillator of Fig. 6.5–6(b) and compare with part (a).

6.24 (Sec. 6.5). Find the exact frequency of oscillation for the SC oscillator of Fig. 6.5–7(b).

6.25 (Sec. 6.5). Show how to make the oscillator of Fig. 6.5–11 have unequal duty cycle, i.e., T_1 different from T_2 of Fig. 6.5–9. Your modifications should be compatible with SC circuit techniques.

6.26 (Sec. 6.5). Show that the frequency of oscillation of Fig. 6.5–12 is given by Eq. (23).

6.27 (Sec. 6.6). (a) Show that the phase shift of Fig. 6.6–1(a) is given by Eq. (1). (b) Find the z-domain transfer function of Fig. 6.6–1(b), and assume that the sampling rate is high, and derive the expression in Eq. (2).

6.28 (Sec. 6.6). Show that the phase shift of Fig. 6.6–2(b) is given by Eq. (2).

6.29 (Sec. 6.6). Show that the phase shift of Fig. 6.6–3(b) is given by Eq. (4).

6.30 (Sec. 6.7). Verify Eq. (16) for the frequency of the VCO of Fig. 6.7–8.

6.31 (Sec. 6.7). Show that the addition of a resistor R_1 in series with both M5 and M6 results in the expression given by Eq. (17).

6.32 (Sec. 6.8). Show that the pole locations of Fig. 6.8–4 are given by

$$r^2 = 1 - \frac{C_u^2}{C_Q C_{f_o}} + \frac{C_u^2}{C_{f_o}^2}$$

and

$$2r \cos \theta = 2 - \frac{C_u^2}{C_Q C_{F_o}}$$

and that the stability condition is given by

$$\frac{\omega_o}{f_c} - \frac{1}{Q} < 0.$$

7 MOS Analog-digital Converters

7.0 INTRODUCTION

The ability to convert between digital and analog signals and vice versa has become very important in many signal-processing applications. The switched capacitor circuits of the previous chapters find applications in systems containing digital circuits, such as microprocessors. Consequently, it is important to consider how to convert between analog and digital signals using technology compatible with switched capacitor circuits, namely MOS technology. It is the objective of this chapter to examine the characterization, architecture, and design of analog-to-digital and digital-to-analog converters that are compatible with single-channel MOS or CMOS.

Figure 7.0–1 illustrates how analog-to-digital (A/D) and digital-to-analog (D/A) converters[1] are used in data systems.[2] In general, an A/D conversion process will convert an analog signal that is sampled and held to a digital word that is a representation of the analog sampled signal. Often, many analog inputs are multiplexed to the A/D converter. The D/A conversion process is essentially the reciprocal of the A/D conversion process. Digital words are applied to the input of the D/A to create from a reference voltage an analog output signal that is a representation of the digital word.

This chapter will characterize analog-digital converters. The characterization will include the static and dynamic specifications of the conversion process. Next, D/A converters will be divided into the categories of binary-weighted, resistive voltage division, and serial conversion techniques. Examples of each of the categories are presented. Last, the chapter discusses A/D converters, which are divided into the categories of serial, successive approximation, and parallel. Each of these categories will be discussed with

[1] A/D and D/A are used as shorthand for analog-to-digital and digital-to-analog converters, respectively.

[2] B. M. Gordon, "Linear Electronic Analog/Digital Conversion Architecture, Their Origins, Parameters, Limitations, and Applications," *IEEE Trans. on Circuits and Systems,* Vol. CAS-25, No. 7, July 1978, pp. 391–418.

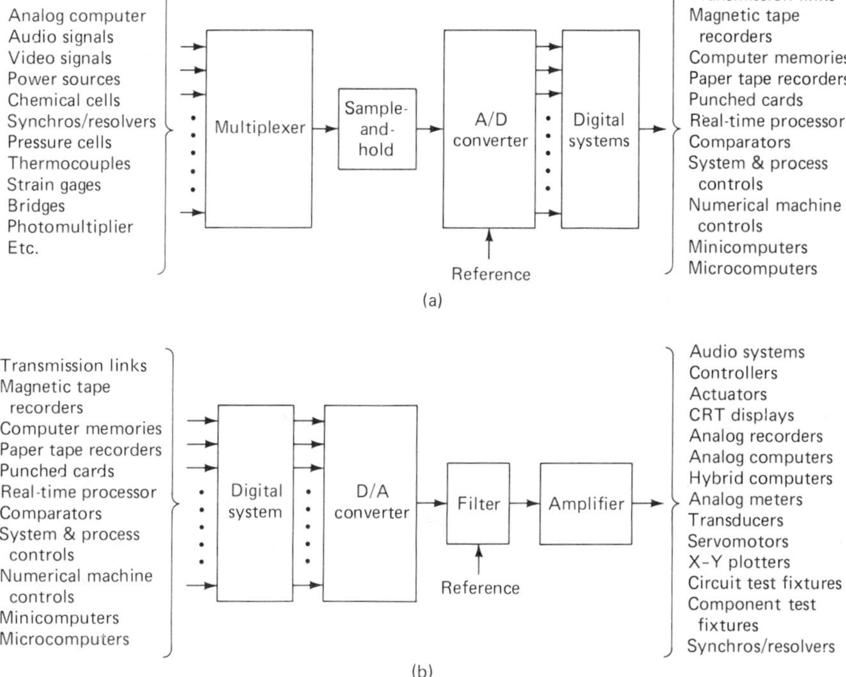

Fig. 7.0–1. (a) A/D converters and (b) D/A converters in data systems.

examples presented. This chapter will complete the presentation of switched capacitor circuits in this text. The remaining chapter, Chapter 8, will be devoted to giving the background necessary to bring the integrated circuit MOS technology together with the design techniques of this and the previous chapters.

7.1 CHARACTERIZATION OF ANALOG-DIGITAL CONVERTERS

In this section we wish to develop the basic definitions by which A/D and D/A converters can be characterized. Because of the complexity of most converters, it is not possible to characterize their performance simply. Because an A/D converter is the reciprocal of a D/A converter, they can be characterized simultaneously. We shall characterize the converters from their static and dynamic properties. Other properties include noise and temperature.

Consider the diagram of a basic D/A converter shown in Fig. 7.1–1. In this figure a digital word is used to control a switch S that applies an analog

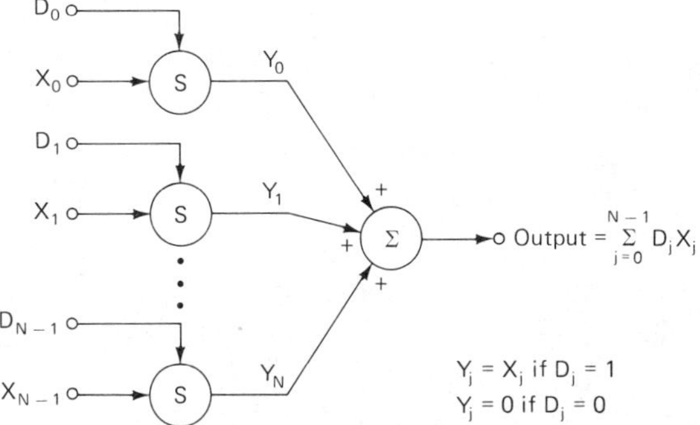

Fig. 7.1–1. Block diagram of a basic D/A converter.

signal to a summing circuit to produce an analog output signal. The analog signals may be either voltage or current. The ideal static behavior of a D/A converter is shown in Fig. 7.1–2. The input is on the horizontal axis and consists of all possible combinations of the digital input word. Figure 7.1–2 is for a three-bit digital word. The vertical scale is the analog output of the ideal D/A converter. The maximum analog output signal is designated

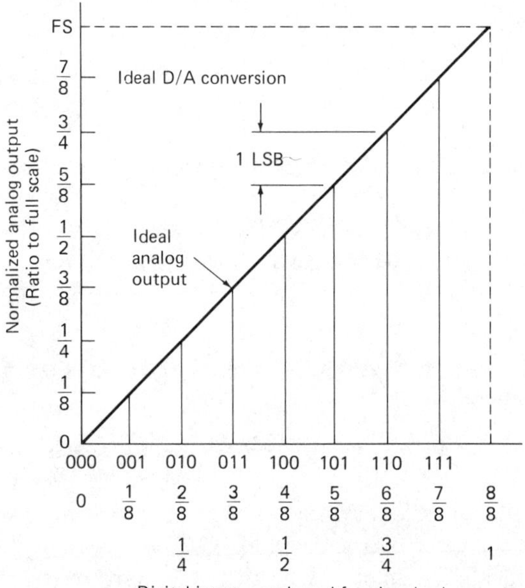

Fig. 7.1–2. Ideal input-output characteristics for a three-bit D/A converter. (By permission from D. H. Sheingold, *Analog-Digital Conversion Handbook,* Analog Devices, Inc., 1972.)

as the full scale (FS) value. For each digital word, there should be a unique analog output signal. Any deviations from Fig. 7.1–2 fall into the category of static conversion errors.

The static conversion errors include integral linearity, differential linearity, absolute linearity, resolution, zero and full-scale error, and monotonicity. *Integral linearity* is the maximum deviation of a D/A's output with any given input code from a straight line drawn from its ideal minimum to its ideal maximum output. Integral linearity can be expressed as a percentage of the full-scale range or in terms of the least significant bit (LSB). Integral linearity has several subcategories, which include absolute, best-straight-line, and end-point linearity.[3] *Absolute linearity* is measured by assuming that the output of a D/A will begin at zero and end at full scale and then comparing the actual outputs to a line drawn between these two points (see Fig. 7.1–3(a). Absolute linearity emphasizes the zero and full-scale errors. The *zero error* is the difference between the actual output and zero when the digital word for a zero output is applied. The *full-scale error* is the difference between the actual output and the reference voltage when the digital word for a full-scale output is applied to the D/A. Another error closely related to full-scale error is gain error. *Gain error* is the difference between the gains of the actual static and ideal input-output characteristics. Gain errors exist when the slope of the actual characteristic is not parallel to the slope of the line indicative of the ideal characteristic. Thus, a disadvantage of the absolute linearity measurement is that zero and full-scale errors can contribute to the nonlinearity of an otherwise linear D/A. Another subcategory of integral linearity is the best-straight-line linearity. *Best-straight-line linearity* defines the accuracy of a D/A in terms of the deviation from the ideal output range without regard to zero or full-scale errors. Figure 7.1–3(b) shows the characteristics of a D/A converter that has a large full-scale error but good linearity. The last subcategory of integral linearity includes end-point linearity and is illustrated in Fig. 7.1–3(c). *End-point linearity* uses a straight line through the *actual* end points, rather than the ideal end points.

Differential linearity is a measure of the separation between adjacent levels. Differential linearity measures bit-to-bit deviations from ideal output steps, rather than along the entire output range. If V_{cx} is the actual voltage change on a bit-to-bit basis and V_s is the ideal change, then the differential linearity can be expressed as

$$\text{Differential linearity} = [(V_{cx} - V_s)/V_s] \times 100\% \tag{1}$$

[3] D. Gilbert, *Electronic Products,* Vol. 24, No. 3, July 1981, pp. 61–63.

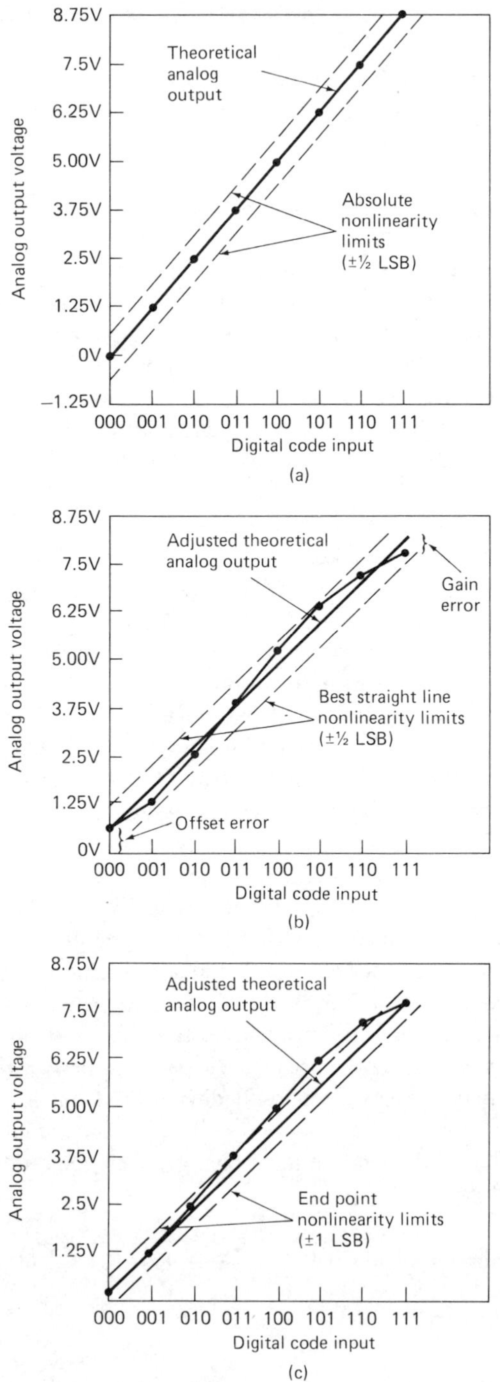

Fig. 7.1–3. Examples of various types of linearity for a three-bit D/A converter. (a) Absolute linearity. (b) Best-straight-line linearity. (c) End-point linearity. (By permission from D. Gilbert, "Understanding D/A Accuracy Specs.," *Electronic Products,* July 1981, pp. 61–63.)

We note that for an N-bit converter and a full-scale voltage range of V_{FSR} that V_s is

$$V_s = V_{FSR}/(2^{N-1}) \tag{2}$$

It is apparent that the differential linearity can also be specified as a fraction of the *least significant bit* (LSB). Figure 7.1–4 shows how differential linearity differs from integral linearity. Figure 7.1–4(a) is for a 12-bit D/A converter having a ±2-LSB integral linearity and a ±½-LSB differential linearity. Figure 7.1–4(b) is for a 12-bit D/A converter having a ±½-LSB integral linearity and a ±1-LSB differential linearity.

In a D/A converter, *monotonicity* means that as the digital input to the converter increases over its full-scale range the analog output never exhibits a decrease between one conversion step and the next. In other words, the slope of the transfer characteristic is never negative in a monotonic converter. Figure 7.1–5 gives an example of a D/A converter that is not monotonic. Another definition used to characterize a D/A converter is resolution. *Resolution* is the smallest input digital code for which an analog output level is produced. Theoretically the resolution is equal to the number of bits. However, the actual resolution will be less if internal noise excursions exceed the quantization level or if component drift causes switching errors (i.e., at major carry points). These types of phenomena will effectively reduce the available resolution of the converter.

The inverse operation of the D/A converter is the A/D converter. Figure 7.1–6 shows a basic diagram of an A/D converter. The input to the A/D is from a sample-and-hold circuit. This input, along with a reference input, is used to determine the digital word that best represents the sampled analog input signal. The means by which the conversion is accomplished may be different from that suggested in Fig. 7.1–6, where a set of fixed reference levels are applied to comparators whose outputs are then decoded into an output digital word. Regardless of the means of conversion, the A/D converter is a device that converts a continuous range of input amplitude levels into a discrete finite set of digital words.

The above definitions also hold for the A/D converter. If the input and output of the D/A converter definitions are interchanged, then the A/D converter definitions follow. Figure 7.1–7 shows the ideal input-output characteristics of an A/D converter. Figure 7.1–8 shows various examples of the previous definitions applied to A/D converters. Figure 7.1–8(a) shows the offset error in an A/D converter. Figure 7.1–8(b) gives an example of gain error. Figure 7.1–8(c) shows an example of absolute linearity error. Finally, Fig. 7.1–8(d) indicates differential linearity.

The above considerations are all considered static errors. Dynamic errors

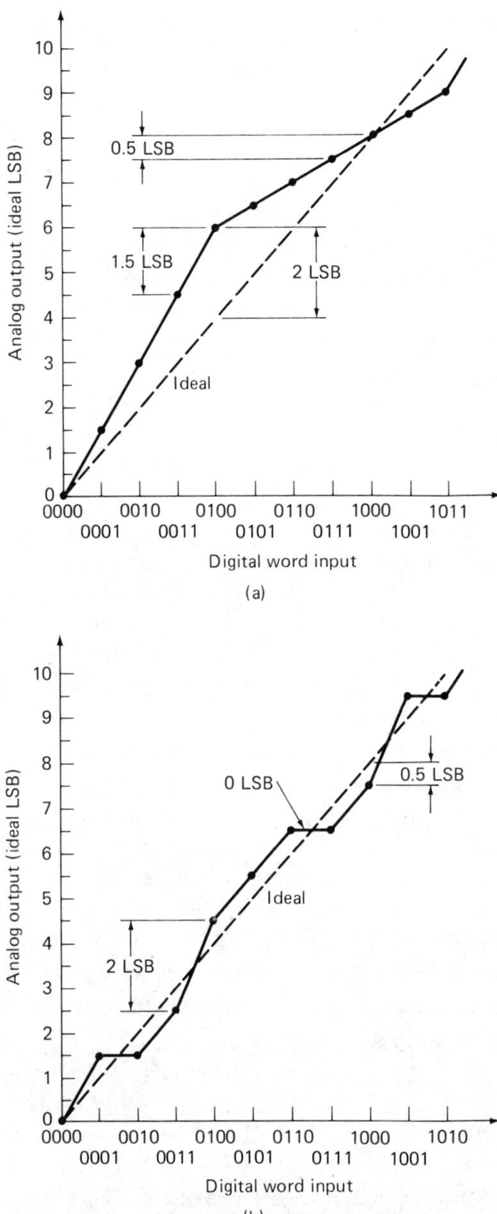

Fig. 7.1–4. Illustration of the difference between integral and differential linearity for a D/A converter. (a) D/A converter with ±2 LSB integral linearity and ±1/2 LSB differential linearity. (b) D/A converter with ±1/2 LSB integral linearity and ±1 LSB differential linearity.

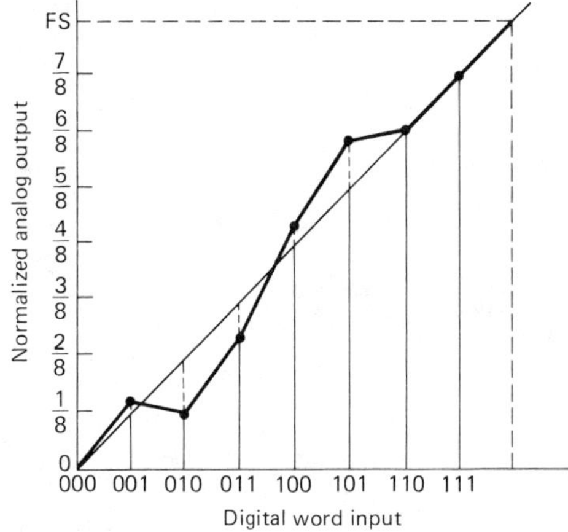

Fig. 7.1–5. Example of a three-bit D/A converter that is not monotonic.

have to do primarily with the speed of operation. Because either the input or the output of a converter is a digital word, sampling or discrete time signals are inherent. Therefore, the rate at which the converter can operate is of interest. *Conversion time* is the time from the application of the start convert signal to the availability of the completed output signal (digital or analog). Typically, a D/A converter can provide an analog output signal as soon as the digital signal is applied. The primary delay would be the settling time of the operational amplifiers or other electronic devices used in the converter scheme. On the other hand, A/D converters typically require a number of clock cycles after the application of the analog signal before the output of the digital word. The reason for this will be seen in the following sections.

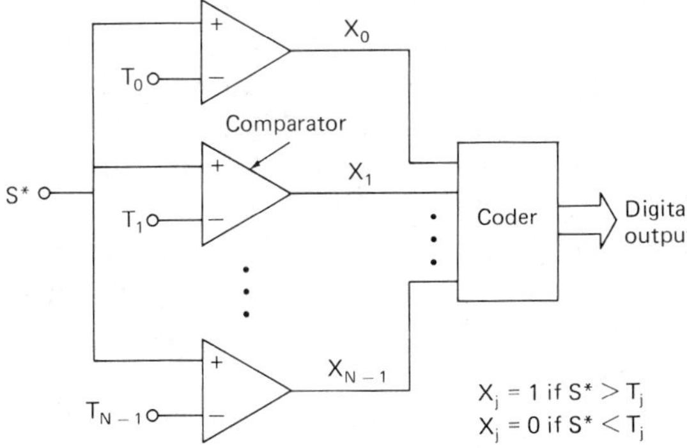

Fig. 7.1–6. Block diagram of the general A/D converter operation.

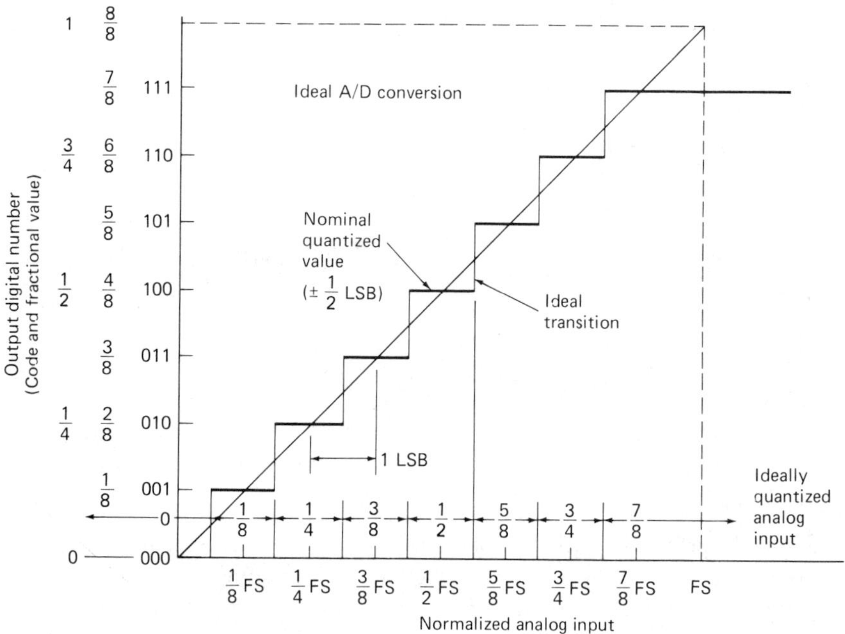

Fig. 7.1–7. Ideal input-output characteristics for a three-bit A/D converter. (By permission from D. H. Sheingold, *Analog-Digital Conversion Handbook,* Analog Devices, Inc., 1972.)

Because a sample-and-hold circuit is used in conjunction with the A/D converter, it is of interest to mention some of the definitions that characterize its performance. A sample-and-hold circuit has two modes of operation. The sample mode applies the input to the memory device. The hold mode disconnects the input from the memory device and ideally provides to the external circuitry the exact value of the input which existed at the first instant of the hold mode. Figure 7.1–9 shows the timing effects in a sample-and-hold circuit. The *acquisition time,* indicated as t_a, is the time during which the sample-and-hold circuit must remain in the sample mode to ensure that the subsequent hold mode output will be within a specified error band of the input level that existed at the instant of the sample-and-hold conversion, after gain and offset effects have been removed. The *settling time,* indicated by t_s, is the time interval between the sample-and-hold transition command and the time when the output transients and subsequent ringing have settled to within a specified error band. The minimum sample time would be

$$T_{\text{sample}} = t_s + t_a \qquad (3)$$

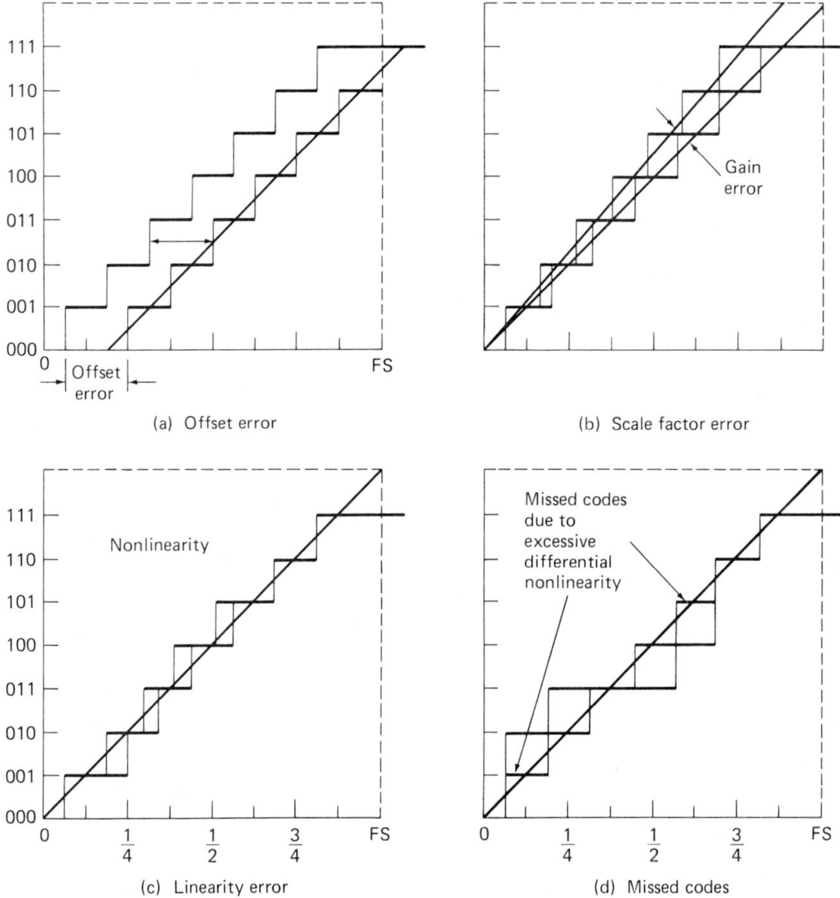

Fig. 7.1–8. Characterization of a three-bit A/D converter. (a) Offset error. (b) Scale factor error. (c) Integral linearity. (d) Differential linearity. (By permission from D. H. Sheingold, *Analog-Digital Conversion Handbook,* Analog Devices, Inc., 1972.)

The minimum conversion time for an A/D converter would be equal to T_{sample}. The maximum sample rate is

$$f_{\text{sample}} = 1/(T_{\text{sample}}) \tag{4}$$

The dynamic performance of the converter will depend largely on the dynamic characteristics of the op amps and converters. Therefore the slew rate, settling time, and overload recovery time of these circuits will be of importance.

In addition to the dynamic characteristics of converters, there are character-

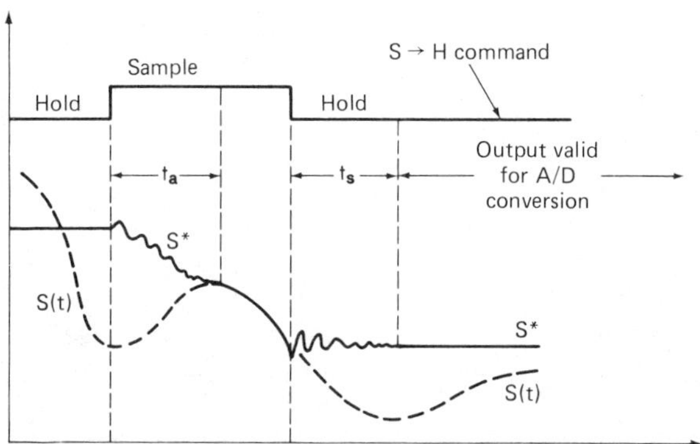

Fig. 7.1–9. Waveforms for a sample-and-hold circuit.

istics having to do with stability. These characteristics define the immunity of the converter to time, temperature, power supplies, and component aging. These characteristics are typically expressed in the change in the converter performance parameter per unit change in the affecting influence. These characteristics include the *temperature coefficient of linearity, temperature coefficient of gain,* and *temperature coefficient of differential nonlinearity.* Also of importance in the stability of a converter is the voltage reference. The voltage reference can be supplied externally or internally to the converter. It is important that the reference provide the stability required for the proper operation of the converter.

It is of interest to consider the testing of D/A and A/D converters. Testing is divided into static and dynamic tests for D/A and A/D converters. Most test configurations demand the ability to resolve the analog signal to within ±½ LSB, which can be very demanding if the number of bits is large. Techniques for testing both types of converters can be found in more detail in the literature.[4,5] Often, a microprocessor is used to perform the tests on the converter. One simple means of testing both the D/A and A/D converter is shown in Fig. 7.1–10. In this test the A/D converter is cascaded with the D/A converter. The output of the D/A converter is compared to the input of the A/D converter, resulting in an error voltage that can be plotted on an *X-Y* recorder as a function of the amplitude of the analog input signal. A typical portion of the error voltage for a 12-bit configuration is shown in Fig. 7.1–11.

[4] D. H. Sheingold, *Analog-Digital Conversion Handbook,* Analog Devices, Norwood, MA 02062, 1972.

[5] J. R. Naylor, "Testing Digital/Analog and Analog/Digital Converters," *IEEE Trans. on Circuits and Systems,* Vol. CAS-25, No. 7, July 1978, pp. 527–538.

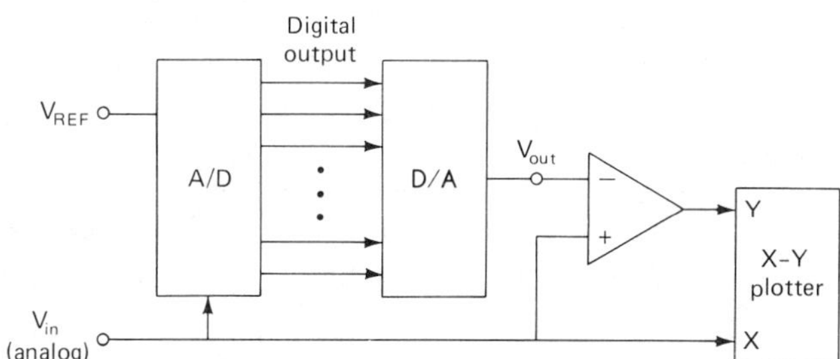

Fig. 7.1–10. A simple method of testing A/D and D/A converters.

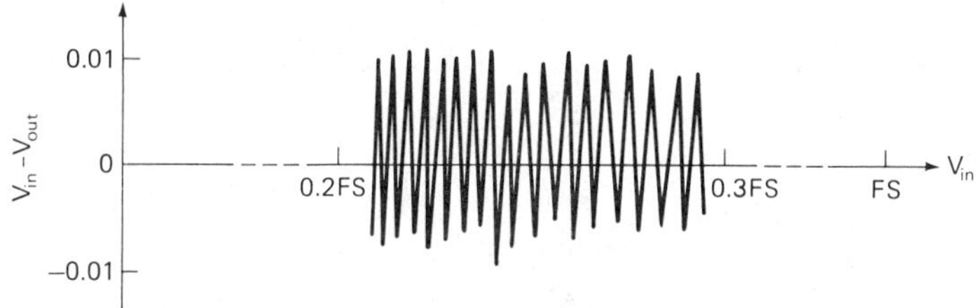

Fig. 7.1–11. Typical portion of the error voltage for Fig. 7.1–10.

In our study of D/A and A/D converters, it is necessary to classify them. One convenient method of characterizing converters is by their speed. Table 7.1–1 shows the characterization of A/D and D/A converters by speed. A/D and D/A converters are divided into ultrahigh-speed, very high-speed, high-speed, intermediate-speed, and low-speed. Our considerations in the re-

Table 7.1–1 Classification of Converters Based on Speed Range

A/D CONVERTERS	CONVERSION RATE	CONVERSION TIME
Ultrahigh-speed	Greater than 3 MHZ	Less than 330 nanoseconds
Very high-speed	300 kHz to 3 MHz	330 nanoseconds to 3.3 microseconds
High-speed	30 kHz to 300 kHz	3.3 to 33 microseconds
Intermediate speed	3 kHz to 30 kHz	33 to 333 microseconds
Low-speed	Less than 3 kHz	Greater than 33 microseconds

D/A CONVERTERS	SETTLING TIME
Ultrahigh-speed	Less than 100 nanoseconds
Very high-speed	100 nanoseconds to 1 microsecond
High-speed	1 to 10 microseconds
Intermediate speed	10 to 100 microseconds
Low-speed	Greater than 100 microseconds

maining sections of this chapter will be divided according to this classification.

In this section we have presented a characterization of the A/D and D/A converter from the viewpoint of static, dynamic, and other characteristics. This information will help us to compare various converters and to quantify their performance. We shall restrict our considerations only to those converters that are compatible with MOS technology. Further information on converters not covered in the following sections can be found in the literature.[6]

7.2 CHARGE-BALANCE/REDISTRIBUTION D/A CONVERTERS

The ability to convert between digital and analog signals and vice versa has become very important in the area of microprocessors and digital signal-processing. We saw in Section 7.1 that the conversion from analog to digital and then from digital back to analog is a major part of a signal-processing system. Though progress is being made in integrating more converter functions onto the same chip with the microprocessor or digital signal-processor, the problems become very difficult at the 12-bit level and beyond. Fundamental limitations, such as the stability of the converter and its voltage reference, are becoming important. As a result, monolithic converters in the 8- to 10-bit range are being improved both in performance and by having more functions associated with the conversion process. The high-speed converters with a large number of bits are relegated to hybrid techniques at the present time.

Our objective in the remainder of this chapter is to examine converter techniques that are compatible with single-channel MOS or CMOS. In so doing, we restrict ourselves from a class of very competitive converters that are implemented by bipolar technology. If a general background on converters is desired, then one should investigate other references.[7,8,9]

D/A conversion is the simpler of the two conversions to implement. Also, because D/A converters are used in the implementation of A/D converters, we will examine them first. If we constrain ourselves to MOS technology, then there are three basic methods of D/A conversion: charge-balance, or

---

[6] D. J. Dooley, "Data Conversion Integrated Circuits," *IEEE Press,* John Wiley & Sons, New York, 1980.

[7] D. Sheingold, *Analog-Digital Conversion Handbook,* Analog Devices, Inc., Norwood, MA 02062, 1972.

[8] E. Hnatek, *A User's Handbook for D/A and A/D Converters,* John Wiley and Sons, Inc., New York, 1976.

[9] R. Allan, "The Inside News on Data Converters," *Electronics,* July 17, 1980, pp. 101–112.

charge redistribution;[10,11,12] resistive voltage division[13] and combinations of these two techniques;[14] and serial D/A converters. This section and the next two will deal with each of these three basic methods.

In a D/A converter, the output variable, voltage or current, is equal to the product of a number represented by the digital input word (code) and an analog reference voltage or current. For example

$$V_{out} = D\, V_{ref} \tag{1}$$

for a D/A converter having a reference voltage V_{ref}, an output voltage V_{out}, and a fractional binary input having the value D given by the

$$D = b_1 2^{-1} + b_2 2^{-2} + b_3 2^{-3} + \cdots + b_N 2^{-N} \tag{2}$$

where b_i are the individual bits and are 1 or 0 as determined by the digital word, b_1 is the *most significant bit* (MSB), and b_N is the *least significant bit* (LSB). V_{ref} may be fixed internally or supplied externally. If V_{ref} is provided externally, the D/A converter is called a *multiplying D/A converter* (MDAC). Unfortunately, it is difficult to build a precise voltage reference using single-channel MOS technology. We shall show in Chapter 8 that reasonably good voltage references are available using CMOS. Consequently, many MOS D/A converters are of the multiplying type. An MDAC can be used with an external reference voltage for ratiometric conversion and for digitally controlled attenuation. This is one of the best methods of achieving a programmable SC filter, because the MDAC placed in series with an SC integrator will vary the integrator time constant.

The MDAC can be characterized in terms of the possible polarities of the input and output. If we consider a plane that consists of the digital word, D, versus V_{ref}, then we can define a *two-quadrant MDAC* as one

---

[10] J. L. McCreary and P. R. Gray, "All-MOS Charge Redistribution Analog-to-Digital Conversion Techniques—Part I," *IEEE J. of Solid-State Circuits,* Vol. SC-10, No. 6, December 1975, pp. 371–379.

[11] Y. S. Lee, L. M. Terman, and L. G. Heller, "A Two-Stage Weighted Capacitor Network for D/A – A/D Conversion," *IEEE J. Solid-State Circuits,* Vol. SC-14, No. 4, August 1979, pp. 778–781.

[12] K. B. Ohri and M. J. Callahan, Jr., "Integrated PCM Code," *IEEE J. of Solid-State Circuits,* Vol. SC-14, No. 1, February 1979, pp. 38–46.

[13] A. R. Hamade, "A Single Chip All-MOS 8-Bit A/D Converter," *IEEE J. of Solid-State Circuits,* Vol. SC-13, No. 6, December 1978, pp. 785–791.

[14] B. Fotouhi and D. A. Hodges, "High-Resolution A/D Conversion in MOS/LSI," *IEEE J. of Solid-State Circuits,* Vol. SC-14, No. 6, December 1979, pp. 920–926.

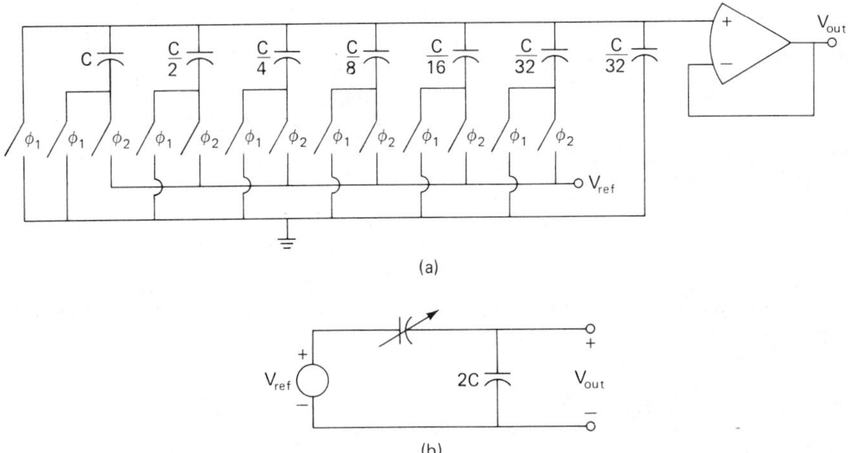

Fig. 7.2–1. (a) A six-bit binary-weighted MDAC. (b) Equivalent circuit of (a). (Note that whether or not the ϕ_2 switches close depends upon b_i.)

whose operating input range consists of only two quadrants of this plane. In this case, either D is unipolar and V_{ref} is bipolar, or V_{ref} is unipolar and D is bipolar. If both D and V_{ref} are bipolar, then the MDAC has four-quadrant operation.

Probably the simplest MDAC is a single-stage, binary-weighted MDAC. Figure 7.2–1(a) shows a simple six-bit MDAC using capacitors, switches, and an op amp as a buffer. This converter works on a charge redistribution principle.[15] The first step in a conversion cycle is to discharge all capacitors (this corresponds to the ϕ_1 phase period). During the ϕ_2 phase period, the binary switches are closed or opened depending on whether the bit is a 1 or 0, respectively. An equivalent circuit for the converter at this point, shown in Fig. 7.2–1(b), is simply a capacitive attenuator. Because it is an attenuator, some or all of the capacitors may be connected to V_{ref}. This is the variable capacitance of Fig. 7.2–1(b). The $2C$ comes from the fact that all the capacitors sum to give the denominator term of the attenuator. Therefore, we may write an expression for V_{out} as

$$V_{out} = \frac{\sum_{i=0}^{n} b_i 2^{-i} C}{2C} V_{ref} = \frac{1}{2} \sum_{i=0}^{n} b_i \, 2^{-i} \, V_{ref} \tag{3}$$

[15] J. L. McCreary and P. R. Gray, *op. cit.*

where any unused inputs (i.e., $b_i = 0$) are grounded. Typically the bottom plate of the capacitors is connected to the binary switches, and the top plate is connected in common. It is seen that the bottom plate parasitic is eliminated from consideration, and, through the charge-balancing technique, the influence of the top plate parasitic can be minimized.[16] The MDAC can have bipolar operation if the largest capacitor is connected to V_{ref} during the ϕ_1 phase period. Then during the ϕ_2 phase period, it is switched to ground rather than to V_{ref}. As a result Eq. (3) becomes

$$V_{out} = \left(-\frac{b_0}{2} + \frac{1}{2}\sum_{i=0}^{n} b_i\, 2^{-i}\right) V_{ref} \tag{4}$$

With the proper scaling this will give us a bipolar behavior for D. If V_{ref} is also bipolar then we have a four-quadrant MDAC.

The accuracy of the capacitor and the area required are both factors that limit the number of bits used. The accuracy is seen to depend upon the capacitor ratios. The accuracy of the capacitor ratios for an MOS technology can be as low as 0.1%. If the capacitor ratios were able to have this accuracy, then the MDAC of Fig. 7.2–1(a) should have the resolution of 10 bits. However, this means that the capacitor ratios will be 1024 in the extreme case, which is undesirable from an area viewpoint. Also, the 0.1% capacitor ratio accuracy is applicable only for ratios in the neighborhood of unity.

The above considerations have led to the development of a two-stage MDAC. Figure 7.2–2(a) shows the schematic of a 13-bit MDAC with bipolar capability for V_{ref}.[17] It is seen that the 1.016-pF capacitor acts as a 64:1 divider, which in effect scales up the last six bits by a factor of 64. An equivalent circuit of the 13-bit MDAC is shown in Fig. 7.2–2(b). Here we have two voltage sources that depend upon the state of the switches. The right-hand voltage source is given as

$$V_r = \sum_{i=0}^{6} \frac{\pm b_i\, V_{ref}}{127} \tag{5}$$

where the polarity of V_{ref} depends upon polarity of the digital word. The left-hand voltage source is given as

$$V_l = \sum_{i=7}^{12} \frac{\pm b_i\, V_{ref}}{64} \tag{6}$$

[16] J. L. McCreary and P. R. Gray, *op. cit.*

[17] K. B. Ohri and M. J. Callahan, Jr., *op. cit.*

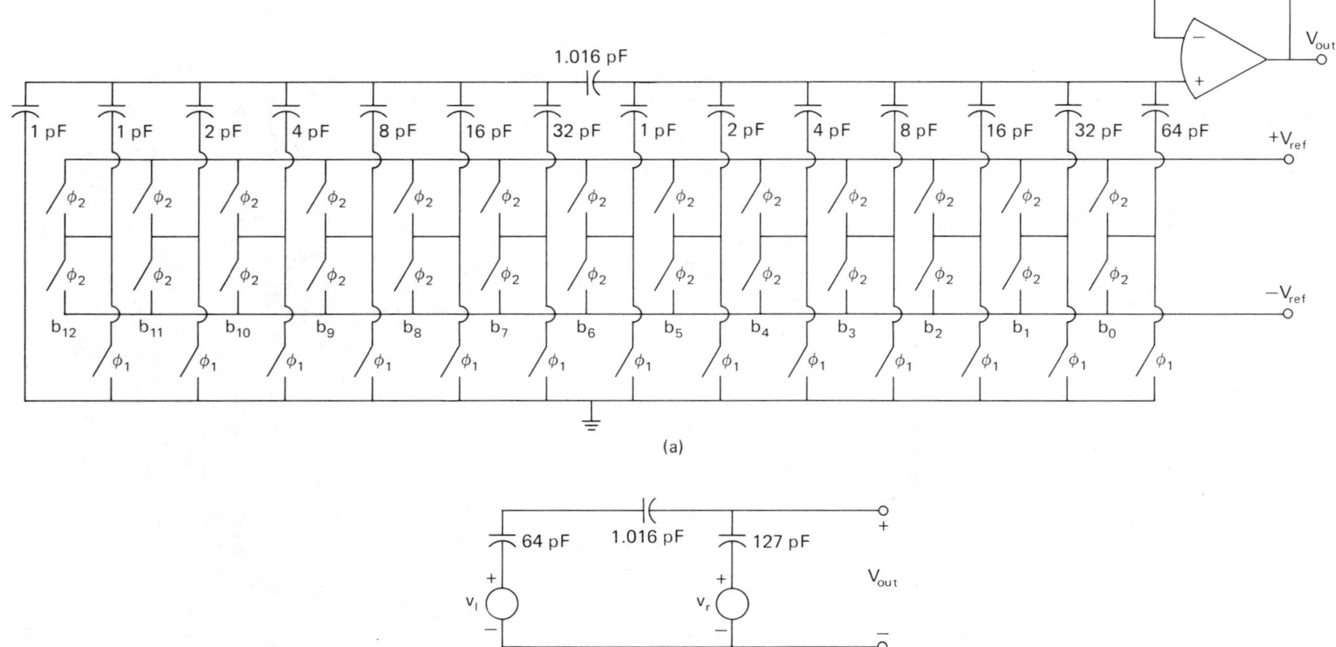

Fig. 7.2–2. (a) A 13-bit two-stage MDAC. Note that whether or not the ϕ_2 switches close depends upon the state of the binary variable, b_i. (b) An equivalent circuit of (a).

where in Eqs. (5) and (6) any unused inputs are assumed to be grounded. The output of the overall MDAC can be written as

$$V_{out} = \frac{\pm V_{ref}}{128}\left[\sum_{i=0}^{6} b_i C_i + \sum_{i=7}^{12} b_i \frac{C_i}{64}\right] \tag{7}$$

One might expect this converter to be influenced by the nonidealities of the system, such as the stray capacitance. The stray capacitance loading on the output, C_L, causes Eq. (7) to be modified accordingly as

$$V_{out} = \left(1 - \frac{C_L}{128}\right)\frac{V_{ref}}{128}\left[\sum_{i=0}^{6} b_i C_i + \sum_{i=7}^{12} b_i \frac{C_i}{64}\right] \tag{8}$$

where we see that C_L has caused an error of $[1-(C_L/128)]$. If C_L is 1.28 pF, then a 1% error is introduced.

A deviation of the 1.016-pF divider capacitance from the desired ratio of 1.016:1 introduces both gain and linearity errors. Assuming a $\pm\Delta C$ variation in the value of the 1.016:1 capacitance ratio modifies the output of the MDAC as

$$V_{out} = \left(\frac{V_r}{128}\right)\left(1 - \frac{\Delta C}{128}\right)\left[\sum_{i=0}^{6} b_i C_i + (1+\Delta C)\sum_{i=7}^{12} b_i \frac{C_i}{64}\right] \tag{9}$$

If we assume that $\Delta C = \pm 0.016$ (1.6% error), then the gain term has an error of

$$\text{Gain error term} = [1 - (\Delta C/128)] = 1 - [1/(64)(128)] \tag{10}$$

which is negligible. The linearity error term is given by

$$\text{Linearity error term} = \Delta C \sum_{i=7}^{12} \frac{b_i C_i}{64} \tag{11}$$

The worst-case error occurs for all $b_i = 1$ and is essentially ΔC. It is also important to keep $+V_{ref}$ and $-V_{ref}$ stable and equal in amplitudes. This will influence the long-term stability and gain tracking of the MDAC.

A different two-stage eight-bit MDAC is shown in Fig. 7.2-3.[18] It has the advantage of having no floating terminals that require charge-balancing.

---

[18] Y. S. Lee, L. M. Terman, and L. G. Heller, *op. cit.*

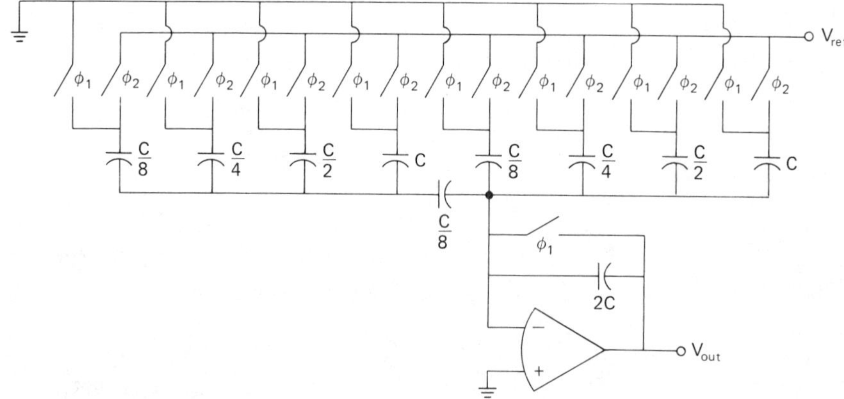

Fig. 7.2–3. A two-stage, weighted capacitor MDAC.

An op amp is connected in its inverting configuration with the $2C$ capacitor fed back from the output to the inverting input of the op amp. Because the input node is a virtual ground during operation, the parasitics associated with the input node are eliminated.

This section has illustrated several MDACs that use the charge-balance/redistribution mode of operation. This method has been used to a large extent in designing D/A converters using MOS technology.[19] It has the advantage of being compatible with SC technology and relating the performance of the D/A converter to the capacitor ratios, which can be very precise. The method has the disadvantage of loss of accuracy as the number of bits become large. Combining the method of this section with the method presented in the next will lead to means of obtaining a larger number of bits.

7.3 RESISTIVE DIVISION D/A CONVERTERS

Thus far we have not made use of the resistors available from the MOS process. These resistors include diffusion resistors and polysilicon resistors and will be discussed in more detail in Chapter 8. Diffusion resistors are made at the same time as the drain or source of the MOS device. Polysilicon resistors are essentially the same conductors as the gates of MOS devices. The typical resistance of diffused resistors is 10 to 100 ohms/square, and that of polysilicon resistors is 30 to 200 ohms/square. The absolute accuracy of these resistors can be as much as ±50%. The relative accuracy is in the

[19] P. R. Gray, D. A. Hodges, and R. W. Brodersen, *Analog MOS Integrated Circuits, IEEE Press,* John Wiley & Sons, Inc., New York, 1980.

neighborhood of ±2%, if the ratio of resistors does not vary widely from unity. We have not used these resistors in filter applications because the accuracy is not sufficient. However, in the D/A converter, the ±2% relative accuracy is in some cases sufficient. We shall illustrate in this section D/A converters that use the resistive division method and combine this method with that of the previous section to achieve high-resolution A/D converters compatible with MOS technology.

Figure 7.3–1 shows a three-bit resistive division D/A converter. A voltage reference, V_{ref}, is applied to the top of a resistor string. The resistor string has taps that correspond to the quantization of the analog signal levels. The designer must choose where he wants the discrete jumps to occur. In Fig. 7.3–1, V_{ref} is divided into $\frac{1}{16}$, $\frac{3}{16}$, $\frac{5}{16}$, etc. of V_{ref}. Each tap is connected to a switching tree whose switches are controlled by the bits of the digital word. If the bit b_i is 1, then the switches controlled by b_i are closed. If the bit b_i is 0, then the switches controlled by $\overline{b_i}$ are closed. It is seen that this structure is very regular and thus well suited for MOS technology. An advantage of this structure is that it guarantees monotonicity, for the voltage at each tap cannot be greater than the tap below it. The area required is not

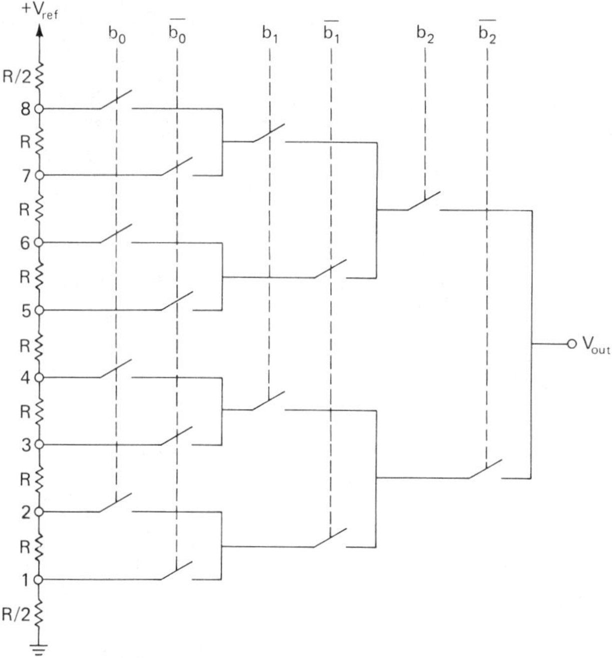

Fig. 7.3–1. Illustration of a three-bit resistive division D/A converter.

large if the number of bits is eight or fewer. The converter is fast unless there are large capacitances to be charged by V_{ref}.

Figure 7.3–1 works as follows. Suppose that the digital word to be converted is $b_0 = 1$, $b_1 = 0$, and $b_2 = 1$. Following the sequence of closed switches, we see that V_{out} is equal to $^{11}\!/_{16}$ of V_{ref}. In general, the voltage at any tap, n, can be expressed as

$$V_n = \frac{V_{ref}}{8}(n - 0.5) \tag{1}$$

It may be desirable to connect the bottom tap to ground, so that a well-defined output is available when the digital word is all zeros. Figure 7.3–2 shows the input-output characteristics of the D/A converter of Fig. 7.3–1. The integral linearity of this converter will depend upon the resistor ratios and as such will not be as good as the integral linearity of the charge-balance/redistribution D/A converters.

The charge-balance/redistribution and resistive division D/A converters can be combined to result in D/A converters having 12-bit resolution using 8-bit techniques.[20] An M-bit resistor string and a K-bit binary-weighted capacitor array are used to achieve an $N = M + K$-bit conversion. Figure 7.3–3 gives an example of such a converter where $M = 4$ and $K = 8$. The resistor string, R_1 through R_{2M}, provides inherently monotonic division of V_{ref} into 2^M nominally identical voltage segments. The binary-weighted capacitor array, C_1–C_{K+1}, is used to subdivide any one of these voltage segments into 2^K levels. This is accomplished by the following sequence of events. First, the switches S_F, S_B, and S_{1B} through $S_{K,B}$ are closed, connecting the top and bottom plates of the capacitors C_2 through C_{K+1} to ground. If the output of the D/A converter is applied to any circuit having a nonzero threshold, switch S_B could be connected to this circuit rather than ground to cancel this effect of a threshold. After opening switch S_F, the buses A and B are connected across the resistor string in such a manner as to achieve the new reference voltage, given as

$$V'_{ref} = V_{ref}[b_1 2^{-1} + b_2 2^{-2} + \cdots + b_{M-1} M^{-(M-1)} + b_M 2^{-M}] \tag{2}$$

Although the resistor string could take the form of Fig. 7.3–1, the string used in Fig. 7.3–3 has the advantage of switching both buses A and B. This causes any switch nonidealities, such as clock feedthrough, to be canceled.

[20] B. Fotouhi and D. A. Hodges, *op. cit.*

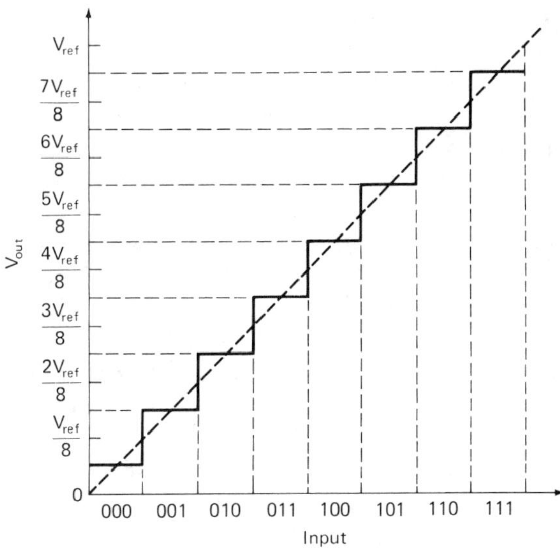

Fig. 7.3–2. Input-output characteristics of Fig. 7.3–1.

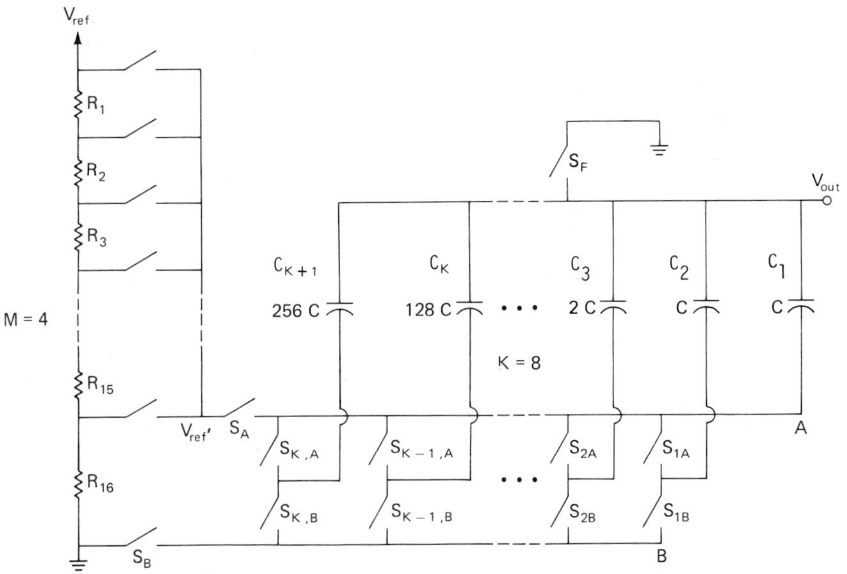

Fig. 7.3–3. A D/A converter using resistive division and charge-balance/redistribution techniques. The four MSB's are accomplished by resistive division, and the eight LSB's are accomplished by charge-balance/redistribution methods.

After V'_{ref} is applied between buses *A* and *B*, we have the equivalent circuit of Fig. 7.3–4(a). The final step is to decide whether or not to connect the bottom plates of the capacitors to bus *A* or bus *B*. This is determined by the *K* bits of the digital word being converted. The equivalent circuit for the analog output voltage of the converter is shown in Fig. 7.3–4(b).

The D/A converter of Fig. 7.3–3 has the advantage that, because the resistor string is inherently monotonic, the first *M* bits will be monotonic regardless of any resistor mismatch. This implies that the capacitor array has to be ratio-accurate to only *K* bits and still provide $(M+K)$ bit monotonic conversion. The conversion speed for the resistors is faster than the capacitors because no precharging is necessary. Therefore some interesting tradeoffs can be made between area and speed of conversion. Such techniques as trimming the resistor string using polysilicon fuses can help to improve the integral linearity of this technique.

Other combinations of resistive string and charge-balance, or charge redistribution, techniques are also possible in D/A converter applications. Instead of using resistive string techniques for the MSBs, one can use weighted capacitor techniques to achieve the MSBs and a resistive divider to achieve the LSBs. It would be necessary to trim the resistors to maintain absolute linearity. Alternatively, one could use a capacitor divider consisting of 2^N identical capacitors along with a resistive string for the LSBs. This approach would be monotonic and should have good linearity. Figure 7.3–5 illustrates how such an A/D converter could be implemented. The three to four MSB's will probably have to be trimmed, using a technique such as polysilicon

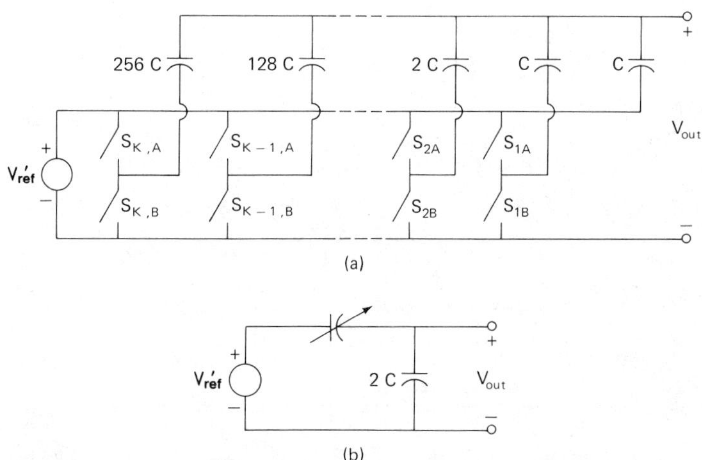

Fig. 7.3–4. (a) Equivalent circuit of Fig. 7.3–3 for resistive division part. (b) Equivalent circuit for entire Fig. 7.3–3.

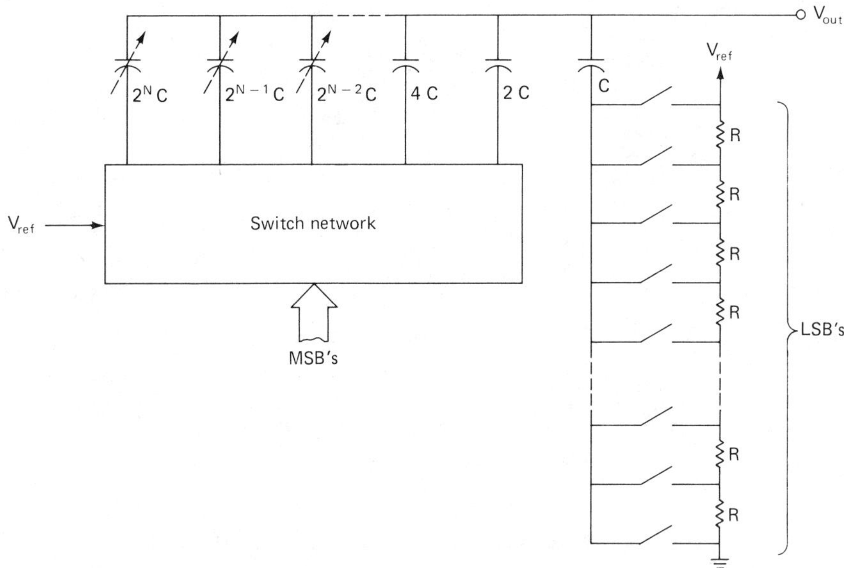

Fig. 7.3–5. An illustration of a weighted current-resistive division D/A converter.

fuses. This is indicated by the dashed arrows through the capacitors in Fig. 7.3–5.

Another approach that could be used to implement D/A converters compatible with MOS technology is the current-switched approach. Figure 7.3–6 shows one possible configuration. Equal voltages are applied to the gate and source of identical transistors to create unit currents. These unit currents are applied to a switch array controlled by the digital word to be converted to derive an output current. Because the current of an MOS transistor is directly proportional to the W/L ratio (width of the gate/length of the gate), the geometric properties of the MOS transistor can be used to create binary-weighted currents. The current-switched D/A converter is easily cascaded as illustrated in Fig. 7.3–6.

The advantages of a current-switched D/A converter are that it is inherently monotonic, as the currents are all positive, and also very compatible with MOS technology and requires very little area. Zero offset current can be achieved, which removes threshold problems. Further, the converter is high-speed and is easily expanded. Some of the disadvantages of the current-switched D/A converter are that the integral linearity is dependent upon the MOS transistor matching and that a large number of current sources will be required. Also, this converter technique uses more power and requires a means of converting I_{out} to V_{out}.

In this section we have examined D/A converters that use combinations

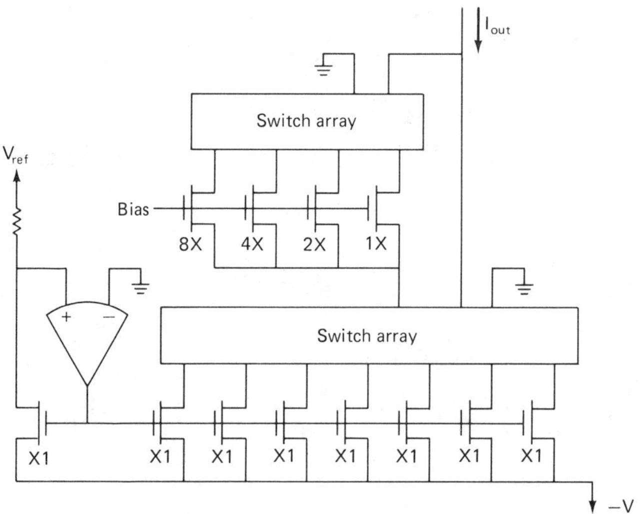

Fig. 7.3–6. A current-switched D/A converter.

of both charge-balance/redistribution and resistive division techniques. It has been shown how MOS D/A converters having a resolution of up to 12 bits can be implemented. The circuits and techniques presented are only representative of a very fast-growing area, and the literature should be consulted for practical examples of these techniques and others that improve upon the performance of the methods illustrated here.

7.4 SERIAL D/A CONVERTERS

In this section we conclude our considerations of D/A converters by considering two different types of converters compatible with MOS technology. These two types of converters fall under the category of serial D/A converters. A serial D/A converter is one in which the conversion is done sequentially. Typically, a clock pulse is required to convert one bit. Thus, N clock pulses would be required for the typical serial, N-bit D/A converter. The two types of converters to be examined here include the serial, charge redistribution and the algorithmic D/A converters.

Figure 7.4–1 shows the simplified schematic of a serial, charge redistribution D/A converter.[21] We see that this converter consists, in its simplest form,

[21] R. E. Suarez, P. R. Gray, and D. A. Hodges, "All-MOS Charge Redistribution Analog-to-Digital Conversion Techniques-Part II," *IEEE J. of Solid-State Circuits,* Vol. SC-10, No. 6, December 1975, pp. 379–385.

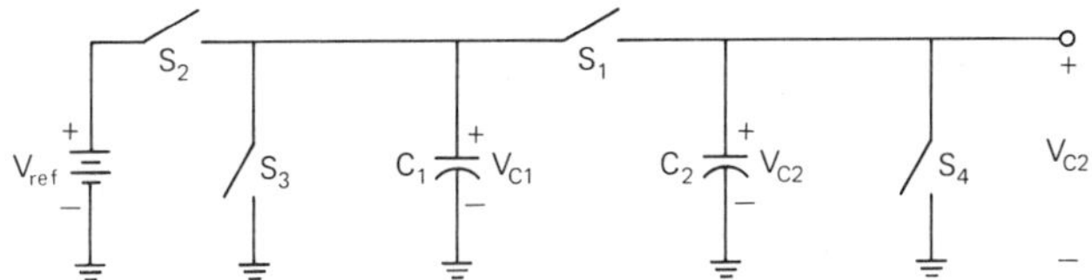

Fig. 7.4–1. Simplified schematic of a serial, charge redistribution D/A converter.

of four switches, two equal-valued capacitors and a reference voltage. The function of the switches is as follows. S_1 is called the *redistribution switch* and places C_1 in parallel with C_2 causing their voltages to become identical through charge redistribution. Switch S_2 is used to precharge C_1 to V_{ref}, if the ith bit, b_i, is a 1, and switch S_3 is used to precharge C_1 to zero volts if the ith bit is 0. Switch S_4 is used at the beginning of the conversion process to initially discharge C_2. An example is used to illustrate the operation of this D/A converter.

Example 7.4–1 *Operation of the serial D/A converter.* Assume that $C_1 = C_2$ and that the digital word to be converted is given as $b_0 = 1$, $b_1 = 1$, $b_2 = 0$, and $b_3 = 1$. The conversion begins with the LSB which is assumed to be b_3. The conversion starts with the closure of switch S_4, so that $V_{C2} = 0$. Since $b_3 = 1$, then switch S_2 is closed causing $V_{C1} = V_{ref}$. Next, switch S_1 is closed causing $V_{C1} = V_{C2} = 0.5\ V_{ref}$. This completes the conversion of the LSB bit. Figure 7.4–2 illustrates the waveforms across C_1 and C_2 during this example. Going to the next most LSB, b_2, switch S_3 is closed, discharging C_1 to ground. When switch S_1 closes, the voltage across both C_1 and C_2 is 0.25 V_{ref}. Because the remaining two bits are both 1, C_1 will be connected to V_{ref} and then connected to C_2 two times in succession. The final voltage across C_1 and C_2 will be $(13/16)V_{ref}$. This sequence of events will require nine sequential switch closures to complete the conversion.

From the above example it can be seen that the serial D/A converter requires considerable supporting external circuitry, most likely in the form of a microprocessor. Although the circuit for the conversion is extremely simple, several sources of error will limit the performance of this type of D/A. These sources of error include the capacitor parasitic capacitances, the switch parasitic capacitances, and the clock feedthrough errors. This converter has the advantage of monotonicity and very small area. An eight-bit converter using this technique has been fabricated and demonstrated a conversion time of 13.5/μsec.[22]

[22] R. E. Suarez, et al., *op. cit.*

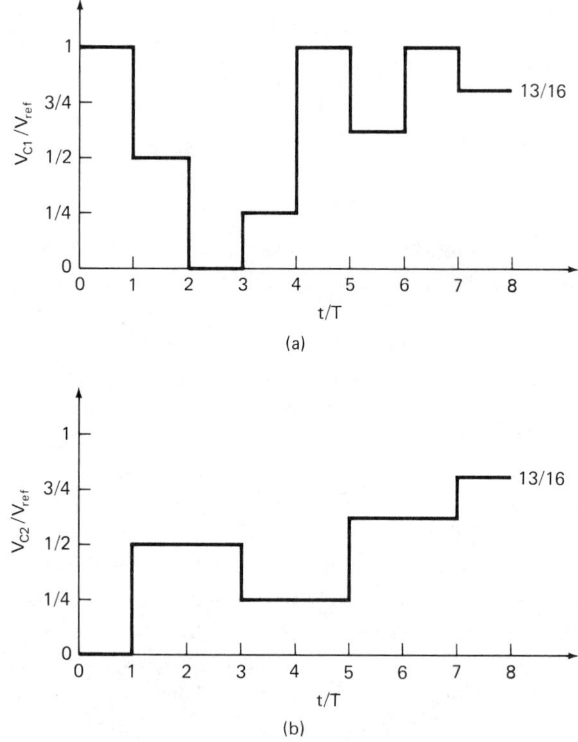

Fig. 7.4–2. Waveforms of Fig. 7.4–1 for the conversion of 1101. (a) Voltage across C_1. (b) Voltage across C_2.

The second approach to serial D/A conversion is called *algorithmic.*[23] Figure 7.4–3 illustrates the pipeline approach of implementing an algorithmic D/A converter. We see that Fig. 7.4–3 consists of unit delays and weighted summers. It can be shown that the output of this circuit can be written as

$$V_{out}(z) = [b_0 z^{-1} + 2^{-1} b_1 z^{-2} + 2^{-2} b_2 z^{-3} + \cdots + 2^{-n} b_n z^{-(n+1)}]\, V_{ref} \quad (1)$$

where b_i is either ± 1. Figure 7.4–3 shows that it takes $n + 1$ clock pulses for the digital word to be converted to an analog signal, even though a new digital word is converted on every clock pulse.

The complexity of Fig. 7.4–3 can be avoided by using two approaches

[23] R. H. McCharles, V. A. Saletore, W. C. Black, Jr., and D. A. Hodges, "An Algorithmic Analog-to-Digital Converter," *IEEE Int. Solid-State Circuits Conf. Digest of Tech. Papers,* February 1977, pp. 96–97.

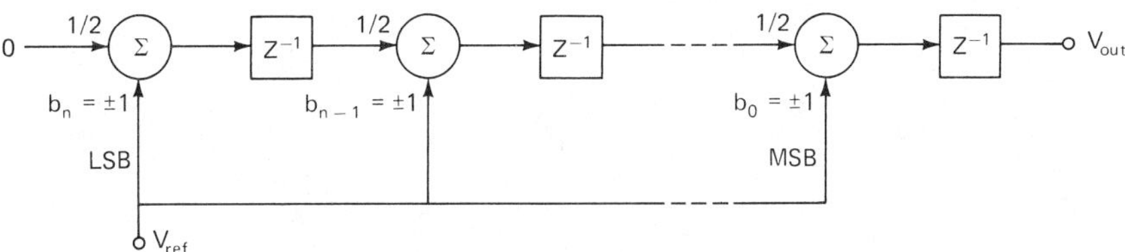

Fig. 7.4–3. Pipeline approach to implementing an algorithmic D/A converter.

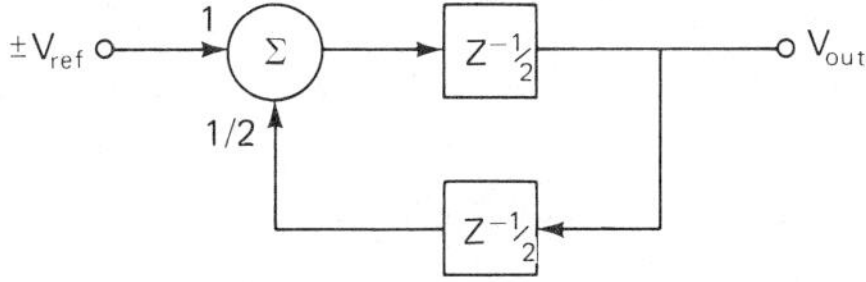

Fig. 7.4–4. Equivalent realization of Fig. 7.4–3, using iteration techniques.

called *replication* and *iteration.*[24] Here we shall consider only the iteration approach. It can be shown that Eq. (1) can be rewritten as

$$V_{out} = b_i z^{-1}[1 - 0.5z^{-1}]^{-1} V_{ref} \tag{2}$$

where we have assumed that the b_i are all identical. The fact that each b_i is either ± 1 will be determined in the following realization. It can be seen that Eq. (2) is realizable by Fig. 7.4–4. An SC implementation of Fig. 7.4–4 is shown in Fig. 7.4–5. In this realization, the unit delay has been implemented by means similar to the SC amplifiers of Section 2.4. We see that there is no delay between $\pm V_{ref}$ and the output, which means the z^{-1} term in the numerator of Eq. (2) would not be realized. This is circumvented by not looking at the output until one full cycle has been completed. The following example illustrates how Fig. 7.4–5 accomplishes the conversion.

Example 7.4–2. *D/A conversion using the algorithmic method.* Let us assume that Fig. 7.4–5 is to be used to convert the digital word 11001 into an analog output. We shall further assume that the first digit is the MSB and that the last digit is the LSB. Figure 7.4–6 shows the timing that is necessary in this example. A cycle consists of ϕ_1 followed by ϕ_2. For the first cycle, switches B and C are closed. This causes $+V_{ref}$ to appear at the output during the ϕ_2 phase. During the remaining five cycles, switch D is closed,

---

[24] R. H. McCharles and D. A. Hodges, "Charge Circuits for Analog LSI," *IEEE Trans. Circuits and Systems,* Vol. CAS-25, No. 7, July 1978, pp. 490–497.

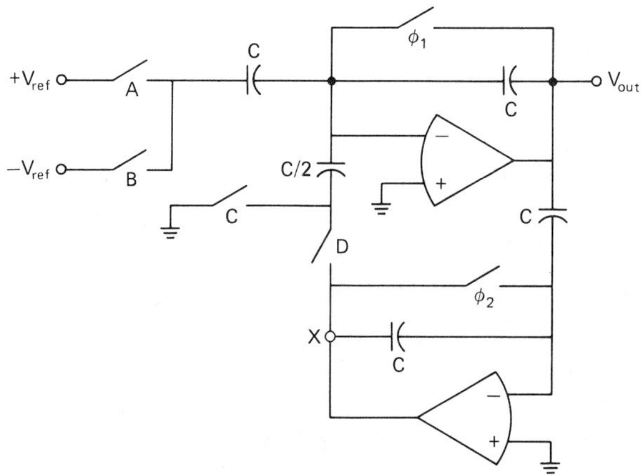

Fig. 7.4–5. An SC realization of Fig. 7.4–4.

and either switch A or B is closed, depending on whether the current bit is a zero or one, respectively. The LSB is input first. On the second cycle, the LSB is 1, so that switch B is closed. During ϕ_2 the output is $+V_{ref}$, due to the closure of switch B and the output of the second stage (point X) is $-V_{ref}$ due to the $+V_{ref}$ applied from the first stage. For the third cycle, the bit is 0, so that switch A is closed. During the ϕ_2 phase period, the output is equal to $-V_{ref}$ plus ½ V_{ref} fed back from point X. On the fourth cycle, the bit is 0, so that switch A is closed. During the ϕ_1 phase period, the voltage at point X is ½ V_{ref}. For the ϕ_2 phase period, the output is $-V_{ref}$ plus $-$¼ V_{ref} or $-$5⁄4 V_{ref}. For the fifth cycle, the voltage at point X during the ϕ_1 phase period is 5⁄4 V_{ref}. As the fourth bit is 1, switch B is closed, and the voltage at the output during the ϕ_2 phase period becomes $+V_{ref}$ minus 5⁄8 V_{ref} or 3⁄8 V_{ref}. Finally, the sixth cycle converts the MSB, which is 1, so that switch B is closed. During the ϕ_1 phase period the voltage at point X becomes $-$3⁄8 V_{ref}. During the ϕ_2 phase period, the final output voltage is V_{ref} plus 3⁄16 V_{ref} or 19⁄16 V_{ref}. As the actual V_{ref} of this example is $\pm V_{ref}$ or 2 V_{ref}, the analog value of the digital word 11001 is 19⁄32 times 2 V_{ref} or 19⁄16 V_{ref}.

The above example has illustrated the operation of an algorithmic D/A converter using an iterative implementation. Many other schemes for implementing an algorithmic converter are possible. The algorithmic converters have the advantages of minimum area and simplicity of implementation. They have the disadvantage of slow speed, in that $n + 1$ clock cycles are necessary to complete the conversion.

This section has presented two approaches to serial D/A conversion. The

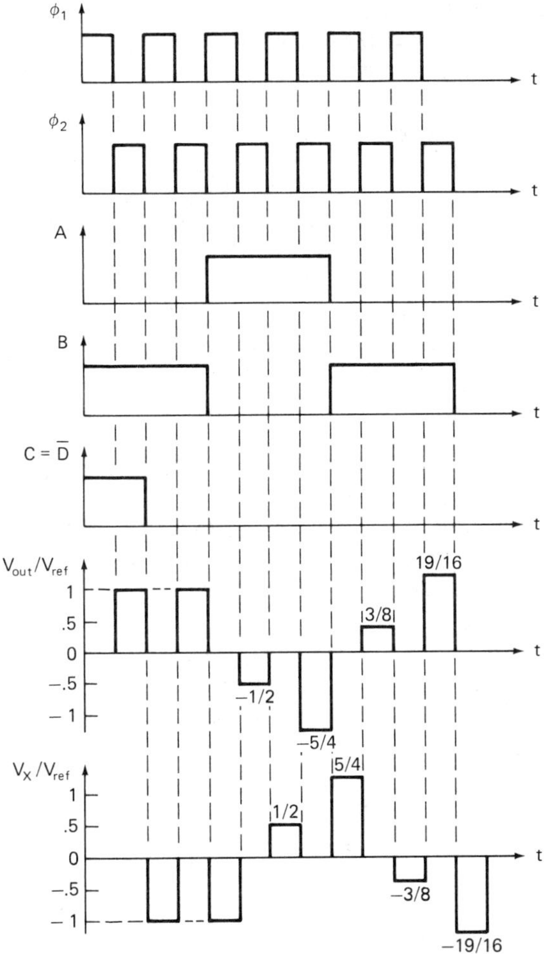

Fig. 7.4–6. Waveforms of Fig. 7.4–5 for Example 7.4–2.

serial converter is seen to be very simple but to require a longer time for conversion. In some applications, these characteristics would be advantageous. This section also completes the presentation of D/A converters suitable for MOS technology. Table 7.4–1 gives a summary of the D/A converters presented thus far. Also included in this table is the primary advantage and disadvantage of each of the types of converters. In the following sections, we shall examine the complementary subject of A/D converters. Many of the converters that will be presented use the D/A converters illustrated in this and the previous sections.

Table 7.4–1 Comparison of the D/A Converters Compatible with MOS Technology

D/A CONVERTERS	FIGURE	ADVANTAGE	DISADVANTAGE
Binary-weighted *C*	7.2–1	Fast, no S/H required	Nonmonotonic
Split-weighted *C*	7.2–2	Minimum area	Complex logic
R string	7.3–1	Monotonic	Large area required
R string-weighted *C*	7.3–3	Monotonic, simple	Trim for abs. accuracy
Weighted *C*-*R* string	7.3–5	Monotonic	Trim for abs. accuracy
Current switch	7.3–6	Monotonic	Large no. of switches
Serial	7.4–1	Simple, min. area	Slow
Algorithmic	7.4–5	Simple	Requires 2 op amps

7.5 SERIAL A/D CONVERTERS

We have seen from the previous sections that the objective of an A/D converter is to convert an analog signal to an equivalent digital word. A variety of methods have been developed to perform A/D conversion. These methods provide various tradeoffs in accuracy, speed, and area. In Fig. 7.1–6 we saw one form of an A/D converter. This type of converter can be classified as an *open loop converter.* It also called a *flash* or *parallel* A/D converter. Another form of an A/D converter can be classified as closed loop, for the converter compares the analog signal generated from the proposed digital word with the actual analog input signal and makes the necessary changes in the digital word, so that the two analog signals match to within the system resolution. The closed loop A/D converters are further divided into serial and successive approximation A/D converters. A good quality comparator is always at the heart of any A/D converter. It is also apparent that the A/D converter cannot continuously convert an analog signal, so that the analog signal must be sampled. Thus the A/D converter is truly an analog sampled data system. Our considerations of A/D converters compatible with MOS technology will be divided into the categories of serial, successive approximation, and open loop converters. In this section we shall consider the serial A/D converters.

The serial A/D converter is much like the serial D/A converter in that it performs serial operations until the conversion is complete. Figure 7.5–1 shows the block diagram of a *single-slope A/D converter.* This type of converter consists of a ramp generator (integration of a constant), an interval counter, a comparator, an AND gate and a counter that generates the output digital word. At the beginning of a conversion cycle, the analog input is sampled and held and applied to the plus terminal of the comparator. The counters are reset, and a clock is applied to both the interval counter and the AND

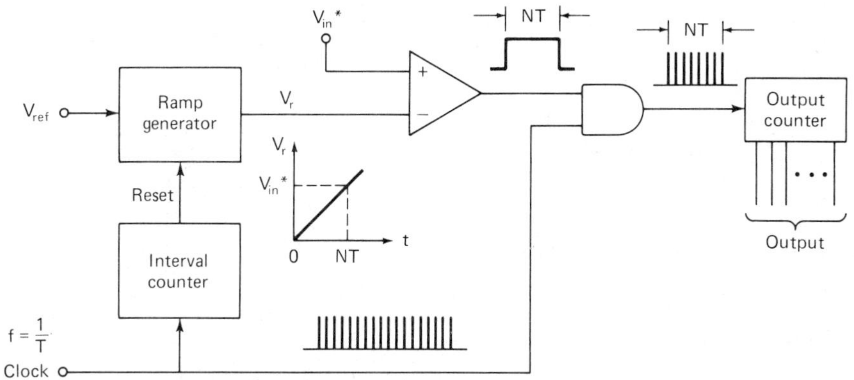

Fig. 7.5–1. Block diagram of a single-slope, serial A/D converter.

gate. On the first clock pulse, the ramp generator begins to integrate the reference voltage, V_{ref}. If V_{in}^{*}[25] is greater than the initial output of the ramp generator, then the output of the ramp generator, which is applied to the minus terminal of the comparator, begins to rise. Because V_{in}^{*} is greater than the output of the ramp generator, the output of the comparator is high, and each clock pulse applied to the AND gate causes the counter at the output to count. Finally, when the output of the ramp generator is equal to V_{in}^{*}, the output of the comparator goes low, and the output counter is now inhibited. The binary number representing the state of the output counter can now be converted to the desired digital word format.

The serial A/D converter can have many different forms. For example, the interval counter can be replaced by some logic to detect the state of the comparator output and reset the ramp generator when its output has exceeded V_{in}^{*}. The serial A/D converter has the advantage of simplicity of operation. A disadvantage of the serial converter is that it is subject to error in the ramp generator and is unipolar. Another disadvantage of the serial converter is that a long conversion time is required if the input voltage is near the value of V_{ref}.

A slightly different form of the single-slope, serial A/D converter is shown in Fig. 7.5–2(a). This serial A/D converter uses a *constant slope* approach.[26] The converter works in the following manner. First switch R is closed, charging the capacitor C to V_{ref}. The current source, I, then discharges C in a

[25] The symbol * will be used to represent a sample-and-hold signal.

[26] G. Smarandoiu, K. Fukahori, P. R. Gray, and D. A. Hodges, "An All-MOS Analog-to-Digital Converter Using a Constant Slope Approach," *IEEE J. of Solid-State Circuits,* Vol. SC-11, No. 4, June 1976, pp. 408–410.

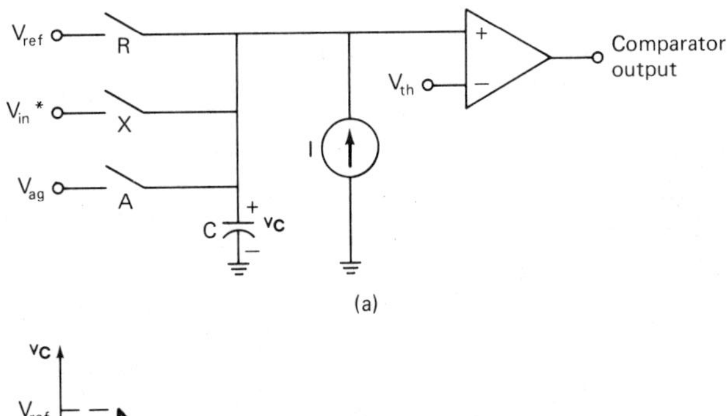

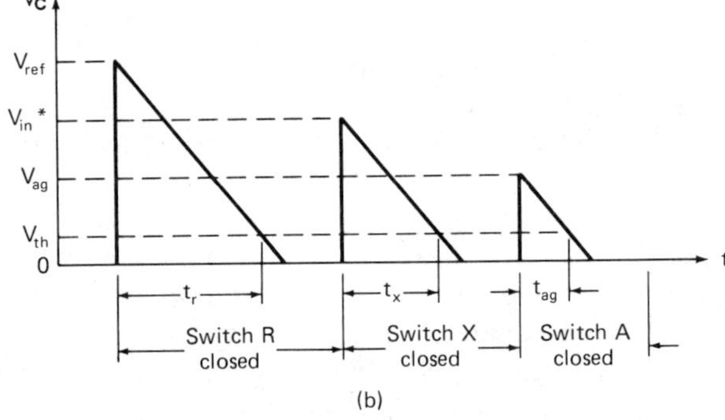

Fig. 7.5–2. (a) Serial A/D converter using constant slope. (b) Waveforms of (a).

linear manner. When the voltage across C is equal to the threshold of a comparator, V_{th}, then the time interval, t_r, is recorded. It can be seen that t_r is given as

$$t_r = (V_{ref} - V_{th})/(dv_C/dt) = (V_{ref} - V_{th})C/I \tag{1}$$

Next, the switch X is closed placing V_{in}^* across C. The current source I once again is used to discharge C in a linear fashion. When v_C reaches the value of V_{th}, the time t_x is recorded; t_x is given as

$$t_x = (V_{in}^* - V_{th})/(dv_C/dt) = (V_{in}^* - V_{th})C/I \tag{2}$$

Finally, switch A is closed placing a voltage across C, which is called the *analog ground reference.* When v_C reaches V_{th}, the time t_{ag} is recorded; t_{ag} is given as

$$t_{ag} = (V_{ag} - V_{th})/(dv_C/dt) = (V_{ag} - V_{th})C/I \tag{3}$$

If we define N as

$$N = (t_x - t_{ag})/(t_r - t_{ag}) \tag{4}$$

then it can be seen that

$$N = (V_{ref} - V_{ag})/(V_{in}^* - V_{ag}) \tag{5}$$

The waveforms of the A/D converter of Fig. 7.5–2(a) are shown in Fig. 7.5–2(b). Consequently, by subtracting the count resulting for V_{ag} from the counts obtained for V_{ref} and V_{in}^*, and dividing the results, a digital representation of the unknown input voltage is obtained.

This A/D converter has the advantage of not requiring an op amp; however, a microprocessor is necessary to do the required counting and digital division. The linearity of the slopes is important to the conversion process, and therefore the output resistance of the current source I must be very large (i.e., 10^9 ohms with $C = 4$ nF). The capacitance voltage coefficient is another source of error. To achieve 12 bits resolution requires a capacitance voltage coefficient of less than 100 ppm/V. The A/D converter of Fig. 7.5–2(a) was implemented as an 11-bit converter using MOS technology and having a conversion time of 20 samples/second.

Another type of serial A/D converter is the *dual-slope converter* shown in Fig. 7.5–3. This converter has the advantage of eliminating the dependence of the conversion process on the linearity and accuracy of the slope. Initially, V_{int} is zero, and the input is sampled and held. In this scheme, it is necessary for V_{in}^* to be greater than zero. The conversion process begins by reseting the positive integrator with switch 1 until the output of the integrator is equal to the threshold, V_{th}, of the comparator. Next, switch 1 is closed,

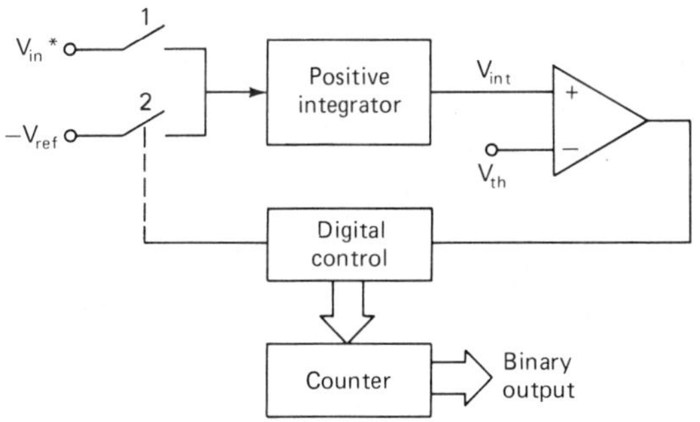

Fig. 7.5–3. Serial, dual-slope A/D converter.

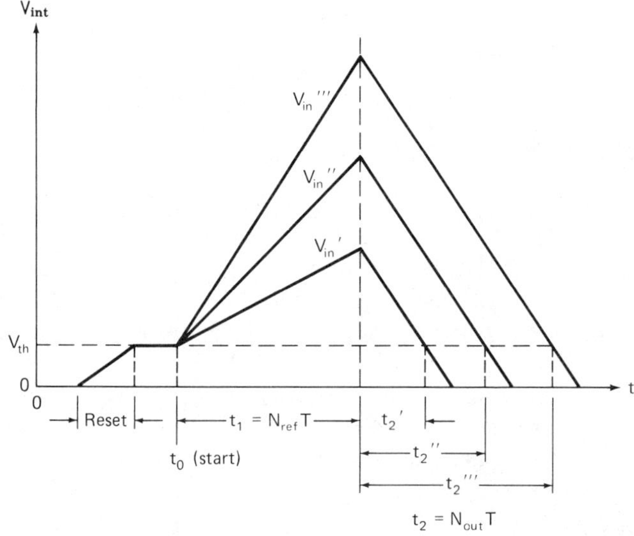

Fig. 7.5–4. Waveforms for the conversion process of Fig. 7.5–3.

and V_{in}^* is integrated for N_{ref} number of clock cycles. Figure 7.5–4 illustrates the conversion process. We see that the slope of the voltage at V_{int} is directly proportional to the amplitude of V_{in}^*. The voltage $v_{int}(t)$ during this time is given as

$$v_{int}(t) = K \int_0^{N_{ref} T} V_{in}^* \, dt + v_{int}(0) = K N_{ref} T \, V_{in}^* + V_{th} \tag{6}$$

where T is the clock period. At the end of the N_{ref} counts, the carry output of the counter is applied to switch 2 and causes $-V_{ref}$ to be applied to the integrator. Now the integrator integrates negatively with a constant slope, as V_{ref} is constant. When v_{int} becomes less than the value of V_{th}, the counter is stopped and the binary count can be converted into the digital word. This can be seen by considering $v_{int}(t)$ during the time designated as t_2 on Fig. 7.5–4. This voltage is expressed as

$$v_{int}(t) = v_{int}(0) + K \int_0^{N_{out} T} (-V_{ref}) \, dt \tag{7}$$

However, when $t = N_{out} T$, then Eq. (7) becomes

$$v_{int}(N_{out} T) = [K N_{ref} T V_{in}^* + V_{th}] - K N_{ref} N_{out} T \tag{8}$$

Because $v_{int}(N_{out}T) = V_{th}$, then Eq. (8) can be solved for N_{out} giving

$$N_{out} = N_{ref}(V_{in}^*/V_{ref}) \tag{9}$$

We see that the output of the dual-slope A/D converter is not a function of the threshold of the comparator, of the slope of the integrator, or of the clock rate. Therefore, it is an extremely accurate method of conversion. The only disadvantage is that it takes a worst case time of $2(2^N)T$ for a conversion where N is the number of bits of the A/D converter. The positive integrator of this scheme can be replaced by any of the integrators developed in Section 2.5, resulting in a straightforward implementation of the converter of Fig. 7.5–3.

The above examples of serial A/D converters are representative of the techniques that can be used. Other forms of serial conversion have been implemented using MOS technology but are not included here.[27] The serial A/D converter can be expected to be slow but to provide a high resolution. Typical values for serial A/D converters are conversion frequencies of less than 100 Hz and greater than 12 bits.[28]

7.6 SUCCESSIVE APPROXIMATION A/D CONVERTERS

In this section we consider a second category of closed loop A/D converters called *successive approximation A/D converters.* This class of A/D converters is the most widely used of all the techniques now available. It has higher speed than the serial converter with almost the same resolution capability. It makes optimum use of chip area and is adaptable to a large number of different internal components.

Figure 7.6–1 illustrates a block diagram of a general successive approximation A/D converter. This diagram consists of a comparator, a D/A converter, and digital control logic. To illustrate the conversion process, let us assume that the converter is unipolar (only positive analog signals can be applied). The conversion cycle begins by sampling the analog input signal to be converted. Next, the digital control circuit assumes that the MSB is 1 and all the other bits are zero. This digital word is applied to the D/A converter, which generates an analog signal of ½ V_{ref}, which is compared to the sampled

---

[27] G. F. Landsburg, "A Charge-Balancing Monolithic A/D Converter," *IEEE J. Solid-State Circuits,* Vol. SC-12, No. 6, December 1977, pp. 662–673.

[28] E. Masuda, C. Sato, T. Lida, Y. Suzuki, Y. Agawa, and T. Shima, "A Single-Chip C²MOS A/D Converter for Microprocessor Systems," *IEEE Int. Solid-State Conf. Digest of Technical papers,* February 1978, pp. 134–135 and 271.

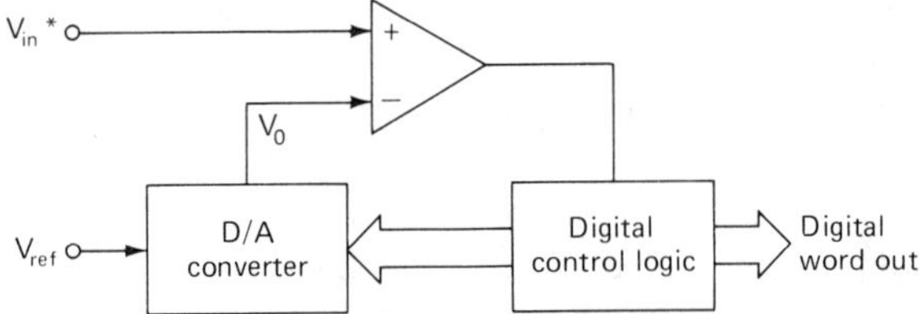

Fig. 7.6–1. Block diagram of a general successive approximation D/A converter.

analog input. If the comparator output is high, then the digital control logic makes the next MSB 1. If the comparator output is low, then the digital control logic makes the next MSB 0. This completes the first step in the approximation sequence. At this point the value of the first MSB is known. The approximation process continues by again applying a digital word to the D/A converter, with the first MSB having its proven value, the second MSB a "guess" of 1, and all other remaining bits having the value of 0. Again, the sampled input is compared to the output of the D/A converter with this digital word applied. If the comparator is high, the second MSB is proven to be 1. However, if the comparator is low, the second MSB is 0. The process continues in this manner until all bits of the digital word have been decided by the successive approximation process. Figure 7.6–2 shows how the successive approximation sequence works in converging to the closest analog output of the D/A converter to the sampled analog input. The digitally generated voltage converges to the sampled analog input in a

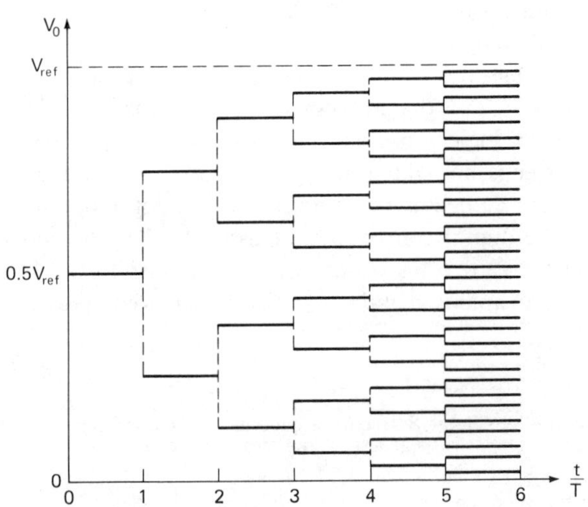

Fig. 7.6–2. The successive approximation process.

series of successive approximations, with each step being half the magnitude of the previous step. It is noted that the number of steps in the conversion process equals the number of bits in the digital word. Thus, the long conversion times required for the serial A/D converters are avoided. Bipolar A/D converters can be implemented by using a sign bit to choose either + or − V_{ref}.

The general successive approximation A/D converter described above can be implemented with any of the D/A converters of the three previous sections. A/D converters have been built that use weighted-capacitor[29] and resistive division[30] D/A converters. Figure 7.6–3 shows an example of a resistor division, weighted-capacitor D/A used to implement a successive approximation A/D converter.[31] The only extra components are a comparator and what is called the *successive approximation register* (SAR). The SAR essentially implements the switch control logic necessary to accomplish the successive approximation process. From the concepts of Section 6.2, we see that the comparator should have a gain greater than $(V_L 2^{M+K}/V_{ref})$, where V_L is the minimum output swing of the comparator required by the logic circuitry.

The conversion operation is described as follows. With S_F closed, the bottom plates of C_2 through C_{K+1} are connected through S_B to V_{in} to sample and hold the unknown input. The voltage stored on the capacitor array at the end of the sampling period is actually V_{in} minus the threshold voltage of the comparator, so that this threshold is removed as a source of offset error. After S_F is opened, a successive approximation search among the resistor string taps is performed to find the segment within which the stored sample lies. Then buses A and B are switched to the ends of the resistor defining this segment. Finally, capacitor bottom plates are switched in a successive approximation sequence until the comparator input voltage converges back to the threshold voltage. The sequence of comparator output is a digital code corresponding to the unknown analog input signal.

The A/D converter shown in Fig. 7.6–3 is capable of 12-bit, monotonic conversion with a differential linearity of less than ±½ LSB and a conversion time of 50 msec. The comparator used in this converter employed a clever biasing technique that avoided the use of op amps and permitted simple inverters to be used.[32] The comparator consisted of a cascade of inverters whose inputs were connected by a switch to the output. During this mode,

---

[29] J. L. McCreary et al., *op. cit.*

[30] A. R. Hamade, "A Single Chip All-MOS 8-Bit A/D Converter," *IEEE J. Solid-State Circuits,* Vol. SC-13, No. 6, December 1978, pp. 785–791.

[31] B. Fotouhi and D. A. Hodges, *op. cit.*

[32] B. Fotouhi and D. A. Hodges, *op cit.*

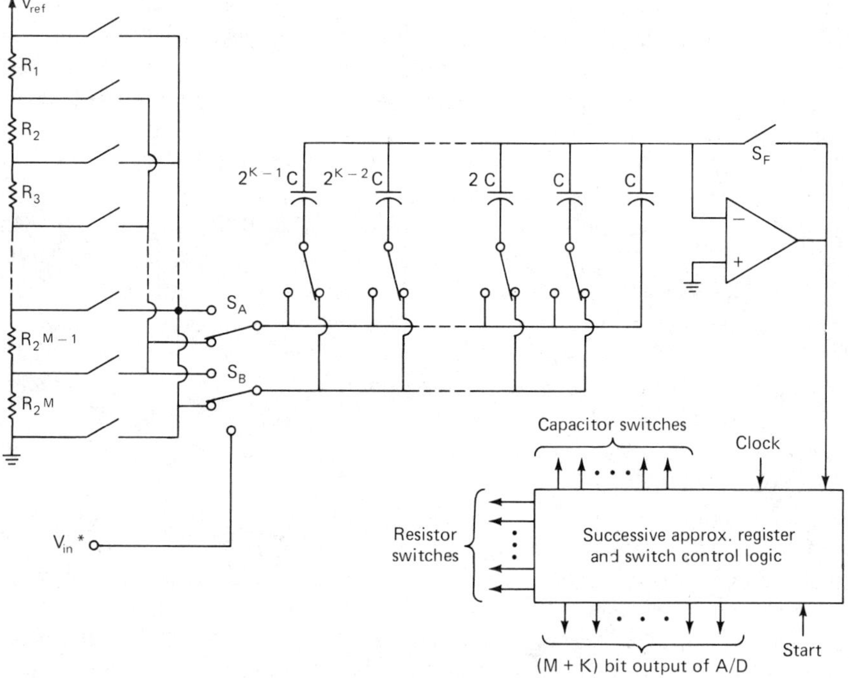

Fig. 7.6–3. A resistor division, weighted-capacitor, successive approximation A/D converter.

the inverters biased themselves. When the comparator was to be used, the inputs would be disconnected from the output and connected in series to form a high-gain amplifier. The inherent capacitances held the biasing point long enough for the comparator to operate. Thus a reasonably simple comparator resulted and used minimum power, for it operated only when required.

Figure 7.6–4 shows an example of using the serial D/A converter of Fig. 7.4–1 in a successive approximation A/D converter.[33] The converter works by converting the MSB, a_N, first. The jth bit is denoted d_j in a D/A conversion and a_j in an A/D conversion. The control logic takes a very simple form, for the D/A input string at any given point in the conversion is just the previously encoded word taken LSB first. For example, consider a point during the A/D conversion in which the first K MSB's have been decided. To decide the $(K + 1)$th MSB, a $(K + 1)$ bit word is formed in the D/A control register by adding a 1 as the LSB to the K bit word already encoded in the data storage register. A $(K + 1)$ bit D/A conversion then establishes

[33] R. E. Suarez et al., *op. cit.*

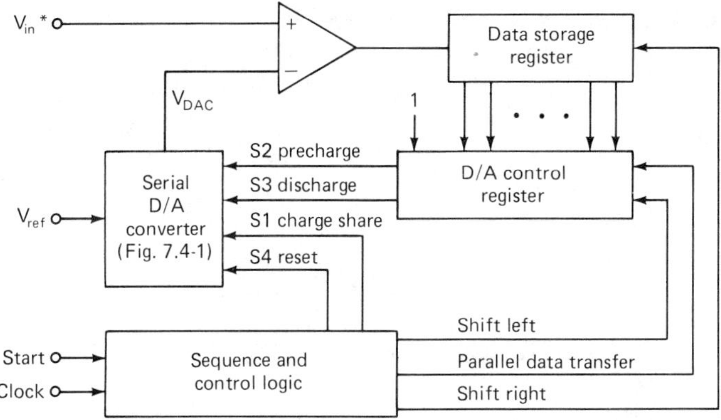

Fig. 7.6–4. Serial A/D converter using the serial D/A converter of Fig. 7.4–1.

the value of a_{N-K} by comparison with the unknown voltage V_{in}^*. The bit is stored in the data storage register, and the next serial D/A conversion is initiated. The conversion sequence is shown in detail in Table 7.6–1. Figure 7.6–5 illustrates a four-bit A/D conversion for $V_{in}^* = (13/16)V_{ref}$. Altogether, $N(N + 1)$ charging steps are required for an N-bit A/D conversion. The total A/D conversion time also includes N comparator settling times.

An algorithmic A/D converter patterned after the algorithmic D/A converter of Section 7.4 is shown as a discrete-time domain block diagram in Fig. 7.6–6.[34,35] This N-bit A/D converter consists of N stages and N comparators for determining the signs of the N outputs. Each stage takes its input, multiplies it by 2, and adds or subtracts the reference depending on the sign of the previous output. The comparator outputs form an N-bit digital representation of the bipolar analog input to the first stage. Figure 7.6–7 shows an implementation of Fig. 7.6–6 using iterative techniques. A sample timing sequence for the A/D converter is shown in Fig. 7.6–8. During the first cycle, switch C is open and switch D is closed, so that there is no feedback from the previous cycle. At the same time, switch E is closed during ϕ_1 then opened during ϕ_2, while switch B is closed. This causes a change in voltage of V_{in}^* at the inverter input between the trailing edges of ϕ_1 and ϕ_2. During the ϕ_1 phase of the second cycle, the comparator will show the sign of the input. This is the most significant bit. During the remaining N cycles, the closing of switches A and B is phased to either add or subtract the reference to reduce the magnitude of the feedback signal. The

---

[34] R. H. McCharles, V. K. Saletore, W. C. Black, Jr., and D. A. Hodges, *op. cit.*

[35] R. H. McCharles and D. A. Hodges, *op. cit.*

Table 7.6-1 Conversion Sequence for the Serial D/A Converter of Fig. 7.6-4

D/A CONVERSION NUMBER	D/A INPUT WORD						COMPARATOR OUTPUT	NUMBER OF CHARGING STEPS
	d_1	d_2	d_3	$\cdots$	d_{N-1}	d_N		
1	1	—	—	$\cdots$	—	—	a_N	2
2	1	a_N	—	$\cdots$	—	—	a_{N-1}	4
3	1	a_{N-1}	a_N	$\cdots$	—	—	a_{N-2}	6
⋮	⋮	⋮	⋮	⋮	⋮	⋮	⋮	⋮
N	1	a_2	a_3	$\cdots$	a_{N-1}	a_N	a_1	$2N$

Total number of charging steps $= N(N+1)$

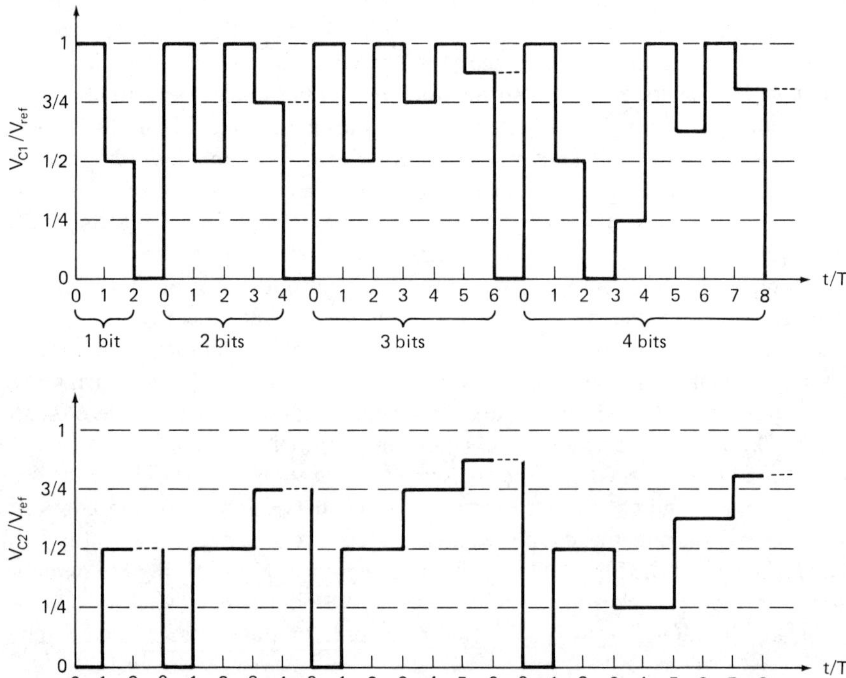

Fig. 7.6–5. Illustration of the operation of the serial A/D converter of Fig. 7.6–4 in the conversion of the sampled analog input voltage of $(13/16)V_{\text{ref}}$. The digital word out is $b_0 = 1$, $b_2 = 1$, and $b_3 = 1$, where b_0 is the MSB.

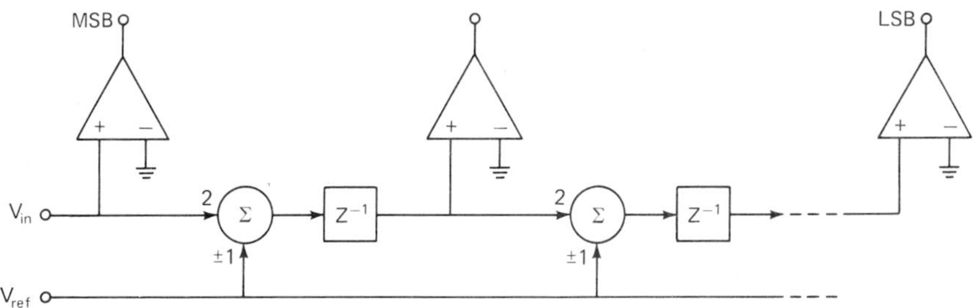

Fig. 7.6–6. Implementation in the discrete-time domain of an algorithmic A/D converter.

sequence of the comparator outputs gives a digital representation of the analog signal, V_{in}^*. An example is given to illustrate the above process.

Example 7.6–1. *Conversion process of an algorithmic A/D converter.* The algorithmic A/D converter of Fig. 7.6–7 is to be used to convert an analog signal of 0.8 V_{ref}. The following sequence of events is illustrated in Fig. 7.6–8. Notice that the phasing of switches *A* or *B* and *C* determine whether V_{in}^* and V_{ref} are positive or negative. As before, a cycle consists of the ϕ_1 and ϕ_2 phase periods. During the first cycle, V_{in}^* is applied in a noninverting manner to the inverter. During the ϕ_2 phase period, this signal appears at the output as V_{in}^*. At the end of the ϕ_1 phase period during the second cycle, $V_X = -V_{in}^*$. At the end of the ϕ_2 phase period during the second cycle, $V_{out} = 2V_{in}^* - V_{ref}$. The comparator at the output now determines if V_{out} is greater or less than zero. If it is greater, then V_{ref} will be subtracted from the next cycle. If not, V_{ref} will be added to the input inverter at that

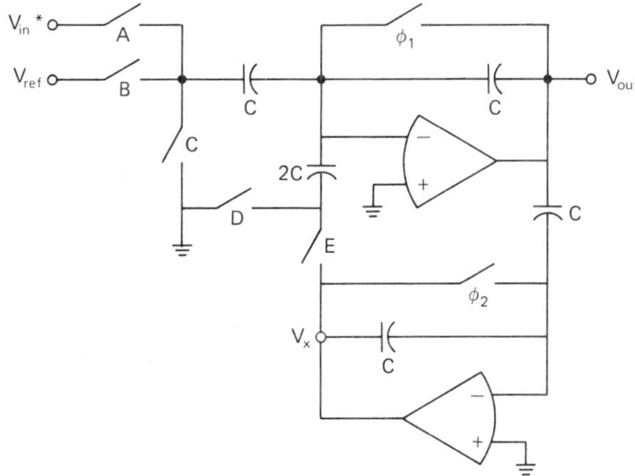

Fig. 7.6–7. An SC implementation of Fig. 7.6–6 using the iterative technique.

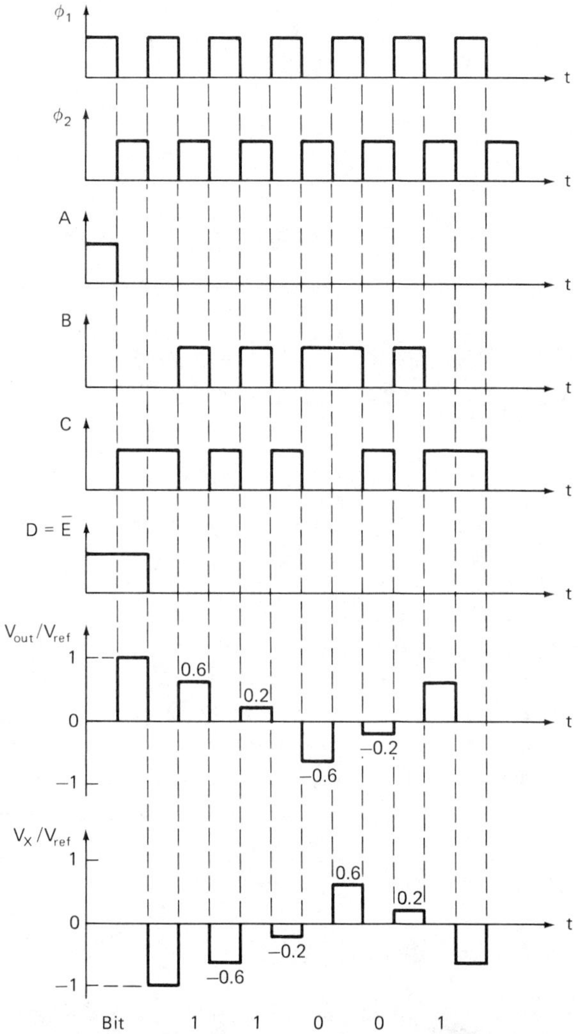

Fig. 7.6–8. Waveforms for the A/D converter of Example 7.6–1.

time. Because V_{in}^* is 0.8 V_{ref}, 2 V_{in}^* will be greater than V_{ref}, so that V_{ref} will be subtracted in the next cycle. This also means that the first MSB is 1. At the end of the ϕ_2 phase period during the third cycle, the output is $V_{out} = 4\ V_{in}^* - 2\ V_{ref} - V_{ref}$ or $4\ V_{in}^* - 3\ V_{ref}$. Again, V_{out} is still positive, so that the next bit will be 1 and V_{ref} will again be subtracted during the next cycle. During the ϕ_2 phase period of the fourth cycle, the output is $V_{out} = 8\ V_{in}^* - 7\ V_{ref}$, which this time is not greater than zero. Thus the third bit is 0, and V_{ref} will be added during the next cycle. During the ϕ_2 phase period of the fifth cycle, the output is $V_{out} = 16\ V_{in}^* - 13\ V_{ref}$, which is not greater than zero. Thus, the fourth bit is 0, and V_{ref} will be added during the next cycle. Finally, during the ϕ_2 phase period of the sixth cycle the output is $V_{out} = 32\ V_{in}^* - 25\ V_{ref}$, which is slightly greater than zero. Thus, the last bit is 1 and if the conversion were to proceed, V_{ref} would be subtracted during the next cycle.

Unfortunately, the above scheme gets the RMS sum of the noise voltage in the op amps during the conversion time, which will limit the resolution of the algorithmic A/D converter. Also, the algorithmic A/D is much slower than the successive approximation techniques presented in this chapter.

This section has presented a general scheme for converting analog signals into digital words. The method uses a successive approximation technique that reduces the conversion time compared with serial converters. It was seen that any of the D/A converters previously discussed could be used in this scheme. The algorithmic method also produced a means to convert analog signals quicker than serial converters using an iterative technique. In general, successive approximation A/D converters can have conversion times that fall within the 10^4 to 10^5 range. They are also capable of 8 to 10 bits untrimmed accuracy, 12 to 14 bits untrimmed monotonicity, and 12 to 14 bits trimmed accuracy. In the next section, we present a method of A/D conversion that provides the highest conversion speeds possible.

7.7 HIGH-SPEED A/D CONVERTERS

We have seen that the A/D converter must sample the analog input signal while the conversion process takes place. Therefore, the A/D converter is inherently a sampled data system. From the concepts of Chapter 1, we know that a signal must be sampled at a frequency of at least twice the highest frequency in the analog signal. Thus, the A/D converters of the previous two sections would be suitable only for analog signals in the audio frequency range or less. However, in many applications it is necessary to process analog signals that contain video frequencies. In this case, one must turn to A/D converters that are capable of very short conversion times. This class of A/D converters is the subject of this section; they are called *high-speed A/D converters.* This class of A/D converters also falls into the category of open loop A/D converters. Typically, the conversion time involves several clock cycles only.

Figure 7.1–6 is a general block diagram of a high-speed A/D converter known as the *flash,* or *parallel,* A/D converter. An example of how this converter works is illustrated in Fig. 7.7–1. Figure 7.7–1 is a three-bit A/D converter. V_{ref} is divided into eight values, indicated on the figure. Each of these values is applied to the positive terminal of a comparator. The outputs of the comparators are taken to a digital decoding network that determines the digital output word from the comparator outputs. For example in Fig. 7.7–1, if V_{in}^* is 0.7 V_{ref}, then the top two comparator outputs are 1, and the bottom five are 0. The digital decoding network would identify 101 as the corresponding digital word. A lot of variations can occur around the basic idea of Fig. 7.7–1. For example, one may wish to have the voltage at the taps to be in multiples of $V_{ref}/16$ with $V_{ref}/8$ voltage difference between taps. Also, the resistor string can be connected between $+V_{ref}$ and $-V_{ref}$ to achieve bipolar conversion.

The parallel A/D converter of Fig. 7.7–1 converts the analog signal to a digital word in essentially one clock cycle. The clock cycle will generally have two phase periods. During the first phase period, the analog input voltage is sampled and applied to the comparator inputs. During the second phase period the digital decoding network determines the correct output digital word and stores it in a register/buffer. Therefore, the conversion time would

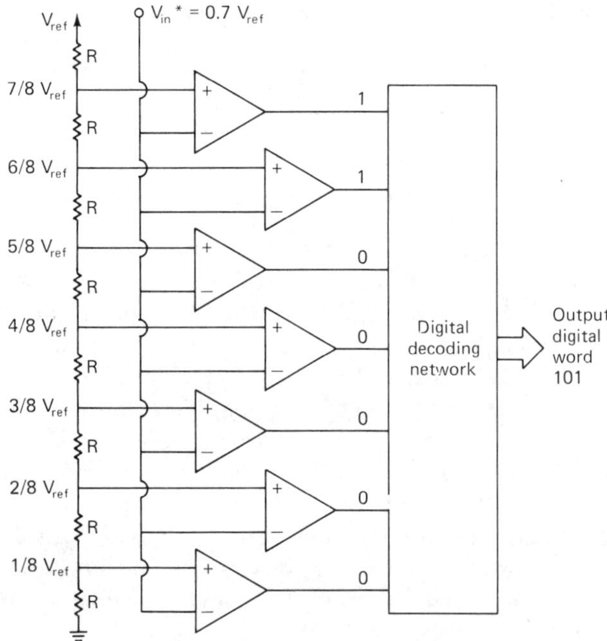

Fig. 7.7–1. A three bit flash A/D converter.

be limited by how fast this sequence of events can occur. Typical clock frequencies can be as high as 10 MHz, which gives a theoretical conversion time of 100 nanoseconds. Unfortunately, the parallel A/D converter requires 2^N comparators, which becomes impractical for N greater than about 6. Other available methods give almost the same conversion times with much more efficient utilization of chip area.

One method to achieve smaller system conversion times using the A/D converters of the previous sections would be to time-interweave them, as indicated in Fig. 7.7–2. Here, M successive approximation A/D converters are used in a parallel, pipeline configuration to convert an analog signal to an N-bit digital word. The sample-and-hold circuits consecutively sample and apply the input analog signal to their respective A/D converters. N cycles later, the A/D converter provides a digital word out. If $M = N$, then a digital word is output at every cycle. If one examines the chip area used in the technique of Fig. 7.7–1 compared with Fig. 7.7–2 with $M = N$, the minimum area is likely to be somewhere in between these limits.

A technique called *flash sequence*[36] is very attractive in minimizing the area and conversion time for A/D converters with more than six bits. This technique is illustrated in Fig. 7.7–3 for a 2-M bit A/D converter. The method

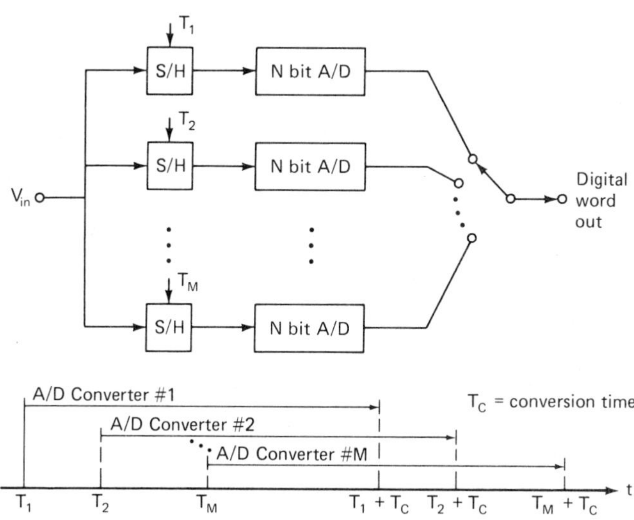

Fig. 7.7–2. A Time-interweaved A/D converter array.

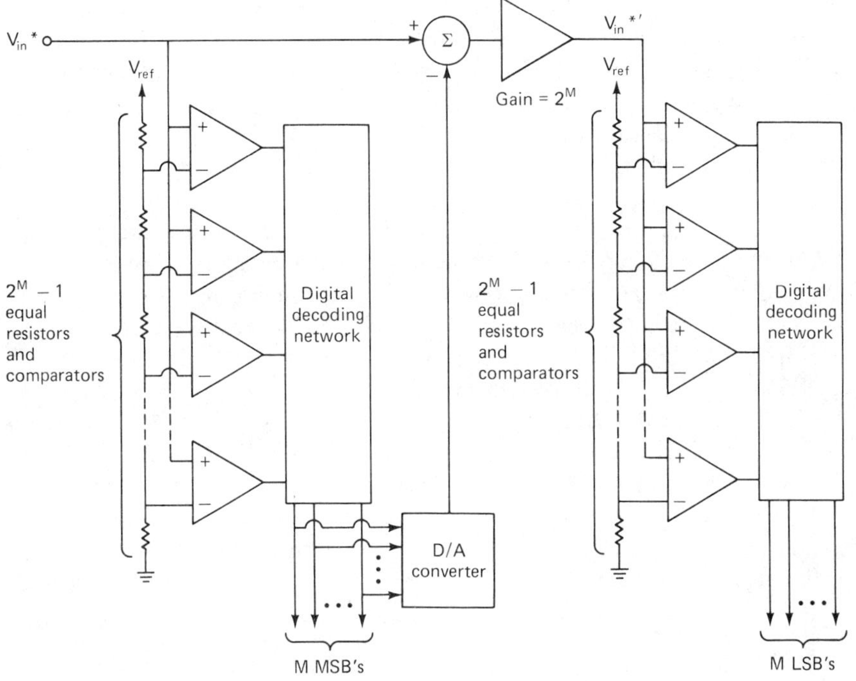

Fig. 7.7–3. A 2 *M* bit flash sequence (two-step) A/D converter.

basically first converts the *M* MSB's followed by the conversion of the *M* LSB's. As a consequence, only 2^M (really $2^M - 1$) comparators and two clock cycles are required to convert an analog input signal to a 2-*M* bit digital word. In the technique shown in Fig. 7.7–3, during the first clock cycle, the analog input is applied to the string of $2^M - 1$ comparators to determine the *M* MSB's. The *M* MSB's are used to create a modified analog signal to be used in the next cycle. The modified analog signal is given as

$$V_{in}^{*\prime} = (V_{in}^{*} - KV_{ref})2^M \tag{1}$$

[36] B. M. Gordon, "Linear Electronic Analog/Digital Conversion Architectures, Their Origins, Parameters, Limitations, and Applications," *IEEE Trans. on Circuits and Systems,* Vol. CAS-25, No. 7, July 1978, pp. 391–418.

where KV_{ref} is the voltage of the resistor tap applied to the highest comparator that was triggered. The value of KV_{ref} can be derived by several methods, one of which is an M-bit D/A converter driven by the M MSB's. At the end of the first cycle, the M MSB's are stored in a register/buffer, and the value of $V_{in}^{*\prime}$ is applied to the previous string of $2^M - 1$ comparators. The outputs of these comparators are used to decode the M LSB's, which are stored in a register/buffer. At the end of the second cycle, the 2-M bit digital word is available for further processing.

The photomicrograph of Fig. 7.7–4 shows an eight-bit A/D converter built using the above principles. The value of KV_{ref} was achieved using the method illustrated in Fig. 7.7–5. The comparators have both a true and a false output available. The false output of the upper comparator was ANDed with the true output of the lower comparator. The AND gate will have only a high output when the lower comparator has triggered (true output was high) and the upper comparator has not (true output low). The output

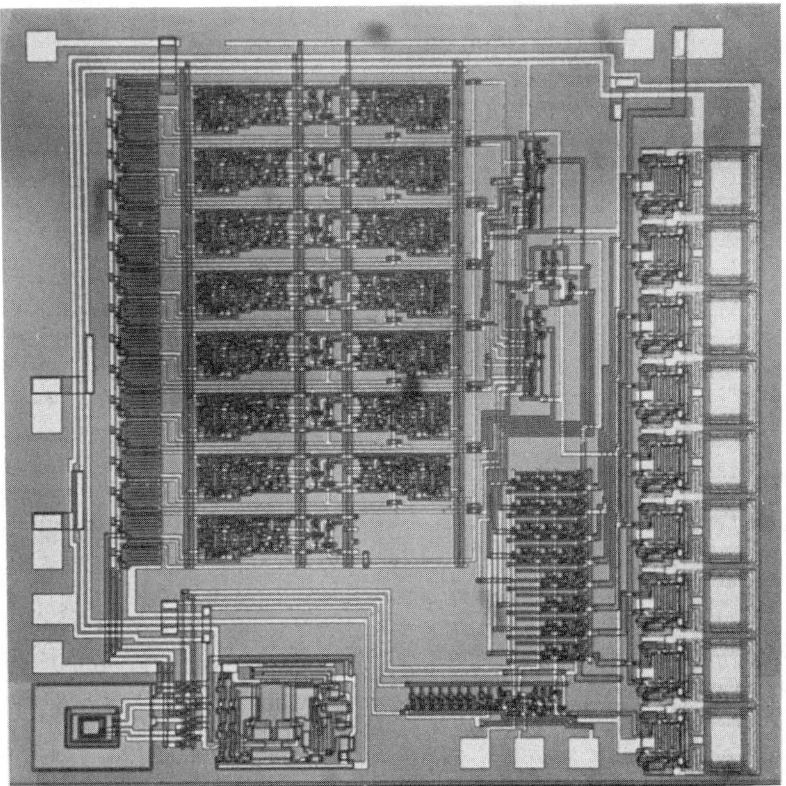

Fig. 7.7–4. Photomicrograph of an eight-bit two-step flash A/D converter.

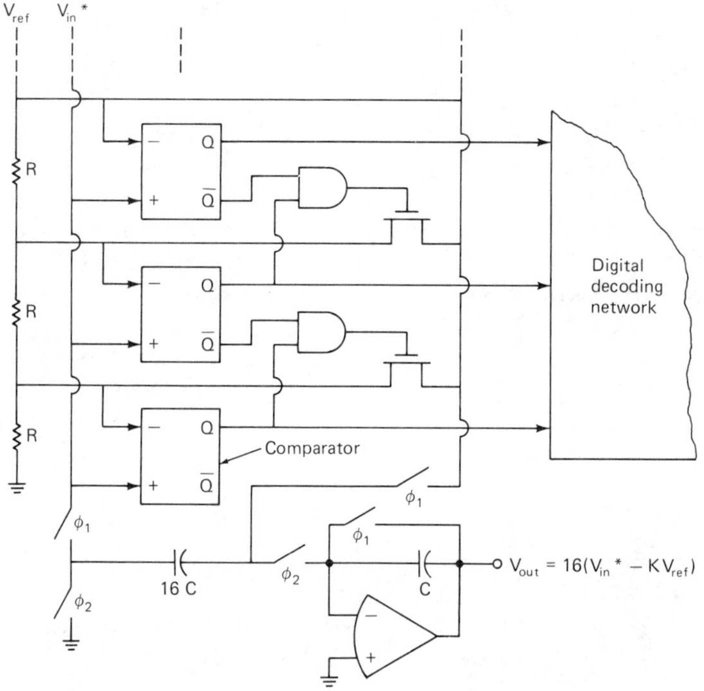

Fig. 7.7–5. A method of generating $V_{in}*'$.

of the comparator is used to control a switch that connects the voltage from the resistor string applied to the negative input of the highest comparator that triggered to one end of the capacitor designated as 16*C*. The voltage across the capacitor designated as 16*C* is now equal to $V_{in}^* - KV_{ref}$. During the ϕ_2 phase period, this voltage will be amplified by 16 times and applied on the next ϕ_1 phase period to the comparators as a new analog input, $V_{in}^{*\prime}$. At the ϕ_2 phase period of the second cycle, the conversion will be complete. The conversion time for this converter was in the range of 2 μsec and was primarily limited by the settling times of the logic.

The flash sequence method described above is essentially a two-step flash A/D converter. Higher speed or less conversion time can be achieved at the cost of 15 more comparators. During the ϕ_2 phase period of the first cycle, the new analog signal, $V_{in}^{*\prime}$, is applied to the second bank of comparators, so that the conversion is complete in 1.5 clock cycles. Figure 7.7–6 shows a photomicrograph of a 1.5-step flash converter using this concept. The conversion time was slightly better than that of Fig. 7.7–4.

Many possible approaches to achieving a high-speed A/D converter exist that combine the techniques presented in this chapter. For example, consider

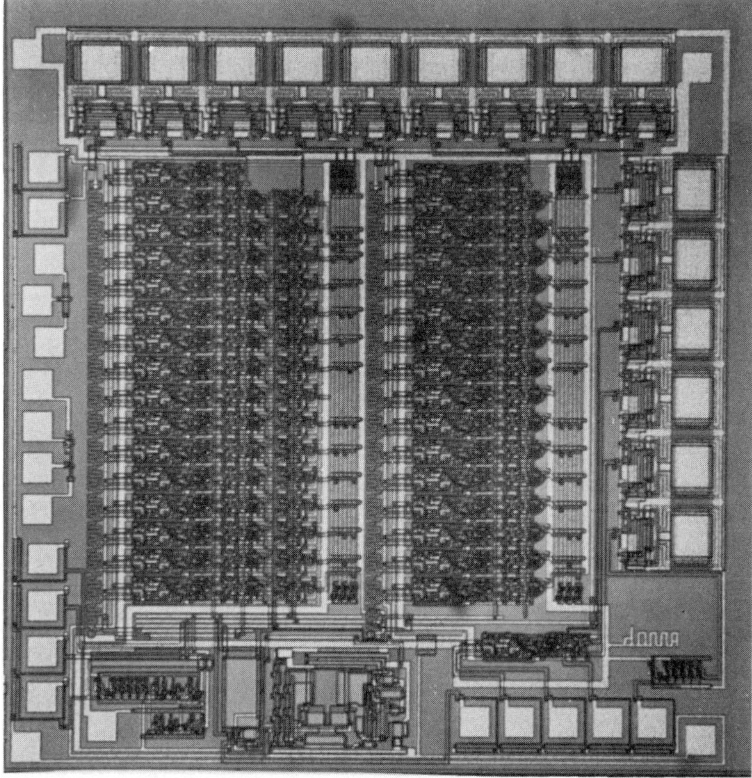

Fig. 7.7–6. A photo micrograph of an eight-bit 1- and ½-step A/D converter.

the block diagram in Fig. 7.7–7, which is an eight-bit ripple A/D converter. This method has the potential of very high speed conversion. The approach uses a resistor string of 256 equal-valued resistors between V_{ref} and ground. The 256-resistor string is divided into 16 equal parts each consisting of 16 series resistors. The voltage across each segment of 16 series resistors is applied to a bank of 15 comparators that will provide the information necessary to decode the 4 MSB's. The remaining 240 taps in the resistor string are taken to a switching network controlled by the 4 MSB's. The switching network selects which of the 16 series resistor segments the analog signal is contained within. The taps of the selected segment are taken to another bank of 15 comparators that supplies the informating necessary to decode the 4 LSB's. As one does not have to generate $V_{in}^{*\prime}$, only a single clock is necessary to drive the circuit. The conversion takes place as fast as the signals can propagate through the comparators and logic circuitry. Some care needs to be taken, so that the state of the LSB's will not influence the determination of the

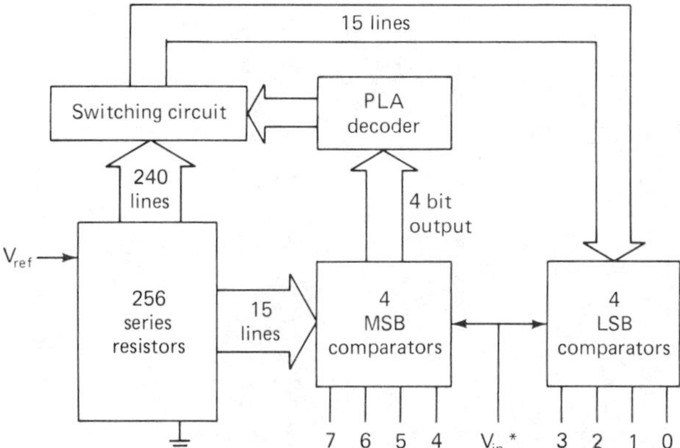

Fig. 7.7–7. Block diagram of an eight-bit ripple A/D converter.

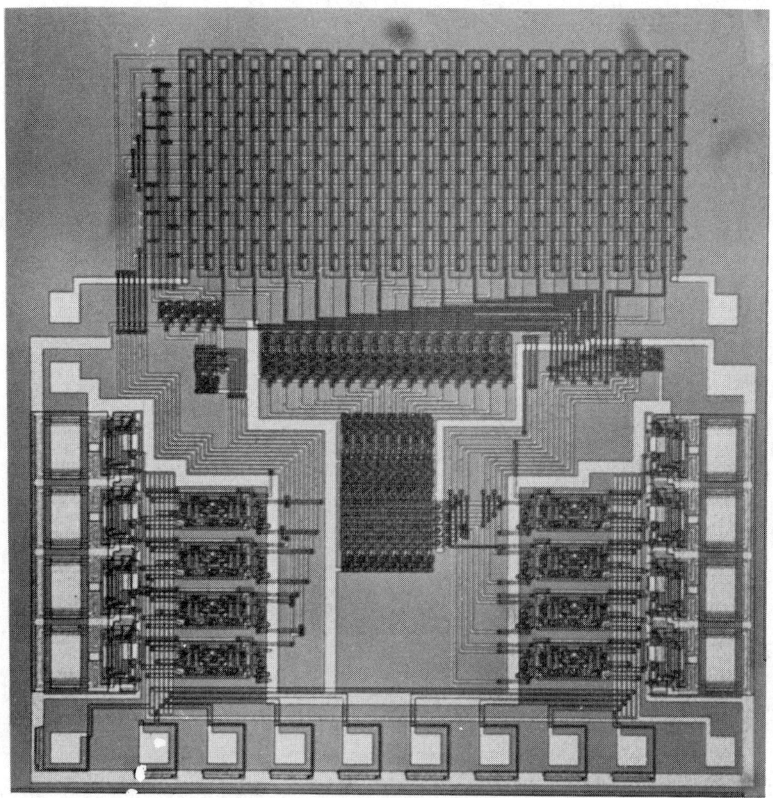

Fig. 7.7–8. A photomicrograph of an eight-bit ripple A/D converter.

Table 7.7–1 Comparison of the Performance of the Various Techniques of Implementing A/D Converters

A/D CONVERTERS	PERFORMANCE
Serial	1–100 conversions/second 12–14 bit absolute accuracy Requires no element-matching Stable V_{ref} necessary
Successive approximation	10^4–10^5 conversions/second 8–10 bits untrimmed accuracy 12–14 bits untrimmed monotonicity 12–14 bit trimmed accuracy
High-speed	10^6–2×10^7 conversions/second 7–8 bit accuracy

MSB's, which then determine the LSB's. The conversion time is in the range of 150 nanoseconds, which corresponds to a maximum clock rate of approximately 6 MHz. Figure 7.7–8 shows a photomicrograph of an eight-bit A/D converter using this technique.

This section has presented A/D conversion methods that provide fast conversion times. It has been seen that the area requirements increase rapidly as the conversion time is reduced. The parallel, or flash, method was presented, along with several modifications including the flash sequence and a ripple method that provided various compromises between speed and area. Typically, the high-speed converters will have a conversion rate of over 10^6 conversions/second with accuracy of seven to eight bits. High-speed converters implemented from CMOS technology should be expected to have a higher conversion rate than those implemented from NMOS technology. High-speed CMOS A/D converters can have conversion rates up to 15 MHz.[37] Table 7.7–1 summarizes the various A/D converters and their performance capabilities that have been presented in this and the previous two sections.

7.8 SUMMARY

This chapter has presented digital-to-analog and analog-to-digital conversion techniques compatible with MOS technology. We have seen that the ability to convert between digital and analog formats is very important in signal-processing applications. In Section 7.1, D/A and A/D converters were charac-

[37] S. Ohr, "6-bit a-d converter pulls only 200 mW at 15 MHz," *Electronic Design,* Vol. 28, December 6, 1980.

terized in terms of their performance. These characteristics divided into static and dynamic characteristics. Sections 7.2 through 7.4 presented various techniques of realizing D/A converters that can be implemented using MOS technology. These techniques included charge-balance/redistribution, resistive division, combinations of the previous two, and serial D/A converters. In all cases, tradeoffs between area and accuracy or resolution were possible. Sections 7.5 through 7.7 presented various techniques of realizing A/D converters compatible with MOS technology. These techniques included serial, successive approximation, and high-speed A/D converters. The important tradeoff in A/D converters was speed versus area.

The development of D/A and A/D converters is fast-growing, and many improvements are expected over the circuits presented in this chapter. The circuits and techniques presented here represent the implementation of the basic principles that exist in the design of converters. In many cases one may be involved in the specialized use of a converter that uses only part of the circuits presented here. Regardless of the application, the information of this chapter should be a good starting point in the design of either an A/D or D/A converter.

PROBLEMS

7.1 (Sec. 7.1). A linear A/D converter is to have a full-scale input of +10 volts and a zero value of 0 volts with a resolution of at least 5 mV. What is the minimum number of bits that can be used to implement this design?

7.2 (Sec. 7.1). Plot the analog output versus the digital word input for a three-bit D/A converter that has ±1 LSB differential linearity and ±1 LSB integral linearity. Assume an arbitrary analog full-scale value.

7.3 (Sec. 7.1). Repeat problem 7.2 for ±1.5 LSB differential linearity and ±0.5 LSB integral linearity.

7.4 (Sec. 7.1). Repeat problem 7.2 for ±0.5 LSB differential linearity and ±1.5 LSB integral linearity.

7.5 (Sec. 7.2). Rederive Eq. (3) if a capacitance of value C is connected from the plus terminal of the op amp of Fig. 7.2–1 to ground.

7.6 (Sec. 7.2). Derive Eqs. (5) and (6) for Fig. 7.2–2(a).

7.7 (Sec. 7.2). Design an eight-bit, two-stage MDAC similar to that of Fig. 7.2–2(a) using two four-bit sections with a capacitive attenuator between the stages. Give all capacitances in terms of C, which is the smallest capacitor of the design.

7.8 (Sec. 7.3). What is V_{out} of Fig. 7.3–1 if $b_0 = 1$, $b_1 = 1$, and $b_2 = 0$?

7.9 (Sec. 7.3). (a) Redesign Fig. 7.3–1 so that the outputs are equal to $iV_{ref}/8$ where $i = 1, 3, \ldots, 8$. (b) Repeat (a) if $i = 0, 1, \ldots 7$.

7.10 (Sec. 7.3). If R of Fig. 7.3–1 is 10 kΩ and a capacitance of 10 pF is attached to the output, calculate the risetime for the cases where $b_0 = 1$, $b_1 = 1$, and $b_2 = 1$, and where $b_0 = 0$, $b_1 = 0$, and $b_2 = 1$ where the risetime is defined as four time constants.

7.11 (Sec. 7.4). What is V_{C1} in Fig. 7.4–1 after the following sequence of operations? S_4 closes, S_3 closes, S_1 closes, S_2 closes, S_1 closes, S_3 closes, S_1 closes, S_2 closes, and S_1 closes.

7.12 (Sec. 7.4). Repeat Problem 7.11 if $C_1 = 1.05C_2$.

7.13 (Sec. 7.4). Show how to derive Eq. (2) from Eq. (1).

7.14 (Sec. 7.4). Repeat Example 7.4–2 for the digital word 10101.

7.15 (Sec. 7.5). Estimate the conversion time in terms of T, the clock period, for an N-bit A/D converter using (a) the single-slope techniques (Fig. 7.5–1(a)), (b) the constant-slope technique (Fig. 7.5–2(a)), and (c) the dual-slope technique (Fig. 7.5–3).

7.16 (Sec. 7.5). Show how the cascade of an A/D converter with a D/A can realize a multiplier/divider if the reference voltages to both converters are themselves analog signals.

7.17 (Sec. 7.6). A switched capacitor successive approximation A/D converter is shown in Fig. P 7–17. The converter starts by resetting V_{out} to zero and connecting

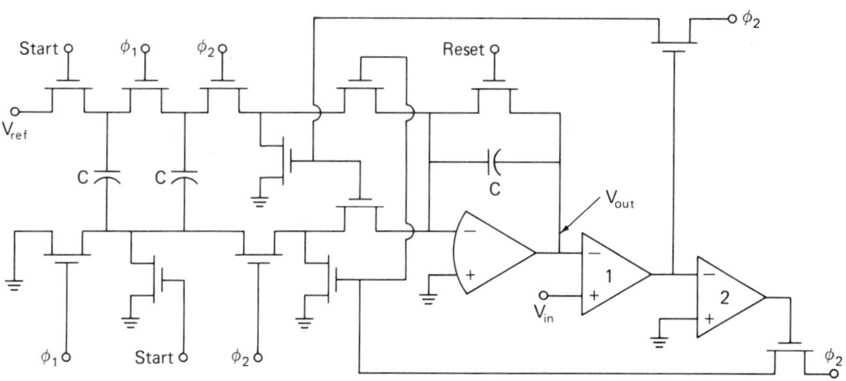

Fig. P 7-17.

the left-most capacitor C to V_{ref}. During the first cycle, the right-hand C samples V_{ref} and applies it to the SC integrator, so that V_{out} is 0.5 V_{ref}. During the ϕ_2 phase period, the number 1 comparator compares V_{out} with V_{in}. If V_{in} is greater than V_{out}, then a 1 is stored for the MSB. If V_{in} is less than 0.5 V_{ref}, a 0 is stored for the MSB. The process continues until a sufficient number of successive approximations have been made. If V_{ref} is 10 V and V_{in} is 4.2 V, plot V_{out} for six clock cycles, and give the equivalent digital output word.

7.18 (Sec. 7.6). Repeat problem 7.17 if V_{in} is 7.8 V.

7.19 (Sec. 7.6). Develop a plot of V_{C1}/V_{ref} and V_{C2}/V_{ref} as a function of time similar to Fig. 7.6–5 for the serial A/D converter of Fig. 7.6–4 if $V_{in} = (7/16)V_{ref}$.

7.20 (Sec. 7.6). Repeat Example 7.6–1 if $V_{in} = 0.3\ V_{ref}$.

7.21 (Sec. 7.7). Figure P 7–21 shows a stage-by-stage A/D technique. (a) Show how this stage converts an analog signal into an N-bit digital output word. (b) Use the SC capacitor techniques developed in the previous chapters to yield an SC implementation of the stage shown in Fig. P 7–21.

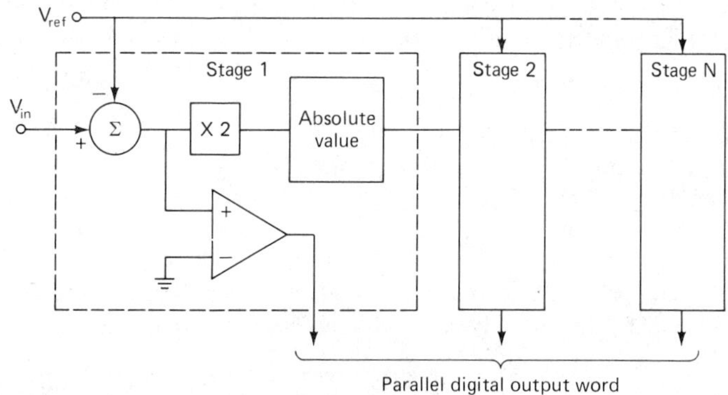

Fig. P 7-21.

7.22 (Sec. 7.7). Compare the serial, successive approximation, and flash-type A/D converters.

8
MOS Technology and Analog Circuit Design

8.0 INTRODUCTION

The previous chapters of this book have shown how to apply the concepts of switched capacitor circuits to analog signal-processing applications. The switched capacitor implementation of analog circuits has resulted in a very general and flexible method of circuit realization. This development did not happen by accident, but was stimulated by its extreme compatibility with integrated circuit fabrication techniques. Technology has had a tremendous influence upon the design of analog sampled data circuits. Although analog switched capacitor circuits can be implemented by using discrete components (op amps, capacitors, and switches) *the real advantage of these circuits is the adaptability to existing IC technology.* As a consequence, it is important to complete our study of switched capacitor circuits by considering MOS technology and the circuit design techniques that are compatible with this technology.

Practical integrated circuit technology can be divided into bipolar and field effect technologies. The field effect technology will be referred to as metal-oxide-semiconductor (MOS) technology. Often polysilicon is used in place of the metal, but the MOS notation is still retained. Each technology has various advantages. The bipolar technology has high values of transconductance, is very fast, and has low turn on values (0.5 volts for silicon). The MOS technology makes an almost ideal switch although it has less transconductance, is slower, and typically has a larger turn on voltage than bipolar technology. Because of the switch realization and because it is used extensively in digital design, MOS technology is typically selected for switched capacitor circuit implementation. This does not imply that bipolar technology is unsuited for switched capacitor circuit realizations. However, we shall consider only MOS technology in this text.

A summary of the component performance of NMOS technology corresponding to a minimum feature dimension of 6–7 microns is shown in Table

Table 8.0–1 NMOS Component Performance Summary

COMPONENT TYPE	RELATIVE ACCURACY	TEMPERATURE COEFFICIENT	VOLTAGE COEFFICIENT	ABSOLUTE ACCURACY	RANGE OF VALUES
Poly-to-poly capacitor	.1%	+25 ppm/°C (less than 100°C)	−30 ppm/V	Few percent (edge-dependent)	.15 to .2 pF/mil²
MOS capacitor	.1% (L = 10 microns)	+25 ppm/°C	−20 ppm/V	Few percent (edge-dependent)	.25 to .3 pF/mil²
Diffused resistor	2% (5 microns) 0.23% (50 microns)	+1500 ppm/°C	−200 ppm/V	±50%	10 ± 2 ohms/sq.
Polysilicon resistor	2% (5 microns)	1500 ppm/°C		±50%	60 ohms/sq.
Ion-implanted resistor	2% (5 microns) 0.15% (50 microns)	400 ppm/°C	−800 ppm/V		.5 k to 20 ohms/sq.
FET resistor (pinched resistor)	Few %	$(T/T_0)^{3/2}$	Voltage-dependent	±25%	10 k to 100 kohms

8.0–1.[1] In continuous analog circuits, the performance depends upon the RC products, which can be seen to have a worst-case variation of ± 50% if implementation is by NMOS technology. For switched capacitor circuits, the circuit performance has been shown to depend upon capacitor ratios. From Table 8.0–1 we see that the worst-case ratios of capacitors compatible with the NMOS process are as small as ± 0.1%. This is a significant result and one of the primary reasons switched capacitor circuits have found such compatibility with IC technology. This result also places switched capacitor circuits in a competitive position with digital techniques. Signal-processing functions that can be accomplished with 0.1% accuracy are equivalent to a 10-bit digital implementation of the signal-processing function.

Another factor of integrated circuit implementation of switched capacitor circuits is that large resistors are replaced by small capacitors (see Eq. (11) of Chapter 2, Section 2), so that the area requirements are reduced. As switched capacitor circuits require only capacitors, switches, and operational amplifiers, they are ideal for MOS technology.

In this chapter we shall introduce the concepts of MOS technology, including single- and double-channel MOS. Double-channel MOS is often called *complementary* metal-oxide-semiconductors (CMOS). This will be followed by methods of modeling MOS components that are compatible with present methods of computer analysis. This is a very important step, for most integrated switched capacitor circuits cannot be properly represented using breadboarding techniques. Next, the MOS transistor will be examined in its application as a switch. Analog circuit techniques that are compatible with MOS technology will be presented, including biasing techniques, voltage references, and single-stage amplifiers. This section will be followed by a discussion of the general approach to the design of MOS op amps. Finally the last section will consider the design of three types of op amps. These types include op amps that use only enhancement single-channel devices, both enhancement and depletion single-channel devices, and enhancement complementary devices. The objective of this chapter is to give the reader a brief background on how the circuits of this book can be implemented as MOS integrated circuits.

8.1 MOS TECHNOLOGY

In this section MOS technology is described in sufficient detail to allow the reader to become familiar with the various MOS processes and how the different components are fabricated. Circuit layout techniques and component accuracy will also be examined. A detailed description of the device electronics

[1] Minimum feature dimension is typically the smallest allowable gate dimension.

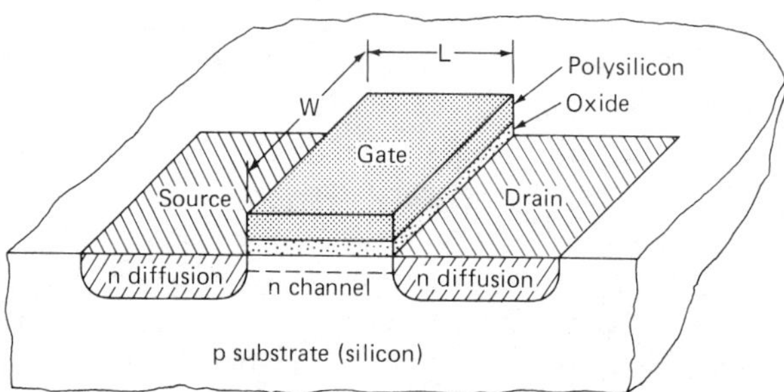

Fig. 8.1–1. Illustration of a n channel MOS device.

can be found in other texts.[2–5] Our discussion will include both MOS and CMOS technology. An idealized n-channel MOS device is shown in Fig. 8.1–1. The two dimensions, L and W, are the only parameters that a circuit designer can vary after a process has been specified. Consequently, these two dimensions and their effect upon the electrical characteristics of the MOS device become very important. In most FET's the drain and source are symmetrical, so that their designation is arbitrary. We recall that the insulated-gate FET can operate in one of two modes. The first mode is called *enhancement.* In this mode the gate potential (the gate consists of the conducting layer of polysilicon shown in Fig. 8.1–1) is made positive with respect to the substrate. The result is to deplete the region under the gate and between the drain and source of majority positive carriers and to form an n-channel. The gate voltage at which the channel begins to conduct current is called the *threshold voltage,* V_T. This behavior is illustrated in Fig. 8.1–2. I_D is the drain current flowing from the drain to the source (it is assumed that the source and substrate are connected to the same potential).

The second mode is called *depletion.* In this mode, an n-channel is formed during the fabrication of the MOS device by ion implantation. Therefore a channel exists for negative values of V_{GS}. An example of the depletion mode behavior is also illustrated in Fig. 8.1–2. In either case, current begins to conduct in the n-channel when V_{GS} is equal to or more positive than V_T. The various relationships implied by Fig. 8.1–2 and the dependence of the

[2] R. S. Muller and T. I. Kamins, *Device Electronics for Integrated Circuits,* John Wiley & Sons, New York, 1977.

[3] W. M. Penny and L. Lau, eds., *MOS Integrated Circuits,* Van Nostrand Reinhold, New York, 1972.

[4] A. S. Grove, *Physics and Technology of Semiconductor Devices,* John Wiley & Sons, New York, 1967.

[5] Engineering Staff of American Microsystems, *MOS Integrated Circuits—Theory, Fabrication, Design and Systems Application MOS/LSI,* Van Nostrand Reinhold, New York, 1972.

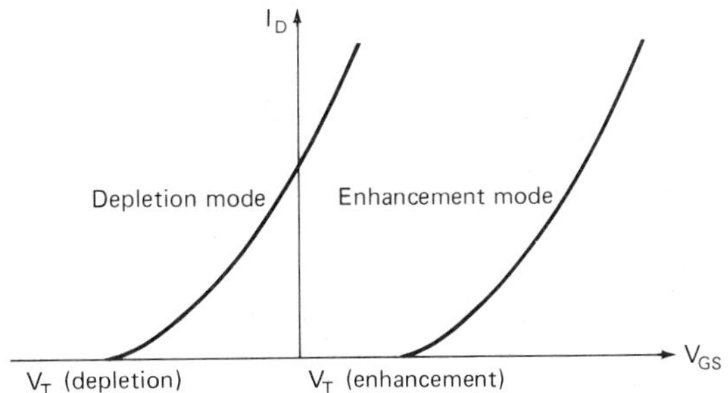

Fig. 8.1–2. Transconductance characteristics for an enhancement mode and a depletion mode MOS device.

drain current upon the terminal voltages of the MOS device will be developed in the next section.

A *p*-channel MOS transistor is similar to that of Fig. 8.1–1 except that the substrate is *n*-type material, and the drain and source diffusions are *p*-type. In CMOS only one type of substrate can be used, so that extra processing steps will be required to make both *n*-channel and *p*-channel devices in the same substrate. Figure 8.1–3 shows how this might be accomplished. In the CMOS process, typically both types of MOS devices are enhancement devices. It is easily seen that the processing steps are more complex for double-channel than for single-channel MOS technology.

Let us now consider the various sequences of steps required to achieve either Fig. 8.1–1 or 8.1–3. Figure 8.1–1 is typical of a silicon gate, *n*-channel MOS process (*p*-channel would be similar except that *n* and *p* are interchanged). To illustrate the process steps, consider the circuit of Fig. 8.1–4. This circuit is a simple MOS inverter and consists of a lower enhancement device called the *pulldown* transistor and an upper depletion device called

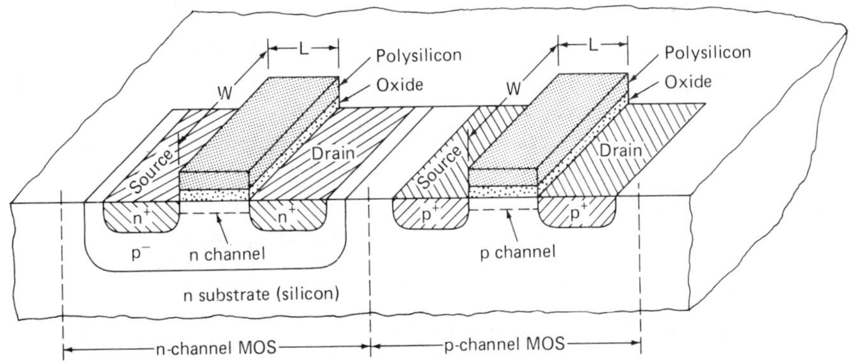

Fig. 8.1–3. Illustration of a double channel MOS (CMOS) process.

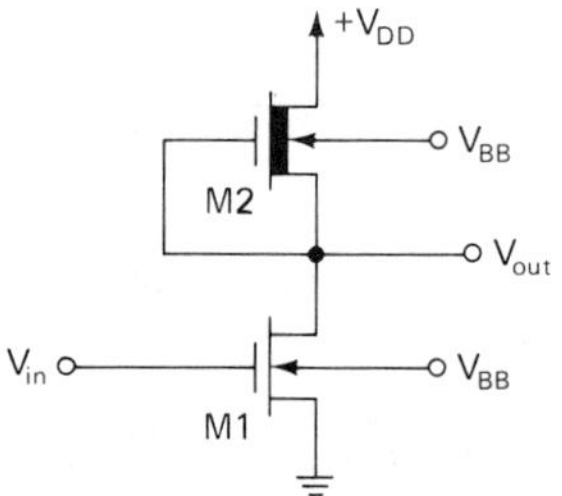

Fig. 8.1–4. An inverter used to illustrate the processing steps of an n-channel MOS process shown in Fig. 8.1–5.

the *pullup* transistor. The pullup transistor has the gate connected to the source, which causes it to function like a current source.

The symbols for the enhanced and depletion MOS devices are indicated in Fig. 8.1–4. These symbols can be simplified when the application of the device is obvious. Often the substrate contact is dropped when one is using a single-channel technology. In CMOS, the substrate is almost always connected to the source. In this case it is convenient to introduce a slightly modified symbol, which will be done later.

Although there are many steps involved in a given process, we shall focus only on the major steps. This means we will be ignoring such steps as application of photoresist, exposure of the photoresist, etching, cleaning, etc. The major steps in the process of Fig. 8.1–1 are given as follows.

1. definition of active device areas, called the *moat*
2. implantation for depletion devices
3. gate oxide
4. polysilicon
5. diffusion of drain and source
6. contact cuts
7. metallization

The basic process has many frequently used modifications. One example is to include two polysilicon layers separated by a thin oxide. This is very convenient in fabricating capacitors and charge-coupled devices.[6]

Each of the above steps is briefly described with the help of Fig. 8.1–5. In this figure we illustrate the fabrication of Fig. 8.1–4 using a silicon gate, *n*-channel MOS process. Figure 8.1–5(a) shows a top view of the inverter

---

[6] W. L. Eversole, D. J. Mayer, P. W. Bosshart, M. deWit, C. R. Hewes, and D. D. Buss, "A Fully-Integrated 32-point Chirp Z-Transform IC," *Proc. ISSCC,* February 1978, San Francisco, CA, pp. 86–87.

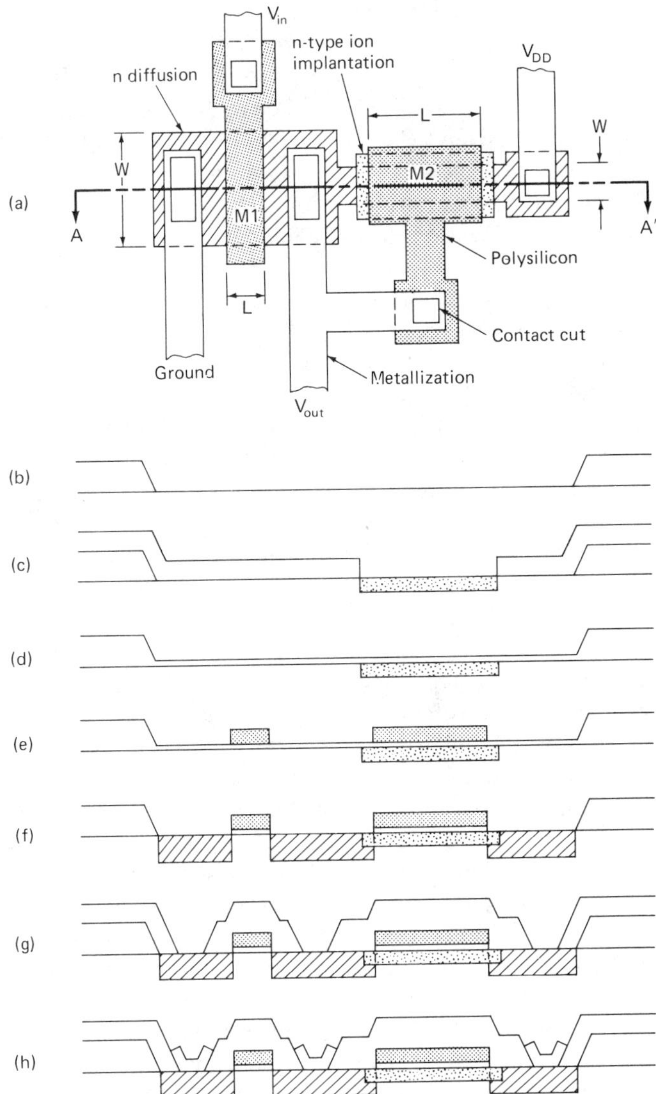

Fig. 8.1–5. Illustration of a silicon gate n-channel MOS process used to fabricate Fig. 8.1–4. (a) Top view. All remaining views are taken looking at the cross-section AA′. (b) Definition of the active areas (moat). (c) Implantation on n-type material creating the M2 depletion device. (d) Gate oxide step. (e) Polysilicon gates. (f) Diffusion of the n^+ drain and source. (g) Contact cuts. (h) Metallization.

as it might appear on an integrated circuit. Note that the channel area ($L \times W$) is defined by the overlapping areas of the diffusion, or moat, and the polysilicon gate. In Fig. 8.1–5(a), W/L of M1 is 3, and W/L of M2 is ⅓. Figure 8.1–5(a) is assumed to be located on a p-type substrate, as suggested by Fig. 8.1–1. The active device area, or moat, is indicated by cross-hatching, the polysilicon by a dark shading, and the ion implantation by random dots. The metallization and oxides are all plain. Because the contact cuts are always beneath the metallization, they are not designated but are easy to identify.

Looking at the cross section AA', we can describe the major steps of the silicon gate, n-channel process. Before these processes occur, the silicon dioxide has been removed everywhere except where the moats are located. This exposed area is called the *field* and is doped slightly more than the moat to increase the parasitic field oxide threshold, V_{TF}. The first step indicated in Fig. 8.1–5(b) is to define the moat by removing all the silicon dioxide in this area. The moat will eventually contain the drain, source, and channel of all devices. The second step shown in Fig. 8.1–5(c) is to implant an n-type material over the area that will become the channel of the depletion devices. This area is slightly larger than the actual channel area. The next step in the process is the gate oxide indicated in Fig. 8.1–5(d). Everything is then covered with a layer of polysilicon that is removed except where indicated in Fig. 8.1–5(a). This step is shown in Fig. 8.1–5(e). The polysilicon is conducting and can serve both as a gate and as a circuit interconnection. It can also form a crossunder beneath metal, because these two layers are isolated by a silicon dioxide layer. The next step is the n^+ diffusion of the drain and source. This step is shown in Fig. 8.1–5(f). This is accomplished by removing all the gate oxide not covered by the polysilicon layer. The source and drain diffusions occur in these exposed areas of the substrate. The major advantage of the silicon gate process is that it does not require a critical alignment between the mask that defines the gate area and the mask that defines the drain and source area. The result is an automatic self-aligned gate. Note that the overlapping of the ion implantation area and the source and drain diffusion has no significance, for both have n-type impurities. Finally, the circuit is covered by another layer of silicon dioxide, and contact cuts are made as indicated in Fig. 8.1–5(g). Metal is then deposited over the entire circuit and removed everywhere except in those areas where metal is to be located, as shown in Fig. 8.1–5(h). Figure 8.1–6 shows scanning electron photomicrographs of a single NMOS device.

The above process is conservative with respect to the fact that the chip area is not necessarily minimized. The process also continues beyond that described above. After metallization, the entire chip is covered with another coat of silicon dioxide. This step is called *overglassing* and provides physical

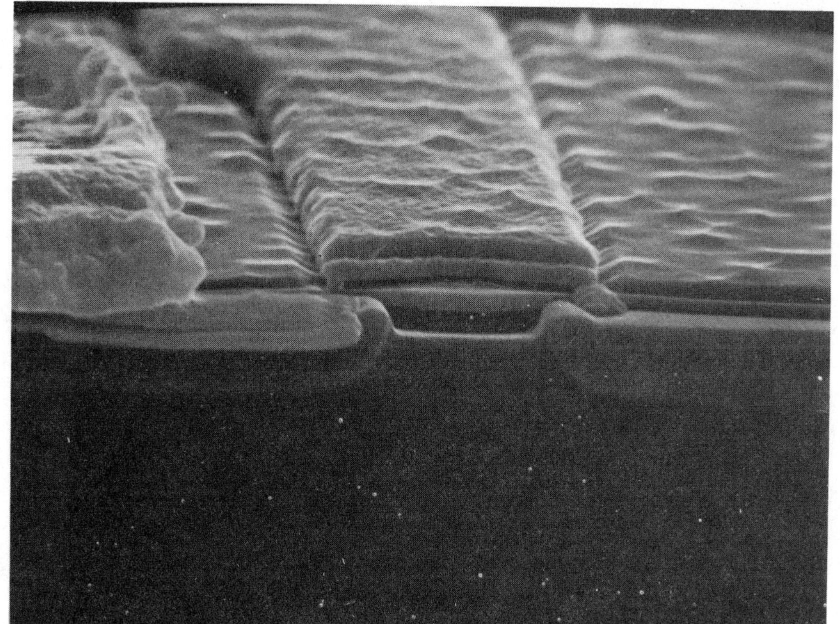

(a)

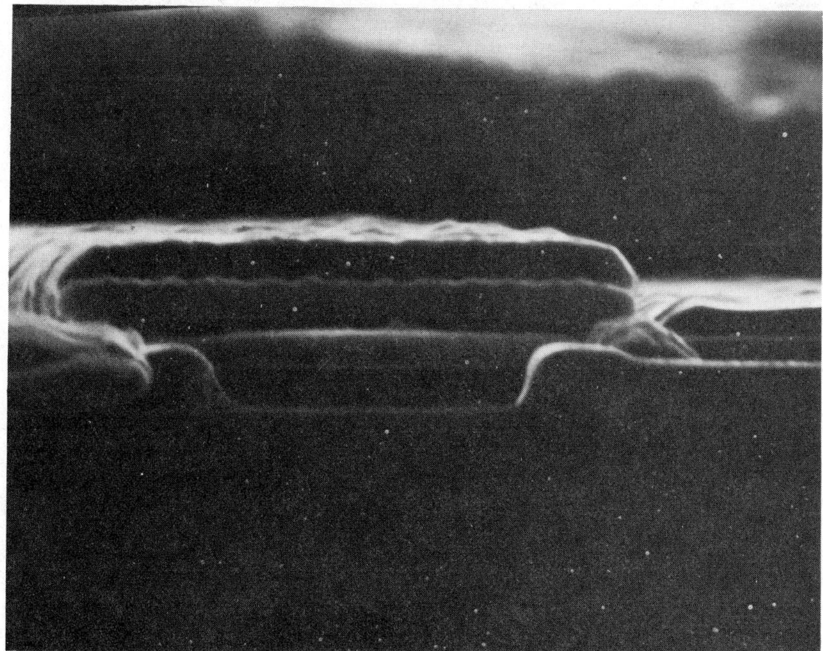

(b)

Fig. 8.1–6. Scanning electron microscope photographs of an NMOS transistor.

protection for the circuit. This step is followed by a removal of the overglassing on top of the bonding pads, which are large metal areas to which electrical contact to the circuit will be made. After testing, each wafer, which contains many chips, will be separated and individually mounted in a package. Fine wire is used to connect from the bonding pads of the chip to the pads in the package that are connected to external pins. The package is then covered and sealed and is ready for testing and subsequent usage.

The double-channel or CMOS silicon gate process uses the same steps as for the single-channel MOS process plus an additional diffusion or implantation of a "well" to build the devices whose channel has the same type of impurity as the substrate. Let us consider the CMOS process illustrated by Fig. 8.1–3. The major processing steps are given as follows.

1. diffusion or implantation of the p^- well
2. gate oxide
3. polysilicon
4. diffusion or implantation of the p^+ drain and source
5. diffusion or implantation of the n^+ drain and source
6. contact cuts
7. metallization

Because no depletion implantation is employed, both the p-channel and n-channel devices are of the enhancement mode. We have neglected other steps which include guard rings and are important in preventing latch-up. An illustration of the CMOS silicon gate process is given in Fig. 8.1–7(a), and a CMOS metal gate process is shown in Fig. 8.1–7(b).

Since we are concerned with switched capacitor circuits, the capacitor and its fabrication are very important. The capacitor must be compatible with the process steps indicated in Figs. 8.1–5 and 8.1–7. There are two types of capacitors that have the desired characteristics for analog sampled data applications. These are the capacitors formed between the gate and the channel and between two layers of polysilicon separated by a thin oxide layer. Figure 8.1–8(a) shows an example of the gate-oxide-channel capacitor. It is seen that the upper plate of this capacitor consists of the polysilicon (or metal) gate and that the lower, or bottom, plate is realized by a region or n-type impurities in the p-type substrate. To be compatible with the steps of the single-channel MOS process, this n-region beneath the gate oxide must be the n-implantation for the n-channel of depletion devices. If the p-substrate is less positive than the n-region beneath the gate, then the bottom plate is isolated from the substrate. However, the reversed biased pn junction leads to a voltage-dependent parasitic capacitance that could have significant influence on the use of the gate-oxide-channel capacitor. The capacitance per area of the gate-oxide-channel capacitor is typically 0.25 to 0.3 pF/mil^2 (0.4 to 0.48 $\times$ 10^{-3} pF/micron^2).

The polysilicon-oxide-polysilicon capacitor is illustrated in Fig. 8.1–8(b). It is seen that a double polysilicon process is required to implement this capacitor. Both the upper and lower plates of this capacitor are formed by polysilicon layers. The dielectric is formed by a thin oxide layer that can be produced only by several additional process steps over the usual single-polysilicon, single-channel process. One of the advantages of this capacitor is a smaller dependence upon the substrate capacitance and a voltage-independent parasitic capacitance. However, even though there is a thick layer of oxide isolating the polysilicon-oxide-polysilicon capacitance, the parasitics are still not negligible. The typical value of capacitance per area of the polysilicon-oxide-polysilicon capacitor is about 0.15 to 0.2 pF/mil^2 (0.24 to 0.32 $\times$ 10^{-3} pF/micron^2). This smaller value is due primarily to a slightly larger oxide thickness than the gate oxide.

We have seen that the performance of analog sampled data circuits can be directly related to the MOS capacitors used in the implementation. Therefore it is important to investigate the important characteristics of these capacitors. From the standpoint of analog sampled data applications the most important characteristics of the MOS capacitor are (1) the ratio accuracy, (2) voltage and temperature coefficient, and (3) parasitic capacitances.[7-9] We shall briefly discuss each of these characteristics.

The value of integrated MOS capacitors is given by

$$C = (\epsilon_{ox} A)/t_{ox} = C_{ox} A \tag{1}$$

where ϵ_{ox} is dielectric constant of the silicon oxide (approximately 3.54 $\times$ 10^{-5} pF/micron), t_{ox} is the thickness of the oxide, and A is the area of the capacitor. The accuracy of the capacitor is seen to depend upon the area A and the oxide thickness t_{ox}. If the thickness has a linear variation that could cause an error, then a common centroid geometry can be used to eliminate its effects.[8,9] The more serious variation is that due to the geometrical definition of the capacitor area. There are several reasons for the area to vary. One is

---

[7] R. W. Broderson, P. R. Gray, and D. A. Hodges, "MOS Switched-Capacitor Filters," *Proceedings of the IEEE,* Vol. 67, No. 1, pp. 61–75, January 1979.

[8] J. L. McCreary and P. R. Gray, "All-MOS Charge Redistribution Analog-to-Digital Conversion Techniques—Part I," *IEEE J. of Solid-State Circuits,* Vol. SC-10, No. 6, December 1975, pp. 371–379.

[9] J. L. McCreary, "Matching Properties, and Voltage and Temperature Dependence of MOS Capacitors," *IEEE J. of Solid-State Circuits,* Vol. SC-16, No. 6, December 1981, pp. 608–616.

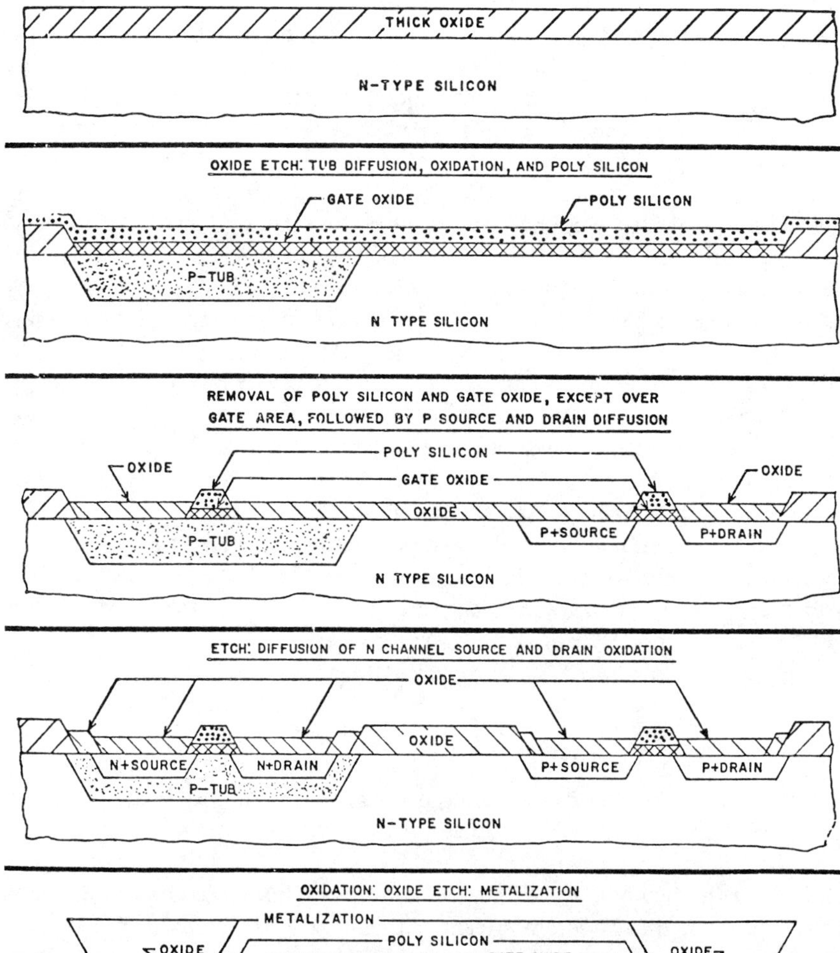

Fig. 8.1–7. (a) CMOS silicon gate process. (Courtesy of Motorola, Inc.)

BASIC SUBSTRATE

THICK OXIDE

N-TYPE SILICON

OXIDE ETCH: TUB DIFFUSION AND OXIDATION

OXIDE

P-TUB

OXIDE

N- TYPE SILICON

OXIDE ETCH: SOURCE AND DRAIN DIFFUSION AND OXIDATION FOR P-CHANNEL DEVICE

OXIDE

OXIDE

OXIDE

P-TUB

P+SOURCE

P+DRAIN

N-TYPE SILICON

ETCH: DIFFUSION OF N-CHANNEL SOURCE AND DRAIN OXIDATION

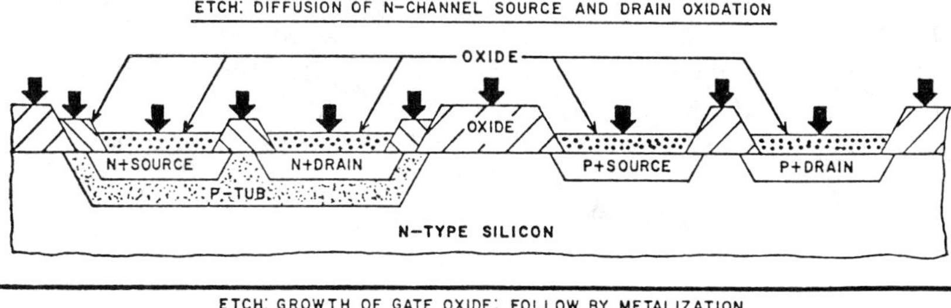

ETCH: GROWTH OF GATE OXIDE; FOLLOW BY METALIZATION

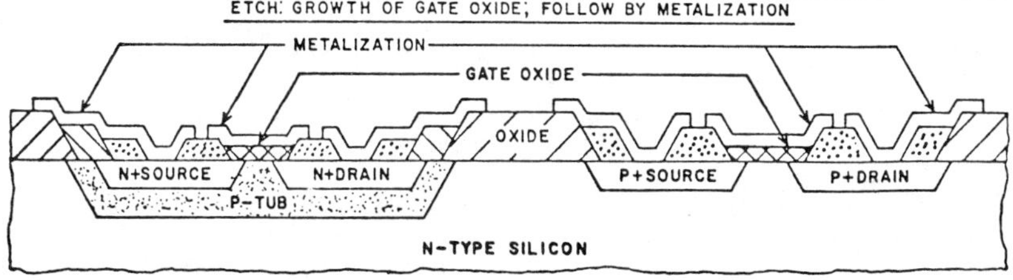

Fig. 8.1–7. (b) CMOS metal gate process. (Courtesy of Motorola, Inc.)

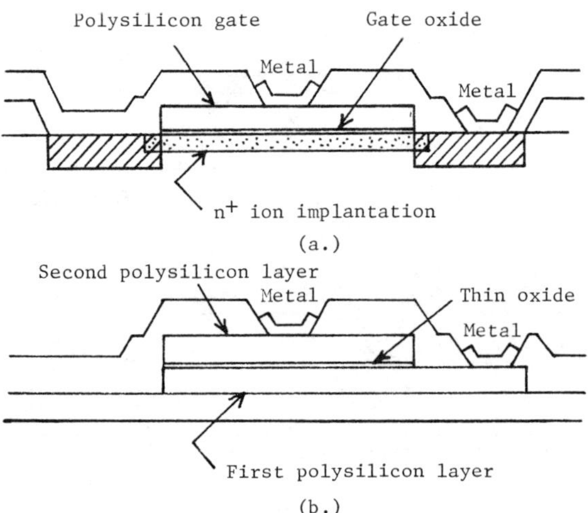

Fig. 8.1–8. MOS capacitors. (a) Polysilicon-oxide-channel and (b) Polysilicon-oxide-polysilicon types.

due to undercutting in the oxide in the photolithography process. Another is due to the resolution of the masks used for fabrication. Both these errors are associated with the periphery of the capacitor area. If the periphery of a given area is minimized, then the errors associated with this area should be minimized. More importantly, in switched capacitor circuits, we have seen that the ratio of two capacitors is the factor determining the circuit performance. Therefore it is desirable to use two capacitors with the same accuracy if we are to make the capacitor ratio as accurate as possible. For example, let C_1' be defined as

$$C_1' = C_1 \pm \Delta C_1 \tag{2}$$

and C_2' be defined as

$$C_2' = C_2 \pm \Delta C_2 \tag{3}$$

where C_1 and C_2 are the ideal values of C_1' and C_2'. The ratio of C_2' and C_1' can be expressed as

$$C_2'/C_1' = (C_2 \pm \Delta C_2)/(C_1 \pm \Delta C_1) \cong (C_2/C_1)[1 \pm (\Delta C_2/C_2) \mp (\Delta C_1/C_1)] \tag{4}$$

where $\Delta C_1/C_1$ and $\Delta C_2/C_2$ are the errors of C_1 and C_2, respectively, and are assumed to be small. If the error of C_1 is the same as C_2, then $C_2'/C_1' = C_2/C_1$. Because ΔC is proportional to the periphery of C, and C is propor-

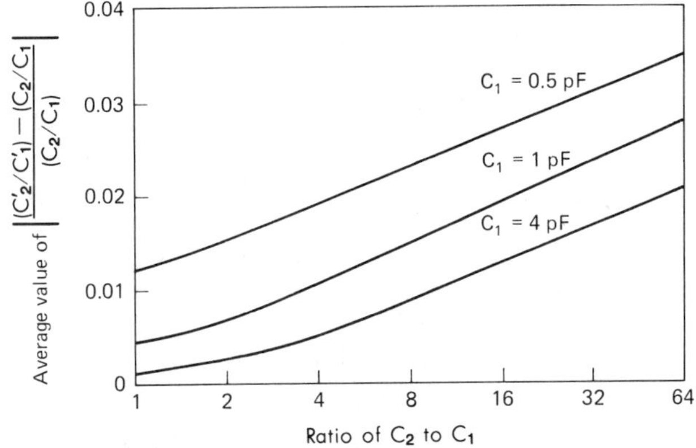

Fig. 8.1–9. Capacitor ratio error for a 128 unit binary weighted capacitor array.

tional to the area, then the periphery-to-area ratio of C_1 and C_2 should be identical. This is typically accomplished by making the smaller capacitor a square and the larger capacitor an integer number of these squares plus some smaller fraction of the square. As the ratio of C_2/C_1 becomes greater than unity, the errors tend to accumulate, as shown in Fig. 8.1–9. It is always desirable to keep the ratio of capacitors as close to unity as possible to maintain accuracy and also to minimize area requirements. In general, the absolute value of MOS capacitors can be several percent and is inversely proportional to the size of the capacitor. The matching or relative accuracy of two MOS capacitors also depends on the size of the capacitors. As the ratio of the two capacitors differs from 1:1, the accuracy becomes worse.

Many design techniques are based on experience; they are not immediately obvious and fall into the category of tricks. One of these tricks becomes important when etching the material forming the top plate. The presence of adjacent similar material will tend to influence the etch rate. Consequently, if a capacitor is made from a number of adjacent squares, the sides of the exterior squares that form the perimeter of the capacitor will etch differently than the sides of a square that are adjacent to another square. Therefore one must surround the capacitor with a strip of top plate material to equalize the etch rates of the outside edges of the exterior squares.

The voltage coefficient of MOS capacitors is negative for n-channel plates and falls within the range of −10 to −100 ppm/V. Temperature coefficients of MOS capacitors vary from 20 to 50 ppm/C°. Both these coefficients become much smaller when the ratio of two capacitors is considered, and they generally have insignificant influence upon the analog sampled data circuit.[10]

---

[10] J. L. McCreary, *op. cit.*

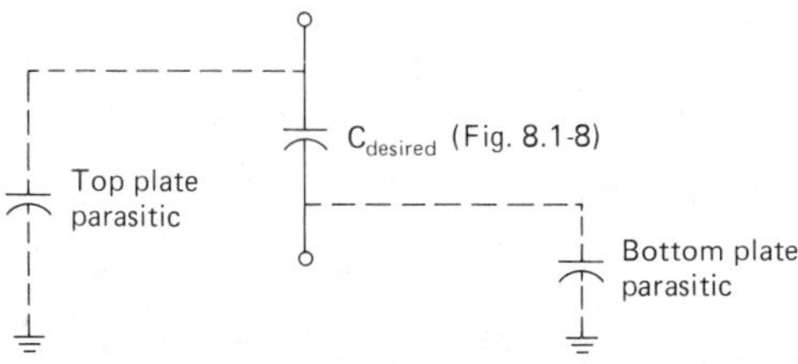

Fig. 8.1–10. A model for the capacitors of Fig. 8.1–8.

The parasitic capacitors associated with the capacitors of Fig. 8.1–8 represent a major source of error in analog sampled data circuits. The upper plate of the capacitor will be designated as the top plate and the lower capacitor plate as the bottom plate. The parasitic capacitor associated with the top plate is due primarily to the circuits connected to the capacitor. This includes the capacitance from the metallization (or polysilicon) to the substrate, the overlap capacitors of the drain or source to the bulk of switches attached to the top plate, and other circuit parasitics. Figure 8.1–10 shows the top plate parasitic capacitance modeled simply as a capacitor to ground. The value of this capacitance is typically in the range of 0.01 to 0.001 of the desired capacitor and is independent of voltage. For the polysilicon-oxide-channel capacitor, the bottom plate capacitor is the capacitance from the channel to the substrate. This capacitor is seen to be a voltage-dependent capacitance, and it can have a value of 0.05 to 0.2 of the desired capacitance. The bottom plate parasitic of the polysilicon-oxide-polysilicon is an oxide capacitor similar to the gate capacitance, but reduced because of the thicker oxide. The bottom plate parasitic capacitance of the polysilicon-oxide-polysilicon capacitor is approximately 0.1 of the desired capacitor. The bottom plate parasitic capacitance is modeled as capacitor to ground, as shown in Fig. 8.1–10. All these parasitic capacitances depend upon the capacitor size, the layout, and the technology and are unavoidable.

The remaining component of MOS technology to be considered is the resistance. Even though we shall primarily use circuits that contain only MOS active devices and capacitors, there are some applications, such as digital-to-analog conversions and anti-aliasing filters, that use the resistor. Resistors compatible with the MOS technology of this section include the diffused, polysilicon, ion implanted, and the pinched resistor.

A diffused resistor is formed using the source-drain diffusion and is shown in Fig. 8.1–11(a). The sheet resistance of such resistors is usually in the range of 10 to 100 ohms per square. The fact that the source-drain diffusion is needed as a conductor in integrated circuits conflicts with its use as a resistor. The diffused resistor is found to have a voltage coefficient of resistance

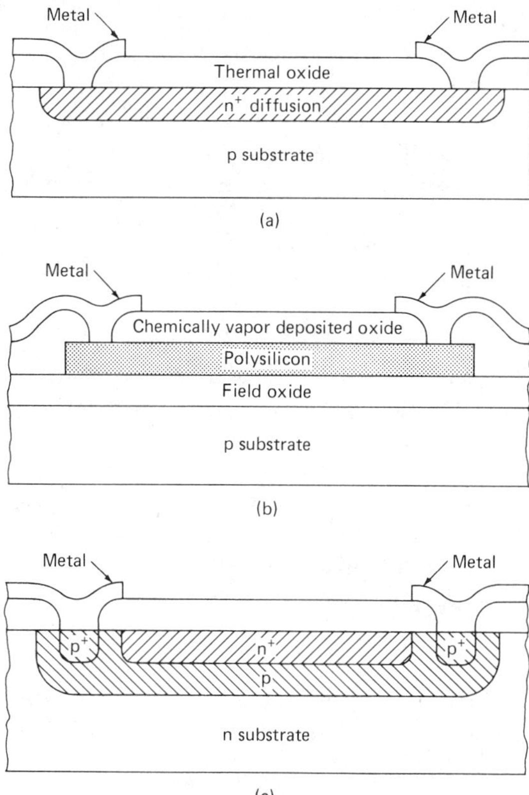

Fig. 8.1–11. (a) Diffused resistor structure. (b) Polysilicon resistor structure. (c) Pinched resistor structure.

in the 100 to 500 ppm/V range. The parasitic capacitance to ground is also voltage-dependent in this type of resistor.

A polysilicon resistor is shown in Fig. 8.1–11(b). This resistor is surrounded by thick oxide and has a sheet resistance in the range of 30 to 200 ohms per square. The parasitics of this resistor are very small and voltage-independent. The sheet resistance of polysilicon can be increased by a factor of 2 to 3 if it is covered by a second polysilicon layer to avoid the source-drain doping, as is available in a double-polysilicon process. Another advantage of the polysilicon resistor is that it can be trimmed by blowing links, either by the application of an electrical current or a laser.

An ion implanted resistor can be obtained at the expense of an additional mask step. The resistor is similar to that of the diffused resistor of Fig. 8.1–11(a). The practical sheet resistances of the ion implanted resistor fall in the range of 500 to 1000 ohms per square. Unfortunately, the ion implanted resistor has a large voltage coefficient, voltage-dependent parasitics, and is subject to errors caused by piezoresistance effects induced by nonuniform residual strain in the chip after packaging.

The pinched resistor is a highly nonlinear resistor capable of sheet resistances in the range of 1000 to 5000 ohms per square. It is constructed as shown in Fig. 8.1–11(c). The pinched resistor is compatible only with the CMOS process; it would require an additional masking step in a single-channel MOS process. The voltage coefficient is very large because the current-voltage characteristics are essentially those of an MOS device with a constant gate-source voltage.

A summary of the performance of single-channel, MOS passive components was given in Table 8.0–1. We have included both resistors and capacitors to further illustrate the differences between the two types of passive components. This summary is based on a typical *n*-channel MOS technology.

The last subject of this section is that of design rules. Design rules are the existing geometrical constraints that guarantee the proper operation of the MOS circuit implemented by a given MOS process. The design rules consist of a set of minimum allowable values for certain widths, separations, extensions, and overlapping of geometrical objects patterned in the various layers of the system. In every process, there is a fundamental resolution capability. This resolution is the distance by which a geometrical feature on any layer may vary from another geometrical feature on the same or different layer, with all processing factors considered and an appropriate safety factor added. All dimensions of the MOS circuit are usually given in terms of this elementary distance unit, which will be designated as λ. As time progresses this unit should decrease, as new technology is developed and perfected. In 1979, this unit was 0.05 mils or about 1.3 microns.[11]

Figure 8.1–12 illustrates the design for a silicon gate, single-channel, MOS process. Each of these figures illustrates various minimum values between different layers of the single-channel, silicon gate MOS process. Each layer represents various levels of the mask used in the fabrication procedure. The following rules are illustrated by Fig. 8.1–12(a).

A. Minimum width of the moat is 4 λ.
B. Minimum spacing between moats is 6 λ.
C. Minimum FET width is 6 λ.
D. Minimum FET length is 6 λ.

[11] Normal design rules are usually designated by the minimum gate dimension, which in the following is 6 λ or 0.3 mils (7.6 microns).

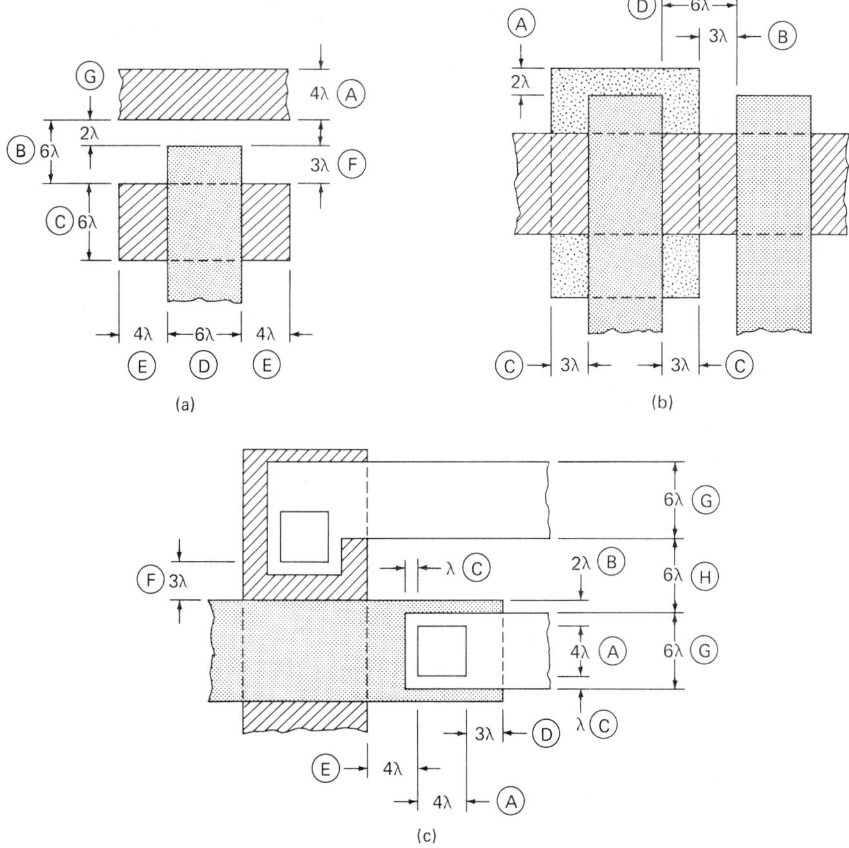

Fig. 8.1–12. Illustration of the design rules for a single-channel, polysilicon gate, MOS process. (a) Polysilicon and moat layers. (b) Depletion, polysilicon, and moat layers. (c) Polysilicon, moat, contact, and metallization layers.

E. Minimum size of source or drain moat is 4 λ.
F. Minimum overlap of the polysilicon over the channel area is 3 λ.
G. Minimum space from polysilicon to another moat is 2 λ.

The following design rules are illustrated by Fig. 8.1–12(b).

A. Minimum overlap of the depletion layer over the polysilicon gate is 2 λ.
B. Minimum separation between the depletion layer and the polysilicon used to form an enhancement mode device is 3 λ.

C. The minimum overlap of the depletion layer over the polysilicon gate in the moat is 3 λ.
D. Minimum spacing between adjacent areas of polysilicon is 6 λ.

The following design rules are illustrated by Fig. 8.1–12(c).

A. The minimum contact size is 4 λ by 4 λ.
B. Minimum distance from the polysilicon contact to substrate is 2 λ.
C. Minimum overlap of metal on a contact is λ (high-current contacts should be 2 λ).
D. Minimum distance from contact on polysilicon to the substrate in the direction of the metal is 3 λ.
E. Minimum distance from contact on polysilicon to a moat is 4 λ.
F. Minimum distance from a contact on the moat to the channel stop or polysilicon gate is 3 λ.
G. Minimum width of metal is 6 λ.
H. Minimum separation between adjacent areas of metal is 6 λ.

In addition to these rules, metal bonding pads should be a square whose side is 4 mils plus 4 λ, with a minimum distance between bonding pad openings of 2 mils. Also, each bonding pad has an area that is 4 λ smaller than the pad used to define the overglassing removal, so that electrical contact may be made to the bonding pads, which will be a square, 4 mils on a side.

Most of the material of this section is only representative of what is actually done in industry. However, the principles developed here should remain unaltered when translated to a specific process. This section has introduced MOS technology from the viewpoint of its use to implement analog sampled data circuits. This leads naturally to the next section, where we develop circuit models to be used in analyzing and designing analog sampled data circuits.

8.2 MODELING OF MOS COMPONENTS

In this section we develop methods for modeling and analyzing MOS transistors and capacitors. This is an important step if the MOS circuit is to be fabricated as an integrated circuit. It is very difficult to properly breadboard such circuits and to experimentally predict their performance. The only means of verifying the design before it is built is to analyze the circuit by hand, calculator, or computer methods. For large circuits, computer analysis becomes the only practical method of analyzing the circuit. To use these methods of analysis it is necessary to be familiar with models for the MOS components. The models presented here are consistent with those used in the computer

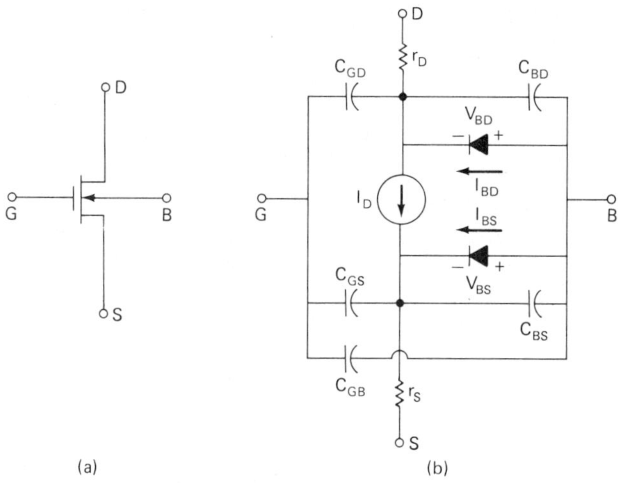

Fig. 8.2–1. (a) n-channel MOS transistor and (b) its large signal model.

program SPICE2.[12] Computer programs such as SPICE2 are suitable for continuous systems, but are difficult to use for analyzing analog sampled data systems in the frequency domain.[13]

The first step will be to develop the large-signal model for an MOS transistor. Figure 8.2–1 shows a modified version of the *n*-channel MOS transistor model presented by Shichman and Hodges.[14] We shall discuss all the parameters of this model, some of which are illustrated in Fig. 8.2–1(b). The most important aspect of this model is the functional relationship of I_D upon the terminal voltages, including the bulk voltage, V_B. The development of these relationships can be found in many references and will not be presented here.[15,16,17] If the charge in the channel can be assumed to be negligible or

[12] L. W. Nagel, "SPICE2: A Computer Program to Simulate Semiconductor Circuits," Electronics Research Laboratory, Memorandum No. ERL-M520, College of Engineering, Univ. of Calif., Berkeley, CA, May 9, 1975.

[13] B. D. Nelin, "Analysis of Switched-Capacitor Networks Using General-Purpose Simulation Programs," *IEEE Trans. on Circuits and Systems,* Vol. CAS-30, No. 1, Jan. 1983, pp. 43–48.

[14] H. Shichman and D. A. Hodges, "Modelling and Simulation of Insulated-Gate Field-Effect Transistor Switching Circuits," *IEEE J. of Solid-State Circuits,* Vol. SC-3, No. 3, September 1968, pp. 285-289.

[15] W. M. Penny and L. Lau, *MOS Integrated Circuits—Theory, Fabrication, Design, and Systems Application of MOS LSI,* Chap. 2, Van Nostrand Reinhold Co., New York, 1972.

[16] R. S. Muller and T. I. Kamins, *Device Electronics for Integrated Circuits,* Chap. 8, John Wiley and Sons, Inc., New York, 1977.

[17] D. R. Alexander, R. J. Antinone, and G. W. Brown, "SPICE2 Modelling Handbook," *Report BDM/A-77-071-TR,* BDM Corporation, 2600 Yale Blvd., Albuquerque, NM 87106.

constant, then a simplified expression for I_D has been developed by Sah[18] and is given as

$$I_D = (\mu_n C_{ox} W)/2L\,[2(V_{GS} - V_T) - V_{DS}]\,V_{DS}\,(1 + \lambda V_{DS}) \tag{1}$$

where the various parameters are defined as

μ_n = mobility of the n-channel (cm²/volt-sec)
$C_{ox} = \epsilon_{ox}/t_{ox}$ = capacitance/unit area of the gate oxide (F/cm²)
λ = channel length modulation parameter $(\text{volt})^{-1}$

The threshold voltage can be expressed as

$$V_T = V_{TO} + \gamma\,[\sqrt{\phi - V_{BS}} - \sqrt{\phi}\,] \tag{2}$$

where

$$V_{TO} = V_{FB} + 2\phi_f + 2\,(\sqrt{q\,\epsilon_{si}\,N_{SUB}\,\phi_f})/C_{ox} \tag{3}$$

γ = bulk threshold parameter $(\text{volt}^{-1/2})$

$$= \sqrt{2\,\epsilon_{si}\,q\,N_{SUB}}/C_{ox} \tag{4}$$

$\phi = 2\phi_f$ = strong inversion surface potential (volt)

$$= (2kT/q)\ln(N_{SUB}/n_i) \tag{5}$$

and

$$V_{FB} = \text{flatband voltage (volt)} = \phi_{ms} - Q_{ss}/C_{ox} \tag{6}$$

These new parameters and constants are defined as

ϕ_{ms} = metal semiconductor work function $= \phi_m - \phi_{so} - (E_g/2q) - \phi_f$
ϕ_m = metal gate work function[19]
Q_{ss} = oxide charge $= N_{ss}q$
C_{ox} = oxide capacitance
ϕ_{so} = $Si - SiO_2$ work function[19]
E_g = silicon bandgap[19]
k = Boltzmann's constant (1.381×10^{-23} J/°K)
T = temperature °K
n_i = intrinsic carrier concentration[18]

---

[18] C. T. Sah, "Characteristics of the Metal-Oxide-Semiconductor Transistor," *IEEE Trans. Electron Devices,* ED-11, July 1964, pp. 324–345.

[19] These values are specified internally in most computer programs.

The effective threshold of the MOS transistor is shown by Eq. (2) to be a function of the voltage between the bulk and the source. A typical dependence of the threshold upon V_{BS} for an n-channel MOS transistor is shown in Fig. 8.2–2. Equation (1) is different from that developed by Sah, in that the channel length modulation parameter is included. The expression for the drain current in Eq. (1) may be simplified as

$$I_D = K' \ W/L \ [(V_{GS} - V_T) - (V_{DS}/2)] \ V_{DS} \ (1 + \lambda \ V_{DS}) \tag{7}$$

where the transconductance parameter K' is given as

$$K' = \mu_n \ C_{\text{ox}} \ (\text{amps/volt}^2) \tag{8}$$

Equation (1) describes the dependence of I_D upon the terminal voltages for an n-channel MOS transistor. This same expression can be used for p-channel MOS transistors by assuming that all terminal voltages are those of an n-channel device and simply reversing the sign on the drain current. This means that V_T is considered positive for enhancement and negative for depletion, regardless of whether the device is n-channel or p-channel.

There are various regions of operation of the MOS transistor that depend upon the value of $V_{GS} - V_T$. If V_{GS} is negative, then we have the *cutoff* region, and Eq. (1) becomes

$$I_D = 0, \qquad V_{GS} - V_T \leq 0 \tag{9}$$

This region is indicated on the output characteristics given in Fig. 8.2–3. A plot of Eq. (1) with $\lambda = 0$ as a function of V_{DS} is shown in Fig. 8.2–4 for various values of $V_{GS} - V_T$. At the maximum of these curves the MOS

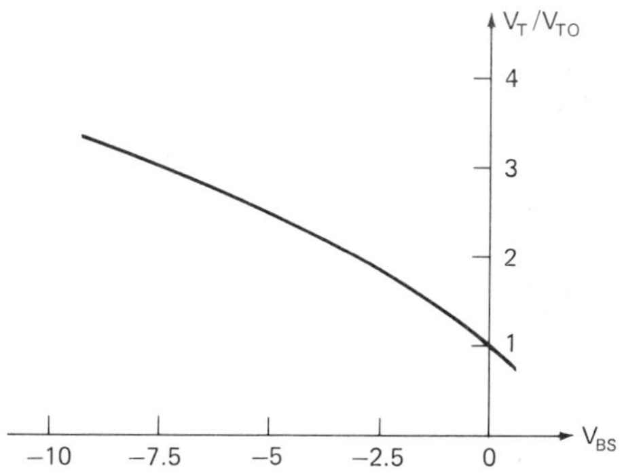

Fig. 8.2–2. Typical dependence of the effective threshold voltage, V_T, upon the bulk-source voltage, V_{BS}.

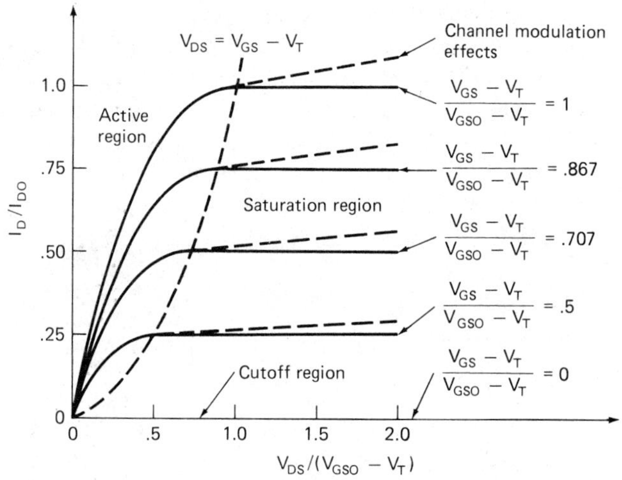

Fig. 8.2–3. Illustration of the regions of operation for the MOS transistor.

transistor is said to *saturate.* The value of V_{DS} at which this occurs is given as

$$V_{DS}(\text{sat}) = V_{GS} - V_T \tag{10}$$

Therefore the second region of operation for the MOS transistor is called the *active,* or *nonsaturated,* region and occurs for $V_{GS} - V_T$ greater than V_{DS}. In this case, Eq. (1) becomes

$$I_D = [(\mu C_{ox} W)/(2L)][2(V_{GS} - V_T) - V_{DS}]V_{DS}(1 + \lambda V_{DS}), \quad 0 < V_{DS} < V_{GS} - V_T \tag{11}$$

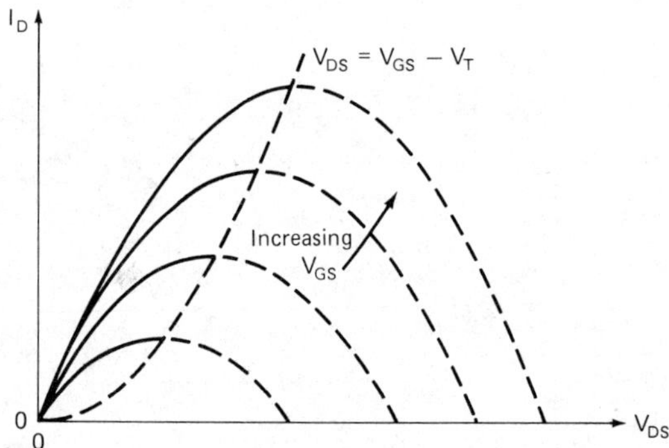

Fig. 8.2–4. Plot of eq. (1) illustrating the saturation effect.

This region is indicated on Fig. 8.2–3 as the area where the curves are nonlinear or the solid portion of the curves in Fig. 8.2–4. The third region occurs when $V_{GS} - V_T$ is less than or equal to V_{DS} and is called the *saturation region.* V_{DS} in Eq. (1) is replaced by V_{DS} (sat) of Eq. (10) to give

$$I_D = [(\mu C_{ox} W)/(2L)](V_{GS} - V_T)^2(1 + \lambda V_{DS}), \quad 0 < V_{GS} - V_T \leq V_{DS} \qquad (12)$$

where the V_{DS} of the $(1 + \lambda\ V_{DS})$ factor has not been replaced by Eq. (10), in order to model the channel length modulation behavior. This region is shown by the horizontal curves of Fig. 8.2–3 when $\lambda = 0$. If $\lambda \neq 0$, then the curves are shown by the dashed lines. Figure 8.2–3 has been normalized to the gate-source voltage of the upper curve. V_{GS} of this curve has been designated as V_{GSO}, which causes a drain current of I_{DO} in the saturation region.

Another important characteristic of the MOS transistor can be obtained by plotting Eq. (12) when λ is zero or V_{DS} is small. Figure 8.2–5 shows this result. This portrayal of the MOS transistor is called the *transconductance characteristic.* We note that the transconductance characteristic in the saturation region can be obtained from Fig. 8.2–3 by drawing a vertical line to the right of the dashed line and plotting values of I_D versus V_{GS}. The effect of the bulk-source voltage, V_{BS}, can be demonstrated very clearly. As V_{BS} becomes large and negative, Fig. 8.2–5 confirms Eq. (2), showing that the threshold voltage, V_T, increases.

Because the MOS transistor is symmetrical, there may be occasions when V_{DS} is negative for the n-channel device. We designate this condition as the *inverse mode.* It becomes necessary to re-express Eqs. (9), (11), and (12) for the inverse mode. The cutoff region in the inverse mode is given by

$$I_D = 0, \qquad V_{GD} - V_T \leq 0 \qquad (13)$$

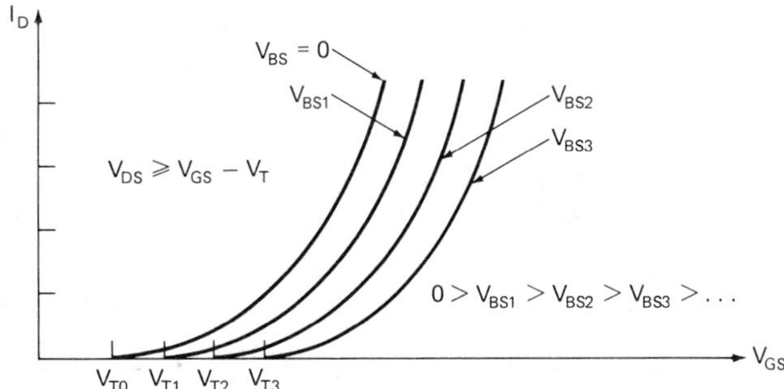

Fig. 8.2–5. Transconductance characteristics of the MOS transistor as a function of V_{BS}.

In comparing Eq. (13) with Eq. (9), we note that V_{GS} has been replaced by V_{GD}. The drain current of the inverse mode in the active region can be expressed as

$$I_D = [(\mu C_{ox} W)/(2L)][2(V_{GD} - V_T) + V_{DS}](-V_{DS})(1 - \lambda V_{DS}),\quad 0 < -V_{DS} < V_{GD} - V_T \tag{14}$$

We see from Eq. (14) that the source is acting like the drain and that current will flow from the source through the channel to the drain. The drain current for the inverse mode in the saturation region is given as

$$I_D = [(-\mu C_{ox} W)/(2L)](V_{GD} - V_T)^2(1 - \lambda V_{DS}),\quad 0 < V_{GD} - V_T \le - V_{DS} \tag{15}$$

If a p-channel transistor is used in the inverse mode, the terminal voltages are considered as having the same polarity as n-channel terminal voltages, but the drain current is reversed.

The diodes of Fig. 8.2–1(b) represent the pn junctions between the source and substrate and drain and substrate. The state of the channel determines whether or not these diodes can be considered in parallel. I_{BD} and I_{BS} are the currents that flow from the bulk to the drain and from the bulk to the source, respectively. These currents are expressed as

$$I_{BD} = I_S[\exp(V_{BD}/V_T) - 1] \tag{16}$$

and

$$I_{BS} = I_S[\exp(V_{BS}/V_T) - 1] \tag{17}$$

which are the standard diode equations for a pn junction. I_S is chosen to model the leakage characteristics of the substrate junctions.

The resistors r_D and r_S represent the ohmic resistance of the drain and source, respectively. Typically, these resistors may be 50 to 100 ohms and have little influence upon the MOS transistor.

The capacitance of Fig. 8.2–1(b) can be separated into three types. The first type includes capacitors C_{BD} and C_{BS}, which are associated with the back-biased depletion region between the drain and substrate and source and substrate. The second type includes capacitors C_{GD}, C_{GS}, and C_{GB}, which are all common to the gate and are dependent upon the operating state of the transistor. The third type includes parasitic capacitors that are independent of the operating state.

The depletion capacitors are a function of the voltage across the pn junction.

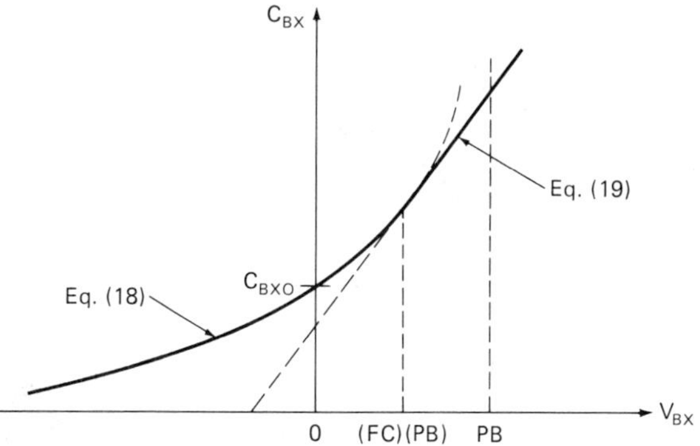

Fig. 8.2–6. Illustration of the modelling of the bulk junction capacitance using eqs. (18) and (19).

The expression of this junction capacitance is divided into two regions. The first is called the *reverse bias region* and is given as

$$C_{BX} = C_{BXO}\, A_{BX}[1 - (V_{BX}/PB)]^{-0.5}, \qquad V_{BX} \leq (FC)(PB) \tag{18}$$

where

$X = D$ for C_{BD} or $X = S$ for C_{BS}
A_{BX} = junction area
$C_{BXO} = C_{BX}$ when $V_{BX} = 0$
PB = bulk junction potential and
FC = forward bias nonideal junction capacitance coefficient (≤ 1)

The second region is called the *forward bias region* and is

$$C_{BX} = (C_{BXO}\, A_{BX})/(1 - FC)^{1.5}\, [(1 - 1.5FC) + (V_{BX}/2PB)], \qquad V_{BX} > (FC)(PB) \tag{19}$$

Figure 8.2–6 illustrates how the junction capacitance of Eqs. (18) and (19) are used together to model the large-signal capacitances C_{BD} and C_{BS}. It is seen that Eq. (19) prevents C_{BX} from approaching infinity as V_{BX} approaches PB.

The large-signal, charge-storage capacitors of the MOS device consist of gate-to-source (C_{GS}), gate-to-drain (C_{GD}), and gate-to-bulk (C_{GB}) capacitances.[20,21] In addition, C_{GS} and C_{DS} can be divided into intrinsic volt-

[20] W. M. Penny and L. Lau, *op. cit.*

[21] R. S. Muller and T. I. Kamins, *op. cit.*

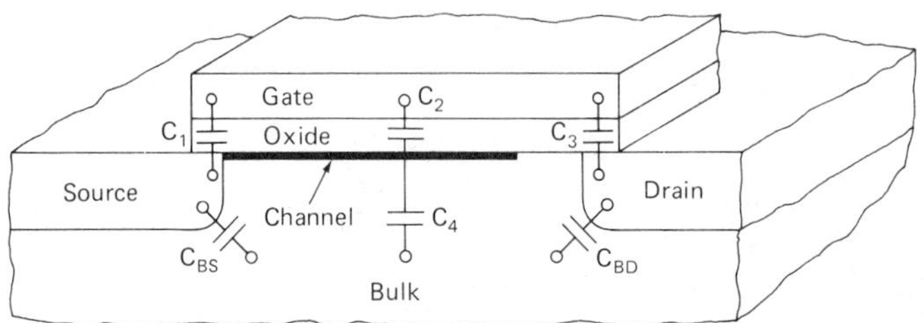

Fig. 8.2–7. Large signal, charge storage capacitors of the MOS device.

age-dependent capacitances (C'_{GS} and C'_{GD}) and extrinsic overlap capacitances (C''_{GS} and C''_{GD}). Figure 8.2–7 shows a cross section of the various capacitances that constitute the charge-storage capacitors of the MOS device. C_{BS} and C_{BD} are the source-to-bulk and drain-to-bulk capacitors discussed above. C_1 and C_3 are overlap capacitances and are due to an overlap of two conducting surfaces separated by a dielectric. The overlapping capacitors are shown in more detail in Fig. 8.2–8. This overlapping is due to lateral diffusion of the source and drain. For typical MOS transistor diffusion depth, the overlap

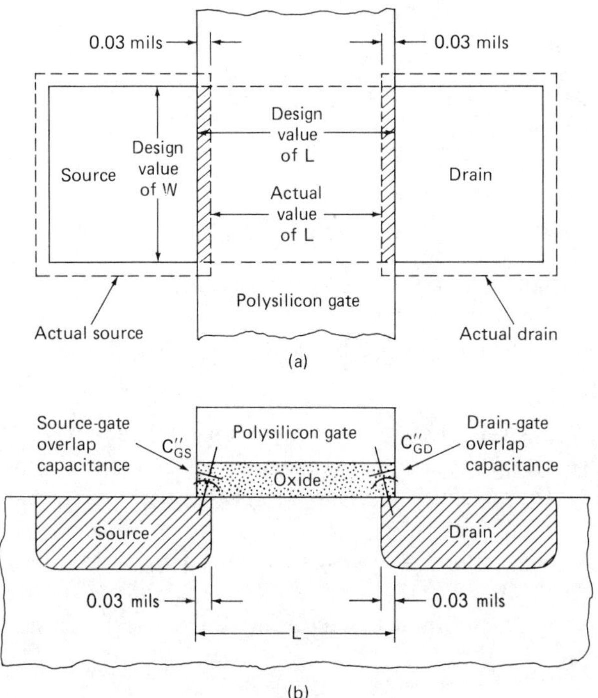

Fig. 8.2–8. Example of capacitances in the MOS transistor. (a) Top view showing the overlap between the source or drain and the gate. (b) Side view.

is approximately 0.03 mils. Therefore, the overlap capacitances can be approximated as

$$C_1 = C_3 \cong 0.03\, WC_{ox}, \qquad (W \text{ in mils}) \tag{20}$$

The channel of Fig. 8.2–7 is shown for the saturated state and would extend completely to the drain if the MOS device were in the active state. C_2 is the gate-to-channel capacitance and is given as

$$C_2 = C_{ox}\, WL \tag{21}$$

C_4 is the channel-to-bulk capacitance, which is a depletion capacitance and will vary with voltage, similar to C_{BS} or C_{BD}.

It is of interest to examine C_1, C_2, C_3, and C_4 as V_{DS} is held constant and V_{GS} is increased from zero. To understand the results, one can imagine following a vertical line upward on Fig. 8.2–3, at say $V_{DS} = 0.5\,(V_{GSO} - V_T)$, as V_{GS} increases from zero. The MOS device will first be off until V_{GS} reaches V_T. Next, it will be in the saturated region until V_{GS} is large enough so that the MOS device will be in the active or triode region. The variation of these capacitances under these conditions is shown in Fig. 8.2–9. In cutoff there is no channel and C_{GB} is approximately equal to C_2. As V_{GS} approaches V_T from the cutoff region, a thin depletion layer is formed, creating a large value of C_4. Since C_4 is in series with C_2, little effect is observed. As V_{GS} increases, this depletion region widens, causing C_4 to decrease and reducing

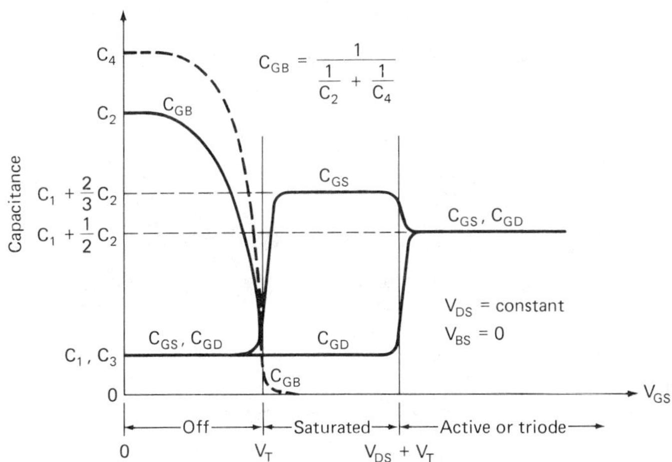

Fig. 8.2–9. Voltage dependence of C_{GS}, C_{GD}, and C_{GB} as a function of V_{GS} with V_{DS} constant and $V_{BS} = 0$.

C_{GB}. When $V_{GS} = V_T$, an inversion layer is formed that prevents further decreases of C_4 (and thus C_{GB}).

C_1, C_2, and C_3 constitute C_{GS} and C_{GD}. The problem is how to allocate C_2 to C_{GS} and C_{GD}. The approach used is to assume in saturation that approximately ⅔ of C_2 belongs to C_{GS} and none to C_{GD}. This is of course an approximation, but it has found to give good results. Figure 8.2–9 shows how C_{GS} and C_{GD} change values in going from the off region to the saturation region. Finally, when V_{GS} is greater than $V_{DS} - V_T$, the device enters the active or triode region. In this case the channel extends from the drain to the source, and C_2 is simply divided evenly between C_{GS} and C_{GD}, as shown in Fig. 8.2–9.

As a consequence of the above considerations, we shall use the following formulas for the charge-storage capacitances of the MOS device in the indicated regions

Off

$$C_{GB} = C_2 = C_{ox} W L \tag{22a}$$

$$C_{GS} = C_1 \cong 0.03\ C_{ox}\ W(\text{mils}) \tag{22b}$$

$$C_{GD} = C_3 \cong 0.03\ C_{ox}\ W(\text{mils}) \tag{22c}$$

Saturation

$$C_{GB} = (C_2 C_4)/(C_2 + C_4) \cong 0 \tag{23a}$$

$$C_{GS} = C_1 + (2/3) C_2 = [0.03 + .67L(\text{mils})]\ C_{ox}\ W(\text{mils}) \tag{23b}$$

$$C_{GD} = C_3 \cong 0.03\ C_{ox}\ W(\text{mils}) \tag{23c}$$

Active or triode

$$C_{GB} = (C_2 C_4)/(C_2 + C_4) \cong 0 \tag{24a}$$

$$C_{GS} = C_1 + 0.5\ C_2 = [0.03 + 0.5L(\text{mils})]\ C_{ox}\ W(\text{mils}) \tag{24b}$$

$$C_{GD} = C_3 + 0.5\ C_2 = [0.03 + 0.5L(\text{mils})]\ C_{ox}\ W(\text{mils}) \tag{24c}$$

Equations that provide a smooth transition among the three regions can be found in the literature.[22]

Other parasitics include the thick oxide sandwiched between polysilicon

[22] D. R. Alexander, R. J. Antinone, and G. W. Brown, *op. cit.*

(or metal) and the field (substrate) and the thick oxide between metal and the first polysilicon or the second polysilicon layers. This type of capacitance typically constitutes the major portion of C_{GB} in the active and saturation regions. Gate connections to other places in the integrated circuit will cause significant parasitic capacitance to occur between the gate and the substrate. All these parasitics are very important and must be considered in the design process of MOS circuits.

The parameters of the large-signal model for the MOS transistor of Fig. 8.2–1(b) are summarized in Table 8.2–1. Typical values are given for each parameter. These values must be adapted to a particular process by carefully measuring the performance of MOS transistors fabricated by the process. In addition to the modeling capability indicated by Fig. 8.2–1, shot and flicker noise of the MOS transistor can be modeled by a current source from the drain to the source.[23] This current source has a spectral density given by

$$\bar{I}_{NM} = [8kTg_m(1+\eta)/3 + (KFI_D^{AF})/(fC_{ox}L^2)]^{-1/2} \text{ (Amps}/\sqrt{\text{Hz}}) \quad (25)$$

where

k = Boltzmann's constant
T = temperature
g_m = small signal transconductance [see Eq. (32)]
$\eta = g_m/g_{mbs}$
g_{mbs} = small signal transconductance of the bulk [see Eq. (33)]
KF = flicker noise coefficient
AF = flicker noise exponent
f = frequency (Hertz)

KF and AF are unitless and have typical values of 10^{-14} and 1.0, respectively.

The large-signal model of the MOS transistor presented by Fig. 8.2–1(b) neglects many important second-order effects. Some of these include charge variation over the length of the channel, short channel effects, weak inversion or subthreshold effects, and many other effects. Several available reports on the simulation of MOS integrated circuits explain these second-order effects in detail.[24,25] Our objective here is to briefly describe some of the more impor-

---

[23] W. M. Penny and L. Lau, *op. cit.*

[24] D. R. Alexander, R. J. Antinone, and G. W. Brown, *op. cit.*

[25] A. Vladmirescu and S. Liu, "The Simulation of MOS Integrated Circuits Using SPICE2," Memorandum No. UCB/ERL M80/7, February 1980, Univ. of California, Berkeley, CA (excellent list of references).

Table 8.2–1 Model Parameters of Fig. 8.2–1 for the *n*-channel or *p*-channel MOS Transistor

PARAMETER NAME	DESCRIPTION	DEFINITION	TYPICAL VALUE	UNITS
V_{TO}	Threshold voltage. Positive (negative) for enhancement and negative (positive) for depletion mode *n*-channel (*p*-channel)	Eq. (3)	± 1	V
ϕ	Surface potential	Eq. (5)	0.5	V
K'	Transconductance parameter	Eq. (8)		A/V^2
	(*n*-channel)		2.5×10^{-5}	
	(*p*-channel)		1.25×10^{-5}	
γ	Bulk threshold parameter	Eq. (4)	0.5	$V^{-1/2}$
λ	Channel length modulation parameter	–	.01	V^{-1}
r_D	Drain ohmic resistance	–	100	Ω
r_S	Source ohmic resistance	–	100	Ω
C_{GS}	Gate-source capacitance	Eqs. (22), (23), and (24)	1	pF
C_{GD}	Gate-drain capacitance	Eqs. (22), (23), and (24)	1	pF
C_{GB}	Gate-bulk capacitance	Eqs. (22), (23), and (24)	1	pF
C_{BDO}	Zero bias bulk-drain junction capacitance per junction area	Eqs. (18) and (19)	5×10^{-8}	pF/cm^2
C_{BSO}	Zero bias bulk-source junction capacitance per junction area	Eqs. (18) and (19)	5×10^{-8}	pF/cm^2
PB	Bulk junction potential	–	0.7	V
FC	Foward bias nonideal junction capacitance coefficient	–	0.5	–
I_S	Bulk junction saturation current	–	10^{-14}	A

tant second-order effects, so that the reader can keep them in mind as they apply to the work that follows.

In the development of Eq. (1) it was assumed that the charge in the depletion region was either constant or zero. If the charge varies over the length of the channel, then it is necessary to use a distributed analysis.[26,27] The result of this analysis is the following relationship of I_D

$$I_D = [(\mu C_{ox} W)/L_{eff}]\{V_{DS}[V_{GS} - 2\phi_f - V_{FB} - (V_{DS}/2)] - (2\gamma/3)[(V_{DS} + 2\phi_F - V_{BS})^{1.5} - (2\phi_f - V_{BS})^{1.5}]\} \tag{26}$$

The various parameters of this relationship have been previously defined. The channel length modulation effect is included in Eq. (26) by defining an effective length

[26] W. M. Penny and L. Lau, *op. cit.*

[27] R. S. Muller and T. I. Kamins, *op. cit.*

$$L_{\text{eff}} = L_0 (1 - \lambda V_{DS}) \tag{27}$$

where L_0 is the value of L for $V_{DS} = 0$, and λ is the channel length modulation parameter. V_{DS} in Eq. (26) should be replaced by $V_{DS}(\text{sat})$ for the saturation region (except for the V_{DS} in Eq. (27)).

Another important second-order effect is that due to short channels. This effect occurs when the channel length approaches the magnitude of the vertical dimensions of the device (the channel depth). This situation occurs in most MOS devices for $L \leq 1$ mil. As a consequence of the short channels, the threshold voltage decreases, the body effect decreases, and the channel length modulation increases. There are several physical reasons for these results. One of these is the fact that the amount of charge underneath the gate depleted by the drain junction field, rather than the gate-to-bulk area, decreases, and the gate-to-bulk field increases with V_{DS}, thus lowering V_T. This effect is referred to as the *electrostatic effect.* Not only is V_T decreased by this effect but λ is increased. Because the gate-to-bulk area decreases, the gate-to-bulk field has less effect on the channel, and thus γ is reduced as well. A typical variation of V_T with L is shown in Fig. 8.2–10. An empirical expression for the dependence of λ upon L for typical substrate concentrations is given as

$$\lambda \cong (0.01)/L(\text{mils}) \qquad \text{for } L \leq 1 \text{ mil} \tag{28}$$

For values of L greater than 1 mil, λ can be considered a constant having a typical value given in Table 8.2–1 of 0.01 volt^{-1}. Most of the existing models for the MOS device start with Eq. (1) or Eq. (26) and account for these additional effects by various modifications of the basic equations.

Another second-order effect of importance is the weak inversion or subthreshold effect. The MOS transistor is not an ideal switch that starts conduction abruptly at $V_{GS} = V_T$. There is a current flowing in the device below the threshold voltage, which is known as the *weak inversion,* or *subthreshold* current. The simulation of this effect is important for present-day MOS circuits

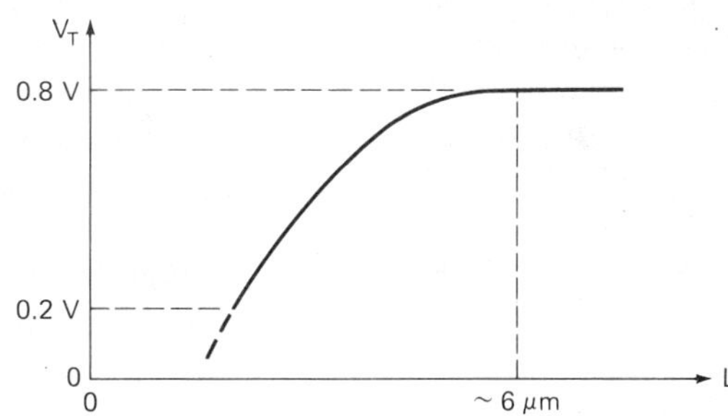

Fig. 8.2–10. Variation of V_T with channel length L for short channel devices.

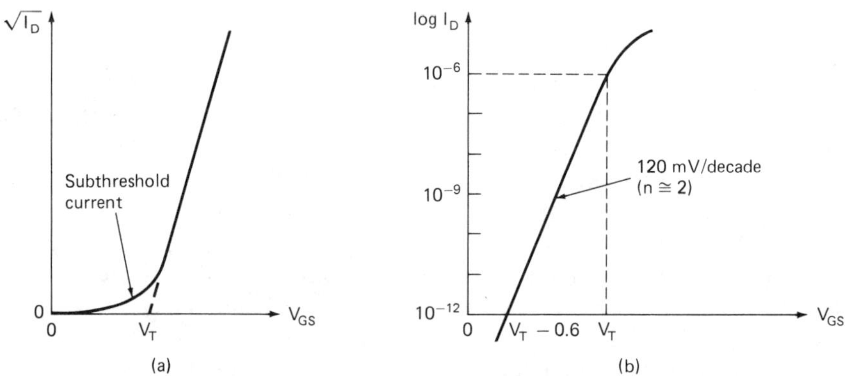

Fig. 8.2–11. (a) Illustration of subthreshold current in a MOS device. (b) Semilog plot of I_D versus V_{GS}.

that are often designed to operate in the weak inversion region. The model used for the drain current of an MOS transistor in weak inversion is[28]

$$I_D = (\mu C_{ox} W/mL))(nkT/q)^2 \exp\left[(q/nkT)(V_{GS} - V_T - (nkT/q))\right] \left[1 - \exp\left(-mqV_{DS}/(nkT)\right)\right] \quad (29)$$

where k is Boltzmann's constant, and m and n are process-independent parameters. If $V_{DS} >> (nkT/mq)$, then the last term of Eq. (29) can be neglected, and an expression very similar to bipolar junction device currents is obtained. Figure 8.2–11 shows the weak inversion current on the MOS device voltage-current characteristics. The result of the weak inversion effect is that it is harder to turn off the MOS device. In fact, the enhancement device is not really turned off until $V_{GS} = 0$. The depletion device never really turns off, even though large negative voltages are applied between the gate and source. An advantage of the weak inversion effect is that the MOS device approximates a bipolar junction device at lower current levels and has a comparable transconductance.

Many other effects of the MOS device become important in specific applications. We shall describe these effects when it becomes necessary in the discussion. We shall try to stay with the simplified model because that is easier for teaching the principles. When using the simulation programs available for MOS circuits, it will become necessary for the reader to understand in more detail some of the second-order effects not presented here.

Up to this point, the considerations in this section have been devoted to the large-signal model of the MOS transistor shown in Fig. 8.2–1(b). Although

[28] R. M. Swanson and J. D. Meindl, "Ion-Implanted Complementary MOS Transistor in Low-Voltage Circuits," *IEEE J. of Solid-State Circuits,* Vol. SC-7, No. 2, April 1972, pp. 146–153.

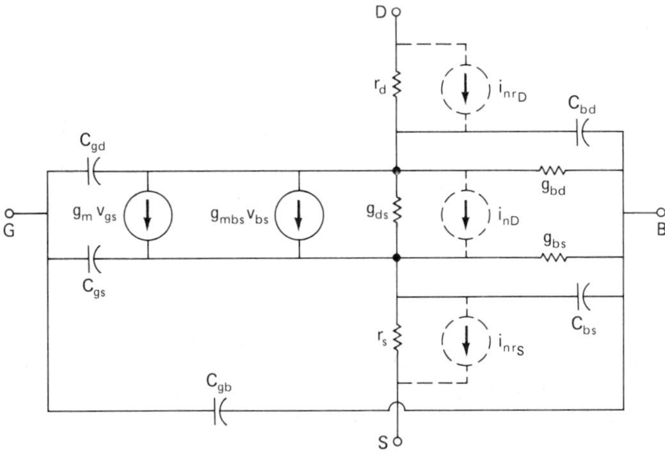

Fig. 8.2–12. Small-signal model of the MOS transistor.

the large-signal model is useful in describing the operation and characteristics of the MOS transistor, its value to the designer is that it provides a means for developing a small-signal model that can be used for linear analysis. Figure 8.2–12 shows a linearized small-signal model for the MOS transistor. The parameters of the small-signal model will be designated by lower case subscripts. The various parameters of this small-signal model are all related to the parameters and DC variables of the large-signal model. The normal relationship between these two models is to assume that the small-signal parameters are defined in terms of the ratio of small perturbations of the large-signal variables or as the partial differentiation of one large-signal variable with respect to another.

The conductances g_{bd} and g_{bs} are the equivalent conductances of the bulk-to-drain and bulk-to-source junctions. Because these junctions are normally reverse-biased, the conductances are very small. They are defined as

$$g_{bd} = (\partial I_{BD}/\partial V_{BD})\Big|_Q \cong 0 \tag{30}$$

$$g_{bs} = (\partial I_{BS}/\partial V_{BS})\Big|_Q \cong 0 \tag{31}$$

where Q implies that these derivatives are evaluated at the DC operating or quiescent point. The channel conductances g_m, g_{mbs}, and g_{ds} and defined as

$$g_m = (\partial I_D/\partial V_{GS})\Big|_Q \tag{32}$$

$$g_{mbs} = (\partial I_D/\partial V_{BS})\Big|_Q \tag{33}$$

and

$$g_{ds} = (\partial I_D / \partial V_{DS}) \Big|_Q \tag{34}$$

The values of these small-signal parameters depend on which region the quiescent point occurs in. For example, in the saturated region, g_m can be found from Eq. (12) as

$$g_m = \sqrt{2K'W/L}\sqrt{|I_D|} = (K'W/L)(V_{GS} - V_T) \tag{35}$$

which emphasizes the dependence of the small-signal parameters upon the large-signal operating conditions. The small-signal channel transconductance due to V_{BS} is found by rewriting Eq. (33) as

$$g_{mbs} = (\partial I_D / \partial V_{BS}) = (\partial I_D / \partial V_T)(\partial V_T / \partial V_{BS}) \tag{36}$$

Using Eq. (2) and noting that $\partial I_D / \partial V_T = -\partial I_D / \partial V_{GS}$, we get

$$g_{mbs} = g_m\ \gamma / 2\sqrt{\phi - V_{BS}} = \eta g_m \tag{37}$$

This transconductance will become important in our small-signal analysis of the MOS transistor when the AC value of the bulk-source potential, v_{bs}, is not zero. The small-signal channel conductance g_{ds} is given as

$$g_{ds} = \lambda\ I_D \tag{38}$$

We can see that g_{ds} will be dependent upon L through Eq. (28) for short-channel devices. We have assumed that the MOS transistor is in saturation for the results given by Eqs. (36), (37), and (38).

The values of r_d and r_s are assumed to be the same as r_D and r_S of Fig. 8.2–1. Likewise, for small signal conditions, C_{gs}, C_{gd}, C_{gb}, C_{bd}, and C_{bs} are assumed to be the same as C_{GS}, C_{GD}, C_{GB}, C_{BD}, and C_{BS}. A very useful approximation for finding C_{bs} or C_{bd}, knowing C_{gs} or C_{gd}, is given as[29]

$$C_{bs} \cong (g_{mbs}/g_m)C_{gs} = \eta C_{gs} \tag{39}$$

---

[29] Y. P. Tsividis, "Relation Between Incremental Intrinsic Capacitances and Transconductances in MOS Transistors," *IEEE Trans. on Electron Devices,* Vol. ED-27, No. 5, May 1980, pp. 946–948.

and

$$C_{bd} \cong (g_{mbs}/g_m)C_{gd} = \eta C_{gd} \tag{40}$$

If the noise of the MOS transistor is to be modeled, then three additional current sources are added to Fig. 8.2–9, as indicated by the dashed lines. The current sources have a spectral density of

$$\bar{i}_{nr_d} = \sqrt{4kT/r_D}\ (\text{Amps}/\sqrt{\text{Hz}}) \tag{41}$$

$$\bar{i}_{nr_S} = \sqrt{4kT/r_S}\ (\text{Amps}/\sqrt{\text{Hz}}) \tag{42}$$

and

$$\bar{i}_{nD} = [8kTg_m(1+\eta)/3 + (KF)I_D^{AF}/(fC_{ox}L^2)]^{1/2}\ (\text{Amps}/\sqrt{\text{Hz}}) \tag{43}$$

The various parameters for these equations have been previously defined. With the noise-modeling capability, the small-signal model of Fig. 8.2–12 is a very general model.

The modeling presented in this section will be sufficient for the purposes of MOS analog sampled data circuit design and analysis. Further sophistication can be achieved by building on these concepts. The most important aspect of modeling is to be able to determine the model parameters that best fit the MOS process being used.

The large- and small-signal modeling concepts presented in this chapter have been applied to various computer and calculator programs whose purpose is to analyze MOS circuits. Computer simulation programs such as SPICE2 provide a very powerful means of modeling complex MOS circuits. The increased capability of programmable calculators has provided the means to analyze simple MOS circuits. Such programs are helpful in designing the circuit that will eventually be analyzed by a more powerful program. Appendix C illustrates a calculator program based on a more sophisticated model of the MOS transistor called the Shichman-Hodges-Allen model. This program was originally developed by D. Mayer and M. de Witt of Texas Instruments and is implemented on a TI-59 with or without the PC 100 printer. This program has also been converted to an HP-41 calculator by John Bradbury and Steve Erickson, students at Texas A&M University. The programs include the ability to store the model parameters of up to nine different MOS devices in a library. Both programs are described in more detail in Appendix C.

8.3 MOS SWITCHES

One of the most important components of switched capacitor circuits is the switches used to connect the capacitors in the various configurations. It is very fortunate that the MOS device makes one of the best switch implementations now available. It is the objective of this section to examine some of the switching characteristics of MOS devices and in particular their application to analog sampled data circuits.

Let us begin with the characteristics of a voltage-controlled switch. Figure 8.3–1 shows a model for a voltage-controlled switch. The voltage V_C is assumed to control the switch and to determine whether the switch is in the ON or OFF state. The voltage-controlled switch is really a three-terminal network, with terminals A and B comprising the switch and terminal C providing the means of applying the control voltage V_C. The most important characteristics of a switch are its ON resistance, R_{ON}, and its OFF resistance R_{OFF}. Ideally R_{ON} is zero and R_{OFF} is infinite. Also R_{ON} should be linear to avoid harmonic distortion. Most switches have some form of voltage offset, which is modeled by V_{OS}, of Fig. 8.3–1. V_{OS} represents the fact that when there is no current through the switch in the ON state, a voltage may exist between terminals A and B. I_{OFF} represents the leakage current that may flow in the OFF state of the switch. This current is primarily due to the subthreshold current between terminals A and B, which flows for small values of V_{GS}. A plot of $\sqrt{I_D}$ versus V_{GS} at small values of V_{GS} was shown in Fig. 8.2–11(a). The polarities of the offset sources are not known and have arbitrarily been assigned the directions indicated in Fig. 8.3–1. The parasitic capacitors are an important consideration in the application of analog sampled

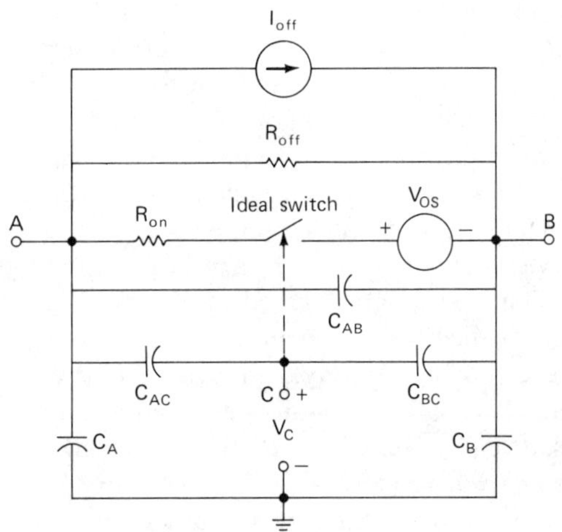

Fig. 8.3–1. Model for a nonideal switch.

data circuits. Capacitors C_A, C_B, and C_{AB} are the parasitic capacitors associated with the switch terminals A and B and ground. Capacitors C_{AC} and C_{BC} are parasitic capacitances that may exist between the voltage control terminal C and the switch terminals A and B. These capacitors contribute to a problem called *feedthrough,* where a portion of the control voltage appears at the switch terminals A and B. There are some other characteristics of switches that are not shown in Fig. 8.3–1. One of these characteristics is called the *commutation rate.* The commutation rate of a switch is the inverse of the time it takes to complete one open-and-close cycle, given in Hertz. Other switch characteristics that will be of interest are linearity and maximum analog signal swing.

In MOS technology, a single MOS transistor makes a very good switch realization. Figure 8.3–2 shows a single-channel NMOS transistor that will be considered as a realization of a switch. The performance of this realization can be determined by comparing Fig. 8.3–1 with Fig. 8.2–1(b), the large-signal model for the MOS transistor. We see that terminals A and B can be either the drain or the source of the NMOS transistor. The ON resistance is seen to consist of r_D, r_S and whatever channel resistance that exists. To find an expression for this resistance, let us consider Eq. (1) of Section 8.2. In the ON state of the switch, the voltage across the switch should be small, and V_{GS} should be large. Therefore, the NMOS device is in the active region. The linear part of the active region is sometimes called the *ohmic* region. Either Eq. (11) or Eq. (14) of Section 8.2 will apply, depending upon the polarity of V_{DS}. Furthermore, let us assume that the channel length modulation effects can be ignored, which results in

$$I_D = (\mu\, C_{ox}\, W)/(2L)\, [2(V_{GS} - V_T)\, V_{DS} - V_{DS}^2] \tag{1}$$

if V_{DS} is less than $V_{GS} - V_T$, but greater than zero. Assuming that there is no offset voltage the large-signal channel resistance is

$$R_{ON} = 1/(\partial I_D/\partial V_{DS}) = [(\mu_n\, C_{ox}\, W/L)\,(V_{GS} - V_T - V_{DS})]^{-1} \tag{2}$$

Figure 8.3–3 illustrates Eq. (2) in the ohmic region where V_{DS} is small. When $V_{GS} = V_T$, R_{ON} is infinite if $V_{DS} = 0$. Figure 8.3–3 is plotted for

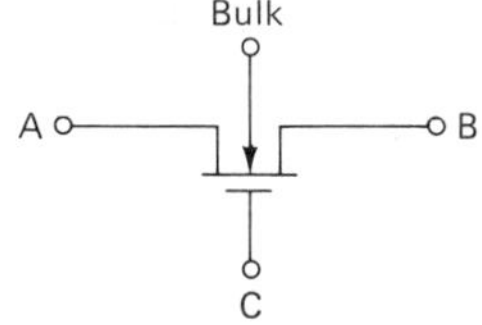

Fig. 8.3–2. NMOS transistor used as a switch.

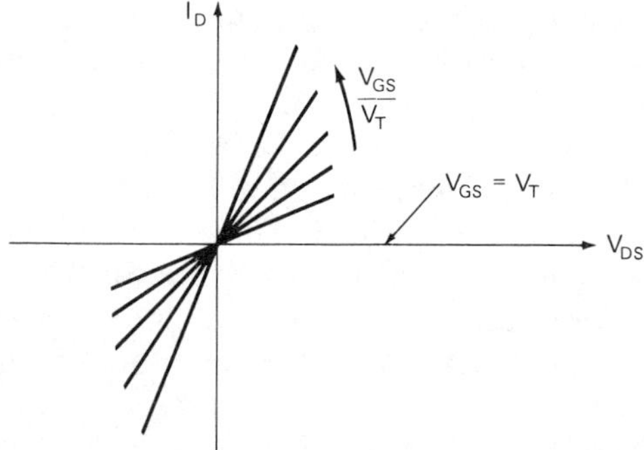

Fig. 8.3–3. Illustration of the ON state of a NMOS switch. V_{GS}/V_T are increasing in equal increments.

equal increasing steps of V_{GS}/V_T. Eventually the curves of Fig. 8.3–3 will start to decrease in slope for increasing V_{DS}. A plot of R_{ON} as a function of V_{GS}/V_T is shown in Fig. 8.3–4 for various values of W/L. It is seen that a lower value of R_{ON} is achieved for larger values of W/L. V_{DS} has been assumed to be zero in Fig. 8.3–4.

When the switch is OFF, V_{GS} is less than or equal to V_T, and the transistor is always in the saturation region. R_{OFF} can be found by differentiating I_D of Eq. (12) of Section 8.2 with respect to V_{DS}. However, when $V_{GS} = V_T$, this resistance is ideally infinite, as illustrated in the cutoff region of Fig. 8.2–3. R_{OFF} is typically in the range of 10^{12} ohms. Therefore, the leakage current is a more important OFF parameter. This leakage current is a combination of the subthreshold current, the surface leakage current, and the package leakage current. Typically this leakage current is in the 10 pA range and doubles every 10°C increase.

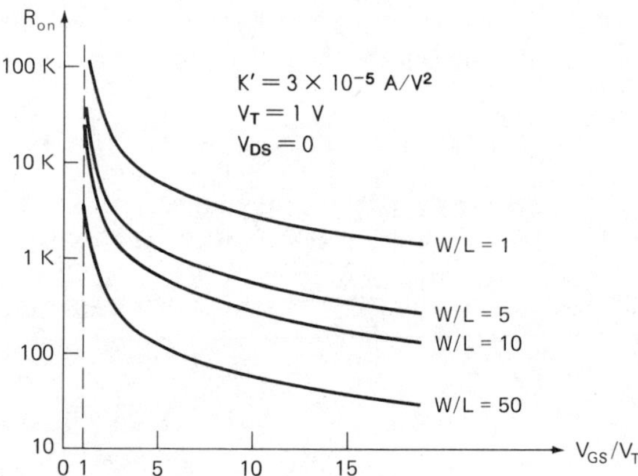

Fig. 8.3–4. Illustration of the R_{ON} of an NMOS transistor with W/L as a parameter.

The offset voltage of the MOS device is zero and does not influence the switch performance. The capacitors C_A, C_B, C_{AC}, and C_{BC} of Fig. 8.3–1 correspond directly to the capacitors C_{BS}, C_{BD}, C_{GS}, and C_{GD} of Fig. 8.2–1(b). C_{AB} is a combination of the capacitors of Fig. 8.2–1(b) and is usually negligible. These capacitors have a great deal of influence upon analog sampled data networks, because the desired capacitors are typically very small.

The commutation rate of the MOS switch is determined primarily by the capacitors of Fig. 8.3–1 and the external resistances. Commutation rates for CMOS switches can be as large as 20 MHz.

One important aspect of the switch is the range of voltages on the switch terminals compared with the control voltage. In the NMOS transistor, we see that the gate voltage must be considerably larger than either the drain or source voltage to ensure that the MOS transistor is ON. Typically the bulk is taken to the most negative potential for the NMOS switch. The problem of dynamic range can be illustrated as follows. Suppose that the ON voltage of the gate is the positive power supply V_{DD}. With the bulk to ground this should keep NMOS switch ON until the analog signal on the switch terminals (which should be approximately identical at the source and drain) approaches $V_{DD} - V_T$. As the analog signal approaches $V_{DD} - V_T$, the switch begins to turn OFF. This introduces an undesired nonlinear distortion in the signal. This problem is often solved by allowing the control signal, which is the clock, to have a value greater than V_{DD}. If this is not done, then the analog signal level must be limited to several V_T less than V_{DD}. The typical voltages on an NMOS switch are shown in Fig. 8.3–5, where the switch is connected between two networks. In the next section, we shall see that it is very difficult to generate clock signals that approach V_{DD}, so that the dynamic range becomes an important consideration.

To illustrate the influence of the switches upon analog sampled data circuits, let us consider the switched capacitor integrator of Fig. 8.3–6. M1 and M2 are the switching devices controlled by the nonoverlapping clock waveforms, ϕ_1 and ϕ_2. The ON resistance of the switches can become important during

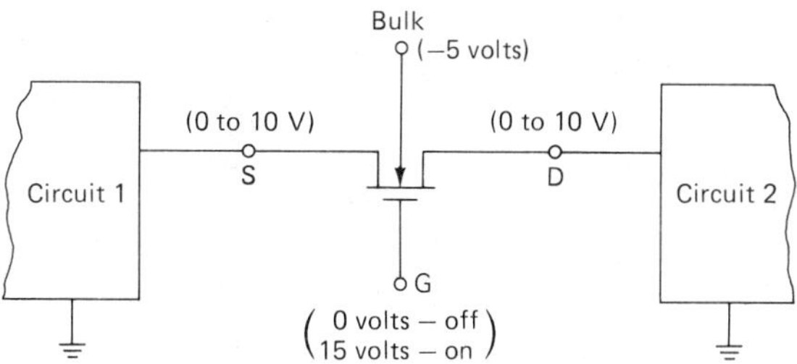

Fig. 8.3–5. Application of an NMOS transistor as a switch with typical terminal voltages indicated.

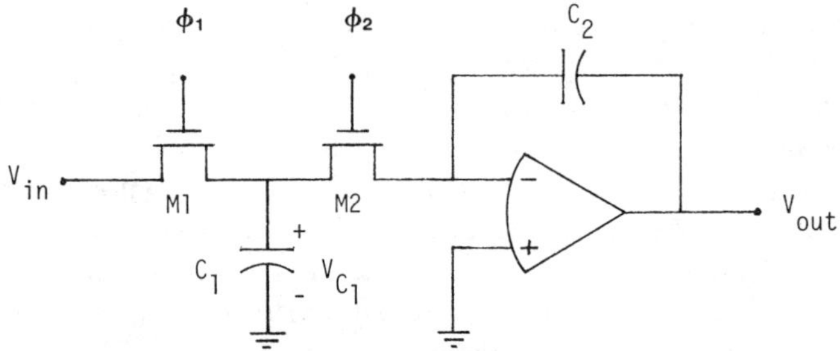

Fig. 8.3–6. An analog sampled data inverting integrator.

the charge-transfer phase of this circuit. For example, when ϕ_1 goes high and ϕ_2 is low, M1 connects C_1 to the voltage source V_{in}. The equivalent circuit at this time is shown in Fig. 8.3–7(a). It is seen that C_1 will charge to V_{in} with the time constant of $R_{ON}C_1$. For successful operation, $R_{ON}C_1 << T$, where T is the period of the clock. It is of interest to determine the value of R_{ON}, for this will determine the size of M1 and M2, as illustrated by Fig. 8.3–4. Typical values of C_1 are less than 10 to 20 pF, because the area that would be required to implement these capacitors would be too

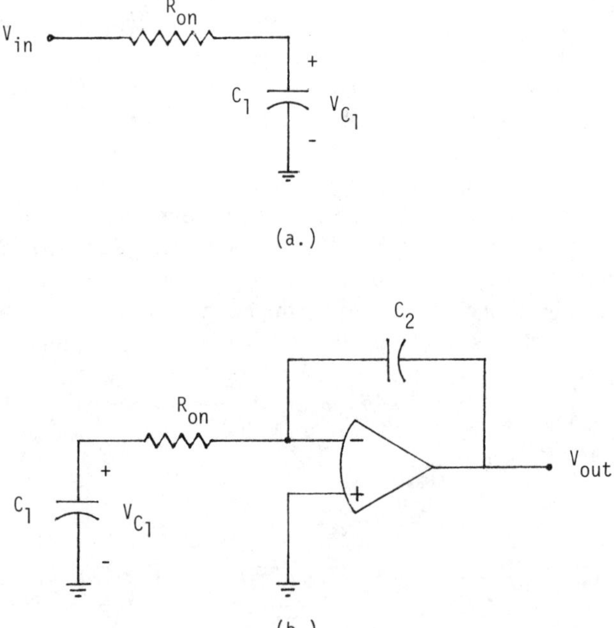

Fig. 8.3–7. Effects of R_{on} on the analog sampled data inverting integrator of Fig. 8.3–6. (a) Charging of C_1 and (b) Discharging of C_1.

large. If $T = 10\ \mu s$, then $R_{ON} << 0.5\ M\Omega$. Since small capacitors are used, R_{ON} can be very large and still perform satisfactorily. As a result the MOS devices used for switching typically use minimum geometry and are square ($W = L$). For a clock of 15 volts, the MOS device of Fig. 8.3–4 with $W = L$ gives R_{ON} of approximately 3 kΩ, which is sufficiently small to transfer the charge in the desired time. The minimum size switches will also help to reduce parasitic capacitances. Similar considerations hold during the transfer of the charge on C_1 to C_2, as shown in Fig. 8.3–7(b). If either the clock frequency or the capacitances is increased, the W/L may have to be increased in order to completely transfer the charge in the necessary time period.

The OFF state of the switch has little influence upon the performance of Fig. 8.3–6 except for the leakage current. Figure 8.3–8 shows two cases where the leakage current can create serious problems. Figure 8.3–8(a) is a sample-and-hold circuit. If C_H is not large enough, then in the hold mode where the MOS switch is OFF, the leakage current can charge or discharge C_H a significant amount. Figure 8.3–8(b) shows the switched capacitor integrator of Fig. 8.3–6 during the ϕ_2 phase period. The leakage current can cause the circuit to integrate in a continuous mode, which can lead to offset unless there is an external feedback path from V_{out} back to the input of the operational amplifier. This feedback must contain switched capacitor resistors and may go through other circuits before returning to the input of Fig. 8.3–6.

Probably the most serious problem caused by switches is the clock feedthrough. Figure 8.3–9 shows all the parasitic capacitors associated with Fig. 8.3–6. Because the clock signal is making very large transitions, it can easily couple from gate to source or drain through C_{GS} or C_{GD}, respectively. An example of the effects of clock feedthrough can be seen in Fig. 8.3–10 during the ϕ_1 phase period. The 0.02-pF capacitors represent the C_{GD} and C_{GS} capacitances. Let us assume that the clock varies from 0 to 15 volts. During the turn-on portion of the ϕ_1 phase period, the gate starts from 0 volts and heads toward 15 volts. During the 0 to $V_{in} + V_T$ transition, M1 is off. Consequently, this portion of the clock waveform can couple to C_1 via C_{GD} or C_{GS}. The remainder of the rising clock waveform is not coupled to C_1,

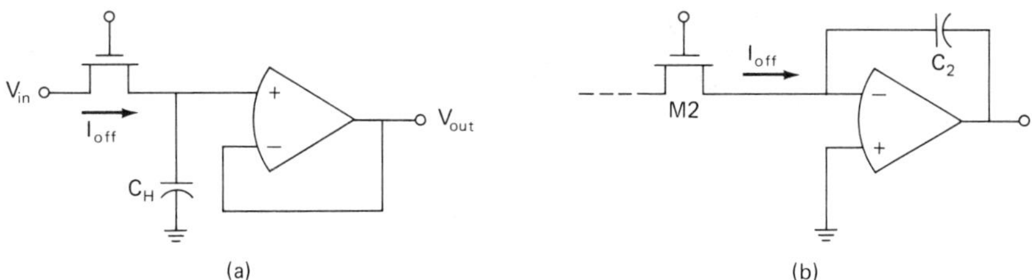

Fig. 8.3–8. Examples of effects of Ioff, (a) Sample-and-hold. (b) Integrator of Fig. 8.3–6.

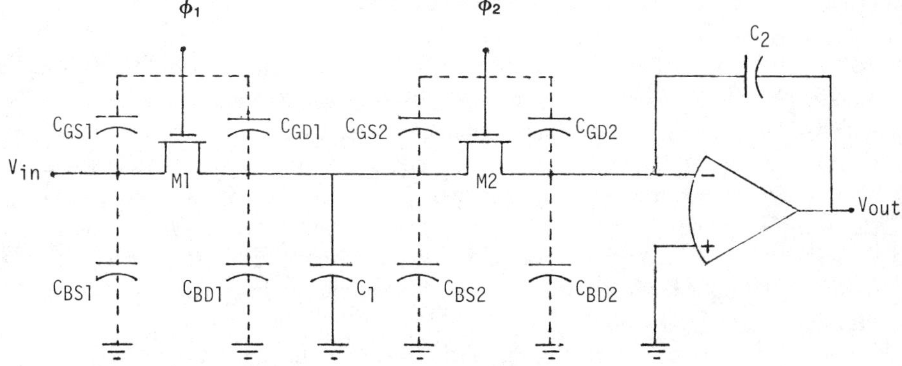

Fig. 8.3–9. Illustration of parasitics of Fig. 8.3–6 due to M1 and M2.

because M1 turns on and connects C_1 to the low-impedance voltage source V_{in}. As a result

$$\Delta V_{C1}^{+} = [C_{GD}/(C_1 + C_{GD})](V_{in} + V_T) \cong 0.02\ V_{in} \tag{3}$$

is coupled onto C_1, which has been assumed to be 1 pF. However, when M1 turns on, C_1 is charged to V_{in}, and it makes no difference what was previously coupled onto C_1. When the clock turns off, feedthrough occurs once more as the clock goes from $V_{in} + V_T$ to zero volts. This feedthrough is given as

$$\Delta V_{C1}^{-} = -[C_{GD}/(C_1 + C_{GD})](V_{in} + V_T) \cong -0.02\ V_{in} \tag{4}$$

If V_{in} is 5 volts, then a feedthrough of −100 mV to C_1 occurs during the ϕ_1 phase period. As ϕ_2 turns on, it should theoretically cancel ΔV_{C1}^{-}. It can be seen that the cancellation between the charge removed from C_1 by ϕ_1 and that added to C_1 by ϕ_2 depends upon the amplitude of V_{in}. Another source of feedthrough-induced offset comes from the incomplete cancellation

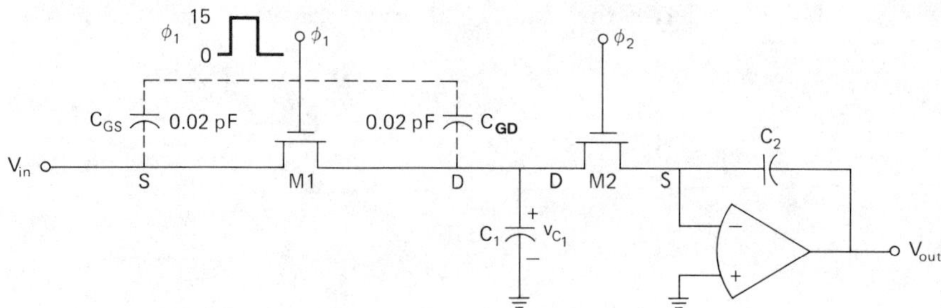

Fig. 8.3–10. Illustration of offset caused by clock feedthrough.

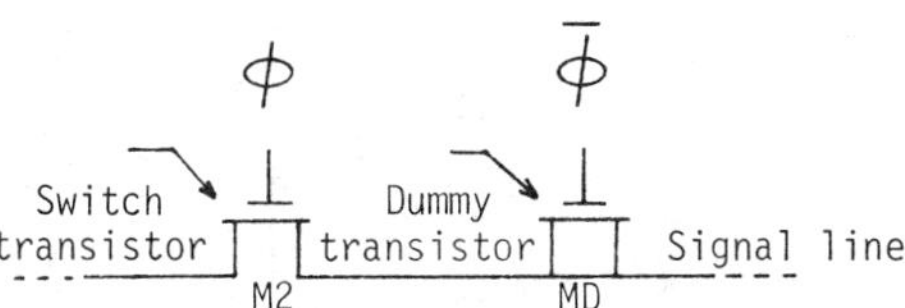

Fig. 8.3–11. Use of a dummy transistor to cancel the clock feedthrough.

between the charge added to and subtracted from C_2 as ϕ_2 turns on then off. Other feedthrough occurs during the clock described above where a switch is connected between a nonzero voltage source and a capacitor. In general, the feedthrough will be dependent upon the switch configuration and the size of the capacitors in the circuit.[30]

It is possible to partially cancel some of the feedthrough effects using the technique illustrated in Fig. 8.3–11. Here a dummy MOS transistor, *MD,* with source and drain both attached to the signal line and the gate attached to the inverse clock, is used to apply an opposing clock feedthrough due to M2. The area of *MD* can be designed to provide minimum clock feedthrough. Unfortunately, this method never completely removes the feedthrough and in some cases may worsen it. Also it is necessary to generate an inverse clock that is applied to the dummy switch. In some cases, the dummy switch is an inverter with the gate attached to the source of M2 and the source and drain of the dummy switch connected to the inverse clock. This avoids charge-pumping of the substrate, which can defeat the purpose of the dummy switch. The best technique to reducing the clock feedthrough is to use the largest capacitors possible, use minimum geometry switches, and keep the clock swings as small as possible. Typically, these solutions will create problems in other areas and will call for compromise. Some recent efforts to eliminate feedthrough use a completely differential signal.[31] This method has greatly reduced feedthrough, but at the cost of additional complexity and more capacitance.

Another important nonideal characteristic of the MOS switch is the capacitance C_{BD} and C_{BS} of Fig. 8.2–1(b). These capacitors cause parasitic capacitances to ground and must be considered. If the parasitic capacitors are connected to a voltage source or to a virtual ground of an operational amplifier, then they do not affect the circuit. Therefore, only C_{BD1} and C_{BS2} of Fig. 8.3–9 are important. Along with these capacitors, we should include the top and bottom plate capacitors due to C_1. As one side of C_1 is grounded,

[30] L. T. Nguyen, First Order Switched-Capacitor Building Blocks for Analog Circuits, MS Thesis, May 1983, Texas A&M University, College Station, TX.

[31] P. R. Gray, R. W. Brodersen, D. A. Hodges, T. C. Choi, R. Kaneshiro, and K. C. Hsieh, "Some Practical Aspects of Switched Capacitor Filter Design," *Proc. of IEEE Int. Conf. on Circuits and Systems,* Chicago, IL, April 1981, pp. 419–422.

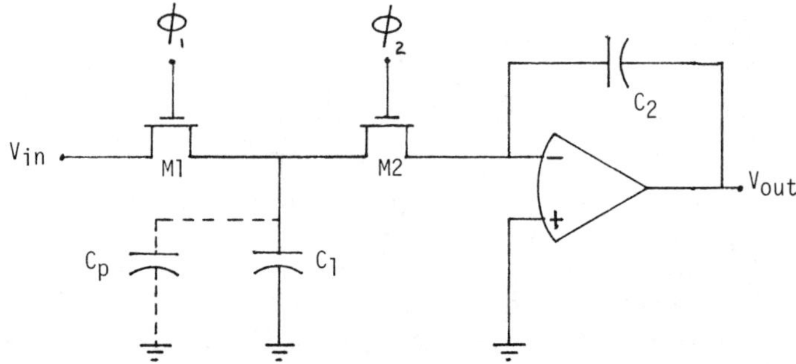

Fig. 8.3–12. Model for the important parasitics of Fig. 8.3–6 effecting the signal transfer.

the parasitic on that side can be eliminated. Because the top plate parasitic capacitor is smaller than the bottom plate parasitic capacitor, the lower polysilicon layer or channel is always connected to ground when possible. Therefore Fig. 8.3–12 has a parasitic capacitor C_p in parallel with C_1, that consists of the top plate parasitic capacitance, C_{BD1}, and C_{BS2}. This parasitic capacitor can cause serious error, since C_1 is normally small. Furthermore, the parasitic capacitance can be voltage-dependent, creating a source of nonlinear distortion. If the top plate parasitic is 0.1 C_1 and $C_{BD1} = C_{BS2} = 0.1$ pF, then an error of 30% may be realized if C_1 is designed to be 1 pF. The presence of this parasitic has caused the development of integrators that are insensitive to these stray capacitances.[32] These stray-insensitive realizations were developed in Chapter 2 and are illustrated in Fig. 8.3–13. It can be seen that the parasitics associated with the terminals of C_1 have no effect on the circuit. For example, the parasitics in Fig. 8.3–13(a) are connected across the input voltage and the virtual ground during the ϕ_1 phase and shorted out during the ϕ_2 phase. The parasitics in Fig. 8.3–13(b) are either shorted or connected to a voltage source or a virtual ground during its operation.

It is also important in the operation of Fig. 8.3–13(b), that the voltage momentarily applied to the op amp not be less than V_{BB}. This can occur when V_{in} is large and positive, for C_1 is reversed and applied to the inverting input. If $V_{in} > |V_{BB}|$, then the terminals of the switch between C_1 and the inverting input of the op amp will be forward-biased with respect to the substrate, and charge will be lost. When the frequency of the integrator is much less than the clock frequency, these circuits behave as ideal integrators. However, it has been shown in Chapter 2 that they have different z-domain transfer functions, which becomes important only as the integrator frequency approaches the clock frequency. One of the undesirable configurations in

[32] K. Martin and A. Sedra, "Stray-insensitive Switched-capacitor Filters Based on Bilinear Z-Transform," *Electronics Letters,* Vol. 15, No. 13, June 21, 1979, pp. 365–366.

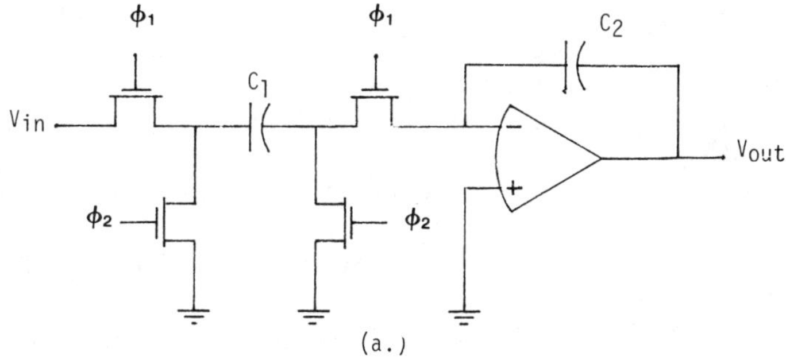

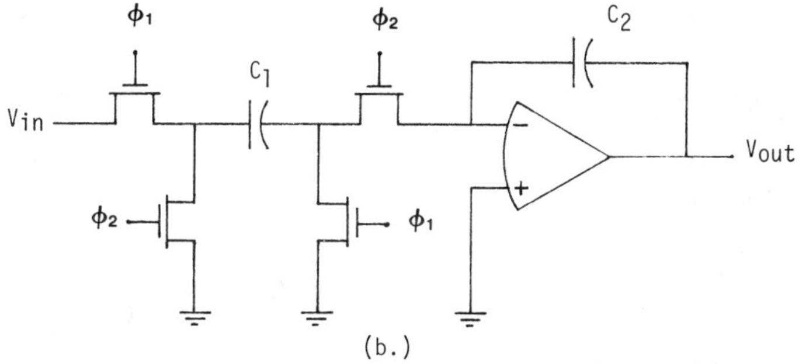

Fig. 8.3–13. Stray insensitive integrator. (a) Inverting and (b) Noninverting.

switched capacitor circuits is where the node of a capacitor is not connected to a defined voltage. This occurs when two capacitors are placed in series.

Some of the problems associated with single-channel MOS switches can be avoided with the CMOS switch shown in Fig. 8.3–14. Using CMOS technology, a switch is usually constructed by paralleling a p-channel and n-channel enhancement transistors, and therefore when ϕ is low, both transistors are

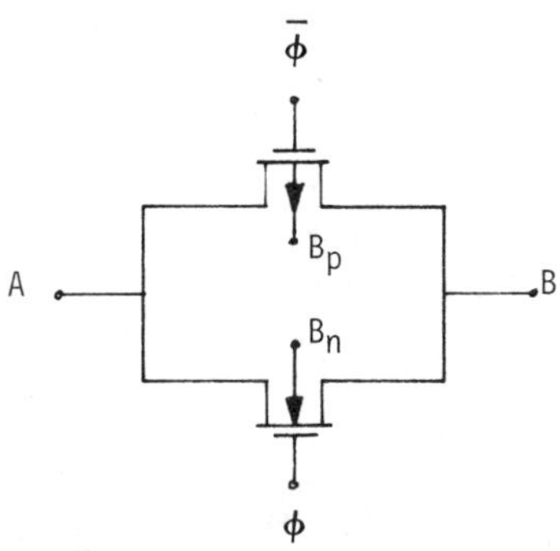

Fig. 8.3–14. A CMOS switch.

OFF, creating an effective open circuit. When ϕ is high, both transistors are ON, giving a low impedance state. The ON resistance of the CMOS switch can be lower than 1 kΩ while the OFF leakage current is in the 10-pA range. The bulk potentials of the *p*-channel (V_{Bp}) and the *n*-channel (V_{Bn}) are taken to the highest and lowest potentials, respectively. It is possible to apply the clocks to the bulks of the MOS devices to improve their switching characteristics.

The CMOS switch has two advantages over the single-channel MOS switch. The first advantage is that the dynamic analog signal range in the ON state is greatly increased. The second is that, because the *n*-channel and *p*-channel devices are in parallel and require opposing clock signals, the feedthrough due to the clock will be diminished through cancellation. The CMOS switch has the disadvantages of requiring more area and a complementary clock signal.

The increased dynamic range of the analog signal can be seen to be a direct result of using complementary devices. When one of the transistors is being turned OFF because of a large analog signal on the drain and source, this large analog signal will be causing the other transistor to be fully ON. As a consequence, both transistors of Fig. 8.3–14 are ON for analog signal amplitudes less than the clock magnitude, and at least one of the transistors is ON for analog signal amplitudes equal to the magnitude of the clock signal.

The feedthrough cancellation of the CMOS switch in Fig. 8.3–14 is not complete for two reasons. One is that the feedthrough capacitances of the *n*-channel device are not necessarily equal to the feedthrough capacitances of the *p*-channel device. The second reason is that the turn-on delay of each type of transistor is not equal, and so the channel conductancess do not necessarily track each other during turn-on and turn-off.

CMOS switches are generally used in place of single-channel switches when the technology and space permits. Although the CMOS switches have larger parasitics than single-channel switches, these parasitics can be minimized through the use of circuit techniques, such as those shown in Fig. 8.3–13. The clock circuitry for CMOS switches is more complex because of the requirement for a complementary clock. A CMOS switch cell or transmission gate including the control circuitry is shown in Fig. 8.3–15(a). It is assumed that V_{control} and $\overline{V_{\text{control}}}$ are generated from the clock waveform. When V_{control} is HIGH, the transmission gate is ON. The equivalent circuit in this condition is shown in Fig. 8.3–15(b). Here we see that the *n*-channel gate is taken to the high state and the *p*-channel gate is the low state. Also the substrates have been connected together by means of the switches M3, M4, and M5. M3 and M4 are OFF while M5 is ON. This helps to keep the switch ON resistance from being a function of the analog signal potential. When V_{control} is LOW, the transmission gate is OFF and has the equivalent circuit shown

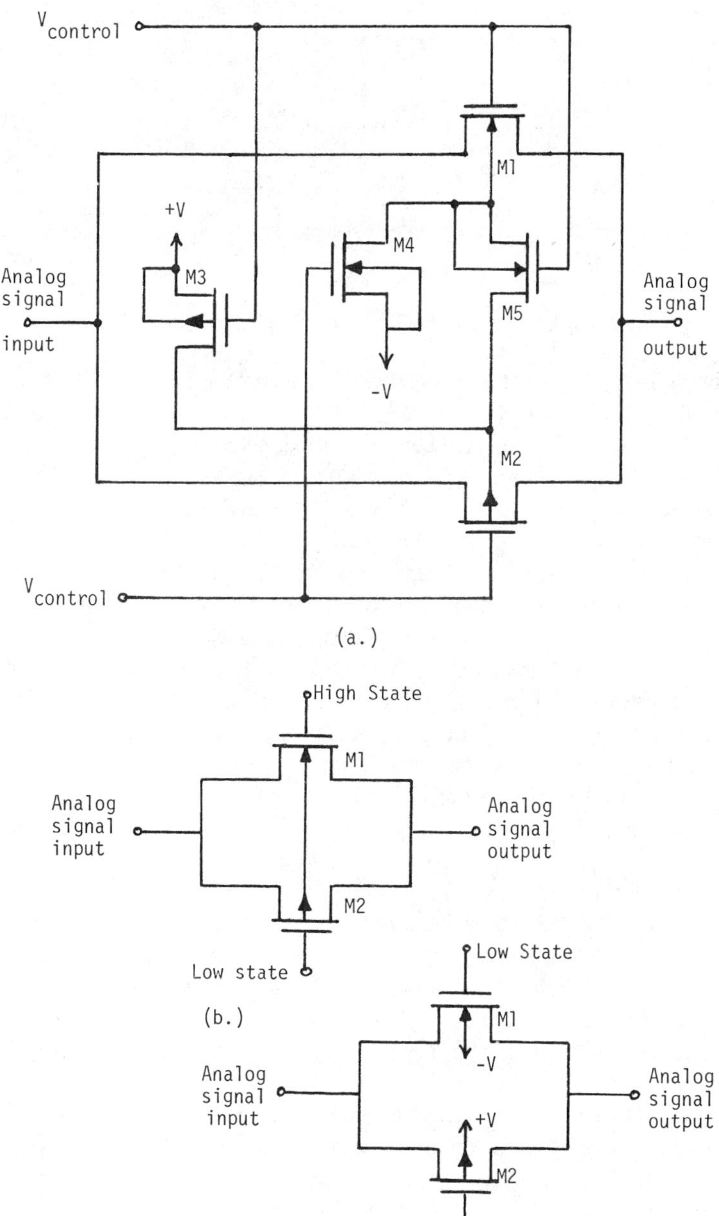

Fig. 8.3–15. CMOS switch with control circuitry. (a) Circuit. (b) ON state. (c) OFF state.

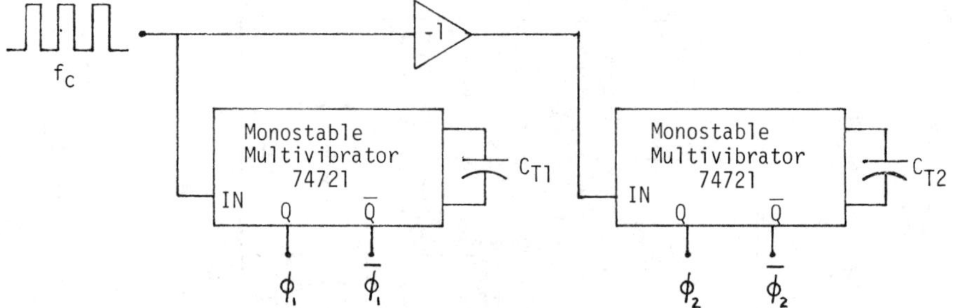

Fig. 8.3–16. A simple two phase clock scheme.

in Fig. 8.3–15(c). The bulks of the n- and p-channel have been taken to $-V$ and $+V$, respectively, for M3 and M4 are ON and M5 is OFF. This ensures that the OFF state will be maintained, because V_{BS} is strongly reverse-biased and causes a large V_T as indicated in Fig. 8.2–2.

The clock circuits are perhaps the most important part of an analog sampled data system. They should not be overlapping and yet should have a duty cycle as large as possible to permit charge transfer. Figure 8.3–16 shows a two-phase clock scheme that uses TTL circuits and a square wave input. The capacitor on the monostable multivibrators determines the duty cycle. The inverter causes the monostables that are triggered by the falling edge to be alternately triggered. Both the clock phase and its complement are available at the Q and $\overline{Q}$ outputs of the monostables. This clock has the disadvantage that if the square wave input frequency is varied, each of the monostable time constants needs to be varied to maintain a constant duty cycle. A clock circuit that avoids this problem is shown in Fig. 8.3–17.[33] This circuit is very general and uses a minimum of components. Another clock circuit that provides complementary as well as nonoverlapping clocks is shown in Fig. 8.3–18.[34] Throughout this study our considerations will generally include only two-phase clock schemes. If clocks phases higher than two are required, then a much more complex scheme is necessary. One method would be to divide a frequency f_0 by n and apply this waveform through a shift register clocked at the f_0 rate. The output of each cell of the shift register would be one of the clock phases.

If the MOS transistor is used as a switch, then the gate is controlled by the clock. If the clock is off chip, then the gate of the MOS switch must be externally connected to the clock circuitry. However, the gate oxide is sensitive to static electricity and should be protected when connected to external circuits if appropriate. Figure 8.3–19(a) shows one possible protection

[33] This method was suggested by Dr. R. L. Geiger of the Dept. of Electrical Engineering, Texas A&M University.

[34] S. Domenik, "On-Chip Clock Buffers," *Lambda,* Vol. II, No. 1, pp. 49–51, 1981.

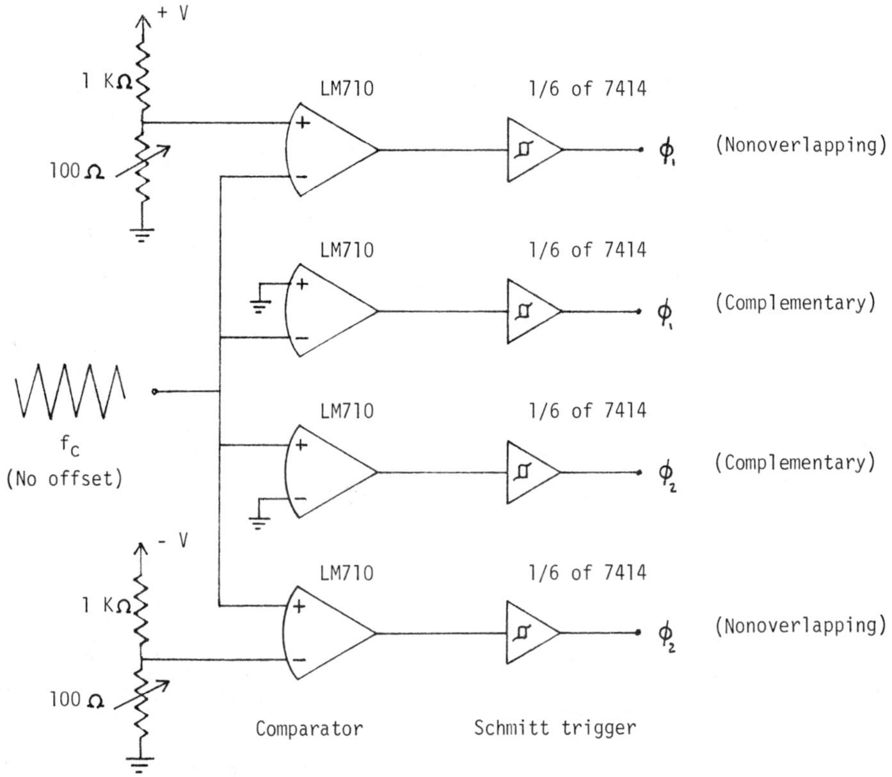

Fig. 8.3–17. A general two phase clock scheme.

circuit against static electricity. The drain of the protection transistor is in common with the gate to be protected. The source and gate of an enhancement resistor (see next section) with a W/L of about 5 is connected to the substrate. Figure 8.3–19(b) illustrates how the protection device works. For large positive voltage spikes, the PN diode formed by the drain and substrate breaks down. Punch through from the drain to the source may also occur. If the voltage spikes are large and negative, then the protection device acts like an inverted

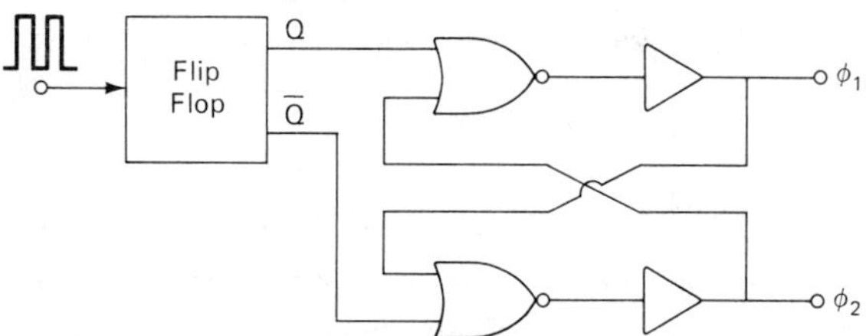

Fig. 8.3–18. A method to generate nonoverlapping clocks.

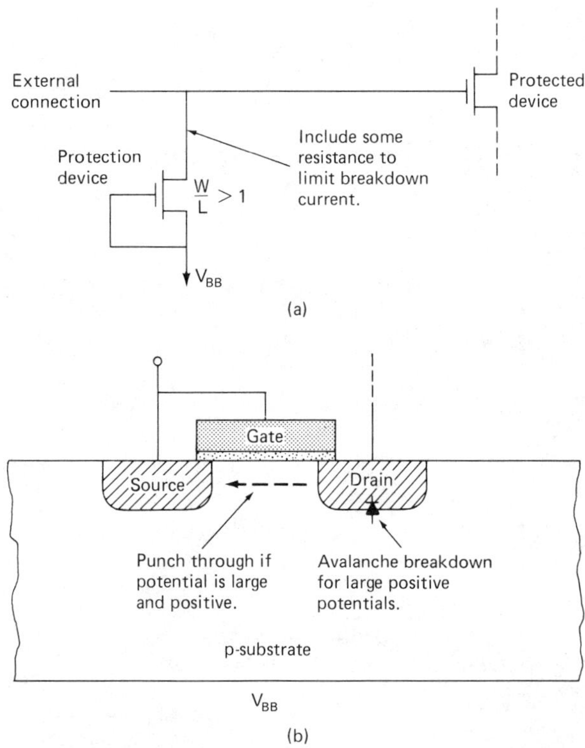

Fig. 8.3–19. (a) Protection scheme for externally connected gates or capacitors. (b) Illustration of the protective behavior.

enhancement resistor and places a finite resistance from the gate of the protected device to the substrate. Other forms of protection devices can be used, and the type quite often depends upon the space available on the chip.

In summary, the MOS transistor makes one of the best switch realizations available in integrated circuit form. If used correctly in the design of switched capacitor circuits, they should function as desired. They take minimum area, dissipate essentially no power, and provide reasonable values of R_{ON} and R_{OFF} for the analog sampled data circuits considered in this study. It is important that the clocks be nonoverlapping and high for a period of time sufficient to transfer charge. Strange behavior in a switched capacitor circuit can often be attributed to a failure to meet these requirements.

8.4 MOS ANALOG CIRCUITS

In this section the basic analog circuit building blocks will be developed. These building blocks will be used to design the more complex circuits of the following sections. The basic circuits presented in this section include the active resistor, or load; current sinks and sources; current mirrors; and voltage references

Active Resistor

The active resistor is used in place of a polysilicon or diffused resistor to produce a DC voltage drop or to provide a small signal resistance that is linear over a small range. There are many cases where the area required to obtain a small signal resistance is more important than the linearity. Thus a small MOS device can simulate the resistance of an equivalent polysilicon or diffused resistor in much less area.

The active resistor is achieved by simply connecting the gate to the drain, as shown in Fig. 8.4–1(a) and (b). For the *n*-channel device, the source should be placed at the most negative power supply voltage, V_{SS}, if possible so that the body effect is eliminated. Similarly, the source of the *p*-channel device should be at the most positive power supply voltage, V_{DD}. Because

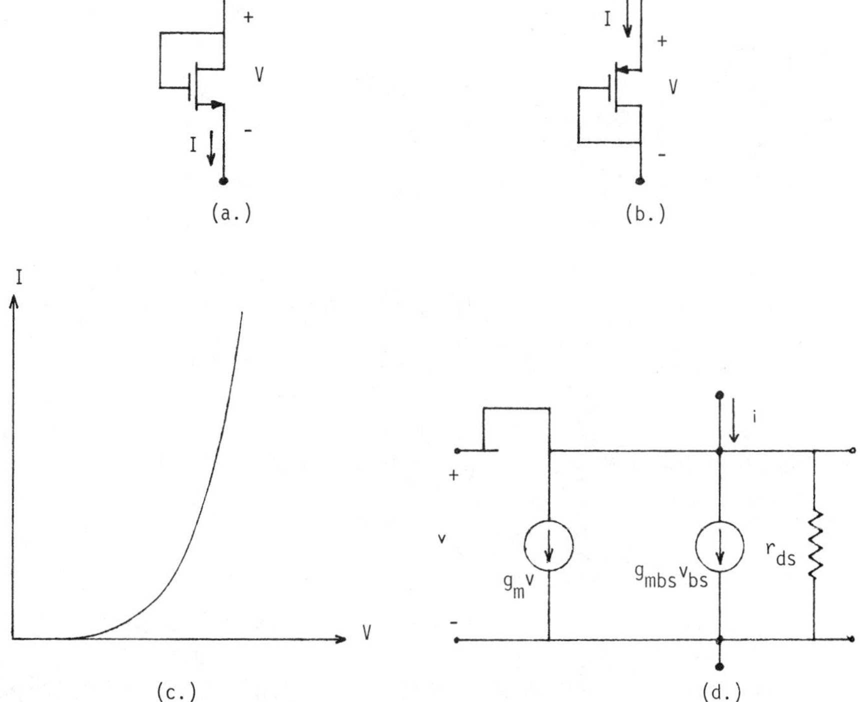

Fig. 8.4–1. Active resistor. (a) n-channel. (b) p-channel. (c) I-V characteristics. (d) AC model.

V_{GS} is now V_{DS}, the transconductance curve of Fig. 8.2–5 characterizes the large signal behavior of the active resistor. Figure 8.4–1(c) illustrates the V-I characteristics of Fig. 8.4–1(a) and (b). Because the connection of the gate to the drain guarantees operation in the saturation region, the V-I characteristics can be written as

$$I = I_D = (K'W/2L)[(V_{GS} - V_T)^2] = \beta(V_{GS} - V_T)^2 \tag{1}$$

or

$$V = V_{GS} = V_{DS} = V_T + (I/\beta)^{0.5} \tag{2}$$

where β is defined as

$$\beta = (\mu_o C_{ox} W/2L) = (K'W/2L) \tag{3}$$

If either V or I is defined, then the remaining variable can be found by using either Eq. (1) or Eq. (2).

Connecting the gate to the drain means that V_{DS} controls I_D, and therefore the channel transconductance becomes a channel conductance. The small-signal model of Fig. 8.2–12 for either the n-channel or p-channel active resistor is shown in Fig. 8.4–1(d). It is easily seen that the small-signal resistance of either of these circuits is

$$r_{out} = \frac{v}{i} = [r_{ds}/(1 + g_m r_{ds})] \cong g_m^{-1} \tag{4}$$

where $g_m r_{ds}$ is greater than unity. Equation (35) of Section 8.2 shows how this small-signal resistance is a function of the large-signal value of V and I.

An illustration of the application of the active resistor is shown in Fig. 8.4–2, where a voltage V_{out} has been derived from V_{DD} and V_{SS}. Both sources are connected so that the body effect has no influence. Because the currents through both transistors must be the same, V_{DS1} can be related to V_{DS2} as follows

$$V_{DS1} = (\beta_2/\beta_1)^{0.5}[V_{DS2} - V_{T2}] + V_{T1} \tag{5}$$

where it is recalled that $V_{GS1} = V_{DS1}$, and $V_{GS2} = V_{DS2}$. A typical situation might be where V_{DD}, V_{SS}, V_{out}, and the current through the devices, I, are specified. In that case, β_1 and β_2 can be found from Eq. (1). For example, if $V_{DD} = 5$ V, $V_{SS} = -5$ V, $V_{out} = 1$ V, and $I = 50$ μA, then using the

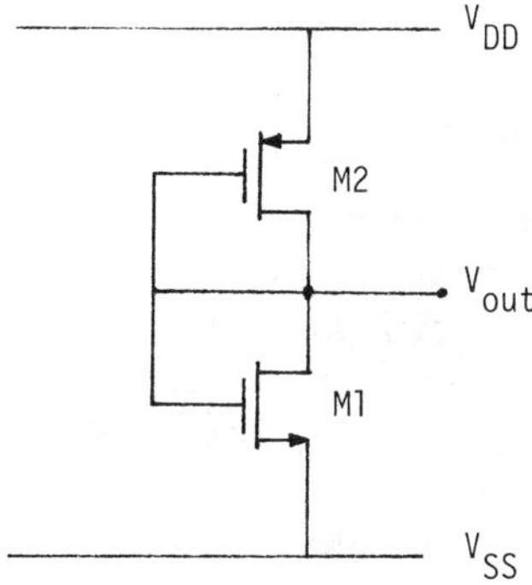

Fig. 8.4–2. Voltage division using active resistors.

model parameters for the MOS transistor of Table 8.2–1, we get $\beta_1 = 2.00 \times 10^{-6}\ A/V^2$ and $\beta_2 = 5.56 \times 10^{-6}\ A/V^2$. This gives a ratio of 0.16 for M1 and 0.89 for M2.

Large DC voltages dropped across a gate-drain connected transistor require large currents or W/L ratios much less than unity. This can be circumvented by cascading devices, as shown in Fig. 8.4–3. The voltages V_1, V_2, and V_3 are given by Eq. (2), where I is from a known current source. If the voltages are specified, then the W/L ratio of each device can be determined from Eq. (2) if the effects of the bulk-to-source voltage are taken into account. If the W/L ratios are given, the voltages V_1 through V_3 can be calculated starting with V_1. Using more than one device to drop the voltage will result in W/L ratios closer to unity and smaller DC currents.

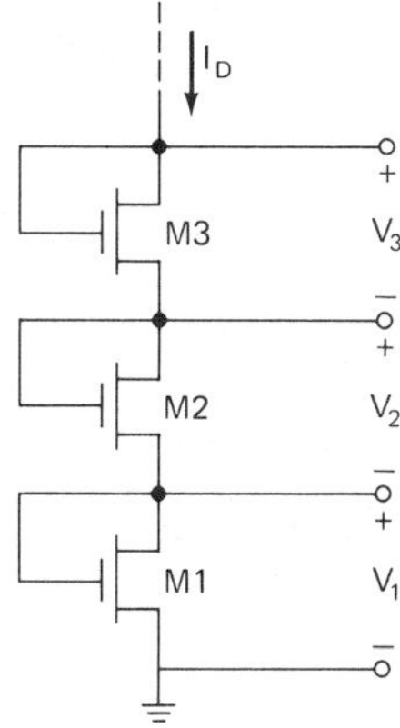

Fig. 8.4–3. Illustration of voltage division using enhancement resistor of Fig. 8.4–1(a).

Current Sinks and Sources

Figure 8.4–4(a) shows an implementation of current sink. The gate is taken to whatever voltage is necessary to create the desired value or current. For example, the voltage divider of Fig. 8.4–2 can be used to provide this value of voltage. If the gate-source voltage is held constant, then the large signal characteristics of Fig. 8.4–4(a) are given by the output characteristics of the MOS transistor, as illustrated in Fig. 8.4–4(b). If the source and bulk are both connected to V_{SS}, then the small-signal output resistance is given by

$$r_{out} = (g_{ds})^{-1} \tag{6}$$

If the source and bulk are not connected to the same potential, the characteristics will not change as long as V_{BS} is a constant. This is an advantage of CMOS over NMOS or PMOS, because the current can always be taken from the drain while the source can be kept at a constant potential.

Figure 8.4–5(a) shows an implementation of a current source. Again, the gate is taken to a constant potential, as is the source. With the value of V_{GS} held constant, the V-I characteristics of the current source are found

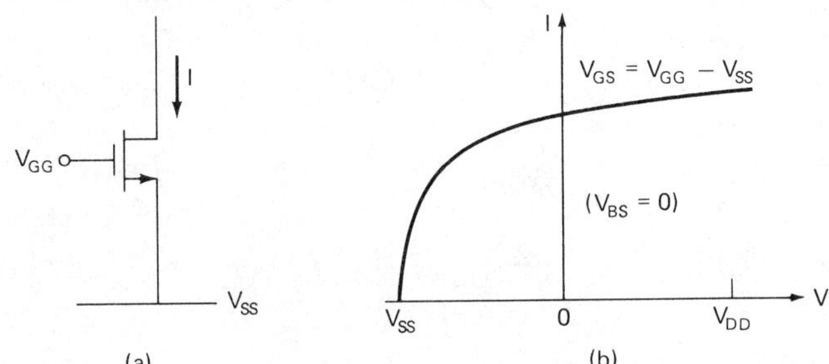

Fig. 8.4–4. (a) Current sink. (b) I-V characteristics of (a).

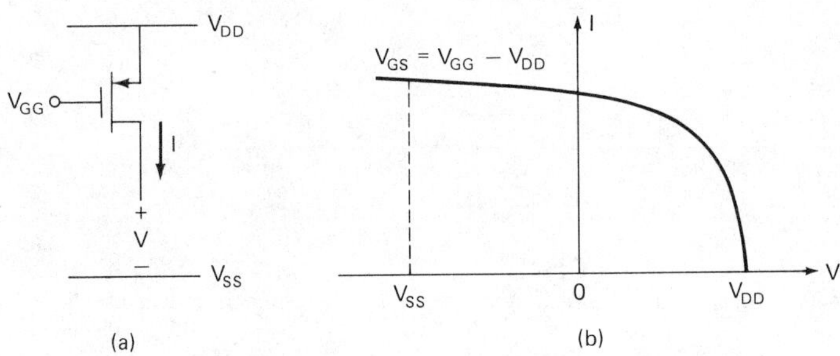

Fig. 8.4–5. (a) Current source. (b) I-V characteristics of (a).

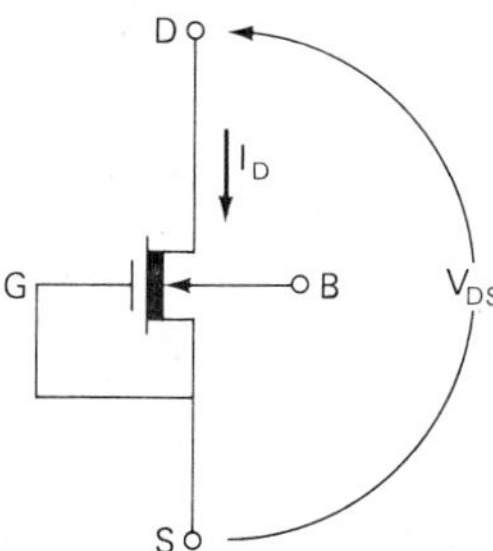

Fig. 8.4–6. An n channel, depletion MOS transistor used as a current source.

from Fig. 8.2–3. With the definition of V and I of the source, as shown in Fig. 8.4–5(a), the large-signal V-I characteristic is shown in Fig. 8.4–5(b). The small-signal output resistance of the current source is given by Eq. (6).

Figure 8.4–6 shows a depletion mode, n-channel MOS transistor connected with its gate to source. This configuration is analogous to the enhancement resistor and could be called the *depletion* resistor. However, it turns out that the depletion device connected as shown is best thought of as a current source, rather than a resistor, because of its large resistance. The graphical characteristics of the depletion device as a current source (or resistor) are shown in Fig. 8.4–7. In this figure we have assumed that $V_{BS} = 0$.

It is interesting to note that Fig. 8.4–6 can serve either as a current source or a current sink. If the source is grounded, then the drain acts as a current sink. This configuration is particularly attractive because the value of V_{BS} is zero, resulting in very large AC output resistance. Figure 8.4–8 shows the gate-source shorted depletion device used as a current source. The output current can be written from Eq. (12) of Section 8.2 using the appropriate substitutions for V_{BS}. This expression is given below, assuming the device is in saturation

$$I_0 = \beta[-V_{TO} - \gamma\sqrt{\phi + V_o} - \gamma\sqrt{\phi}]^2 (1 + \lambda V_{DD} - \lambda V_o) \qquad (7)$$

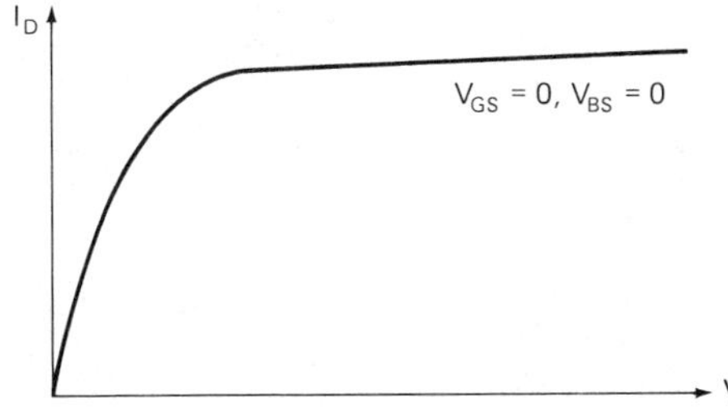

Fig. 8.4–7. Voltage-current characteristics of Fig. 3.6–5.

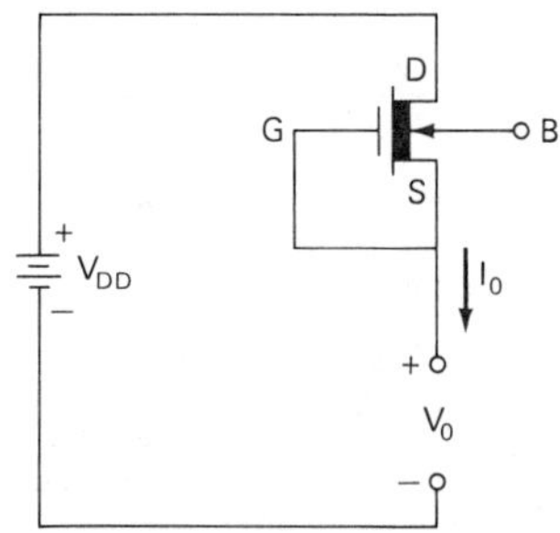

Fig. 8.4–8. An n-channel, depletion MOS transistor used as a current source.

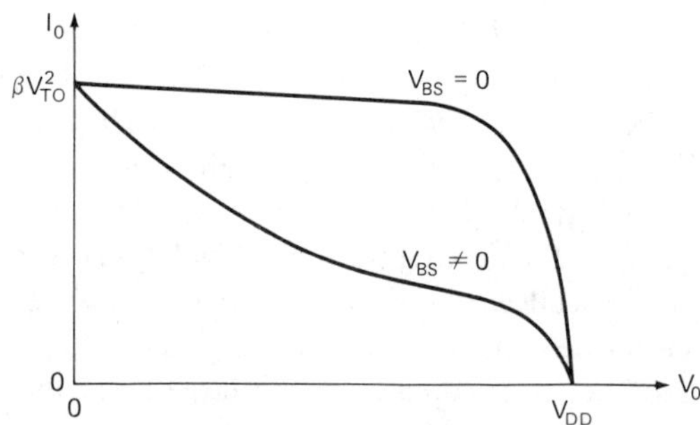

Fig. 8.4–9. Voltage-current characteristics of the current source of Fig. 8.4–8.

The voltage-current characteristic of this current source is shown in Fig. 8.4–9. For comparison, the curve for $V_{BS} = 0$ is shown. However, since V_{TO} in Eq. (7) is negative for an n-channel depletion device, the actual curve shown makes a very poor current source.

The small-signal output resistance of the curve with $V_{BS} = 0$ is equal to r_{ds}. The small-signal output resistance for the case where $V_{BS} \neq 0$ is found to be

$$r_{\text{out}} \cong r_{ds}/(1 + g_{mbs}r_{ds}) \cong 1/g_{mbs} \tag{8}$$

Although the AC output resistance of Fig. 8.4–8 is approximately $1/g_{mbs}$, it is still larger than the AC output resistance of the enhancement resistor. The small-signal output resistance of Fig. 8.4–8 can be improved through the use of additional circuitry (see Problem 8.22).[35]

[35] E. Toi, "An NMOS Operation Amplifier," *Proc. IEEE ISSCC,* Paper THAM 11.5, Philadelphia, PA, 1979, pp. 134–135.

Current Mirrors

A very useful building block in MOS analog circuit design is called the *current mirror.* This building block has been well developed for bipolar IC technology.[36] The current mirror uses the principle that if the gate-source potentials of two identical MOS transistors are equal then the channel currents should be equal. Figure 8.4–10(a) shows the implementation of a simple n-channel current mirror. I_{in} is assumed to be defined by a current source or some other means, and I_{out} is the output or "mirrored" current. M1 is in saturation because $V_{DS} = V_{GS}$. Assuming that V_{DS2} is greater than V_{T2} allows us to use the equation for the saturation region of the MOS transistor. I_{out} may be solved in terms of I_{in} using Eq. (12) of Section 8.2 as

$$I_{out} = [(W_2L_1/W_1L_2)(1 + \lambda\ V_{DS2})/(1 + \lambda\ V_{DS1})]\ I_{in} = A_iI_{in} \qquad (9)$$

where it is assumed that μ_o, C_{ox}, and λ are identical for both n-channel devices. Ideally, if $W_2/L_2 = W_1/L_1$, then $I_{out} = I_{in}$. However, Eq. (9) shows that if V_{DS2} is not equal to V_{DS1}, then I_{out} is not equal to I_{in} unless λ is equal to zero. Because $V_{DS1} = V_{GS1}$, and $V_{DS2} = V_{out}$ is typically much larger, this current mirror does not reproduce I_{in} precisely. Figure 8.4–10(b) shows the experimental results of the current mirror in Fig. 8.4–10(a). Both devices had an identical W/L ratio of 0.3 mils/0.3 mils. It is clearly seen that the value of I_{out} is dependent upon V_{out}. The small-signal output resistance of the current mirror is inversely proportional to the slope of the curves in Fig. 8.4–10(b). Using the small-signal model of Fig. 8.2–12, we can express the output resistance as

$$r_{out} = r_{ds2} = (\lambda\ I_D)^{-1} \qquad (10)$$

In many applications, this resistance is too low and must be increased using the configurations that are described below.

A configuration that causes A_i to be less dependent upon V_{out} is shown in Fig. 8.4–11(a). In this current mirror the current gain A_i is determined by the matching of M1 and M2. M3 and M4 force the value of V_{DS1} to be equal to the value of V_{DS2}, causing the dependence of I_{out}/I_{in} on λ to be eliminated. Thus the current gain of Fig. 8.4–11(a) is written as

$$A_i = I_{out}/I_{in} = [(W_2/L_2)/(W_1/L_1)] \qquad (11)$$

[36] P. R. Gray and R. G. Meyer, *Analysis and Design of Analog Integrated Circuits,* Chap. 4, John Wiley & Sons, Inc., New York, 1977.

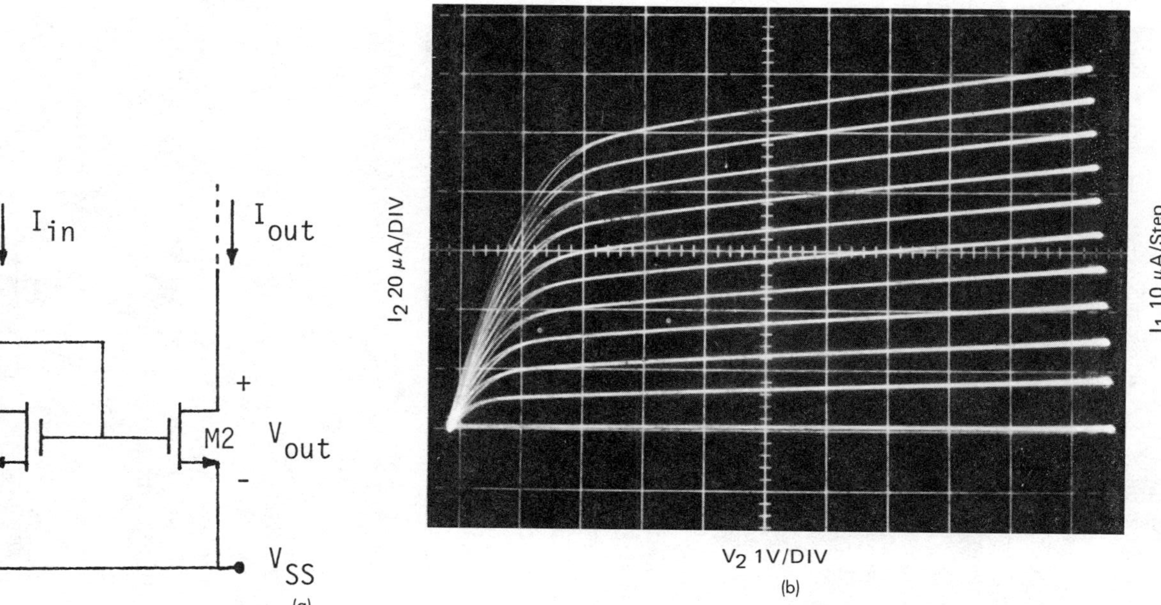

Fig. 8.4–10. (a) Simple current mirror. (b) Experimental results of (a) with $W_2/L_2 = W_1/L_1 = 0.3\text{mils}/0.3\text{mils}$. Vertical scale is 20μA/div. and horizontal scale is 1 volt/div. $V_{BS} = 0$ and step size is 10μA/step.

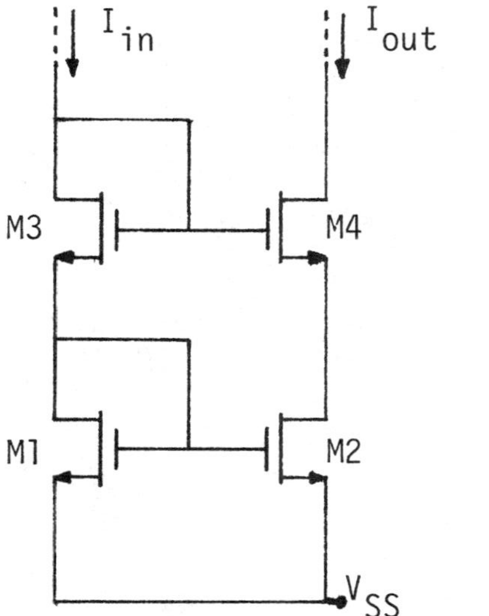

(a.)

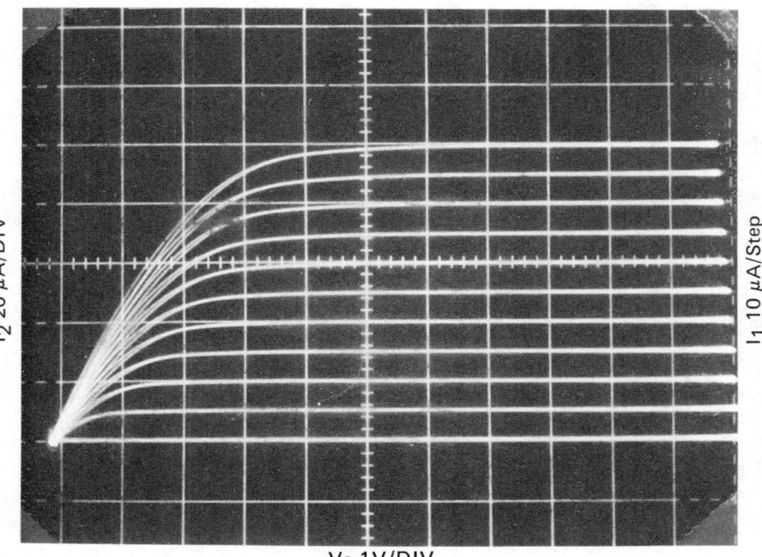

(b.)

Fig. 8.4–11. (a) An improved current mirror using n-channel MOS transistors. (b) Input-output characteristics of (a) with $V_{BS} = -5$ volts and $W_1/L_1 = W_2/L_2 = 0.5/0.5$ and $W_3/L_3 = W_4/L_4 = 0.3/0.5$. Vertical scale is 20μA/div. and the horizontal scale is 10μA/step.

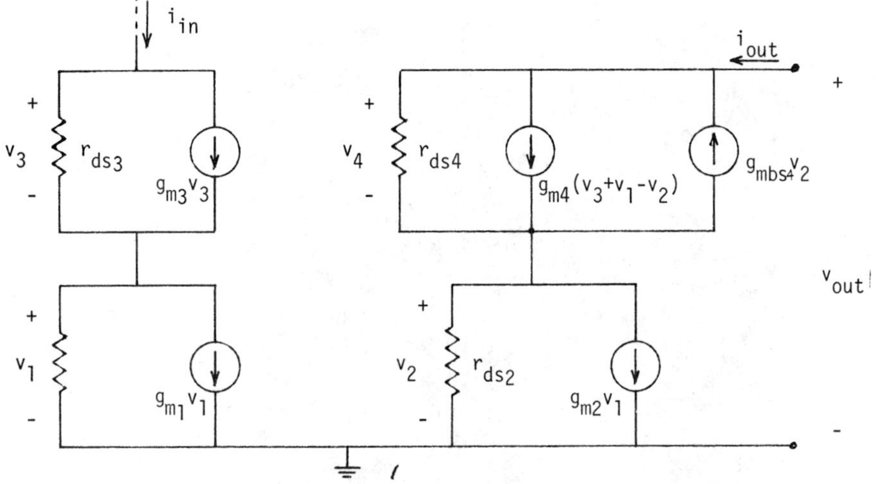

Fig. 8.4–12. Small signal model of the current mirror of Fig. 8.4–11.

Figure 8.4–11(b) shows the experimental performance and the resulting improvement in both A_i and the increased output resistance. The W/L ratios for M1 and M2 were 0.5 mils/0.5 mils and for M3 and M4 were 0.3 mils/0.5 mils; V_{BS} was −5 volts.

The small-signal output resistance of Fig. 8.4–11(a) can be calculated from the small-signal model for this circuit shown in Fig. 8.4–12. V_{out} may be expressed as

$$v_{out} = v_4 + v_2 = r_{ds4}[i_{out} + g_{mb4}\, v_2 - g_{m4}(v_3 + v_1 - v_2)] + v_2 \quad (12)$$

Also

$$v_2 = r_{ds2}\,(i_{out} - g_{m2}\, v_1) \quad (13)$$

Because $i_{in} = 0$ under small signal conditions, v_1 and v_3 are also zero. Therefore, substitution of Eq. (13) into Eq. (12) yields

$$v_{out} = i_{out}\,[r_{ds2} + r_{ds4} + r_{ds2}r_{ds4}\,(g_{m4} + g_{mbs4})] \quad (14)$$

Solving for the output resistance, r_{out}, results in

$$r_{out} = r_{ds2} + r_{ds4} + r_{ds2}r_{ds4}g_{m4}\,(1 + \eta_4) \quad (15)$$

where η was defined in Eq. (37) of Sec. 8.2. Comparing Eq. (15) with Eq. (10) shows that Fig. 8.4–10(a) has significantly larger output resistance than Fig. 8.4–11(a).

Another current mirror is shown in Fig. 8.4–13(a). This circuit is an NMOS implementation of the well-known Wilson current mirror.[37] The output resistance of the Wilson current mirror is increased through the use of negative feedback. Figure 8.4–13(b) shows the experimental results of the Wilson current mirror of Fig. 8.4–13(a) where the W/L ratios for M1 and M2 are 1.0 mils/0.3 mils and 5.0 mils/0.5 mils for M3. The bulk voltage was −5 volts.

It can be shown that the small-signal output resistance of Fig. 8.4–13(a) is given as (see Problem 8.23)

$$r_{out} = r_{ds3} + \frac{r_{ds2}[1 + r_{ds3}g_{m3}(1 + \eta_3) + g_{m1}r_{ds1}g_{m3}]}{[1 + g_{m2}r_{ds2}]} \tag{16}$$

The output resistance of Fig. 8.4–13(a) is found to be comparable with that of Fig. 8.4–11(a). Unfortunately, both these circuits require at least 2 V_T across the input before they behave as described above. The influence of the W/L ratios upon the range of V_{out} before the curves become flat should be noted.

A practical consideration in matching the devices that are mirroring the current is to keep the channel lengths of both devices equal. Since the channel length, L, is susceptible to lateral diffusion and the channel width is not (see Fig. 8.2–8(a)), then the channel width, W, can be used to achieve nonunity current gains. Each of the current mirrors discussed above can be implemented using p-channel devices. The circuits perform in an identical manner and exhibit the same small-signal output resistance.

Voltage References

An ideal voltage reference is independent of power supply and temperature. Many applications in analog circuits require such a building block. Examples are analog-to-digital and digital-to-analog converters. A voltage reference that approaches the ideal requirements has been very difficult to achieve using single-channel MOS technology. One approach that has been used with reasonable success is shown in Fig. 8.4–14. Here we see that M2 is a depletion mode device with $V_{GS} = 0$ acting like a current source and M1 is an enhancement device using this current to create a voltage, V_{ref}. Assuming both devices are saturated leads to an expression for V_{ref} given as

$$V_{ref} = V_{T1} - \sqrt{(\mu_2 W_2/L_2)/(\mu_1 W_1/L_1)}\; V_{T2}\sqrt{(1 + \lambda V_{DS2})} \tag{17}$$

---

[37] P. R. Gray and R. G. Meyer, *op. cit.*

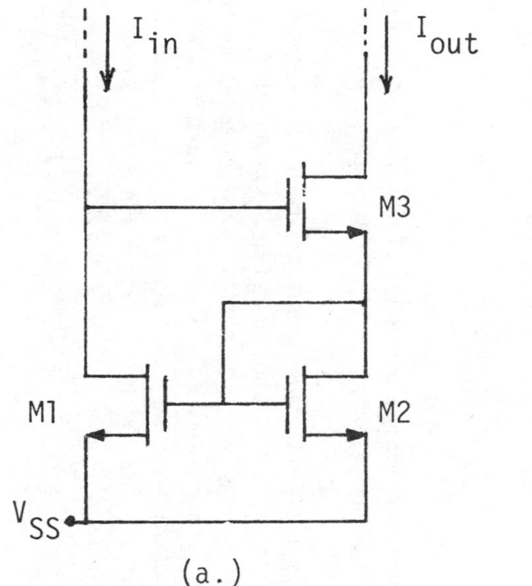

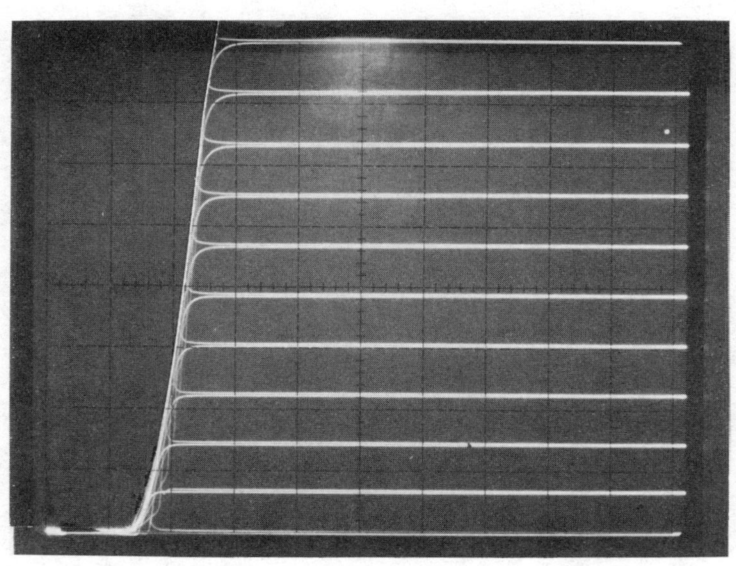

Fig. 8.4–13. (a) Wilson current mirror using n-channel MOS transistors. (b) Experimental characteristics of (a) with $V_{BS} = -5$ volts, and $W_1/L_1 = W_2/L_2 = 1/.3$ and $W_3/L_3 = 5/0.5$. Vertical scale is $10\mu A$/div. Horizontal scale is 1 volt/div. Step size is $10\mu A$/step.

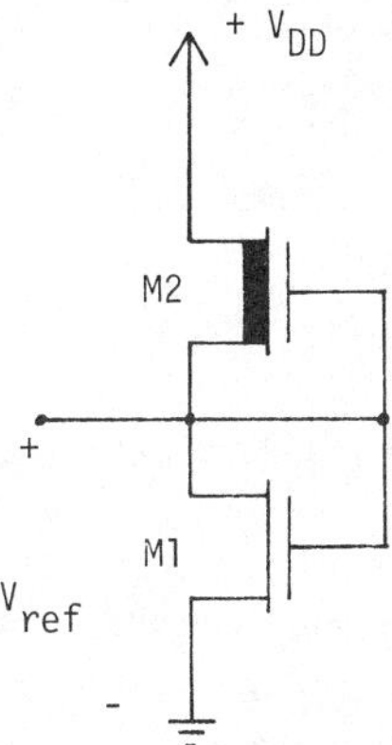

Fig. 8.4–14. A simple voltage reference using n-channel MOS transistor.

Because λ is typically very small, Eq. (17) reduces to

$$V_{\text{ref}} \cong V_{T1} - \sqrt{(\mu_2 W_2/L_2)/(\mu_1 W_1/L_1)}\, V_{T2} \tag{18}$$

Unfortunately, V_{T1} and V_{T2} are process-dependent, and V_{T2} also depends upon the bulk source potential, which should be a constant. Using the values in Table 8.2–1 and assuming equal W/L ratios gives $V_{\text{ref}} \cong 2$ volts. This reference has eliminated the dependence upon V_{DD}, though it is still dependent upon temperature through V_{T1} and V_{T2}.

Let us examine the temperature dependence of V_{GS} of an MOS device to see how a voltage reference depends upon temperature. If the MOS transistor is in saturation, then from Eq. (2) we have

$$dV_{GS}/dT = dV_T/dT + \sqrt{(2L)/(W C_{\text{ox}})}\; d\sqrt{I_D/\mu}/dT \tag{19}$$

The terms in Eq. (2) that cause appreciable temperature dependence are the threshold voltage, V_T, the device current, I_D, and the channel mobility, μ. The threshold voltage has been shown to have a temperature variation given by

$$V_T = V_{T0} - \alpha T \tag{20}$$

where V_{T0} and α are different for enhancement and depletion devices. These values are strongly dependent upon the substrate dopant density and the implant dosages used during fabrication. The channel mobility has a first-order dependence on temperature given as $T^{-1.5}$. This dependence can be ignored compared with that of V_T given by Eq. (20). For the present we shall assume that I_D does not depend upon temperature. Therefore, the variation of V_{GS} with respect to temperature is approximately

$$dV_{GS}/dT \cong dV_T/dT = -\alpha \tag{21}$$

Thus, the temperature dependence of Fig. 8.4–14 can be expressed as

$$dV_{\text{ref}}/dT = -\alpha_1 + \sqrt{(W_2/L_2)/(W_1/L_1)}\,\alpha_2 \tag{22}$$

Unfortunately a zero temperature coefficient implies a zero value of V_{ref} if $V_{T1} = |V_{T2}|$. Other single-channel, MOS voltage references have been developed that permit a nonzero value of V_{ref} when $dV_{\text{ref}}/dT = 0$.[38] Such circuits do not have a low-temperature coefficient over a wide range of temperatures, because the temperature characteristics of enhancement and depletion transistors do not track each other.

Voltage references similar to the bipolar bandgap voltage reference[39] can be achieved using CMOS technology. The principle of the bandgap voltage reference is illustrated in Fig. 8.4–15, where a diode voltage, V_{BE}, with a temperature coefficient of approximately −2.5 mV/°C is summed with a voltage $V_T = kT/q$, which has a temperature coefficient of approximately 0.085 mV/°C. Zero temperature coefficient can be achieved if V_T is multiplied by a gain, G, so that

$$V_o = V_{BE} + GV_T \tag{23}$$

Differentiating Eq. (23) with respect to temperature, T, gives

$$dV_o/dT = dV_{BE}/dT + G\,dV_T/dT = (-2.5 + 0.085\,G)\;mV/°C \tag{24}$$

Therefore if $G \cong 29.4$, then dV_o/dT should be approximately zero. This gives a bandgap reference voltage of approximately 1.2 to 1.5 volts, depending upon the current levels in the diodes. If the current also varies as a function of temperature, i.e., $I_s(T)$, then the above analysis must be repeated.

A CMOS bandgap voltage reference is shown in Fig. 8.4–16. $Q1$ and $Q2$ are *npn* substrate transistors. $Q1$ is assumed to have an emitter area of

[38] R. A. Blauschild, P. Tucci, R. S. Muller, and R. G. Meyer, "A NMOS Voltage Reference," *Proc. of 1978 Int. Solid-State Circuits Conference,* February 1978, San Francisco, CA, pp. 50–51.

[39] P. R. Gray and R. G. Meyer, *Analysis and Design of Analog Integrated Circuits,* John Wiley & Sons, New York, 1977, pp. 254–261.

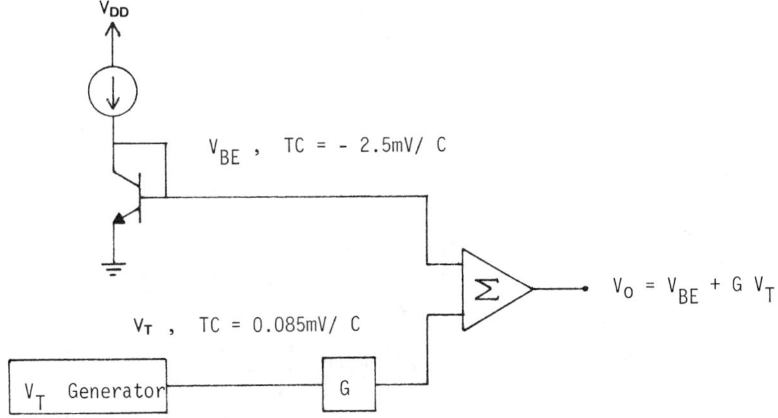

Fig. 8.4–15. Block diagram of the bandgap voltage reference.

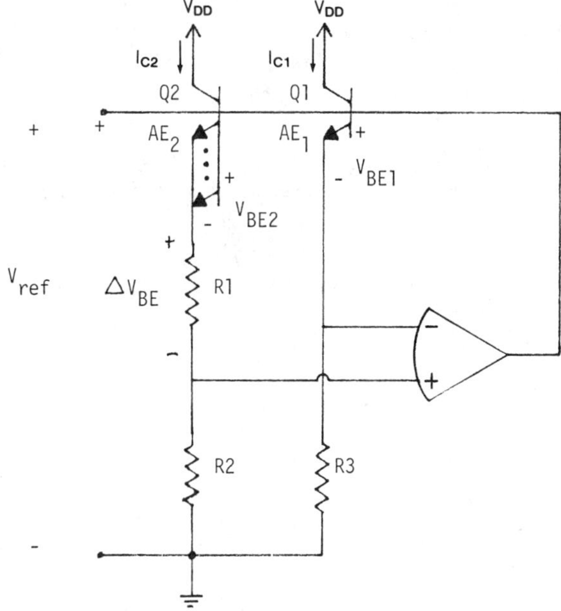

Fig. 8.4–16. CMOS bandgap voltage reference circuit.

A_{E1} and $Q2$ an emitter area of A_{E2}. The output voltage, V_{ref}, can be found in the following manner. The voltage across R_1 is

$$\Delta V_{BE} = V_T \ln[(I_{C1}/I_{C2})(I_{S2}/I_{S1})] = V_T \ln[(I_{C1}/I_{C2})(A_{E2}/A_{E1})] \quad (25)$$

However, the op amp forces the relationship

$$I_{C1}R_3 = I_{C2}R_2 \tag{26}$$

Thus

$$\Delta V_{BE} = V_T \ln\left[(R_2/R_3)(A_{E2}/A_{E1})\right] \tag{27}$$

The reference voltage, V_{ref}, can be written as

$$V_{\text{ref}} = V_{BE1} + V_{R3} = V_{BE1} + V_{R2} = V_{BE1} + (\Delta V_{BE}R_2)/R_1 \tag{28}$$

Substituting Eq. (27) into Eq. (28) gives

$$V_{\text{ref}} = V_{BE1} + (R_2/R_1)\, V_T \ln[A_{E2}R_2/A_{E1}R_3] \tag{29}$$

We note that $(R_2/R_1)\ln[A_{E2}R_2/A_{E1}R_3]$ is equal to G of Eq. (23). The area ratio and the resistors can be used to adjust the temperature coefficient to be zero. Exact cancellation will not be possible because of component tolerances and second-order effects that have been neglected. Problem 8.24 shows that the influence of the op amp offset voltage is multiplied by $(1 + R_2/R_1)$ and can become a significant source of error and temperature-dependence.

The $Q2$ diode can be replaced by an MOS device operating in the weak inversion region if the current levels are small. Such a voltage reference uses very little power. In the weak inversion region, the V-I characteristics of the MOS transistor are exponential and can be used to generate V_T. An equation describing the MOS transistor in weak inversion is given as[40]

$$I_D = S\, I_{DO}\, e^{V_G/nV_T}(e^{-V_S/V_T} - e^{-V_D/V_T}) = S\, I_{DO}\, e^{V_G/nV_T}(1 - e^{-V_{DS}/V_T}) \tag{30}$$

where

$S = W/L$
$V_T = kT/q$
$n =$ slope factor
$I_{DO} =$ characteristic current

Equation (30) holds only for the weak inversion region where $V_{DS} < 3V_T$. It is found that n is controllable, but has a strong temperature-dependence.[41]

[40] E. Vittoz and J. Fellrath, "CMOS Analog Integrated Circuits Based on Weak Inversion Operation, "*IEEE J. of Solid-State Circuits,* Vol. SC-12, June 1977, pp. 224–231.

[41] G. Tzanateas, C. Salama, And Y. Tsividis, "A CMOS Bandgap Voltage Reference," *IEEE J. of Solid-State Circuits,* Vol. SC-12, No. 3, June 1979, pp. 655–657.

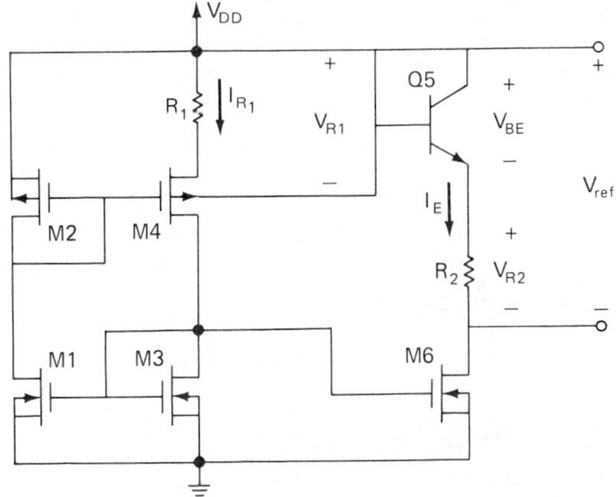

Fig. 8.4–17. A CMOS bandgap voltage reference with no dependence on n.

I_{DO} is difficult to control and also has a strong temperature-dependence. A voltage reference using the above concepts was proposed by Tsividis and Ulmer.[42] Though the reference voltage of this circuit was independent of I_{DO}, it was dependent upon n, which prevented satisfactory reduction of the temperature coefficient. A CMOS bandgap voltage reference that eliminated the dependence on n is shown in Fig. 8.4–17. The current mirrors M1 and M4 form a closed loop with an initial loop gain greater than unity. Therefore, the current in both branches increases until equilibrium is achieved when the loop gain is reduced to 1 by the voltage V_{R1} across R_1. If we assume that M1 through M4 operate in the weak inversion region and that V_{DD} is high enough to ensure drain current saturation of M1 and M4, then V_{R1} can be expressed as[43]

$$V_{R1} = V_T \ln[(S_4 S_1)/(S_2 S_3)] \tag{31}$$

Note that V_{R1} depends only on the thermal voltage and the ratio of the geometry of the devices and is independent of n. I_{R1} and I_E are related as

$$I_{R1}/I_E = S_3/S_6 \tag{32}$$

Solving for I_E gives

$$I_E = (S_6/S_3) V_T \ln[(S_4 S_1)/(S_2 S_3)] \tag{33}$$

[42] Y. P. Tsividis and R. W. Ulmer, "CMOS Reference Voltage Source," *ISSCC Dig. Tech. Papers*, February 1978, pp. 49–50.

[43] E. Vittoz and J. Fellrath, *op. cit.*

V_{ref} is given as

$$V_{\text{ref}} = V_{BE} + I_E R_2 = V_{BE} + (R_2/R_1)(S_6/S_3) V_T \ln[(S_4 S_1)/(S_2 S_3)] \quad (34)$$

An expression for V_{BE} that emphasizes its temperature dependence is given by[44]

$$V_{BE}(T) = V_{GO}\left(1 - \frac{T}{T_o}\right) + V_{BEO}(T/T_o) + (mkT/q)\ln(T_o/T) \quad (35)$$

where

V_{GO} = the extrapolated bandgap voltage of silicon
V_{BEO} = V_{BE} of a diode connect transistor at $T = T_o$
m = constant dependent on the diode fabrication and temperature characteristics

It can be shown that the condition for $dV_{\text{ref}}/dT = 0$ is given by

$$(R_2/R_1)(S_6/S_3)\ln[(S_4 S_1)/(S_2 S_3)] = q/kT_o\,(V_{GO} - V_{BEO}) + m \quad (36)$$

This results in a reference voltage of

$$V_{\text{ref}} = V_{GO} + (mkT/q)\left(1 + \ln\frac{T_o}{T}\right) \quad (37)$$

To ensure proper operation of the voltage reference described, the following precautions must be taken. First the devices must be in weak inversion even at the highest temperature of operation. Secondly, leakage currents, particularly in the n-channel devices, must be minimized to prevent these currents from becoming a major source of error at higher temperature. Lastly, the output resistance of the devices must be large enough to ensure proper operation of the devices as current mirrors. This can be accomplished by long devices or by use of the various current mirrors presented earlier in this section.

8.5 MOS AMPLIFIERS

This section uses the basic analog building blocks from the previous section to realize MOS amplifiers. The amplifiers considered here include the inverter,

[44] J. S. Brugler, "Silicon Transistor Biasing for Linear Collector Current Temperature Dependence," *IEEE J. of Solid-State Circuits,* Vol. SC-2, June 1967, pp. 57–58.

cascode, source follower, and differential amplifiers. These amplifiers will in turn be used to realize the MOS op amps in the remaining sections of this chapter. The large-signal and small-signal characteristics of each of the amplifiers will be discussed briefly.

Inverter

The inverter is the simplest and most basic gain stage in analog circuits. It consists of a common source transistor and a load. The load can be an active resistor or a current source/sink. Four types of MOS inverters are shown in Fig. 8.5–1. Figure 8.5–1(a) shows an all-enhancement NMOS inverter with the active resistor of Fig. 8.4–1(a) as the load. Figure 8.5–1(b) shows an NMOS enhancement transistor with the depletion MOS current source of Fig. 8.4–8. Figure 8.5–1(c) shows a CMOS inverter using a

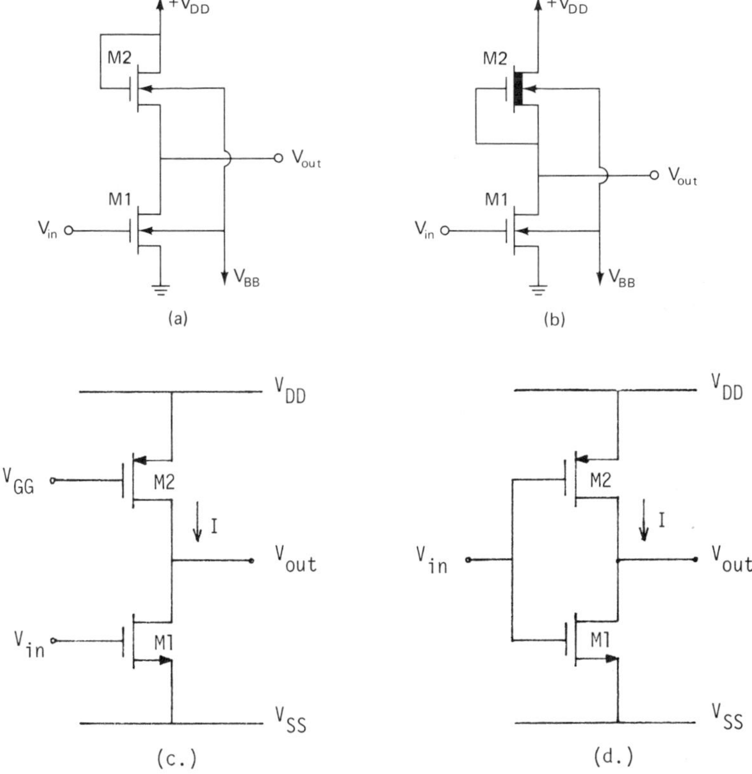

Fig. 8.5–1. Inverter using (a) enhancement load and (b) a depletion load. (c) Inverter with an active load. (d) Push pull inverter.

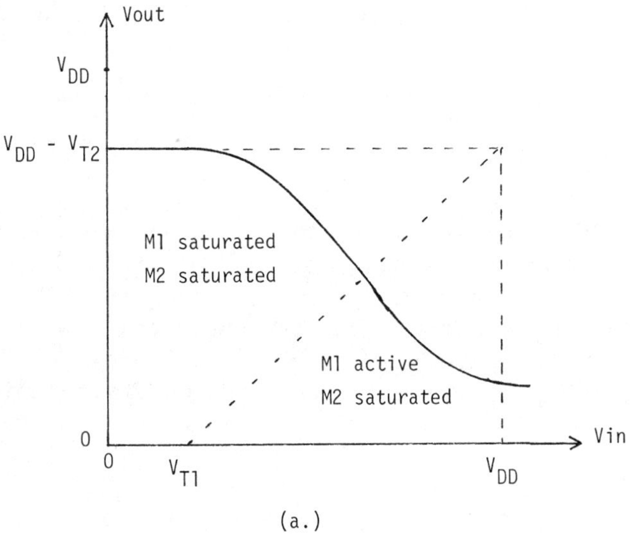

(a.)

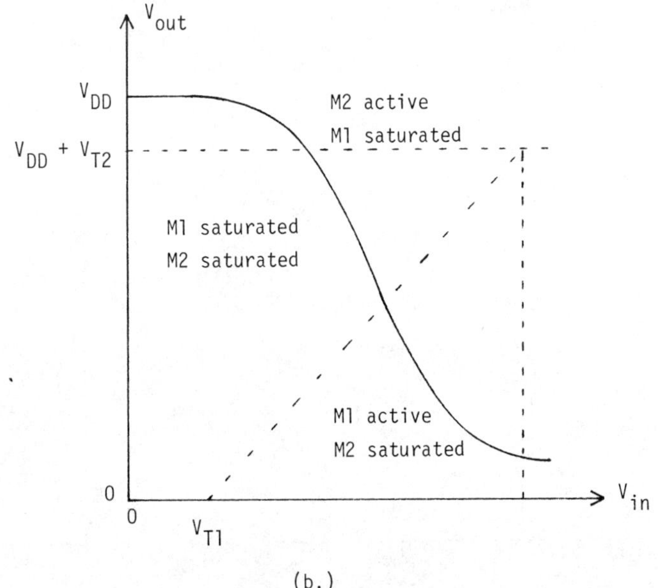

(b.)

Fig. 8.5–2. Large signal characteristics of NMOS inverters. (a) Fig. 8.5–1(a). (b) Fig. 8.5–1(b).

p-channel current source as the load. Figure 8.5–1(d) shows a CMOS push-pull inverter.

The large-signal characteristics can be demonstrated by graphical means, as illustrated in Fig. 8.5–2, for the NMOS inverters, and Fig. 8.5–3, for the CMOS inverters. These characteristics are obtained by superimposing the drain-source characteristics of M2 on the drain-source characteristic of M1 and then plotting values of V_{out} as a function of V_{in}. We have also shown the regions of operation for each device, assuming typical values of V_T.

If the inverter is to be used as an amplifier, it should have a small-signal gain that is as large as possible. Because the slope of the transfer characteristic at any point is equal to the small-signal gain at that point, it becomes necessary to bias the inverter in the region of steepest slope. The quiescent point of the inverter can be determined from the large-signal model developed in Section 8.2. The procedure is to first determine the region of operation for each device. Typical regions of operation have been indicated in Figs. 8.5–2 and 8.5–3. Next the DC currents in M1 and M2 are written as a function

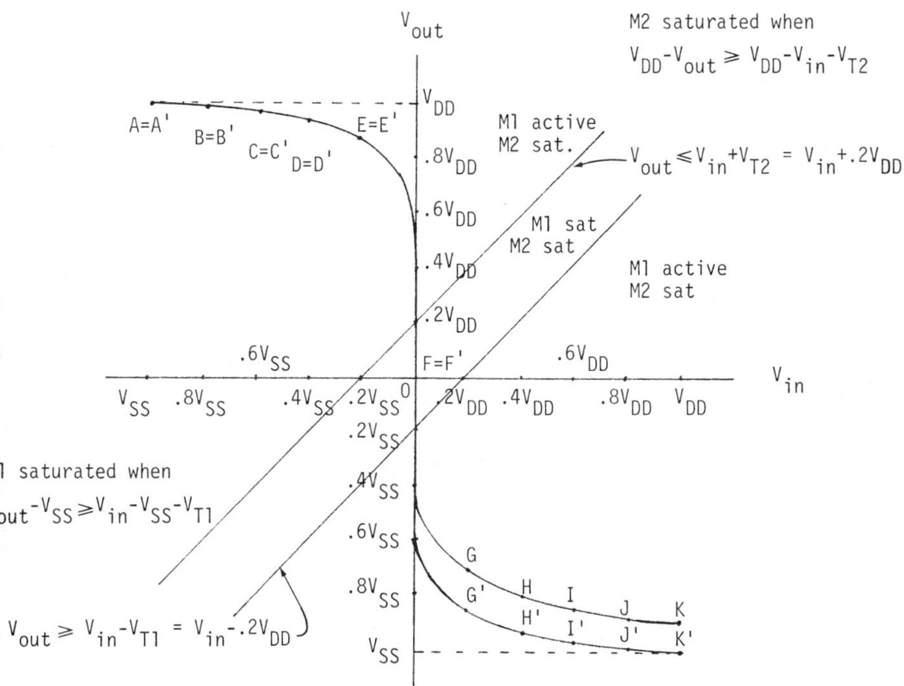

Fig. 8.5–3. Transfer characteristics of the inverters of Fig. 8.5–1(c) and 8.5–1(d).

of the DC terminal voltages. These two currents are equated giving generally a complex expression from which one can design the W/L ratios of M1 and M2, given that all the terminal voltages usually known.

If we assume that M1 and M2 are both operating in the saturated region (note this is typically the region of steepest slope), we may express the geometrical parameter of Fig. 8.5–1(a) and Fig. 8.5–1(b) in terms of the model parameters and terminal voltages as

$$\frac{W_1/L_1}{W_2/L_2} = \frac{[V_{DD} - V_{out}(DC) - V_{TO2} - \gamma\,[\sqrt{\phi - V_{BB} + V_{out}(DC)} - \sqrt{\phi}]}{V_{in}(DC) - V_{TO1} - \gamma\,[\sqrt{\phi - V_{BB}} - \sqrt{\phi}]} \tag{1}$$

and

$$\frac{W_1/L_1}{W_2/L_2} = \frac{[\gamma\sqrt{\phi} - \gamma\sqrt{\phi - V_{BB} + V_{out}} - V_{TO2}}{V_{in} - V_{TO1} - \gamma\sqrt{\phi - V_{BB}} + \gamma\sqrt{\phi}} \tag{2}$$

respectively. If the assumption of the operating regions is correct, then one can use the above equations to design the W/L ratios to achieve a desired operating point. Because the small signal gain is a function of the W/L ratios, we will find that the choice of an operating point will influence the small-signal performance, necessitating a compromise. Normally, CMOS inverters are biased by an external DC negative feedback path, so that we will not give relationships equivalent to Eqs. (1) and (2) for CMOS inverters.

The small-signal gain of the NMOS inverters of Fig. 8.5–1(a) and (b) can be calculated in one of two ways. One may differentiate the large-signal model equations or use the small-signal circuit model developed in Section 8.2. We shall illustrate both approaches. Assuming that both M1 and M2 of Fig. 8.5–1(a) are in saturation allows us to write

$$I_{D1} = \beta(V_{in} - V_{T1})^2 \tag{3}$$

and

$$I_{D2} = \beta_2(V_{DD} - V_{out} - V_{T2})^2 \tag{4}$$

Equating Eqs. (3) and (4) and solving for V_{out} results in

$$V_{out} + \gamma\sqrt{\phi + V_{out} - V_{BB}} = -\sqrt{\beta_1/\beta_2}\,(V_{in} - V_{TO1}) + V_{DD} - V_{TO2} + \gamma\sqrt{\phi} \tag{5}$$

Differentiating V_{out} with respect to V_{in} results in

$$A_{VE} = \partial V_{out}/\partial V_{in} = \frac{-\sqrt{\beta_1/\beta_2}}{1 + \dfrac{\gamma}{2\sqrt{V_{out} + \phi - V_{BB}}}} \tag{6}$$

It can be shown that the connection of the gate of M2 to its drain results in the maximum gain for the enhancement load inverter, A_{VE}.

An alternative and perhaps easier way to calculate the small-signal voltage gain of the enhancement inverter is to use the small-signal circuit model for the MOS device given in Section 8.2. The small-signal model of a general MOS inverter is shown in Fig. 8.5–4. For Fig. 8.5–1(a) $v_{gs2} = -v_{out}$, so that the small signal voltage gain is

$$\frac{v_{out}}{v_{in}} = \frac{g_{m1}}{g_{ds1} + g_{ds2} + g_{m2} + g_{mbs2}} = -\frac{g_{m1}}{g_{ds1} + g_{ds2} + g_{m2}(1 + \eta_2)} \tag{7}$$

where η was defined in Eq. (37) of Section 8.2. If we assume that the output conductances are negligible, then Eq. (7) reduces to

$$A_{VE} \cong (-g_{m1})/[g_{m2}(1 + \eta_2)] \tag{8}$$

Note that Eqs. (8) and (6) are identical.

The small-signal voltage gain of the depletion load inverter of Fig. 8.5–1(b) can be found by equating the large signal expressions of Section 8.2 for

$$I_{D2} = \beta_2(-V_{T2})^2 \tag{9}$$

and

$$I_{D1} = \beta_1(V_{in} - V_{T1})^2 \tag{10}$$

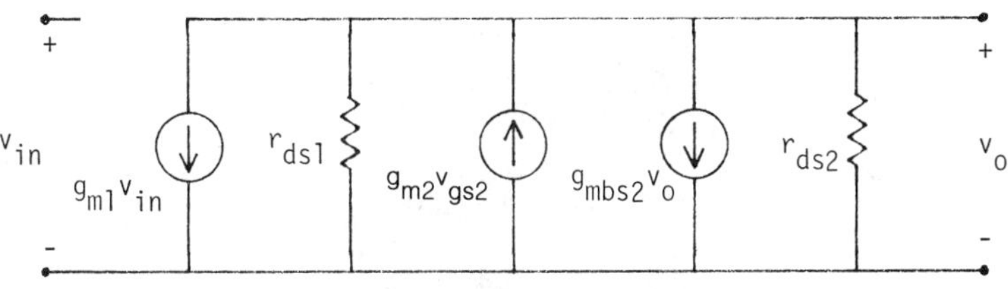

Fig. 8.5–4. A general MOS small signal model.

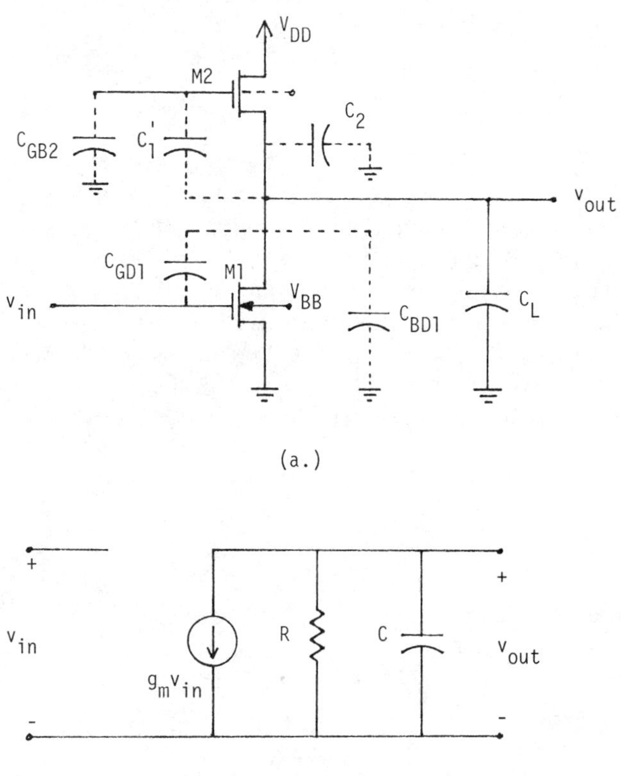

Fig. 8.5–5. Frequency response of the inverter. (a) Parasitic capacitance of the inverter. (b) Small signal model.

to get

$$V_{TO} - \gamma[\sqrt{\phi + V_{out} - V_{BB}} - \sqrt{\phi}] = -\sqrt{\beta_1/\beta_2}\,(V_{in} - V_{TO1}) \quad (11)$$

assuming both M1 and M2 are in saturation. Differentiating V_{out} with respect to V_{in} results in

$$A_{VD} = (\partial V_{out})/(\partial V_{in}) = -(2/\gamma)\sqrt{\beta_1/\beta_2}\sqrt{\phi + V_{out} - V_{BB}} \quad (12)$$

As with Fig. 8.5–1(a), we can calculate the small-signal gain using the small-signal circuit model of Section 8.2 with $v_{gs2} = 0$ gives

$$A_{VD} = v_{out}/v_{in} = -g_{m1}/(g_{ds1} + g_{ds2} + g_{mbs2}) \quad (13)$$

If the output conductances are negligible, then Eq. (13) becomes

$$A_{VD} \cong -g_{m1}/\eta_2 g_{m2} \tag{14}$$

Again note that Eqs. (14) and (12) are identical. Because η is normally small, we see that

$$A_{VD} = (1+\eta_2)/\eta_2 \, A_{VE} \cong A_{VE}/\eta_2 \tag{15}$$

Equation (15) indicates that the small-signal gain of the depletion load inverter is $1/\eta_2$ times the small-signal gain of the enhancement load inverter. Since η_2 is approximately 0.1, the depletion load inverter will have a gain of roughly 10 times the enhancement load inverter. Using the values of Table 8.2–1 and 10 μA gives $A_{VE} \cong -2.87$ and $A_{VD} \cong -28.11$ if $W_1/L_1 = 10(W_2/L_2)$ and $V_{BS} = -5.5$ volts.

As we did not develop the large-signal equations for the CMOS inverter, we shall use only the small-signal circuit model method of calculating the small-signal performance of Figs. 8.5–1(c) and (d). For the CMOS inverters, the g_{mbs2} term is not present because M2 has its source on AC ground. For Fig. 8.5–1(c), $v_{gs2} = 0$, so that the small-signal voltage gain is

$$A_V = v_{out}/v_{in} = -g_{m1}/(g_{ds1}+g_{ds2}) = -2\frac{\sqrt{K'_n/I_D}\,\sqrt{W_1/L_1}}{\lambda_1+\lambda_2} \tag{16}$$

We note with interest that as I_D becomes small A_V increases. Assuming that $W_1/L_1 = 10$ and $I_D = 10\ \mu$A and using the values of Table 8.2–1, Eq. (16) gives a small-signal gain of approximately −500, which is at least an order of magnitude more than the NMOS inverters.

For the push-pull inverter of Fig. 8.5–1(d), $g_{mbs} = 0$ and $v_{gs2} = -v_{in}$, so that the small-signal voltage gain is

$$\begin{aligned} A_V = v_{out}/v_{in} &= -(g_{m1}+g_{m2})/(g_{ds1}+g_{ds2}) \\ &= \frac{-2}{\sqrt{I_D}}\,\frac{(\sqrt{K'_n\,W_1/L_1}+\sqrt{K'_p\,W_2/L_2})}{\lambda_1+\lambda_2} \end{aligned} \tag{17}$$

If $W_1/L_1 = 10$, $W_2/L_2 = 1$, and $I_D = 10\ \mu$A, then the small-signal voltage gain of Fig. 8.5–1(d) is approximately −612.

In both CMOS inverters, the gain is inversely proportional to the square root of I_D. This occurs because the output conductance is proportional to I_D, whereas the transconductance is proportional to the square root of I_D. This of course assumes that the simple relationship for the drain current as a function of gate-source voltage given in Eq. (12) of Section 8.2 is valid. The increase of gain as I_D decreases holds true until this current reaches the subthreshold region of operation, where weak inversion occurs. At this point the transconductance becomes proportional to the bias current, and the small-signal voltage gain becomes a constant as a function of bias current. Thus the small-signal gain is independent of I_D. According to Fig. 8.2–11, this transition would occur in the vicinity of 0.1 to 1 μA.

Next we want to examine the small-signal frequency response of the inverter. The frequency response of the inverter is determined primarily by the capacitance to AC ground at the output mode, assuming the inverter is driven by a voltage source. A generalized inverter representing any of the four configurations of Fig. 8.5–1 is shown in Fig. 8.5–5(a). Here, whether M2 is an enhancement or depletion NMOS or a PMOS transistor is undesignated. C_1', C_2, and C_{GB2} represent the parasitic capacitances of M2. Also the pertinent parasitic capacitances of M1 are designated. For example, if M2 is an enhancement NMOS transistor, then the gate and drain are connected to V_{DD}, and the source is connected to V_{out}. Thus C_{GB2} is not important, and C_1' and C_2 become C_{GS2} and C_{BS2}, respectively. Probably the most significant capacitance is the load capacitance, C_L. C_L consists of the load capacitance seen by the inverter and includes the gate or gates of the next devices and any parasitics due to connections.

The small-signal model of Fig. 8.5–5(a) is shown in Fig. 8.5–5(b). The frequency response of this circuit is

$$V_{\text{out}}/V_{\text{in}} = -g_m R\,[\omega_1/(s+\omega_1)] \tag{18}$$

where

$$\omega_1 = (RC)^{-1} \tag{19}$$

and

$$R = [g_{ds1} + g_{m2} + g_{mbs2} + g_{ds2}]^{-1} \tag{20}$$

$$C = C_{gd1} + C_{bd1} + C_1 + C_2 + C_L \tag{21}$$

where C_1 is either C_1' or C_{gb2}, depending upon the type of inverter. The presence of various terms in Eqs. (20) and (21) depends upon the type of inverter. The above analysis shows that ω_1, which is the −3-dB frequency of the inverter, is proportional to the square root of the drain current for the NMOS inverters and proportional to the drain current for the CMOS inverters. As the drain current increases, the frequency response will also

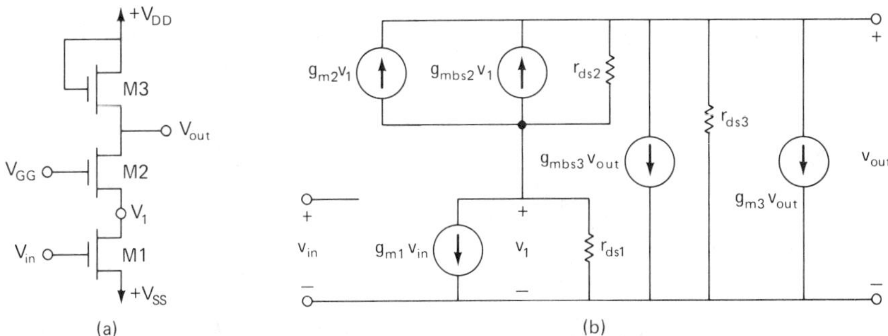

Fig. 8.5–6. (a) Cascode configuration using n-channel enhancement devices. All bulk potentials are assumed to be at ground or a negative potential V_{BB}. (b) AC small signal model for (a).

increase. A more detailed analysis of the frequency response of the inverter can be found in the literature.[45]

Cascode

The cascode stage is a very useful component in analog circuit design using MOS technology. The primary advantage of the cascode stage is that it reduces the capacitance loading on the preceding stage. The cascode stage also has higher output impedance than an inverting stage. Figure 8.5–6(a) shows a cascode stage using all-enhancement n-channel devices. The bulks are assumed to be taken to a potential designated as V_{BB}. Normally V_{BB} will be ground or negative. M1 is operating in a common source configuration, while M2 is in a common gate configuration. M2 presents a small load resistance to the drain of M1, which keeps the AC voltage swing at the drain of M1 small. This feature prevents the capacitance, C_{GD1}, from becoming a load capacitance on the previous stage through the Miller effect. M3 serves as a load resistance for M2. All the voltage gain of the cascode configuration comes from the common gate device, M2. A finite value of DC voltage, V_{GG}, is necessary to bias the gate of M2 so that M1 will operate in saturation. This voltage can be derived using the methods discussed in the previous section.

Figure 8.5–6(b) shows the small-signal model for the cascode stage of Fig. 8.5–6(a). M2 and M3 have a nonzero value of v_{bs}, so that g_{mbs2} and g_{mbs3} are included in the model. The voltage gain of the cascode configuration can be found as

[45] Y. P. Tsividis, "Design Considerations in Single-Channel MOS Analog Integrated Circuits—A Tutorial," *IEEE J. of Solid-State Circuits,* Vol. SC-13, No. 3, June 1978, pp. 383–391.

$$\frac{v_{out}}{v_{in}} = \frac{-g_{m1}\,[(1+\eta_2)\,g_{m2} + g_{ds2}]}{[(1+\eta_3)\,g_{m3} + g_{ds3}]\,[(1+\eta_2)\,g_{m2} + g_{ds1} + g_{ds2}] + g_{ds1}\,g_{ds3}} \tag{22}$$

Assuming that the g_m's are greater than the g_{ds}'s allows Eq. (22) to be simplified to

$$v_{out}/v_{in} \cong -g_{m1}/[(1+\eta_3)g_{m3}] = -\,g_{m1}/g_{mbs3} = \frac{-1}{1+\eta_3}\sqrt{\frac{W_1/L_1}{W_3/L_3}} \tag{23}$$

Equation (23) illustrates the manner in which the geometry can determine the gain of the cascode stage. Obviously, for high gain we would like W_1/L_1 to be much larger than W_3/L_3. However, we must also consider the gain from v_{in} to v_1 of Fig. 8.5–6(b). If this gain is large, then the Miller effect becomes significant, and the previous stage will have a significant capacitive load. The gain from v_{in} to v_1 is exactly like the inverter of Fig. 8.5–1(a). If we assume that $v_{out} = 0$, then from Eq. (8) we get

$$\frac{v_1}{v_{in}} \cong \frac{-1}{1+\eta_2}\frac{g_{m1}}{g_{m2}} = \frac{-1}{1+\eta_2}\sqrt{\frac{W_1/L_1}{W_2/L_2}} \tag{24}$$

Now we see that to reduce the Miller effect on C_{GD1} that W_1/L_1 should be less than W_2/L_2. This cannot be carried to the extreme because if M2 becomes large then it will contribute significant capacitive loading in the cascode stage. Typically W_1/L_1 is approximately equal to W_2/L_2, and W_3/L_3 is much smailer than W_1/L_1. This gives a gain close to unity for Eq. (24) and eliminates the Miller effect. Thus the cascode circuit finally has about the same gain as an inverter. Under normal conditions, the frequency response of the cascode circuit will be due to the resistance $1/g_{mbs3}$ and the capacitance associated with the output node. This capacitance would include C_{GS3} plus C_L.

One way the gain of the cascode can be increased is illustrated in Fig. 8.5–7. The bulks of the *n*-channel devices are assumed to be either at ground or at a negative potential V_{BB}. M2 is a depletion device that permits us to directly ground the gate and still have M1 operating in the saturation region. M3 was selected as a depletion device to increase the gain of the cascode. Unfortunately, the fact that the source is the output node detracts from this objective due to g_{mbs3}. Finally, the current source provided by the depletion device, M4, allows the DC current in M1 to be different from the DC current in M2 or M3. This also permits an increase in gain, as seen in the following small-signal analysis. Because the gain of the depletion load cascode

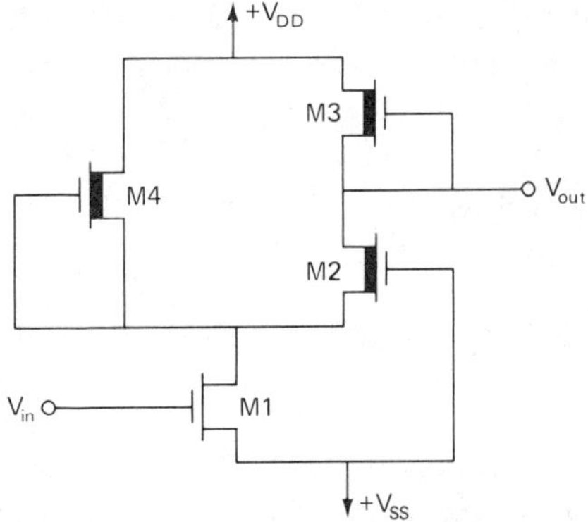

Fig. 8.5–7. Cascode using depletion devices and a current source to increase the voltage gain.

is the same as the depletion load inverter, we may use Eq. (14) to solve for the gain of Fig. 8.5–7 as

$$\frac{v_{out}}{v_{in}} \cong -\frac{g_{m1}}{g_{mbs3}} = -\frac{1}{\eta_3}\frac{g_{m1}}{g_{m3}} = -\sqrt{\frac{\mu_1(W_1/L_1)\,I_{D1}}{\mu_3(W_3/L_3)\,I_{D3}}}\,\frac{2}{\gamma}\sqrt{\phi + V_{out} - V_{BB}} \qquad (25)$$

where we have assumed that V_{BS} of M4 is approximately zero and have used the relationships expressed in Eqs. (35) and (37) of Section 8.2. M4 now allows the freedom to make I_{D1} greater than I_{D3}, so that the gain of Fig. 8.5–7 can be increased. Unfortunately, I_{D3} must be large enough to charge any capacitance attached to the output node, so that to get much increase in gain would result in M1 becoming a large device. Again, a compromise is in order, and although some improvement in gain is possible, large increases will cause a sacrifice in performance elsewhere.

The cascode also can be implemented using CMOS technology. The load device would be replaced by a p-channel device giving an order of magnitude or more gain over the cascode in Fig. 8.5–7.

Source Followers

A source follower is useful for providing a low-output resistance. Unfortunately, because the source is the output node, the MOS device becomes dependent upon the body effect. The body effect causes the threshold voltage, V_T, to increase as the output voltage is increased, creating a situation where the maximum output is substantially lower than V_{DD}. Figure 8.5–8(a) shows a general configuration for the MOS source follower. Either or both devices may be replaced by depletion NMOS transistors. The primary difference will be the available output signal swing. Although the gate of M2 can be taken to the output to form an active resistor, we will consider only the source follower that uses a current source load.

Figure 8.5–8(b) shows the small-signal model for the source follower. The effect of the bulk is implemented by the g_{mbs1} transconductance. The small-signal voltage gain can be found as

$$v_{out}/v_{in} = g_{m1}/[g_{ds1} + g_{ds2} + g_{m1}(1+\eta_1)] \cong 1/(1+\eta_1) \qquad (26)$$

If we assume that $V_{DD} = 5$ V, $V_{SS} = -5$ V, $V_{out} = 0$ V, and $I_D = 100$ μA, then, using the parameters of Table 8.2–1, we find that $\eta_1 = 0.10545$, and the small-signal voltage gain is 0.872. If the bulk effect were not present, $\eta_1 = 0$, the small-signal voltage gain would become 0.96. Because of the bulk effect, the small-signal voltage gain of MOS source followers cannot be assumed to be unity and must be taken into account in most applications.

The frequency response of the source follower is determined by the two capacitances designated as C_1 and C_2, shown in the small-signal model of Fig. 8.5–8(b). C_1 consists of the capacitance connected between the input and the output of the source follower, which is primarily C_{gs1}. C_2 consists

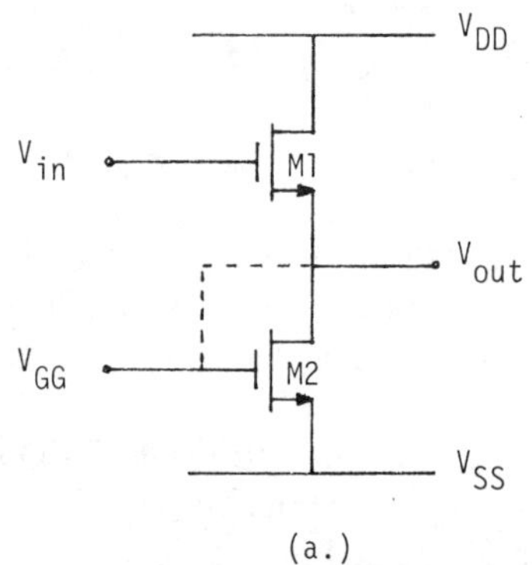

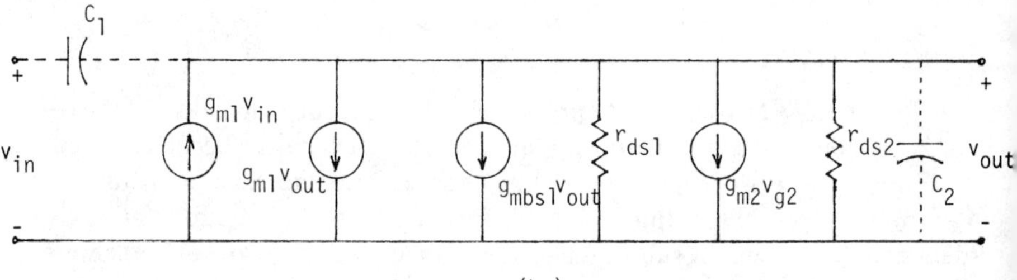

Fig. 8.5–8. (a) Source follower. (b) Small signal model of (a).

of capacitances connected from the output of the source follower to ground. This includes C_{gd2}, C_{bd2}, C_{bs1}, and C_L of the next stage. The small-signal frequency response can be found as

$$v_{out}/v_{in} = (g_{m1} + sC_1)/[g_{ds1} + g_{ds2} + g_{m1}(1 + \eta_1) + s(C_1 + C_2)]^{-1} \quad (27)$$

The presence of a zero leads to the possibility that in most cases the pole and zero will provide some degree of cancellation, leading to a broadband response.

Differential Amplifiers

The differential amplifier is one of the more versatile circuits and is highly compatible with integrated circuit technology. The objective of the differential amplifier is to amplify only the difference between two signals regardless of the common-mode value. Thus, a differential amplifier can be characterized by its common-mode rejection ratio, which is the ratio of the differential gain to the common-mode gain. In addition, the input common-mode range specifies over what range of common-mode values the differential amplifier continues to sense and amplify the difference signal. Another characteristic that affects the performance of the differential amplifier is offset. In MOS differential amplifiers, the most serious offset is the voltage offset. If the input terminals of the differential amplifier are connected together, the offset voltage is the voltage that appears at the output of the differential amplifier. If this voltage is divided by the differential voltage gain of the differential amplifier, then the offset is expressed as the *input-voltage offset.* Typically, the input-voltage offset of an MOS differential amplifier is 10 to 30 mV.

We shall consider both the large- and small-signal characteristics of MOS differential amplifiers. Let us consider first the all-enhancement MOS differential amplifier shown in Fig. 8.5–9(a). M1 and M2 form the differential pair, M3 and M4 are the loads, and M5 is the current sink for the differential amplifier. The large-signal characteristics can be developed by assuming that the differential pair, M1 and M2, is always in saturation. This condition is reasonable in most cases and illustrates the behavior even when this assumption is not valid. The pertinent relationship describing the large-signal behavior is given as

$$V_{ID} = V_{GS1} - V_{GS2} = (I_{D1}/\beta)^{0.5} - (I_{D2}/\beta)^{0.5} \quad (28)$$

and

$$I_{SS} = I_{D1} + I_{D2} \quad (29)$$

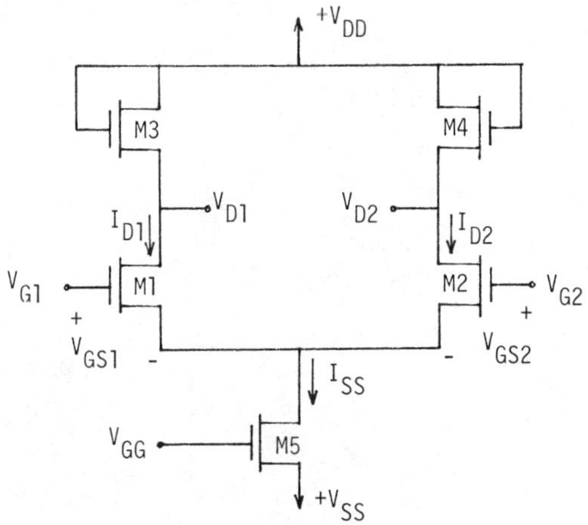

Fig. 8.5–9. Differential amplifier using n-channel enhancement devices. The bulk is assumed to be taken to a negative potential V_{BB}, $V_{BB} \leq V_{SS}$.

where it has been assumed that M1 and M2 are matched. Substituting Eq. (29) into Eq. (28) and forming a quadratic allows the solution for I_{D1} and I_{D2} as

$$I_{D1} = (I_{SS}/2) + (I_{SS}/2)[(2\beta\ V_{ID}^2/I_{SS}) - (\beta^2\ V_{ID}^4/I_{SS})]^{0.5} \tag{30}$$

and

$$I_{D2} = (I_{SS}/2) - (I_{SS}/2)\ [(2\beta\ V_{ID}^2/I_{SS}) - (\beta^2\ V_{ID}^4/I_{SS})]^{0.5} \tag{31}$$

where these relationships are valid only for $V_{ID} < (2\sqrt{2}\ I_{SS}/\beta)$. Figure 8.5–10 shows a plot of the normalized drain current versus the normalized differential-input voltage. The dotted portions of the curves are meaningless and are ignored. The above analysis has illustrated the relationship between the drain current and the differential input voltage, V_{ID}.

It is of interest to determine the slope of this curve that leads to a definition of g_m of the differential amplifier. Differentiating Eq. (30) with respect to V_{ID} and setting $V_{ID} = 0$ gives the differential transconductance of the differential amplifier as

$$g_{md} = \partial I_{D1}/\partial V_{ID} = [\beta I_{SS}/2]^{0.5} \tag{32}$$

We note, when comparing this result with Eq. (35) of Section 8.2, that a difference of 2 exists (assuming $I_{SS}/2 = I_D$). The reason for this difference is that only half of V_{ID} is being applied to M1.

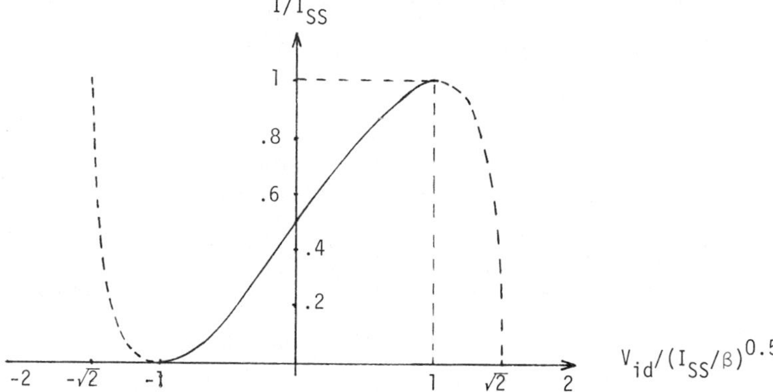

Fig. 8.5–10. Large signal transfer characteristics of a MOS differential amplifier.

The small-signal characteristics can be determined by decomposing the input voltage into a differential voltage, v_d, and a common-mode voltage, v_c. For example, if v_1 and v_2 are two different voltages, then their difference value is defined as

$$v_{id} = v_1 - v_2 \tag{33}$$

and the common-mode value is defined as

$$v_{ic} = (v_1 + v_2)/2 \tag{34}$$

The small-signal analysis of the differential amplifier can be divided into an analysis of the differential input and an analysis of the common-mode input.

Figure 8.5–11 shows two equivalent models of the differential amplifier of Fig. 8.5–9 suitable for the differential-mode analysis. The gain of v_{d1}/v_{id} is the same as that found in Eq. (8) for the inverter using enhancement devices.[46] Therefore

$$\frac{v_{d1}}{v_{id}} = \frac{-g_{m1}}{2(1+\eta_3)g_{m3}} = -\frac{1}{2}\sqrt{\frac{W_1/L_1}{W_3/L_3}}\,\frac{1}{1+\dfrac{\gamma}{2\sqrt{\phi + V_{D1} - V_{BB}}}} \tag{35}$$

[46] Note that because the input is $v_{id}/2$, the gain of Fig. 8.5–11(a) is half the enhancement load inverter of Fig. 8.5–1(a).

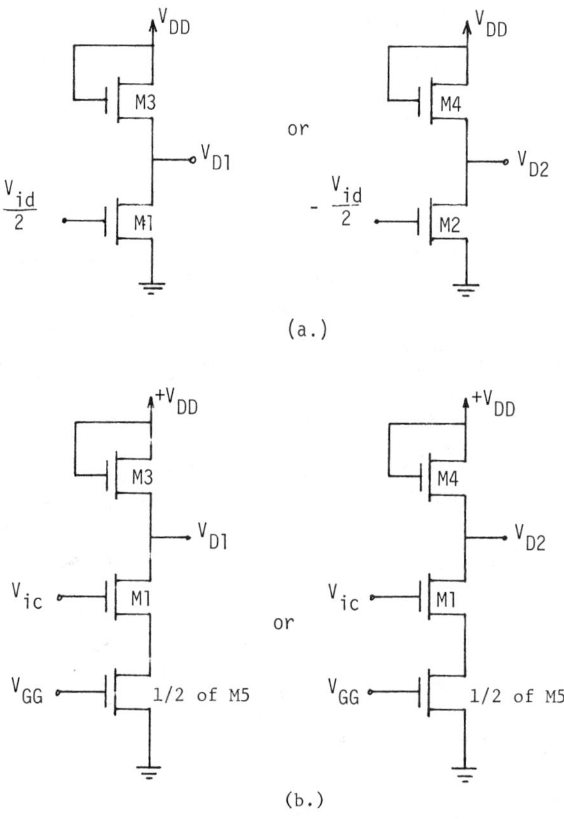

Fig. 8.5–11. (a) Differential mode equivalent of Fig. 8.5–9. (b) Common mode equivalent of Fig. 8.5–9.

The differential gain is found by using $v_o = v_{d1} - v_{d2}$ to get

$$\frac{v_o}{v_{id}} = -\frac{g_{m1}}{(1+\eta_3)g_{m3}} = -\sqrt{\frac{W_1/L_1}{W_3/L_3}}\frac{1}{1+\dfrac{\gamma}{2\sqrt{\phi+V_{D1}-V_{BB}}}} \tag{36}$$

Next, it is of interest to calculate the common-mode gain. In the common mode, the differential amplifier of Fig. 8.5–9 reduces to that shown in Fig. 8.5–11(b). The term "½ of M5" in Fig. 8.5–11(b) means currents are halved and resistances are doubled. The small-signal model for Fig. 8.5–11(b) is

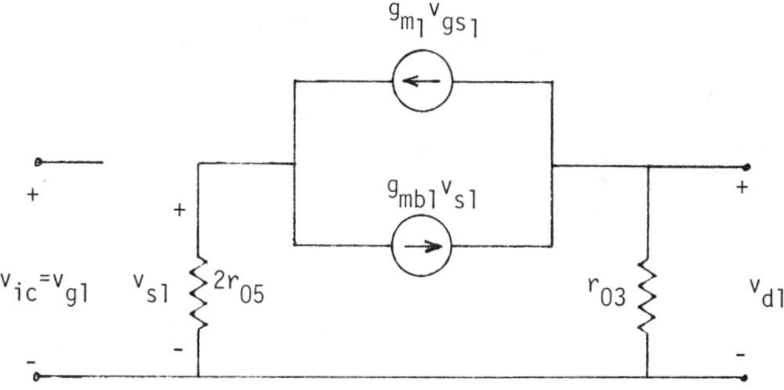

Fig. 8.5–12. Small signal model for Fig. 8.5–11(b).

shown in Fig. 8.5–12. It can be shown that the voltage gain v_{d1}/v_{ic} can be expressed as

$$\begin{aligned} v_{d1}/v_{ic} &= (-g_{m1}\, r_{03})/[1 + 2\, g_{m1}\, r_{05}(1+\eta_1)] \\ &\cong -1/[2 r_{05}\, g_{m3}(1+\eta_1)(1+\eta_3)] \end{aligned} \tag{37}$$

where

$$r_{03} \cong 1/[(1+\eta_3) g_{m3}] \tag{38}$$

It is desirable that this gain be as small as possible. As a figure of merit we define the common-mode rejection ratio (CMRR) as

$$CMRR = |(v_{d1}/v_{id})/(v_{d1}/v_{ic})| \tag{39}$$

From Eqs. (34) and (35), we find that the CMRR of Fig. 8.5–9 is given as

$$CMRR = 2\, g_{m1}\, r_{05}(1+\eta_1) \tag{40}$$

The importance of having a large value of r_{05} for large CMRR is evident from Eq. (40). Consequently, M5 should be a long and narrow device or the current source output resistance could be increased by any of the methods given in Section 8.4.

A depletion load differential amplifier can be achieved by replacing M3 and M4 of Fig. 8.5–9 by depletion devices. The resulting differential amplifier should perform identical to that in Fig. 8.5–9, but have a larger differential-mode gain. When M3 and M4 are depletion devices, it becomes necessary to derive V_{GG} from V_{D1} or V_{D2}, so that the operating point does not become

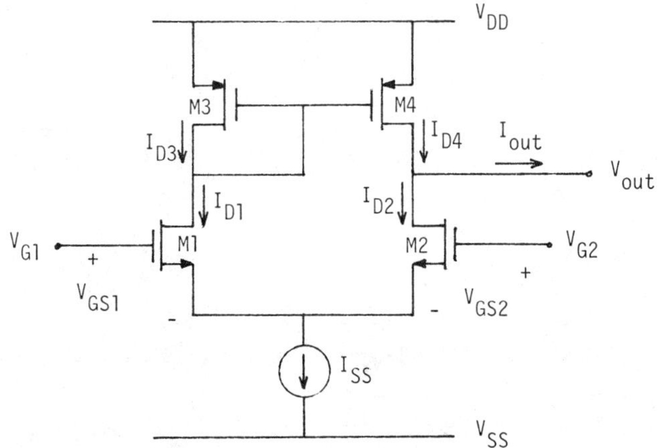

Fig. 8.5–13. CMOS differential amplifier using n-channel input devices.

indeterminate (i.e., I_{D5} may not equal $I_{D3} + I_{D4}$). Other techniques, such as increasing the CMRR through common-mode feedback, have been developed and resulted in reasonably good NMOS differential amplifiers.[47]

Figure 8.5–13 shows a CMOS differential amplifier that uses *n*-channel MOS devices, M1 and M2, as the differential pair. I_{SS} is a current source that can be any of the types previously considered. The loads for M1 and M2 are obtained from a simple *p*-channel current mirror. If M3 and M4 are matched, then the current of M1 will determine the current in M3. This current will be mirrored in M4. If $V_{GS1} = V_{GS2}$, then the currents in M1 and M2 are equal. Thus the current that M4 sources to M2 should be equal to the current that M2 sinks, causing I_{out} to be zero. If $V_{GS1} > V_{GS2}$, then I_{D1} increases with respect to I_{D2}, since $I_{SS} = I_{D1} + I_{D2}$. This increase in I_{D1} implies an increase in I_{D3} and I_{D4}. However, I_{D2} decreased when V_{GS1} became greater than V_{GS2}. Therefore, the only way that circuit equilibrium can be established is for I_{out} to become positive. It can be seen, if $V_{GS1} < V_{GS2}$, that I_{out} becomes negative. This is a simple way in which the differential-output signal of the differential amplifier can be converted back to a single-ended signal, i.e., one referenced to AC ground.

Figure 8.5–14 shows a CMOS differential amplifier that uses *p*-channel MOS devices, M1 and M2, as the differential pair. The circuit operation is identical to that of Fig. 8.5–13, but there is a difference in the bulk effect. If we assume that the CMOS technology uses an *n*-substrate, then the bulks of the *p*-channel devices can be taken only to the most positive potential,

[47] Y. P. Tsividis, *op. cit.*

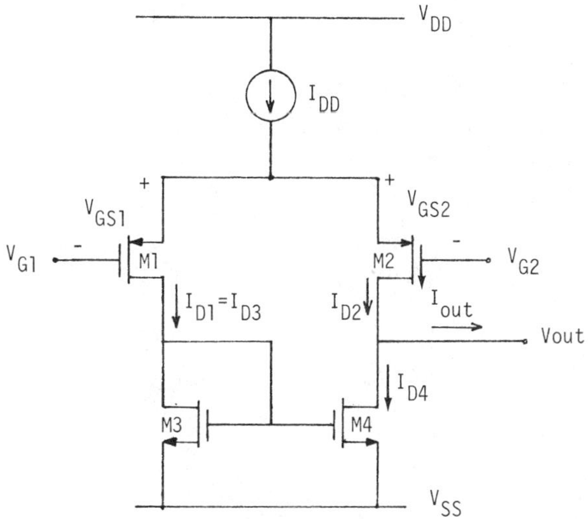

Fig. 8.5–14. CMOS differential amplifier using p-channel input devices.

V_{SS}. However, the n-channel bulks are not necessarily constrained to be connected to the most negative potential, V_{DD}. This permits the n-channel differential amplifier of Fig. 8.5–13 to have the sources of M1 and M2 connected to their bulks. Thus, M1 and M2 would be made in a p-tub, which is allowed to float. Therefore, Fig. 8.5–13 is insensitive to the bulk effects, whereas Fig. 8.5–14 is not.

The common-mode input-voltage range for the p-channel differential transconductance amplifier will be examined next. Consider the differential transconductance amplifier shown in Fig. 8.5–15. Here the current source has been implemented by M5, which is a part of a current mirror. The lowest possible input voltage at the gate of M1 (or M2) is found to be

$$V_{GS1}(\text{Low}) = V_{SS} + V_{T3} + V_{DS1} - V_{T1} \tag{41}$$

If the input devices, M1 and M2, are large, then V_{DS1} and V_{GS2} can be very small as V_{GS1} is taken negative. Therefore, Eq. (41) can be reduced to

$$V_{GS1}(\text{low}) \cong V_{SS} + V_{T3} - V_{T1} \tag{42}$$

In a similar manner, the highest possible input voltage can be written as

$$V_{GS1}(\text{High}) = V_{DD} - V_{DS5} - V_{T1} \cong V_{DD} - V_{T1} \tag{43}$$

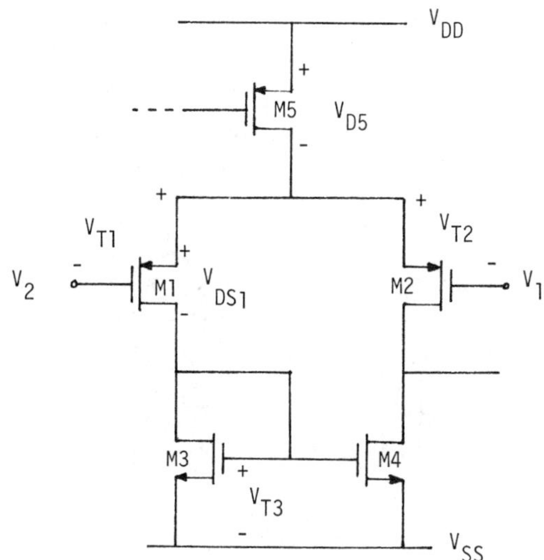

Fig. 8.5–15. Circuit for common-mode voltage input range considerations.

If we assume that V_{DD} varies from 4 to 6 volts and that $V_{SS} = 0$, then using the values of Table 8.2–1 under worst-case conditions shows that V_{GS1}(Low) is 0.4 volts and that V_{GS1}(High) is 2.8 volts. These results will be aggravated by the dependence of the threshold voltages upon the bulk-source voltage. The common-mode input range for the n-channel input differential transconductance amplifier can be found in a similar manner, but will not be influenced by the bulk-source voltage.

The large-signal analysis of the CMOS differential amplifier is identical to the MOS differential amplifier of Fig. 8.5–9. Because of the current mirrors, the output current, I_{out}, is the difference between I_{D1} and I_{D2}. Although the differential transconductance of the CMOS differential amplifier is approximately equal to the MOS differential amplifier, the voltage is much larger, because there is no bulk effect and the equivalent load resistance is much higher.

The small-signal analysis of the differential amplifier of Fig. 8.5–13 can be accomplished with the assistance of the model shown in Fig. 8.5–16. This model is appropriate for differential analysis where, if both sides of the amplifier are assumed to be matched, the point where the two sources of M1 and M2 are connected is an AC ground. If we assume that the differential stage is unloaded, then the unloaded differential transconductance gain can be expressed as

$$i'_{out} = [g_{m1}\, g_{m3}\, r_{ds1}/(1 + g_{m3}\, r_{ds1})]\, v_{gs1} - g_{m2}\, v_{gs2} \tag{44}$$

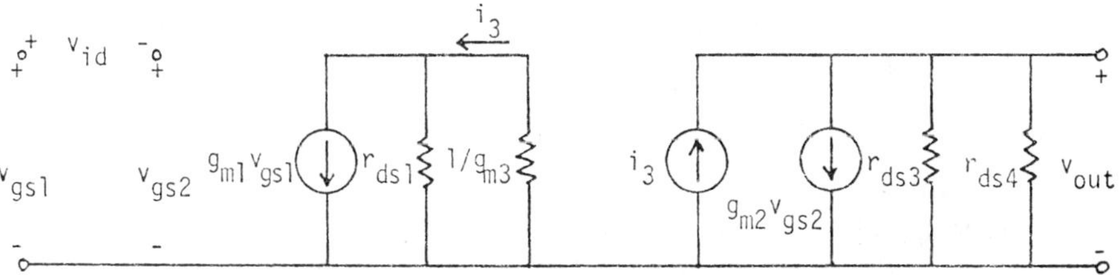

Fig. 8.5–16. Small signal model for CMOS differential amplifier.

or

$$i'_{\text{out}} \cong g_{m1}\, v_{gs1} - g_{m2}\, v_{gs2} = g_m\, v_{id} \tag{45}$$

where $g_{m1} = g_{m2} = g_m$. The unloaded differential voltage gain can be expressed as

$$v_{\text{out}}/v_{id} = [g_m/(g_{ds2} + g_{ds4})] \tag{46}$$

Using Eqs. (35) and (37) of Section 8.2 in Eq. (46) gives the following expression

$$v_{\text{out}}/v_{id} = [2/(g_2 + g_4)](\beta_1/I_{SS})^{0.5} \tag{47}$$

Again we note that dependence of the small-signal gain on the inverse of $I_{SS}{}^{0.5}$ similar to that of the inverter. This relationship is in fact valid until I_{SS} approaches subthreshold values. Assuming that $W_1/L_1 = 1$ and that $I_{SS} = 10\ \mu\text{A}$, the small-signal gain of the differential amplifier is found to be 157. The small-signal gain of the *p*-channel differential amplifier under the same conditions is 100. This difference is due to the difference in mobilities.

The common-mode gain of the CMOS differential amplifiers shown in Figs. 8.5–13 and 8.5–14 is ideally zero. The reason is that the current mirror loads reject any common-mode signal. The fact that a common-mode response exists is due to the mismatches in the differential amplifier.

The slew rate performance of an MOS differential amplifier depends upon the value of I_{SS} and the capacitance from the output node to AC ground. Slew rate is defined as *the maximum voltage rate, either positive or negative.* Because the slew rate in the MOS differential amplifier is determined by the amount of current that can be sourced or sunk into the output capacitor, we find that the slew rates of the MOS differential amplifiers are given by

$$\text{Slew rate} = I_{SS}/C_L \tag{48}$$

For example, if $I_{SS} = 10\ \mu\text{A}$ and $C_L = 5$ pF, the slew rate is found to be 2 volts/microsecond. The value of I_{SS} must be increased to increase the slew rate capability of the differential amplifier.

The frequency response of the MOS differential amplifier is due to the various parasitic capacitors at high resistance nodes. Figure 8.5–17 shows the location of various parasitic capacitors in the differential amplifier of Fig. 8.5–13. C_T is called the *tail capacitor* and is not important to the differential-mode response, because the node to which it is connected acts as an AC ground in the differential mode. C_M is the mirror capacitance. This capacitance consists primarily of $C_{gs3} + C_{gs4}$ and can become significant when M3 and M4 are made large to reduce the $1/f$ noise. C_{out} is the capacitance connected from the output node to ground. The dominant pole will be due to the output capacitor because of the very high output resistance of the differential amplifier. These capacitors will be examined in more detail in the next section, where compensation is considered.

The noise performance of the CMOS differential amplifier can be due to both thermal and $1/f$ noise. Depending upon the frequency range of interest, one can neglect one of the sources in favor of the other. At low frequencies, $1/f$ noise is important, whereas at high frequencies/low currents, thermal noise is important. Figure 8.5–18 shows the n-channel differential amplifier with equivalent noise voltage sources shown at the input of each device. These equivalent noise voltage sources have been obtained from the equivalent

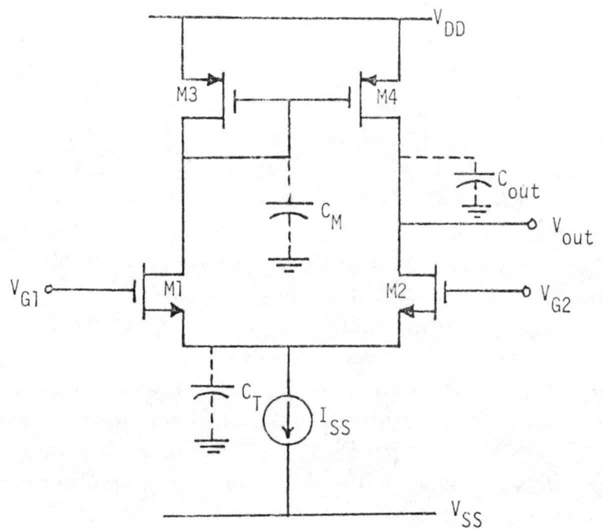

Fig. 8.5–17. Parasitic capacitance in the CMOS differential amplifier.

drain current noise sources of Fig. 8.2–12 via g_m^2 and ignoring the thermal noise sources or r_D and r_S. For example, assuming that $1/f$ noise is the more significant source of noise, Eq. (43) of Section 8.2 can be expressed as

$$\bar{i}_{nD}^2 \cong [(KF)/(fC_{ox}\,L^2)]\,I_D^{AF}\ (\text{Amps}^2/\text{Hz}) \tag{49}$$

Thus the equivalent noise voltage source at the gate can be expressed as

$$\bar{v}_n^2 = \bar{i}_{nD}^2/g_m^2 = [(KF)/(f\mu_0\,C_{ox}\,ZL)]\,I_D^{(AF-1)}\ (\text{Volts}^2/\text{Hz}) \tag{50}$$

Because we desire the total equivalent noise voltage source at the input of the CMOS differential amplifier of Fig. 8.5–18(a), let us solve for the total output noise current, $\bar{i}_{od}^2$ at the output of the circuit. Furthermore, let us assume that the output is shorted to ground to simplify calculations. The total output noise current is found by summing each of the noise current contributions to get

$$\bar{i}_{od}^2 = g_{m1}^2\,\bar{v}_{n1}^2 + g_{m2}^2\,\bar{v}_2^2 + g_{m3}^2\,\bar{v}_{n3}^2 + g_{m4}^2\,\bar{v}_{n4}^2 \tag{51}$$

Because the equivalent output noise current is expressed in terms of the equivalent input noise voltage, we may use

$$\bar{i}_{od}^2 = g_{m1}\,\bar{v}_{neg}^2\ (\text{Amps}^2/\text{Hz}) \tag{52}$$

to get

$$\bar{v}_{neg}^2 = \bar{v}_{n1}^2 + \bar{v}_{n2}^2 + (g_{m3}/g_{m1})[\bar{v}_{n3} + \bar{v}_{n4}^2]\ (\text{Volts}^2/\text{H}_3) \tag{53}$$

We have assumed that $g_{m1} = g_{m2}$ and $g_{m3} = g_{m4}$ in the above.

It is seen that the input MOS devices contribute directly to the noise, while the transistors of the current mirror load contribute in an amount related to the ratio of the transconductances. The transconductances of the devices in the mirror can be made much less than the transconductances of the input devices by simply increasing the channel length. This approach works well until the channel resistance of M3 or M4 becomes large enough that thermal noise is important or the frequency response of the differential amplifier is deteriorated. To reduce the noise further, it is necessary to reduce the contribution due to the input transistors, M1 and M2. We see from Eq. (53) that increasing the size of the input transistors will decrease the noise. Also, larger devices imply better matching and lower offsets.

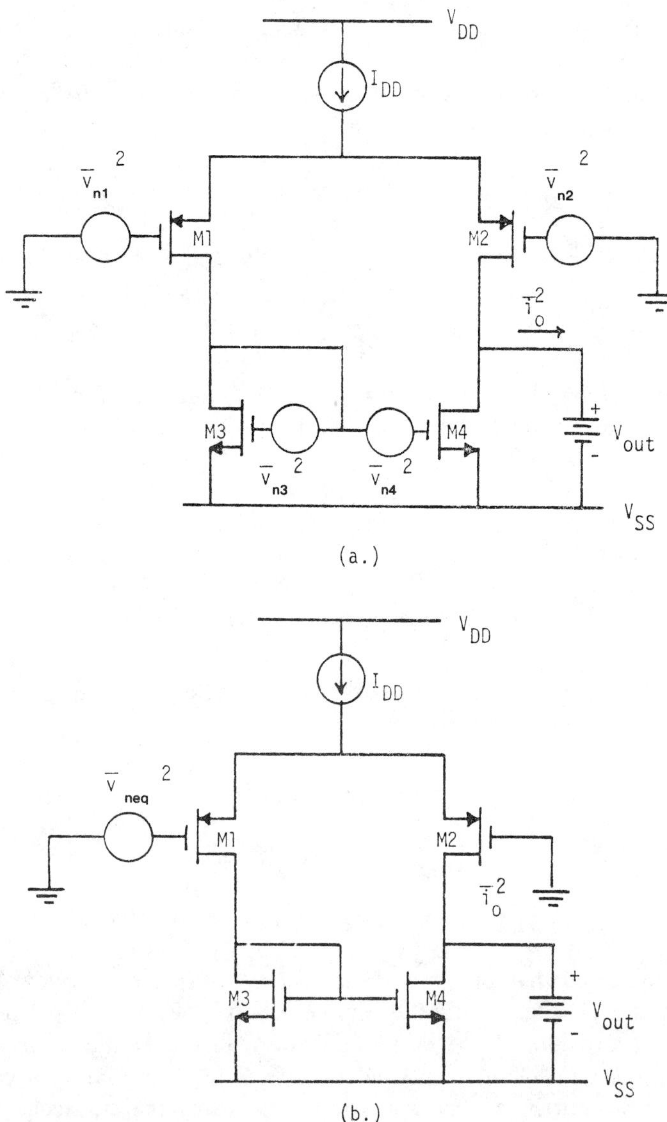

Fig. 8.5–18. Noise analysis of a CMOS differential amplifier. (a) Noise model of each device. (b) Equivalent noise model of (a).

Another important consideration is the empirical observation that *p*-channel devices seem to have significantly lower $1/f$ noise than *n*-channel devices, all else being equal. The reasons for this are not well understood, but generally the best noise performance can be achieved by using the *p*-channel differential amplifier of Fig. 8.5–14.

8.6 MOS OP AMP DESIGN

The operational amplifier, called the *op amp,* has become one of the most important components in analog circuit design. The op amp is a versatile and flexible building block. The objective of this section is to present the preliminary ideas and concepts necessary to design MOS op amps. This objective will be supported by the circuits examined in the previous two sections. In the next section we shall examine several types of MOS op amps in more detail. Further information can be found in several excellent references on the subject of integrated circuit op amps.[48,49,50,51]

Figure 8.6–1 shows a general block diagram of an MOS op amp. This figure is very general and yet can be used to represent and characterize some of the important aspects of MOS op amps. The differential stage is the input stage of the op amp. The purpose of this stage is to amplify the differential-input signal and to provide an output referenced to AC ground (called single-ended) to the second, or high-gain, stage. The differential stage may consist of several stages in the single-channel MOS op amps because the conversion from differential to single-ended may require an additional stage. Also the differential stage must provide the output signal at the desired DC level for the high-gain stage. This may require a DC level-shifting circuit.

The high-gain stage serves two functions. The first is to provide sufficient low-frequency gain for the op amp, and the second is to permit the op amp to be compensated for closed-loop stability. We shall discuss the function of this stage in more detail in this section. The output stage must provide the desired signal swing to the external load. Typically the output stage has unity gain and provides a high input and low output impedance. The output stage usually consumes the majority of the quiescent power of the op amp. It will be seen that in switched capacitor applications it is not necessary to have low output impedance. Therefore if an op amp does not have to drive a resistive load, the output stage can be very simple and in some cases may not even exist.

The remaining blocks of the op amp block diagram of Fig. 8.6–1 are the bias and compensation circuits. Typically these circuits are an inherent part of the previous circuits. The compensation circuit has the objective of stabiliz-

---

[48] J. E. Solomon, "The Monolithic Op Amp, A Tutorial Study," *IEEE J. of Solid State Circuits,* Vol. SC-9, No. 6, December 1974, pp. 314–332.

[49] R. G. Meyer, *IEEE Press-Integrated Circuit Operational Amplifiers,* John Wiley & Sons, New York, 1978.

[50] Y. P. Tsividis, "Design Considerations in Single-Channel MOS Analog Integrated Circuits—A Tutorial," *IEEE J. of Solid-State Circuits,* Vol. SC-B, No. 3, June 1978, pp. 383–391.

[51] P. R. Gray, *IEEE Press-Analog MOS Integrated Circuits,* John Wiley & Sons, New York, 1980, pp. 28–49.

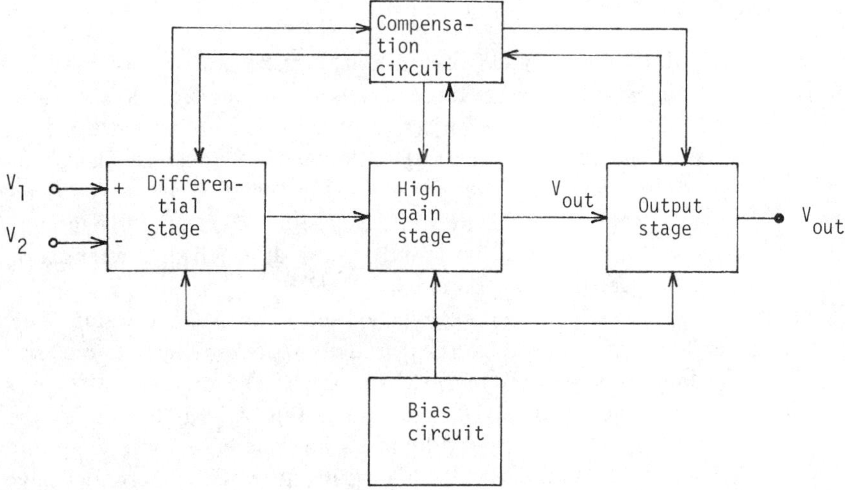

Fig. 8.6–1. General block diagram of a MOS op amp.

ing the op amps using small capacitors to avoid large area requirements. The bias network must be independent of process variations if possible. This means that the performance of the DC conditions should not depend on the value of V_T. Techniques of achieving this goal will be presented in the next section. The bias network should also make the DC conditions independent of the power supply and temperature.

Op Amp Parameters

An op amp ideally has infinite input resistance, infinite differential gain, and zero output resistance. An actual op amp only approaches these three values. The most important of these three values is the large gain. The symbol for an op amp is shown in Fig. 8.6–2 (see also Fig. 2.4–2). In the nonideal case, the output voltage, V_0 can be expressed as

$$V_0 = A_v(V_1 - V_2) \tag{1}$$

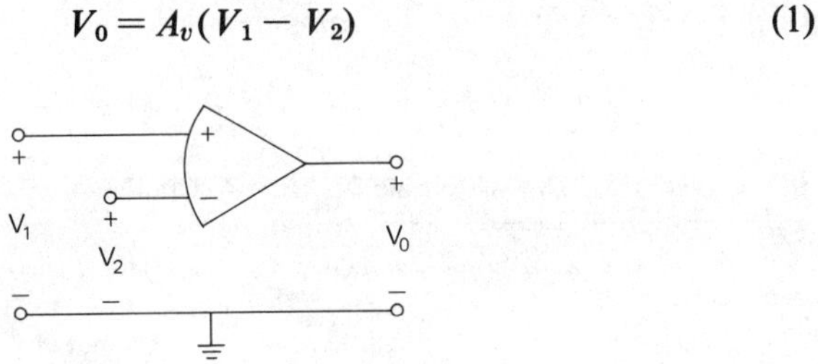

Fig. 8.6–2. Symbol for an operational amplifier realized by a voltage-controlled voltage source.

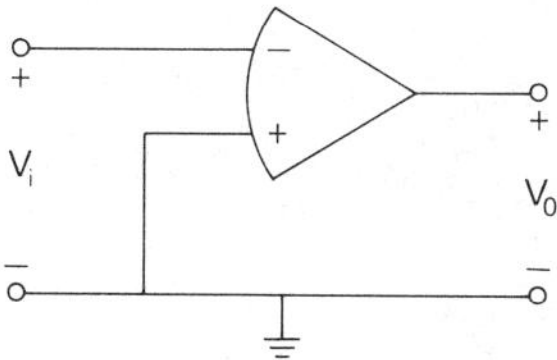

Fig. 8.6–3. Typical configuration of the op amp for analog sampled data circuit applications.

where upper-case variables with lower-case subscripts indicate a transform variable, such as $V_0(s)$ or $V_0(j\omega)$. V_1 and V_2 are the input voltages applied to the noninverting and inverting terminals, respectively. The difference between these two voltages is amplified by A_v and appears at the output of the op amp. In analog sampled data circuits, the op amp is almost always used with the noninverting terminal connected to AC ground. Therefore, the typical configuration of the op amp in analog sampled data circuits is given in Fig. 8.6–3. Here, the input voltage is defined as V_i and causes an output voltage V_0, given as

$$A_0 = -A_v V_i \tag{2}$$

In most of the cases we can assume that A_v approaches infinity. If this assumption is valid, then external feedback connected from V_0 back to V_i (sometimes through one or more op amp circuits) will cause the input of the op amp to simulate a virtual ground. Therefore, V_i can be assumed to be zero. The current flowing into the op amp is also zero because of the infinite input impedance and because of the zero voltage. If the virtual ground concept illustrated in Section 2.4 is valid, then op amp circuits are very easy to analyze.

Unfortunately, the operational amplifier only approaches the ideal, infinite-gain voltage amplifier in practice. Some of its other nonideal characteristics are illustrated in Fig. 8.6–4. This model will be used to define the various characteristics. The finite *differential input impedance* is modeled by R_{id} and C_{id}. The *output resistance* is modeled by R_{out}. The *common-mode input resistances* are resistances, R_{icm}, connected from each of the inputs to ground. V_{OS} is the *input offset voltage* necessary to make the output voltage zero if $V_1 = V_2$. I_{OS} (not shown) is the *input offset current,* which is necessary to make the output voltage zero if the op amp is driven from two identical current sources. Therefore, I_{OS} is defined as the magnitude of the difference between the two *input bias currents* I_{B1} and I_{B2}. The *common-mode rejection ratio* (CMRR) is modeled by the voltage-controlled voltage-source, indicated as V_1/CMRR. This is an approximate model for the effects of a common-

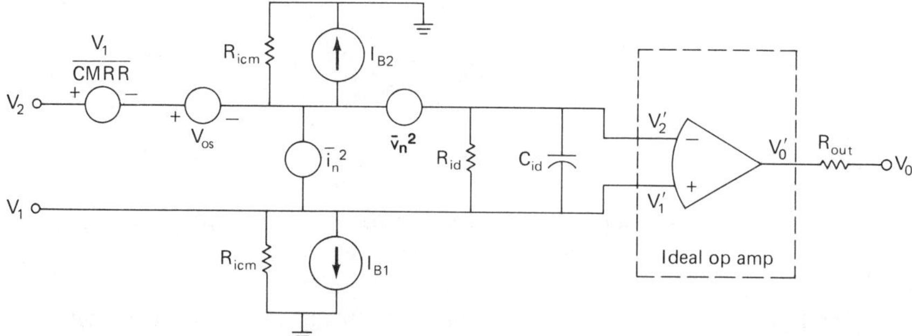

Fig. 8.6–4. A model for a nonideal op amp showing some of the nonideal characteristics.

mode input signal on the op amp. The two sources designated as $\bar{v}_n^2$ and $\bar{i}_n^2$ are used to model the op amp noise and are called *voltage and current noise spectral densities.* Their units are mean square volts per Hertz and mean square amperes per Hertz, respectively. These noise sources have no polarity and are always assumed to add.

Not all the nonideal characteristics of the op amp are illustrated in Fig. 8.6–4. Other pertinent characteristics of the op amp will now be defined. The output voltage of Fig. 8.6–2 can be defined as

$$V_0(s) = A_d(s)[V_1(s) - V_2(s)] + A_c(s)\,[(V_1(s) + V_2(s))/2] \qquad (3)$$

where the first term on the right is the differential portion of $V_0(s)$, and the second term is the common-mode portion of $V_0(s)$. The *differential frequency response* of the op amp is given as $A_d(j\omega)$, and the *common-mode frequency response* is given as $A_c(j\omega)$. A typical differential frequency response of an op amp is given as

$$A_d(s) = (A_{0d}\,\omega_1\,\omega_2\,\omega_3\ldots)/(s+\omega_1)(s+\omega_2))s+\omega_3)\ldots \qquad (4)$$

where $\omega_1, \omega_2, \ldots$ are poles of the operational amplifier. Though the operational amplifier may have zeros, they will be ignored at present. A_{0d} (or simply A_0 where the use is understood) is the *low-frequency gain* of the op amp. Figure 8.6–5 shows a typical frequency response of the magnitude of $A_d(j\omega)$. In this case we see that ω_1 is much lower than the rest of the poles, causing ω_1 to be the dominant influence in the frequency response. The intersection of the −6 dB/octave slope from the dominant pole and the 0 dB axis is designated as *GB* and is defined as the *unity-gain bandwidth* of the op amp. Even if the next higher-order poles are lower than *GB,* we shall still continue to use the unity-gain bandwidth as defined above.

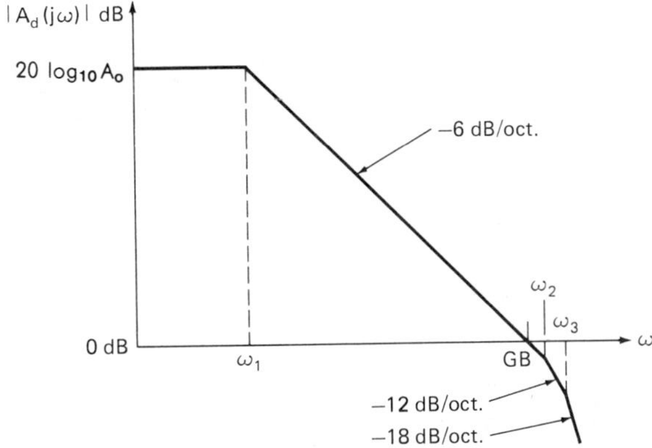

Fig. 8.6–5. Typical frequency response of the magnitude of $A_d(j\omega)$ for an op amp.

Other nonideal characteristics of the op amp not defined by Fig. 8.6–4 include the *power supply rejection ratio* (PSRR). The PSRR is defined as the product of the ratio of the change in the supply voltage to the change in the output voltage of the op amp caused by the change in the power supply, and the open-loop gain of the op amp; PSRR $= (\Delta V_{DD}/\Delta V_{\text{out}}) \times A_d$. An ideal op amp would have an infinite PSRR. The *common-mode input range* is the voltage range over which the input common-mode signal can vary. Typically, this range is several volts less than the maximum and several volts more than the minimum power supply voltages. The output of the op amp has several limits that are important. One of these limits is the *maximum output current sourcing and sinking* capability. Another limit of the output of the op amp is the *output voltage swing.* Like the input common-mode range, the output voltage swing is typically limited to within several volts of the maximum and minimum power supply voltages. The output also has a voltage rate limit called the *slew rate.* The slew rate is the maximum rate at which the output voltage can rise or fall and is generally determined by the maximum current available to charge a capacitance. The last characteristic that happens to be of great interest to analog sampled data circuit applications is the *settling time.* The settling time is the amount of time the op amp takes in a given negative-feedback configuration to respond to an input step and to settle within a given portion of the final value of the step response. The importance of the settling time to analog sampled data circuits is illustrated by Fig. 8.6–6. If the output has not settled to within a few tenths of a percent, then the accuracy of the charge transfer as determined by the capacitor ratios will not be realized.

Fortunately, the MOS op amp does not suffer all the nonideal characteristics previously discussed. Because of the extremely high input resistance of the

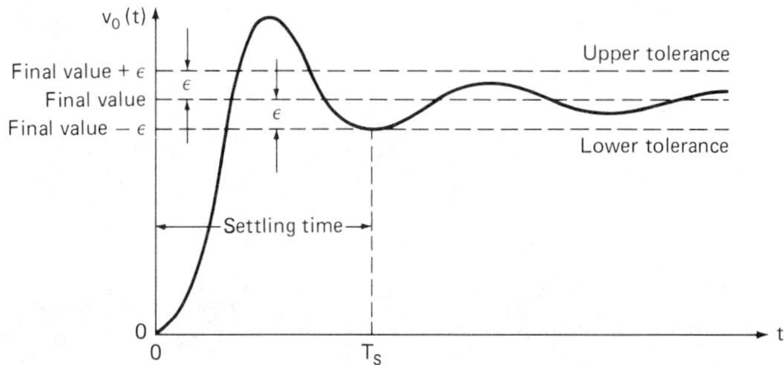

Fig. 8.6–6. Transient response of an op amp with negative feedback illustrating settling time, T_s. ϵ is the tolerance to the final value which is used to define the settling time.

MOS devices, both R_{id} and I_{OS} (or I_{B1} and I_{B2}) are not of importance. A typical value of R_{id} is in the range of 10^{14} ohms. Also R_{icm} is extremely large and can be ignored. If the op amp is used in the configuration of Fig. 8.6–3, then all common-mode characteristics can be ignored.

General Approach to the Design of MOS Op Amps

The first question in designing any op amp concerns the specifications. These specifications involve any of the previously defined parameters used to describe the characteristics of op amps. In analog circuit applications the following considerations may be pertinent.

1. How much gain is desired?
2. How much output voltage swing?
3. How much power consumption can be provided?
4. How much capacitance does the op amp need to drive?
5. What levels of power supply are available?
6. Is a substrate (back-gate) bias available?
7. Is substantial common-mode input range needed?
8. How about power supply rejection ratio?

In addition to resolving these questions, one would like to be able to design an op amp that is insensitive to the processes used to build it. Table 8.6–1 lists specifications for a typical MOS op amp used in switched capacitor circuits.

The block diagram of Fig. 8.6–1 is useful for guiding the MOS op amp design. The compensation method has a large influence on the design of each block. Two basic methods of compensation are suggested by the opposite

Table 8.6–1 Specifications of a Typical High Output Impedance MOS op amp

SPECIFICATION	REQUIREMENT
Area	Less than or equal to 200 mils square
Power	Less than or equal to 1 milliwatt
Programmable	Changing the bias will change the frequency response
Unity-gain bandwidth	Greater than 1 MHz
Slew rate	Greater than or equal to 1 volt/microsecond
Settling time	Less than or equal to 2 microseconds
Phase margin for 10pF load	Greater than 60°

parallel paths into the compensation block of Fig. 8.6–1. These two methods are feedback and feedforward compensation, and they will be discussed below.

In designing an op amp, one can begin at many points. The design procedure must be iterative, inasmuch as it is almost impossible to relate all specifications simultaneously. As a rule, all MOS transistors should be operated in the saturation region if possible, which is defined as

$$V_{DS} \geq V_{GS} - |V_T| \tag{5}$$

The reason for this is that larger values of g_m and voltage swings are obtained in the saturation region. For a typical MOS op amp design, the following steps may be appropriate.

1. Block diagram. From the specifications determine a suitable block diagram similar to Fig. 8.6–1 that shows how the gains will be distributed throughout the op amp. Although the actual circuit is not identified at this point, the type and size of compensation should be determined as well as possible. The size of the compensation capacitor will have a direct influence on the design of the op amp.

2. Input stage. This stage is differential and determines the common-mode input range, CMRR, PSRR, and noise performance. The output of this stage will generally have a low-frequency pole that will be used by the Miller compensation technique to achieve the dominant pole of Fig. 8.6–5.

3. Gain stage. This stage should provide most of the gain of the op amp. This gain will permit Miller compensation techniques to be applied using small capacitances. The poles of this stage should be moved outside *GB* where possible.

4. Output stage. The output stage should have the desired current and voltage output capability. Because the output stage will consist of large devices, the frequency response is very important. The poles should be moved

above GB wherever possible. One must watch out for complex poles and pole-zero pairs, both of which will lead to poor settling times.

5. Bias network. At this point the requirements of the bias network should be identifiable. The bias network should be independent of power supply, temperature, and process if possible.

6. Compensation. The preliminary considerations of compensation in step 1 can now be followed through in more detail. The typical compensation is to apply a capacitor around a high-gain, inverting stage with high input-output resistance. This will create a dominant pole, as will be seen in the following. It may also be necessary to provide feedforward compensation to obtain a larger phase margin.

If the specifications cannot be satisfied after the designer has gone through the above steps, then it will be necessary to iterate, making changes that will meet the specifications. Once a preliminary design has been developed so that W/L ratios of the MOS transistors have been estimated, then the design is taken to the computer for further simulation. In some cases, it may be necessary to computer-simulate portions of the circuit before considering the entire circuit. The approach to designing MOS op amps will be illustrated in the next section, using various types of MOS technologies.

Compensation Methods

Although there are many compensation methods applicable to the op amp, we shall consider only some of the simpler ones. Obviously, the high open-loop gain of the MOS op amp makes it necessary to examine the stability of the circuit in a closed-loop mode. A good op amp should have a phase margin of 60° in the closed-loop unity gain configuration. Smaller phase margins will cause the step response to "ring" and contribute to a longer settling time. The basic approach to compensation consists of two steps. The first is to determine all the poles in the circuit, including the loading capacitance. One should consider all parasitic capacitances in this step as well. These parasitic capacitances include those from the moat bottom to the substrate, the moat sidewall to the substrate, the first poly to the substrate, the metal to the substrate, and the second poly (if used) to the substrate. These capacitances are illustrated in Fig. 8.6–7(a) and (b). Usually one must also take into account the lateral diffusion in calculating the moat bottom capacitance. The next step is to redesign, if possible, any stage that has a pole inside the unity-gain bandwidth, except the pole intended to be used for the dominant pole (this is usually the pole between the input and the gain stage). Once these two steps are complete, then one is ready to apply one or more of several compensation techniques.

An AC model of the operational amplifier suitable for compensation pur-

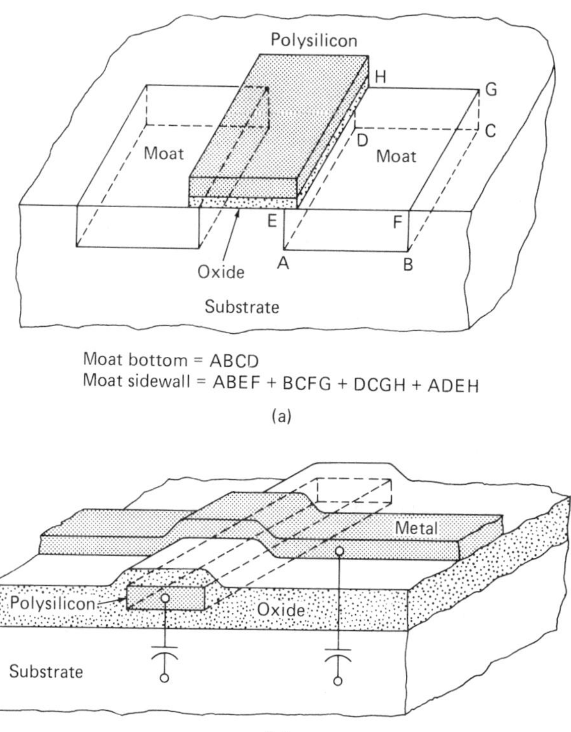

Fig. 8.6–7. Illustration of various parasitic capacitances. (a) Moat bottom and moat sidewall to substrate capacitances. (b) Stray capacitances between the polysilicon and substrate and metal and substrate.

poses is shown in Fig. 8.6–8.[52,53] This model represents two voltage-gain stages with a capacitor, C_c, being placed around the second voltage-gain stage. If the gain of the second voltage stage is high, then this capacitor uses the Miller effect to accomplish compensation. g_{m1} is the transconductance of the first stage, and R_1 is the combined resistance to ground of the output of the first stage and the input of the second stage. C_1 represents the capacitance to ground at this node. g_{m2} is the transconductance of the second stage, and R_2 and C_2 include the output and load resistances and capacitances, respectively.

[52] J. E. Solomon, "The Monolithic Op Amp—A Tutorial Study," *IEEE J. of Solid-State Circuits,* Vol. SC-9, No. 6, December 1974, pp. 314–332.

[53] Y. P. Tsividis, "Design Considerations in Single-Channel MOS Analog Integrated Circuits—A Tutorial," *IEEE J. of Solid-State Circuits,* Vol. SC-13, No. 3, June 1978, pp. 383–391.

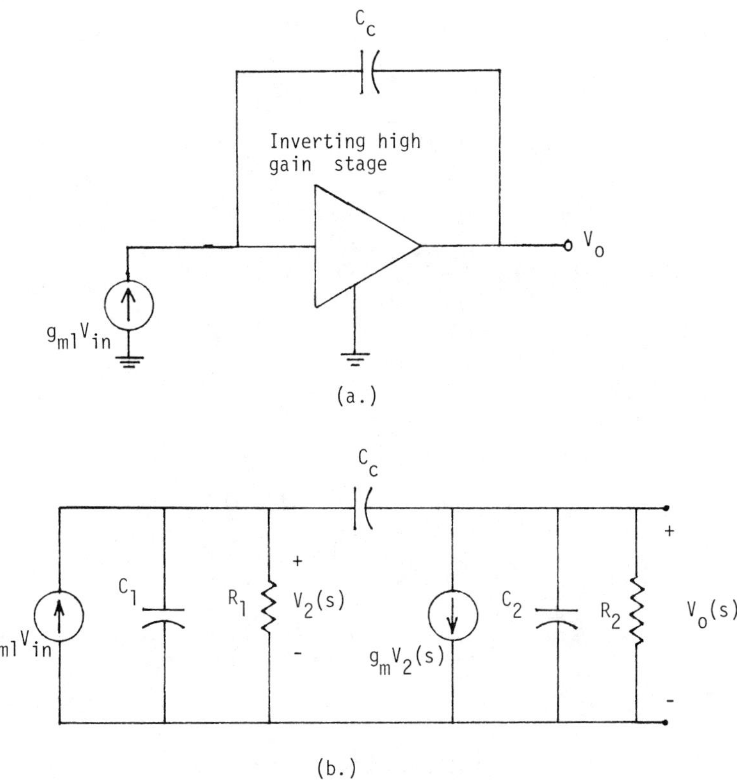

Fig. 8.6–8. Miller compensation scheme. (a) General circuit. (b) Equivalent model.

If C_c is zero, then the roots of this case correspond to

$$p_1' = -1/R_1C_1 \tag{6}$$

and

$$p_2' = -1/R_2C_2 \tag{7}$$

These poles are on the negative real axis and are shown on Fig. 8.6–9(a). When C_c is not zero, then the transfer function $V_o(s)/V_{in}(s)$ can be found as

$$\frac{V_o(s)}{V_{in}(s)} = \frac{g_{m1}\,g_{m2}\,R_1R_2(1 - sC_c/g_{m2})}{1 + s[R_1(C_1 + C_c) + R_2(C_2 + C_c) + g_{m2}\,R_1R_2C_c] + s^2R_1R_2[C_1C_2 + C_c(C_1 + C_2)]} \tag{8}$$

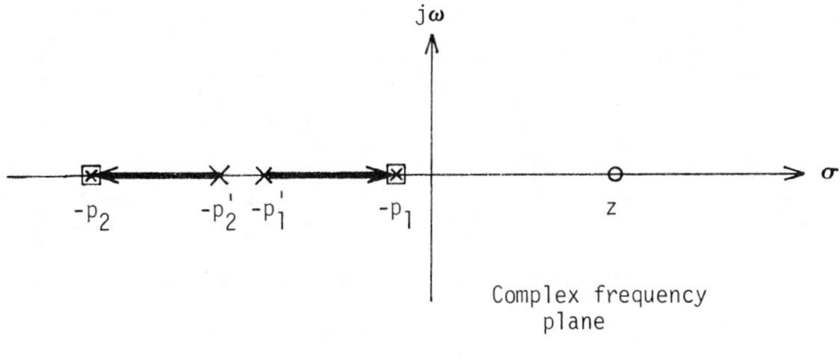

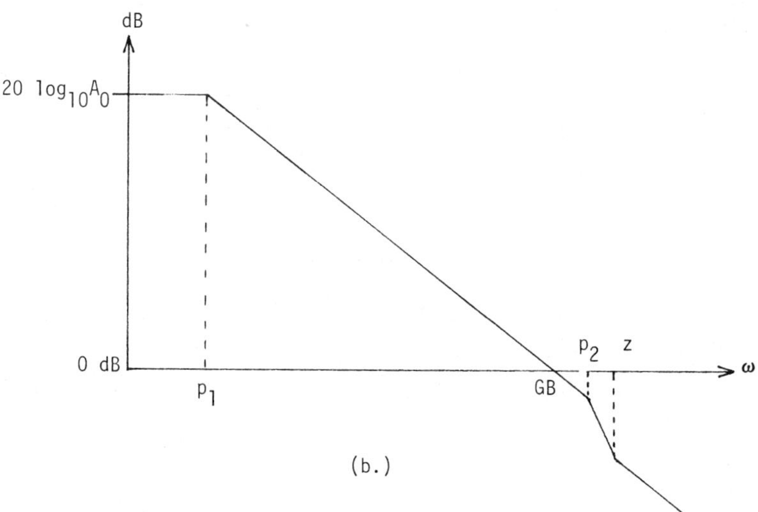

Fig. 8.6–9. (a) Open and closed loop roots of Fig. 8.6–8(a). (b) Closed loop frequency response of Fig. 8.6–8(a).

Assuming that the Miller capacitance, C_c, will split p_1' and p_2' by a significant amount allows the use of approximations that give the roots of Eq. (8) as

$$p_1 \cong -1/(g_{m2} R_1 R_2 C_c) \tag{9}$$

$$p_2 \cong (-g_{m2}\, C_c)/(C_1 C_2 + C_2 C_c + C_c C_1) \tag{10}$$

and

$$z = g_{m2}/C_c \tag{11}$$

Figure 8.6–9(a) shows the compensated root locations. It is clear how C_c has split p_1' and p_2'. However, in the process of applying the Miller compensation, a zero located on the positive real axis has appeared.

The frequency response of the compensated op amp is shown in Fig. 8.6–9(b). The objective of compensation is to have a single dominant pole, p_1, and then to have all other poles, including p_2 and higher-frequency poles that have not been considered here, above the unity-gain bandwidth, *GB*. However, the right-hand plane zero can destroy attempts to compensate the op amp if g_{m2} is not large. This zero not only boosts the magnitude of the op amp gain, but also causes phase lag, which will cause the op amp to become unstable when negative feedback is applied around it. The desired objective of compensation is to have sufficient phase margin in a given application. Typically this value of phase margin is between 45° to 60° for satisfactory performance. This performance will generally be achieved when p_2 is equal to or greater than *GB*. It can be shown that *GB* is given as

$$GB = A_0\, p_1 \cong (g_{m1}\, R_1\, g_{m2} R_2)/(g_{m2}\, R_1 R_2 C_c) = g_{m1}/C_c \qquad (12)$$

The ratio of z to *GB* is given as

$$z/GB = g_{m2}/g_{m1} \qquad (13)$$

Obviously for successful compensation, z must be greater than *GB*, which means g_{m2} must be greater than g_{m1}. Unfortunately, MOS op amps are usually operated with comparable bias currents in both the first and second stages. Also the transconductance varies as the square root of the bias current, so that very large current differences are necessary to achieve g_{m2} greater than g_{m1}.

Because the zero of Fig. 8.6–8(a) is caused by the feedforward path through C_c, the above situation can be avoided by the use of a buffer in the feedback path, as shown in Fig. 8.6–10(a). A model for Fig. 8.6–10(a) is shown in Fig. 8.6–10(b). R_1, C_1, R_2, and C_2 are associated with the inverting high-gain stage and have already been described. $AV_0(s)$ and R_0 represent the buffer and its output resistance. If R_0 is small, then the transfer function is found to be

$$\frac{V_0(s)}{V_{\text{in}}(s)} = \frac{-g_{m1}\, g_{m2}\, R_1 R_2\, \omega_1\, \omega_2}{(s+\omega_1)(s+\omega_2)} \qquad (14)$$

where

$$\omega_1 = \frac{1}{g_{m2} A R_1 R_2 C} \qquad (15)$$

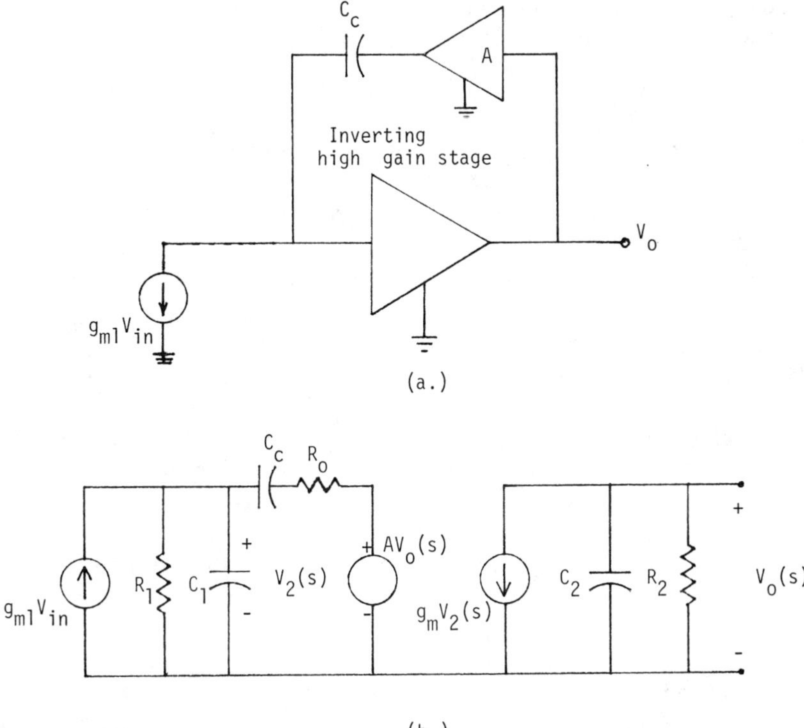

Fig. 8.6–10. (a) Modified Miller compensation scheme. (b) Model for gain calculation.

and

$$\omega_2 \cong \frac{g_{m2}C_c}{C_2(C_1 + C_c)} \tag{16}$$

Unfortunately, the value of R_0 is not small enough to be neglected and must be incorporated into the above analysis. To simplify our considerations, let us assume that $C_1 = C_2 = 0$. This assumption is justified for the buffer isolates C_2 from the input, and, if the Miller effect is strong enough, then C_1 has little effect on the dominant pole. The transfer function of Fig. 8.6–10(b) therefore becomes

$$\frac{V_0(s)}{V_i(s)} = \left(\frac{-g_{m1}\, g_{m2}R_0R_1R_2}{R_1 + R_0 + g_{m2}AR_1R_2}\right)\left[\frac{s + \dfrac{1}{R_0C_c}}{s + \dfrac{1}{C_c(R_1 + R_0 + g_{m2}AR_1R_2)}}\right] \tag{17}$$

We see that R_0 causes a zero on the left-hand positive axis that may effect the compensation efforts. If R_0 becomes small, then Eq. (17) approaches

$$\lim_{R_0 \to 0} \frac{V_0(s)}{V_i(s)} = g_{m1} g_{m2} R_1 R_2 \left(\frac{\dfrac{1}{C_c(1+g_{m2}AR_2)R_1}}{s + \dfrac{1}{C_c(1+g_{m2}AR_2)R_1}} \right) \tag{18}$$

This is seen to be approximately the expression in Eq. (14) if ω_2 approaches infinity and C_1 is equal to zero.

As an example let us assume that $R_0 = 35.71$ kΩ, $C_c = 32.3$ pF, $R_1 = 134.23$ kΩ, and $g_{m2}AR_2 = 44$. Substituting these values into Eq. (17) gives a zero at 138 kHz and a pole at 810 Hz. The zero at 138 kHz gives a leading phase shift, which is desirable, but also boosts the open-loop gain, which can be undesirable. Unless the output impedance of the buffer is small, Eq. (17) should be used for the modified Miller compensation scheme. If R_1, R_0, g_{m2}, A, and R_2 are known, then C_c may be designed to realize the desired dominant pole, providing the zero is on the negative real axis. This zero could also be incorporated into the compensation scheme using the concepts of lead compensation.[54]

Another means of eliminating the effect of the right-half plane zero of Eq. (11) is to insert a nulling resistor in series with C_c. Figure 8.6–11 shows the small-signal equivalent model for the resistor nulling method. This circuit has the following node voltage equations

$$g_{m1} V_{in} + V_2/R_1 + V_2 s C_1 + (V_2 - V_0)[sC_c/(1+sC_cR_z)] = 0 \tag{19}$$

$$g_{m2} V_2 + V_0/R_2 + V_0 s C_2 + (V_0 - V_2)[sC_c/(1+sC_cR_z)] = 0 \tag{20}$$

These equations can be solved to give

$$\frac{V_0(s)}{V_{in}(s)} = \frac{a\left[1 - s\left(\dfrac{C_c}{g_{m2}} - R_z C_c\right)\right]}{1 + bs + cs^2 + ds^3} \tag{21}$$

where

$$a = g_{m1} g_{m2} R_1 R_2 \tag{22}$$

[54] J. K. Roberge, *Operational Amplifiers—Theory and Practice,* Chaps 5 and 11, John Wiley & Sons, Inc., New York, 1975.

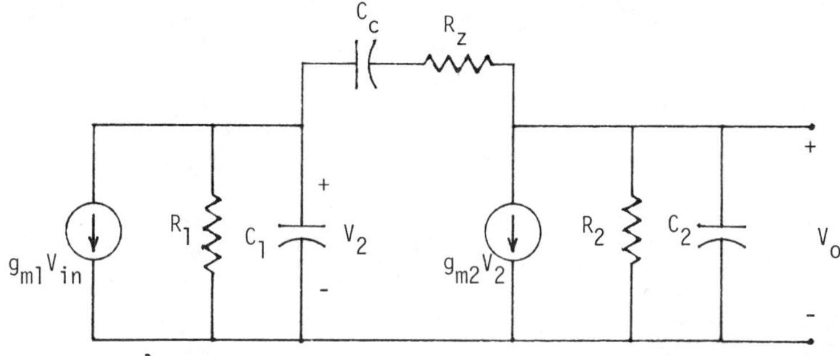

Fig. 8.6–11. Model for the nulling resistor compensation.

$$b=[(C_2+C_c)R_2+(C_1+C_c)R_1+g_{m2}R_1R_2C_c+R_zC_c] \tag{23}$$

$$c=[R_1R_2(C_1C_2+C_cC_1+C_cC_2)+R_zC_c(R_1C_1+R_2C_2)] \tag{24}$$

and

$$d=R_1R_2R_z\,C_1C_2C_c \tag{25}$$

If we assume that R_z is less than R_1 or R_2, and that the poles will be widely spaced, then we may approximate the roots as

$$p_1 \cong \frac{-1}{(1+g_{m2}R_2)R_1C_c} \tag{26}$$

$$p_2 \cong \frac{-g_{m2}C_c}{C_1C_2+C_cC_1+C_cC_2} \tag{27}$$

and

$$z=\frac{1}{C_c\left(\dfrac{1}{g_{m2}}-R_z\right)} \tag{28}$$

If $R_z = 1/g_{m2}$, then the zero is at infinity, which eliminates its influence on the circuit. One may wish to make R_z slightly larger than $1/g_{m2}$ to take advantage of lead compensation. If R_z is realized by MOS transistors, then Eq. (26) will be satisfied over a wider range of operating conditions.

Another compensation technique that has found use in MOS op amps is the feedforward scheme in Fig. 8.6–12(a). In this circuit we see that again a buffer is used to get directionality in the signal path. Figure 8.6–12(b)

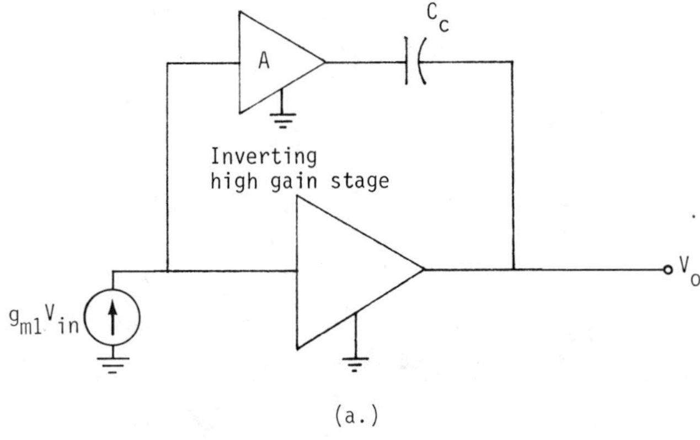

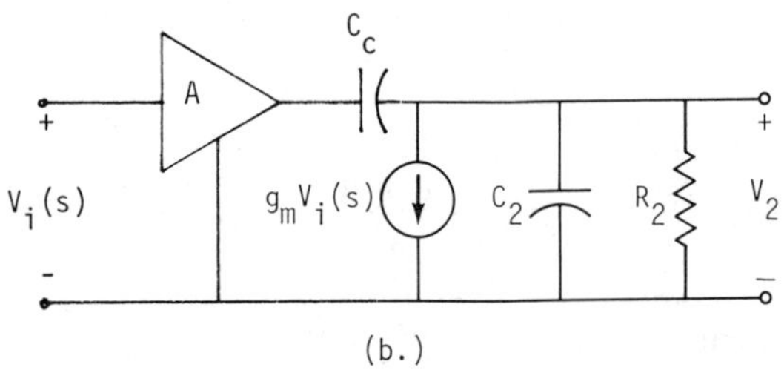

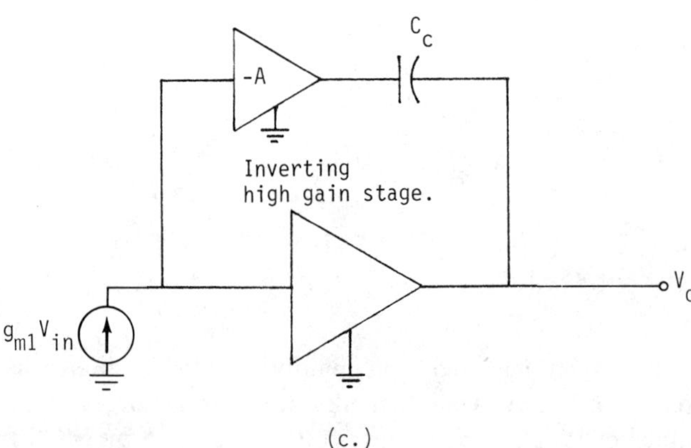

Fig. 8.6–12. (a) Feedforward compensation scheme. (b) Model for gain calculation. (c) Proper polarity for feedforward compensation.

can be used as a model for this circuit. The voltage transfer function, $V_0(s)/V_i(s)$, can be found as

$$\frac{V_0(s)}{V_i(s)} = \left(\frac{A\,C_c}{C_c + C_2}\right)\left(\frac{s - g_m/AC_c}{s + \dfrac{1}{R_2(C_c + C_2)}}\right) \tag{29}$$

We see that this scheme results in a slight shift of the original pole, $-1/R_2C_2$, when C_c was zero. However, we note that the zero is located on the positive real axis, which causes phase lag rather than the desired phase lead. Consequently it is necessary to use an inverting buffer with a gain of $-A$. Alternatively, one could feedforward around a noninverting stage. Figure 8.6–12(c) gives the desired compensation scheme using feedforward techniques. The voltage transfer function is

$$\frac{V_0(s)}{V_i(s)} = \left(\frac{-A\,C_c}{C_c + C_2}\right)\left(\frac{s + g_m/AC_c}{s + \dfrac{1}{R_2(C_c + C_2)}}\right) \tag{30}$$

To use Fig. 8.6–12(c) to achieve compensation, it is necessary to place the zero located at g_m/AC_c above the value of GB, so that boosting the magnitude will not cancel the phase lead of the zero. Fortunately, the phase effects extend over a much broader frequency range than the magnitude effects, so that this method will provide additional phase margin over that provided by the feedback compensation technique. It is quite possible that several zeros can be generated in the transfer function of the op amp. These zeros should all be placed above GB and should be as close to a pole as possible to avoid large settling times caused by a pole-zero or doublet in the transient response.[55]

The load of the op amp greatly affects its compensation requirements. In many switched capacitor applications, the op amp drives only internal capacitance nodes and supplies no DC current to the load. In this case it becomes possible to omit the buffer stage of the op amp and to drive the capacitive load directly from the high-impedance output of the second stage. This results in substantial savings in power and in chip area. The key factor in determining

---

[55] Y. P. Tsividis and P. R. Gray, "An Integrated NMOS Operational Amplifier with Internal Compensation," *IEEE J. of Solid-State Circuits,* Vol. SC-11, No. 6, December 1976, pp. 748–753.

the amount of capacitance that can be attached to the output is the location of the first nondominant pole. This pole was designated as p_2 and is

$$p_2 = \frac{-g_{m2}\,C_c}{C_1C_2 + C_2C_c + C_cC_1} \tag{31}$$

If C_2 and C_1 are greater than C_1, then Eq. (31) simplifies to

$$p_2 \cong \frac{-g_{m2}\,C_c}{C_2C_c} = \frac{-g_{m2}}{C_2} \tag{32}$$

Therefore, the first nondominant pole is located at approximately $-g_{m2}/C_2$. However, to ensure adequate phase margin, the magnitude of this pole must be greater than *GB*. Therefore, we must satisfy

$$g_{m2}/C_2 > g_{m1}/C_c \tag{33}$$

Thus the load capacitance, C_2, must be less than

$$C_2 < (g_{m2}/g_{m1})C_c \tag{34}$$

Because g_{m2} is generally larger than g_{m1}, most op amps can drive a load capacitor on the order of the compensating capacitor. This is entirely adequate for many applications.

8.7 MOS OP AMPS

This section uses the previously developed concepts and principles to illustrate the design and analysis of MOS op amps. The general types of MOS architectures will be examined first. A general design approach that illustrates the use of process-insensitive bias techniques will follow. An example of designing an MOS op amp will be presented, to illustrate the principles of the design process. Several other types of MOS op amps will be briefly compared to this design. Finally, considerations for testing MOS op amps will be presented.

MOS Op Amp Architecture

The general form of an NMOS op amp is shown in Fig. 8.7–1. Whether this op amp uses enhancement or depletion devices is not of concern at this point. This figure illustrates one method of converting from a differential to a single-ended signal that we have not considered previously. The loads

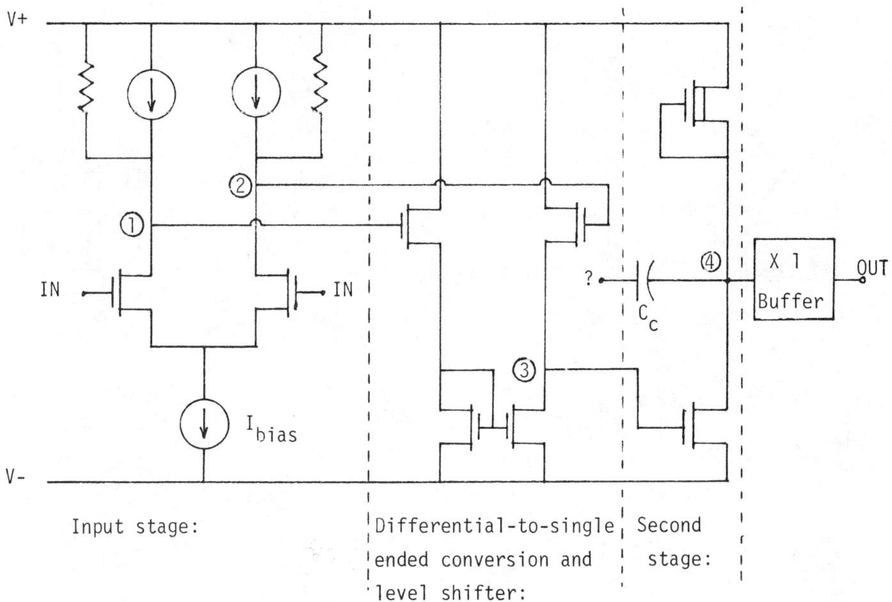

Fig. 8.7–1. High impedance level shifter NMOS op amp.

are considered as low-impedance current sources, which serves as a model for either the enhancement or depletion load NMOS differential amplifiers. The difference signal at points 1 and 2 is applied differentially to the gates of the source followers. The current through the left-hand source follower is mirrored and subtracted from the current in the right-hand source follower. The difference in the two currents is applied to the gate of the common source transistor of the second stage (point 3). Because the currents are differential, the gain of the differential-to-single-ended converter and level-shifter is approximately 2. The output of the second stage (point 4) can be used directly or applied to a buffer to lower the output resistance of the op amp.

As was noted in the previous section, the Miller compensation method splits two poles, making one dominant and driving the other away from the origin. The most obvious place to use the Miller compensation is around the second stage, where the pole at 4 is driven toward infinity and the pole at 3 becomes dominant. Unfortunately this approach does not work well with Fig. 8.7–1 because there is a pole at each of the points 1 and 2 that cannot be ignored. These poles cannot be ignored because the high impedance at points 1 and 2 makes the parasitic capacitances to ground significant. This problem is usually solved in part by placing a small feedforward capacitor

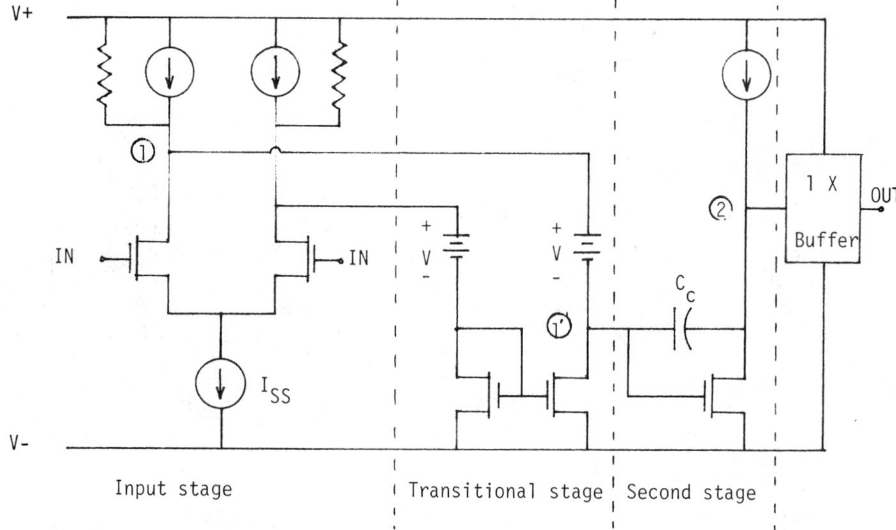

Fig. 8.7–2. Low impedance level shifter NMOS op amp.

from each gate to the source of both source followers. This reduces the impedance level at points 1 and 2 as frequency increases and permits C_c to be connected to point 3. This method is satisfactory although the transient response is more complex and the phase margin is minimum.

To eliminate the problems of too many high-impedance points, the NMOS op amp of Fig. 8.7–2 has been used.[56] The objective behind this architecture is clear in that points 1 and 2 of Fig. 8.7–1 are no longer high-impedance. The only high-impedance points are designated as 1 and 2 in Fig. 8.7–2. The batteries can be implemented by an active resistor in series with a current sink, as shown in Fig. 8.7–3(a). Unfortunately the impedance level of this device becomes large as the DC voltage across it increases. Thus a more practical configuration is shown in Fig. 8.7–3(b). This is a network that uses shunt, negative feedback to reduce the resistance across M2. The W/L ratio of M2 can be small, resulting in a large DC voltage drop, while the value of the small signal impedance is small.

Figure 8.7–4 shows a general architecture for a p-channel input CMOS op amp. The availability of both n- and p-channel devices not only makes

[56] D. Senderowicz, D. A. Hodges, and P. R. Gray, "High-Performance NMOS Operational Amplifier," *IEEE J. of Solid-State Circuits,* Vol. SC-13, December 1978, pp. 760–766.

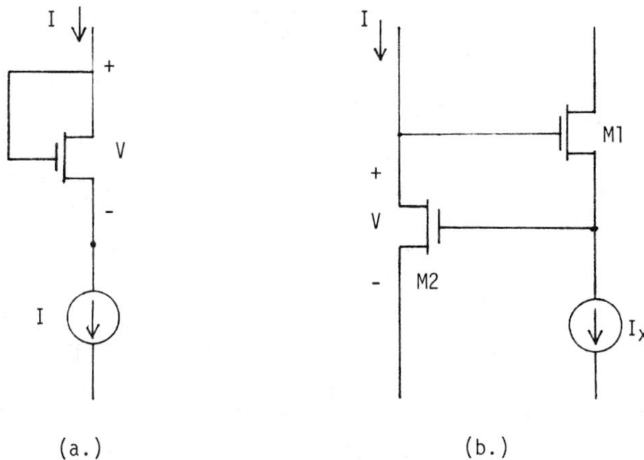

Fig. 8.7–3. Level shifters (a) Active resistor approach. (b) shunt feedback approach.

the gain higher, but avoids the need for a differential-to-single-ended converter. The second stage consists of a simple inverter. Because the mobility of the *n*-channel device is larger than the mobility of the *p*-channel device, the second stage should have more gain than the differential stage, all other things being equal. Because of the additional gain, the compensation capacitor

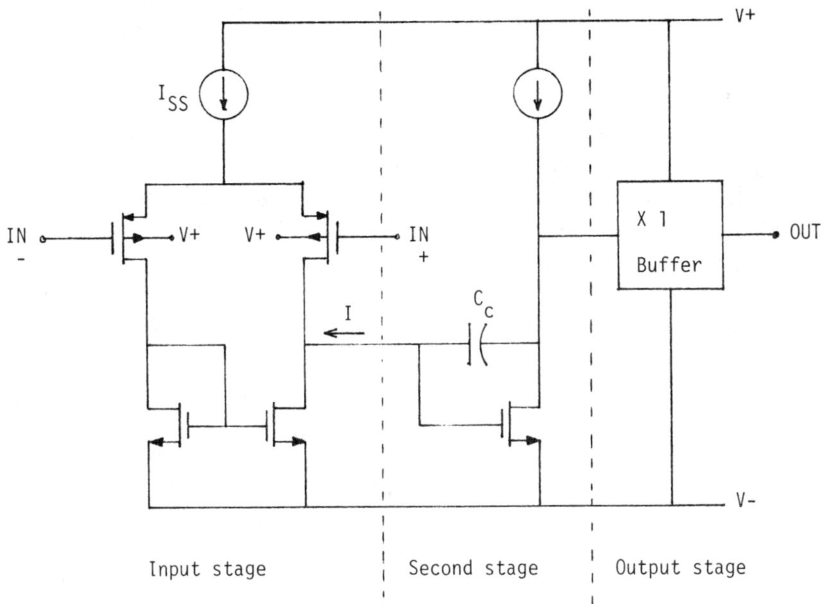

Fig. 8.7–4. CMOS op amp architecture.

can be smaller. The use of feedforward or nulling resistor techniques can help achieve the desired stability. If the CMOS op amp is to drive resistive loads, then a buffer stage must be added to provide the voltage swing and current capability.

Design Procedure for MOS Op Amps

In designing an op amp, one can begin at many points. The design procedure must be iterative, inasmuch as it is almost impossible to relate all specifications simultaneously. If the specifications cannot be satisfied after going through the steps given in Section 8.6, then it will be necessary to iterate, making changes that will meet the specifications. Once a preliminary design has been developed so that W/L ratios of the MOS transistors have been estimated, then the design is taken to the computer for further simulation. In some cases, it may be necessary to compute-simulate portions of the circuit before considering the entire circuit.

The objective of process-independent biasing is to develop geometrical constraints that will guarantee that each device remains in saturation regardless of how V_T varies. Equation (5) of Section 8.6 can be rewritten as

$$V_{DG} \geq -V_T \tag{1}$$

Fortunately, this constraint can be guaranteed for an enhancement NMOS device if the drain is connected to the gate, resulting in $V_{DG} = 0$. As V_T is always positive unless the process is out of control, Eq. (1) is always satisfied, regardless of the terminal voltages. The same considerations hold for the enhancement PMOS device.

Now we show how the gate-source voltages of various devices can be interrelated, allowing one to ensure saturation. Let us define the large-signal expression for the ith MOS device in saturation as

$$I_i = K'S_i\,(V_{GSi} - V_{Ti})^2 \tag{2}$$

where

$$S_i = W_i/L_i \tag{3}$$

Let us consider four MOS devices, $i = 1, 2, 3$, and 4, that are of the same type, i.e., enhancement NMOS, depletion NMOS, etc. If we form the ratio

$$S_1/S_2 = S_3/S_4 \tag{4}$$

then we can solve for S_i from Eq. (2) and show that Eq. (4) is equivalent to

$$\frac{V_{GS2} - V_{T2}}{V_{GS1} - V_{T1}} = \frac{V_{GS4} - V_{T4}}{V_{GS3} - V_{T3}} \tag{5}$$

If V_{GSi} is greater than V_{Ti} for $i = 1, 2, 3$, and 4, then Eq. (5) may be written as

$$\frac{V_{GS2}/V_{GS1}}{V_{GS4}/V_{GS3}} \cong \frac{1 + \dfrac{V_{T3}}{V_{GS3}} - \dfrac{V_{T4}}{V_{GS4}} - \dfrac{V_{T3}V_{T4}}{V_{GS3}V_{GS4}}}{1 + \dfrac{V_{T1}}{V_{GS1}} - \dfrac{V_{T2}}{V_{GS2}} - \dfrac{V_{T1}V_{T2}}{V_{GS1}V_{GS2}}} \approx 1 \tag{6}$$

Therefore, Eq. (4) implies

$$V_{GS2}/V_{GS1} \cong V_{GS4}/V_{GS3} \tag{7}$$

or

$$V_{GS1}/V_{GS2} \approx V_{GS3}/V_{GS4} \tag{8}$$

This is an important result. For example, if the ratio of V_{GS2} to V_{GS1} is known, then the geometric ratios of the devices can be used to establish the ratio of V_{GS4} to V_{GS3}. Many applications of this principle will follow.

MOS Enhancement Op Amp

The first op amp considered will use only enhancement devices and falls into the category of Fig. 8.7–1. We shall be attempting to meet the general specifications for an MOS op amp given in Table 8.7–1. The block diagram for our first op amp is shown in Fig. 8.7–5. This block diagram is based

Table 8.7–1 General Specifications for an MOS Operational Amplifier

PARAMETER	VALUE
Gain	60 dB
V_{OS}	±20 mV
Output swing	$V_{DD}-2$ V to $V_{SS}+2$ V
Common-mode input range	$V_{DD}-2$ V to $V_{SS}+2$ V
CMRR	60 dB
PSRR from any supply	60 dB
GB	2 MHz
Slew rate	± 2V/μS
Power consumption	50 mW
Area	< 200 mils²
Input referred noise	300 nV/$\sqrt{\text{Hz}}$ at 1 kHz

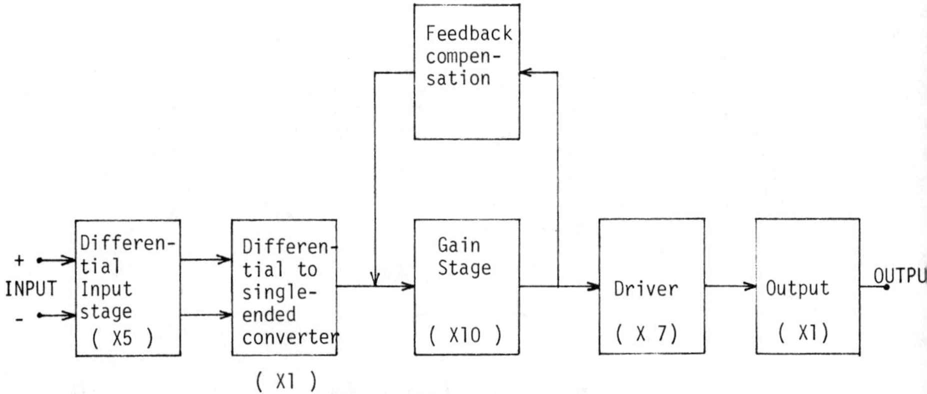

Fig. 8.7–5. Block diagram of an all-enhancement op amp design.

on experience, and some best guesses and will be modified as necessary. The next step is to fill in the blocks of this diagram with circuits of the previous sections. One possible result is shown in Fig. 8.7–6. It is at this point that the design begins. We shall try to design this circuit to meet the specifications in Table 8.7–1, modifying or iterating the design where necessary. In the case we are considering, the op amp has already been designed.[57]

[57] Y. P. Tsividis and P. R. Gray, "An Integrated NMOS Operational Amplifier with Internal Compensation," *IEEE J. of Solid-State Circuits,* Vol. SC-11, No. 6, December 1976, pp. 748–753.

However, it will be satisfactory for our purposes to follow through the design of this op amp, to understand the design process.

A closer look at Fig. 8.7–6 shows, with the exception of the output stage, that all the component circuits of this op amp have been previously discussed. The bias current consists of two strings of devices M1, M2, and M3 and M15, M16, and M17. These devices are used as voltage dividers. The input stage, consisting of M6, M7, M8, M9, and M10, is the enhancement differential amplifier of Section 8.5. The differential-to-singled-ended conversion is accomplished by M4, M5, M11, and M12, as illustrated in Fig. 8.7–1. The high-gain stage consists of M18, M19, and M20 and is recognized as a cascode stage. M21 and M22 provide a DC level shift from the drain of M19 to the gate of M24. The output stage consists of M23, M24, M25, and M26. M24 provides voltage gain, while M26 and M23 form a voltage-shunt feedback that reduces the output resistance and, unfortunately, the load resistance seen by M24. M25 serves the function of a "pullup" resistor. Last, the frequency compensation is accomplished by the source follower M13 and M14 and the capacitor C_c.

Our first step is to design the DC conditions of the circuit so that all transistors are in saturation. This will result in constraints on the W/L ratios that must be considered in the design of each block. Transistors that are in saturation because the drain is connected to the gate include M1, M2, M3, M5, M6, M9, M15, M16, M17, M18, and M25. Transistors that are in saturation because V_{DG} is greater than zero include M4, M11, M13, M21, and M22. Symmetry is assumed in the differential amplifier, so that $S_7 = S_{10}$ and $S_6 = S_9$. The differential-to-single-ended converter is also assumed to be symmetric, so that $S_4 = S_{11}$ and $S_5 = S_{12}$. S_8 and S_3 are chosen to be equal, so that $I_8 = I_3 = I_1$.[58] Therefore, a current equal to $I_1/2$ flows through M6, and if we choose $S_6 = S_9 = S_1/2$, we have $V_{GS1} = V_{GS6} = V_{GS9}$. We therefore have $V_{GS2} + V_{GS3} = V_{GS4} + V_{GS5}$. If we now choose $S_2/S_3 = S_4/S_5 = S_{11}/S_{12}$, then from Eq. (8) we have $V_{GS3}/V_{GS2} = V_{GS5}/V_{GS4} = V_{GS12}/V_{GS11}$. From symmetry $V_{GS4} = V_{GS11}$, so we can conclude that $V_{GS5} = V_{GS12} = V_{GS3}$ and $V_{GS4} = V_{GS11} = V_{GS2}$. Thus, the drain voltages of M2, M7, and M10 are all equal and are designated V_A. Likewise the drain voltages of M3, M5, and M12 are equal and are designated as V_B. V_A and V_B are designer-selected voltages chosen to optimize signal swing and to ensure V_{DG} greater than zero.

The status of M7, M8, M10, and M12 is not yet established. M7, M8, and M10 depend upon the common-mode input and are in saturation if

[58] I_i is used to designate the current of the ith MOS device.

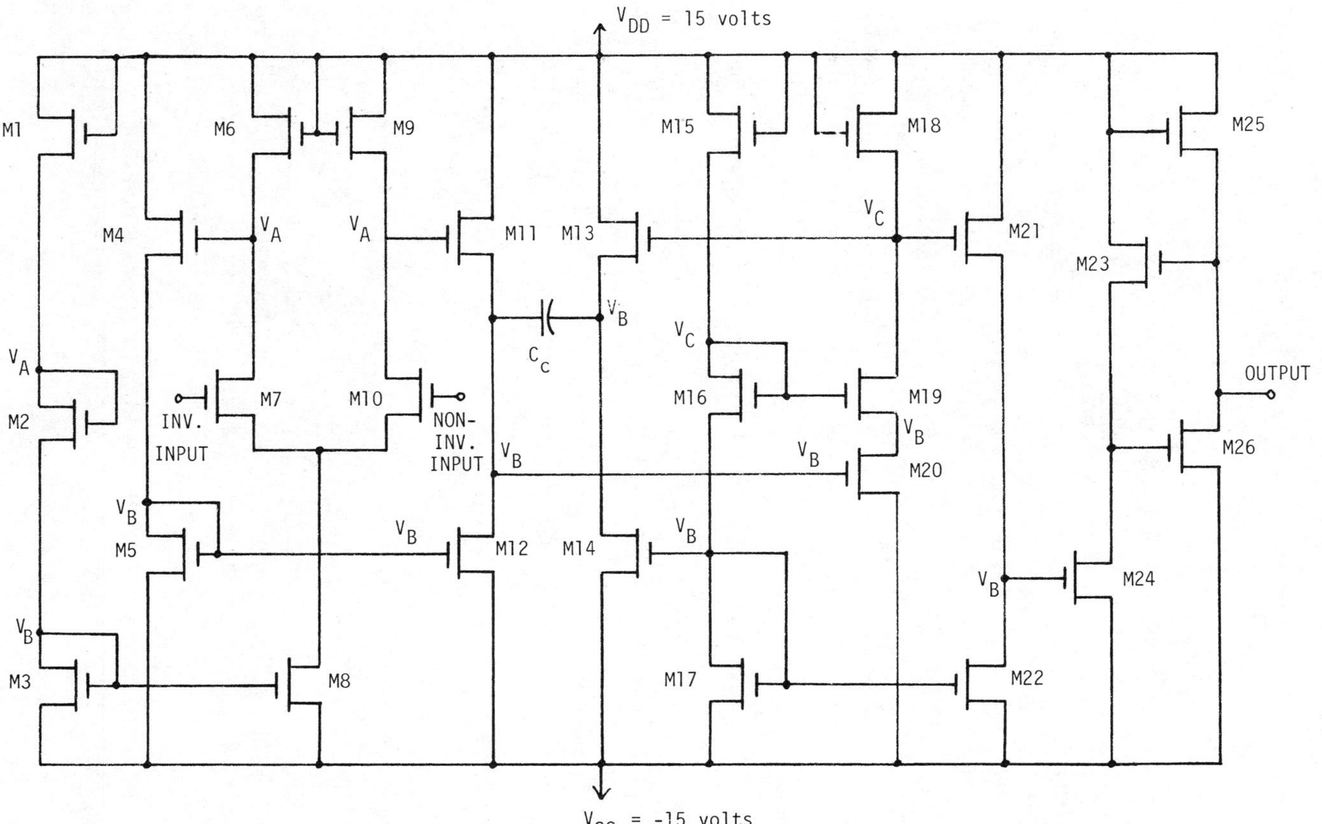

Fig. 8.7–6. All-enhancement MOS operational amplifier.

$$V_{SS} + V_{GS\,8} + V_T \leq V_{\text{in}} \leq V_{DD} - V_{GS\,6} \tag{9}$$

Because $V_{GS\,4} = V_{GS\,11}$, then $V_{DG\,12} = 0$ and M12 is also in saturation.

The second stage consists of M13 through M22. The two bias strings M1-M2-M3 and M15-M16-M17 have been designed so that $V_{GS\,17} = V_{GS\,3}$ for a typical value of V_T. Computer simulation shows that these two voltages remain nearly equal even if V_T changes, so we will assume that $V_{GS\,17} = V_{GS\,3}$ independently of V_T. We have chosen $S_{16}/S_{17} = S_{19}/S_{20}$, and because $V_{GS\,20} = V_{GS\,17}$, this guarantees that $V_{GS\,16} = V_{GS\,19}$. Therefore $V_{GS\,20} = V_{DS\,20}$, so that M20 is kept in the saturation region for any $V_T > 0$. In addition, we have $S_{15}/S_{16} = S_{18}/S_{19}$, and therefore $V_{GS\,18} = V_{GS\,15}$. This guarantees that $V_{DS\,19} = V_{GS\,19}$, so that M19 is also kept in the saturation region. Finally $S_{21}/S_{22} = S_{16}/S_{17}$, and because $V_{GS\,21} + V_{GS\,22} = V_{GS\,16} + V_{GS\,17}$, we get $V_{GS\,21} = V_{GS\,19}$, and so $V_{DS\,22} = V_{GS\,22}$, keeping M22 in saturation. The same considerations hold for M13 and M14. The previous design results in equal voltages at various places in the op amp of Fig. 8.7–6. All voltages indicated by V_A are equal, as is the case with the voltages indicated by V_B and V_C. As all devices are kept in saturation independently of V_T, the gain from the input to the gate of M24 remains high, and any DC imbalance will correspond to a small input offset voltage when reflected back to the input of the op amp. Table 8.7–2 summarizes the various constraints developed on the W/L ratios.

The various voltages V_A, V_B, and V_C, will now be designed. We shall choose $V_A = 8$ volts, $V_B = -11.5$ volts, and $V_C = -7$ volts. This selection assures that all devices are in saturation. Assuming that the current through M1, M2, and M3 is 100 μA allows us to design the W/L ratio of these devices. Using the parameters of Table 8.2–1 with $V_T = 2$ and $K_n = 10^{-5}$

Table 8.7–2 Design Constraints for Process-Independence of Fig. 8.7–6

CONSTRAINTS FOR PROCESS-INDEPENDENCE OF FIG. 8.7–6
$S_4 = S_{11}$
$S_5 = S_{12}$
$S_6 = S_9$
$S_7 = S_{10}$
$\dfrac{S_{15}}{S_{18}} = \dfrac{S_{16}}{S_{19}} = \dfrac{S_{17}}{S_{20}}$
$\dfrac{S_{13}}{S_{14}} = \dfrac{S_{16}}{S_{17}} = \dfrac{S_{21}}{S_{22}}$

A/V^2, the W/L ratio of M3 is 4.4. The bulk effect must be accounted for in the design of M2 and M1. The W/L ratio for M2 is 0.035 and for M1 is 1.2. Assuming that the current through M15, M16, and M17 is 150 μA leads to $W_{15}/L_{15} = 0.042$, $W_{16}/L_{16} = 4.4$, and $W_{17}/L_{17} = 6.7$.

Now let us consider in more detail the individual stages, beginning with the output stage. Our objective is to design the values of S_i consistent with the constraints developed for biasing. It will be assumed that the output should be able to charge a 70 pF load to $\pm$ 5 volts in 1 μs. The value of M25 can be readily determined from this specification. For example, the current required is given by

$$I_{25} = C_L \; (dV_{\text{out}}/dt) \tag{10}$$

Equating Eq. (10) to the current of M25 in saturation results in

$$C_L \; (dV_{\text{out}}/dt) = \frac{\mu \; C_{\text{ox}}}{2} S_{25} \, [V_{DD} - V_{\text{out}} - V_T]^2 \tag{11}$$

Integrating Eq. (11) results in

$$\frac{t_c}{C_L} = \frac{V_{\text{out}_2} - V_{\text{out}_1}}{\dfrac{\mu \; C_{\text{ox}}}{2} S_{25} \, [V_{DD} - V_{\text{out}_2} - V_T][V_{DD} - V_{\text{out}_1} - V_T]} \tag{12}$$

where we have assumed that V_T is constant and has its worst-case value, and t_c is the time required to charge C_L to $\pm$ 5 volts ($t_c = t_2 - t_1$). Assuming a worst-case value of V_T of 5 volts and a value of $\mu C_{\text{ox}}/2$ of 10^{-5} $\mu A/V^2$ gives $S_{25} = 0.7$. Because we prefer integers when convenient, $S_{25} = 1$ is selected.

The ratios of M23, M24, M25, and M26 are designed next. There are two concerns with these devices. The first is that M26 can sink the necessary current to discharge C_2 to -5 volts in 1 μs, and the second is that the poles of the output stage are greater than GB. Let us consider more closely the output stage used in Fig. 8.7–6 and shown in Fig. 8.7–7(a). Figure 8.7–7(a) also shows the capacitances present in this circuit. A small-signal model of this circuit is shown in Fig. 8.7–7(b). We have assumed that the bulk

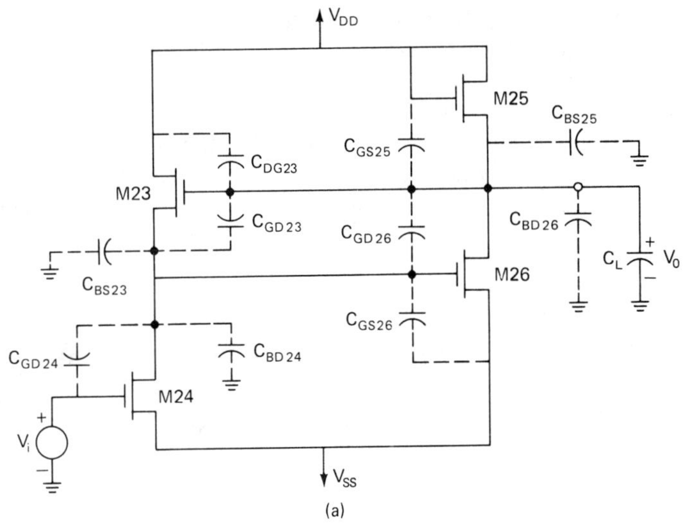

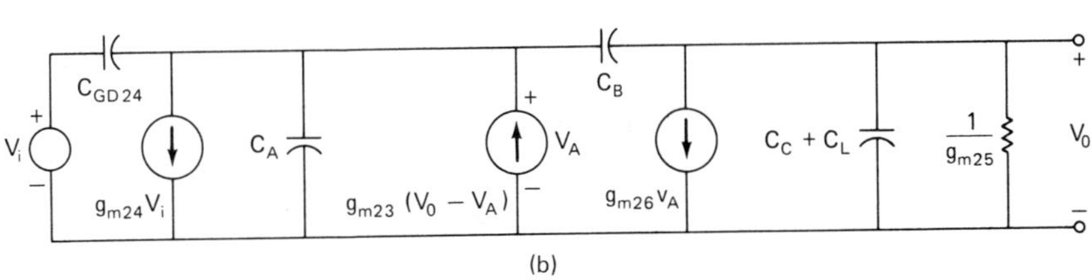

Fig. 8.7–7. (a) Output stage of Fig. 8.7–6 showing the location of parasitic capacitors. (b) Small signal model for Fig. 8.7–7(a).

effects are negligible in this model. The capacitors C_A, C_B, and C_C are expressed as

$$C_A = C_{BS\,23} + C_{BD\,24} + C_{GS\,26} \tag{13}$$

$$C_B = C_{GS\,23} + C_{GD\,26} \tag{14}$$

and

$$C_C = C_{GS\,25} + C_{BS\,25} + C_{BD\,26} \tag{15}$$

The low-frequency gain of Fig. 8.7–7(b) can be calculated as

$$\frac{V_0(0)}{V_i(0)} = \frac{g_{m\,24}}{g_{m\,23}}\left(\frac{g_{m\,26}/g_{m\,25}}{1 + g_{m\,26}/g_{m\,25}}\right) \tag{16}$$

Unfortunately Fig. 8.7–7(b) cannot be simplified to make hand calculations because of the small values of g_m. Therefore, one must turn to a computer-assisted analysis of this circuit to locate the poles. A tedious but workable approach is to choose W/L values of the devices and to iterate these choices until the poles are satisfactory. The results of such an analysis for two different choices of W/L values with C_L as a parameter are shown in Fig. 8.7–8(a) and 8.7–8(b). Because the poles of the output stage will not be affected by

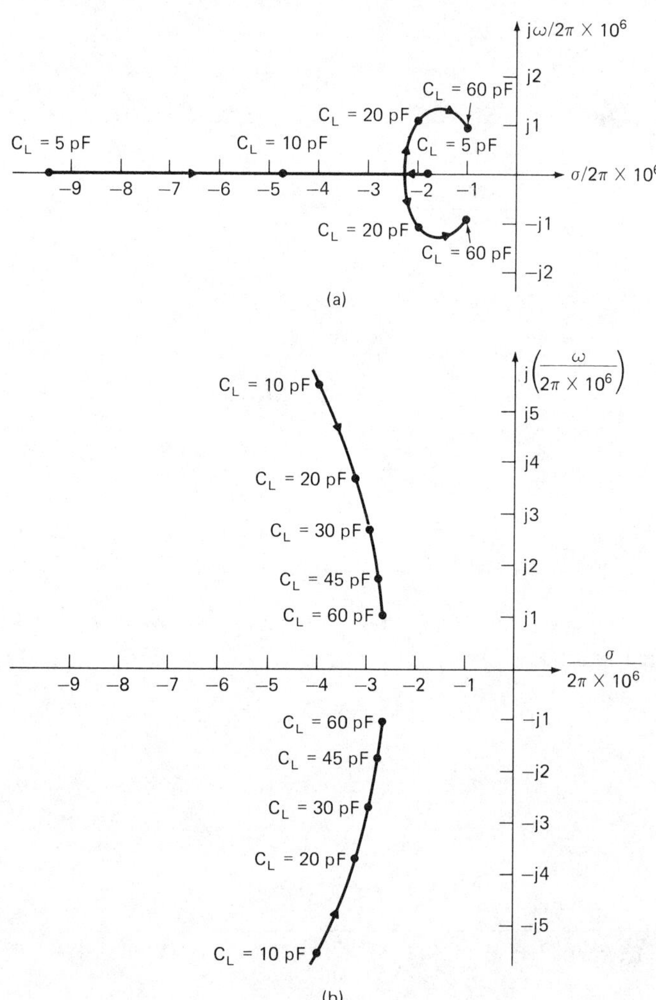

Fig. 8.7–8. Root locus plots of the output stage of Fig. 8.7–7(a) as a function of C_L. (a) $S_{23} = 0.1$, $S_{24} = 10$, $S_{25} = 1$, and $S_{26} = 40$. (b) $S_{23} = 0.53$, $S_{24} = 53$, $S_{25} = 1$, and $S_{26} = 16$.

the compensation of the op amp, they will have a strong influence on the transient response of the circuit. The exact influence will be determined by the negative feedback applied around the amplifier. Consequently, the output stage must be designed in view of the intended application of the op amp. For the op amp of Fig. 8.7–6, the W/L ratios corresponding to Fig. 8.7–8(b) were selected after extensive computer analysis. The gain of Eq. (16) becomes 8 in this case.

The next stage is the source follower that drives the output stage. As a starting point in the design of this stage, we shall assume that $S_{21} = S_{22}$. Because the voltage at the gate of M21 is low, the value of W/L of M21 and M22 will be larger than unity if sufficient current is desired in these devices. The primary concern of this stage is to make the poles occur above *GB*. The transfer function of the source follower of M21 and M22 is given in Eq. (27) of Section 8.5. The pole of this circuit can be expressed as

$$\omega_p \cong g_{m21}/(C_1 + C_2) \tag{17}$$

where we have assumed that g_{m21} is greater than $g_{mb21} + g_{ds21} + g_{ds22}$. From Eq. (23b) of Section 8.2, we have

$$C_1 \cong C_{GS21} \cong (2/3) C_{ox} W_{21} L_{21} \tag{18}$$

and from Eq. (23c) of Section 8.2

$$C_2 \cong C_L + C_{GD22} = C_{24} + 0.03\, W_{22}\, C_{ox} \tag{19}$$

C_{24} will be considered constant for this analysis. It is seen from Eqs. (18) and (19) that $C_1 < C_2$, so that cancellations of the pole and zero of the source follower cannot occur. In this case we will select a design that makes ω_p as large as possible. If we assume that $W_{21} = W_{22}$, then Eq. (8) becomes

$$\omega_p = \frac{\sqrt{2\,\mu\, C_{ox}}\sqrt{W_{21}/L_{21}}}{.67\, W_{21} L_{21} C_{ox} + C_{24}} = \frac{\sqrt{2\,\mu\, C_{ox}}\sqrt{S_{21}}}{0.67\, L_{21}^2 C_{ox} S_{21} + C_{24}} \tag{20}$$

A value for S_{21} that maximizes ω_p of Eq. (20) is

$$S_{21} = \frac{W_{21}}{L_{21}} = 1.5\frac{C_{24}}{W_{21}^2 C_{ox}} = \left(\frac{L_{24}}{L_{24}}\right)^2 S_{24} \tag{21}$$

If $L_{21} = L_{24} = 0.5$ mils, then $S_{21} = 53$. This value of M21 will result in a large capacitance at the cascode output and will increase the frequency response of the driver stage at the expense of the frequency response of the cascode stage. To increase the frequency response of the cascode stage, M21 and M22 were reduced to approximately 30. The gain of the stage is approximately 0.94, assuming $V_{BS\,21} = -15$ volts.

The design of the cascode stage, consisting of M18, M19, and M20, requires a compromise between power dissipation and frequency response. Assuming a current in M18, M19, and M20 of 1 μA and letting M18 have most of the voltage drop between V_{DD} and V_{SS} results in $W_{18}/L_{18} \cong 0.3$ $W_{19}/L_{19} = 30$ and $W_{20}/L_{20} = 45$. The gain of the cascode stage is found from Eq. (23) of Section 8.5 as approximately 9.25, where $V_{BS\,18} \cong -7$ volts.

The differential-to-single-ended converter consists of M4, M5, M11, and M12. For symmetry purposes we want $S_4 = S_{11}$ and $S_5 = S_{12}$, as shown in Table 8.7–2. However, to reduce the common-mode gain, it can be shown that the following expression must be satisfied.[59]

$$(g_{m\,4}\, g_{m\,12})/[g_{m\,11}(g_{m\,4} + g_{m\,5})] \cong 1 \tag{22}$$

If $g_{m\,5} >> g_{m\,4}$ and $g_{m\,12} >> g_{m\,11}$, then Eq. (22) can be satisfied. Therefore, W/L ratios of 10 were chosen for M5 and M12 and 0.1 for M4 and M11. The current through these devices is 250 μA. The differential gain of this stage is seen to be −0.99.

The input differential amplifier consists of M6, M7, M8, M9, and M10. To have a large, positive, common-mode input signal range, the voltage across M6 and M9 must be small. This voltage is given by V_{DD} and V_A and results in a W/L ratio of approximately 0.6 for M6 and M9. To increase the common-mode swing and improve the frequency response, S_6 and S_9 were increased to 2. From Eq. (35) of Section 8.5, we see that a choice of 60 for the W/L ratios of M7 and M10 results in a gain of approximately 5. The channel length of M8 should be as long as possible, to decrease the common-mode gain. However, long channels will lead to a restriction on the negative common-mode input signal. Therefore, a compromise of $W/L = 4$ was selected for M4.

Finally the frequency compensation must be designed. The total gain up to this point in the amplifier is given as

$$A_{v0} = 5 \times .99 \times 9.25 \times .94 \times 8 = 330 = 50 \text{ dB} \tag{23}$$

Although this gain is low, we shall continue. If the gain bandwidth, *GB*, of the amplifier is to be 2 MHz, then the dominant pole must be at 6 kHz.

[59] Y. P. Tsividis, "Design Considerations in Single-Channel MOS Analog Integrated Circuits—A Tutorial," *IEEE J. of Solid-State Circuits*, Vol. SC-13, No. 3, June 1978, pp. 383–391.

Using the feedback method of Fig. 8.6–10(a) results in the dominant pole given by Eq. (15) of Section 8.6, which can be approximated as

$$\omega_1 \cong \frac{1}{R_1 g_{m2} A R_2 C} \tag{24}$$

R_1 consists of the resistance seen to ground between M11 and M12 and is approximately $1/g_{m11}$. Because $g_{m2}\, R_2$ is the gain of the cascode stage and A is the gain of the source follower consisting of M13 and M14 (assumed to be 0.95), the value of C can be found from Eq. (24) as 67.5 pF. This rather large value of a capacitor is overconservative and could probably be reduced without significantly decreasing the phase margin.

The design of the source follower consisting of M13 and M14 is based on the ability to drive the compensation capacitor C_c. Therefore, M13 should be as large as possible without loading the output of the cascode stage. A W/L value of 20 was chosen for this design. Therefore, from Table 8.7–2 we get $S_{13} = S_{14} = 20$.

At this point all W/L ratios have been approximately designed and will be used as a starting point in a more detailed computer-assisted design. Now we must specify the actual sizes of the devices. Typically, the smallest dimension is selected as the minimum size allowable by the design rules. If matching is of concern, then the minimum size may be increased beyond this. Table 8.7–3 shows how the W/L ratios previously designed for Fig. 8.7–6 can be decomposed into actual values for W and L. Here we have assumed that 0.5 mils is the minimum channel dimension. It will probably be necessary to account for lateral diffusion in the final design of the W/L ratios. This information will depend upon the process used and should be readily available. Another example of an all-enhancement NMOS op amp design can be found in the literature.[60]

Space will not permit us to consider the design of other types of NMOS op amps in the detail of the previous design. However, we shall try to give a very brief review of one other type of NMOS op amp. Figure 8.7–9 shows the circuit schematic of an NMOS op amp that uses both enhancement and depletion mode devices.[61] In many respects this design is similar to the design

[60] I. A. Young, "A High-Performance All-Enhancement NMOS Operational Amplifier," *IEEE J. of Solid-State Circuits,* Vol. SC-14, No. 6, December 1979, pp. 1070–1077.

[61] A. W. T. Ismail, "Design of a Low Power, Internally Compensated, NMOS Operational Amplifier," Master of Engineering Report, Dept. of Electrical Engineering, Texas A&M University, College Station, TX 77843, May 1981.

Table 8.7–3 Channel Dimensions and *W/L* Ratios for Operational Amplifier of Fig. 8.7–6

DEVICE	*W* (mils)	*L* (mils)	*W/L*
M1	1.20	1.00	1.2
M2	0.50	14.30	.035
M3	4.40	1.00	4.4
M4	0.50	5.00	.1
M5	5.00	0.50	10
M6	2.00	1.00	2
M7	30.00	0.50	60
M8	4.00	1.00	4
M9	2.00	1.00	2
M10	30.00	0.50	60
M11	0.50	5.00	.4
M12	5.00	0.50	10
M13	10.00	0.50	20
M14	10.00	0.50	20
M15	0.50	11.90	.042
M16	2.20	0.50	4.4
M17	3.35	0.50	6.7
M18	0.50	1.67	.3
M19	15.00	0.50	30
M20	15.00	0.50	30
M21	15.00	0.50	30
M22	15.00	0.50	30
M23	0.75	1.42	.53
M24	26.50	0.50	53
M25	1.50	1.50	1
M26	16.00	1.00	16

of Fig. 8.7–6, except that the higher gain of the depletion devices permits the use of a source follower output stage. The use of M11 to increase the gain of the cascode is also a difference between the designs. This was done so that the compensation capacitor, which was 6 pF, would be small. The block diagram from which Fig. 8.7–9 was designed is shown in Fig. 8.7–10. In addition to C_c, two feedforward capacitors were also used. Figure 8.7–11 shows the open-loop frequency response of Fig. 8.7–9 for loads of 10 and 20 pF. The sensitivity of the stability of the loop to the capacitive load is apparent in these results. Later versions of this op amp also included a nulling resistor, R_z, in series with C_c to increase the phase margin. Tables 8.7–4, 8.7–5, 8.7–6, and 8.7–7 give pertinent information and the results of

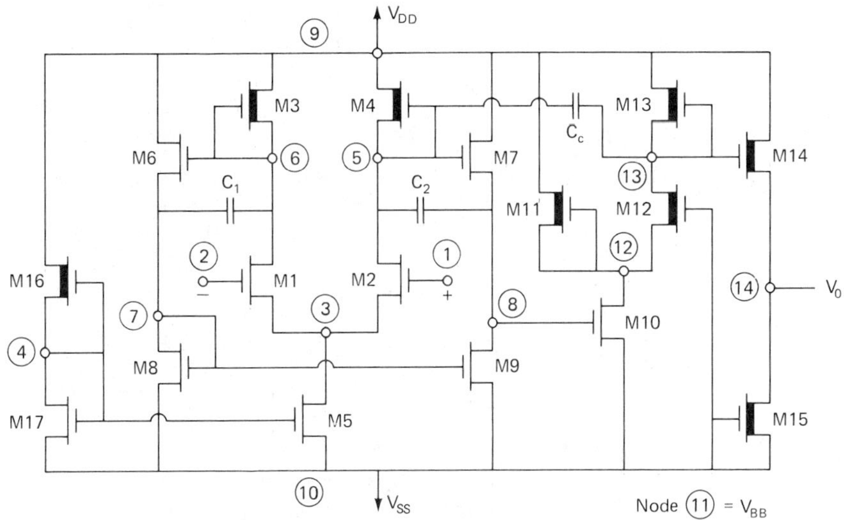

Fig. 8.7–9. NMOS op amp using enhancement and depletion devices.

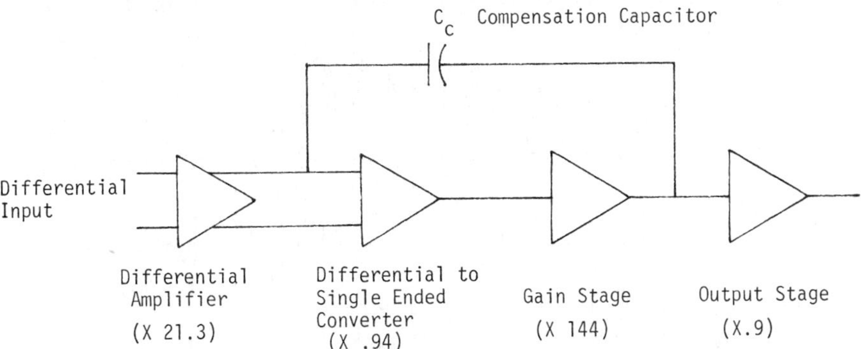

Fig. 8.7–10. Block diagram of the op amp of Fig. 8.7–9.

this design. Other examples of enhancement and depletion NMOS op amps can be found in the literature.[62–65]

---

[62] A. W. T. Ismail, "Design of a Low Power, Internally Compensated, NMOS Operational Amplifier," Master of Engineering Report, Dept. of Electrical Engineering, Texas A&M University, College Station, TX 77843, May 1981.

[63] B. J. Hosticka, R. W. Brodersen, and P. R. Gray, "MOS Sampled Data Recursive Filters Using Switched Capacitors Integrators," *IEEE J. of Solid-State Circuits,* Vol. SC-12, No. 6, December 1977, pp. 600–608.

[64] W. L. Eversole, D. J. Mayer, P. W. Bosshart, M. deWit, C. R. Hewes, and D. D. Buss, "A Fully Integrated 32-point Chirp *z*-transform IC," *1978 IEEE Int. Solid-State Circuits Conference Digest of Technical Papers,* Philadelphia, February 1978, pp. 86–87, 269.

[65] Y. P. Tsividis, D. L. Fraser, Jr., and J. E. Dziak, "A Process-Insensitive High-Performance NMOS Operational Amplifier," *IEEE J. of Solid-State Circuits,* Vol. SC-15, No. 6, December 1980, pp. 921–928.

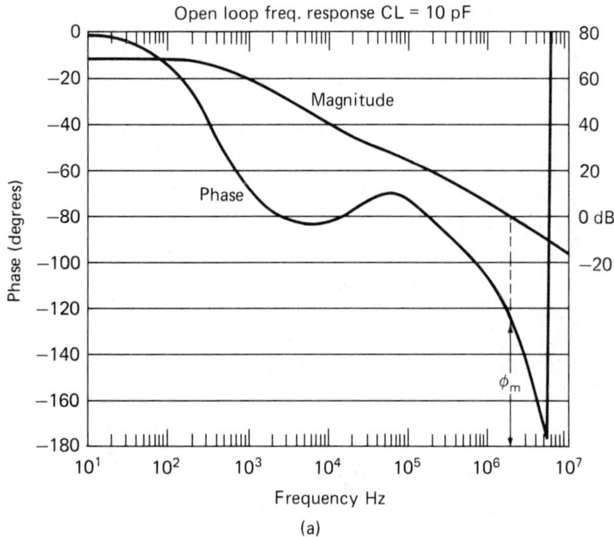

Fig. 8.7–11. (a) Open loop frequency response with $C_c, C_1, C_2, C_L = 10\text{pF}$ for Fig. 8.7–9.

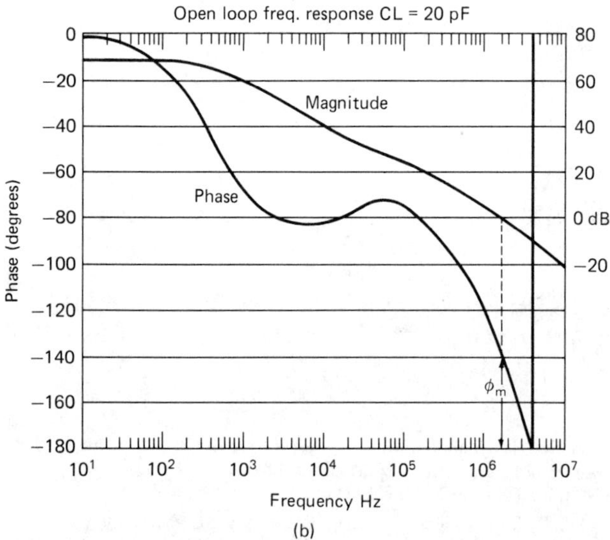

Fig. 8.7–11. (b) Open loop frequency response with $C_c, C_1, C_2, C_L = 20\text{pF}$ for Fig. 8.7–9.

Table 8.7–4 Device *W/L*, Current, Transconductance for the Op Amp of Fig. 8.7–9

DEVICE	TYPE	$\frac{W}{L}$	W, mil	L, mil	I_D, µA	g_m, µA/V	g_{mbs}, µA/V
M1	Enh.	1.7	0.85	0.5	20.8	46.1	2.46
M2	Enh.	1.7	0.85	0.5	20.8	46.1	2.46
M3	Dep.	.35	.35	1.0	20.8	20.5	1.32
M4	Dep.	.35	.35	1.0	20.8	20.5	1.32
M5	Enh.	1.0	1.0	1.0	41.6	49.3	3.25
M6	Enh.	.125	0.3	2.4	50.3	19.9	1.15
M7	Enh.	.125	0.3	2.4	50.3	19.9	1.15
M8	Enh.	8.0	2.4	0.3	50.3	152	10
M9	Enh.	8.0	2.4	0.3	50.3	152	10
M10	Enh.	10.	5.0	0.5	63.0	190	12.5
M11	Dep.	0.5	0.4	0.8	48.2	38.9	3.39
M12	Dep.	2.0	1.0	0.5	14.8	41.7	3.63
M13	Dep.	0.2	0.3	1.5	14.8	13.4	0.97
M14	Dep.	4.0	2.0	0.5	314	275	20
M15	Dep.	3.0	1.5	0.5	314	238	23.8
M16	Dep.	.464	0.65	1.4	41.4	34.6	2.83
M17	Enh.	1.0	1.0	1.0	41.4	49.1	3.24

$C_c = 6$ pF
$C_1 = C_2 = 3$ pF

N-Channel, CMOS Op Amp

A circuit of an *n*-channel, unbuffered CMOS op amp is shown in Fig. 8.7–12. The bias is provided by R_B and M6. The differential amplifier consists of M1, M2, and the loads M3 and M4, with M7 acting as a current source. M5 and M8 form the gain stage, and C_c is the Miller compensating capacitor. This circuit has the advantage of simplicity and is expected to perform very

Table 8.7–5 Operating Point Node Voltages of the Open-Loop Amplifier of Fig. 8.7–9

NODE	VOLTAGE	NODE	VOLTAGE
1	0.0	8	−3.336
2	0.0	9	5.0
3	−2.076	10	−5.0
4	−2.311	11	−10.0
5	2.824	12	−3.235
6	2.824	13	0.004
7	−3.336	14	−0.0562

Table 8.7–6 SPICE MOSFET Parameters for the Op Amp of Fig. 8.7–9

TYPE	ENHANCEMENT	DEPLETION	
V_{To}	0.5	−3.4	V
Kp	14	14	μA/V^2
γ	0.31	0.47	V$^{0.5}$
ϕ	0.54	0.54	V
λ	0.015	0.015	1/V
C_{BD}	2.0E-14	2.0E-14	F
C_{BS}	2.0E-14	2.0E-14	F
P_B	0.7	0.7	V
C_{GSO}	3.3E-10	3.3E-10	F/m
C_{GDO}	3.3E-10	3.3E-10	F/m
C_{GBO}	1.5E-9	1.5E-9	F/m
T_{ox}	8.0E-8	8.0E-8	m
F_c	0.5	0.5	–

well in applications with small capacitive loads, e.g., switched capacitor circuits. Also R_B and M6 can be shared with other op amps.

The design of the DC conditions of the op amp is the first step. The objective is to make sure all devices are saturated regardless of process variations. For $V_{DD} = 5$ volts and $V_{SS} = -5$ volts, and since $|V_T| = 1$ volt, choose the voltage at node 2 in Fig. 8.7–12 as −2 volts. This will give ample current with moderately sized devices. Selecting R_B as 100 kΩ gives $I_6 =$

Table 8.7–7 Expected Performance of the Op Amp of Fig. 8.7–9 as Predicted by SPICE

Power supply	$V_{DD} = 5$ volt
	$V_{SS} = -5$ volt
	$V_{BB} = -10$ volt
Power consumption	5.6 mW
Open-loop low-frequency gain	68 dB
Unity-gain bandwidth	2 MHz
Phase margin (no load)	70°
(20-pF load)	40°
Input common-mode voltage range	±3.5 volt
Common-mode rejection ratio (CMRR)	60 dB
Power supply rejection ratio (PSRR)	for V_{DD} 79 dB
	V_{SS} 59 dB
Output voltage swing	±3.7 volt
Capacitive load (phase margin > 45 degrees)	20 pF
Slew rate	±4 volt/μsec
1% Settling time (C_L = 20 pF)	1 μsec

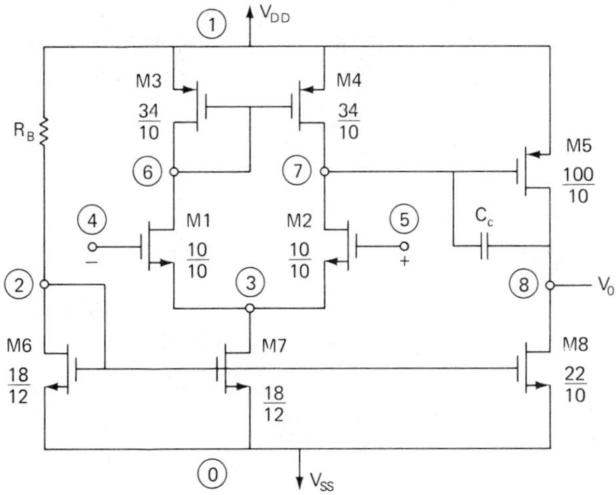

Fig. 8.7–12. Circuit schematic of an unbuffered CMOS op amp with an n-channel input pair.

70 μA. To simplify the notation, we define S_i as the W_i/L_i ratio of the ith device. Now $S_6 = W_6/L_6$ is found from

$$W_6/L_6 = \frac{I_6}{K'_n(V_{GS6} - V_{Tn6})^2} = \frac{70}{48} \tag{25}$$

where K'_n is assumed to be 12.375 $\mu A^2/V$. Thus, select S_6 = 18 microns/12 microns. To apply a current of 70 μA to the differential n-channel stage, select $S_7 = S_6$. Because the gate source voltages of M8 and M6 are equal, we have

$$S_8 = S_6 \times (I_8/I_6) \tag{26}$$

For M4 to remain in saturation

$$V_{GS6} = V_{GS3} \quad \text{or} \quad S_3 = S_5 \times (I_3/I_5) \tag{27}$$

Choosing $S_3 = S_4$ and $S_1 = S_2$, and because $I_5 = I_8$, Eq. (27) becomes $S_4 = S_5 \times (I_4/I_8)$, but $I_4 = ½\ I_6$, and from Eq. (26), the condition for M4 to remain in saturation becomes

$$S_5 = 2\,(S_4 S_8)/S_7 \tag{28}$$

Therefore, all devices in the op amp are guaranteed to be in saturation under normal operating conditions. The differential amplifier gain can be seen to be

$$A_d = \sqrt{\frac{K'_n S_1}{I_1 \lambda^2}} \tag{29}$$

where $\lambda_n = \lambda_p = 0.01$ (all channel length modulation parameters are assumed equal because all lengths are either 10 or 12 microns), and $I_1 = 35\ \mu\text{A}$. Therefore

$$A_d = 59.46\sqrt{S_1} \tag{30}$$

The inverter gain, A_1, is given by

$$A_I = \sqrt{\frac{K'_p S_5}{I_5 \lambda^2}} \tag{31}$$

where $K'_p = 5.065\ \mu A/V^2$, $I_5 = 100\ \mu\text{A}$, and $\lambda = 0.01$. Therefore

$$A_I = 22.5\sqrt{S_5} \tag{32}$$

Choosing $S_1 = 1 = 10$ microns/10 microns and $S_5 = 10S_1 = 100$ microns/10 microns, we find $A_d = 59.46$ and $A_I = 71.15$. From Eq. (28), $S_3 = S_4$ is found to be 3.41. Let $S_3 = S_4 = 34$ microns/10 microns. Therefore, the overall gain becomes $A_d \times A_I = 4230$ (73 dB), and the output resistance $= 1/(2\lambda\ I_8) = 500\ \text{k}\Omega$.

The AC considerations are the next step in the op amp design procedure. The transconductance of the first stage is given by

$$g_{m1} = 2\sqrt{S_1 K'_n I_1} = 41.6\ \mu\mho \tag{33}$$

and that of the second stage is

$$g_{m2} = 2\sqrt{S_5 K'_p I_5} = 142\ \mu\mho \tag{34}$$

If a 1 MHz unity-gain bandwidth is required, then from Eq. (12) in Section 8.6, C_c is found to be 6.6 pF. Unfortunately, a right-half-plane zero appears due to signal feedforward through C_c at high frequencies. It is located at

$$\omega_0 = g_{m2}/C_c \tag{35}$$

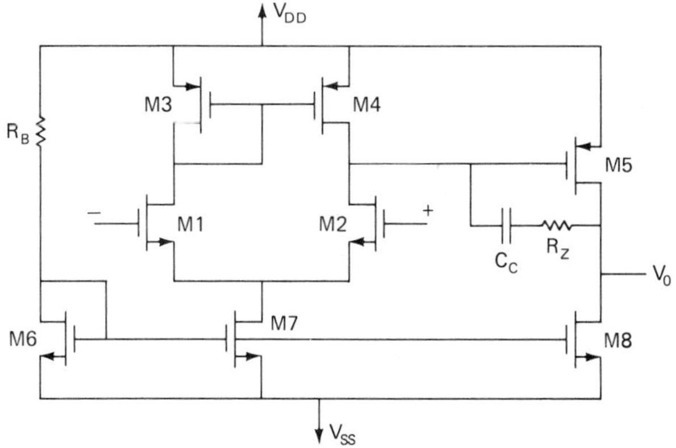

Fig. 8.7–13. The use of nulling resistance R_z to improve the phase margin.

and for $C_c = 6.6$ pF, the zero is at 3.41 MHz. Also, a nondominant pole is located approximately at

$$\omega_{p2} = -g_{m2}/C_L \tag{36}$$

where C_L is the capacitive load at node 8. From Eqs. (35) and (36), it is seen that in order to minimize any resulting phase degradation, g_{m2} should be designed to be greater than g_{m1}.

One way to reduce the effects of the right-half-plane zero is by connecting a nulling resistance R_z in series with C_c, as shown in Fig. 8.7–13. In this case, Eq. (28) of Section 8.6 gives the zero location as

$$\omega'_0 = \frac{1}{C_c\left(\dfrac{1}{g_{m2}} - R_z\right)} \tag{37}$$

By proper choice of R_z, the zero can be moved away from the unity-gain bandwidth or even to the left-half plane, where it would improve the phase

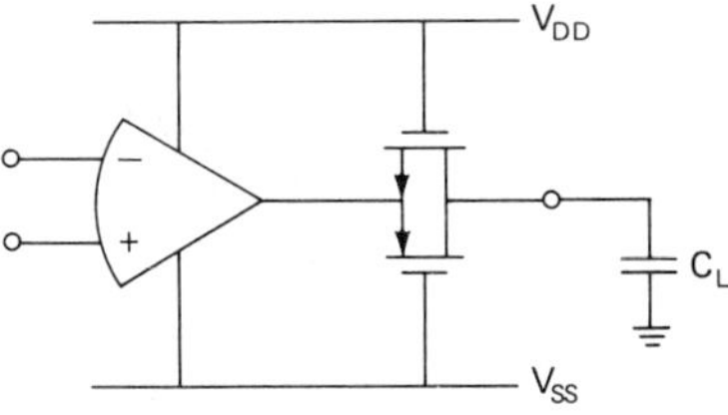

Fig. 8.7–14. Improvement of response time by adding a resistance between op amp and C_L.

Table 8.7–8 Dependence of the Performance of Fig. 8.7–13 upon DC Current, *W/L* Ratios, and the Compensating Capacitor

	DRAIN CURRENT		DIFFERENTIAL PAIR, M1, M2		DIFFERENTIAL LOAD M3, M4		INVERTER		INVERTER LOAD		COMP. CAP.
	I_7	I_8	W	L	W	L	W_5	L_5	W_8	L_8	C_c
Increase DC gain	$(\downarrow)^{1/2}$	$(\downarrow)^{1/2}$	$(\uparrow)^{1/2}$	$(\downarrow)^{1/2}$		$\uparrow$	$(\uparrow)^{1/2}$	$(\downarrow)^{1/2}$		$\uparrow$	
Increase gain BW	$(\uparrow)^{1/2}$		$(\uparrow)^{1/2}$	$(\downarrow)^{1/2}$							$\downarrow$
Increase RHP zero f		$(\uparrow)^{1/2}$					$(\uparrow)^{1/2}$	$(\downarrow)^{1/2}$			$\downarrow$
Increase slew rate	$\uparrow$	$\uparrow$									$\downarrow$
Increase capacitive load without buffer											$\uparrow$

margin of the amplifier rather than degrade it. For example, if $g_{m2} = 142$ $\mu℧$ or $1/g_{m2} = 7.04$ kΩ and $R_z \cong 7$–8 kΩ, then the zero is in the left-half plane, with subsequent improvement in phase margin.

It is also possible to improve the AC performance of the CMOS op amp and hence its settling time by using external devices in series with the output of the op amp. Figure 8.7–14 illustrates this situation. These devices act as ON switches (pass transistors) and reduce the effective capacitive load seen at node 8 of Fig. 8.7–12.

The slew rate of the op amp is given by Eq. (48) of Section 8.5 as I_1/C_c or 5.3 volts/microsecond. This slew rate limitation occurs only for input voltage changes of greater than $|V_T|$.

Table 8.7–8 summarizes the dependence of the performance of the op amp with respect to current, *W/L* ratios, and the compensating capacitor. An up arrow means increase and a down arrow means decrease. It is seen that the design represents the compromise of several factors. Table 8.7–9 gives the results of a SPICE2 analysis of the CMOS op amp in Fig. 8.7–12 using the model parameters of Table 8.7–10. The influence of R_B upon this op amp is illustrated in Table 8.7–11. If an external resistor is used for R_B, then the op amp can be programmed for various operating conditions.

A *p*-channel CMOS op amp is shown in Fig. 8.7–15. It is designed in the same manner as the one in Fig. 8.7–13. The primary difference is that in a bulk CMOS technology using an N^- substrate, the bulks of the *p*-channel input devices must be taken to V_{DD}, so that the body effects cannot be ignored. The primary influence of the body effect will be in limiting the negative

Table 8.7–9 Results of the Computer Analysis of Fig. 8.7–12

Power supplies	$V_{DO} = +5$	
	$V_{SS} = -5$	
Open-loop low-frequency gain	$5037 = 74$ dB	
Output resistance	497 kΩ	
Power consumption	2.46 mW	
Power supply rejection ratio	81.76 dB V_{DD}	
	95.23 dB V_{SS}	
Unity-gain Bandwidth	1 MHz	No load
Phase margin	75°	No load
Input CMR	$-4 \rightarrow +3.8$V	
Output voltage swing	$-4.75 \rightarrow +4.75$V	
CMRR	81 dB (at CMI − 0)	
	60 dB (at CMI = −0.8)	
Slew rate	for $C_L = 10$ pF	+10 V/μs
		−7 V/μs
Settling time	for $C_L = 10$ pF	≤ 1.5 μs

Table 8.7–10 MOS Model Parameters for SPICE2

TYPE	NMOS	PMOS	UNITS
LEVEL	1	1	
VTO	1.000	−1.000	V
KP	2.40D-05	1.20D-05	A/V^2
LAMBDA	1.00D-02	1.00D-02	V^{-1}
CBD	2.00D-14	2.00D-14	F
CBS	2.00D-14	2.00D-14	F
PB	0.700	0.700	V
CGSO	3.30D-10	3.30D-10	F/m
CGDO	3.30D-10	3.30D-10	F/m
CGBO	1.50D-09	1.50D-09	F/m
TOX	8.00D-08	8.00D-08	m
FC	0.500	0.500	

potential swing of the common-mode input range. Another difference between Figs. 8.7–13 and 8.7–15 is that the second stage should have more gain for the p-channel input op amp because the n-mobility is greater than the p-mobility.

The CMOS op amps of Figs. 8.7–13 and 8.7–15 represent a very basic architecture. Many more CMOS op amp architectures have been developed. Some of these approaches take advantage of the substrate bipolar junction transistor, which is available in the bulk CMOS process. Further information can be found in the literature.[66-74]

---

[66] D. Herbst, B. Hoefflinger, K. Schumacher, R. Schweer, A. Fettweis, K. A. Owenier, and J. Pandel, "MOS Switched Capacitor Filters with Reduced Number of Operational Amplifiers," *IEEE J. of Solid-State Circuits,* Vol. SC-14, December 1979, pp. 1010–1019.

[67] B. J. White, G. M. Jacobs, and G. F. Landsburg, "A Monolithic Dual Tone Multifrequency Receiver," *IEEE J. of Solid-State Circuits,* Vol. SC-14, December 1979, pp. 991–997.

[68] Y. Suzuki, E. Masuda, C. Stoh, and T. Iida, "A New Single-Chip C2MOS A/D Converter for Microprocessor Systems-Penta Phase Integrating C2MOS A/D Converter," *IEEE J. of Solid-State Circuits,* Vol. SC-13, December 1978, pp. 779–785.

[69] R. Gregorian and W. E. Nicholson, "CMOS Switched Capacitor Filters for a PCM Voice Codec," *IEEE J. of Solid-State Circuits,* Vol. SC-14, December 1979, pp. 970–980.

[70] W. C. Black, D. J. Allstot, and R. A. Reed, "A High Performance Low Power CMOS Channel Filter," *IEEE J. of Solid-State Circuits,* Vol. SC-15, December 1980, pp. 929–938.

[71] G. Samarandoiu, D. A. Hodges, P. R. Gray, G. F. Landsburg, "CMOS Pulse-Code Modulation Voice Codec," *IEEE J. of Solid-State Circuits,* Vol. SC-13, August 1978, pp. 504–510.

[72] Y. Haque, R. Gregorian, R. W. Blasco, R. A. Mao, and W. E. Nicholson, "A Two Chip PCM Voice Codec with Filters," *IEEE J. of Solid-State Circuits,* Vol. SC-14, December 1979, pp. 961–969.

[73] W. J. Parrish, An Ion Implanted CMOS Amplifier for High Performance Active Filters, Ph.D. dissertation, Univ. of California, Santa Barbara, June 1976.

[74] F. H. Musa and R. C. Huntington, "A CMOS Monolithic 3½-Digit A/D Converter," *IEEE Int. Solid State Circuits Conf. Digest of Tech. Papers,* 1976, pp. 144–145.

Table 8.7-11 Influence of R_B upon the Performance of Fig. 8.7-12

R_B, kΩ	POWER mW	GAIN	R_o, kΩ	GB, MHz	I_7, μA	I_8, μA	PHASE MARGIN (NO LOAD)	g_{m1}, m℧
50	4.21	347.8	33.43	1.2	115 (active)	178 (active)	≃ 75°	54
75	3.15	3950	389	1.1	89.5	135	≃ 75°	47.6
100	2.46	5036	497	1	70	106	≃ 75°	42.1
200	1.33	9266	918	0.79	38	57.2	≃ 75°	31
500	0.57	21,480	2136	0.5	16.3	24.6	≃ 75°	20.3
1000	0.295	41,310	4118	0.39	8.47	12.7	≃ 74°	14.6

We may use the p-channel CMOS op amp of Fig. 8.7–15 to study the influence of the power supply upon the performance of the op amp. Figure 8.7–16 shows a model of Fig. 8.7–15 that includes a small-signal ripple, $V_s^+(\omega)$ and $V_s^-(\omega)$, in series with V_{DD} and V_{SS}, respectively. Power supply rejection ratio, PSRR, can be defined using this model and is

$$\text{PSRR}^+ = [V_0(\omega)/V_i(\omega) \mid V_s^+(\omega) = 0]/[V_0(\omega)/V_s^+(\omega) \mid V_i(\omega) = 0] \quad (38)$$

and

$$\text{PSRR}^- = [V_0(\omega)/V_i(\omega) \mid V_s^-(\omega) = 0]/[V_0(\omega)/V_s^-(\omega) \mid V_i(\omega) = 0] \quad (39)$$

Because the numerator of Eqs. (38) and (39) is the compensated open-loop frequency response of the op amp, then all one must do is calculate the denominators.

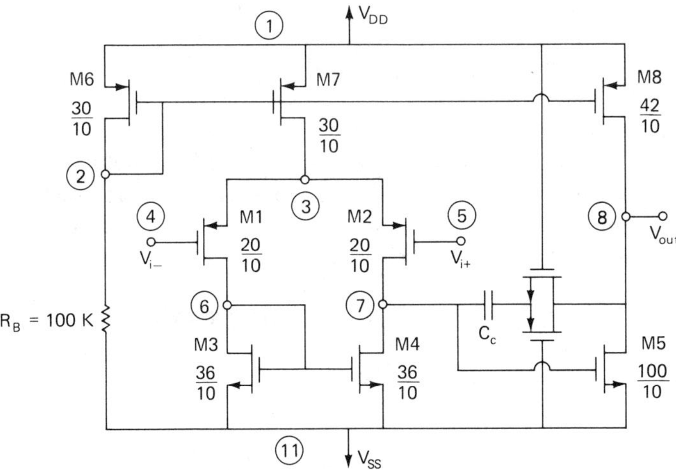

Fig. 8.7–15. Unbuffered p-channel input pair, CMOS op amp.

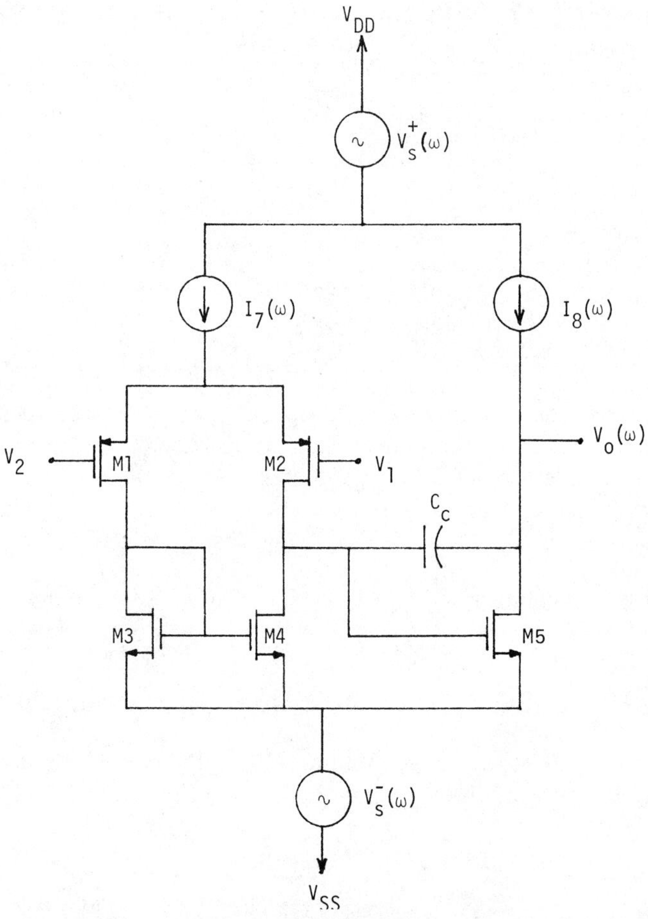

Fig. 8.7–16. Model for PSRR for Fig. 8.7–15.

$V_s^+(\omega)$ or $V_s^-(\omega)$ will cause the current sources of Fig. 8.7–16 to have an AC ripple. However, because $V_{id} = 0$, the differential-input stage will reject this common-mode input signal. Consequently, the primary source of supply ripple is from the current source of M5. Assuming that $V_{gs5}(\omega) \cong 0$ and that $I_8(\omega)$ is given approximately as

$$I_8(\omega) \cong (10/7)\,[V_s^+(\omega) - V_s^-(\omega)/R_B] \tag{40}$$

which leads to

$$V_0(\omega) \cong (10/7)\,\frac{r_{ds5}\,r_{ds8}}{R_B(r_{ds5} + r_{ds8})}\,[V_s^+(\omega) - V_s^-(\omega)] \tag{41}$$

For the values of our example, we get

$$V_0(\omega) \cong 7.145\,[V_s^+(\omega) - V_s^-(\omega)] \tag{42}$$

Therefore, we see that the PSRR magnitude curve will have the shape of $|A_V(\omega)|$ but be reduced by a factor of 1/7.145, or −17.1 dB.

The noise performance of Fig. 8.7–12 or 8.7–15 can be calculated using the results of noise analysis of the differential amplifier of Section 8.5. In Fig. 8.5–18, the equivalent input noise voltage was calculated. If we assume that the input differential stage has sufficient gain, then the noise of the second stage will not be important when referred to the input. Therefore, the equivalent input noise voltage of Eq. (53) of Section 8.5 is approximately the equivalent input noise voltage of Fig. 8.7–12 or Fig. 8.7–15.

This section has attempted to use the design principles of the previous section and to illustrate the design method for MOS op amps. Two MOS op amp designs were considered in detail. Many topics and alternative architectures have been omitted and should be examined to complete the subject.

8.8 SUMMARY

In this chapter we have presented MOS technology and MOS analog circuit design that can be used to implement the switched capacitor circuits of the previous chapters as integrated circuits. It is difficult to provide the necessary depth in a single chapter, but a brief overview of MOS technology and MOS analog circuit design has been presented. Of necessity many important topics, such as simulation and testing of MOS circuits, have not been included. At least, the underlying principles applicable to future problems have been identified here in this chapter.

PROBLEMS

8.1 (Sec. 8.1). If the *n* substrate and *p* well of Fig. 8.1–3 are taken to V_{DD} and V_{SS}, respectively, show how a parasitic lateral *pnp* and a vertical *npn* bipolar transistor can be formed that is the equivalent of an SCR. V_{DD} is the most positive circuit voltage, and V_{SS} is the most negative circuit voltage.

8.2 (Sec. 8.1). Show how a bipolar junction transistor can be achieved from the CMOS process shown in Fig. 8.1–3. Is this transistor *npn* or *pnp,* and are the terminals constrained?

8.3 (Sec. 8.1). Use the design rules given and lay out on a grid where the smallest resolution is λ, the circuit of Fig. 8.1–4 where the W/L ratio of M1 is 6λ/6λ and that of M2 is 6λ/24λ.

8.4 (Sec. 8.1). Assume that λ is equal to 0.05 mils. Lay out the three-stage ring oscillator shown in Fig. P8.3 to minimize the two dotted rectangles shown. The W/L values are given in terms of mils in the circuit. The objective of this problem is to maximize the value of the SCORE defined as

$$\text{SCORE} = \frac{500}{\begin{array}{l}\text{Minimum rectangular}\\ \text{area that just encom-}\\ \text{passes your active cir-}\\ \text{cuit not including}\\ \text{bond pads in sq. mils}\end{array}} + \frac{5000}{\begin{array}{l}\text{minimum rectangular}\\ \text{area that just encom-}\\ \text{passes your entire cir-}\\ \text{cuit including the}\\ \text{bond pads in sq. mils}\end{array}} - 10 \times \begin{array}{c}\text{number of}\\ \text{design rule}\\ \text{violations}\end{array}$$

The inner rectangle must not touch anything except the four metal runs, which must cross it orthogonally. The outer rectangle can be coincident with the outer edge of the bond pad metal. (If a design rule checker exists, it can be used to determine the number of design rule violations.)

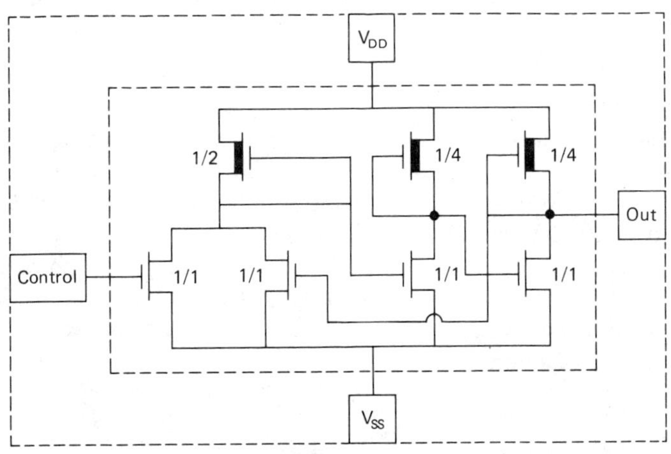

Fig. P8.3.

8.5 (Sec. 8.2). Show that Eq. (10) corresponds to the maximum of the curves in Fig. 8.2–4.

8.6 (Sec. 8.2). Use the values given in Table 8.2–1 to find the drain current in an n-channel enhancement transistor, assuming the following terminal voltages: (a) $V_D = 10$ volts, $V_G = 1.5$ volts, $V_S = 0$ volts, and $V_B = -5$ volts, (b) $V_D = 1$ volt, $V_G = 2$ volts, $V_S = 0$ volts, and $V_B = -5$ volts.

8.7 (Sec. 8.2). Repeat problem 8.6 if $V_B = 0$ volts.

8.8 (Sec. 8.2). Use the values given in Table 8.2–1 to find the drain current of an n-channel depletion transistor, assuming that the terminal voltages are $V_D = 10$ volts, $V_G = V_S = 0$ volts, and $V_B = -5$ volts. The threshold voltage for the n-channel depletion transistor is assumed to be −3.5 volts.

8.9 (Sec. 8.2). Use the parameters in Table 8.2–1 to estimate the values of C_1, C_2, C_3, C_4, C_{BS}, and C_{BD} of Fig. 8.2–7 for an MOS device with a length of 1 mil and a width of 1 mil for the active region.

8.10 (Sec. 8.2). Repeat Problem 8.9 for the saturated region.

8.11 (Sec. 8.2). Repeat Problem 8.9 for the cutoff region.

8.12 (Sec. 8.2). Find the conductances and transconductances of the small-signal model of Fig. 8.2–12, assuming the values in Table 8.2–1 and that (a) $I_D = 10$ microamperes and (b) $I_D = 100$ microamperes.

8.13 (Sec. 8.2). A depletion current source is shown in Fig. P8.13. The parameters of the transistor are assumed to be given by Table 8.2–1. Assume that V_{T0} is − 3.5 volts. Find the value of W/L that will give a current of 50 microamperes.

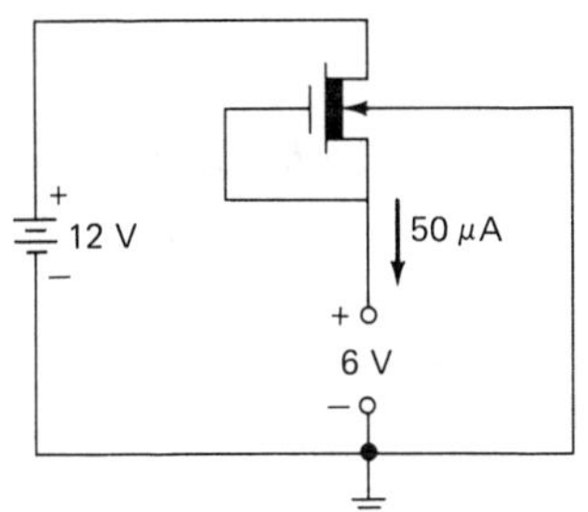

Fig. P8.13.

8.14 (Sec. 8.2). Fig. P8.14 shows an enhancement load inverter. If the W/L ratios of both devices are equal to 17, find the value of V_{out}. Assume that the parameters for the devices are given in Table 8.2–1.

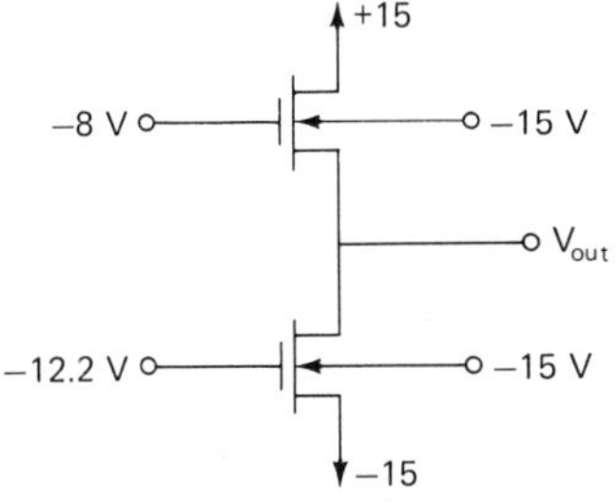

Fig. P8.14.

8.15 (Sec. 8.3). Develop Eq. (2) from Eq. (1).

8.16 (Sec. 8.3). What is the ON resistance of an enhancement switch using the values of Table 8.2–1 if $V_S = 0$ volts, $W/L = 1$, and (a) $V_G = 10$ volts and (b) $V_G = 2$ volts? Assume V_D is greater than or equal to V_S.

8.17 (Sec. 8.3). Repeat Problem 8.16 if $V_S = 5$ volts.

8.18 (Sec. 8.3). If C_1 in Fig. 8.3–6 is 5 pF, what value of W/L ratio is required for M1 to charge C_1 to 99.9% of V_{in}, if the amplitude of the clock is 15 volts, V_{in} is 5 volts, and the clock frequency is 100 kHz with clock pulses of 40% duty cycle?

8.19 (Sec. 8.3). Calculate the feedthrough of the clock in Fig. 8.3–10 if C_1 is 5 pF.

8.20 (Sec. 8.4). Two depletion devices are used, as shown in Fig. P8.20, to divide V_{DD}. (a) Assume that the body and channel modulation effects are negligible, and find an algebraic expression for the value of V_{out}. (b) If $V_{DD} = 12$ volts and $V_{TO} = -3.4$ volts, what value of $(W_2/L_2)/(W_1/L_1)$ gives $V_{out} = 6$ volts, assuming that the values of Table 8.2–1 are valid for these devices.

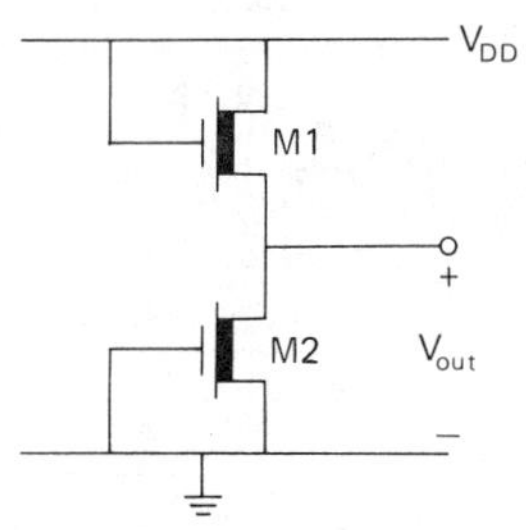

Fig. P8.20.

8.21 (Sec. 8.4). Find the W and L values of M1 and M2 if the minimum device dimensions are 10 microns and minimum-size devices are to be used. Assume that $V_{TO1} = 1$ volt, $V_{TO2} = -3$ volts, $\gamma = 0.5\ V^{1/2}$, $\phi = 0.5$ volts, $\mu = 1000$ cm²/volt.second, $C_{ox} = 0.5 \times 10^{-7}$ F/cm², and $\gamma = 0.01\ V^{-1}$.

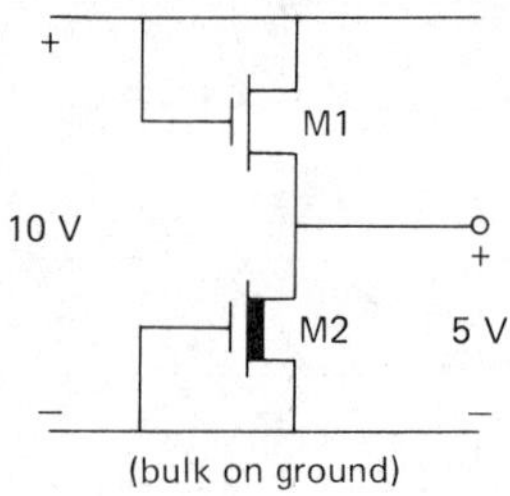

Fig. P8.21.

8.22 (Sec. 8.4). Solve for I_0 of Fig. P8.22 and differentiate with respect to V_0 to find an expression for the output resistance of this current source. Compare to Eq. (8).

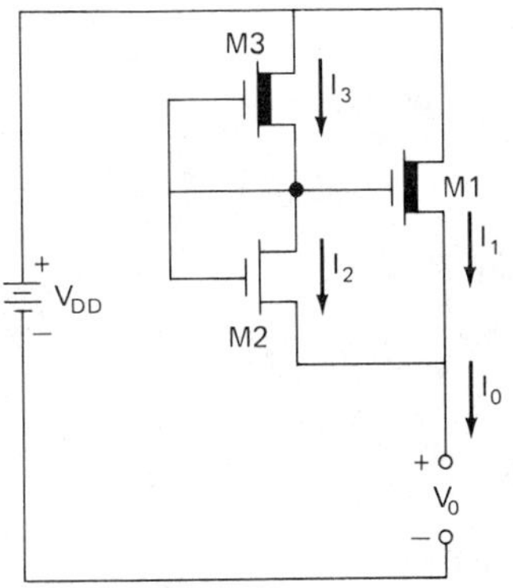

Fig. P8.22.

8.23 (Sec. 8.4). Develop the expression for the small-signal output resistance of Fig. 8.4–13(a) given in Eq. (16).

8.24 (Sec. 8.4). Assume that the op amp of Fig. 8.4–16 has an input offset voltage of V_{OS}. (a) Develop an expression for V_{ref} including V_{OS} of the op amp. (b) If $A_{E2} = 10A_{E1}$, $R_2/R_3 = 25$, and $V_{BE1} = 0.7$ volts, find R_2/R_1, and V_{ref} if $V_{OS} = 0$. (c) Find V_{ref} if $V_{OS} = 20$ mV.

8.25 (Sec. 8.4). Assume that M1 and M2 are matched and that the parameters of Table 8.2–1 are valid for the devices in Fig. P8.25. Find I_{out}. State and justify all assumptions necessary.

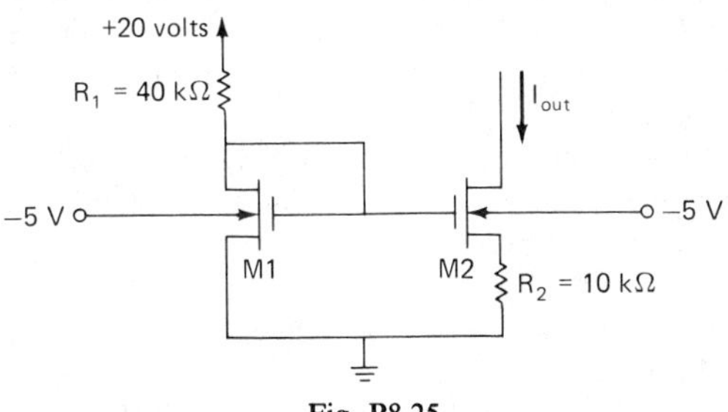

Fig. P8.25.

8.26 (Sec. 8.5). Show how Figs. 8.5–2(a) and 8.5–2(b) would change if the bulk effects were taken into account.

8.27 (Sec. 8.5). Use the small-signal model of the MOS device to verify the voltage gains of Figs. 8.5–1(a) and 8.5–1(b) given in Eqs. (7) and (13).

8.28 (Sec. 8.5). Find the value of the small-signal voltage gain, v_{out}/v_{in}, of the circuit shown in Fig. P8.28 if the DC value of V_{in} is such that the DC value of V_{out} is 0 volts. Assume that $K' = 20 \times 10^{-6} V/A^2$, $\gamma = 0.5\ V^{1/2}$, $\phi = 0.5$ V, and $\lambda = 0.01\ V^{-1}$.

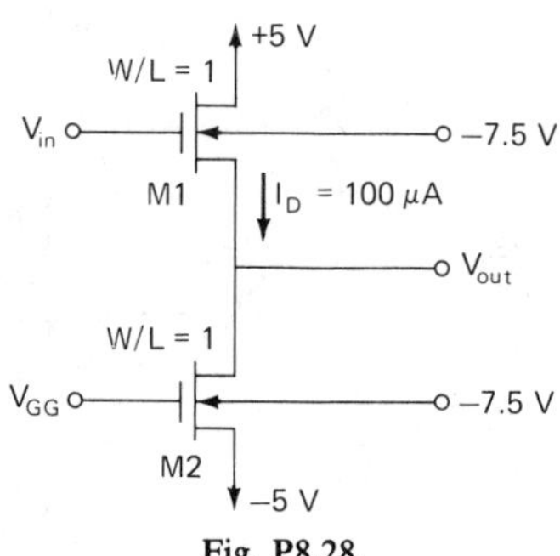

Fig. P8.28.

8.29 (Sec. 8.5). Find the AC voltage gain of the CMOS inverter shown in Fig. 8.5–1(c). The parameter values are: $K'_p = 4\ \mu A/V^2$, $K'_n = 10\ \mu A/V^2$, $V_{T01} = 0.8$ V, $V_{T02} = -0.8$ V, $\lambda_n = \lambda_p = 0.01\ V^{-1}$, and $I = 10\ \mu$A.

8.30 (Sec. 8.5). Use Fig. 8.5–6 to derive Eq. (22).

8.31 (Sec. 8.5). Assume that $V_{DD} = 5$ V, $V_{SS} = -5$ V, $V_{out} = 0$ V, and $I_D = 100$ microamperes of Fig. 8.5–8(a). Find the gain of the source follower of Fig. 8.5–8 if (a) $V_{BB} = -5$ V, (b) $V_{BB} = -10$ V, and (c) $V_{BB} = -15$ V.

8.32 (Sec. 8.5). Ignore the bulk and channel modulation effects, and obtain an expression for V_{D1} and V_{D2} of Fig. 8.5–9 as a function of the differential-input voltage, V_{id}. Give a sketch of V_{D1} and V_{D2} versus V_{id} similar to Fig. 8.5–10.

8.33 (Sec. 8.5). Develop an expression for the small-signal differential and common-mode voltage gain of Fig. 8.5–9 if M3 and M4 are depletion devices with the gate connected to the source.

8.34 (Sec. 8.5). If all the devices in the differential amplifier shown in Fig. 8.5–9 are saturated, find the worst-case offset voltage, V_{OS}, *if* $V_{Ti} = 1.0 \pm 0.01$ volts and $\beta_i = 10^{-5} \pm 5 \times 10^{-7}$ amps/volt². Assume that $\beta_1 = \beta_2 = 10\beta_3 = 10\beta_4$ and $\Delta\beta_1/\beta_1 = \Delta\beta_2/\beta_2 = \Delta\beta_3/\beta_3 = \Delta\beta_4/\beta_4$. State any assumptions that you make beyond those above in working this problem.

8.35(Sec. 8.5). If a CMOS differential amplifier uses p-channel devices as the differential pair with n-channel active loads, and another CMOS differential amplifier uses n-channel devices as the differential pair with p-channel active loads, which has the most gain, assuming identical sizes?

8.36 (Sec. 8.6). (a) Derive Eq. (8) from Fig. 8.6–8(b). (b) Show how to get the approximations used for the roots of Eqs. (9), (10), and (11).

8.37 (Sec. 8.6). Derive Eq. (17) from Fig. 8.6–10(b).

8.38 (Sec. 8.6). A block diagram of an NMOS op amp suitable for designing the compensation is shown in Fig. P8.38. Find the value of C that will give a single-pole rolloff to the unity-gain bandwidth of 1 MHz. Assume that all other poles of the amplifier except for $1/(R_1C_1)$ are above 1 MHz. What value of current through C corresponds to a slew rate of 2 volts/microsecond?

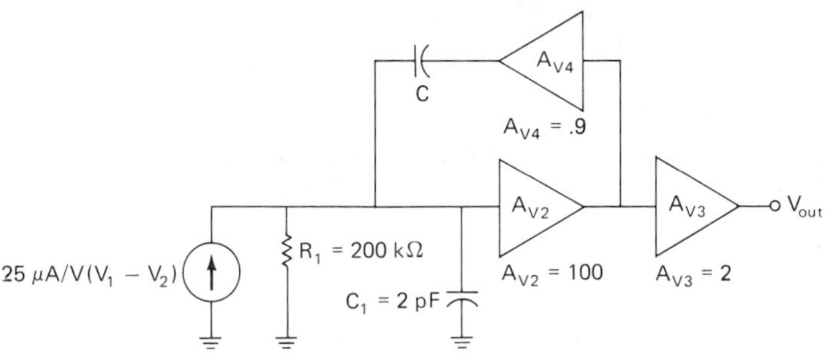

Fig. P8.38.

8.39 (Sec. 8.7). List the steps in the design of a CMOS op amp starting with the specifications and ending with digitization.

8.40 (Sec. 8.7). What relationship between the sizes of M1 through M4 of Fig. P 8.40 will ensure that V_{DG4} is zero, thus ensuring that M4 is saturated? If $S_4 = 5$, $S_3 = 0.2$, and $S_1 = 1$, what is the value of S_2?

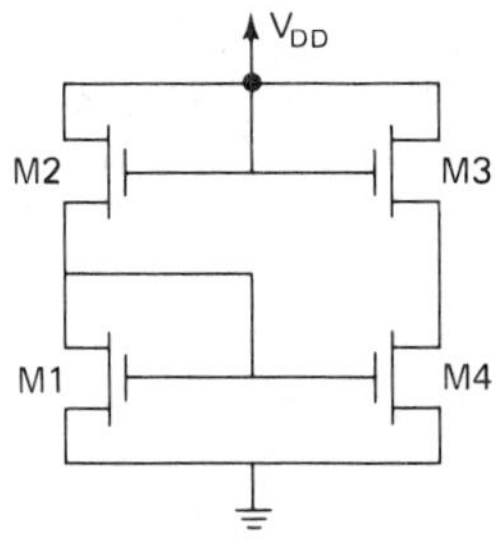

Fig. P8.40.

8.41 (Sec. 8.7). Draw a small-signal model of the output circuit shown in Fig. P 8.41, and prove that the AC voltage gain of this circuit is approximately

$$\frac{v_{out}}{v_{in}} \cong \frac{g_{m2}}{g_{m1}}\left(\frac{g_{m4}/g_{m3}}{1 + g_{m4}/g_{m3}}\right)$$

if the effect of the bulk is neglected.

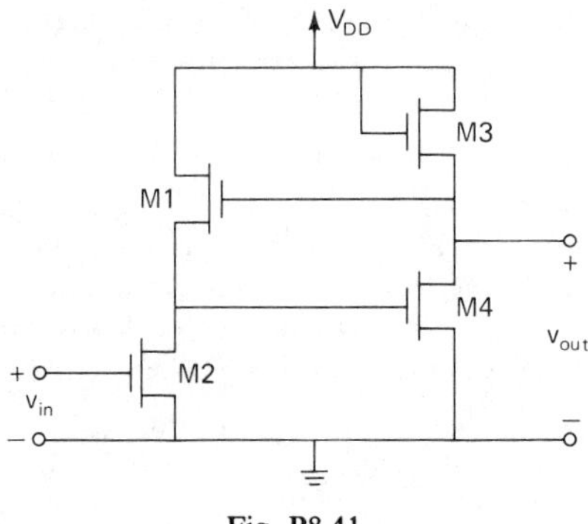

Fig. P8.41.

8.42 (Sec. 8.7). Identify in Fig. 8.7–6 places where feedforward compensation could be applied.

8.43 (Sec. 8.7). Calculate the slew rate of Fig. 8.7–6.

8.44 (Sec. 8.6). Using a computer program such as SPICE, obtain an open-loop frequency response of Fig. 8.7–6 using the values in Table 8.7–3 for C_L = 0 pF, 10 pF, and 20 pF. What is the phase margin of this amplifier under these conditions?

8.45 (Sec. 8.7). Design the values of W/L and the bias currents of Fig. 8.7–9 to obtain the performance given in Table 8.7–7, and compare with the values in Table 8.7–4.

8.46 (Sec. 8.7). For the p-channel input-pair, CMOS op amp of Fig. 8.7–15, calculate the open-loop, low-frequency gain, the output resistance, the power consumption, the power supply rejection ratio, the input common-mode range, the output voltage swing, the slew rate, the common-mode rejection ratio, and the unity-gain bandwidth. Use SPICE to find the phase margin for no load and for a 20-pF load and the 1% settling time for no load and for a 20-pF load.

8.47 (Sec. 8.7). Show how to measure (a) open-loop gain, (b) input common-mode range, (c) output voltage swing, (d) open-loop frequency response, and (e) the settling time. Your measurement scheme should be practical and consider the nonideal performance of the op amp.

8.48 (Sec. 8.7). Design the values of W and L for each transistor of the CMOS op amp in Fig. 8.7–13 which will give a low frequency, differential voltage gain of 4000. Assume that $K'_n = 10^{-5}$ amps/volt², $K'_p = 5 \times 10^{-6}$ amps/volt², $V_{TN} = 1$ volt, $V_{TP} = -1$ volt, and $\lambda = 0.01$ volts^{-1} for both PMOS and NMOS devices. Also, assume the minimum device dimension is 10 microns and chose the smallest devices possible. Design C_c and R_z to give a unity-gainbandwidth of 1 MHz and to eliminate the influences of the RHP zero. How much load capacitance should this op amp be capable of driving without suffering a degradation in the phase margin? What is the slew rate of this op amp?

8.49 (Sec. 8.7.). Use the electrical model parameters of Prob. 8.48 to design W_3, L_3, W_4, L_4, W_5, L_5, C_c, and R_z of Fig. P8.49 if $W_1 = L_1 = W_2 = L_2 = 10$ microns to obtain a low frequency, differential voltage gain of 5000 and a unity gainbandwidth of 1 MHz. All devices should be in saturation under normal operating conditions and the effect of the RHP should be cancelled. How much load capacitance should this op amp be able to drive before suffering a degradation in the phase margin? What is the slew rate of this op amp?

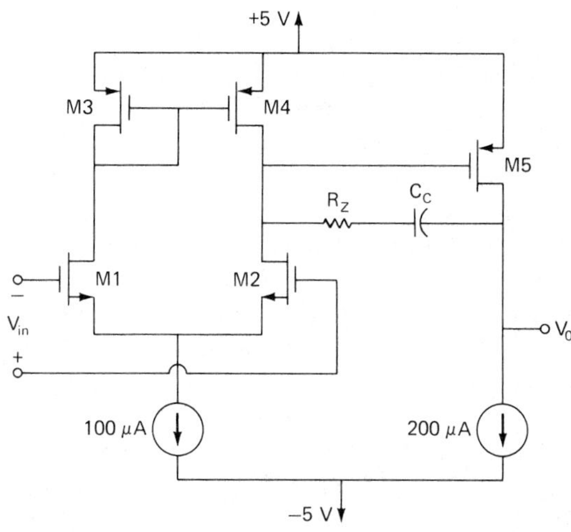

Fig. P8.49.

Appendix A
Passive, Low-Pass, Prototype Filter Realizations

In this appendix we present tables for passive filter realizations of various types of low-pass network functions. The first type of filter is the single-resistance, terminated, lossless ladder network. The form of the realization for a voltage-source excitation is shown in Figs. A1(a) (for even-order functions) and A1(b) (for odd-order functions). For a current-source excitation the realizations have the form shown in Figs. A2(a) (for even-order functions) and A2(b) (for odd-order functions). The element values for (normalized) Butterworth, Chebyshev (½ and 1-dB ripple), and Thomson filters are given in Table A1.

A second type of filter realization is the double-resistance, terminated, lossless ladder one. The form of the realization is shown in Figs. A3(a) (for even-order functions) and A3(b) (for odd-order functions). An alternative realization form is shown in Figs. A4(a) (for even-order functions) and A4(b) (for odd-order functions). The element values for (normalized) Butterworth, Chebyshev (½ and 1-dB ripple), and Thomson network functions are given in Table A2 for the equal-resistance case. In this case, no solution exists for even-order Chebyshev functions. Table A3 gives the element values for the case where the load and source resistances are related by a factor of 2. In this case no solution exists for even-order 1-dB ripple Chebyshev functions.

The form of the filter for double-resistance, terminated, lossless ladder realizations of elliptic low-pass functions is given in Fig. A5. An alternative realization form is shown in Fig. A6. The element values for various values of passband ripple and for various orders are given in Tables A4 and A5. Table A4 is for equal-resistance terminated realizations for even-order (Case C) and odd-order filters (0.1 and 1.0-dB passband ripple). Table A5 is for Case B (unequal-resistance terminations) for even-order filters (0.1 and 1.0-dB passband ripple).

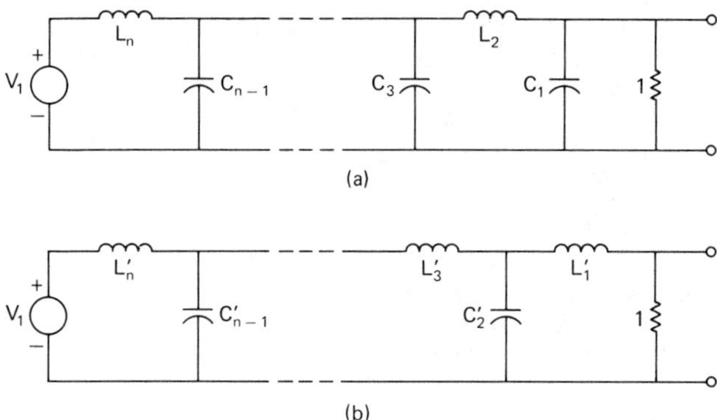

Fig. A-1. Network configuration for Table A1 for voltage-source excitation. (a) Even. (B) Odd.

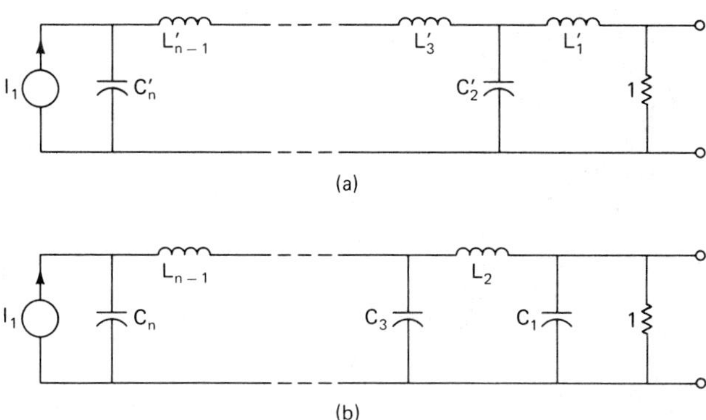

Fig. A-2. Network configuration for Table A1 for current-source excitation. (a) Even. (b) Odd.

Table A1 Element Values for Low-pass Single-Resistance Terminated Lossless Ladder Realizations

	Elements in Figs. A1(a) (even) and A2(b) (odd)									
n	C_1	L_2	C_3	L_4	C_5	L_6	C_7	L_8	C_9	L_{10}
2	0.7071	1.4142				Butterworth				
3	0.5000	1.3333	1.5000			(1 rad/sec bandwidth)				
4	0.3827	1.0824	1.5772	1.5307						
5	0.3090	0.8944	1.3820	1.6944	1.5451					
6	0.2588	0.7579	1.2016	1.5529	1.7593	1.5529				
7	0.2225	0.6560	1.0550	1.3972	1.6588	1.7988	1.5576			
8	0.1951	0.5776	0.9370	1.2588	1.5283	1.7287	1.8246	1.5607		
9	0.1736	0.5155	0.8414	1.1408	1.4037	1.6202	1.7772	1.8424	1.5628	
10	0.1564	0.4654	0.7626	1.0406	1.2921	1.5100	1.6869	1.8121	1.8552	1.5643
2	0.7014	0.9403				1/2-dB ripple Chebychev				
3	0.7981	1.3001	1.3465			(1 rad/sec bandwidth)				
4	0.8352	1.3916	1.7279	1.3138						
5	0.8529	1.4291	1.8142	1.6426	1.5388					
6	0.8627	1.4483	1.8494	1.7101	1.9018	1.4042				
7	0.8686	1.4596	1.8675	1.7371	1.9712	1.7254	1.5982			
8	0.8725	1.4666	1.8750	1.7508	1.9980	1.7838	1.9571	1.4379		
9	0.8752	1.4714	1.8856	1.7591	2.0116	1.8055	2.0203	1.7571	1.6238	
10	0.8771	1.4748	1.8905	1.7645	2.0197	1.8165	2.0432	1.8119	1.9816	1.4539
2	0.9110	0.9957				1-dB ripple Chebychev				
3	1.0118	1.3332	1.5088			(1 rad/sec bandwidth)				
4	1.0495	1.4126	1.9093	1.2817						
5	1.0674	1.4441	1.9938	1.5908	1.6652					
6	1.0773	1.4601	2.0270	1.6507	2.0491	1.3457				
7	1.0832	1.4694	2.0437	1.6736	2.1192	1.6489	1.7118			
8	1.0872	1.4751	2.0537	1.6850	2.1453	1.7021	2.0922	1.3691		
9	1.0899	1.4790	2.0601	1.6918	2.1583	1.7213	2.1574	1.6707	1.7317	
10	1.0918	1.4817	2.0645	1.6961	2.1658	1.7306	2.1803	1.7215	2.1111	1.3801
2	0.3333	1.0000								
3	0.1667	0.4800	0.8333			Thomson (1 second delay at DC)				
4	0.1000	0.2899	0.4627	0.7101						
5	0.0667	0.1948	0.3103	0.4215	0.6231					
6	0.0476	0.1400	0.2246	0.3005	0.3821	0.5595				
7	0.0357	0.1055	0.1704	0.2288	0.2827	0.3487	0.5111			
8	0.0278	0.0823	0.1338	0.1806	0.2227	0.2639	0.3212	0.4732		
9	0.0222	0.0660	0.1077	0.1463	0.1811	0.2129	0.2465	0.2986	0.4424	
10	0.0182	0.0541	0.0886	0.1209	0.1549	0.1880	0.2057	0.2209	0.2712	0.4161
n	L'_1	C'_2	L'_3	C'_4	L'_5	C'_6	L'_7	C'_8	L'_9	C'_{10}
	Elements in Figs. A1(b) (odd) and A2(a) (even)									

Reprinted by permission from L. Weinberg, *Network Analysis and Synthesis,* McGraw-Hill Book Co., New York, 1962; reprinted by R. E. Krieger Publishing Co., Huntington, New York, 1975.

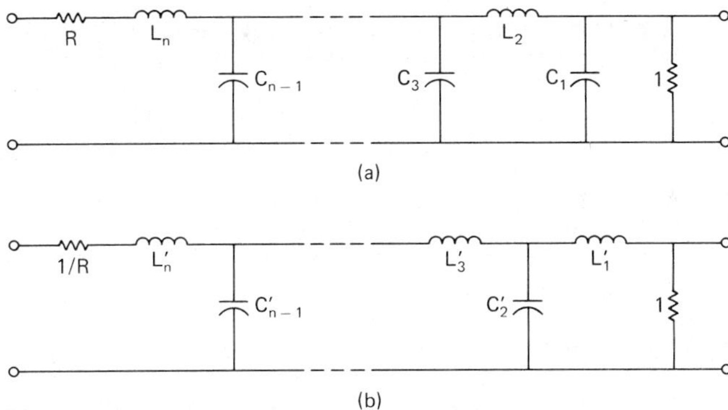

Fig. A-3. Network configuration with a shunt C as the left-most reactive element for Tables A2 and A3. (a) Even. (b) Odd.

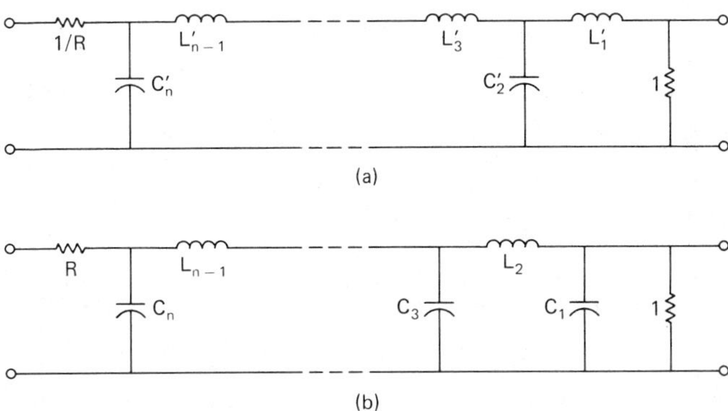

Fig. A-4. Network configuration with a series L as the left-most reactive element for Tables A2 and A3. (a) Even. (b) Odd.

Table A2 Element Values for Low-pass Double-Resistance Terminated Lossless Ladder Realizations with $R = 1$ ohm

	Elements in Figs. A3(a) (even) and A4(b) (odd)									
n	C_1	L_2	C_3	L_4	C_5	L_6	C_7	L_8	C_9	L_{10}
2	1.4142	1.4142				Butterworth				
3	1.0000	2.0000	1.0000			(1 rad/sec bandwidth)				
4	0.7654	1.8478	1.8478	0.7654						
5	0.6180	1.6180	2.0000	1.6180	0.6180					
6	0.5176	1.4142	1.9319	1.9319	1.4142	0.5176				
7	0.4450	1.2470	1.8019	2.0000	1.8019	1.2470	0.4450			
8	0.3902	1.1111	1.6629	1.9616	1.9616	1.6629	1.1111	0.3902		
9	0.3473	1.0000	1.5321	1.8794	2.0000	1.8794	1.5321	1.0000	0.3473	
10	0.3129	0.9080	1.4142	1.7820	1.9754	1.9754	1.7820	1.4142	0.9090	0.3129
3	1.5963	1.0967	1.5963			1/2-dB ripple Chebychev				
5	1.7058	1.2296	2.5408	1.2296	1.7058	(1 rad/sec bandwidth)				
7	1.7373	1.2582	2.6383	1.3443	2.6383	1.2582	1.7373			
9	1.7504	1.2690	2.6678	1.3673	2.7239	1.3673	2.6678	1.2690	1.7504	
3	2.0236	0.9941	2.0236			1-dB ripple Chebychev				
5	2.1349	1.0911	3.0009	1.0911	2.1349	(1 rad/sec bandwidth)				
7	2.1666	1.1115	3.0936	1.1735	3.0936	1.1115	2.1666			
9	2.1797	1.1192	3.1214	1.1897	3.1746	1.1897	3.1214	1.1192	2.1797	
2	1.5774	0.4226				Thomson (1-second delay at DC)				
3	1.2550	0.5528	0.1922							
4	1.0598	0.5116	0.3181	0.1104						
5	0.9303	0.4577	0.3312	0.2090	0.0718					
6	0.8377	0.4116	0.3158	0.2364	0.1480	0.0505				
7	0.7677	0.3744	0.2944	0.2378	0.1778	0.1104	0.0375			
8	0.7125	0.3446	0.2735	0.2297	0.1867	0.1387	0.0855	0.0289		
9	0.6678	0.3203	0.2547	0.2184	0.1859	0.1506	0.1111	0.0682	0.0230	
10	0.6305	0.3002	0.2384	0.2066	0.1808	0.1539	0.1240	0.0911	0.0557	0.0187
	L'_1	C'_2	L'_3	C'_4	L'_5	C'_6	L'_7	C'_8	L'_9	C'_{10}
	Elements in Figs. A3(b) (odd) and A4(a) (even)									

Reprinted by permission from L. Weinberg, *Network Analysis and Synthesis*, McGraw-Hill Book Co., New York, 1962; reprinted by R. E. Krieger Publishing Co., Huntington, New York, 1975.

Table A3 Element Values for Low-pass Double-Resistance Terminated Lossless Ladder Realizations with $R = 1/2$ ohm

Elements in Figs. A3(a) (even) and A4(b) (odd)

n	C_1	L_2	C_3	L_4	C_5	L_6	C_7	L_8	C_9	L_{10}
2	3.3461	0.4483				Butterworth				
3	3.2612	0.7789	1.1811			(1 rad/sec bandwidth)				
4	3.1868	0.8826	2.4524	0.2175						
5	3.1331	0.9237	3.0510	0.4955	0.6857					
6	3.0938	0.9423	3.3687	0.6542	1.6531	0.1412				
7	3.0640	0.9513	3.5532	0.7512	2.2726	0.3536	0.4799			
8	3.0408	0.9558	3.6678	0.8139	2.6863	0.5003	1.2341	0.1042		
9	3.0223	0.9579	3.7426	0.8565	2.9734	0.6046	1.7846	0.2735	0.3685	
10	3.0072	0.9588	3.7934	0.8864	3.1795	0.6808	2.1943	0.4021	0.9818	0.0825
2	1.5132	0.6538				1/2-dB ripple Chebychev				
3	2.9431	0.6503	2.1903			(1 rad/sec bandwidth)				
4	1.8158	1.1328	2.4881	0.7732						
5	3.2228	0.7645	4.1228	0.7116	2.3197					
6	1.8786	1.1884	2.7589	1.2403	2.5976	0.7976				
7	3.3055	0.7899	4.3575	0.8132	4.2419	0.7252	2.3566			
8	1.9012	1.2053	2.8152	1.2864	2.8479	1.2628	2.6310	0.8063		
9	3.3403	0.7995	4.4283	0.8341	4.4546	0.8235	4.2795	0.7304	2.3719	
10	1.9117	1.2127	2.8366	1.2999	2.8964	1.3054	2.8744	1.2714	2.6456	0.8104
3	3.4774	0.6153	2.8540			1-dB ripple Chebychev				
5	3.7211	0.6949	4.7448	0.6650	2.9936	(1 rad/sec bandwidth)				
7	3.7916	0.7118	4.9425	0.7348	4.8636	0.6757	3.0331			
9	3.8210	0.7182	5.0013	0.7485	5.0412	0.7429	4.9004	0.6797	3.0495	
2	2.6180	0.1910				Thomson				
3	2.1156	0.2613	0.3618			(1 second delay at DC)				
4	1.7893	0.2461	0.6127	0.0530						
5	1.5686	0.2217	0.6456	0.1015	0.1393					
6	1.4102	0.1999	0.6196	0.1158	0.2894	0.0246				
7	1.2904	0.1821	0.5797	0.1171	0.3497	0.0542	0.0735			
8	1.1964	0.1676	0.5395	0.1135	0.3685	0.0683	0.1684	0.0142		
9	1.1202	0.1558	0.5030	0.1081	0.3580	0.0744	0.2195	0.0336	0.0453	
10	1.0569	0.1460	0.4710	0.1024	0.3586	0.0763	0.2456	0.0450	0.1100	0.0925
n	L'_1	C'_2	L'_3	C'_4	L'_5	C'_6	L'_7	C'_8	L'_9	C'_{10}

Elements in Figs. A3(b) (odd) and A4(a) (even)

Reprinted by permission from L. Weinberg, *Network Analysis and Synthesis,* McGraw-Hill Book Co., New York, 1962; reprinted by R. E. Krieger Publishing Co., Huntington, New York, 1975.

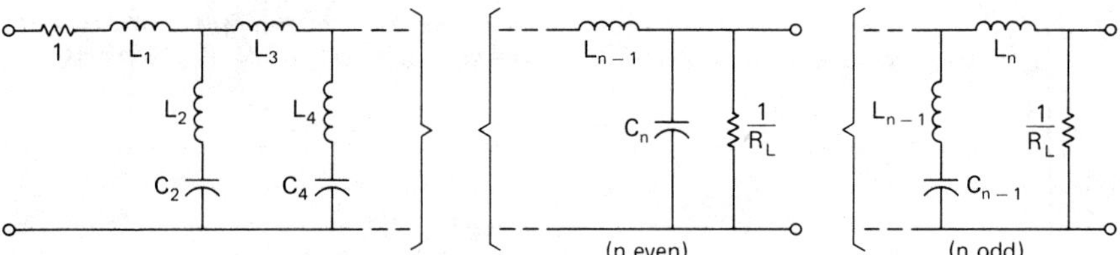

Fig. A-5. Network configuration with a series L as the left-most reactive element for Tables A4 and A5.

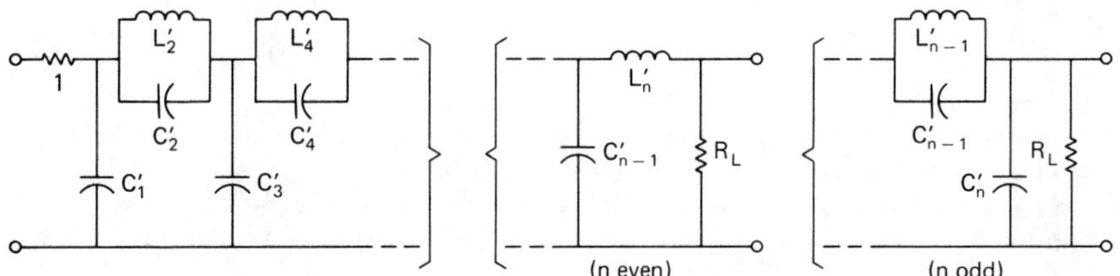

Fig. A-6. Network configuration with a shunt C as the left-most reactive element for Tables A4 and A5.

Table A4 Element Values for Elliptic Low-pass Double-Resistance Terminated Lossless Ladder Realizations with $R_L = 1$ ohm (Case C for even cases)[a]

0.1-dB passband ripple

			Elements in Fig. A5													
n	ω_s	K_P	L_1	C_2	L_2	L_3	C_4	L_4	L_5	C_6	L_6	L_7	C_8	L_8	L_9	C_{10}
3	1.05	1.748	.35550	.15374	5.39596	.35550										
	1.10	3.374	.44626	.26993	2.70353	.44626										
	1.20	6.691	.57336	.44980	1.30805	.57336										
	1.50	14.848	.77031	.74561	.47797	.77031										
	2.00	24.010	.89544	.93759	.20697	.89544										
4	1.05	3.284	.00442	.17221	4.93764	1.01224	.84445									
	1.10	6.478	.17279	.32758	2.30986	1.04894	.89415									
	1.20	12.085	.37139	.56638	1.09294	1.11938	.92440									
	1.50	23.736	.62815	.94009	.40730	1.24711	.93518									
	2.00	36.023	.77554	1.17646	.17957	1.33473	.93382									
5	1.05	13.841	.70813	.76630	.73572	1.12761	.20138	4.38116	.04985							
	1.10	20.050	.81296	.92418	.49338	1.22445	.37193	2.13500	.29125							
	1.20	28.303	.91441	1.06516	.31628	1.38201	.60131	1.09329	.52974							
	1.50	43.415	1.02789	1.21517	.15134	1.63179	.93525	.44083	.81549							
	2.00	58.901	1.08758	1.29322	.07317	1.79387	1.14330	.20038	.97720							
6	1.05	18.727	.44177	.71651	.90905	.83142	.36274	2.44680	.80463	.99857						
	1.10	26.230	.57630	.88798	.61282	.97304	.59060	1.35666	.94305	1.01381						
	1.20	36.113	.70984	1.06266	.39136	1.15974	.87407	.76185	1.09176	1.02462						
	1.50	54.202	.86595	1.27403	.18554	1.43106	1.27235	.33007	1.28253	1.03317						
	2.00	72.761	.95131	1.39297	.08926	1.60132	1.51866	.15421	1.39521	1.03621						
			C'_1	L'_2	C'_2	C'_3	L'_4	C'_4	C'_5	L'_6	C'_6	C'_7	L'_8	C'_8	C'_9	L'_{10}
			Elements in Fig. A6													

Table A4 *Cont.*

			Elements in Fig. A5													
n	ω_s	K_P	L_1	C_2	L_2	L_3	C_4	L_4	L_5	C_6	L_6	L_7	C_8	L_8	L_9	C_{10}
7	1.05	30.470	.91937	1.07659	.34220	1.09623	.40518	2.20850	.84335	.50342	1.51827	.41098				
	1.10	39.357	.98821	1.16726	.24374	1.27743	.59720	1.35681	1.04029	.67881	.96669	.58282				
	1.20	50.963	1.05029	1.24872	.16124	1.48377	.82869	.81542	1.28723	.87428	.58918	.75395				
	1.50	72.129	1.11593	1.33554	.07857	1.75687	1.15174	.37160	1.63827	1.12502	.26822	.95588				
	2.00	93.809	1.14910	1.37979	.03822	1.92026	1.35221	.17692	1.85664	1.27023	.12694	1.06720				
8	1.05	36.268	.68105	1.01040	.47466	.91103	.48838	1.83667	.67613	.70502	1.12616	.98462	1.04901			
	1.10	46.399	.77921	1.13273	.33839	1.08668	.70667	1.15081	.86562	.92698	.74468	1.09627	1.06140			
	1.20	59.639	.87290	1.25079	.22404	1.28897	.97154	.69915	1.09601	1.17713	.46698	1.21321	1.07163			
	1.50	83.807	.97779	1.38443	.10930	1.56051	1.34331	.32099	1.41679	1.50420	.21807	1.35777	1.08093			
	2.00	108.575	1.03311	1.45551	.05321	1.72464	1.57463	.15328	1.61419	1.69718	.10439	1.44022	1.08484			
9	1.05	47.276	1.02597	1.21654	.20583	1.29803	.60675	1.36728	.76114	.44746	2.01083	.94136	.74312	.84407	.63916	
	1.10	58.707	1.07226	1.27741	.14773	1.46403	.79046	.92320	1.00154	.63575	1.28473	1.14956	.89544	.57633	.77015	
	1.20	73.629	1.11295	1.33139	.09815	1.64257	.99964	.58858	1.29050	.86538	.78945	1.39229	1.05143	.36877	.89697	
	1.50	100.842	1.15493	1.38761	.04800	1.86765	1.27611	.27927	1.69055	1.18960	.36523	1.72037	1.23694	.17437	1.04130	
	2.00	128.717	1.17576	1.41568	.02339	1.99761	1.44055	.13471	1.93638	1.39225	.17487	1.91900	1.33887	.08371	1.11842	
10	1.05	53.576	.82096	1.17319	.30389	1.09174	.66563	1.26609	.64760	.54466	1.65398	.76756	.97842	.69517	1.11421	1.08154
	1.10	66.262	.89544	1.26213	.21933	1.25856	.87195	.85540	.86007	.76733	1.06638	.95500	1.18228	.48206	1.20554	1.09096
	1.20	82.830	.96461	1.34441	.14648	1.44189	1.10918	.54607	1.11622	1.03951	.65893	1.17011	1.39629	.31255	1.29689	1.10032
	1.50	113.056	1.03982	1.43372	.07199	1.67758	1.42540	.25986	1.47272	1.42505	.30603	1.45794	1.65959	.14962	1.40489	1.10919
	2.00	144.023	1.07857	1.47977	.03516	1.81552	1.61465	.12563	1.69286	1.66661	.14673	1.63110	1.80819	.07224	1.46428	1.11311
			C'_1	L'_2	C'_2	C'_3	L'_4	C'_4	C'_5	L'_6	C'_6	C'_7	L'_8	C'_8	C'_9	L'_{10}
			Elements in Fig. A6													

Elements in Fig. A5

1.0-dB passband ripple

n	ω_s	K_P	L_1	C_2	L_2	L_3	C_4	L_4	L_5	C_6	L_6	L_7	C_8	L_8	L_9	C_{10}
3	1.05	8.134	1.05507	.25223	3.28904	1.05507										
	1.10	11.480	1.22525	.37471	1.94752	1.22525										
	1.20	16.209	1.42450	.52544	1.11977	1.42450										
	1.50	25.176	1.69200	.73340	.48592	1.69200										
	2.00	34.454	1.85199	.85903	.22590	1.85199										
4	1.05	11.322	.63708	.35277	2.41039	1.11522	1.39953									
	1.10	15.942	.80935	.54042	1.40015	1.18107	1.45001									
	1.20	22.293	1.00329	.77733	.79634	1.26621	1.49217									
	1.50	34.179	1.25675	1.11431	.34362	1.38981	1.53225									
	2.00	46.481	1.40677	1.32367	.15960	1.46762	1.55071									
5	1.05	24.135	1.56191	.67560	.83449	1.55460	.26584	3.31881	.88528							
	1.10	30.471	1.69691	.77511	.58827	1.79892	.39922	1.98907	1.12109							
	1.20	38.757	1.82812	.87005	.38720	2.09095	.56347	1.16672	1.38094							
	1.50	53.875	1.97687	.97694	.18824	2.49161	.79362	.51950	1.71889							
	2.00	69.360	2.05594	1.03392	.09152	2.73567	.93561	.24486	1.91939							
6	1.05	29.133	1.07458	.80116	.81300	.92735	.51753	1.71498	.92186	1.60511						
	1.10	36.680	1.22059	.94235	.57746	1.10900	.75718	1.05819	1.01676	1.64682						
	1.20	46.571	1.37146	1.08633	.38284	1.32610	1.05110	.63354	1.12484	1.68498						
	1.50	64.661	1.55425	1.25876	.18779	1.62529	1.46557	.28655	1.26961	1.72482						
	2.00	83.221	1.65661	1.35450	.09179	1.80860	1.72376	.13586	1.35729	1.74424						
7	1.05	40.926	1.82156	.86343	.42668	1.67632	.34381	2.60271	1.23696	.46779	1.63392	1.22362				
	1.10	49.816	1.91040	.92662	.30705	1.93579	.48016	1.68753	1.55276	.59277	1.10699	1.41994				
	1.20	61.422	1.99168	.98474	.20446	2.22804	.64444	1.04856	1.92724	.73012	.70551	1.62539				
	1.50	82.588	2.07882	1.04761	.10016	2.61372	.87393	.48973	2.44021	.90483	.33349	1.87717				
	2.00	104.268	2.12329	1.07993	.04884	2.84446	1.01638	.23538	2.75306	1.00567	.16034	2.01924				
			C'_1	L'_2	C'_2	C'_3	L'_4	C'_4	C'_5	L'_6	C'_6	C'_7	L'_8	C'_8	C'_9	L'_{10}

Elements in Fig. A6

Elements in Fig. A5

n	ω_s	K_P	L_1	C_2	L_2	L_3	C_4	L_4	L_5	C_6	L_6	L_7	C_8	L_8	L_9	C_{10}
8	1.05	46.727	1.34673	1.00922	.47521	1.08540	.54692	1.64007	.70773	.86533	.91752	1.00315	1.72154			
	1.10	56.858	1.46597	1.10092	.34817	1.28842	.75883	1.07170	.90030	1.11345	.61997	1.07811	1.75961			
	1.20	70.098	1.58346	1.18748	.23599	1.52224	1.01386	.66997	1.12753	1.39664	.39359	1.15904	1.79429			
	1.50	94.266	1.71869	1.28337	.11791	1.83710	1.36932	.31489	1.43764	1.77189	.18513	1.26147	1.83033			
	2.00	119.034	1.79131	1.33358	.05807	2.02800	1.58946	.15185	1.62645	1.99561	.08878	1.32075	1.84787			
9	1.05	57.736	1.95471	.95672	.26172	1.94887	.46951	1.76694	1.12605	.35392	2.54224	1.40979	.63099	.99407	1.47897	
	1.10	69.167	2.01503	.99976	.18875	2.18048	.60062	1.21501	1.47339	.48929	1.66927	1.71691	.73764	.69962	1.63779	
	1.20	84.089	2.06867	1.03833	.12585	2.43022	.74985	.78464	1.88531	.65376	1.04499	2.06810	.84652	.45803	1.79580	
	1.50	111.302	2.12470	1.07895	.06173	2.74601	.94696	.37634	2.45079	.88531	.49076	2.53625	.97584	.22103	1.97957	
	2.00	139.176	2.15275	1.09942	.03012	2.92871	1.06415	.18236	2.79680	1.02979	.23642	2.81748	1.04688	.10706	2.07916	
10	1.05	64.036	1.52461	1.11270	.32041	1.31260	.68239	1.23499	.69671	.59634	1.51065	.77127	1.18327	.57482	1.06945	1.79994
	1.10	76.722	1.62236	1.17191	.23621	1.51411	.87139	.85595	.92029	.82679	.98969	.94303	1.42284	.40056	1.12735	1.83423
	1.20	93.289	1.71564	1.22500	.16075	1.73787	1.08551	.55798	1.18808	1.10906	.61761	1.13757	1.68092	.25963	1.18655	1.86522
	1.50	123.515	1.81947	1.28114	.08056	2.02843	1.36690	.27098	1.55972	1.50976	.28886	1.39429	2.00434	.12388	1.25749	1.89727
	2.00	154.482	1.87378	1.30964	.03972	2.19972	1.53360	.13227	1.78908	1.76138	.13884	1.54729	2.18935	.05966	1.29688	1.91286
			C'_1	L'_2	C'_2	C'_3	L'_4	C'_4	C'_5	L'_6	C'_6	C'_7	L'_8	C'_8	C'_9	L'_{10}

Elements in Fig. A6

[a] Computed using a program written by David Jose Miguel Baezlopez, Sensitivity and Synthesis of Elliptic Functions, Ph.D Dissertation, University of Arizona, 1978.

Table A5 Element Values for Elliptic Low-pass Double-Resistance Terminated Lossless Ladder Realizations for Case B[a]

Elements in Fig. A5

n	ω_s	K_P	L_1	C_2	L_2	L_3	C_4	L_4	L_5	C_6	L_6	L_7	C_8	L_8	L_9	C_{10}
4	1.05	4.485	.15780	.18091	4.73822	1.20743	.82637									
	1.10	8.308	.33411	.33438	2.28333	1.26881	.84827									
	1.20	14.387	.53773	.55478	1.12558	1.36980	.85261									
	1.50	26.320	.79962	.88310	.43628	1.53672	.84068									
	2.00	38.697	.95051	1.08631	.19517	1.64684	.83004									
6	1.05	20.307	.57153	.65752	1.01346	.92972	.32584	2.72744	1.03524	.88809						
	1.10	27.889	.70783	.81703	.67992	1.10484	.51890	1.54640	1.19779	.88523						
	1.20	37.827	.84244	.98082	.43111	1.32791	.75659	.88144	1.37708	.87992						
	1.50	55.966	.99836	1.17887	.20248	1.64500	1.08849	.38623	1.61158	.87198						
	2.00	74.548	1.08280	1.28970	.09690	1.84134	1.29301	.18123	1.75160	.86710						
8	1.05	37.529	.79899	.92996	.52853	1.02243	.42917	2.09084	.78906	.59418	1.34317	1.24821	.89367			
	1.10	47.686	.89538	1.04944	.37264	1.22051	.61878	1.31484	1.0703	.77697	.89307	1.38901	.89115			
	1.20	60.949	.98606	1.16509	.24409	1.44606	.84957	.79988	1.29323	.98124	.56277	1.53864	.88795			
	1.50	85.138	1.08612	1.29596	.11768	1.74572	1.17448	.36725	1.67684	1.24564	.26411	1.72539	.88347			
	2.00	109.915	1.13830	1.36543	.05694	1.92553	1.37705	.17531	1.91249	1.40030	.12672	1.83237	.88075			
10	1.05	54.608	.92833	1.09111	.33387	1.21271	.58502	1.44313	.74643	.46770	1.92642	.91319	.80993	.84618	1.41635	.89491
	1.10	67.307	.99839	1.18097	.23833	1.39449	.76963	.97094	.99183	.65631	1.24697	1.14180	.97201	.59039	1.53570	.89383
	1.20	83.887	1.06237	1.26448	.15755	1.59212	.98322	.61709	1.28687	.88676	.77257	1.40608	1.14191	.38429	1.65727	.89178
	1.50	114.123	1.13092	1.35538	.07660	1.84367	1.26943	.29213	1.69639	1.21247	.35973	1.76038	1.34833	.18475	1.80184	.88893
	2.00	145.095	1.16590	1.40227	.03721	1.98983	1.44131	.14083	1.94870	1.41617	.17269	1.97377	1.46362	.08940	1.88151	.88719
			C'_1	L'_2	C'_2	C'_3	L'_4	C'_4	C'_5	L'_6	C'_6	C'_7	L'_8	C'_8	C'_9	L'_{10}

Elements in Fig. A6

0.1-dB passband ripple
$R_L = 0.73781\ \Omega$

Table A5 *Cont.*

n	ω_s	K_P	L_1	C_2	L_2	L_3	C_4	L_4	L_5	C_6	L_6	L_7	C_8	L_8	L_9	C_{10}
			Elements in Fig. A5													
4	1.05	13.243	.95111	.26779	3.20104	1.90749	.80699									
	1.10	18.140	1.16239	.39958	1.91077	2.05228	.80907									
	1.20	24.700	1.40135	.56068	1.11374	2.23453	.80633									
	1.50	36.771	1.71483	.78307	.49201	2.49368	.79924				1.0-dB passband ripple					
	2.00	49.156	1.90048	.91820	.23091	2.65459	.79441				$R_L = 0.37598\ \Omega$					
6	1.05	30.730	1.40432	.58067	1.14761	1.37588	.31837	2.79144	1.79883	.82259						
	1.10	38.342	1.56906	.69149	.80335	1.66832	.45609	1.75937	1.99786	.82076						
	1.20	48.285	1.73631	.80659	.52424	2.01190	.62218	1.07185	2.22816	.81822						
	1.50	66.425	1.93461	.94611	.25229	2.47740	.85305	.49283	2.53990	.81461						
	2.00	85.008	2.04359	1.02402	.12205	2.75884	.99547	.23540	2.72966	.81243						
8	1.05	47.987	1.67197	.76069	.64615	1.55530	.34969	2.56605	1.17579	.48610	1.64181	2.03669	.82473			
	1.10	58.146	1.79301	.84446	.46309	1.83957	.48544	1.67598	1.51273	.61510	1.12807	2.21830	.82355			
	1.20	71.408	1.90884	.92585	.30716	2.16030	.65000	1.04547	1.91110	.75902	.72753	2.41639	.82209			
	1.50	95.597	2.03847	1.01816	.14980	2.58395	.88106	.48955	2.45568	.94522	.34805	2.66846	.82006			
	2.00	120.374	2.10671	1.06721	.07286	2.83733	1.02492	.23554	2.78753	1.05416	.16833	2.81453	.81883			
10	1.05	65.067	1.83096	.86998	.41873	1.81829	.45061	1.87358	1.10022	.35991	2.50341	1.35196	.63152	1.08524	2.24376	.82561
	1.10	77.767	1.92046	.93288	.30171	2.07454	.58273	1.28236	1.45156	.49556	1.65147	1.67847	.74584	.76942	2.40166	.82484
	1.20	94.346	2.00313	.99155	.20092	2.35262	.73518	.82529	1.86993	.66055	1.03714	2.05233	.86549	.50703	2.56480	.82390
	1.50	124.583	2.09263	1.05564	.09835	2.70617	.93910	.39489	2.44690	.89307	.48839	2.55089	1.01109	.24638	2.76130	.82261
	2.00	155.554	2.13863	1.08880	.04792	2.91150	1.06145	.19123	2.80119	1.03824	.23556	2.85037	1.09249	.11976	2.87052	.82181
			C'_1	L'_2	C'_2	C'_3	L'_4	C'_4	C'_5	L'_6	C'_6	C'_7	L'_8	C'_8	C'_9	L'_{10}
			Elements in Fig. A6													

[a] Computed using a program written by David Jose Miguel Baezlopez, Sensitivity and Synthesis of Elliptic Functions, Ph.D dissertation, University of Arizona, 1978.

Appendix B
Design of Sampled Data Filters Using the Bilinear Transformation Method

The design of SC filters can be obtained by several approaches. One is to transform an analog filter function into a sampled data filter using one of the conventional techniques employed in digital filter design.[1] Thus, once the discrete-time equations are obtained, they can be manipulated to form known or new structures that have the same overall transfer function. The final discrete-time filter can then be realized with SC building blocks.

We present here an approach using the bilinear transformation. Note that a resistor substitution using a "bilinear resistor equivalent" is a different technique. The approach discussed here is illustrated in Fig. 3.1–16; see boxes 1, 2, 3, 4, and 5. Before we describe the method, let us define the following notation

$s =$ analog prototype Laplace variable

$\omega =$ steady-state radian frequency corresponding to $s = j\omega$ and $\omega = 2\pi f$

$H(s) =$ analog prototype transfer function

$H(z) =$ discrete time transfer function

$z = e^{j\omega T}$

$\omega_a =$ is the analog radian frequency; ω_d is the discrete time radian frequency

The desired transfer function $H(z)$ can be obtained by replacing s in $H(s)$ by the bilinear transformation, that is

$$s = K\frac{1 - z^{-1}}{1 + z^{-1}} \tag{1}$$

[1] B. Gold and C. M. Radar, *Digital Processing of Signals,* McGraw-Hill Book Co., New York, 1969.

where K is a constant involving a frequency prewarping[2] and is given by

$$K = \omega_a / \tan(\omega_d T/2) \tag{2}$$

Observe that for high sampling rates, $\omega_d T/2 << 1$, K simply becomes

$$K = \frac{2}{T} \quad \text{for } \omega_d T/2 << 1 \tag{3}$$

To illustrate the method, let us consider a second-order filter

$$H(s) = \frac{A_0 + A_1 s + A_2 s^2}{B_0 + B_1 s + B_2 s^2} \tag{4}$$

The corresponding transfer $H(z)$ using Eq. (1) results

$$H(z) = \frac{a_0 + a_1 z^{-1} + a_2 z^{-2}}{1 + b_1 z^{-1} + b_2 z^{-2}} \tag{5}$$

and

$$H(\hat{z}) = \frac{\alpha_0 + \alpha_1 \hat{z}^{-1} + \alpha_2 \hat{z}^{-2}}{1 + \beta_1 \hat{z}^{-1} + \beta_2 \hat{z}^{-2}} \tag{6}$$

where the relations of the coefficients are given next

$$\begin{array}{ll}
K = \omega_a / \tan(\omega_d T/2) & \alpha_0 = a_0 \\
A = B_0 + B_1 K + B_2 K^2 & \alpha_1 = 2a_0 + a_1 \\
a_0 = (A_0 + A_1 K + A_2 K^2)/A & \alpha_2 = a_0 + a_1 + a_2 \\
a_1 = (2A_0 - 2A_2 K)/A & \beta_1 = 2 + b_1 \\
a_2 = (A_0 - A_1 K + A_2 K^2)/A & \beta_2 = 1 + b_1 + b_2 \\
b_1 = (2B_0 - 2B_2 K^2)/A & \\
b_2 = (B_0 - B_1 K + B_2 K^2)/A &
\end{array}$$

$H(\hat{z})$ is obtained for convenience. Actual implementation of SC filters using "accumulators" $\hat{z}^{-1}$ rather than delay z^{-1} is more economical, and at the

[2] William D. Stanley, *Digital Signal Processing*, Reston Publishing Co., Inc., 1975.

same time they have better sensitivity properties.[3] The relation between $\hat{z}$ and z is

$$\hat{z} = z - 1 \tag{7}$$

It should be observed that an nth-order pole polynomial in s will cause an nth polynomial in the denominator *and* numerator of $H(z)$ or $H(\hat{z})$ independently of the order of the zero polynomial in $H(s)$.

A program named S2ZHAT has been implemented to determine $H(z)$ and $H(\hat{z})$ from a given $H(s)$.[4] The orders of the filter can be 1, 2, 3, 4, and 5. The basic idea is illustrated in Fig. 3.1–16.

Example B-1. Consider the following bandpass function,

$$H(s) = V_0/V_{\text{in}} = \frac{-3141.75}{0.39459 \times 10^8 + 3141.75s + 1.0\, s^2} \tag{8}$$

$$= \frac{H_0 \cdot \dfrac{\omega_n}{Q} s}{\omega_n{}^2 + \dfrac{\omega_n}{Q} s + s^2} \tag{9}$$

where

$$\omega_n = 6.282 \times 10^3 = 2\pi f_n$$

$$Q = 2.0$$

$$H_0 = 1.0$$

$$f_n = f_{\text{peak}} = 1000 \text{ Hz}$$

$$|V_0/V_{\text{in}}| = 0 \text{ dB at } 100 \text{ Hz}$$

$$|V_0/V_{\text{in}}| = -25.94 \text{ dB at } 100 \text{ Hz}$$

$$|V_0/V_{\text{in}}| = -25.94 \text{ dB at } 10 \text{ kHz}$$

---

[3] R. C. Agarwal and C. S. Burrus, "New Recursive Filter Structures Having Very Low Sensitivity and Roundoff Noise," *IEEE Trans. on Circuits and Systems,* Vol. CAS-22, December 1975, pp. 921–927.

[4] Eddie R. Klinkovsky, Master's thesis, Dept. of Electrical Engineering, Texas A&M Univ., 1980.

Thus the above s-domain transfer function is a bandpass filter centered at 1kHz. The S2ZHAT program found the following transfer functions in the z and $\hat{z}$ domains. The clock frequency is 128 kHz in all cases. The printout in all cases showed

$$|V_0/V_{in}| = 0 \text{ dB at 1 kHz}$$

$$|V_0/V_{in}| = -25.94 \text{ dB at 100 kHz}$$

$$|V_0/V_{in}| = -26.12 \text{ dB at 10 kHz}$$

Second-order

$$H(z) = V_0/V_{in} = \frac{-.0121 + .0121z^{-2}}{1 - 1.97z^{-1} + .976z^{-2}} \tag{10}$$

$$H(\hat{z}) = V_0/V_{in} = \frac{-.0121 - .024\hat{z}^{-1}}{1 + .0266z^{-1} + .00237\hat{z}^{-2}} \tag{11}$$

Third-order

$$H(z) = V_0/V_{in} = \frac{-.0121 - .0121z^{-1} + .0121z^{-2} + .0121z^{-3}}{1 - .973z^{-1} - .997z^{-2} + .976z^{-3}} \tag{12}$$

$$H(\hat{z}) = V_0/V_{in} = \frac{-.0121 - .0485\hat{z}^{-1} - .048\hat{z}^{-2}}{1 + 2.03\hat{z}^{-1} + .055\hat{z}^{-2} + .0047\hat{z}^{-3}} \tag{13}$$

Fourth-order

$$H(z) = \frac{-.021 - .024z^{-1} + .024z^{-2} + .0121z^{-3}}{1 + .026z^{-1} - 1.97z^{-2} - .022z^{-3} + .976z^{14}} \tag{14}$$

$$H(\hat{z}) = \frac{-.012 - .073\hat{z}^{-1} - .145\hat{z}^{-2} - .096\hat{z}^{-3}}{1 + 4.03\hat{z}^{-1} + 4.11\hat{z}^{-2} + .116\hat{z}^{-3} + .0095\hat{z}^{-4}} \tag{15}$$

Fifth-order

$$H(z) = \frac{-.0121 - .036z^{-1} - .024z^{-2} + .024z^{-3} + .036^{-4} + .012z^{-5}}{1 + 1.03z^{-1} - 1.94z^{-2} - 1.99z^{-3} + .954z^{-4} + .976z^{-5}} \tag{16}$$

$$H(\hat{z}) = \frac{-.0121 - .097z^{-1} - .290z^{-2} - .387z^{-3} - .194z^{-4} + .55 \times 10^{-7}z^{-5}}{1 + 6.03z^{-1} + 12.16z^{-2} + 8.33z^{-3} + .2414z^{-4} + .019z^{-5}} \tag{17}$$

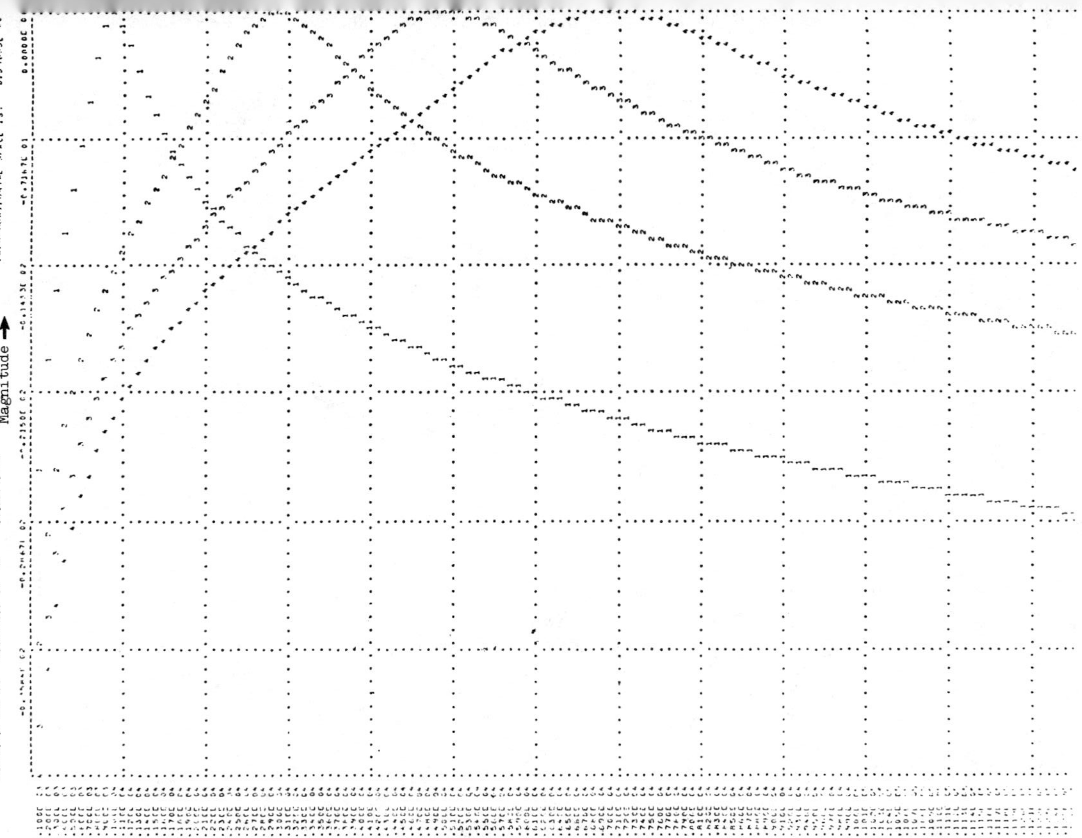

Fig. B-1. Line printer output of the magnitude resonse of Eq. (8) for four different center frequencies.

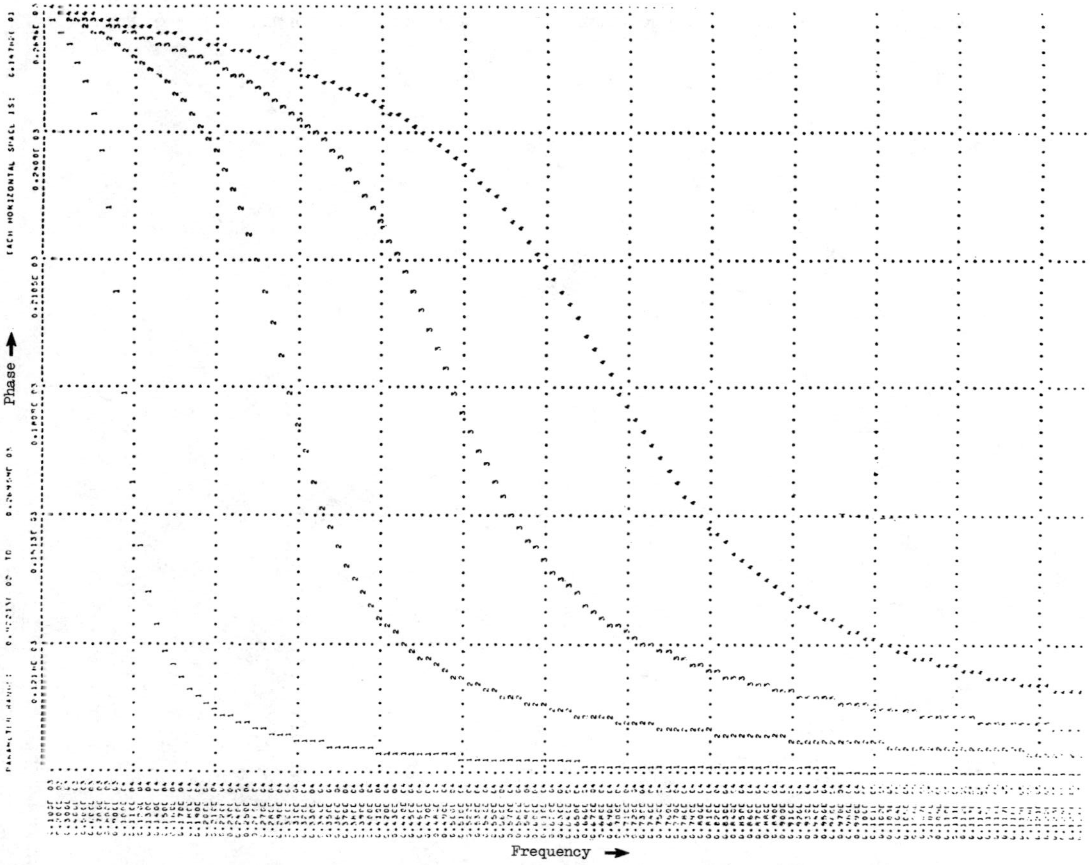

Phase →

Frequency →

Fig. B-2. Line printer output of the phase response of Eq. (8) for four different center frequencies.

Observe that with the same specifications given by Eq. (8) that one can obtain four possible sampled data implementations. The graphical results are shown in Figs. B1 and B2, where different center frequencies were specified for each filter, i.e.

Order	Center
2nd	1 kHz
3rd	3 kHz
4th	5 kHz
5th	7 kHz

S2ZHAT USER'S MANUAL

The program can compute and plot up to nine $H(z)$ at a time. For each data set the order of transfer function, the s-domain coefficients, the sampling (clock) frequency f_c, $\omega_a/2\pi$ the analog frequency f_a of the filter prototype, and $\omega_d/2\pi$ the desired sampled data frequency f_d are specified. Note that all the results will be plotted on the same graph including $H(s)$. Plotting or printing results is completely optional; one can obtain results in only one form or both. Furthermore, the results can be specified in logarithmic or linear scale for the frequency. The data deck setup is as follows.

Card 1. TITLE
i.e., Second-order Bandpass

Card 2. Initial frequency, linear frequency increment, number of points (per decade or for linear), number of decades (just for the logarithmic scale, otherwise 0.0).
i.e., 0.1E3, 0.05E3, 250.0E0, 0
0.1E3, 0.0, 25.0, 10
The first card is for linear scale, and the second is for logarithmic scale.

Card 3. Printing flag (Column 1)
i.e., 0
1
2
Printing flag 0 will provide only numerical results; 2 only the plot is printed, and 1 both graphical and numerical results are printed.

Card 4. Order of filter[5], f_c, f_a, f_d
i.e., 2, .128E6, 0.1E4, 0.32E4

[5] If the order of the filter is specified as 6, the $H(s)$ will be printed. One is still required to provide all the information of Cards 4, 5, and 6. All cards use free-format unless otherwise specified.

Card 5. $A_0, A_1, \ldots, A_n$
i.e., 0.0, −0.31417E4, 0.0

Card 6. $B_0, B_1, \ldots, B_n$
0.39459E8, 0.31417E4, 0.0

Card 7. GO (Column 1 and 2)
This card is included only when another filter data set is given. Note that another data set will include the equivalent of cards 4, 5, and 6.

i.e., 3, 0.128E6, 0.1E4, 0.32E4
0.0, −0,31417E4, 0.0. 0.0
0.39459E8, 0.31417E4, 0.1E1, 0.0

Last card. FN (in Columns 1 and 2)
This is an end card (Final)

Appendix C
TI-59 and HP-41 Programmable Calculator Program to Analyze MOS Devices

MOSFET CURRENT PROGRAM FOR TI-59 WITH PC100[1]

This program uses the Shichman-Hodges model with a modification by J. Allen of Texas Instruments, Inc. The current and other quantities are calculated in the saturated region only.

Formulas

Given the four voltages: V_D, V_G, V_S, and V_B
Given the model parameters: K, V_{TO}, BE, θ, λ, DE, γ, α, WR, TLD, $2\phi_F$, $V_{DS} = V_D - V_S$, $V_{GS} = V_G - V_S$, etc.

Drain current:

$$I_D = I_1 + g_{ds\,1} V_{SAT}$$

Conductances:

$$G_m = \mathrm{I_D}\left(\frac{2}{V_{GST}} - \frac{\theta}{1 + \theta\ V_{GST}}\right) - \frac{g_{ds\,1}}{A_g}\left(1 + \gamma\,(3\,V_{SAT} - V_{DS})\right)$$

$$G_B = \frac{BE \cdot G_m}{2(V_{SB} + 2\phi_F)}\frac{3}{2}$$

$$G_D = g_{ds\,1} + \frac{\mathrm{DE} \cdot G_m}{2\ V_{DS}\ \frac{3}{2}}$$

where

[1] This program was written by D. Mayer and M. de Wit, Texas Instruments, Inc.

$$G_m = \partial I_D/\partial V_{GS},\ G_B = \partial I_D/\partial V_{BS},\ G_D = \partial I_D/\partial V$$

and

$$I_1 = \frac{K(W - WR)\ V_{GST}{}^2}{(L - TLD)\ (1 + \theta \cdot V_{GST})};\ V_{GST} = V_G - V_S - V_T$$

$$V_T = V_{TO} + BE\ ((V_{SB} + 2\phi_F)^{1/2} - (2\phi_F)^{1/2}) - DE\ V_{DS}{}^{1/2}$$

$$g_{ds\,1} = I_1\ \lambda\ \alpha^2/A_g^2$$

$$A_g = \alpha + \gamma \cdot V_{GST}$$

$$V_{SAT} = V_{DS} - V_{DSAT}$$

$$V_{DSAT} = V_{GST}/A_g$$

The model parameters represent the surface mobility (K), the zero back-gate bias threshold (V_{TO}), the body effect (BE), mobility reduction (θ), short-channel effects (λ and DE (Allen's parameter)), and γ. α defines the boundary between the saturated and linear regions.

Program Description

The model parameters are currently structured to use the units: Volts, microamps, micromhos, and mils. There is space for nine different model sets. The parameters for model i are stored in compacted form in memory locations $19 + 4i$ through $19 + 4i + 3$. Given the units above the compacted form is

1st Location	$10^3 \cdot V_{TO} + 10^{-3} \cdot BE$
2nd Location	$10^3 \cdot K + \lambda$
3rd Location	$10^3 \cdot \theta + 10^{-3}\ \alpha$
4th Location	$10^3 \cdot \gamma + 10^{-3}\ DE$

Note that the second parameter in the compacted form must be positive and that the digits of the two parameters must not overlap.

The program has an initialization cycle during the callup of a new set of model parameters (Key E'). The memory is repartitioned to 400 program steps (0–399) and to storage locations (0–399) and to storage locations (0–69). If $V_{DS} \leq$ or $V_{GST} \leq 0$ or $V_{SAT} \leq 0$, the current is not calculated.

The parameters $2\phi_F$, WR, and TLD are the same for all FETs and are in locations 59, 60, and 61, respectively. They are twice the Fermi potential, channel width reduction, and total lateral diffusion.

The quantities W, L, V_G, V_S, V_D, V_B are input via keys A, B, C, D, A′, B′, respectively, and remain unchanged by the I_D calculation (E) or model call (E′)

Labeled results are printed out for

$$I_D,\ G_M,\ G_B,\ G_D,\ V_{GST},\ V_T,\ V_{SAT}$$

Key D prints out the current parameters

$$V_B,\ V_D,\ V_S,\ V_G,\ L,\ W,\ 0,\ V_{TO},\ BE,\ K,\ \lambda,\ \theta,\ \alpha,\ \gamma,\ DE$$

The listing shows seven models in memory for typical values of an NMOS process. The program and data are stored on four card sides.

TITLE SHA (Shichman-Hodges-Allen) PAGE 1 OF 1

TI Programmable Program Record

PROGRAMMER M de Wit DATE 4/16/79

Partitioning (Op 17) 4 7 9 . 5 9 ← NORMAL(1) Library Module NONE Printer YES Cards 4 SIDES

PROGRAM DESCRIPTION

$I_D^{(2)} = I_1 + g_{ds1} V_{SAT}$, $G_m = I\left(\frac{2}{V_{GST}} - \frac{\theta}{1+\theta V_{GST}}\right) - \frac{g_{ds1}}{A_g}\left(1+\gamma\left(3V_{SAT} - V_{DS}\right)\right)$, $G_B = \frac{BE \cdot G_m}{2(V_{SO}+2\phi_F)^{3/2}}$, $G_D = g_{ds1} + \frac{DE \cdot G_m}{2 V_{DS}^{3/2}}$

$I_1 = \frac{KP(W-WR)V_{GST}^2}{(L-TLD)(1+\theta V_{GST})}$, $V_{GST} = V_G - V_S - V_T$, $V_T = V_{TO} + BE\left(\sqrt{V_{SO}+2\phi_F} - \sqrt{2\phi_F}\right) - DE\sqrt{V_{DS}}$, $g_{ds1} = I_1 \lambda \alpha^2 / A_g^2$, $A_g = \alpha + \gamma V_{GST}$

$V_{SAT} = V_{DS} - V_{DSAT}$, $V_{DSAT} = V_{GST}/A_g$

Model #i then STORAGE:
- $VTO(V) \times 10^3 \cdot BE \times 10^{-3}$ in location $19+4i$
- ~~$KP(A/V^2) \times 10^3 \cdot \lambda$ " " $19+4i+1$~~
- $\theta \times 10^3 \cdot \alpha \times 10^{-3}$ " " $19+4i+2$
- $\gamma \times 10^3 \cdot DE \times 10^{-3}$ " " $19+4i+3$

USER INSTRUCTIONS

STEP	PROCEDURE	ENTER	PRESS			DISPLAY
1	Initialize(1) and call up model (model # 1 thru 9)	model #	E'			V_{TO}
2	Enter W, L, V_G, V_S, V_D, V_B	VALUE	A, B	C, D,	A', B'	ENTRY
3	Calculate(2) $I_D, G_M, G_B, G_D, V_{GST}, V_T, V_{SAT}$ & PRINT with name		E			ON PRINTER
4	GO TO 1 OR 2 OR 5					
5	PRINT PARAMETERS ($V_B, V_D, V_S, V_G, L, W, \theta, V_{TO}, BE, KP, \lambda, \theta, \alpha, \gamma, DE$)		D'			ON PRINTER

(2) If $V_{DS} \leq 0$ or $V_{GST} \leq 0$ or $V_{SAT} \leq 0$ it will be printed.

I_D is <u>not</u> calculated for $V_{SAT} \leq 0$

USER DEFINED KEYS		DATA REGISTERS (INV List)				LABELS (Op 08)
A	ENTER W	0 CTR, V_{DS}	10 W	20 V_{GST}	60 WR	INV _ lnx _ CE _ CLR _ x⇌t _ x² _
B	" L	1 DE	11 L	21 V_T	61 TLD	√x _ 1/x _ STO _ RCL ✓ SUM _ yˣ _
C	" V_G	2 γ	12 V_G	22 V_{SAT}	62 V_{DS} TAG	EE _ (_) _ ÷ ✓ GTO _ X ✓
D	" V_S	3 α	13 V_S	23 Models	63 I_D "	SBR _ − ✓ RST _ + _ R/S _ • _
E	Calculate I_D, etc	4 θ	14 V_D	4 ↓	64 GM "	+/− ✓ = _ CLR _ INV _ log _ CP _
A'	ENTER V_D	5 λ	15 V_B	5 thru	65 GB "	tan _ Pgm _ P→R _ sin _ cos _ CMs _
B'	" V_B	6 KP	16 I_D, I_1	6 58	66 GD "	Exc _ Prd _ |x| _ Eng _ Fix _ Int _
C'	—	7 BE	17 Gm, $(1+BV_{GST})$	7	67 VGST "	Deg _ Pause _ x=t _ Nop _ Op _ Rad _
D'	PRT Parameters	8 VTO	18 G_B, $\sqrt{V_{SGT}2\phi_F}$	58 ↓	68 VT "	Lbl _ x≥t _ Σ+ _ x̄ _ Grad _ St flg _
E'	Initialize and model call	9 Address, flag Gd G_D, g_{ds1}		59 $2\phi_F$	69 VSAT "	If flg _ D.MS _ π _ List _ Write _ Dsz _
						Adv ✓ Prt _

FLAGS	0	1	2	3	4	5	6	7	8	9

GO	000	76	LBL
CALCULATE	001	15	E
ID, ETC.	002	61	GTO
	003	00	00
	004	35	35
STORE	005	76	LBL
W	006	11	A
	007	42	STO
	008	10	10
	009	92	RTN
	010	76	LBL
STORE	011	12	B
L	012	42	STO
	013	11	11
	014	92	RTN
	015	76	LBL
STORE	016	13	C
V_G	017	42	STO
	018	12	12
	019	92	RTN
	020	76	LBL
STORE	021	14	D
V_S	022	42	STD
	023	13	13
	024	92	RTN
	025	76	LBL
STORE	026	16	A'
V_D	027	42	STO
	028	14	14
	029	92	RTN
	030	76	LBL
STORE	031	17	B'
V_B	032	42	STO
	033	15	15
	034	92	RTN
	035	29	CP
CALCULATE	036	22	INV
ID, ETC.	037	58	FIX
	038	43	RCL
	039	12	12
	040	75	-
	041	43	RCL
	042	13	13
	043	75	-
	044	53	(
	045	53	(
	046	43	RCL
	047	08	08
	048	85	+
	049	43	RCL
	050	07	07

	051	65	×
	052	53	(
	053	53	(
	054	43	RCL
	055	13	13
	056	75	-
	057	43	RCL
	058	15	15
	059	85	+
	060	43	RCL
	061	59	59
	062	54	)
STORE	063	34	$\sqrt{x}$
$\sqrt{V_{SB} + 2\phi_F}$	064	42	STO
	065	18	18
	066	75	-
	067	43	RCL
	068	59	59
	069	34	$\sqrt{x}$
	070	54	)
	071	75	-
	072	43	RCL
	073	01	01
	074	65	×
	075	53	(
	076	43	RCL
	077	14	14
	078	75	-
	079	43	RCL
	080	13	13
	081	54	)
STORE	082	42	STO
V_{DS}	083	00	00
	084	32	X⇄T
IS $V_{DS} \leq 0$?	085	77	GE
	086	75	- → YE
NO ↓	087	32	X⇄T
	088	34	$\sqrt{x}$
	089	54	)
STORE	090	42	STO
V_T	091	21	21
	092	95	=
STORE	093	42	STO
V_{GEST}	094	20	20
IS	095	32	X⇄T
$V_{GST} \leq 0$?	096	77	GE
	097	55	÷ → YE
NO ↓	098	32	X⇄T
	099	43	RCL
	100	00	00

	Step	Code	Key
	1Ø1	75	-
	1Ø2	53	(
	1Ø3	43	RCL
	1Ø4	2Ø	2Ø
	1Ø5	55	÷
	1Ø6	53	(
	1Ø7	43	RCL
	1Ø8	Ø3	Ø3
	1Ø9	85	+
	11Ø	43	RCL
	111	Ø2	Ø2
	112	65	×
	113	43	RCL
	114	2Ø	2Ø
STORE	115	54	)
$\alpha + \gamma$ V_{EST}	116	42	STO
	117	Ø9	Ø9
STORE	118	95	=
V_{SAT}	119	42	STO
	12Ø	22	22
IS $V_{SAT} \leq Ø$?	121	32	X⇄T
	122	77	GE
	123	65	× → YES
NO ↓	124	43	RCL
	125	1Ø	1Ø
	126	75	-
	127	43	RCL
	128	6Ø	6Ø
	129	95	=
	13Ø	65	×
	131	43	RCL
	132	Ø6	Ø6
	133	55	÷
	134	53	(
	135	43	RCL
	136	11	11
	137	75	-
	138	43	RCL
	139	61	61
	14Ø	54	)
	141	55	÷
	142	53	(
	143	Ø1	1
	144	85	+
	145	43	RCL
	146	04	Ø4
	147	65	×
	148	43	RCL
	149	2Ø	2Ø
	15Ø	54	)

	Step	Code	Key
STORE	151	42	STO
I & θ V_{EST}	152	17	17
	153	65	×
	154	43	RCL
	155	2Ø	2Ø
	156	33	X^2
	157	95	=
STORE	158	42	STO
I_1	159	16	16
	16Ø	65	×
	161	43	RCL
	162	Ø5	Ø5
	163	65	×
	164	43	RCL
	165	Ø3	Ø3
	166	33	X^2
	167	55	÷
	168	43	RCL
	169	Ø9	Ø9
	17Ø	33	X^2
	171	95	=
STORE	172	42	STO
g_{dsi}	173	19	19
	174	65	×
	175	43	RCL
	176	22	22
	177	95	=
CALC	178	44	SUM
I_D	179	16	16
	18Ø	43	RCL
	181	19	19
	182	94	+/-
	183	55	÷
	184	43	RCL
	185	09	Ø9
	186	65	×
	187	53	(
	188	Ø1	1
	189	85	+
	19Ø	43	RCL
	191	Ø2	Ø2
	192	65	×
	193	53	(
	194	Ø3	3
	195	65	×
	196	43	RCL
	197	22	22
	198	75	-
	199	43	RCL
	2ØØ	ØØ	ØØ

	Step	Code	Key
	2Ø1	54	)
	2Ø2	54	)
	2Ø3	85	+
	2Ø4	43	RCL
	2Ø5	16	16
	2Ø6	65	×
	2Ø7	53	(
	2Ø8	Ø2	2
	2Ø9	55	÷
	21Ø	43	RCL
	211	2Ø	2Ø
	212	75	-
	213	43	RCL
	214	Ø4	Ø4
	215	55	÷
	216	43	RCL
	217	17	17
	218	54	)
	219	95	=
STORE	22Ø	42	STO
G_M	221	17	17
	222	65	×
	223	43	RCL
	224	Ø7	Ø7
	225	55	÷
	226	02	2
	227	55	÷
	228	43	RCL
	229	18	18
	23Ø	68	NOP
	231	68	NOP
	232	95	=
STORE	233	42	STO
G_B	234	18	18
	235	43	RCL
	236	17	17
	237	65	×
	238	43	RCL
	239	01	Ø1
	24Ø	55	÷
	241	02	2
	242	55	÷
	243	43	RCL
	244	ØØ	ØØ
	245	34	√x
	246	68	NOP
	247	68	NOP
	248	95	=
STORE	249	44	SUM
G_D	25Ø	19	19

	Step	Code	Key
	251	Ø7	7
STORE	252	42	ST
PRINT COUNTER	253	ØØ	ØØ
→	254	22	IN
PRINT OUT LOOP	255	58	FI
	256	Ø7	7
	257	ØØ	Ø
	258	75	-
	259	43	RC
	26Ø	ØØ	ØØ
	261	95	=
	262	42	ST
	263	Ø9	09
	264	73	RC
	265	09	09
	266	69	OP
	267	Ø4	Ø4
	268	04	4
	269	Ø7	7
	27Ø	22	IN
	271	44	SU
	272	09	Ø9
	273	58	FI
	274	02	Ø2
	275	73	RC
	276	Ø9	Ø9
	277	69	OP
	278	Ø6	Ø6
	279	97	DS
	28Ø	ØØ	ØØ
	281	Ø2	Ø2
(loop back to 254)	282	54	54
	283	76	LB
	284	98	AD
	285	98	AD
	286	98	AD
	287	98	AD
DONE!	288	92	RT
INITIALIZE AND CALL MODEL	289	76	LB
	29Ø	10	E'
	291	22	IN
	292	58	FI
	293	32	X⇄
REPARTITION MEMORY	294	Ø7	7
	295	69	OP
	296	17	17
	297	32	X⇄
	298	65	×
	299	Ø4	4
	3ØØ	85	+

Comment	Step	Code	Key
	301	01	1
	302	09	9
STORE	303	95	=
MODEL START	304	42	STO
ADDRESS	305	09	09
	306	08	8
STORE	307	42	STO
COUNTER	308	00	00
	309	76	LBL
	310	43	RCL
→	311	73	RC*
READ	312	09	09
MODEL	313	59	INT
PARAMS,	314	72	ST*
UP COMPACT	315	00	00
AND	316	01	1
STORE	317	52	EE
	318	94	+/-
	319	03	3
	320	64	PD*
	321	00	00
	322	69	OP
	323	30	30
	324	73	RC*
	325	09	09
	326	22	INV
	327	59	INT
	328	50	I×I
	329	72	ST*
	330	00	00
	331	01	1
	332	52	EE
	333	03	3
	334	64	PD*
	335	00	00
	336	69	OP
	337	29	29
	338	97	DSZ
	339	00	00
	340	03	03
	341	11	11
	342	22	INV
	343	49	PRD
	344	05	05
	345	22	INV
	346	52	EE
	347	43	RCL
	348	08	08
DONE	349	92	RTN
	350	76	LBL
PRINT	351	19	D'
CURRENT	352	01	1
PARAMETERS	353	05	5
	354	42	STO
	355	00	00
	356	22	INV
	357	58	FIX
	358	00	0
	359	42	STO
	360	09	09
	361	73	RC*
	362	00	00
	363	99	PRT
	364	97	DSZ
	365	00	00
	366	03	03
	367	61	61
	368	61	GTO
	369	02	02
	370	85	85
PRINT	371	76	LBL
$V_{GST} \leq 0$	372	55	÷
	373	43	RCL
	374	67	67
	375	76	LBL
	376	94	+/-
	377	69	OP
	378	04	04
	379	32	X⇄T
	380	69	OP
	381	06	06
	382	98	ADV
	383	92	RTN
PRINT	384	76	LBL
$V_{DS} \leq 0$	385	75	-
	386	95	=
	387	43	RCL
	388	62	62
	389	61	GTO
	390	03	03
	391	77	77
PRINT	392	76	LBL
	393	65	×
$V_{SAT} \leq 0$	394	43	RCL
	395	69	69
	396	61	GTO
	397	03	03
	398	77	77
	399	00	0

LABELS USED IN PROGRAM

001	15	E
006	11	A
011	12	B
016	13	C
021	14	D
026	16	A'
031	17	B'
284	98	ADV
290	10	E'
310	43	RCL
351	19	D'
372	55	÷
376	94	+/-
385	75	-
393	65	×

FIXED MEMORY CONTENTS WITH TYPICAL NMOS PARAMETERS

600.000275	23	MODEL #1	
16800.029	24	ENHANCEMENT #1	
109.001	25	L=.3 MIL	
35.000097	26		
3200.00035	27	DEPLETION	
14300.035	28		#2
65.001	29	L=.3 MIL	
63.00011	30		
520.00035	31	ENH	
15400.021	32		
58.001	33	L=.5	#2
11.000022	34		
-3240.0004	35	DEPL	
14300.023	36		#4
31.001	37	L=.5	
30.00003	38		
520.00039	39	ENH	
13900.011	40		#5
29.001	41	L=1.0	
0.0000025	42		
3300.00045	43	DEPL	
13700.008	44		#6
7.001	45	L=1.0	
12.0000129	46		
610.00018	47	ENH	
18900.046	48		#7
202.001	49	L=.2	
79.000222	50		
0.	51		
0.	52		
0.	53		
0.	54		
0.	55		
0.	56		
0.	57		
0.	58		
0.54	59	2φF	
0.06	60	WR	
0.06	61	TLD	

PRINT LABELS

42163600.	62	VDS
24160000.	63	ID
22300000.	64	GM
22140000.	65	GB
22160000.	66	GD
42223637.	67	VGST
42370000.	68	VT
42361337.	69	VSAT

AFTER LOADING THE 4 CARD SIDES
CLEAR AND ENTER 1. PRESS: E'
ENTER: W, L, V_D, ETC.

OUTPUT EXAMPLE

PRESS: E

88.41	ID	(MICRO AMP)
139.83	GM	(MICRO MHO)
8.17	GB	
5.18	GD	
1.17	VGST	(VOLT)
Ø.83	VT	
3.87	VSAT	

CURRENT PARAMETERS

PRESS: D'

-5.	V_B
5.	V_D
Ø.	V_S
2.	V_G
Ø.3	L
1.	W
Ø.	-
Ø.6	VTO
Ø.275	BE
16.8	K
Ø.Ø29	λ
Ø.109	θ
1.	α
Ø.035	γ
Ø.097	DE

HP-41 PROGRAM TO ANALYZE MOS DEVICES

This program was originally developed by M. de Wit of Texas Instruments for the TI-59 and converted to the HP-41 by John Bradbury and Steve Erickson, students at Texas A&M University.

SIZE: Ø52

STEP	INSTRUCTIONS	INPUT	FUNCTION	DISPLAY
1	Set Status, Key in Program, Place in User Mode		USER	
2	Load Model Parameters		E	
	Using self-loading models: (Calculate beeps when finished)	Model#	R/S	(TLD)
	Using different model or hand-loading model:		D	DE?
		DE	R/S	GAMMA?
		γ	R/S	ALPHA?
		α	R/S	THETA?
		θ	R/S	LAMBDA?
		λ	R/S	KP?
		KP	R/S	BE?
		BE	R/S	VTO?
		V_{TO}	R/S	2*PHIF?
		$2\phi F$	R/S	WR?
		WR	R/S	TLD?
		TLD	R/S	(TLD)
3	Load Circuit Parameters		A	W?
		W	R/S	L?
		L	R/S	VG?
		VG	R/S	VS?
		VS	R/S	VD?
		VD	R/S	VB?
		VB	R/S	(VB)
4	Calculate I_D, G_m, G_B, G_D, V_{GST}, V_T, V_{sat}		B	
	If $V_{DS} \leq \emptyset$			VDS = (VDS)
	If $V_{GST} \leq \emptyset$			VGST = (VGST)
	If $V_{sat} \leq \emptyset$			VSAT = (VSAT)

SIZE: Ø62

STEP	INSTRUCTIONS		INPUT	FUNCTION	DISPLAY
4 (cont'd)	ELSE calculator beeps when finished				
	Display values:	I_D		R/S	ID = (ID)
		G_M		R/S	GM = (GM)
		G_B		R/S	GB = (GB)
		G_D		R/S	GD = (GD)
		V_{GST}		R/S	VGST = (VGST)
		V_T		R/S	VT = (VT)
		V_{SAT}		R/S	VSAT = (VSAT)
5	Repeat steps 3 and 4 with new circuit parameters if desired				
6	Repeat steps 2, 3, and 4 with new model parameters				

IMPORTANT STATUS:

SIZE = Ø62
DEG
USER MODE
SF 26

2 MEMORY MODULES REQUIRED

MODELS ARE STORED IN FOLLOWING FORMAT, WITH MODEL NUMBER + i:

REGISTER	CONTENTS
22 + 4i	V_{TO}(V) x 1Ø^3 . BE x 1Ø$^{-3}$
22 + 4i + 1	KP(μA/V) x 1Ø^3 . λ
22 + 4i + 2	θ x 1Ø^3 . α x 1Ø$^{-3}$
22 + 4i + 3	γ x 1Ø^3 . DE x 1Ø$^{-3}$

9 MODELS PROVIDED.

REGISTER	CONTENTS
RØØ	Model #, V_{DS}
RØ1	DE
RØ2	γ
RØ3	α
RØ4	θ
RØ5	λ
RØ6	KP
RØ7	BE
RØ8	V_{TO}
RØ9	α + γ · VGST
R1Ø	W
R11	L
R12	VG
R13	VS
R14	VD
R15	VB
R16	ID, I1
R17	GM, (1+θ·VGST)
R18	GB, $\sqrt{VSB+2\phi F}$
R19	GD, g_{ds1}
R2Ø	VGST
R21	VT
R22	VSAT
R23	2φF
R24	WR
R25	TLD

NOTE: 1. LINES 1-4 ARE USED WHEN LOADING THE PROGRAM AND MODEL PARAMETERS FROM MAGNETIC CARDS. USERS WITHOUT CARD READERS MAY DELETE THESE LINES.

2. IN THE INTEREST OF CONSERVING MEMORY SPACE, THE USER MAY WISH TO ELIMINATE THE SELF-LOADING MODEL OPTION AND WITH IT THE ALLOCATION OF REGISTERS 26-61 FOR MODEL PARAMETER STORAGE. IF SO, DELETE LINES 226-255 AND LET SIZE = Ø26. THIS WILL ALLOW THE PROGRAM TO RUN USING A SINGLE MEMORY MODULE.

```
241 STO IND   <------ STORE PARAMETER
ØØ
242 DSE ØØ    <------ DECREMENT PARAMETER COUNTER
243 RCL IND
Ø9
244 FRC
245 ABS
246 1ØØØ
247 *
248 STO IND   <------ STORE PARAMETER
ØØ
249 ISG Ø9
25Ø DSE ØØ    <------ DECREMENT PARAMETER COUNTER, EXIT IF Ø
251 GTO Ø1    <------ LOOP TO LOAD NEXT PARAMETERS
252 1ØØØ
253 ST/Ø5     <------ CORRECTION FOR λ
254 BEEP      <------ FINISHED!
255 RTN
256 .END.
```

Register	Value		Group
R23	Ø.5400	$2\phi_F$	
R24	0.0600	WR	
R25	0.0600	TLD	
R26	600.000275		#1
R27	16,800.0290		#1
R28	109.0010		#1
R29	35.000097		#1
R30	-3,200.00035		#2
R31	14,300.0350		#2
R32	65.0010		#2
R33	63.00011		#2
R34	52Ø.ØØ035		#3
R35	15,4ØØ.0210		#3
R36	58.0010		#3
R37	11.000022		#3
R38	-3,240.0004		#4
R39	14,300.0230		#4
R40	31.0010		#4
R41	30.00003		#4
R42	520.00039		#5
R43	13,9ØØ.0110		#5
R44	29.0010		#5
R45	2.5000-06		#5
R46	-3,300.00045		#6
R47	13,700.0080		#6
R48	7.0010		#6
R49	12.0000129		#6
R50	610.00018		#7
R51	18,9ØØ.0460		#7
R52	2Ø2.ØØ10		#7
R53	79.ØØØ222		#7
R54	Ø.Ø600		#8
R55	42,163,600.00		#8
R56	24,160,000.00		#8
R57	22,300,000.00		#8
R58	22,140,000.00		#9
R59	22,160.ØØØ.00		#9
R60	42,223.637.ØØ		#9
R61	42,370,000.00		#9

MODEL PARAMETER DATA (R26–R61)

```
Ø1*LBL "X"       Used to load
Ø2 23.Ø61        model parameters
Ø3 RDTAX         from cards
Ø4 STOP
Ø5*LBL A         Load circuits parameters
Ø6 "W?"              W
Ø7 PROMPT
Ø8 STO 1Ø
Ø9 "L?"              L
1Ø PROMPT
11 STO 11
12 "VG?"             VG
13 PROMPT
14 STO 12
15 "VS?"             VS
16 PROMPT
17 STO 13
18 "VD?"             VD
19 PROMPT
2Ø STO 14
21 "VB?"             VB
22 PROMPT
23 STO 15
24 RTN
25*LBL B
26 RCL 13         Calculate VGST
27 RCL 15
28 -
29 RCL 23
3Ø +
31 SQRT
32 STO 18         √(VSB + 2φF)
33 RCL 23
34 SQRT
35 -
36 RCL Ø7
37 *
38 RCL 14
39 RCL 13
4Ø -
41 STO ØØ         VDS
42 X < = Ø?
43 GTO 1Ø         ERROR IF VDS ≤ Ø
44 SQRT
45 RCL Ø1
46 *
47 -
48 RCL Ø8
49 +
5Ø STO 21         VT
51 CHS
52 RCL 12
53 +
54 RCL 13
55 -
56 STO 2Ø         VGST
57 X < = Ø?
58 GTO 11         ERROR IF VGST<Ø
59 RCL ØØ         CALCULATE VSAT
6Ø RCL 2Ø
61 RCL Ø2
62 RCL 2Ø
63 *
64 RCL Ø3
65 +
66 STO Ø9         α + γ . VGST
67 /
68 -
69 STO 22         VSAT
7Ø X < = Ø?
71 GTO 12         ERROR IF VSAT≤Ø
72 RCL 1Ø         CALCULATE ID
73 RCL 24
74 -
75 RCL Ø6
76 *
77 RCL 11
78 RCL 25
79 -
8Ø /
81 RCL Ø4
82 RCL 2Ø
83 *
84 1
85 +
86 STO 17         1 + θ(VGST)
87 /
88 RCL 2Ø
89 X ↑ 2
9Ø *
91 STO 16         I1
92 RCL Ø5
93 *
94 RCL 03
95 X ↑ 2
96 *
97 RCL 09
98 X ↑ 2
99 /
1ØØ STO 19        gds1
1Ø1 RCL 22
1Ø2 *
```

Step	Instruction	Comment
1Ø3	ST+ 16	ID
1Ø4	2	CALCULATE GM, GB
105	ENTER↑	
1Ø6	RCL 2Ø	
107	/	
108	RCL Ø4	
110	/	
111	-	
112	RCL 16	
113	*	
114	RCL 22	
115	3	
116	*	
117	RCL 00	
118	-	
119	RCL 02	
12Ø	*	
121	1	
122	+	
123	RCL 19	
124	*	
125	RCL 09	
126	/	
127	-	
128	STO 17	G_m
129	RCL 07	
13Ø	*	
131	2	
132	/	
133	RCL 18	
134	3	
135	Y↑X	
136	/	
137	STO 18	GB
138	RCL 17	CALCULATE GD
139	RCL Ø1	
14Ø	*	
141	2	
142	/	
143	RCL ØØ	
144	SQRT	
145	3	
146	Y↑X	
147	/	
148	ST + 19	
15Ø	FIX 2	
151	"ID="	
152	ARCL 16	
153	PROMPT	VALUE OF ID
154	"GM="	
155	ARCL 17	
156	PROMPT	VALUE OF GM
157	"GB="	
158	ARCL 18	
159	PROMPT	VALUE OF GB
16Ø	"GD="	
161	ARCL 19	
162	PROMPT	VALUE OF GD
163	"VGST="	
164	ARCL 2Ø	
165	PROMPT	VALUE OF VGST
166	"VT="	
167	ARCL 21	
168	PROMPT	VALUE of VT
169	"VSAT="	
17Ø	ARCL 22	
171	PROMPT	VALUE OF VSAT
172	RTN	
173	LBL 1Ø	VDS ≤ Ø
174	"VDS="	
175	ARCL ØØ	
176	BEEP	
177	PROMPT	VALUE OF VDS
178	RTN	
179	LBL 11	
18Ø	"VGST="	
181	ARCL 2Ø	
182	BEEP	
183	PROMPT	VALUE OF VGST
184	RTN	
185	LBL 12	VSAT ≤ Ø
186	"VSAT="	
187	ARCL 22	
188	BEEP	
189	PROMPT	VALUE OF VSAT
19Ø	RTN	
191	LBL D	ENTER MODEL PARAMETERS BY HAND
192	"DE?"	
193	PROMPT	
194	STO Ø1	DE
195	"GAMMA?"	
196	PROMPT	
197	STO Ø2	γ
198	"ALPHA?"	
199	PROMPT	
200	STO Ø3	α
201	"THETA?"	
202	PROMPT	
2Ø3	STO Ø4	θ
204	"LAMBDA?"	
205	PROMPT	
2Ø6	STO Ø5	λ
207	"KP?"	
208	PROMPT	

```
STO Ø6          KP
"BE?"
PROMPT
STO Ø7          BE
"VTO?"
PROMPT
STO Ø8          VTO
"2*PHIF?"
PROMPT
STO 23          2φF
"WR?"
PROMPT
STO 24          WR
"TLD?"
PROMPT
STO 25          TLD
RTN
-----------------------------------------
LBL E           LOAD MODEL PARA-
"MODEL N"         METERS BY CALCULATOR
PROMPT          INPUT MODEL #
4
*
22.9
+
STO Ø9          FIRST REGISTER OF MODEL #
8
STO ØØ          COUNTER
LBL Ø1          BEGIN LOADING LOOP
RCL IND Ø9
INT
.ØØ1
*
```

EQUATIONS USED: (SHICHMAN-HODGES-ALLEN)

$$I_D = I_1 + g_{ds1} \cdot V_{SAT}$$

$$I_1 = \frac{KP(W-WR) \cdot V_{GST}^2}{(L-TLD)(1+\theta \cdot V_{GST})}$$

$$g_{ds1} = \frac{I_1 \cdot \lambda \cdot \alpha^2}{\alpha + \gamma \cdot V_{GST}}$$

$$V_{SAT} = V_{DS} - \frac{I_1 \cdot \lambda \cdot \alpha^2}{\alpha + \gamma \cdot V_{GST}}$$

$$V_{GST} = V_G - V_S - V_T$$

$$V_T = V_{TO} + BE(\sqrt{V_{SB} + 2\phi_F} - \sqrt{2\phi_F}) - DE \cdot \sqrt{V_{DS}}$$

$$G_m = I_1\left(\frac{2}{V_{GST}} - \frac{\theta}{1 + \theta \cdot V_{GST}}\right) - \frac{g_{ds1}}{\alpha + \gamma \cdot V_{GST}} [1 + \gamma(3V_{SAT} - V_{DS})]$$

$$G_B = \frac{BE \cdot G_m}{2(V_{SB} + 2\phi_F)^{3/2}}$$

$$G_D = g_{ds1} + \frac{DE \cdot G_m}{2 \cdot (V_{DS})^{3/2}}$$

EXAMPLE:

(Using Model #7, see p. 738)

MODEL PARAMETERS	CIRCUIT PARAMETERS	OUTPUT
DE = Ø.222	W = 4	
γ = Ø.Ø79	L = Ø.3	ID = 956.76
α = 1	VG = 2	G_m = 87Ø.74

MODEL PARAMETERS	CIRCUIT PARAMETERS	OUTPUT
θ = 0.202	V_S = 0	G_B = 6.01
λ = 0.046	V_D = 10	G_D = 29.06
KP' = 18.9	V_B = -5	V_{GST} = 1.80
BE = 0.18		V_T = 0.20
V_{TO} = 0.61		V_{SAT} = 8.42
$2\phi_F$ = 0.54		
WR = 0.06		
TLD = 0.06		

Bibliography

The books listed below provide additional information on many of the topics treated in this book. These volumes cover the subjects of filter theory, filter tables, analog-sampled data concepts, MOS technology, analog MOS design, and many other subjects.

Bowen, B. A. and W. R. Brown: *VLSI Systems for Digital Signal Processing,* Prentice-Hall Inc., Englewood Cliffs, N.J., 1982.

Bruton, Leonard T.: *RC-Active Circuits-Theory and Design,* Prentice-Hall, Inc., Englewood Cliffs, N.J., 1980.

Budak, Aram: *Passive and Active Network Analysis and Synthesis:* Houghton Mifflin Company, Boston, 1974.

Daryanani, Gobind: *Principles of Active Network Synthesis and Design,* John Wiley & Sons, Inc., New York, 1976.

Dooley, Daniel J., Editor: *Data Conversion Integrated Circuits,* IEEE Press, John Wiley & Sons, Inc., New York, 1980.

Ghausi, M. S., and K. R. Laker: *Modern Filter Design-Active RC and Switched Capacitor,* Prentice-Hall, Inc., Englewood Cliffs, N.J., 1981.

Graeme, Jerald G., Gene E. Tobey, and Lawrence P. Huelsman, *Operational Amplifiers-Design and Application,* McGraw-Hill Book Company, New York, 1971.

Graeme, Jerald G.: *Application of Op Amps-Third Generation Techniques,* McGraw-Hill Book Company, New York, 1973.

Gray, Paul R. and Robert G. Meyer: *Analysis and Design of Analog Integrated Circuits,* John Wiley & Sons, Inc., New York, 1977.

Gray, Paul R., David A. Hodges, and Robert W. Brodersen, Editors: *Analog MOS Integrated Circuits,* IEEE Press, John Wiley & Sons, Inc., New York, 1980.

Grebene, Alan B.: *Analog Integrated Circuit Design,* Van Nostrand Reinhold, New York, 1972.

Grebene, Alan B., Editor: *Analog Integrated Circuits,* IEEE Press, John Wiley & Sons, Inc., New York, 1978.

Hnatek, E. R.: *A User's Handbook of D/A and A/D Converters,* John Wiley & Sons, Inc., New York, 1976.

Hoeschele, D. F.: *Analog-to-Digital and Digital-to-Analog Conversion Techniques,* John Wiley & Sons, Inc., New York, 1968.

Hodges, David A. and Horace G. Jackson: *Analysis and Design of Digital Integrated Circuits,* McGraw-Hill Book Company, New York, 1983.

Huelsman, Lawrence P. and Phillip E. Allen: *Introduction to the Design and Application of Active Filters,* McGraw-Hill Book Company, New York, 1980.

Jong, M. T.: *Methods of Discrete Signal and System Analysis,* McGraw-Hill Book Company, New York, 1982.

Karni, Shlomo and William J. Byatt: *Mathematical Methods in Continuous and Discrete Systems,* Holt, Rinehart and Winston, New York, 1982.

Kuo, Benjamin C.: *Analysis and Synthesis of Sampled-Data Control Systems,* Prentice-Hall, Inc., Englewood Cliffs, N.J., 1963.

Maxwell, James Clerk: *A Treatise on Electricity and Magnetism,* Oxford University Press, London, 1873; Lowe & Brydone, Printers, Ltd., London, 1946.

Mead, Carver, and Lynn Conway: *Introduction to VLSI Systems,* Addison-Wesley Publishing Company, Reading, Massachusetts, 1980.

Mitra, Sanjit K.: *Analysis and Synthesis of Linear Active Networks,* John Wiley & Sons, Inc., New York, 1969.

Moschytz, George S.: *Linear Integrated Networks: Design,* Van Nostrand Reinhold Co., New York, 1975.

Muller, Richard S., and Theodore I. Kamins: *Device Electronics for Integrated Circuits,* John Wiley & Sons, Inc., New York, 1977.

Oppenheim, Alan V., Editor: *Applications of Digital Signal Processing,* Prentice-Hall, Englewood Cliffs, N.J., 1978.

Oppenheim, Alan V., and Ronald W. Schafer: *Digital Signal Processing,* Prentice-Hall, Inc., Englewood Cliffs, N.J., 1975.

Oppenheim, Alan V., Alan S. Willsky, and Ian T. Young: *Signals and Systems,* Prentice-Hall, Inc., Englewood Cliffs, N.J., 1983.

Roberge, James K.: *Operational Amplifiers,* John Wiley & Sons, Inc., New York, 1975.

Schmid, H.: *Electronic Analog/Digital Conversions,* Van Nostrand Reinhold, New York, 1970.

Sedra, Adel S., and Peter O. Brackett: *Filter Theory and Design: Active and Passive,* Matrix Publishers, Inc., Champaign, Ill., 1977.

Sheingold, Daniel H., Editor: *Analog-Digital Conversion Handbook,* Analog Devices, Inc., Norwood, Massachusetts, 1972.

Temes, Gabor C., and Sanjit K. Mitra: *Modern Filter Theory and Design,* John Wiley & Sons, Inc., New York, 1973.

Van Valkenberg, M. E.: *Analog Filter Design,* Holt, Rinehart and Winston, New York, 1982.

Van Valkenberg, M. E.: *Introduction to Modern Network Synthesis,* John Wiley & Sons, Inc., New York, 1960.

Wait, John V., Lawrence P. Huelsman, and Granino A. Korn: *Introduction to Operational Amplifier Theory and Applications,* McGraw-Hill Book Company, New York, 1975.

Weinberg, Louis: *Network Analysis and Synthesis,* McGraw-Hill Book Company, New York, 1962; R. E. Krieger Publishing Co., Huntington, N.Y., 1975.

Zverov, Anatol I.: *Handbook of Filter Synthesis,* John Wiley & Sons, Inc., New York, 1967.

Index